48

Springer Series in Chemical Physics

Edited by Fritz P. Schäfer

Springer Series in Chemical Physics

Editors: Vitalii I. Goldanskii   Fritz P. Schäfer   J. Peter Toennies

Managing Editor: H. K. V. Lotsch

Volume 40 **High-Resolution Spectroscopy of Transient Molecules**
By E. Hirota

Volume 41 **High Resolution Spectral Atlas of Nitrogen Dioxide 559–597 nm**
By K. Uehara and H. Sasada

Volume 42 **Antennas and Reaction Centers of Photosynthetic Bacteria**
Structure, Interactions, and Dynamics
Editor: M. E. Michel-Beyerle

Volume 43 **The Atom-Atom Potential Method.** Applications to Organic Molecular Solids
By A. J. Pertsin and A. I. Kitaigorodsky

Volume 44 **Secondary Ion Mass Spectrometry SIMS V**
Editors: A. Benninghoven, R. J. Colton, D. S. Simons, and H. W. Werner

Volume 45 **Thermotropic Liquid Crystals, Fundamentals**
By G. Vertogen and W. H. de Jeu

Volume 46 **Ultrafast Phenomena V**
Editors: G. R. Fleming and A. E. Siegman

Volume 47 **Complex Chemical Reaction Systems**
Editors: J. Warnatz and W. Jäger

Volume 48 **Ultrafast Phenomena VI**
Editors: T. Yajima, K. Yoshihara, C. B. Harris, and S. Shionoya

Volume 49 **Vibronic Interactions in Molecules and Crystals**
By I. B. Bersuker and V. Z. Polinger

Volumes 1–39 are listed on the back inside cover

Ultrafast Phenomena VI

Proceedings of the 6th International Conference, Mt. Hiei, Kyoto, Japan, July 12–15, 1988

Editors: T. Yajima, K. Yoshihara, C. B. Harris, and S. Shionoya

With 487 Figures

Springer-Verlag Berlin Heidelberg New York
London Paris Tokyo

Professor Tatsuo Yajima
Institute for Solid State Physics, University of Tokyo,
Roppongi, Minato-ku, Tokyo 106, Japan

Professor Keitaro Yoshihara
Institute for Molecular Science, Myodaiji, Okazaki 444, Japan

Professor Charles B. Harris
Department of Chemistry, University of California at Berkeley,
Berkeley, CA 94720, USA

Professor Shigeo Shionoya
Department of Electronics, Tokyo Engineering University,
Katakura, Hachioji, Tokyo 192, Japan

ISBN-13: 978-3-642-83646-6 e-ISBN-13: 978-3-642-83644-2
DOI: 10.1007/978-3-642-83644-2

Softcover reprint of the hardcover 1st edition 1988

Printing: Druckhaus Beltz, 6944 Hemsbach/Bergstr.

Preface

This volume is a collection of papers presented at the Sixth International Conference on Ultrafast Phenomena (ICUP 88) held at Mt. Hiei, Kyoto, Japan, July 12–15, 1988. The conference brought together scientists from a wide range of disciplines, including physics, chemistry, biology, and electronics, who share a common interest in discussing ultrafast processes at picosecond and femtosecond time scales. It was attended by approximately 350 participants from 19 countries.

The ICUP 88 conference was held just ten years after the first conference in the series. The progress of both the technology and applications in the field during this decade is truly remarkable. In addition to the achievement of short pulses far below 10 fs, which is almost the inherent limit in the visible region, extensions in wavelength range, power levels and other performance parameters of ultrashort light pulses are prominent. With these high-performance light sources, ultrafast phenomena in physical, chemical and biological systems and in artificial devices are being studied extensively and earnestly. Ultrafast technology is now becoming one of the basic and common tools and concepts presently entering a wide variety of scientific fields not only for basic research in natural science but also for promoting new applications in various engineering fields. We feel that these proceedings vividly reflect the present status of the field.

Numerous people contributed to make the conference so successful. We would like to thank especially the members of the international program committee and the Japanese organizing committee for their helpful advice and great effort in organizing the conference. Special thanks are due to Mr. S. Ohtani and his colleagues at the secretariat of ICUP 88 for their hard work in implementing the meeting arrangements. We are also indebted to the Japan Society of Applied Physics, the Optical Society of America, the Institute of Electrical Engineers of Japan, the Institute of Electronics, Information and Communication Engineers, the Laser Society of Japan, and the Optoelectronic Industry and Technology Development Association for their sponsorship. In addition, we gratefully acknowledge financial support from the Commemorative Association for the Japan World Exposition 1970, the Optical Society of America, and a number of industrial companies and firms.

One of us (C.B.H.) also acknowledges the National Science Foundation, the Office of Naval Research, and the Air Force Office of Scientific Research for supporting participants from USA.

Tokyo, Okazaki,
Berkeley, Tokyo
August 1988

T. Yajima, K. Yoshihara
C.B. Harris, S. Shionoya

Contents

Part II Generation of Short-Wavelength Ultrashort Light Pulses

Part III Propagation, Control and Measurement of Ultrashort Light Pulses

Part VI Dynamical Processes in Semiconductor Quantum Wells and Microstructures

Part VII Nonlinear Optics, Coherent Spectroscopy and Spectroscopic Methods

Part VIII **Dynamics on Surfaces and at Interfaces**

Part IX **Energy Transfer and Relaxation**

Part X Chemical Reaction and Solvation Dynamics

Part XI **Dynamics of Biological Processes**

Part I

Generation, Amplification and Compression of Ultrashort Light Pulses

Chirped Pulse Amplification: Present and Future

P. Maine[1], *D. Strickland*[1], *M. Pessot*[1], *J. Squier*[1], *P. Bado*[1], *G. Mourou*[1], *and D. Harter*[2]

[1]Laboratory for Laser Energetics, University of Rochester, 250 East River Road, Rochester, NY 14623-1299, USA
[2]Allied-Signal, Inc., P.O. Box 1021R, Morristown, NJ 07960, USA

Short pulses with ultrahigh peak powers have been generated in Nd:glass and Alexandrite using the Chirped Pulse Amplification (CPA) technique. This technique has been successful in producing picosecond terawatt pulses with a table-top laser system. In the near future, CPA will be applied to large laser systems such as NOVA to produce petawatt pulses (1 kJ in a 1 ps pulse) with focused intensities exceeding 10^{+21} W/cm^2. These pulses will be associated with electric fields in excess of 100 e/a_0^2 and blackbody energy densities equivalent to 3 x 10^{10} J/cm^3. This petawatt source will have important applications in x-ray laser research and will lead to fundamentally new experiments in atomic, nuclear, solid-state, plasma, and high-energy density physics.

Three fundamental requirements have to be fulfilled for the amplification of short pulses. First, the bandwidth of the amplifier must be large enough to accommodate the full spectrum of the short seed pulse. Second, to efficiently extract the energy stored in the amplifier the fluence of the input pulse has to be near the saturation fluence of the medium, i.e., $F_s = h\nu/\sigma$, where σ is the stimulated emission cross section. Finally, the intensity within the amplifier must stay below a critical level at which nonlinear effects become significant and begin to distort the spatial and temporal profiles of the pulse. This critical level is a function of the integrated nonlinear index along the optical path and is generally known as the B integral [1]

$$B = \frac{2\pi}{\lambda} \int \frac{\Delta n}{n} dl = \frac{2\pi}{\lambda} n_2 \int_0^L I(z)\, dz\,.$$

The B integral at any position across the beam gives the amount of phase delay experienced by the laser beam. The high spectral frequencies are amplified with a gain factor proportional to exp(B). Above a threshold value of B = 3, high spatial frequencies are amplified to an unacceptable level and must be removed by spatial filtering. This condition limits the beam intensity to below a critical value of 10 GW/cm^2 for solid or liquid amplifiers 10 cm in length. The CPA approach lowers the peak power during the amplification and alleviates this last requirement. Thus it allows the utilization of superior storage energy amplifiers such as Nd:glass, Alexandrite, and Ti:sapphire.

Solid-state media doped with neodymium, chromium, or titanium have saturation fluences on the order of 1-20 J/cm^2 and offer excellent energy storage capabilities. They have some additional features of interest, such as high doping concentration and very large bandwidth. Chromium-doped media have already shown lasing capabilities from 700 to 1100 nm with large bandwidths. One of them, Alexandrite, with a bandwidth covering the 700–800 nm range, has reached a high average power performance of 100 W [2]. Another very promising medium is titanium-sapphire, which has been reported to lase between 700 and 1000 nm. A well-developed and widely used solid-state medium is neodymium-glass which has a bandwidth larger than 20 nm and so can support the amplification of subpicosecond pulses.

However, if a 1 ps pulse is amplified in Nd:glass to the saturation level, the power density will be on the order of 1 TW/cm^2, a value well above the critical non-linear effect threshold.

Springer Series in Chemical Physics, Vol. 48: **Ultrafast Phenomena VI**
Editors: T. Yajima · K. Yoshihara · C.B. Harris · S. Shionoya

Thus the peak intensity limits the amount of energy which can be extracted in solid-state amplifiers by short pulses to a value much lower than the saturation energy. For this reason, solid-state lasers until now have been well suited for long (nanosecond) pulse amplification, but could not be used efficiently for the amplification of picosecond pulses.

A new technique was necessary to amplify short pulses to saturation energies in wide bandwidth solid-state media while maintaining low power levels in the amplifier. Our CPA approach, illustrated in Fig. 1, reaches this goal [3,4]. A short seed optical pulse is frequency-chirped and stretched before being amplified. The long pulse is then amplified to saturation levels, while maintaining a low peak power. After amplification, an optical compressor is used to restore the original short pulse duration. Thus we are able to generate a pulse with short duration and high energy.

Our present CPA system uses phosphate Nd:glass amplifiers and a wavelength compatible cw-pumped mode-locked Nd:YLF oscillator [5] to generate 0.5 TW pulses of 1 ps duration. A schematic of the CPA system is shown in Fig. 2. The oscillator pulses are coupled into a 9 μm core, 1.3 km single-mode optical fiber. The combined self-phase-modulation and group-velocity-dispersion (GVD) produce a 300 ps linearly chirped pulse with a 3.5 nm bandwidth [6]. Because of the low dispersion of silica, a long fiber (1.3 km) is necessary to accumulate sufficient GVD and produce a long pulse with a highly linear chirp (see, for example, [7]). At this point, the pulses could be compressed to 1 ps using a double-pass grating compressor [6,8]. In the CPA technique, the chirped pulses are first amplified and only then compressed. By using this approach, 300 times more energy can be extracted than by directly amplifying a (compressed) 1 ps pulse. The chirped pulses are amplified in phosphate Nd:glass (Kigre Q98). In the absence of gain narrowing, the Nd:glass bandwidth (21 nm) allows for amplification of pulses as short as 100 fs. Energy in excess of 1 J is reached using three flashlamp-pumped amplifiers. The amplification system is described in detail in [9].

In order to fully exploit the potential of CPA, it is desirable to have the largest chirped pulse/compressed pulse ratio. The present embodiment relies on positive group velocity dispersion of the optical fiber and negative group velocity dispersion of the grating pair. The frequency chirp arising from dispersion within the fiber can be canceled by a grating pair only to the first order. Thus this combination allows for a compression ratio of a few hundred at most. To achieve higher compression ratios a new approach is necessary.

The group time delay τ from a grating pair in a double pass can be written to the second order as [10]

$$\tau(\lambda_0 + \Delta\lambda) = \tau_0 + 2\frac{\lambda_0}{c} L \frac{m^2}{a^2 \cos^3\theta}\Delta\lambda + \frac{3Lm^2}{ca^2\cos^3\theta}\left(1 + \frac{\lambda_0 m \sin\theta}{a\cos^2\theta}\right)(\Delta\lambda)^2 \qquad (1)$$

$$= \tau_0 + \tau_1 + \tau_2 \qquad (2)$$

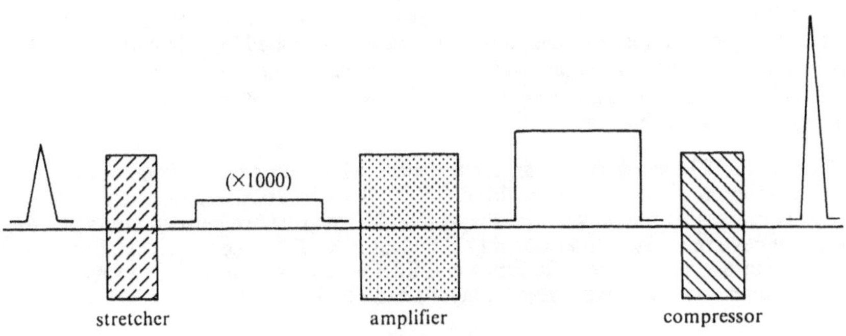

Fig. 1 The concept of Chirped Pulse Amplification

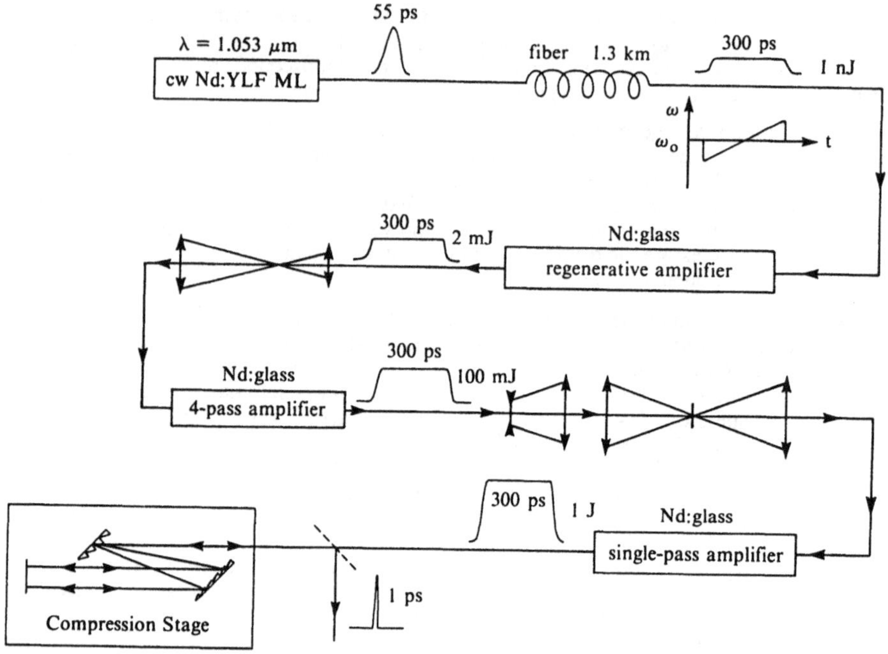

Fig. 2 Diagram of the current laser system

where $\lambda = \lambda_0 + \Delta\lambda$ is the wavelength, λ_0 is the central wavelength of the pulse, c is the speed of light, L is the distance between the planes containing the gratings, m is the diffraction order, a is the grating line spacing, and θ is the diffracted angle.

The ratio between the linear and the quadratic terms is given by

$$\frac{\tau_2}{\tau_1} = \frac{3}{2}\left(1 + \frac{\lambda_0 \, m \sin\theta}{a \cos^2\theta}\right)\frac{\Delta\lambda}{\lambda} \tag{3}$$

where τ_1 is the stretched pulse duration. The quadratic term, τ_2, has to be negligible compared to the Fourier transformed pulse duration τ_p, which could be produced by an ideal compressor. The pulse duration τ_p, and the pulse spectrum $\Delta\nu$ are simply related by the uncertainty expression $\Delta\nu \cdot \tau_p \simeq 0.5$. Due to the quadratic term, τ_p will be stretched by a factor $\sqrt{2}$ when τ_2 equals τ_p. Because the pulse spectrum $\Delta\nu$ is limited by gain narrowing and so can only be a fraction of the gain bandwidth $\Delta\nu_G$, it can be calculated from the expression (3) that for a $\Delta\nu/\Delta\nu_G = 0.2$, the maximum compression ratio τ_1/τ_p that can be obtained with a fiber-grating system is a few hundred.

MARTINEZ [11,12] has described an arrangement which uses a telescope between a grating pair in order to provide a net positive GVD. This offers the possibility of both stretching and compressing the pulses with dispersive devices that are perfectly matched to one another. We have recently demonstrated [13] the validity of this technique by stretching an 80 fs pulse from a dye laser by 1000 times and then recompressing it back to its initial value using this approach. The experimental layout and results are shown in Figs. 3 and 4, respectively.

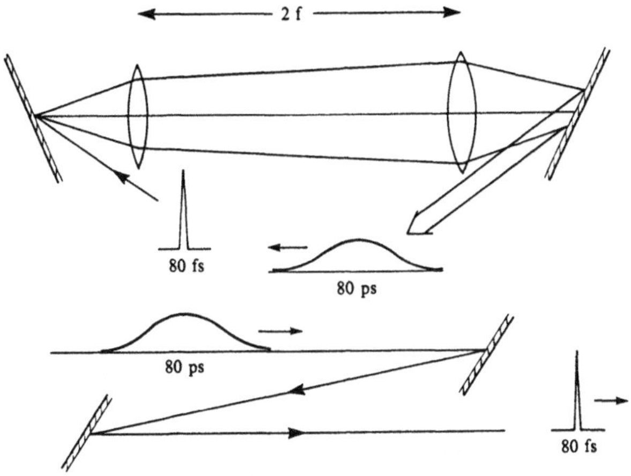

Fig. 3 Expansion-compression setup

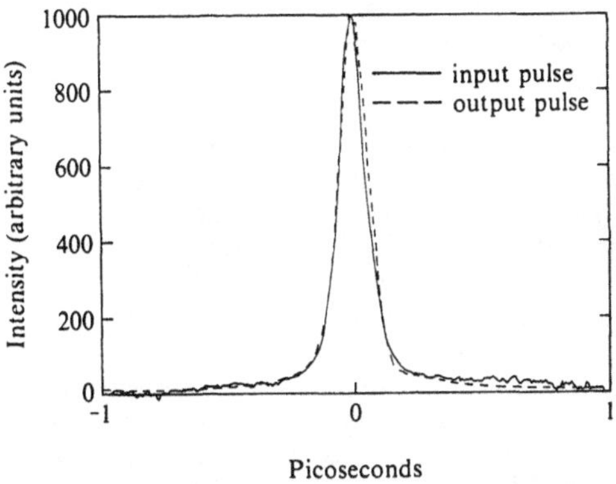

Fig. 4 Comparison of the auto-correlation traces of the input and output pulses after an expansion-compression of 1000

Alexandrite is a very promising material for high power amplifiers. With a saturation fluence of ~20 J/cm^2 [14] (~4 times higher than Nd:glass), it has great energy storage capabilities. In addition, with a lasing bandwidth stretching from 700–800 nm, Alexandrite can potentially support pulses as short as a few femtoseconds. These properties make Alexandrite an excellent choice for amplification of sub-picosecond pulses using expansion/compression techniques.

We have built a CPA source combining a synchronously pumped dye laser with a flash-pumped Alexandrite regenerative amplifier. The dye laser is described elsewhere [15]. The oscillator is tuned by a 6 μm thick uncoated pellicle over a 750–770 nm range. The pulse duration is on the order of 300 fs. The pulse expansion system consists of two 1800 l/mm gratings in an antiparallel configuration separated by 140 cm. Between the gratings are two 500 mm focal length lenses forming a unit magnification telescope. This grating configuration has previously been shown to provide a net positive GVD [12], with an effective grating separation of 60 cm. A 270 fs (FWHM) 2.9 nm input pulse to the expansion system results in a positively chirped pulse of 50 ps duration. The Alexandrite regenerative amplifier has been described elsewhere [16]. The output pulse has an energy as high as 3.5 mJ. The amplifier could be operated at repetition rates as high as 30 Hz. A standard double pass grating compressor is used to compress the pulse to nearly its original duration. The stretched pulse of about 60 ps is measured with a streak camera, while the compressed pulse is measured with a single-shot autocorrelator. The compressed width (FWHM) is 300 fs assuming a sech^2 pulse shape (Fig. 5).

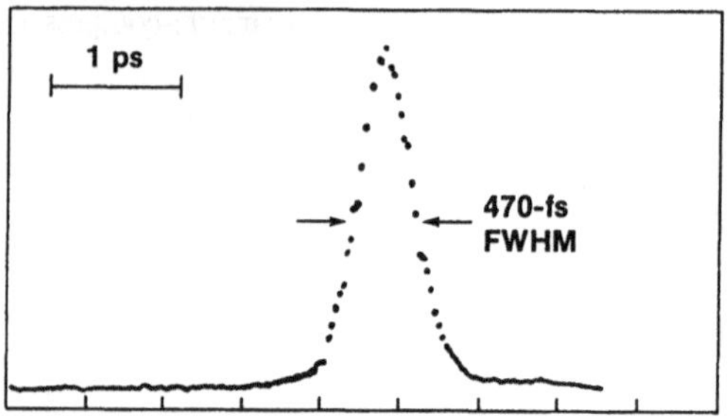

Fig. 5 Autocorrelation of the amplified pulse after Chirped Pulse Amplification

correlation width = 470 fs
pulse width = 305 fs (assuming $sech^2$ profile)

In conclusion, chirped pulse amplification has been used with Nd:glass amplifiers to produce single picosecond pulses at the terawatt level. The beam divergence is 2.2 times the diffraction limit, making the brightness of this source greater than 10^{+18} W/(cm^2 • sr), the highest brightness ever reported. Presently, the technique of CPA uses a combination of fiber and grating pair. This approach is limited by the grating pair higher order dispersion term to a compression ratio of a few hundred. A better embodiment of the chirped pulse amplification technique uses two grating pairs, one for pulse expansion and the other one for compression. With this approach, we have already demonstrated an expansion-compression ratio of 1000 and should be able to obtain a ratio as high as 5000. Although in Nd:glass gain narrowing presently forces us to amplify pulses of ~1 ps, shorter pulses could be obtained by artificially broadening the overall amplifier bandwidth using different types of glass. Even more exciting is the fact that CPA techniques can and will be implemented on very large Nd:glass amplifiers such as the NOVA system at Lawrence Livermore National Laboratory. With such a system, nanosecond pulses with kilojoule energies will be generated before being compressed to a single picosecond producing powers in the petawatt range. These pulses will produce power densities in the range of 10^{21} W/cm^2, corresponding to electric fields in excess of 100 e/a_0^2, and a blackbody energy density equivalent to 3 x 10^{10} J/cm^3. Such a source will have important applications in x-ray laser research and lead to fundamentally new classes of experiments in atomic physics, nuclear, solid-state plasma, and high-energy density physics.

This work was supported by AFOSR-URI, ARO, the U.S. Department of Energy Office of Inertial Fusion under agreement No. DE-FC08-85DP40200 and by the Laser Fusion Feasibility Project at the Laboratory for Laser Energetics which has the following sponsors: Empire State Electric Energy Research Corporation, New York State Energy Research and Development Authority, Ontario Hydro, and the University of Rochester. Such support does not imply endorsement of the content by any of the above parties. We thank the Laboratoire d'Optique Appliquée, E.N.S.T.A., Palaiseau, France, for the loan of the 16 mm amplifier, Allied-Signal for the loan of the Alexandrite laser, and Lawrence Livermore National Laboratory for the loan of the diffraction gratings. We also thank Corning for generously providing us with the single-mode fiber.

1. W. Koechner: In Solid-State Laser Engineering (New York, Springer-Verlag, 1976), p. 581.
2. L. G. Deshazer, Laser Focus 23, 54 (1987).
3. D. Strickland and G. Mourou, Opt. Commun. 56, 219 (1985).
4. D. Strickland, P. Maine, M. Bouvier, S. Williamson. and G. Mourou, In Ultrafast Phenomena V, Ed. by G. R. Fleming and A. E. Sigman, (New York, Springer-Verlag, 1986), p. 38.
5. P. Bado, M. Bouvier, and J. S. Coe, Opt. Lett. 12, 319 (1987).
6. H. Nakatsuka and D. Grischkowsky, Opt Lett. 6, 13 (1981).
7. H. Roskos, A. Seilmeier, W. Kaiser, and J. D. Harvey, Opt. Commun.61, 81 (1987).
8. J. Desbois, F. Gires, and P. Tournois, IEEE J. Quantum Electron. QE-9, 213 (1973).

9. P. Maine, D. Strickland, P. Bado, M Pessot, and G. Mourou, IEEE J. Quantum Electron. QE-24, 398 (1988).
10. J. D. McMullen, Appl. Opt. 18, 737 (1979).
11. O. E. Martinez, J. P. Gordon, and R. L. Fork, J. Opt. Soc. Amer. A 1, 1003 (1984).
12. O. E. Martinez, IEEE J. Quantum Electron. QE-23, 59 (1987).
13. M. Pessot, P. Maine, and G. Mourou, Opt. Commun. 62, 419 (1987).
14. J. C. Walling, D. F. Heller, H. Samelson, D. J. Harter, J. A. Pete, and R. C. Morris, IEEE J. Quantum Electron. QE-21, 1568 (1985).
15. M. Pessot, J. Squire, P. Bado, G. Mourou, and D. Harter, submitted to IEEE. J. Quantum Electron.
16. P. Bado, M. Pessot, J. Squire, G. Mourou, and D. J. Harter, IEEE J. Quantum Electron. QE-24, 1167 (1988).
17. See for example A. Siegman, Laser (University Science Books, Mill Valley, CA, 1986).

Generation of Intense 20-fs Pulses and Their Application in Multiphoton Ionization

P.B. Corkum, C. Rolland, and S.L. Chin***

National Research Council of Canada, Ottawa,
Ontario, K1A 0R6, Canada
*Present address: Bell Northern Research, P.O. Box 3511,
Station C, Ottawa, Ontario, K1Y 4H7, Canada
**Work performed on sabbatical leave from LORL,
Laval University, Quebec, P.Q., Canada

This paper reports the first high power compression of femtosecond pulses using bulk materials.[1] More than 100-μJ of output energy with pulse duration <22-fsec were achieved, an energy increase of more than four orders of magnitude over that available from fiber compression. It also reports an experimental investigation of multiphoton ionization of xenon and krypton with femtosecond pulses.[2]

High power compression is possible because material lengths required for optimum pulse compression have now become considerably shorter than the self-focusing length of typical amplified femtosecond pulses. In that case the beam profile cannot evolve significantly inside the material. Thus an appropriate portion of the beam can be selected after it emerges from the nonlinear medium. The approach is general and can be applied to a number of bandwidth limited ultrashort pulse lasers, notably CO_2 or excimer lasers.

Because of the relative ease of generating femtosecond pulses with dye laser we have chosen to demonstrate high power pulse compression at 625-nm. The amplified output of a colliding pulse dye laser[3] produced a near transform limited ($\Delta t\ \Delta\nu = 0.5$) 90 fsec pulse. This beam was spatially filtered by focusing the pulse through a 500 μm pinhole mounted in vacuum. The resulting ~300 μJ diffraction limited pulse was allowed to expand to a full-width at half-maximum (FWHM) beam diameter of 785 μm. It was then incident on a 1.2 cm thick quartz (CaF_2 was also used) window. An aperture mounted on an XYZ translator was placed immediately after the quartz to select a region of nearly constant intensity from the output beam. Compression was accomplished on a pair of 600 l/mm gratings.

Autocorrelation traces of the recompressed pulse are displayed in Fig. 1(a) and 1(b). Figure 1(b) shows the energy in the wings. The compression ratio of 5 is consistent with the increased bandwidth. The output energy measured immediately after the 200 μm aperture is 11 μJ, three orders of magnitude more than the energy obtained in fiber compression.

For larger apertures one should anticipate a reduction in the measured spectral width and a degradation in the autocorrelation trace since the small spectral broadening generated in the low-intensity

Springer Series in Chemical Physics, Vol. 48: **Ultrafast Phenomena VI**
Editors: T. Yajima · K. Yoshihara · C.B. Harris · S. Shionoya

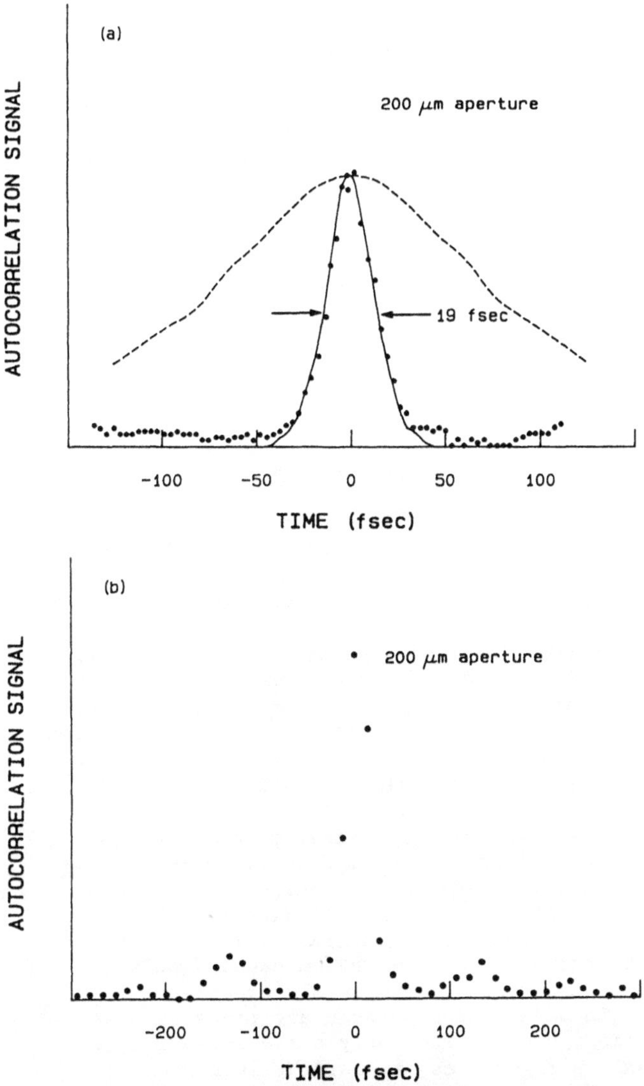

Fig. 1: Autocorrelation traces of a chirped and compressed pulse with a 200 μm aperture. The solid curve shows a $sech^2$ pulse envelope with a FWHM of 19 fsec. The intensity autocorrelation of the incident pulse (dashed line) is shown for comparison.

regions now contributes to the radiation transmitted through the pinhole. For a 700 μm aperture the spectral width is ~25% narrower than that observed through the 200 μm pinhole. In addition some spectral modulation is observed. The reduction in the spectral width is consistent with the slightly longer pulse duration. Pulse widths of 22 fsec are measured with an energy of 110 μJ. This represents 37% of the incident beam energy on the quartz. (With antireflection-coated quartz, the transmission could be increased to 42%).

For the 700 μm and smaller apertures, the beam divergence was dominated by diffraction from the aperture. A slit was placed in front of an OMA to sample a cross section of the beam. The far-field diffraction pattern of the attenuated beam (i.e., with no self-phase modulation) and the 22 fsec beam were compared under otherwise identical conditions. The divergence of the two beams are within 15% of the theoretical Fraunhoffer diffraction value.

With such good beam divergence, it is possible to use 22 fsec pulses to investigate high power phenomena. We have investigated multiphoton ionization of xenon and krypton with 22, 90 and 900 fsec pulses. The 90 fsec and 900 fsec pulses were produced by amplifying respectively, the output of a colliding pulse mode-locked or a synchronously pumped dye laser. Our aim was to investigate the pulse dependence of multiphoton ionization since ultrafast continuum experiments had indicated that atoms could be illuminated by surprisingly high intensity light (relative to extrapolation of long pulse experiments) before significant multiphoton ionization occurs.[4]

An achromatic lens focused the pulse into a vacuum chamber equipped with a time of flight ion spectrometer. The laser intensity was varied by reflecting the beam from a Brewster's angle germanium slab (polarizer) after it has passed through a $\lambda/2$ mica waveplate.

Figure 2 is a graph of the number of ions as a function of the laser intensity for both krpton and xenon (two gases used for continuum generation at higher pressures[4]) and for all three pulse durations. The solid lines were obtained from a modified Keldysh theory.[5] The high intensities that are characteristic of femtosecond multiphoton ionization in Fig. 2 confirm the conclusions in Ref. 4.

It is not clear why the intensities required for multiphoton ionization with femtosecond pulses are so high relative to extrapolations of those observed with 25 psec pulses.[6] Keldysh theories, which adequately fit the experimental data for the first ionization state, ignore the bound state internal structure of the atom. That this approximation is valid with femtosecond pulses should not be surprising. The bound state energy levels experience large and rapid AC Stark shifts. Thus dephasing between any transition and a harmonic of the laser field occurs after only a few optical cycles. Resonances that last only a few cycles are hardly resonances at all and they should contribute only marginally to the overall ionization rate with femtosecond pulses. Approximations that ignore transient resonances are much more uncertain with relatively long pulses.

There is a second issue that deserves comment. It has been noted[7] that if nonsequential ionization were to occur, it is most probable with ultrashort pulses. Thus 22 fsec pulses (only 11 optical cycles) with the associated high saturation intensity ($>10^{14}$ W/cm^2) should produce the most favorable conditions. A necessary, but not sufficient condition for nonsequential ionization is the observation at the same (below saturation) intensity of both the parent and daughter species. Experimentally we see, for example, the presence of both neutral and doubly ionized xenon at the same intensity for both 22 fsec and 90 fsec pulses. However, the modified Keldysh theory (solid curves in Fig. 1)

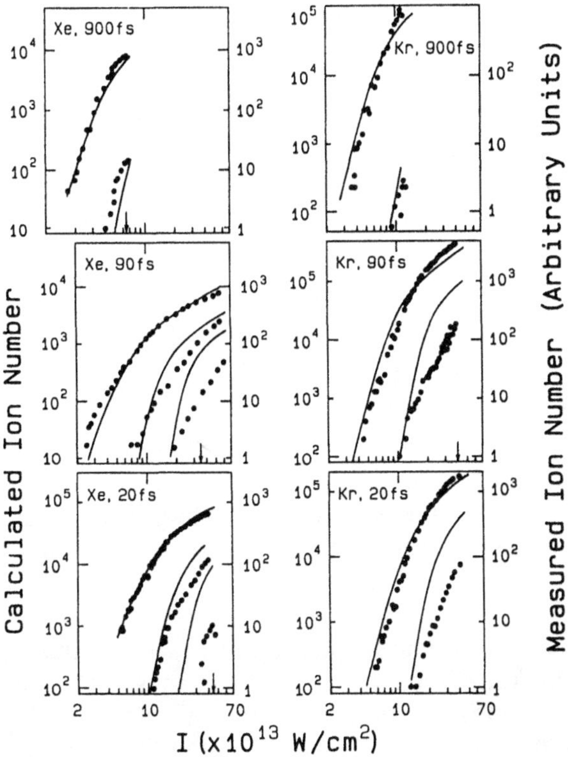

Fig. 2 Ion yield of Xe and Kr. Solid curves are calculated from a modified Keldysh theory.[5] The left curve on each panel is the singly ionized species. The arrows indicate the appearance intensity for ion species that were observed over too small an intensity range to plot.

which assumes sequential ionization, also predicts the simultaneous presence of all three species of these intensities. Although on purely experimental grounds we cannot eliminate the possibility of nonsequential ionization, we do not see significant enhancement of the highly ionized species compared with predictions of a sequential model.

References

1 C. Rolland and P.B. Corkum, J. Opt. Soc. Am. B5, 641 (1988).
2 S.L. Chin, C. Rolland, P.B. Corkum and P. Kelly, Phys. Rev. Lett. 61, 153 (1988).
3 C. Rolland and P.B. Corkum, Opt. Commun. 59, 64 (1986).
4 P.B. Corkum, C. Rolland and T. Srinivasan-Rao, Phys. Rev. Lett. 57, 2268 (1986); Ultrafast Phenomena V, eds. G.R. Fleming and A.E. Siegman, Springer-Verlag, New York (1986), p. 149.
5 A. Szoke, NATO ASI Series - Physics B, V. 171, A. Bandrauk ed., Plenum Press (1988) p. 207.
6 A. l'Huillier, L.A. Lompre, G. Mainfray, and C. Manus, Phys. Rev. A27, 2503 (1983).
7 P. Lambropoulos, Phys. Rev. Lett. 55, 2141 (1985).

Generation of Tunable 9-fs Optical Pulses in the Near Infrared

P.C. Becker, H.L. Fragnito, R.L. Fork, F.A. Beisser, and C.V. Shank

AT&T Bell Laboratories, Crawfords Corner Road,
Holmdel, NJ 07733, USA

We report the generation of tunable 9 femtosecond optical pulses in the near infrared (800-840nm). The pulses are produced by generating a continuum in a flowing jet of ethylene glycol with a focused intense 50 femtosecond pulse at 620 nm; selecting and amplifying a portion of that continuum in the near infrared[1]; then compressing the pulse in a fiber followed by a grating and prism compressor to generate 9 femtosecond pulses in the near infrared at an 8 kHz repetition rate.

The intial amplified 50 femtosecond pulses are produced by a colliding pulse modelocked laser followed by a dye amplifier pumped by a copper vapor laser (Plasma Kinetics model 251) operating at a repetition rate of 8 kHz[2]. The amplified pulses have energies of a few μJ and are centered at 620 nm. These pulses are then focused into a flowing stream of ethylene glycol with a microscope objective to generate a continuum covering the uv to the infrared[3]. A diagram of the experimental apparatus is shown in Figure 1. A prism filter is used to select the near infrared portion of the spectrum and to compensate the linear dispersion[4]. The input and output beams of the prism filter are displaced in the vertical plane. The infrared beam is then sent through a dye amplifier pumped by a copper vapor laser (Plasma Kinetics model 151) which consists mainly of reflective optical components, as shown in Figure 1. The dye used in the infrared amplifier is LDS 821 which amplifies ultrashort pulses in the 800-840 nm range. The saturable absorber is a 5 μm thick piece of GaAs[5]. The infrared pulses are amplified to 0.5 μJ with less than a percent of amplified spontaneous emission. Their duration is measured to be 60 femtoseconds using the second harmonic upconversion technique.

The amplified infrared pulses were compressed by passing the pulses through an 8 mm piece of single mode fiber. The pulse generated in the fiber is then compressed by a grating and

Springer Series in Chemical Physics, Vol. 48: Ultrafast Phenomena VI
Editors: T. Yajima · K. Yoshihara · C.B. Harris · S. Shionoya

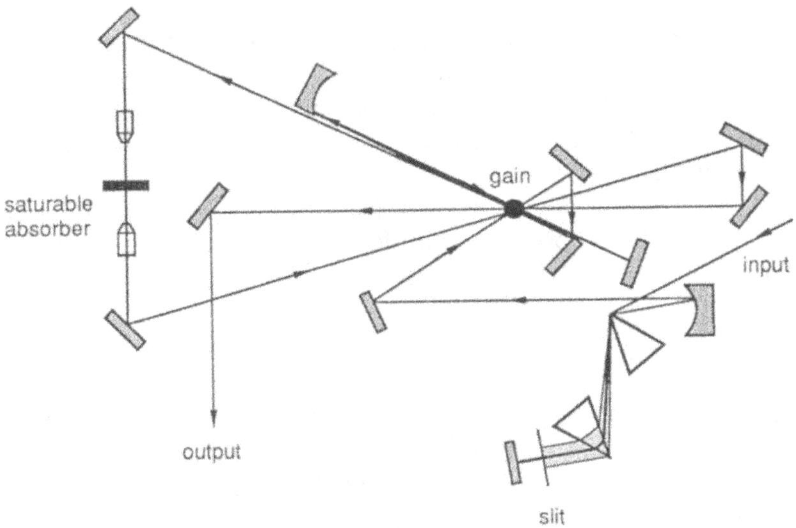

Fig. 1: Experimental arrangement of the prism filter and infrared amplifier.

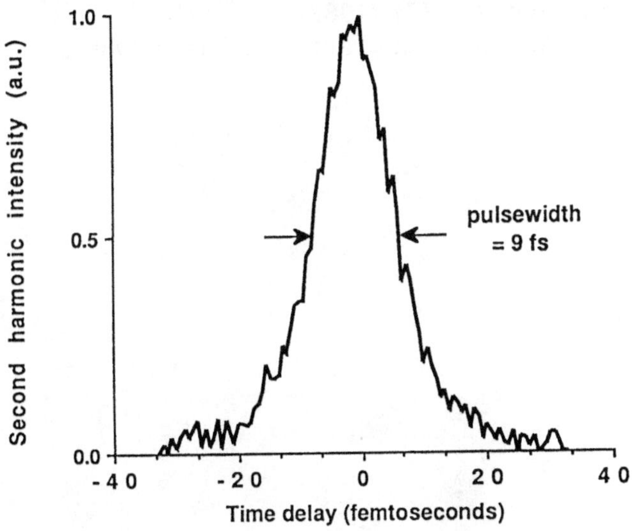

Fig. 2: Autocorrelation of the 9 femtosecond infrared pulse, with a center frequency of 820 nm.

prism pair compressor sequence, used previously to generate 6 femtosecond pulses at 620 nm[6]. The autocorrelation of the pulse, measured by the background free second harmonic upconversion technique, is shown in Figure 2. The pulse duration is determined to be 9 femtoseconds assuming a hyperbolic secant squared intensity profile.

The tuning range achievable with this scheme can be extended to cover the visible and infrared region of the spectrum by selecting the appropriate gain dye. We expect that the wavelength tunable ultrashort pulses produced with the help of this amplifier will find many applications in studying the dynamics of physical systems.

H.L. Fragnito has a postdoctoral fellowship from the Brazilian agency CNPq.

REFERENCES

[1] A. Migus, A. Antonetti, J. Etchepare, D. Hulin, and A. Orszag, J. Opt. Soc. Am. B **2**, 584 (1985).

[2] W.H. Knox, M.C. Downer, R.L. Fork, and C.V. Shank, Opt. Lett. **9**, 552 (1984).

[3] R.L. Fork, C.V. Shank, C. Hirlimann, R. Yen, and W.J. Tomlinson, Opt. Lett. **8**, 1 (1983).

[4]. R.L. Fork, Opt. Lett. **11**, 629 (1986).

[5] W.H. Knox, J. Opt. Soc. Am. **4**, 1771 (1987).

[6] R.L. Fork, C.H. Brito Cruz, P.C. Becker, and C.V. Shank, Opt. Lett. **12**, 483 (1987).

Amplification and Compression of 16-fs Spectrally Broadened Pulses to the Microjoule Level at 10 kHz

P. Georges[2], *J.P. Chambaret*[1], *F. Salin*[2], *G.R. Boyer*[1], *M.A. Franco*[1], *G. Le Saux*[2], *G. Roger*[2], *and A. Brun*[2]

[1]Laboratoire d'Optique Appliquée, Ecole Polytechnique, Ecole Nationale Supérieure de Techniques Avancées, F-91120 Palaiseau, France

[2]Institut d'Optique Théorique et Appliquée, CNRS UA.14, Centre Universitaire d'Orsay, F-91406 Orsay Cedex, France

Rapid advances have taken place in the compression of optical pulses using self-phase modulation in optical fibers. Pulses as short as 6 fs have recently been reported [2] . Unfortunately, the single mode fiber used limits the compressed pulse energy to a few nanojoules. These pulses can only be used as a probe in spectroscopic experiments in which the temporal resolution is still limited around 50 fs by the high power pump pulse duration. In this paper, we report the generation of less than 20 fs pulses with energy in the microjoules range. The principle is the following : an optical pulse is spectrally broadened by self-phase modulation in a single mode fiber ; the output pulse is amplified in a multipass amplifier and then compressed by a sequence of four prisms. The obtained energy level is sufficient to generate non-linear phenomena such as spectral continuum.

We will present consecutively two experiments using this technique. The first one was performed at the "Institut d'Optique" and produces 10 µJ, 20 fs pulses at low repetition rate (50 Hz). The other one was performed at the "Laboratoire d'Optique Appliquée" and leads to 16 fs microjoule pulses at high repetition rate (10 kHz).

The experimental set-up is shown in fig. 1.

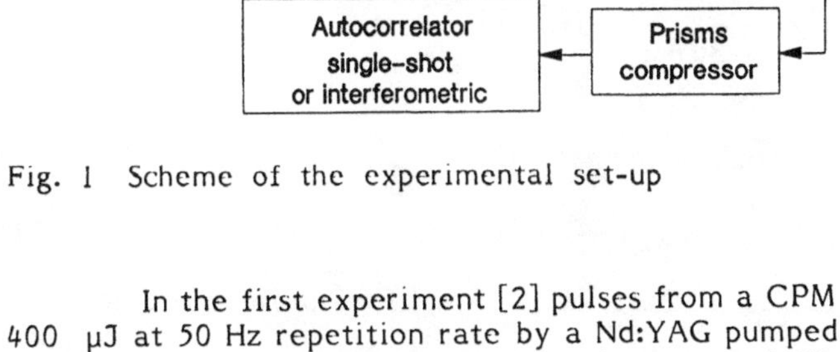

Fig. 1 Scheme of the experimental set-up

In the first experiment [2] pulses from a CPM laser are amplified to 400 µJ at 50 Hz repetition rate by a Nd:YAG pumped amplifier. A few tens of nanojoules are coupled in a 1.5 cm long optical fiber. At the fiber output

Springer Series in Chemical Physics, Vol. 48: **Ultrafast Phenomena VI**
Editors: T. Yajima · K. Yoshihara · C.B. Harris · S. Shionoya

the pulse spectrum has been broadened by self phase modulation ; this broad spectrum (40 nm) is then injected in a multipass amplifier pumped by the Nd:YAG laser. Many dyes were tried in order to obtain the widest gain bandwidth centered on the pulse spectrum. Typical amplification bandwidths were about 30 nm. To increase the width of the gain curve we used several dyes mixture. The "red" dye must have an absorption band which does not overlap the fluorescence band of the "yellow" dye. We have found experimentally that DCM is the best candidate for amplification of the red part of the spectrum. When mixed with Rhodamine B, we have observed gain bandwidth exceeding 40 nm in our Nd:YAG pumped multipass amplified with total energy gain of one thousand [2]. The pulse at the output of the amplifier has an energy of about 10 microjoules and a duration of about 3 ps. We then compress this pulse using high index prisms and measure its duration with a single shot autocorrelator. The amplifier pulses can be compressed down to 20 fs. Figure 2 shows the amplified pulses spectrum and single-shot autocorrelation trace. The pulse peak power (0.25 GW) is sufficient to generate a spectral continuum in a 200 μm thick glass plate.

The second experiment [3] was carried out using optical pulses generated in a colliding-pulse mode locked laser and amplified at a 10 kHz

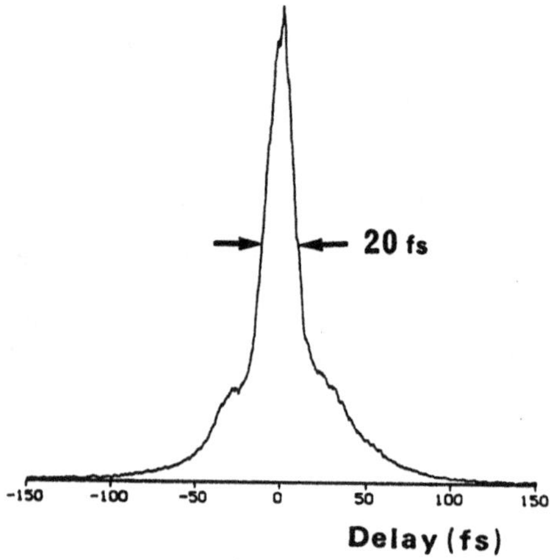

Fig. 2

Single shot autocorrelation

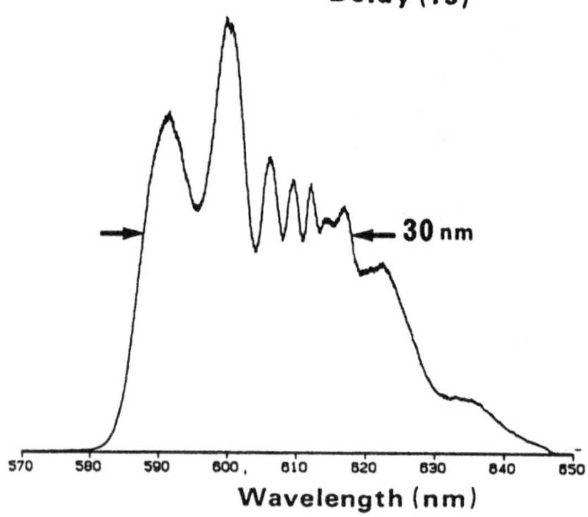

Amplified pulse spectrum

repetition rate in a copper vapor laser-pumped amplifier to energy of about 2 µJ [4]. Group velocity dispersion in the amplifier was compensated using a four prisms arrangement. The amplified pulses had durations of 50 fs and a 10 nm wide spectrum centered at 614 nm. A small fraction of the amplified pulse energy was coupled into a 11 mm long single mode fiber. The pulse spectrum was then broadened by self-phase modulation up to a width of 60 nm (FWHM). The pulse energy at the fiber output was typically 3 nJ and due to the chirp its duration was on the order of .3 ps. The beam was then injected in a multipass amplifier pumped by 5 W of the green line of the copper vapor laser. In this experiment, we also used a mixture of DCM and Rhodamine 610. After 5 passes in the dye jet, the pulse energy has increased to about 1 µJ (x 300 gain). The output pulse spectrum was similar to the input one and widths of 50 nm were typically obtained (Fig. 3). We notice a hole in the gain curve near 600 nm. These amplified chirped pulses were then compressed by a sequence of four high index prisms (SF10 from Schott). An interferometric autocorrelator was used in order to measure the pulse duration. The shortest pulses that could be obtained using these prisms were 16 fs (Fig. 3). The compressed pulse energy was decreased by losses in the compressor to about 0.5 µJ. These pulses are sufficiently powerful to create a spectral continuum when focused in a 1 mm thick glass plate with a 50 mm lens (estimated power density 10^{12} W/cm²).

By replacing the prisms by a two pass grating pair we obtained less than 12 fs pulses but at the expense of much greater losses.

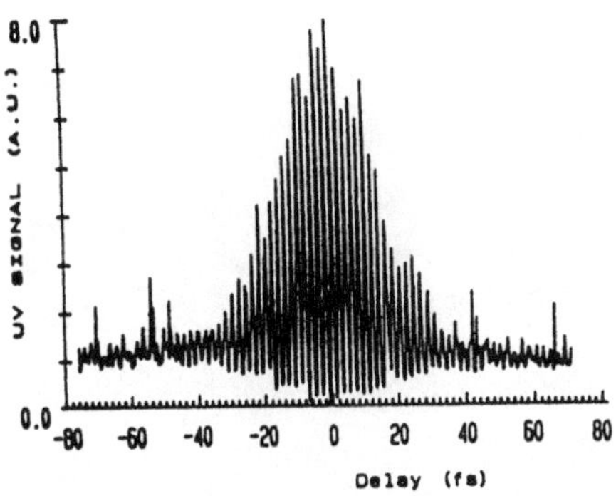

Fig. 3

Interferometric autocorrelation

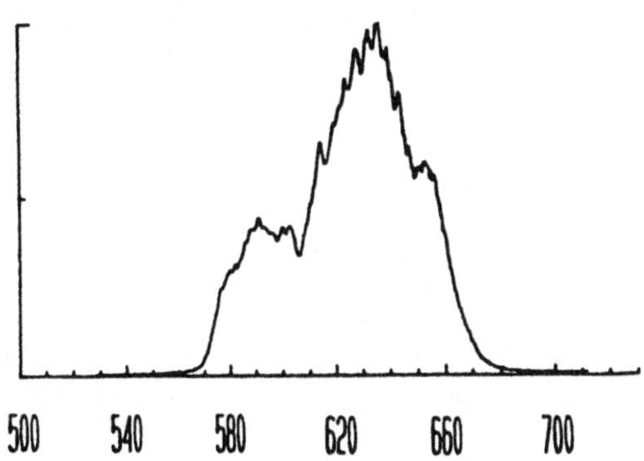

10 kHz amplified pulses spectrum

In both experiments, the amplified beam profile was nearly diffraction limited. This allowed to obtain power density higher than 10^{14} W/cm².

This work was partially supported by Direction des Recherches, Etudes et Techniques (Contract # 86/089 and # 86/0184), Ministère de la Recherche Scientifique and Centre National d'Etude des Télécommunications (Contract # 86/1B/161).

References :

[1] R.L. Fork, C.H. Brito Cruz, P.C. Becker and C.V. Shank. Opt. Lett. 12, 483 (1987).
[2] P. Georges, F. Salin, G. Le Saux, G. Roger and A. Brun. Opt. Comm. (1988) (in press).
[3] G. Boyer, M. Franco, J.P. Chambaret, A. Migus, A. Antonetti, P. Georges, F. Salin and A. Brun. Appl. Phys. Lett. (1988) to be published.
[4] W.B. Knox, M.C. Downer, R.L. Fork and C.V. Shank. Opt. Lett. 9, (1984).

Generation of High-Repetition-Rate 14-fs Amplified Pulses Using a Q-Switched CW YAG Pumping Laser

Y. Ishida, T. Tokizaki, and T. Yajima

Institute for Solid State Physics, University of Tokyo,
Roppongi, Minato-ku, Tokyo 106, Japan

For spectroscopy in the extremely short time region, it is required to have well-controlled ultrashort light pulses, not only in shortening the pulse width but also with regard to stability, tunability, peak power and, in addition, ease in maintenance and handling.

For this purpose, we have developed a relatively simple and practical ultrashort-pulse dye-laser-based system. The stable basic laser source is a cavity-dumped, hybridly mode-locked CW dye laser (rhodamine 6G/DQOCI) pumped by a mode-locked Ar laser coupled with a single-stage fiber compressor, which can generate transform-limited, tunable femtosecond pulses (≃100 fs) with no wings and also clean linear chirped pulses [1,2].

In order to obtain extremely short pulses and largely broadened spectrum with a second optical fiber compressor, the tunable femtosecond pulses were further amplified with a single-stage dye amplifier (20 mm cell length) longitudinally pumped by internal SHG pulses from a CW Nd:YAG laser Q-switched at repetition rates up to 1 kHz. The average power of the SHG pulses was 400 mW at 1 kHz and the pulse width was 45 ns.

A gain of 1×10^3 can be obtained in the wavelength range of 595-630 nm for a rhodamine 640 dye solution (3×10^{-5} M). The maximum peak power of the amplified pulses, without broadening by the use of a mixed saturable absorber (DQOCI and brilliant green), was about 500 kW at 603 nm. The fluctuation of the amplified pulses was about ± 15 %, and was attributed to the time jitter between the dye laser and pump laser pulses. High coupling efficiency of over 50 % for the fiber (4 μm core diameter) was attained due to the good mode pattern of the amplified pulses. The spectrum of the output pulses from the fiber (15 mm length) was efficiently broadened to 50-90 nm for the power density of 1-2 TW/cm^2 in the fiber. For 35 mm fiber length the spectral width became 143 nm.

Figure 1(a) shows a single rapid-scanned SHG autocorrelation trace of the shortest pulse compressed by using a grating-pair (600 ℓ/mm). The figure indicates a trace of only one scanning of the SHG correlator which is operating at 33 Hz and is triggered by the synchronizing clock pulses at a repetition rate of 1 kHz. The trace consists of several discrete signals corresponding to the sampled output of compressed pulses at 1 kHz. The correct correlation trace of the pulse is given by the envelope (dotted curve), and is fitted by a sech^2 shape of 14 fs pulse width accompanied by small wings. This type of rapid scanning is effective for monitoring the instantaneous behavior of the compressed pulses. It was found that the fluctuation of the compressed pulse width (± 25 %) occurs due to the fluctuation of the input pulse intensity which changes the condition of the optimum phase compensation of the compressor. The large spectral broadening of 90 nm shown in Fig. 1(b) (corresponding to that for 4 fs transform-

Springer Series in Chemical Physics, Vol. 48: Ultrafast Phenomena VI
Editors: T. Yajima · K. Yoshihara · C.B. Harris · S. Shionoya

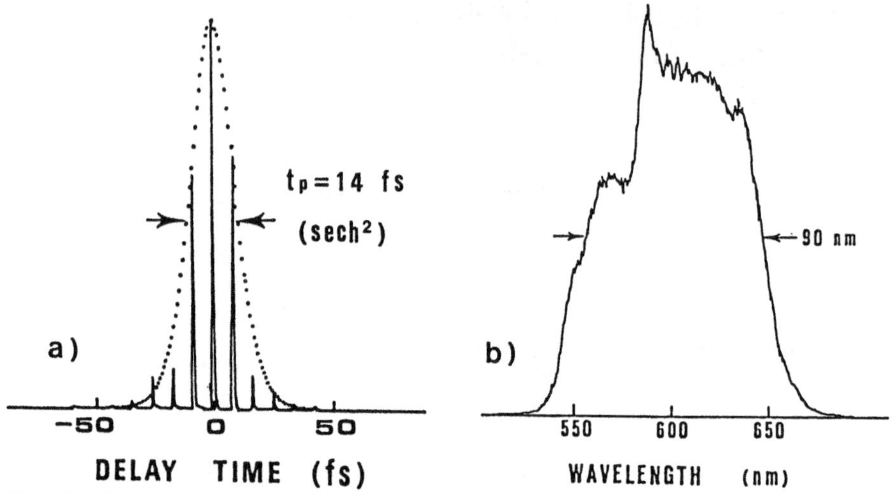

Fig. 1 (a) Single rapid-scanned SHG autocorrelation trace of the shortest compressed pulse. (b) Spectrum of the pulse for (a).

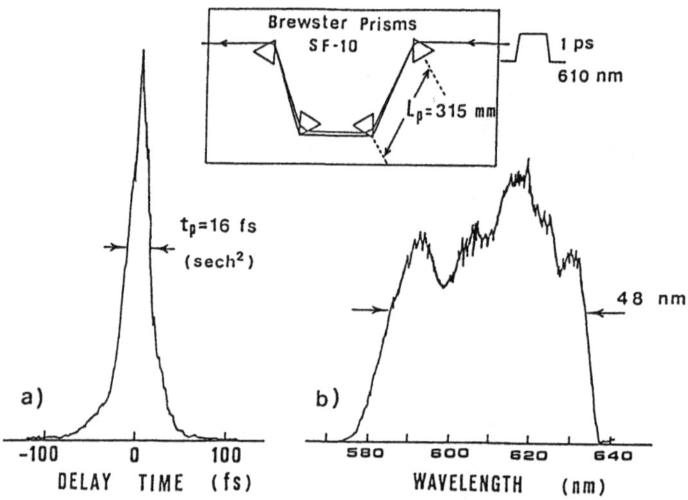

Fig. 2 (a) Slow-scanned SHG autocorrelation trace of the pulse compressed by using prism-pairs. (b) Spectrum of the pulse for (a).

limited $sech^2$ pulse) indicates that the measured pulses were not yet transform-limited even at the shortest width.

As a practical dispersive delay line for pulse compression, we alternatively used two Brewster prism-pairs (SF-10 glass) with a high transmission rate instead of the grating-pair. A slow-scanned SHG autocorrelation trace of the pulse compressed by using the prism-pairs under the conditions of 12 mm fiber length and 1 TW/cm^2 power density in the fiber is shown in Fig. 2(a). The pulse width of 16 fs (assuming a $sech^2$ shape) around 610 nm was obtained with small wing. The spectral width of the pulse was 48 nm (FWHM) as shown in Fig. 2(b). In comparison with the grating-pair compressor, the prism-pair compressor increases the peak power of the compressed pulse about two times due to low energy-loss, and is easy in alignment and in adjustment.

Our numerical calculations by nonlinear Schrödinger equation indicate that the effect of nonlinear chirp in the pulse arising from the cubic-order dispersion and the shock term in the fiber [3] becomes important under our experimental conditions, and, as a result, increases largely the side lobes in the compressed pulse. The asymmetric spectrum as well as the pulse shape measured in our experiment is also explained with these effects. For further shorter pulse generation, it is important to consider the nonlinear chirp in the fiber. However, an SHG autocorrelation trace of the broadened pulse before compression implied that the chirping is not far from linear. It could therefore be expected to realize much shorter pulses by careful phase compensation in consideration of all dispersion effects in the measurement system.

The features of our laser system using a high repetition-rate CW Nd:YAG pumping laser are the stable operation for a long time (e.g., maintenance free), the ease in handling and the compactness in comparison with systems using other pumping lasers. The present system will serve as a practical femtosecond light source for ultrafast spectroscopy around 10 fs region. The broadband linear chirped pulses produced by this system are also useful for a new type of transient nonlinear spectroscopy [2,4], where the time resolution is governed by the light coherence time much shorter than the pulse width.

References

1. Y. Ishida and T. Yajima: Optics commun. 58 (1986) 335.
2. Y. Ishida and T. Yajima: Revue Phys. Appl. 22 (1987) 1629.
3. W. Zhao and E. Bourkoff: IEEE J. Quant. Electron. 24 (1988) 365.
4. T. Yajima and N. Morita: In *Methods of Laser Spectroscopy*, ed. by Y. Prior, A. Ben-Reuven and M. Rosenbluh (Plenum Publishing Corp. 1986) p.75.

10 kHz-Rate Amplification of 40-fs Optical Pulses at Low Pumping Energy

E.V. Khoroshilov, I.V. Kryukov, P.G. Kryukov, and A.V. Sharkov

P.N. Lebedev Physical Institute, USSR Academy of Sciences, SU-117924 Moscow, USSR

One of the most promising approaches to obtaining amplified femtosecond pulses is the use of a multipass jet amplifier pumped by a pulsed copper vapour laser (CVL) /1/, or Nd-YAG laser /2/. The amplification of 2.5 x 10^3 /1/, or about 4 x 10^3 /2/ was achieved at pump pulse energy of 2 mJ. In the paper by KNOX et al. /1/ such a pulse energy corresponds to 10 W of the CVL average power at 5 kHz repetition rate.

To amplify 40-fs pulses at 610 nm of a colliding-pulse mode-locked (CPM) dye laser we have used a two-mirror telescopic optical scheme of an amplifier (Fig. 1), which enabled us to obtain the gain greater than 10^4 at CVL pumping power less than 1 W with pulse repetition rate of 10 kHz.

A 200 µm -thick jet of dye dissolved in ethylene glycol is located in the common focal point between the two mirrors with 120 and 150 mm radii. An amplified pulse, by propagating between the mirrors, illustrates a principle of an unstable cavity. With each next pass the beam is propagated closer to the optical axis of the cavity, and after a certain number of passes leaves the amplifier through the hole in the mirror.

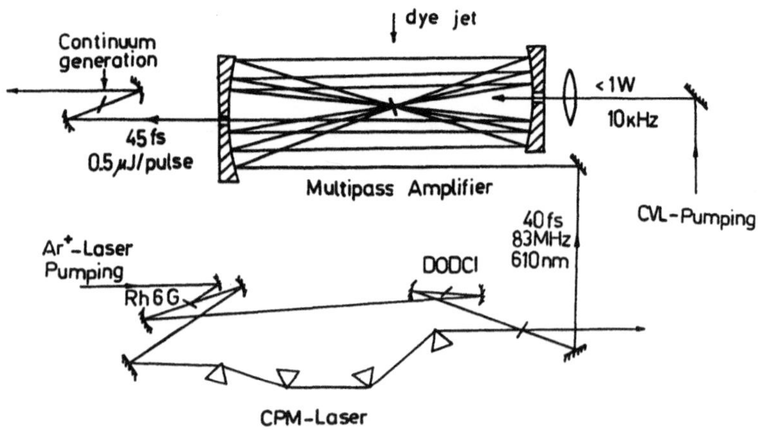

Fig. 1. Schematic diagram of the amplifier.

Springer Series in Chemical Physics, Vol. 48: Ultrafast Phenomena VI
Editors: T. Yajima · K. Yoshihara · C.B. Harris · S. Shionoya

In our experiment we have realized six passes through active medium. An amplifying pulse propagates at a small angle to the pumping pulse. The maximum angle is 12° for the first pass and for subsequent passes the geometrical matching between the gain volume and the focal volume of the amplifying beam is better than for the first one. Utilization of a thin jet is also optimal for a good geometrical matching. Such an optical scheme allowed us to focus the pump radiation into a spot of 0.3 mm diameter by a lens of 25 mm focal length and to obtain high gain at small pumping energy.

The focal diameter of the amplified beam in such a telescopic system is 3 times greater at the sixth pass than that at the first pass (120 and 40 μm, respectively), which seems optimal, provided the amplification saturation takes place.

We have studied amplification in a dye N 160 /3/ absorbing mainly the pump line at 511 nm and having the gain maximum near 600 nm. The gain of 1.2×10^4 was achieved, and the energy of the amplified pulse was 0.5 μJ. Average power of the CVL was 0.7 W at 511 nm (70 μJ/pulse).

Since we used a thin jet and a total path of 1.5 mm the amplifying medium at 6 passes, the pulse duration did not actually change and was equal to 45 ± 5 fs. The beam quality remains close to diffractional. At focusing the amplified pulses to a 2 mm-thick silica plate we have observed the generation of a femtosecond continuum throughout 380-720 nm. The amplification of pulses up to the energy sufficient for a continuum generation, at pump power < 1 W, makes it possible to use the rest of CVL radiation for the amplification in the second analogous amplifier of the chosen spectral part of the continuum from green to infrared. We believe the described method is one of the most simple and effective for the amplification of femtosecond pulses.

1. W.H. Knox, M.C. Downer, R.L. Fork, C.V. Shank: Opt. Lett. 9, 552 (1984).
2. C. Hirlimann, O. Seddiki, J-F. Morhange, R. Mounet, A. Goddi: Opt. Comm. 59, 52 (1986).
3. Yu.A. Matveetz, C.V. Chekalin, A.V. Sharkov: J. Opt. Soc. Am. B 2, 634 (1985).

Generation of 29-fs Pulses from a Synchronously Pumped Dye Laser and Its Cavity-Dumping Technique

M. Nakazawa, H. Kubota, and K. Kurokawa

NTT Transmission Systems Laboratories, Tokai,
Ibaraki-ken 319-11, Japan

With a colliding pulse mode-locking (CPM) technique and an intracavity group velocity dispersion (GVD) control by Brewster-angled prisms, pulse widths of dye lasers have been reduced well into the femtosecond region. [1,2] We have reported 65 fs pulse generation from a synchronously-pumped dye laser with GVD compensation, in which the possibility of generation of femtosecond pulses below 50 fs was discussed.[3] Here, we newly report the generation of 29 fs pulses at 615 nm by adopting Kiton Red S (KRS) instead of Rh6G. We also report the generation of 190 kW, 48 fs pulses with the use of a cavity dumper and a high repetition, high power femtosecond laser system which generates 10-20 MW, 15 fs pulses at a repetition rate of 5 kHz.

The cavity configuration of our synchronously pumped femtosecond dye laser is shown in Fig. 1. The laser has five mirrors, two dye jets, four Brewster-angled prisms, and an adjustable spatial filter. The output mirror is mounted on a precision translation stage, so that the cavity length can be adjusted to a pump repetition rate within 0.1 μm resolution. The laser is pumped by a frequency-doubled cw Nd:YAG laser with 1.3 W average power, 80 ps width, and 76 MHz repetition. The dye concentration in ethylene glycol is about 2×10^{-3} mol/l for KRS, 8×10^{-5} mol/l for DODCI, and 1.6×10^{-4} mol/l for DQOCI. The dye jet thickness for gain is 400 μm and that for the saturable absorber is 100 μm.

In the case of usual combination of Rh6G and DODCI/DQOCI, the typical oscillation wavelength region is around 580 nm, and this short wavelength component is interrupted with a spatial filter to achieve oscillation at 615 nm. For a combination of KRS and DODCI/DQOCI, however, the laser tends to oscillate near 630 nm, and therefore the suppression of longer wavelength component is required.

The shortest pulses thus obtained in a background-free auto-correlation trace are shown in Fig. 2. The full width at half maximum of the pulse

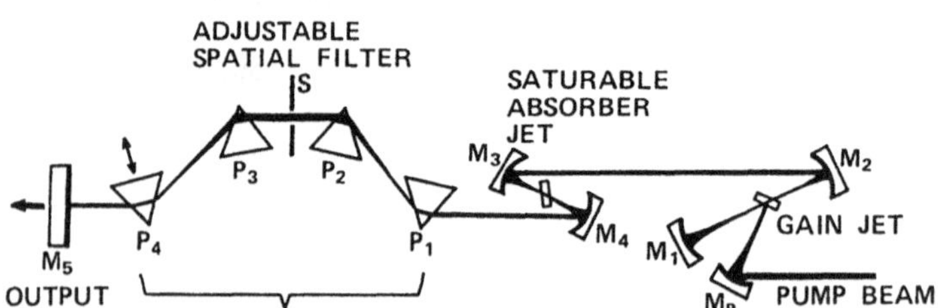

Fig. 1
Laser-cavity configuration incorporating a four-prism sequence and a spatial filter to change the oscillation wavelength.

Springer Series in Chemical Physics, Vol. 48: **Ultrafast Phenomena VI**
Editors: T. Yajima · K. Yoshihara · C.B. Harris · S. Shionoya

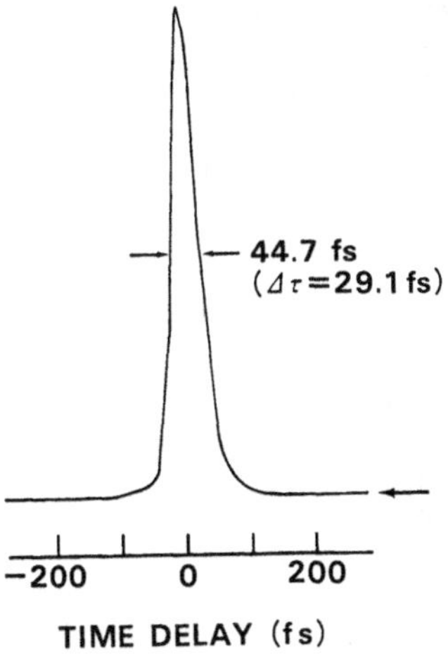

Fig. 2 Autocorrelation trace of the laser output.

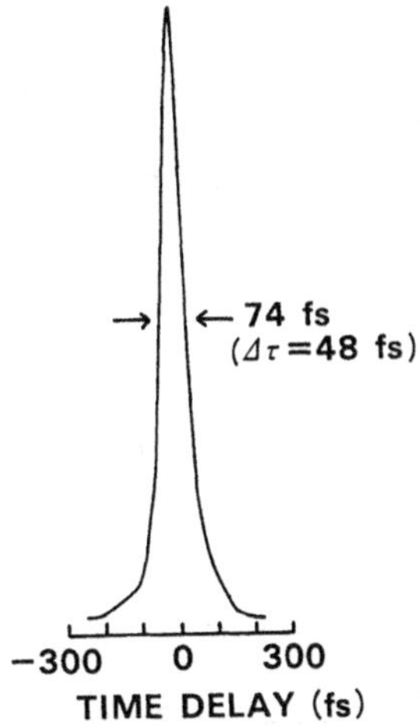

Fig. 3 Autocorrelation trace of the cavity dumped dye laser.

width was 44.7 fs, which corresponded to a pulse width of 29.1 fs for a $sech^2$ pulse shape. The average output power was 60 mW at a repetition rate of 76 MHz, resulting in a peak power as high as 27 kW. The amplitude stability was ±10 % and becomes more stable when the pulse is set to 35 fs.

In the next step, we describe a synchronously pumped, cavity dumped dye laser which can emit high power short pulses with a high repetition rate. Because of its simple configuration which do not employ a dye amplifier, it is very advantageous in studying ultrafast phenomena.

The cavity dumped dye laser is achieved by replacing mirror M_5 with a double-pass cavity dumper. Two kinds of dyes, Rh6G and KRS, were used for the gain medium. A mixture of DODCI and DQOCI was also used for the saturable absorber. In the case of Rh6G, pulses as short as 240 fs were generated with a peak power of 55 kW at 3.8 MHz. The center wavelength of the oscillation was 615 nm. The pulses were compressed to 56 fs with a peak power of 188 kW by using a pulse compressor, which consisted of an optical fiber and a pair of Brewster-angled TeO_2 prisms. In the case of Kiton Red S, 48 fs, 190 kW pulses were generated at 619 nm, as shown in Fig. 3. The pulses were further compressed to 30 fs with a peak power of 130 kW with the same pulse compressor. The duration of 30 fs quite agrees with the minimum pulse width attainable by compression using TeO_2 prisms.

Nonlinear optical phenomena often require large pulse intensities. In a pulse compression technique, high peak power is essential to generate short pulses.[4] Here, we describe a femtosecond laser system capable of emitting 10-100 μJ/pulse with a 5 kHz repetition rate at 620 nm by using the above mentioned cw dye laser. Pulses as short as 10-20 fs have been obtained by a fiber grating-pair compressor.

The femtosecond laser system has three major components which consist of an oscillator, an amplifier, and a compressor, as shown in Fig. 4. The oscillator is a synchronously pumped dye laser with a four-prism GVD controller and a spatial filter as mentioned above. 100 fs, 1 nJ pulses from this dye laser were introduced into the dye amplifier. The dye medium for the laser amplifier was KRS in methanol and this was filled into fused

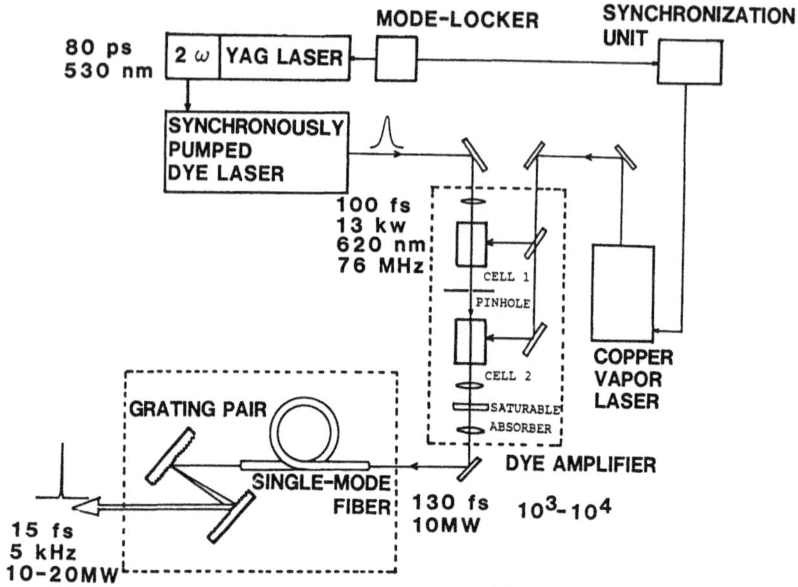

Fig. 4 Schematic view of the femtosecond laser system.

silica vessels which were connected to 5 liter reservoirs. Usually, a dye stream with relatively thick jet (>0.5 mm) is adopted as an amplifier.[5] Here we used dye cells with a high speed flow, i.e. high speed flow cells, to overcome a heat-up problem due to high power from a copper vapor laser. The cuvette length of both cells was 2 cm. They were pumped with the copper vapor laser which produced 30-ns pulses with 5 mJ/pulse at a 5 kHz repetition rate. A Corning glass plate (color filter number CS2-60) was installed after the second stage as a solid state saturable absorber so as to eliminate amplified spontaneous emission. The pulse broadened to 130 fs after passing through the two stage amplifier. This broadening, however, was compensated for by a negative group velocity dispersion effected by a grating pair or a prism pair.

The amplified pulses were focused into a 12 mm long single-mode optical fiber with a core diameter of 4 μm. The chirped pulse from the fiber was collimated with a 20x microscope objective and was introduced into a grating pair with 600 line/mm gratings, where the distance between the gratings was set to 20 mm. Thus, pulses as short as 15 fs with 0.25 μJ/pulse have been achieved at a repetition rate of 5 kHz.

References

1. R. L. Fork, B. I. Greene, and C. V. Shank: Appl. Phys. Lett. 38, 671 (1981)
2. R. L. Fork, O. E. Martinez, and J. P. Gordon: Opt. Lett. 9, 150 (1984)
3. M. Nakazawa, T. Nakashima, H. Kubota, and S. Seikai: Opt. Lett. 12, 681 (1987)
4. W. J. Tomlinson, R. H. Stolen, and C. V. Shank: J. Opt. Soc. Am. B1 139 (1984)
5. W. H. Knox, M. C. Downer, R. L. Fork, and C. V. Shank: Opt. Lett. 9 552 (1984)

Colliding Pulse Mode-Locked Femtosecond Laser Using Binary-Energy-Transfer Gain Dye Mixture

M. Mihailidi, Y. Budansky, X.M. Zhao, Y. Takiguchi, and R.R. Alfano

Institute for Ultrafast Spectroscopy and Lasers,
Photonics Application Laboratory,
Departments of Physics and Electrical Engineering,
The City College of New York, NY 10031, USA

We report on femtosecond pulse generation using binary gain dye mixture of Rhodamine 590 and Kiton Red in an improved cavity design. The absorption spectrum and the emission spectrum of the dyes overlap to ensure energy transfer. The overall fluorescence spectrum of the mixture shifts to the red, thus improving the CPM laser performance characteristics for short pulse generation. The new cavity design consists of a quasilinear CPM ring. This results in better intracavity chirp compensation to routinely generate pulse durations of ~30 fs. A bandwidth of 18.7 nm is achieved which exceeds the best results so far by 17% and a symmetric shape which indicates good chirp compensation [1].

In all recent experimental work the number of dielectric mirrors has been reduced to six to minimize mirrors' contribution to cavity dispersion but the incidence angles have remained considerably large. We solved this problem by a near linear ring cavity configuration for the CPM laser. The mirror incidence angles have been greatly reduced resulting in almost normal incidence. Thus, we have avoided astigmatic distortions introduced by oblique angles of incidence that limit the performance of the system [2]. For multilayer dielectric mirrors the dependence of the reflection coefficient on the angle of incidence is the smallest for normal incidence. Our configuration has minimized this dependence, which is important since small deviations of the incidence angles from the optimum for which the mirror has been coated for, causes chirping of the pulse for frequencies at bandwidth edge. This effect becomes significant when the bandwidth is large [3,4].

We recorded the laser pulse autocorrelation functions and the corresponding spectra as a function of Kiton Red and Rhodamine 590 concentration in the gain medium, while keeping the saturable absorber concentration constant (Fig. 1.). The corresponding fluorescence spectra of the binary-energy-transfer gain dye mixture was also measured (Fig 2.). Routinely achieved operation results, which were obtained when the DODCI concentration was also optimized, are presented in Fig. 3. DODCI was used as the standard saturable absorber

Springer Series in Chemical Physics, Vol. 48: **Ultrafast Phenomena VI**
Editors: T. Yajima · K. Yoshihara · C.B. Harris · S. Shionoya

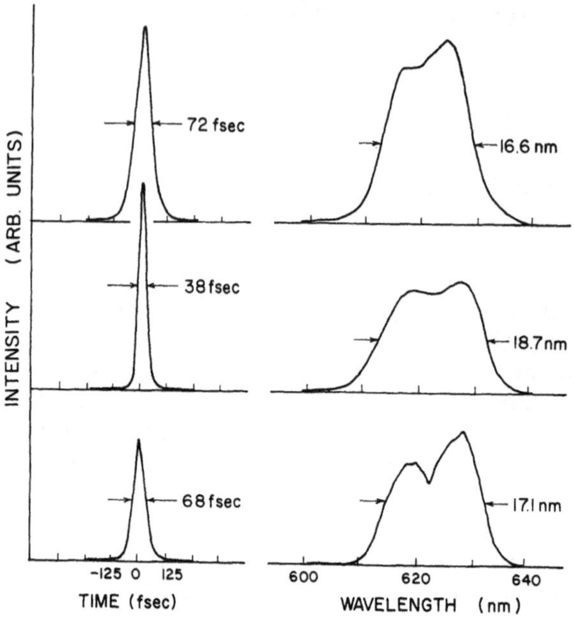

Fig.1. Autocorrelation functions and their corresponding spectra presented for three different Kiton Red concentrations in Rhodamine 590 - Ethylene Glycol solution (Rhodamine 590; 1.6×10^{-3} mol/l): (a) No Kiton Red at all, (b) 3.9×10^{-5} mol/l of Kiton Red, (c) 7.1×10^{-5} mol/l of Kiton Kiton Red

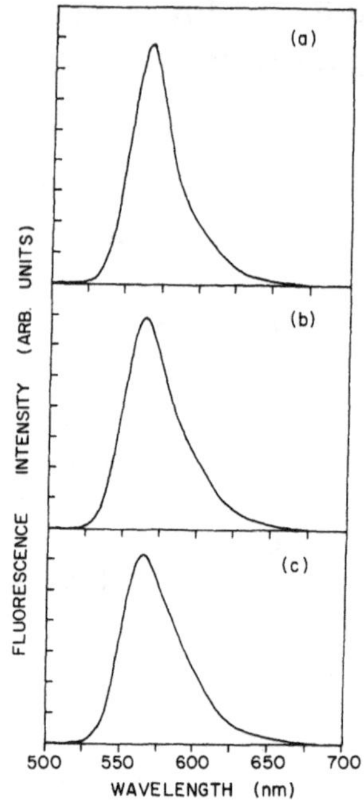

Fig. 2. Fluorescence spectra of the binary energy transfer gain dye mixture corresponding to the results presented in Fig.1.

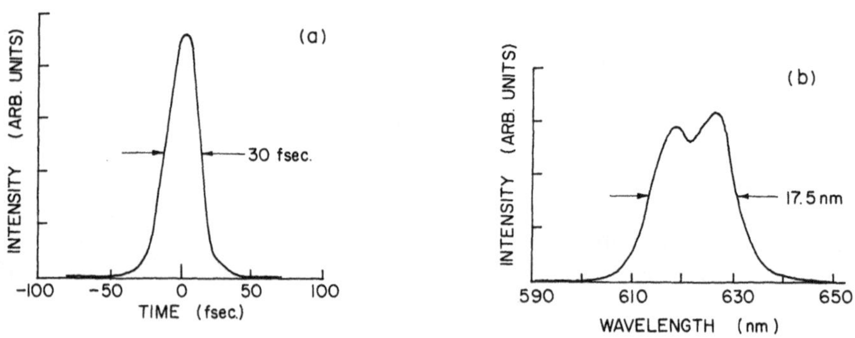

Fig. 3. Autocorrelation function of pulses having 30 fs duration (a) and corresponding spectrum (b) obtained for optimum Kiton Red - Rhodamine 590 solution when DODCI concentration was also optimized (DODCI;1.3×10^{-4} mol/l)

dye in Ethylene Glycol. Typical pump power that gave the shortest output pulses of 27 fs ranged between 2.0 W and 2.3 W depending on the concentration of the gain medium. The intracavity pulse energy is 16 nJ .

It has been shown that an average positive chirp is generated for pulses at energies higher than 10 nJ and shorter than 70 fs. This positive chirp is due to the prevailing contribution of the positive chirp from the nonlinear refractive index of the solvent, over the negative chirp from the absorber saturation. The negative chirp from the absorber saturation shifts to the beginning of the pulse as the saturation parameter $g=E_0/E_S$ increases [5]. This, and the fact that the positive chirp from the solvent increases with pulse energy is considered to have a special effect in our experiment. Our DODCI concentration is 20 times smaller than the optimum concentration used by JACOBOVITZ et al. [6] which increases g and makes the negative chirp considerably less pronounced by shifting it to the very beginning of the pulse, thus affecting an even smaller part of it. At the same time the affect of the positive chirp due to the nonlinear refractive index of the solvent is the same as in the study by the same authors [5]. Since the part of the chirp from the DODCI jet is reduced, this results in a more successful intracavity compensation. This was possible because high concentrations of DODCI were not necessary to shift the lasing spectrum to the red.

Both the design and the gain mixture contribute to reducing the causes of chirp while at the same time considerably broadening the spectral width. The achieved spectral width of 18.7 nm in principle permits generation of even shorter pulses (22 fs). The failure to do so, is attributed to the amount of glass in the output mirror and to the excessive intracavity power as a consequence of the Kiton Red presence. Reducing the amount of glass in the output coupler and increasing its transmission is expected to result in transform limited pulses.

This research was supported by Hamamatsu Photonics K. K.

References

1. J. Valdmanis, R.L. Fork: IEEE J. Quantum electron. **QE-22**, 1 (1986)
2. H.W. Kogelnik, E.P. Ippen, A. Dienes, C.V. Shank: IEEE J. Quantum Electron. **QE-8**, 3 (1972)
3. W. Dietel, E. Dopel, K. Hehl, W. Rudolph , E. Schmidt: Optics Comm. **50**, 3 (1984)
4. S. De Silvestri, P. Laporta, and O. Svelto, IEEE J. Quantum Electron. **QE-20**, 5 (1984)
5. R.S. Miranda , G.R. Jacobovitz, C.H. Brito Cruz, M.A.F. Scarparo: Opt. Lett. **11**, 4 (1986)
6. G.R. Jacobovitz, C.H. Brito Cruz, M.A. Scarparo: Optics Comm. **57**, 2 (1986)

Saturable Amplification Without Pulse Broadening

J.D. Kafka and J.B. Clark

Spectra-Physics, Inc., 1250 West Middlefield Road,
Mountain View, CA 94042, USA

Dye amplifiers are commonly used for amplification of subpicosecond pulses. Higher efficiency can be obtained from a saturated amplifier but at the expense of pulse broadening[1]. In this paper we show that chirped pulse amplification (CPA) can be used to avoid this pulse broadening in a saturated dye amplifier.

In CPA, a pulse is temporally dispersed before amplification and subsequent recompression. Recently, CPA has been used to avoid damage in Nd:Glass amplifiers by reducing the peak power during amplification [2]. Stretching a subpicosecond pulse to match the nanosecond pump pulse in a dye amplifier has also been proposed [3].

When the energy density of the input pulse to a dye amplifier approaches 1 mJ per square cm, the gain in the amplifier will become saturated. This causes pulse broadening because the leading edge of the pulse experiences a larger gain than the peak of the pulse.

We have numerically modeled CPA in a saturated amplifier. First, a fourier transform is taken of the hyperbolic secant squared input pulse. Next, we multiply by a frequency dependent phase factor and then take a second fourier transform. This procedure produces the chirped pulse which we amplify in the saturated amplifier. Finally, we fourier transform this pulse, apply the inverse of the phase factor used previously, and transform back to produce the output pulse. This model does not include spectral hole burning effects or pulse broadening due to spectral narrowing.

In Fig. 1A we demonstrate the effect of saturable amplification without CPA. The hyperbolic secant squared input pulse was amplified just above the saturation level and the output pulse was then normalized.

An optical fiber can be used to broaden the pulse prior to amplification and a prism pair to recompress the pulse after amplification. The input pulse will be stretched by the first dispersive delay line and the carrier frequency will sweep linearly from red to blue. The pulse is now amplified in a saturated amplifier and the leading edge is broadened as before. The amplified pulse has proportionally more red components and less blue than the input pulse. The total bandwidth, however, is slightly larger and shifted to the red.

Springer Series in Chemical Physics, Vol. 48: **Ultrafast Phenomena VI**
Editors: T. Yajima · K. Yoshihara · C.B. Harris · S. Shionoya

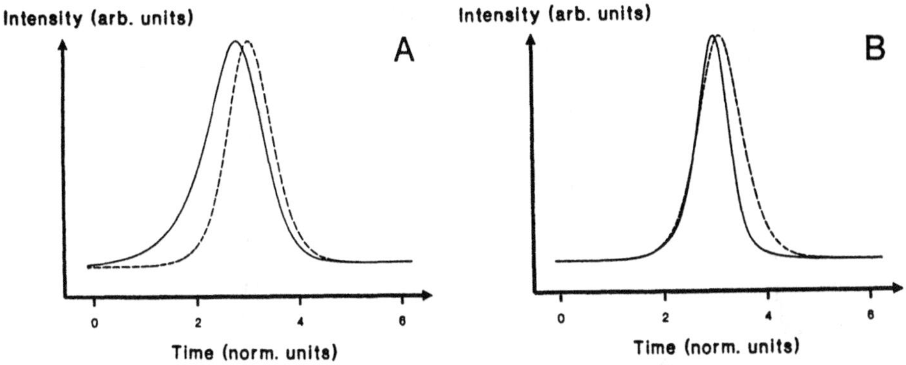

Fig. 1 Theoretical comparison of the input pulse (dashed curve) and the amplified pulse (solid curve) A) without chirped pulse amplification and B) with chirped pulse amplification. The units have been normalized so that the input hyperbolic secant squared pulse has a pulsewidth of 1.

Despite the change in the pulse envelope and spectrum, the amplified pulse is still described by a linear frequency sweep. After traversing the final prism pair, the resulting pulse is slightly shorter in duration than the input pulse. This is shown in Fig. 1B where we demonstrate saturable amplification with CPA.

The oscillator for our system consists of a pulse compressed cw mode-locked Nd:YAG laser which is frequency doubled and used to pump a dye laser [4]. The output of the dye laser is further compressed in fiber-prism compressor to produce tunable pulses of less than 100 fsec [5,6]. Our best result from this system was a pulsewidth of 65 fsec with 100 mW of average power. We then amplified these pulses in a three stage dye amplifier pumped by a frequency doubled Q-switched Nd:YAG laser. At low pump power we obtained 100 fsec pulses at 330 µJ. When the pump power was increased, we obtained 620 uJ pulses but the pulsewidth broadened to 150 fsec and the wings on the pulse became large. The autocorrelation of this pulse is shown in Fig. 2A.

We then divided the fiber-prism pulse compressor into two parts so that the pulse from the fiber was first amplified and then recompressed in the prism pair. The duration of the pulse before the prism pair was 0.9 psec. The pump power to the amplifier was increased as before but no pulse broadening was observed. Even at an output energy of 850 µJ we observed an amplified pulsewidth of 87 fsec. The autocorrelation is shown in Fig. 2B.

In summary, we have proposed and demonstrated the use of CPA in eliminating pulse broadening due to saturation in dye amplifiers. Since the staging of dye amplifiers is predicated on remaining below saturation, the use of CPA may allow some novel amplifier designs in the future.

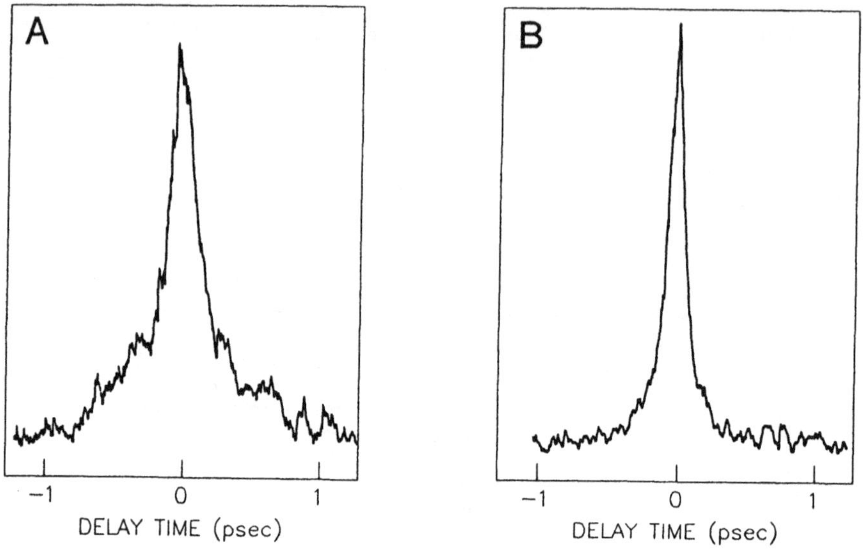

Fig. 2 Amplified pulse autocorrelation measurements for A) prism pair before the amplifier and B) prism pair after the amplifier.

References

1. J. D. Kafka, Ph.D. dissertation, Univ. of Rochester, Rochester, N.Y., 1983.
2. D. Strickland and G. Mourou, Opt. Comm. 56, 219(1985) and P. Maine, D. Strickland, P. Bado, M. Pessot and G. Mourou, IEEE J. Quantum Electron., QE-24, 348 (1988).
3. O. E. Martinez, IEEE J. Quantum Electron., QE-23, 1385 (1987).
4. J. D. Kafka and T. Baer, Proc. SPIE 533, 38 (1985).
5. J. D. Kafka and T. Baer, Opt. Lett. 12, 401 (1987).
6. T. C. Damen and J. Shah, Appl. Phys. Lett. 52, 1291 (1988).

General Analysis of Optical Cavities for Femtosecond Dye Lasers

S. De Silvestri, Liu Yu-Pu, V. Magni, and O. Svelto*

Centro di Elettronica Quantistica e Strumentazione Elettronica del C.N.R., Istituto di Fisica del Politecnico di Milano, Piazza Leonardo da Vinci 32, I-20133 Milano, Italy

Light pulses less than 100 fs in duration have been recently achieved by synchronously pumped hybrid dye lasers at different wavelengths. The duration and stability of the pulses critically depend on the design of the resonator, which includes several optical elements such as two pairs of focusing mirrors and dispersion compensating prisms. Simple three mirror linear resonators with a single focus were first analyzed by KOGELNIK [1], and more recently cavities containing two foci were numerically studied [2-3]. The stability ranges, as well as the ratio of the spot sizes at the foci, were calculated as a function of the focusing mirror separation, to optimize saturation in the absorber and gain jets. An important parameter not yet considered in the literature is the misalignment of the resonator, which determines the sensitivity of the cavity to perturbations.

In this work, we present a simple and general analysis of resonators with multiple foci, which includes the characterization of the misalignment sensitivity. The study is based on a theory recently developed for multi--element resonators [4]. Let us first consider one of the focal regions in a generic linear resonator, whose ray matrix equivalent is shown in Fig. 1. The distance z represents the separation between the two focusing mirrors of a folding configuration. The matrices describe the propagation towards the two end mirrors, which are assumed to be flat, and include the folding mirrors and all the other optical elements contained in the resonator. The misalignment in the optical components is treated by adding a 2×1 vector to that obtained by multiplying the input ray vector by the appropriate ray

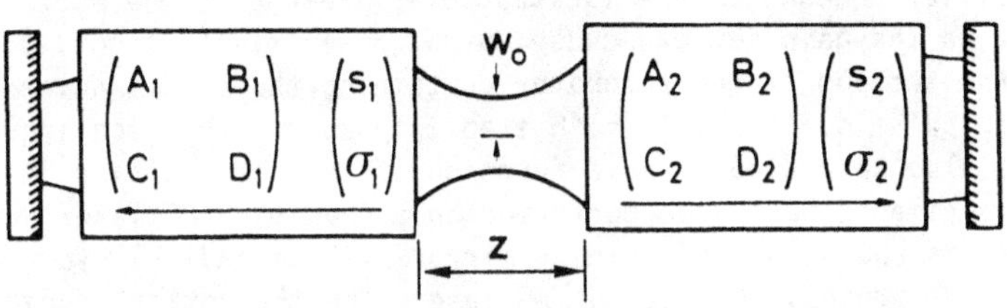

Fig. 1 Schematic of a linear resonator with one folding. The matrices and the misalignment vectors represent any optical system present in the cavity

*Shanghai Institute of Optics and Fine Mechanics, ICTP fellowship.

Springer Series in Chemical Physics, Vol. 48: **Ultrafast Phenomena VI**
Editors: T. Yajima · K. Yoshihara · C.B. Harris · S. Shionoya

transfer matrix [5]. The elements of the misalignment vectors give the position and the slope of the output ray when the input ray coincides with the reference axis of the system, and therefore the input-output relationship is no longer homogeneous. The analysis of such a general resonator shows the existence of two stability zones of equal amplitude, as a function of z. The spot size of the gaussian TEM_{00} mode at the beam waist inside the folding vanishes at the stability limits and, as a function of z, reaches a maximum value w_{00} given by

$$w_{00}^2 = (\lambda/2\pi)\Delta z \quad , \qquad (1)$$

where Δz in the width of the stability zone, and λ the laser wavelength. This simple, but rigorous, relationship is of general validity and does not depend on the particular configuration. Since w_{00} is the same in both zones the laser can operate in each zone without loss in the output power.

The resonator sensitivity to the misalignment provides a criterion to choose the stability zones. The displacement and the slope of the mode axis with respect to that of a perfectly aligned resonator can be determined by applying the ray matrix technique and solving a self consistency matrix equation that expresses the fact that the mode axis is the unique ray that retraces itself after one round trip around the resonator. The misalignment sensitivity in the two zones is markedly different: in fact, at one edge of one stability zone the displacement, x, and the slope, θ, of the mode axis diverge for small misalignment in the optical elements. The stability zone characterized by this limit is called zone II, the other is called zone I. Figure 2 shows a typical behavior of the misalignment sensitivity defined as the ratio between beam displacement (or slope) and the tilting angle of one of the two end mirrors as a function of the folding mirror separation. The mode configuration at the limits of the stability zones is also shown for a simple resonator made by flat end mirrors and a pair of focusing mirrors. In this case the misalignment sensitivity diverges near the confocal arrangement of the two folding mirrors.

The effect of the displacement of the mode axis on the stability of the system becomes prominent in particular planes in the resonator where limiting apertures or peculiar components are located. Displacement of the mode from the pumped area in the gain jet can cause output power fluctuations, as well as that in the section of the resonator containing the prisms determines fluctuations in pulse duration. In both stability zones, the misalignment sensitivity has also been studied as a function of the jet position inside the folding. Experiments have been performed on a Rhodamine 6G laser made by two flat mirrors and a pair of folding mirrors (10 cm radius) synchronously pumped by a frequency doubled Nd:YAG laser. The theoretical curves of misalignment sensitivity have been compared with the reciprocal of the mirror tilting angle that halves the output power. In zone I the misalignment sensitivity is almost independent of the jet position and of the folding mirror distance, whereas in zone II the divergence of the misaglinment sensitivity can be eliminated only if the jet is placed exactly in the focus of one of the two folding mirrors.

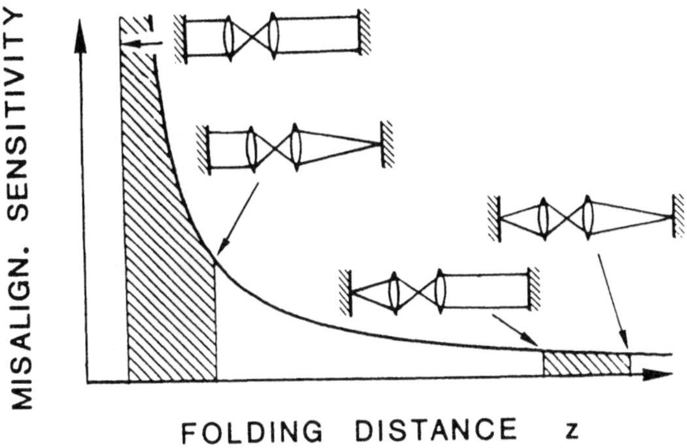

Fig. 2 Theoretical behavior of the misalignment sensitivity of the resonator. The resonators show the mode configurations at the edges of the stability zones

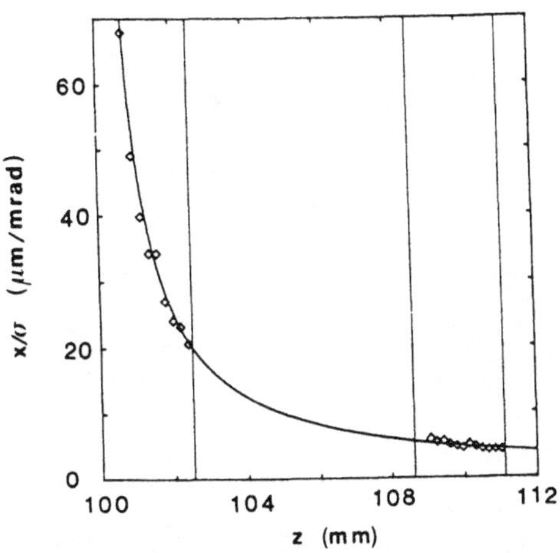

Fig. 3 Ratio of the mode axis displacement in the long arm of the resonator to the tilting angle of one of the end mirrors as a function of the distance between the folding mirrors

To understand the effects of misalignment in a resonator containing prisms we have measured the misalignment sensitivity by placing an aperture in the long arm of the resonator and the gain jet at the focus of one of the folding mirrors. Figure 3 shows the measured mode axis displacement compared with the theoretical prediction. The behavior in zone II indicate a more critical control of the cavity dispersion, which depends on the optical path inside the prisms, and therefore on the point of incidence of the beam on the prisms.

The analysis has been also extended to take into account the presence of a second folding. With the help of a two dimensional diagram, with the two folding distances as axes, several stability areas can be recognized. The misalignment sensitivity diverges along a particular line, which determine the regions that should be avoided for a reliable operation.

References

1. H.W.Kogelnik, E.P.Ippen, A.Dienes and C.V.Shank, IEEE J. Quantum Electr. QE-8 (1972) 373.
2. K.S.Budil, I.A.McIntyre, and C.K.Rhodes, Opt. Commun. 64 (1987) 279.
3. C.E.Wagstaff, M.H.Dunn, A.I.Ferguson, and S.J.Bastow, Opt. Commun. 25, (1978) 379.
4. V.Magni, J. Opt. Soc. Am. A 4 (1987) 1962.
5. A.Gerrard and J.M.Burch: Introduction to Matrix Methods in Optics (Wiley, London, 1975) p. 106 and 286.

Chirp-Compensation Cavity-Mirrors of Small Third-Order Dispersion for a Femtosecond Pulse Laser

M. Yamashita [1], *S. Kaga* [2], *K. Torizuka* [1], *and T. Sato* [1]

[1] Laser Research Section, Electrotechnical Laboratory, 1-1-4 Umezono, Tsukuba, Ibaraki 305, Japan

[2] Research and Development Department, Vacuum Optics Corp. of Japan, 2-29-10 Nishigotanda, Shinagawa, Tokyo 141, Japan

Compensation for chirp arising from self-phase moduration and dispersion in a cavity is essential for the generation of pulses shorter than 100 fs from a colliding-pulse mode-locked (CPM) CW dye laser. There are presently three compensation-techniques: 1) the method employing four prisms [1], 2) the Gires-Tournois interferometers' method [2], and 3) the multilayer dielectric cavity-mirrors' method [3, 4]. The latter has the advantages of a simple cavity-configuration and adjustment-free compensation, but the disadvantage of a large third-order dispersion $\dddot{\phi}(\omega)$ which broadens the pulse duration with subpulses. In this paper, we show that $\lambda_0/4$ multilayer dielectric mirrors providing a small value of $\dddot{\phi}(\omega)$ ($\leq 1\times10^{-42}\,s^3$) at an optimum value of $\ddot{\phi}(\omega)$ ($\sim 2\times10^{-28}\,s^2$ [3, 4]) and small variations of $\ddot{\phi}(\omega)$ and $\dddot{\phi}(\omega)$ around the lasing spectrum for chirp-compensation can be achieved.

Cavity mirrors for chirp compensation should satisfy the following conditions around λ_d=635 nm for an R6G+DODCI CPM laser: 1) the values of the sum of $\ddot{\phi}(\omega)$ and $\dddot{\phi}(\omega)$ of all the cavity mirrors be $\sim 2\times10^{-28}\,s^2$ and $\leq 1\times10^{-42}\,s^3$, respectively, 2) the reflectivity $R(\omega)$ of the mirrors be 99.9 %, and 3) the value of the fourth-order dispersion $\ddddot{\phi}(\omega)$ be zero so that the variations of $\ddot{\phi}(\omega)$ and $\dddot{\phi}(\omega)$ are small around 635 nm.

Using a well-known matrix formulation [5] for the amplitude reflectance $|r(\omega)|\times \exp[i\phi(\omega)]$ ($R(\omega) = |r(\omega)|^2$) of a $\lambda_0/4$ multilayer dielectric mirror, a phase shift $\phi(\omega)$ and its angular-frequency derivatives of a reflected EM-wave were numerically calculated as a function of the frequency with the aid of a computer. Refractive indices of a high refractive layer n_H = 2.25, a low refractive layer n_L = 1.48 and substrate n_S = 1.52 were employed for the calculation. After many calculations of $\ddot{\phi}(\omega)$ and $\dddot{\phi}(\omega)$ at $\ddddot{\phi}(\omega)$ = 0 for different coating structures of double- and triple-stacking mirrors, we found that one can design simple double-stacking mirrors satisfying the above conditions. For a double-stacking mirror, the frequency dependences of the dispersion and reflectivity around the resonance wavelength λ_{B0} of the lower-side stack are affected mainly by those of upper- and lower-side stacks, respectively. In the region of the wavelength longer than the resonance wavelength λ_{A0} of the upper-side stack, the value of $\ddot{\phi}(\omega)$ becomes positive, corresponding to negative group velocity-dispersion, while around the wavelength λ_{B0} the value of $R(\omega)$ becomes about 99.9 %. Therefore, we investigated double-stacking mirrors for which λ_{A0} is in the wavelength-region shorter than λ_d, λ_{B0} is near to λ_d and the layer number L_B of the lower-side stack is 24.

For the practical cavity-configuration as shown in Fig. 1-C, the following three cases were considered: in case A, the chirp is compensated for by one cavity-mirror M_1 ($\theta \cong 0°$, typically 3.5°); in case B, by two mirrors M_1 and M_6 (θ = 45°); and in case C, by three mirrors M_1, M_5 (θ =

Springer Series in Chemical Physics, Vol. 48: **Ultrafast Phenomena VI**
Editors: T. Yajima · K. Yoshihara · C.B. Harris · S. Shionoya

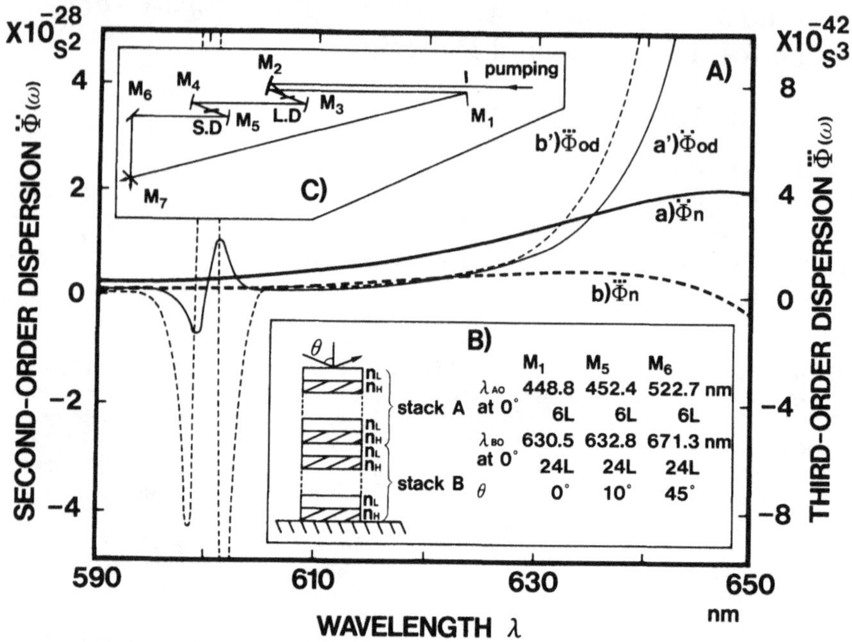

Fig.1 A) Wavelength dependences of the sum of the second- and third-order dispersions ($\ddot{\phi}_n$ (a) and $\dddot{\phi}_n$ (b)) of chirp-compensation mirrors M_1, M_5 and M_6 in case C, and those ($\ddot{\phi}_{0d}$ (a') and $\dddot{\phi}_{0d}$(b')) of compensation-mirror M_1 employed in the previous experiment for 50 fs-pulse generation [3]. B) Doubly stacked layer-structures of new mirrors for case C. C) Cavity configuration of a CPM laser to be chirp-compensated for by cavity mirror(s).

Table. 1 Multilayer structures of double-stacking mirrors obtained.

Case	A	B		C		
M_i	M_1	M_1	M_6	M_1	M_5	M_6
L_A	8	6	8	6	6	6
λ_{A0} (nm) at 0°	489.1	457.9	550.7	448.8	452.4	522.7
L_B	24	24	24	24	24	24
λ_{B0} (nm) at 0°	601.9	598.4	699.4	630.5	632.8	671.3

10°) and M_6. For all cases, suitable mirrors were obtained, which are summarized in Table 1. Here all the resonance wavelengths λ_{A0} and λ_{B0} were defined at the incident angle of 0° but not at the actual angle.

In Fig. 1-A, the dispersion curves $\ddot{\phi}(\omega)$ and $\dddot{\phi}(\omega)$ of the sum of all the mirrors obtained for case C are shown. In all cases (A, B and C), both $\ddot{\phi}(\omega)$ and $\dddot{\phi}(\omega)$ similarly increase with the wavelength up to the 643 ~ 645 and 635 nm, respectively, and then decrease with the further increase of the wavelength. However, the degree of the variations of $\ddot{\phi}(\omega)$ and $\dddot{\phi}(\omega)$ are much different, especially $\dddot{\phi}(\omega)$. Those variations in case C are the smallest. In addition, the maximum values of $\ddot{\phi}(\omega)$ are the same ($\sim 2 \times 10^{-28} s^2$), while those of $\dddot{\phi}(\omega)$ are different and those in case C are once again the smallest ($1.16 \times 10^{-42} s^3$). Therefore, it is concluded that the compensation method of case C is the best. From this result we can deduce that when chirp is compensated for by all (six) cavity-mirrors except for an output mirror the smaller $\dddot{\phi}(\omega)$ and variations of $\ddot{\phi}(\omega)$ and $\dddot{\phi}(\omega)$ will be obtained. It should be noted that, in general, as the layer number L_A of the upper-side stack is small the variations of $\ddot{\phi}(\omega)$ and $\dddot{\phi}(\omega)$ and their maximum values become small.

The wavelength dependences of $\ddot{\phi}(\omega)$ and $\dddot{\phi}(\omega)$ of the compensation mirrors employed previously for 50 fs pulse generation [3] are also shown in Fig. 1-A for the comparison with case C. It is found that the value of $\dddot{\phi}(\omega)$ and the variations of $\ddot{\phi}(\omega)$ and $\dddot{\phi}(\omega)$ around λ_d for the new mirrors are much smaller than those of the previous mirrors. In addition, the maximum values of $\ddot{\phi}(\omega) = 2.05 \times 10^{-28} s^2$ and $\dddot{\phi}(\omega) = 1.16 \times 10^{-42} s^3$ around λ_d of the new mirrors are comparable to those of $\ddot{\phi}(\omega) = 3.12 \times 10^{-28} s^2$ and $\dddot{\phi}(\omega) = 1.19 \times 10^{-42} s^3$ which were estimated for compensation by the four prisms' method [1.6]. The latter values were evaluated from the approximate equations of Ref. 6 with the experimental data of Ref. 1 (where $\ell = 30$ cm and $\beta = 0°$), where the practical effect of positive group-velocity dispersion of prism materials originating from the beam-passing distance in four prisms was neglected. The calculation [7] of the third-order dispersion effect of $\dddot{\phi}(\omega) = 1.16 \times 10^{-42} s^3$ on 30 fs pulses shows that no broadening occurs on reflecton by the new mirrors.

References

1. J.A.Valdmanis and R.L.Fork, IEEE J.Quantum Electron. QE-22, 112 (1987).
2. J.Heppner and J.Kuhl, Appl.Phys.Lett. 47, 453 (1985).
3. M.Yamashita, M.Ishikawa, K.Torizuka and T.Sato, Opt.Lett. 11,504(1986).
4. M.Yamashita, K.Torizuka and T.Sato, IEEE J.Quantum Electron, QE-23, 2005 (1987).
5. H.M.Liddell, *"Computer-Aided Techniques for the Design of Multilayer Filters"* (Hilger, Bristol, England, 1981).
6. R.L.Fork, C.H.Brito Cruz, P.C.Becker and C.V.Shank, Opt.Lett. 12, 483 (1987).
7. M.Miyagi and S.Nishida, Appl.Opt. 18, 678 (1979).

Theoretical and Experimental Study of Synchronously Pumped Dispersion Compensated Femtosecond Fiber Raman Lasers and Amplifiers

E.A. Golovchenko, E.M. Dianov, P.V. Mamyshev, A.M. Prokhorov, and D.G. Fursa

General Physics Institute, Academy of Sciences of the USSR, 38 Vavilov Street, Moscow 117942, USSR

Fiber Raman laser with intracavity pulse compression is investigated both theoretically and experimentally. Smoothly tuned generation inside two Stokes components with minimum pulse width 150/fs/is experimentally obtained. Potentialities of such lasers are discussed. It is shown that they can generate smoothly tuned pulses in all silica fiber transparency spectral region. Pump energy conversion into the energy of Raman generation can reach more than 80 % (30% experimentally obtained) and the average output power - a few watts. Experiments on forward and backward Raman amplification of femtosecond pulses up to the energy of ~/μJ /are described.

One of the most promising laser sources are wavelength-tuned fiber Raman lasers, where fiber acts not only as a Raman active medium with a broad gain line but also as dispersive medium with Kerr nonlinearity. In the positive group velocity dispersion spectral region of fibers (for silica fibers can be 0.2-1.8/μm/) traditional synchronously-pumped fiber Raman lasers generate transform unlimited pulses with durations of the order of pump pulse duration (typically ~100/ps/)[1]. But with intracavity negative dispersion delay line it is possible to obtain femtosecond pulses. That was realized by KAFKA et al. [2] and DIANOV et al. [3] who obtained 0.8/ps/and 0.4/ps/pulses correspondingly.

Figure 1 demonstrates laser schematics. A fiber, a dispersion delay line (consisting of a diffraction grating and a prism) and a spectral filter (slit S_1) are introduced into the laser ring cavity. Synchronous pumping through the dichroic mirror M_1 is produced by cw mode locked laser. The theoretical model of pump and SRS radiation propagation through the fiber is based on a system of two bound Schrodinger type equations describing the effects of SRS, self-phase modulation, cross-phase modulation, group velocity dispersion of fiber and differences between laser parameters for stable high-quality femtosecond pulse generation were found. At optimal conditions the width of generated pulses τ_p can be evaluated as $\tau_p \sim |d^2K/d\omega^2|\, L/\tau_p$ (L is the fiber length, τ_p is the pump pulse width). The minimum pulse width is limited by SRS-gain line width (in silica fibers ~100/fs/). Con-

Springer Series in Chemical Physics, Vol. 48: Ultrafast Phenomena VI
Editors: T. Yajima · K. Yoshihara · C.B. Harris · S. Shionoya

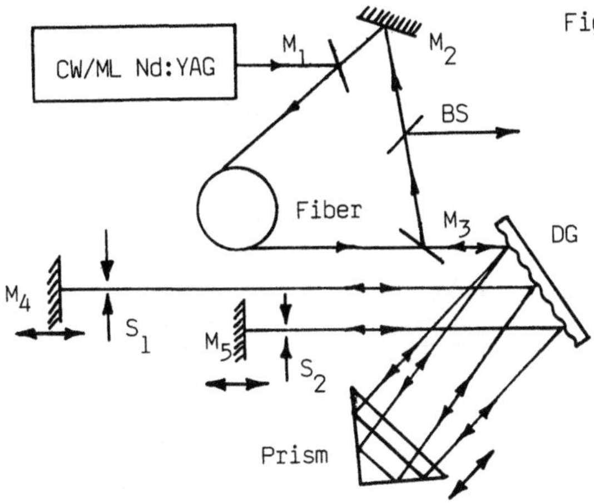

Figure 1.

version efficiency of pump energy into SRS radiation can be >80%. On the other hand the optimum value of feed-back (the ratio between SRS energy at fiber output and energy coupled back into the fiber) is $\sim 10^{-3}$ .This means that practically all energy of SRS radiation can be taken out from the cavity as a laser output. Figure 2a shows the dynamics of passage by passage evolution of laser generation from spontaneous noise level calculated on a computer. It can be seen that the stable generation regime sets in already after ~ 20 cavity round trips. In a steady state generation the SRS pulse at fiber output has a rectangular shape with a width of about a pump pulse width and a linear chirp these pulses are compressed in a dispersive delay line.

CW mode-locked at 125 MHz Nd:YAG laser (λ = 1.06 /μm/, τ_p=60/ps/) was used as a pump. Figure 2b demonstrates obtained results on dynamics of passage by passage setting in of laser generation (fiber length 110 m) after the pump is switched on. In agreement with theory, stable generation of 440/fs/pedestal free pulses sets in after first few tens of cavity round trips. Pump energy conversion into Raman generation was ~30% at average pump power ~1/W/. SRS pulse at fiber output has a rectangular shape with 23/ps/ width. The evaluated feed-back index was $\sim 10^{-3}$. Energy transmission of the dispersive delay line was ~20%, laser output losses (reflection of the beamsplitter BS) was 17% while the average output power was 10-15 /mW/. Having transmission of a delay line ~ 80-90% and feed-back coupling efficiency > 30% the reflection of a beamsplitter can be increased up to >90%. Under these conditions the average output power of our laser will increase up to > 0.2/W/ and at pumping powers 10/W/ the output will be a few watts. Smooth tuning of generated wavelength inside the first Stokes component λ=1.07-1-12/μm/ was carried out by tuning the spectral slit and adjusting the laser cavity length by the mirror M_4. With an additional mirror M_5 and a slit S_2 we had a smoothly- tuned second Stokes generation in the region of λ =1.15-1.18/μm/. Having a 110/m/ long fiber the pulse width was 400/fs/ and 1.5/ps/ on the first and second Stokes components respectively. Comparatively

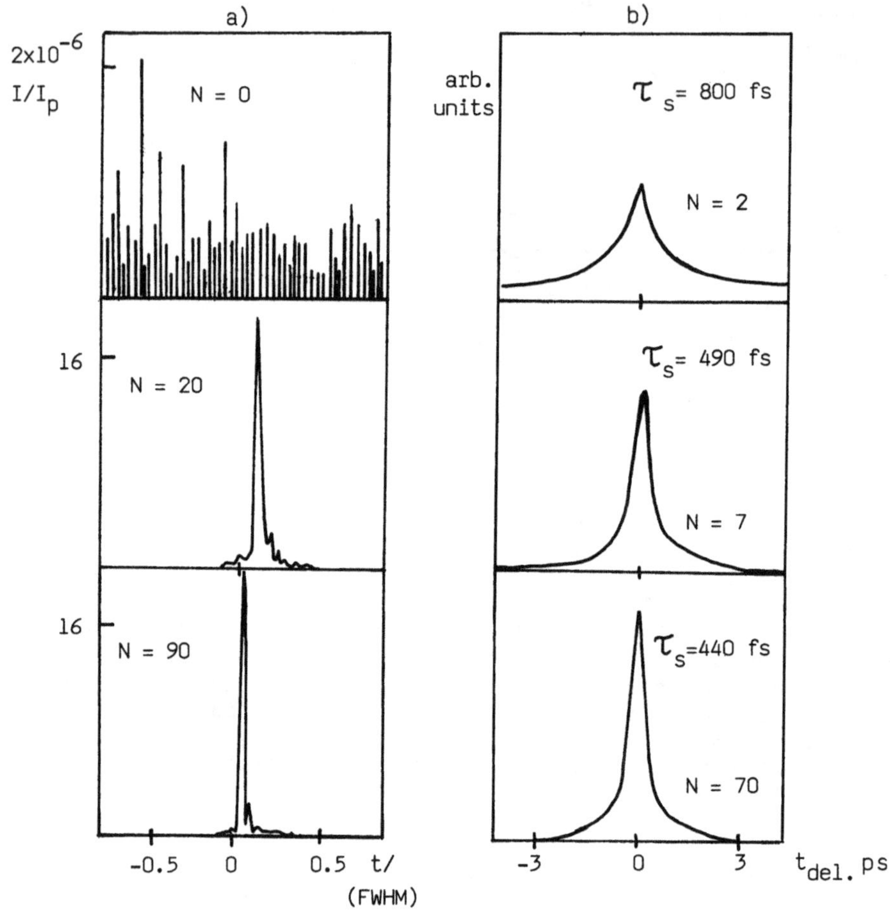

Figure 2. Theoretical (a) and experimental (b) results on dynamics of evolution of laser generation versus N -cavity round trips

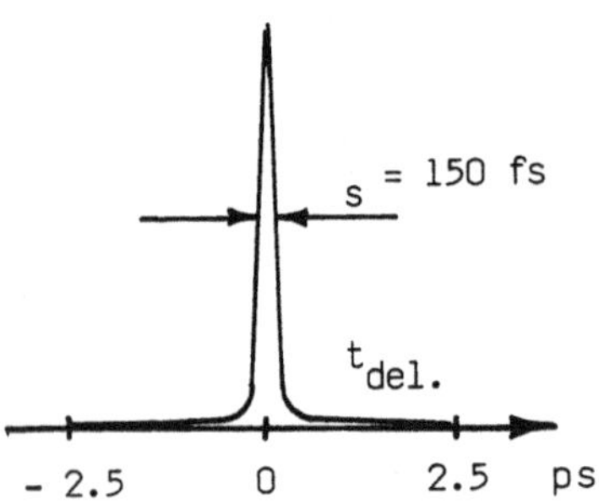

Figure 3.

large pulse width on the second SRS component, in our opinion, can be explained by an insufficient pump power coupled into the fiber (<2/W/). Stable 150/fs/ generation was obtained on the first Stokes component with the 30/m/-long fiber (Figure 3).

In experiments on forward and backward SRS-amplification a cw Q-switched Nd:YAG laser (repetition rate <30/KHz/) and a cw Q-switched and mode-locked Nd:YAG laser were used, respectively. 400/fs/ pulses were preliminarily dispersively broadened in the 100/m/-long fiber up to 20/ps/. After amplification to the energy of about /μJ/ the pulses were compressed in the dispersive delay line to 500-600/fs/ [5].

Literature

1. R.H.Stolen: Fiber and Integrated Optics, 3, 21 (1980).
2. J.D.Kafka, D.F.Head, T.Baer: In Ultrafast Phenomena V.(New York:Springer Verlag,1986) p.51
3. E.M.Dianov, P.V.Mamyshev, A.M.Prokhorov,D.G.Fursa:JETP Lett.,45,599 (1987)
4. E.A.Golovchenko, E.M.Dianov, P.V.Mamyshev, A.M.Prokhorov, D.G.Fursa: submitted to J.Opt.Soc.Am.B.
5. E.M.Dianov, P.V.Mamyshev, A.M.Prokhorov, D.G.Fursa: Pis'ma Zh. Eksp.Teor.Fiz. 46, 383 (1987) [JETP Lett.].

Compression of the High Energy Pulsed Mode-Locked Nd:YLF and Nd:YAG Lasers

P.H. Chiu, P. Pax, and R. Aubert

Quantel International, 3150 Central Expressway,
Santa Clara, CA 95051, USA

1. Introduction

Optical pulse compression,utilizing fiber/grating pairs and/or intracavity prism pairs is the most recent advanced technique in generating ultrashort pulses[1]. The pulse-compression methods have been demonstrated in a number of experiments using CW mode-locked dye lasers and Nd:YAG laser in both picosecond and femtosecond regime[2,3]. However, little work has been done on compressing the optical pulse from the high energy pulsed mode-locked lasers. In this paper, we report compressing the pulsed mode-locked Nd:YLF and Nd:YAG lasers by a combination of compression methods.

2. Compression of Nd:YLF pulses using intracavity Brewster angle prism pair

We have investigated a number of materials capable of generating ultrashort pulses at higher repetition rates. By simply replacing the Q-switching dye #1 (9740) with the fast recovery dye #5 with very little modification of the laser cavity, we have shortened the pulsewidth of the laser by approximately a factor of 2 (see Table 1). We have also reported previously obtaining 7ps (FWHM) from a Nd:YLF laser[4]. Recently when a prism pair is introduced inside the cavity, a near bandwidth limited pulse of 2ps from the same laser was observed. This is attributed to the prism pair creating a negative dispersion inside the cavity to compensate for the positive dispersion due to the thermal effect of the rod. Figure 1 demonstrates that the pulse duration of the Nd:YLF laser decreases as the amount of quartz introduced inside the cavity increases. The thickness of the quartz inserted in the cavity varies from approximately 1mm to 10mm. On the top of Figure 2, the separation of the two main peaks is measured to be 2.5ps which corresponds to the round trip time of the 250 micrometer etalon output coupler of the laser. A small amount of dye # 1 was mixed with dye #5 to achieve a better stability (< ±3%). A similar effect has also been observed in a Nd:YAG laser where a bandwidth limited pulse as short as 7ps is obtained.

Table 1

SUMMARY OF RESULTS ON ACTIVE/PASSIVE MODE-LOCKED LASERS

Material	Lasing length	Fluorescence bandwidth	Dye	Pulsewidth (FWHM)	Stability (peak-to-peak)
Nd:YAG	1.064μm	4cm^{-1}	9740	30ps	±2.0%
			5	15ps	±2.0%
Nd:YLF	1.053/	12cm^{-1}	9740	12-14ps	±2.5%
	1.047μm		5 + 9740	3-4ps	±2.5%
Nd:YAP	1.079μm	10cm^{-1}	5	10ps	±3.0%
Nd:Cr:GSGG	1.061μm	10cm^{-1}	5	8-10ps	----

Springer Series in Chemical Physics, Vol. 48: **Ultrafast Phenomena VI**
Editors: T. Yajima • K. Yoshihara • C.B. Harris • S. Shionoya

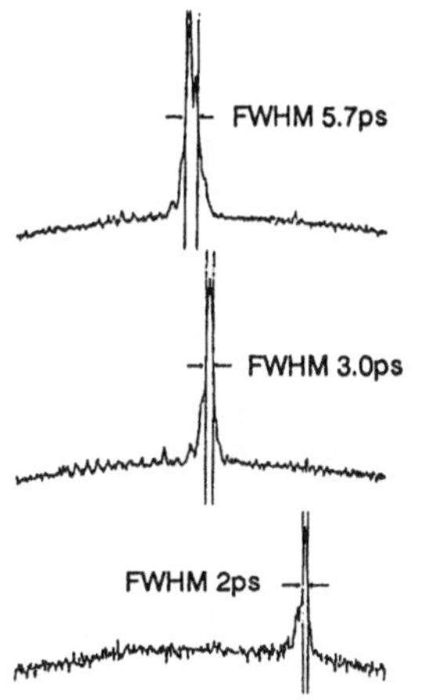

Streak camera measurements of Nd:YLF pulse using mixture of Exciton #5 and Kodak 9740 dye as the saturable absorber. The measurements were taken at 523.5nm.

Figure 1

3. Compression of Second Harmonic Pulse from a mode-locked Nd:YAG laser.

We also report the compression of 20ps (FWHM) second harmonic pulse from a Nd:YAG laser to 5ps (FWHM) using a standard fiber/grating pair compressor. The experimental set-up is shown in Figure 2. Approximately 1% of 30mJ, 532nm single pulse is selected for the pulse compression. The spectral and temporal characteristics of the pulses before and after the fiber are respectively monitored by the grating spectrometer (Mcpherson model 270, resolution 0.3Å) and the Hamamatsu streak camera with 2ps time resolution. The input beam is focused by a 75mm f.l. lens into a approximately 300 micrometers spot. The focused spot is about 10mm away from the entrance face of the fiber in order to avoid damaging it. The input beam is spatially filtered and energy to the fiber is controlled by a set of neutral density filters. We selected a high damaged threshold UV stepped index filter with core diameter of 100 micrometers in our experiment.

This fiber can accept at least 10 times more energy than the graded-index fiber (50mm core) and 2 or more orders of magnitude than can a single mode fiber (5mm core). We have delivered energy as much as 20μJ through this filter without damaging it. Since the length of the fiber used in the experiment is roughly one meter, the group velocity dispersion influence can be neglected. Figure 3 shows the spectrum of the pulse before and after the fiber at the input energy of 10μJ with the time-bandwidth product 2 to 3 times above the transform limit. It was found that the width of the chirped pulse follows linearly with the input energy. Figure 4 shows the input and compressed 532nm pulsewidth under the same condition. The measured temporal variation of the compressed pulses was not more than ±1ps.

1. Flowing dye cell & back mirror
2. Acoustic mode-locker
3. Delay prisms
4. Glan-Taylor polarizer
5. Crystal rod
6. Output coupler
7. Single pulse slicer
8. Quarter wave plate
9. Amplifier
10. Second harmonic generator
11. Lens
12. Fiber optic
13. Lens
14. Grating
15. Prism

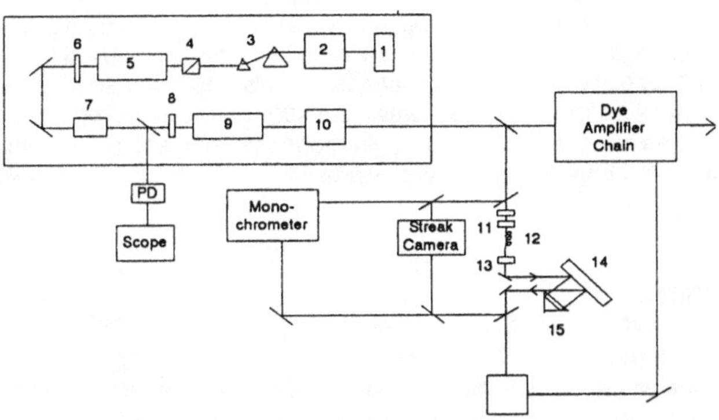

Figure 2 Experimental Set-up

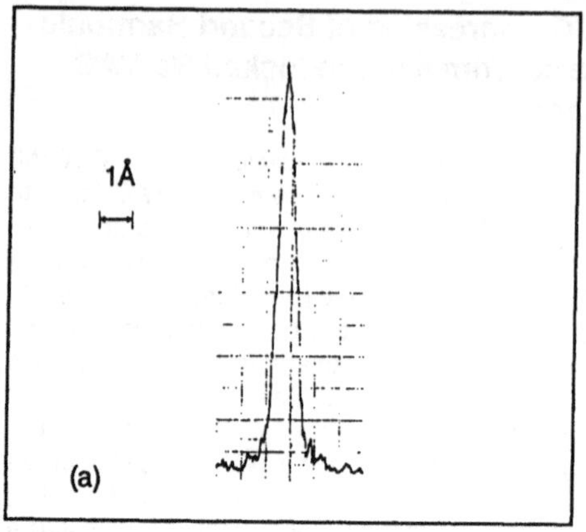

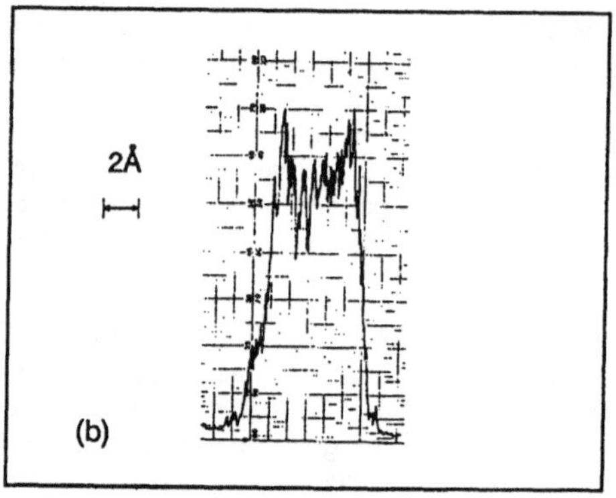

Figure 3 *(a) Spectrum of Input Laser Pulse*
(b) Spectrum of Chirped Pulse

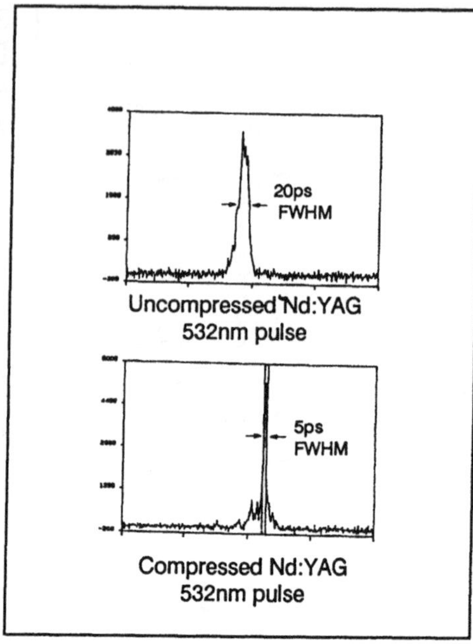

Figure 4

4. Results on Dye Laser Pumping

By using the compressed pulsed of 532nm pulses to pump the newly designed electromagnetically-tuned short cavity dye oscillator and the rest of the uncompressed pulse to pump the four-stage amplifer chain[5] we obtained dye laser pulses less than 2ps with energy as high as 5mJ at 600nm region. Energy as high as 1mJ at 380nm was achieved by mixing the fundamental and the dye laser pulse through a KDP crystal. We also obtained an 8% energy efficiency conversion with the blue and green dyes between the 400nm to 560nm region. Doubling efficiency better than 13% was achieved with input single-mode dye laser energy of 0.5mJ at 490nm region when a thin type I BBO crystal (1.0-1.5mm) was used.

References

1. A.E. Siegman, Lasers (University Science Books, California, 1986)
2 H. Nakatsuka and D. Grischkowsky, Appl. Phys. Lett. 42, 1 (1982)
3 A. M Johnson, R. H. Stolen and W. M. Simpson, Appl. Phys.Lett. 44, 729 (1984)
4. J. Weston, P.H. Chiu and R. Aubert, Opt. Comm. 61,208 (1987)
5. P. H. Chiu, P. Pax, J. Henden, L. Marshall, F. Wilson and R. Aubert, J. of Optics and Lasers in Engineering, to be published.

Active Pulse Compression

B.H. Kolner

Hewlett-Packard Laboratories, P.O. Box 10350, MS 26M16, Palo Alto, CA 94303, USA

Optical pulse compression has become a very important technique in the generation of picosecond and femtosecond pulses. What has now become the standard approach involves launching pulses down single mode optical fibers where they are chirped by self-phase modulation (SPM) and subsequently passing them through a dispersive delay line, such as a diffraction grating pair, where the compression actually takes place [1]. In the sense that there is no power released into the pulse from the fiber or any external source, the process is passive.

In spite of its popularity and effectiveness, passive pulse compression suffers from several disadvantages. First, owing to the nonlinear origin of self-phase modulation, fairly high peak powers are required to obtain sufficient spectral broadening for high compression ratios. This limits its usefulness to high power lasers which tend to be physically large with commensurate power and cooling requirements. Second, the spectral broadening produced by SPM is intensity dependent and therefore amplitude noise on the input pulse stream leads to enhanced output amplitude noise and pulsewidth fluctuations [2]. Finally, in spectral regions where group velocity dispersion is negligible, the induced chirp is linear only over the central part of the pulse and this leads to the formation of a pedestal when compressed by a quadratic dispersive delay line [1,3].

In contrast to the passive interaction of SPM, we have taken the approach of actively chirping the optical pulses with electro-optic phase modulators [4-7]. With the development of high efficiency integrated optical phase modulators, substantial chirp rates can be impressed on optical pulses by using microwaves at modest power levels. Since the process is independent of the optical power, it lends itself well to low power solid state lasers. Also, by adjusting the period of the modulating waveform with respect to the input pulsewidth, a linear chirp can be maintained over the whole pulse envelope. These features, combined with a linear sensitivity to amplitude fluctuations, make active pulse compression an attractive alternative for generating ultrashort light pulses.

In an ideal electro-optic phase modulator, an optical pulse will copropagate with a microwave field in a region where that field varies quadratically. The pulse will then accumulate a quadratic phase shift or retardation which is proportional to both the strength of the field and the duration of the interaction. The total accumulated retardation can be described by an overlap integral that sums the contribution of the electro-optically induced phase shift along the path of propagation. Using a traveling-wave coordinate system centered on the optical pulse allows us to observe the effects on the pulse as a function of the propagation distance x. For a sinusoidal modulating field in one spatial dimension, the integral takes the form

Springer Series in Chemical Physics, Vol. 48: **Ultrafast Phenomena VI**
Editors: T. Yajima · K. Yoshihara · C.B. Harris · S. Shionoya

$$\Gamma(x,\xi) = \frac{\omega_o}{c}\int_0^x \Delta n(x',\xi)\,dx' = A\,\frac{\sin\left(\frac{\Delta\phi}{2}\right)}{\frac{\Delta\phi}{2}}\cos\left(\frac{\Delta\phi}{2} + \omega_m\xi + \theta\right) \tag{1}$$

where $A \equiv \frac{\omega_o}{c}\Delta n_o x = \pi\frac{V}{V_\pi}$, $\Delta\phi = \omega_m x\left(\frac{1}{v_{go}} - \frac{1}{v_{pm}}\right)$,

ω_m is the modulation frequency, and θ is an inital phase offset between the pulse and the modulating field. A is the peak phase deviation or modulation index, $\xi = t - x/v_{go}$ is the traveling wave variable centered on the optical pulse moving at velocity v_{go} and $\Delta\phi$ is the walkoff due to the mismatch between the optical group velocity v_{go} and the microwave phase velocity v_{pm}. The effect of velocity walkoff is twofold. First, the amplitude is reduced in a $\sin x/x$ fashion, characteristic of non-phasematched harmonic generation. Second, the phase of the final retardation is shifted which causes the region of quadratic phase to no longer be centered on the optical pulse. This effect can be compensated for by injecting the pulse into the modulator with an initial offset $\theta = -\Delta\phi/2$. The amplitude reduction, however, cannot. Equation (1) is plotted in Fig. 1 for the case of no velocity mismatch (solid cosine) and two cases of mismatch producing walkoffs of $\Delta\phi = \pm\pi/2$. The Gaussian curve in the center represents the optical pulse.

We can calculate the performance of active pulse compressors by Fourier transforming the chirped pulse and applying the effect of a quadratic dispersive delay line to cancel the imaginary part of the phase. Assuming no velocity walkoff and a Gaussian input pulse, we arrive at the following expression for the pulse compression ratio:

$$\frac{\tau_i}{\tau_o} = \sqrt{1 + \left(\frac{A\pi^2}{\ln 2}\left(\frac{\tau_i}{T}\right)^2\right)^2} \approx Vf^2\,\frac{\pi^{3/2}\tau_i^{\,2}}{V_\pi \ln 2} \tag{2}$$

where τ_i and τ_o are the input and output pulsewidths, respectively, T is the modulation period, and $f = 1/T$. The right hand side of this expression is the form for large modulation index. It is interesting to see that the compresion ratio scales as the square of the modulation frequency for a fixed input pulsewidth. This is a reflection of the de-

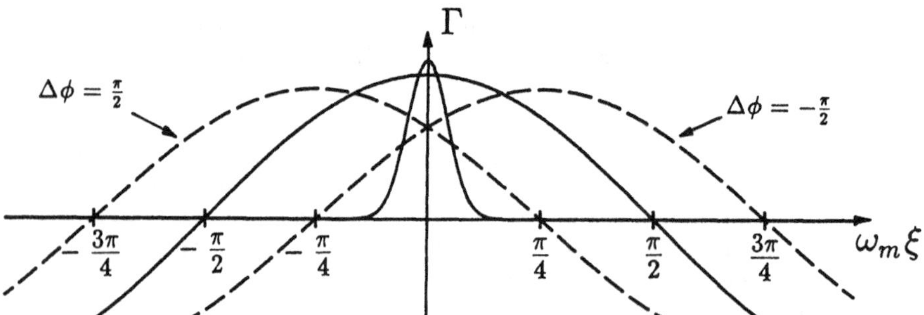

Fig. 1. Retardation as a function of velocity walkoff. Solid cosine: no walkoff; dashed cosines: walkoff of $\pm\pi/2$; solid Gaussian: optical pulse envelope.

pendence of the chirp rate on the 'curvature' of the driving waveform. We also notice in (2) the dependence on T/τ_i, the ratio of the microwave period to the input pulsewidth. Clearly, the best compression ratio is obtained by spreading the pulse over the largest region of quadratic phase. However, when sinusoids are used as the driving waveform, one is limited by the higher order terms far away from the center of the cusps. We have found a practical limit of $T/\tau_i \approx 6$ for producing clean compressed pulses. Input pulsewidths larger than this have enough energy outside the quadratic portion of the RF cycle to induce nonlinear chirp and hence are not effectively compressed by a simple quadratic delay line.

We recently demonstrated this technique by compressing the pulses from a mode-locked Nd:YAG laser from 100 ps to 45 ps [6]. The RF drive waveform had a frequency of 3.365 GHz with a peak power of about 0.5 watt. The average power was reduced to below 10 mW because the higher average power appeared to heat the electrodes and impair waveguiding. Under these conditions, the modulation index was approximately 5 and was limited by the available peak drive power. The delay line was constructed from a single 1800 line/mm diffraction grating used in a four-pass arrangement [8] at grazing incidence. The optimum grating-grating spacing was found to be 1 meter which is equivalent to 2 meters in a single-pass two grating configuration. Using the gratings at grazing incidence resulted in a very low throughput efficiency of several percent but was necessary because of the weak chirp rate. Increasing the modulation index thus has the double advantage of better compression ratio and higher efficiency since the gratings could then be used at less acute angles.

Active pulse compression is very attractive for use with low power solid state and semiconductor diode lasers. It is interesting to consider what is required to generate 1 picosecond pulses using this technique. For the moment let us neglect the problem of inherent chirp in gain-switched diode lasers and assume we can generate 30 ps pulses from such a device. We choose T/τ_i=6 for clean pulses which implies a driving frequency of 5.6 GHz. To get to 1 ps requires a compression ratio of 30:1 and from (2) this implies a modulation index of A=76. With 10 watts peak drive power and a 50Ω transmission line, the modulator needs a half-wave voltage of V_π=1.3 volts. This does not seem unreasonable and we anticipate that such a modulator can be constructed.

References

1. D. Grischkowsky and A. C. Balant, Appl. Phys. Lett., 41, 1 (1982).
2. J. D. Kafka and T. M. Baer, IEEE J. Quant. Elect., 24, 341 (1988).
3. J. P. Heritage, R. N. Thurston, W. J. Tomlinson, A. M. Weiner, and R. H. Stolen, Appl. Phys. Lett., 47, 87 (1985).
4. J. K. Wigmore and D. R. Grischkowsky, IEEE J. Quant. Elect., QE-14, 310 (1978).
5. T. Kobayashi, A. Morimoto, T. Fujita, K. Amano, T. Uemura, and T. Sueta, Ultrafast Phenomena V, (Springer-Verlag, Berlin Heidelberg, 1986), p. 134.
6. B. H. Kolner, Appl. Phys. Lett., 52, 1122 (1988)
7. T. Kobayashi, H. Yao, K. Amano, Y. Fukushima A. Morimoto, and T. Sueta, IEEE J. Quant. Elect., 24, 382 (1988).
8. A. M. Johnson, R. H. Stolen, and W. M. Simpson, Appl. Phys. Lett., 44, 729, (1984).

Two Novel Techniques for Femtosecond Pulse Compression: Utilization of Induced Phase Modulation and Highly Nonlinear Organic-Fibers

M. Yamashita[1], *K. Torizuka*[1], *T. Shiota*[2], *and T. Sato*[1]

[1]Laser Research Section, Electrotechnical Laboratory, Umezono 1-1-4, Tsukuba, Ibaraki 305, Japan

[2]Fujikura Optoelectronics Laboratory, Rokusaki 1440, Sakura, Chiba 285, Japan

1. Femtosecond pulse compression by induced phase-modulation

The technique of pulse compression utilizing self phase-modulation (SPMT) has the disadvantage that it is impossible to control separately the modulation characteristics (i.e. the chirping intensity and linearity, the dispersion's sign and value, and the modulation fluctuation) and the parameters (i.e. the power, the wavelength and spectral shape, the temporal duration and shape, the chirping characteristics and the intensity fluctuation) of the input-pulse to be compressed. In practice the SPMT cannot be applied to compression of an ultrashort pulse at the wavelength where it is difficult to generate a high peak-power pulse. In addition, the SPMT does not allow us to adjust finely and independently the chirping intensity and linearity. Furthermore, the temporal fluctuation of the highly-amplified input pulse in the SPMT causes severe fluctuation of the compressed pulse because of the effect of the multiplication of the modulation fluctuation and the intensity fluctuation of the pulse to be compressed. In order to overcome the disadvantage, we propose a compression technique utilizing induced phase-modulation [1], where the I_1-pulse with a high peak-power inducing the phase-modulation and the I_2-pulse with a low and stable peak-power to be compressed are independently propagated in a single-mode fiber followed, usually but not always, by a grating pair (IPMT). The results of a computer analysis based on two nonlinear Schrödinger equations [2] are presented for typical cases.

In case A (the same 630 nm - wavelength for the I_1- and I_2-pulses), it is found that the 70.7 fs, 100 W I_2-pulse is compressed to 5 fs (or 15 fs) by the modulation induced by the 100 fs, 1 MW (or 100 kW) I_1-pulse followed by a grating pair. This means that a compression ratio comparable to the case of the SPMT is observed. In case B (the wavelengths different in the normal dispersion region for the I_1- and I_2-pulses), the 70.7 fs, 100 W I_2-pulse at 315 (SH pulse), 550 or 800 nm are, respectively, compressed to 15, 12 or 13 fs by the same I_1-pulse with the delay time adjustment of 10, 8.5 or -10 fs. The compressed pulse-duration becomes longer due to the walk-off between those pulses in comparison with case A, which is presumably avoided by using a combination of a phase-matching fiber and two cross-polarization beams. In case C (the wavelength different in the normal (I_1-pulse at 1.06 μm) and anomalous (I_2-pulse at 1.5 μm) regions), the 20 ps, 1 W I_2-pulse is compressed to 1.8 ps (or 1.5 ps) by only the fiber propagation with the modulation by the 20 ps, 100 W (or 10 kW) I_1-pulse in the case of group-velocity matching (or of the walk-off).

From the above analysis, we find the IPMT has the following advantages: (1) one can select, as the I_1 modulation pulse, a pulse of a wavelength so

Springer Series in Chemical Physics, Vol. 48: **Ultrafast Phenomena VI**
Editors: T. Yajima · K. Yoshihara · C.B. Harris · S. Shionoya

that the high peak-power and ultrashort pulse-duration are easily generated (e.g. 630 nm or 1.06 μm), while as the I_2-pulse to be compressed a pulse of low and stable peak-power at any different wavelength can be selected; (2) when the wavelength of the I_2-pulse is in the anomalous-dispersion region, the weak I_2-pulse such as a semiconductor laser pulse is compressed by the fiber propagation only; (3) while monitoring the rapid-scanning autocorrelation traces of the compressed I_2-pulse on an oscilloscope, the fine adjustment for the modulation characteristics can be independently done without changing I_2 input-pulse parameters; (4) the severe fluctuation of the compressed pulse due to the above-mentioned multiplication effect in the SPMT can be avoided since a very stable I_2-pulse of a low peak-power can be utilized. The latter two points are the most important for the compression of ultrashort pulses in the femtosecond time-region where the higher-order dispersion effect and the temporal pulse-fluctuation severely affect the practical compression ratio.

2. Organic fibers for a femtosecond pulse compressor

The compression ratio by the SPMT is determined by the intensity of incident pulses, the nonlinear refractive index (NRI) of the fiber glass and the group velocity dispersion (GVD) [3]. Some kinds of organic materials (e.g. benzene derivatives) are attractive candidates for a compressor of low peak-power ultrashort pulses becauses of its extremely large NRI contributed by π-electrons [4] and moderate GVD. We report an experimental study on the properties of organic fibers in the femtosecond region.

Our fibers consist of a 1 cm long fused-quartz capillary and a 4 μm radius organic-liquid core. Sample (a) is a multimode fiber of pure nitrobenzene and Sample (b) is a single-mode one of a mixture of 2-methyl-4-nitroaniline, nitrobenzene and ethanol. In both of (a) and (b), electronic (fast-response) NRI and GVD are expected to be about 70 times and 5 times of those in a glass fiber, respectively.

Figure 1 shows typical results of the background-free autocorrelation measurements. The peak-power of the incident pulse (i : a colliding pulse mode-locked ring laser) in the fibers is not so strong that self phase-modulation (SPM) occurs. The output-pulse of Sample(a) is broadened by not only material GVD but also additional mode-dispersion. That of Sample(b) is much shorter and is really recompressed (r) by the negative GVD of a grating-pair.

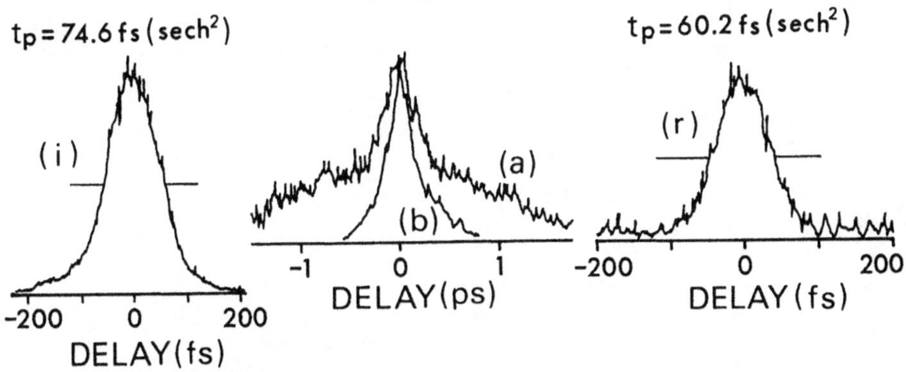

Fig.1 Non-collinear SHG autocorrelation traces of the incident pulse (i), the transmitted pulses through the organic fibers (a, b) and the recompressed pulses (r).

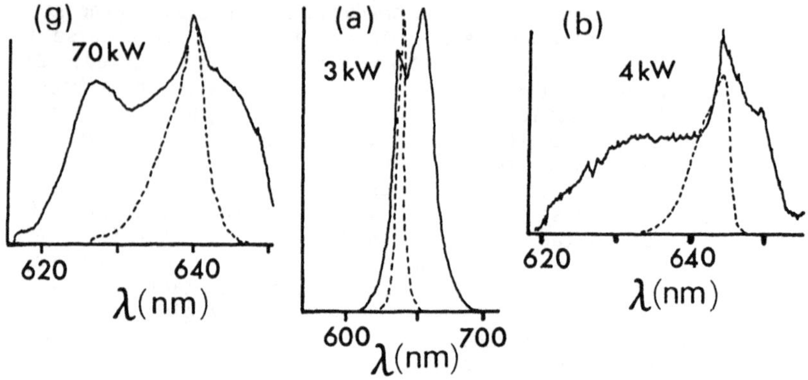

Fig.2 Spectra of the incident pulses (dashed curves) and the modulated pulses (solid curves). Peak-power of the incident pulses is also shown.

The spectrum-broadening by SPM is measured by amplified and GVD-compensated 70 fs pulses (the repetition rate: 10 pps), as shown in Fig.2. Curve (g) is the result of the single-mode glass fiber. In Sample (a), a large asymmetric broadening to the longer-wavelength side occurs. Here, the mode-dispersion enhances a contribution of the slow-response NRI by molecular reorientation. In Sample (b), the observed modulation is similar to that in (g) by the fast-response NRI, where the incident peak-power is lower about one order of magnitude. The pulse of (g) is compressed to 32 fs by a grating-pair. However, the temporal-characters of modulated-outputs of organic fibers cannot be measured because of the lower peak-power. We consider that a high-repetition laser (e.g. a cavity-dumped laser) is a suitable source for organic-fiber compression.

References

1. R.R.Alfano, Q.X.Li, T.Jimbo, J.T.Manassah and P.P.Ho: Opt.Lett. 11, 636 (1986).
2. A.Hasegawa and Y.Kodama: Proc. of IEEE, 69, 1145 (1981).
3. W.J.Tomlinson, R.H.Stolen and C.V.Shank: J.Opt.Soc.Am. B1, 139 (1984).
4. B.F.Levine and C.G.Bethea: J.Chem.Phys. 63, 2666 (1975).

Generation of Sub-100-fs Pulses at 532 nm from Modulation Instability Induced by Cross-Phase Modulation in Optical Fibers

P.L. Baldeck[1], *R.R. Alfano*[1], *and G.P. Agrawal*[2]

[1]The City College of New York, New York, NY 10031, USA
[2]AT&T Bell Laboratories, Murray Hill, NJ 07974, USA

Modulation instability (MI) refers to the sudden break-up of waves propagating in nonlinear dispersive media. MI is a possible method to generate ultrashort pulses with gigahertz and terahertz repetition rates [1]. Tai et al. have observed the generation of 500 fsec pulses at 1.319 um with a 0.3 THz repetition rate [2,3]. Recently, Agrawal suggested that a new kind of modulation instability can occur even in the normal-dispersion regime when two copropagating fields interact with each other through the nonlinearity-induced cross-phase modulation [4].

In this paper, we report the first observation of such a modulation instability initiated by cross-phase modulation in the normal-dispersion regime of silica optical fibers. MI is observed at 532 nm by propagating intense mode-locked picosecond pulses or Q-switched nanosecond pulses in short lengths of a single-mode optical fiber. Modulation-instability sidebands appear in the pump spectrum as a result of cross-phase modulation induced by the simultaneously generated Raman pulses.

Figure 1 shows spectra of intense 25-ps pulses recorded for different fiber lengths. Figure 1-a is the reference spectrum of low-intensity pulses. Figures 1-b and 1-c show spectra measured at about the modulation-instability threshold for fiber lengths of 3 m and 0.8 m, respectively. The latter two figures show modulation-instability sidebands on both sides of the laser wavelength at 532 nm, and the first-order stimulated Raman scattering (SRS) line at 544.5 nm. Notice that the frequency shift of sidebands is larger for the shorter fiber. Secondary sidebands were also observed for pulse energies well above the modulation instability threshold and longer optical fiber lengths. According to the theory, modulational instability at this wavelength is possible only if there is a cross-phase modulation interaction [4]. As shown in Fig. 1, modulation-instability sidebands were observed only in presence of stimulated Raman scattering light. Moreover, it has been recently demonstrated that cross-phase modulation is intrinsic to the SRS process [5,7].

To strengthen our conclusion that the sidebands arise from the modulation instability induced by cross-phase modulation, we have measured and compared with theory the dependence of sideband shifts on the fiber length. For this measurement, we used 10-ns pulses from a Q-switched Nd:YAG laser to ensure the

Springer Series in Chemical Physics, Vol. 48: **Ultrafast Phenomena VI**
Editors: T. Yajima · K. Yoshihara · C.B. Harris · S. Shionoya

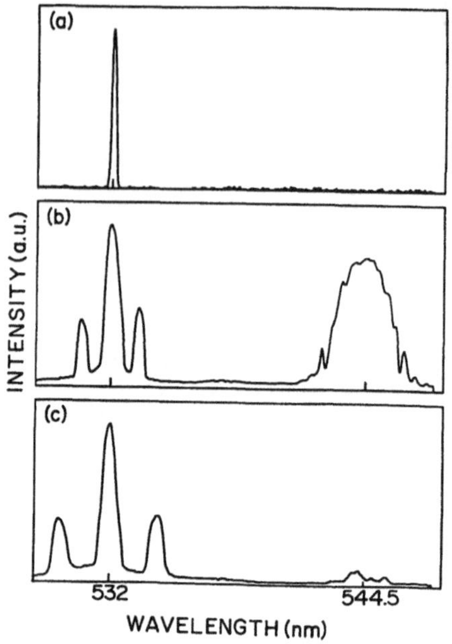

Figure 1. XPM-induced modulation instability sidebands for different fiber lengths. The laser line is at 532 nm and the Raman line at 544.5 nm;
a) reference spectrum;
b) measured spectrum near the MI threshold for L=3 m; c) L=0.8 m

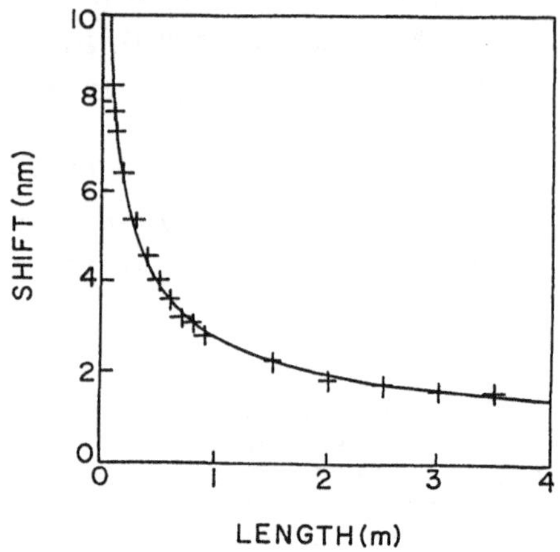

Figure 2. Sideband shift versus fiber length near the modulation instability threshold. The solid line is the theoretical fit

quasi-CW operation. The spectra were similar to those obtained with 25-ps pulses (Fig.1). As shown in Fig.2, the sidelobe separation, defined as the half-distance between sideband maxima, varied from 1.5 nm to 8.5 nm for fiber lengths ranging from 4 m to 0.1 m, respectively. The energy of input pulses was approximately set at the modulation instability threshold for each fiber length. The solid line in Fig.2 corresponds to the theoretical fit. As discussed in [4], the maximum gain of modulation instability sidebands is given by $G=k''w^2$ where w is the sideband frequency shift. Thus, the power of a sideband for an optical fiber length L is given by

$$P(w,L) = Pn \exp(k''w^2 L) \tag{1}$$

where Pn is the initial spontaneous noise and k" is the group velocity dispersion at the laser frequency.
For such amplified spontaneous emission, it is common to define a threshold gain Go by

$$Po(L) = Pn \exp(Go) \tag{2}$$

where Po is the sideband power near threshold such that each sideband contains about 10% of the input energy. A typical value for Go is 16. [2]

From (1) and (2), the dependence of the sideband shift with the fiber length near threshold is given by

$$w = (Go/k''L)^{1/2} . \tag{3}$$

At 532 nm, the group velocity dispersion is $k''=0.06\ ps^2/m$ Using this value and Go=18.1 in (3), we obtain the theoretical fit shown in Fig.2 (solid line). The good agreement between the experimental data and the theory of modulation instability supports our belief that we have observed cross-phase-modulation-induced modulation instability, as predicted in [4].

The maximum sideband shift shown in Fig.2 is 8.5 nm, which corresponds to the generation of sub-100-fsec pulses within the envelope of the 10-ns input pulses with a repetition rate of about 8.5 THz. Even though autocorrelation measurements were not possible, because of the low repetition rate (10Hz) needed to generate pulses with kW peak powers, we believe that we have generated for the first time modulation instability subpulses shorter than 100 fs.

The research at CCNY is supported in part by Hamamatsu Photonics K.K.

References:

1. A. Hasegawa, and W.F.Brinkman: IEEE J. Quantum Electron. QE-11, 100 (1975).
2. K. Tai, A. Hasegawa, and A. Tomita: Phys. Rev. Lett. 56, 135 (1986).
3. K. Tai, A. Tomita, J.L. Jewell, and A. Hasegawa: Appl. Phys. Lett. 49, 236 (1986).
4. Govind P. Agrawal: Phys. Rev. Lett. 59, 880 (1987).
5. M.N. Islam, L.F. Mollenauer, R.H. Stolen, J.R. Simson, H.T. Shang: Opt. Lett. 12, 625 (1987).
6. R.R. Alfano, P.L. Baldeck, F. Raccah, and P.P. Ho: Appl. Opt. 26, 3491-3492 (1987).
7. Jamal T. Manassah: Appl. Opt. 26, 3747-3749 (1987).

Femtosecond Continuum Generation in Fibers Near 1.6 μm

M.N. Islam, G. Sucha, J.P. Gordon, I. Bar-Joseph, and D.S. Chemla

AT&T Bell Laboratories, Room 4D-411, Holmdel, NJ 07733, USA

We report the generation of continuum with $\gtrsim$ 100fsec duration between 1.55 and 1.85μm in a fiber pumped by a color center laser (CCL). Working in the anomalous group velocity dispersion (GVD) regime of fibers, several groups have generated short pulses with extremely broad bandwidth , which they attribute to stimulated Raman scattering [1]. However, our experiments and computer simulations show that the continuum evolves from multi-soliton collisions resulting from modulation instability (MI)[2] and soliton self-frequency shift (SSFS) effects [3]. We demonstrate the utility of the continuum for spectroscopy by using the continuum in pump-probe, absorption saturation experiments in InGaAs/InP multiple quantum wells.

In the anomalous GVD regime, an intense, broad pulse is unstable to perturbations, and will try to break up into a train of solitons. MI amplifies a weak signal or perturbation through the interplay between anomalous GVD and the fiber nonlinearity. As MI starts, both upper and lower sidebands are generated symmetrically about the pump, and the beat frequencies compress through the addition of more sidebands. However, the Raman or SSFS effect translates the envelope of the spectrum to longer wavelengths, while preserving the periodicity of the sidebands [2]. As the solitons compress further, SSFS causes a down shift in the frequency of the pulses, with a consequent shift in the velocity of the solitons. However, the shift will be nonuniform: the solitons near the center of the pump are larger and are generated sooner than the others. As they accelerate toward the rear of the pump pulse, they catch up with their neighbors, so that after a time there forms a short region in which a group of the solitons collide. High peak powers and narrow pulses are generated during these collisions, generally resulting in one dominant soliton.

In the experiments a modelocked NaCl color center laser provides the $\lambda\sim$ 1.5μm pump pulses with $P_{peak} \lesssim$ 100W (in the fiber) and $\tau\sim$ 12psec at 100MHz repetition rate. Following an isolator and a telescoping set of lenses, the light is coupled into a 100m or 500m length of single mode, polarization maintaining fiber with a zero dispersion wavelength of 1.41μm. The output from the fiber is sent to a scanning Fabry-Perot, a spectrometer, and an autocorrelator.

Springer Series in Chemical Physics, Vol. 48: **Ultrafast Phenomena VI**
Editors: T. Yajima · K. Yoshihara · C.B. Harris · S. Shionoya

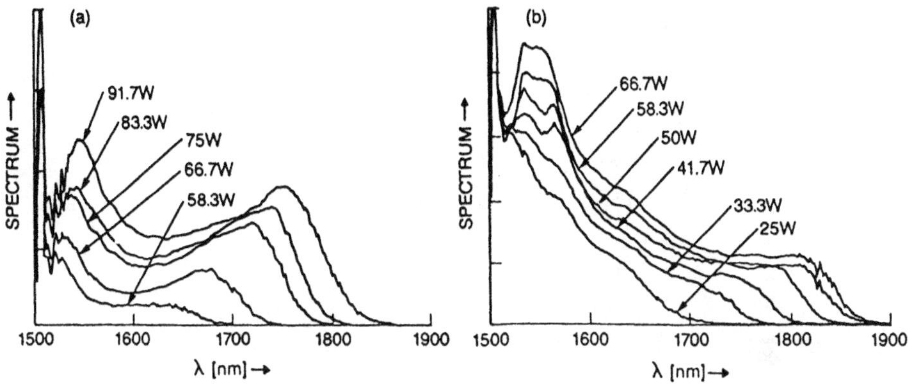

Fig. 1 Continuum spectra versus pump power from (a) 100m and (b) 500m fibers

In Fig. 1a we show the spectra for different pump powers in a 100m length of fiber. As the power is increased, the spectrum reaches to longer wavelengths, but the spectral broadening slows down beyond 1.85μm. The sharp peak at $\sim$ 1.51μm corresponds to the residual pump, and the features between 1.53 and 1.58μm correspond to the MI peaks in the leading half of the pump (which do not collide). At higher pump powers the energy in the continuum shifts to longer wavelengths as SSFS slides more energy to the Stokes side. Note that almost half of the pump energy is in the continuum.

To understand the temporal behavior of various parts of the continuum, we introduced band pass filters before the autocorrelator. In Fig. 2 we show a typical autocorrelation after a 1.7μm filter (band pass of 94nm, 100m fiber, P = 75W). As the pump is varied between 66 W and 100W, the pulse width from the 100m fiber varies between 145 $\leq \tau \leq$ 160 fsec after the 1.7μm filter, and between 116 $\leq \tau \leq$ 132 fsec after a 1.635μm filter (band pass 100nm). However, the level of pedestal increases as we go to shorter wavelengths. For example, for different filters we find (center band pass wavelength, peak-to-pedestal ratio): (a)1.57μm, 13:1 ; (b)1.6μm, 56:1 ; (c)1.635μm, 71:1 ; and, (d) 1.7μm, 153:1. We believe that the randomly timed (since they start from noise) MI solitons in the front half of the pump (which do not collide) contribute significantly to the pedestal. Furthermore, for 83W of pump power, the measured peak powers after the 1.635μm (1.7μm) filters are 1.78kW (1.33kW).

To show the usefulness of the continuum for spectroscopy, we used the fiber output in pump-probe, absorption saturation experiments in InGaAs/InP multiple quantum wells. Since we use the continuum (after appropriate band pass filters) as both the pump and probe, the timing of the femtosecond pulses relative to the color center laser pulses is unimportant. In Fig. 3 we show a typical transmission versus delay for counter-

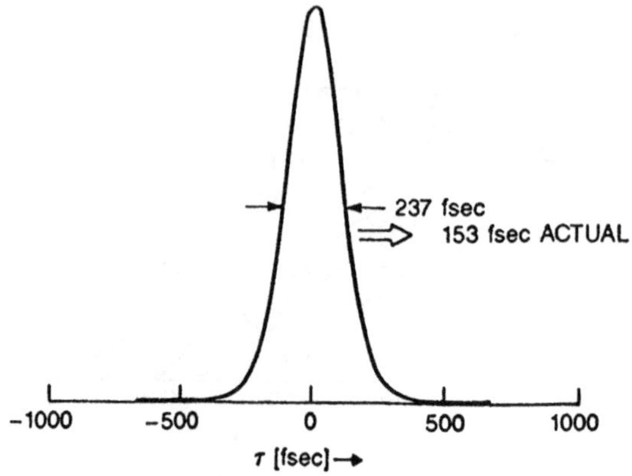

Fig. 2 Autocorrelation of pulses from 100m fiber (P=75W) after 1.7μm filter

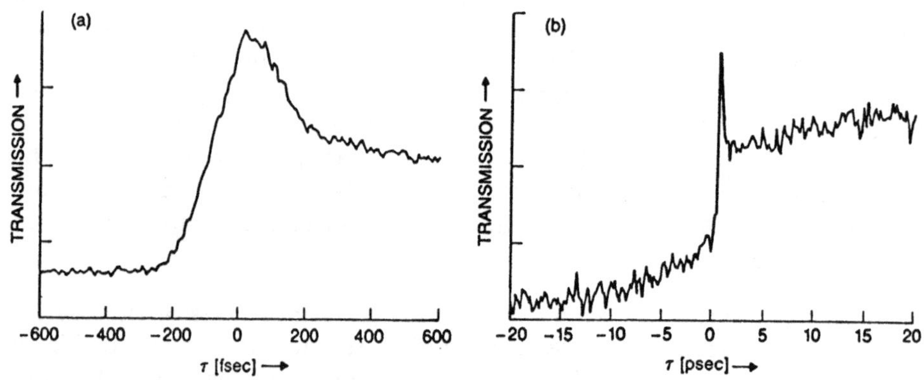

Fig. 3 Transmission versus delay from pump-probe experiments in InGaAs/InP multiple quantum wells (after 1.6μm filter)

propagating pump and probe pulses. We use a 1.6μm filter to pump and probe near the heavy hole exciton feature in the samples. The rise time is $\lesssim$200fsec, and the decay time is initially ~300 fsec, followed by a much slower recovery. The first recovery time results from ionization of excitons, and the second corresponds to the band-to-band recombination time. Although the levels of pedestal are almost immeasurable in the autocorrelations, they show up distinctly in the pump-probe since they are wide and have considerable energy.

To see what happens as the continuum propagates further in the fiber, we studied the output of a 500m length of identical fiber. The spectra (Fig. 1b) reach out as far as 1.88μm, which is slightly further than in the 100m fiber. The threshold for the continuum is lower: even at 25W the spectrum reaches beyond 1.7μm. Again, we see the pump and MI peaks, but the MI peaks are broader. This broadening is attributable to SSFS: a 270fsec pulse at $\lambda\sim$ 1.5μm down shifts by ~18nm in 500m of fiber because of SSFS. Unlike the 100m fiber results, the spectra are almost flat at the long wavelength side before they roll off. However, the spectra show a second feature between 1.6 and 1.65μm, which may be due to Raman amplification ($\sim$ 100nm is peak Raman gain) by the MI solitons.

Unlike the 100m results, in 500m at P$\sim$ 83W there is considerable pedestal for $\lambda \lesssim$ 1.65μm. Although the pedestal is as broad as the pump ($\tau\sim$ 12psec) for the 100m fiber, in 500m the pedestal reaches out to beyond 100 psec. On the other hand, at and above 1.7μm the pulses from the 500m fiber show less than 200:1 peak-to-pedestal ratio in autocorrelation (this is the flat part of the continuum in Fig. 1b). After the 1.7μm filter, the pulse that is 153 fsec at 100m broadens to $\sim$262 fsec in 500m: this is more than the broadening expected from linear loss for a fundamental soliton. The width varies between 262 fsec at 1.7μm and 320 fsec at 1.82 μm.

The short pulse must be a soliton to persist for 500m, but a soliton should be transform limited (for 100fsec $\Delta\lambda\sim$ 20nm). We hypothesize that the broad spectra in Fig.1 represent an ensemble average of narrow $\sim$100fsec pulses which have SSFS'd differently. A 100 fsec pulse rapidly SSFS's to longer wavelengths in a fiber. For example, if the 100 fsec pulse width does not broaden, then the pulse shifts by $\Delta\lambda\sim$150nm in $\sim$30m of fiber. Since the process starts from noise, each pulse will be formed at a different distance in the fiber, and will shift by a different amount. After we ensemble average over many such pulses, we obtain the smooth, continuum spectra of Fig. 1.

SSFS acts like a loss mechanism as the non-soliton part separates from the main soliton. A fundamental soliton pulse adiabatically broadens to compensate for the lower pulse power. Consequently, our pulses broaden as we go to longer fibers, as we increase the pump power, or as we go to longer wavelengths. Furthermore, the spectral broadening does not continue indefinitely because SSFS is a self-limiting process. As the pulse broadens, SSFS slows down since the shift $\Delta\nu \sim 1/\tau^4$. Also, for $\lambda >$ 1.7μm the fiber losses increase, preventing broadening much beyond 1.85μm.

In summary, we find that MI and SSFS play crucial roles in starting the continuum generation process: MI initiates the process by providing the short pulses, and SSFS causes relative velocity shift between pulses and amplification when they overlap. The short pulses generated through this multi-soliton collision then SSFS to give the broad spectral widths. The high peak powers ($>$ 1kW) and short pulses ($\tau \gtrsim$ 100fsec) combined with the 100MHz repetition rate make the broad continuum a simple, yet powerful, tool for spectroscopy.

[1] K.L. Vodop'yanov, A.B. Grudinin, E.M. Dianov, L.A. Kulevskii, A.M. Prokhorov and D.V. Khaidarov, Sov. J. Quantum Electron. **17** , 1311 (1987) ; A.S. Gouveia-Neto, A.S.L. Gomes and J.R. Taylor, IEEE J. Quantum Electron. **24** , 332 (1988).
[2] M.N. Islam, S.P. Dijaili and J.P. Gordon, Opt. Lett. **13** , 518 (1988) ; A. Hasegawa and W.F. Brinkman, IEEE J. Quantum Electron. **QE-16** , 694 (1980).
[3] J.P. Gordon, Opt. Lett. **11** , 662 (1986).

Picosecond Pulse Generation by Two-Photon Induced Amplified Spontaneous Emission

*A. Penzkofer and P. Qiu**

Naturwissenschaftliche Fakultät II – Physik,
Universität Regensburg, D-8400 Regensburg, Fed. Rep. of Germany
*On leave from Shanghai Institute of Optics and Fine Mechanics,
Academia Sinica, Shanghai, People's Republic of China

1. Introduction

Two-photon pumped travelling-wave dye lasers [1,2] and semiconductor lasers [3,4] generate frequency tunable ultrafast light pulses at the anti-Stokes side of the pump laser. Here the two-photon induced amplified spontaneous emission (ASE) in a dye generator cell, the two-photon induced seeding pulse amplification (SPA) of a picosecond light continuum and the signal amplification (AMP) in a two-photon pumped dye amplifier are studied.

2. Experimental

The experimental setup is depicted in Fig.1. Single picosecond light pulses of a passively mode-locked Nd-phosphate glass laser (pulse duration $\Delta t_L \simeq$ 5 ps) are used for two-photon pumping. The dye rhodamine B dissolved in hexafluoroisopropanol serves as gain medium in the generator cell and the amplifier cell (concentrations 0.01 mol/dm³, cell lengths 2 cm). In the case of two-photon pumped seeding pulse amplification a picosecond light continuum is generated in a 2 cm long cell filled with D_2O [5]. Spectral tuning of the generated signal is possible by putting a tuning element behind the generator cell.

3. Results

The absorption and emission cross-section spectrum of rhodamine B in hexafluoroisopropanol is shown in Fig.2. The spectral widths at half maximum (FWHM) and one tenth height of the ASE signal are indicated. Efficient two-photon pumped amplified spontaneous emission requires remarkable two-

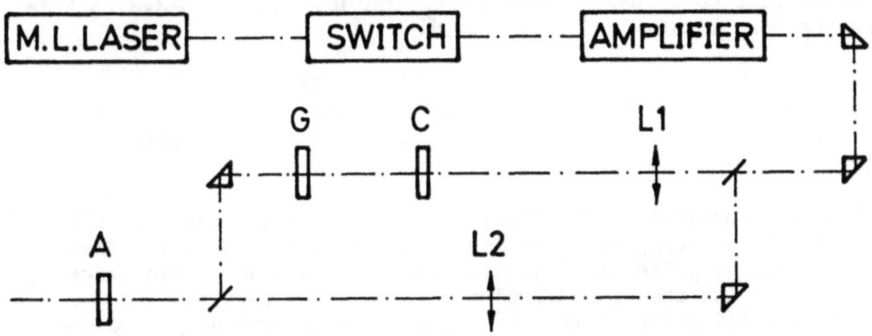

Fig.1 Experimental setup. L1, L2, lenses. C, picosecond continuum generation in D_2O cell. G, generator cell. A, amplifier cell.

Springer Series in Chemical Physics, Vol. 48: Ultrafast Phenomena VI
Editors: T. Yajima · K. Yoshihara · C.B. Harris · S. Shionoya

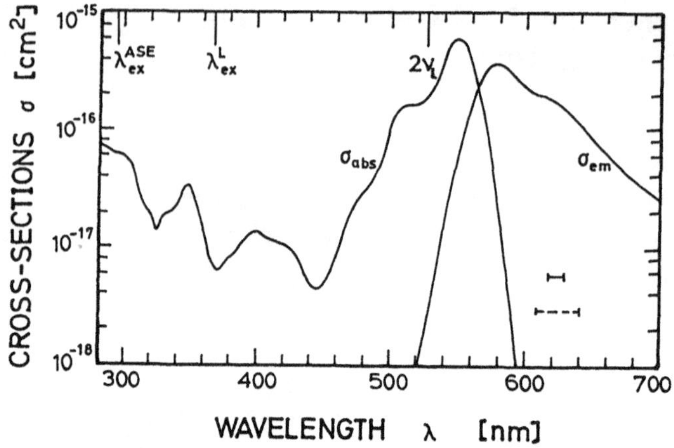

Fig.2 Absorption and emission spectrum of rhodamine B in hexafluoroisopropanol. Horizontal bars indicate spectral widths of ASE signal. Solid bar, FWHM; dashed bar, width at one tenth height.

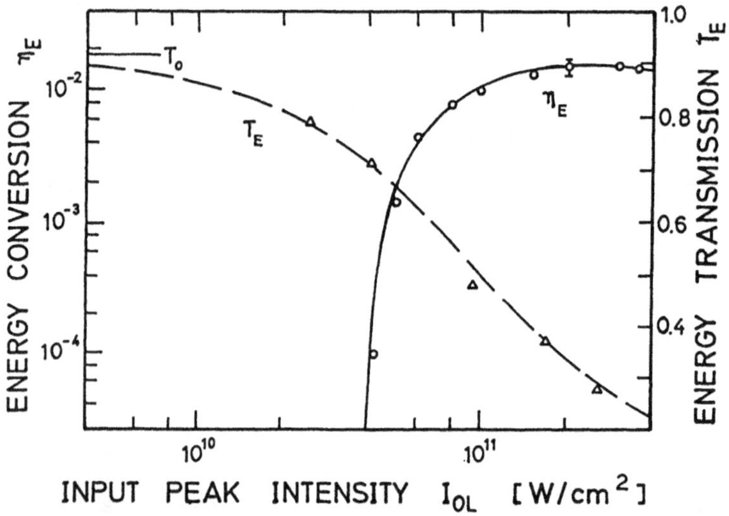

Fig.3 ASE energy conversion efficiency η_E and pump pulse energy transmission T versus pump pulse peak intensity I_{0L} for 0.01 molar rhodamine B in hexafluoroisopropanol. Sample length ℓ = 2 cm.

photon absorption (Fig.3), large stimulated emission cross-sections (Fig.2) and weak excited-state absorption cross-sections for the pump laser (λ_{ex}^{L}) and the generated ASE signal (λ_{ex}^{ASE}).

The ASE energy conversion efficiency versus pump laser peak intensity is depicted in Fig.3. The pump pulse transmission is included. The two-photon pumped seeding pulse amplification reduces the signal divergence $\Delta\Theta$ considerably while the energy conversion efficiency is only reduced slightly. The amplification of the ASE signal in a two-photon pumped dye amplifier cell also reduces strongly the divergence and increases the energy conversion efficiency. The results are summarized in Table 1.

Table 1 Two-photon pumped light generation. The pump pulse parameters are $I_{0L} = 1.5\times10^{11}$ W/cm², $\lambda_L = 1.054$ µm, $\Delta t_L = 5$ ps. ASE-AMP, amplification of ASE signal. SPA-AMP, amplification of seeding-pulse-amplification signal.

Parameter		ASE	SPA	ASE-AMP	SPA-AMP
η_E		0.013	0.003	0.018	0.035
λ_{max}	[nm]	622	617	623	620
$\Delta\lambda$ (FWHM)	[nm]	13	20	19	15
$\Delta\Theta$ (FWHM)	[rad]	2×10^{-2}	2.5×10^{-3}	7×10^{-4}	7×10^{-4}

The two-photon pumped amplified spontaneous emission may be extended to other laser dyes and saturable absorbers [6]. Other picosecond pump lasers may be applied in order to enlarge the frequency range. Computer simulations indicate that at high pump pulse intensities the ASE signals become considerably shorter than the pump pulses, especially for saturable absorbers with fast absorption recovery time [6].

References

1. V.I. Prokhorenko, E.A. Tikhonov: Sov. J. Quant. Electron. 16, 1214 (1986).
2. A. Penzkofer, W. Leupacher: Opt. Quant. Electron. 19, 327 (1987).
3. W.L. Cao, A.M. Vaucher, C.H. Lee: Appl. Phys. Lett. 38, 306 (1981).
4. A. Penzkofer, A.A. Bugayev: Opt. Quant. Electron., submitted.
5. A. Penzkofer, W. Kaiser: Opt. Quant. Electron. 9, 315 (1977).
6. P. Qiu, A. Penzkofer: Appl. Phys. B, submitted.

Mode-Locked Continuous Wave Titanium Sapphire Laser

J.D. Kafka, A.J. Alfrey, and T. Baer

Spectra-Physics, Inc., 1250 West Middlefield Road,
Mountain View, CA 94042, USA

We have actively mode-locked a continuous wave (cw) Titanium doped Sapphire laser and obtained tunable pulses as short as 6 psec. At the peak of the tuning curve (790 nm) we obtained 750 mW of average power. We have also frequency doubled the output of this laser with 15% efficiency using Potassium Niobate.

Recently, Sanchez et al. [1] reported 1.6 W of output power from the Titanium doped Sapphire laser pumped by an Argon ion laser. Several groups have actively mode-locked this laser and obtained pulses as short as 100 psec and powers of a few hundred mW [2-4].

We have used a cw Argon laser running all lines (15 W) to pump the Titanium doped Sapphire laser. Using a 4 mirror astigmatically compensated cavity and a 20 mm long Brewster cut crystal we have obtained 3.5 W of average power. The cavity was then extended to a length of 1.8 meters and we placed a high Q, high modulation depth, acousto-optic mode-locker close to one end mirror.

We have generated mode-locked pulses as short as 6 psec and with 750 mW of average power at 790 nm. An autocorrelation of the pulse is shown in Figure 1. The bandwidth of these pulses was 150 Ghz which was 3 times the transform limit. We used a three plate birefringent tuner to control the wavelength and were able to tune the laser from 770 nm to 860 nm with the

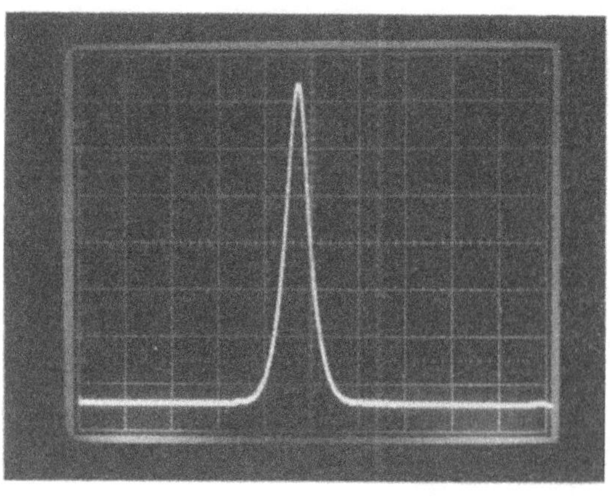

Fig. 1 Autocorrelation of a pulse from the mode-locked Titanium doped Sapphire laser at 790 nm. The FWHM is 6 psec assuming a hyperbolic secant squared pulseshape.

Springer Series in Chemical Physics, Vol. 48: **Ultrafast Phenomena VI**
Editors: T. Yajima · K. Yoshihara · C.B. Harris · S. Shionoya

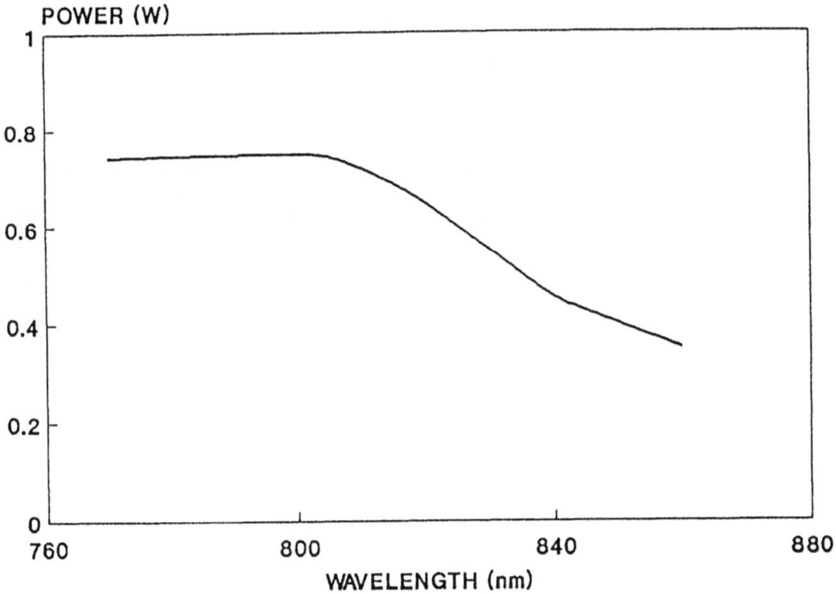

Fig. 2 A graph of the output power versus wavelength for the mode-locked Titanium doped Sapphire laser. The tuning range was limited by intracavity coatings.

pulsewidth remaining below 10 psec. This tuning range was limited by the intracavity coatings. In Fig. 2 we show the power as a function of wavelength. We anticipate that it will be possible to generate pulses of less than 10 psec duration across the entire tuning range of the laser (680 nm to 1000 nm).

Short pulses were obtained over a 20 micron range of cavity lengths. The shortest pulses were generated when the cavity was mismatched slightly from the mode-locker driver frequency. One consequence of this mismatch is a small regular oscillation in the time domain. The oscillation had a depth of a few percent and a period of several microseconds.

An excellent frequency doubling crystal for the wavelength range between 840 nm and 940 nm is temperature tuned Potassium Niobate. At room temperature, this crystal is 90 degree phase matched for doubling 860 nm light and we obtained 350 mW of power at this wavelength. We used a 5X microscope objective to focus into the 3 x 3 x 5 mm crystal and generated 50 mW at 430 nm.

In conclusion, we have actively mode-locked and frequency doubled the cw Titanium doped Sapphire laser. The fundamental and second harmonic of this laser cover two regions of the spectrum where the existing dye lasers have limited lifetime. As a result, we believe that the Titanium doped Sapphire laser is a viable solid state alternative to dye lasers as a picosecond source.

References

1. A. Sanchez, R. E. Fahey, A. J. Strauss, and R. L. Aggarwal, Opt. Lett. 11, 363 (1986).
2. R. Roy, P. A. Schultz and A. Walther, Opt. Lett. 12, 672 (1987).
3. J. T. Darrow and R. K. Jain, in Digest of Conference on Lasers and Electro-Optics, Paper ME1, (1988).
4. P. A. Schultz, IEEE J. Quantum Electron., vol. QE-24, 1039 (1988).

Non-soliton Modelocking of an F-Centre Laser with a Nonlinear External Cavity

K.J. Blow, D.S. Forrester, and B.P. Nelson

British Telecom Research Laboratories, Ipswich, IP5 7RE, UK

The soliton laser [1] was the first example of a nonlinear external cavity being used to improve the modelocked performance of a laser. The laser cavity was coupled to an external cavity which contained an optical fibre with positive group delay dispersion thus supporting solitons. When the returned pulse is synchronous with the pulse in the fundamental laser cavity then pulses with a much shorter duration than those produced by the laser alone are observed. The first explanation of this phenomenon was in terms of the properties of the $N=2$ soliton. However, a theory of the soliton laser [2] and further experimental studies [3] have both indicated the existence of many operating points which were not always related to the $N=2$ soliton [1].

A more recent theory [4] has considered the coupling of a modelocked laser to an external cavity containing other nonlinear elements. The two nonlinearities considered were a saturable absorber, giving pulse compression, and a saturable amplifier, which leads to pulse broadening. The surprising result was that both nonlinear responses can lead to the production of shorter pulses. The proposed explanation for this effect is that the nonlinearity in the external cavity causes coupling between the longitudinal modes of the laser which enables the better communication of phase information between the modes with a consequent increase in the modelocked bandwidth. Thus the behaviour of the soliton laser should be understood as part of a more general class of nonlinear external cavity modelocked lasers.

Here we report on some experiments which will elucidate the role of the external cavity. A Burleigh Tl:KCl colour centre laser was end-fire coupled to an optical fibre. The external cavity was formed either by retro-reflection from a plane mirror or by using an integrated fibre loop mirror [5]. The crucial difference between our experiment and the soliton laser is that the fibre had negative group delay dispersion and does not support (bright) solitons. In this dispersion regime the combined effects of the nonlinear response of the fibre and its dispersion lead to enhanced pulse broadening [6], [7] as we will later show.

Springer Series in Chemical Physics, Vol. 48: **Ultrafast Phenomena VI**
Editors: T. Yajima · K. Yoshihara · C.B. Harris · S. Shionoya

In figure 1 we show (normalised) background free autocorrelation traces of the laser alone (a) and the laser coupled to the external cavity (b). The reduction in pulse autocorrelation width can clearly be seen, from 30*ps* to 5*ps*, and is quite similar to results obtained with the soliton laser [1]. This result was obtained with a 20*m* length of optical fibre at an average launched power of 5*mW* which was close to the threshold power for pulse shortening to be observed. Similar results were also obtained with shorter fibre lengths with the pulse width generally decreasing with fibre length. The implication of this result is that it is not essential for the optical fibre to support solitons merely that it have some nonlinear response.

The group delay dispersion of this fibre was measured on a 1*km* length to be $-10 ps/nm/km$. However, in order to be quite sure that the actual length of fibre used in this experiment had negative dispersion and was not positive due to some longitudinal inhomogineity in the fibre we performed the following test. We set up the standard soliton laser configuration to produce pulses with an autocorrelation FWHM of 2ps. These pulses were then launched into the same 20*m* length of fibre used in the first experiment and figure 2 shows the input (a) and output (b) autocorrelation traces at a launched power of 5*mW*. The output pulse is broader by a factor of about three and the shape is somewhat triangular which is consistent with the pulse shaping effects normally observed in the negative group delay dispersion regime. This result is also consistent with the solution of the Nonlinear Schrodinger Equation for the experimental parameters used. These two measurements clearly indicate that the fibre used in the modelocking experiment did not support soliton solutions.

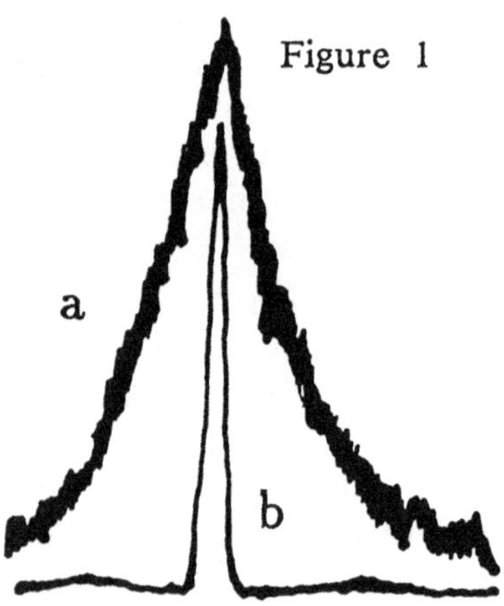

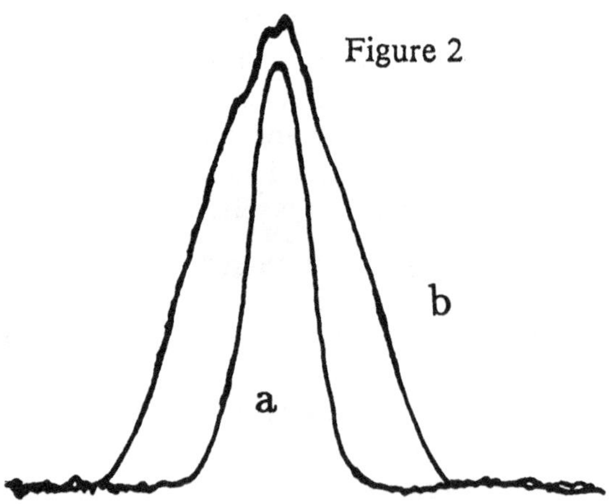

Figure 1. Laser output without optical feedback (a) and with optical feedback (b).

Figure 2. Input pulse to non-soliton fibre (a) and output pulse (b).

In conclusion we have observed behaviour similar to the soliton laser using an optical fibre which does not support solitons at $1.5\mu m$.

We acknowledge the director of Research and Technology for permission to publish.

REFERENCES

1. L. F. Mollenauer and R. H. Stolen, The Soliton Laser, Optics Lett 9 13-15 (1984)
2. K.J.Blow and D.Wood, Stability and Compression of Pulses in the Soliton Laser, IEEE JQE 22 1109 (1986)
3. F.M.Mitschke and L.F.Mollenauer, Stabilizing the Soliton Laser, IEEE JQE 22 2242-2250 (1986)
4. K.J.Blow and D.Wood, Modelocked Lasers with Nonlinear External Cavities, JOSA B 5 629-632 (1988)
5. D.B.Mortimore, Optical Fibre Loop Mirrors, IEEE JLT 6 (1988)
6. B P Nelson, D Cotter, K J Blow and N J Doran, Large Nonlinear Pulse Broadening in Long Lengths of Monomode Fibre, Opt Commun 48 292-294 (1983)
7. K.J.Blow and B.P.Nelson, Observation of Stimulated Raman Scattering and Nonlinear Pulse Broadening at $1.32\mu m$ in Monomode Optical Fibres,IEE PROC J 134 161-162 (1987)

Significance of Enhanced Differential Gain for Short Pulse Generation in Semiconductor Lasers

T. Sogawa[1], *Y. Arakawa*[1], *and T. Kamiya*[2]

[1]Research Center for Advanced Science and Technology, University of Tokyo, Komaba, Meguro-ku, Tokyo 155, Japan
[2]Faculty of Engineering, University of Tokyo, Hongo, Bunkyo-ku, Tokyo 113, Japan

1. INTRODUCTION

Picosecond pulse generation technology in semiconductor lasers is important for ultrahigh bit rate optical communication systems and ultrafast optical information processing systems. To this end, mode-locking [1], Q-switching [2], and gain switching [3] have been investigated. Among these schemes the gain switching method is the most popular, since this scheme requires no external mirror and no special laser structure. In this method, the use of short cavity structures and high peak injected current have been investigated for short pulse generation. Recently, we have been discussing the effect of enhanced differential gain on short pulse generation by the gain switching method, where the differential gain g' is defined as the ratio of the derivative of the bulk gain to the derivative of the carrier concentration n (i.e. $g'=\partial g/\partial n$). Our theoretical calculation predicted that the pulse duration is reduced with the increase of g'. In fact, the generation of a short pulse as narrow as 2ps was successfully observed in a quantum-well laser in which g' is enhanced compared to double heterostructure lasers due to two-dimensional properties of carriers [4]. However, more systematic experiment is required in order to confirm the high g' effect. In this paper, we investigate short pulse generation from distributed feedback (DFB) lasers, in which g' can be enhanced by detuning Bragg wavelength from the gain peak to a shorter wavelength [5], demonstrating significance of high g' for the short pulse generation.

2. EXPERIMENT

In our experiment we used InGaAsP FBH (flat-surface buried heterostructure) DFB lasers (Fujitsu) [5]. Two samples were prepared to examine the detuning effect in the gain switching method: (i) the Bragg wavelength λ_B is tuned to the gain peak; and (ii) λ_B is detuned to a shorter wavelength by 100Å. The threshold currents under CW operation at 25°C are 8.5 and 11mA, respectively. The structures of both lasers are the same except for the Bragg wavelength. Gain switching is realized by using current pulse trains (1GHz, 100ps FWHM, 100mAp-p,) which are generated by a comb generator (HP33005C) with DC bias. The temporal characteristic of the optical pulses is measured by a synchronously scanning streak camera (Hamamatsu C1587+M1955). Fig.1 shows the measured pulse duration, plotted against the

Springer Series in Chemical Physics, Vol. 48: **Ultrafast Phenomena VI**
Editors: T. Yajima • K. Yoshihara • C.B. Harris • S. Shionoya

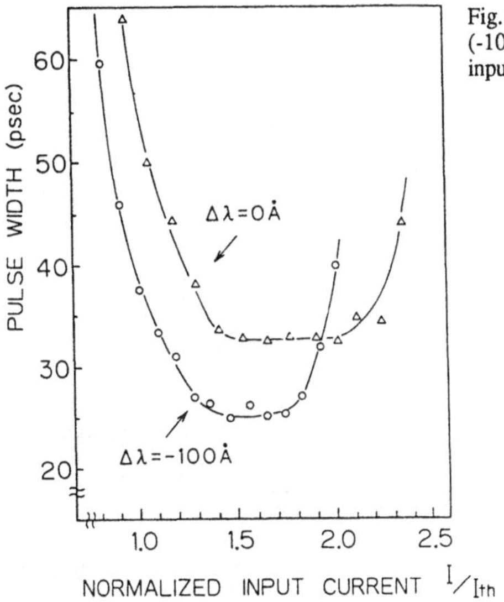

Fig.1 Measured pulse duration of tuned and detuned (-100Å) DFB lasers, plotted against the average input current normalized by each threshold current.

average input current normalized by each threshold current. In this experiment the modulation amplitude is fixed and the direct current is varied. The circles and triangles represent the experimental data for the detuned and tuned lasers, respectively. In both lasers the pulse duration becomes shorter with increasing input current. The shortest pulse is 25ps in the detuned laser, while 33ps is obtained in the tuned laser. This result demonstrated that enhancement of g' through the detuning leads to reduction of the pulse duration. Note that the increase in the pulse duration at higher injected current level is due to the onset of the second pulse through the relaxation effect.

3. DISCUSSION

Figure 2 shows the calculated bulk gain against carrier concentration in InGaAsP lasers at wavelengths of 1300, 1310, and 1320nm. When the bulk gain necessary for lasing is supposed to be 250cm^{-1}, the optimum Bragg wavelength $\lambda_B{}^{opt}$ for minimizing the threshold current is about 1320nm. On the other hand, g' which corresponds to the slope of the bulk gain curve is enhanced by detuning λ_B from $\lambda_B{}^{opt}$ to 1310nm or 1300nm. This indicates that g' can be controlled in DFB lasers by detuning λ_B.

Figure 3 shows the calculated pulse duration of lasers tuned to 1300, 1310, and 1320nm against the averaged input current normalized by each threshold, using the gain profile in Fig.2. The parameter values used in the calculation were as follows: internal loss α = 30cm^{-1}, confinement factor Γ= 0.25, spontaneous emission factor β = 10^{-4}, carrier lifetime τ_s = 3ns, and photon lifetime τ_p = 3ps. The injection current pulse is a raised-cosine pulse (150ps FWHM, five times as large as the threshold current) added to the direct current. As shown in

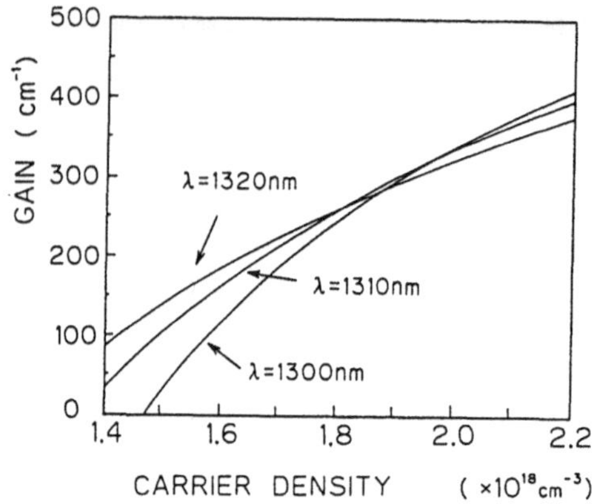

Fig.2 The bulk gain plotted as a function of carrier concentration at various wavelengths of 1300nm, 1310nm, and 1320nm.

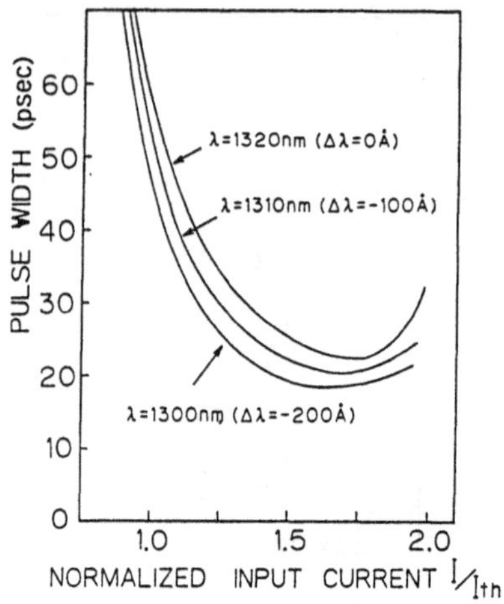

Fig.3 Calculated pulse duration of tuned and detuned (-100Å and -200Å) lasers against average input current normalized by each threshold current.

Fig.3, this calculation explains our experimental result, indicating reduction of the pulse duration due to the detuning effect as well as the increase in the pulse duration in the region of high bias current.

4. CONCLUSION

The effect of detuning the Bragg wavelength in picosecond pulse generation in InGaAsP distributed feedback lasers was investigated using the gain switching method. We observed that the pulse duration is reduced in a detuned DFB laser, in which the differential gain is enhanced through detuning the Bragg wavelength from the gain peak to a shorter wavelength by 100Å. This result demonstrates the significance of the differential gain for short pulse generation.

We thank Dr.Imai, Fujitsu Co.Ltd., for supplying excellent samples and fruitful discussions, and also thank Prof.Y.Fujii and Prof.H.Sakaki for their encouragement.

References

1. J. P. van der Ziel, R. A. Logan, and R. M. Mikulyak, Appl. Phys. Lett. 39, 867 (1981)
2. Y. Arakawa, A. Larsson, J. Paslaski, and A. Yariv, Appl. Phys. Lett. 48, 561 (1986)
3. G. J. Aspin, J. E. Carroll, and R. G. Plumb, Appl. Phys. Lett. 39, 860 (1981)
4. Y. Arakawa, T. Sogawa, M. Nishioka, M. Tanaka, and H. Sakaki, Appl. Phys. Lett. 51, 1295 (1987)
5. K. Kamite, H. Sudo, M. Yano, H. Ishikawa, and H. Imai, 10th IEEE int. semicond. laser conf.,Tokyo(1986)

Characteristics of Picosecond Pulse Amplification by a Traveling-Wave InGaAsP Optical Amplifier

*J.M. Wiesenfeld, G. Eisenstein, R.S. Tucker, G. Raybon, and P.B. Hansen**

AT&T Bell Laboratories, Crawford Hill Laboratory,
P.O. Box 400, Holmdel, NJ 07733, USA

An understanding of the properties of semiconductor optical amplifiers when used with ultrashort optical pulses is important for many applications, including high-speed optical time-division multiplexing [1], optical switching, and optical logic operations. The physics underlying the response of semiconductor optical amplifiers is also of fundamental interest, and includes such processes as hole-burning, carrier thermalization and cooling [2], and gain recovery. The time scale of the pulses used in the experiment will determine which processes are significant. In this paper, we report on the amplification of optical pulses of duration 3 - 25 ps, derived from semiconductor lasers, by 1.3-μm InGaAsP traveling-wave (TW) amplifiers, at pulse repetition rates from 200 kHz to 6 GHz. We observe gain compression (saturation) in the amplifier that depends on pulse energy and repetition rate. By treating the amplifier as a two-level, homogeneously broadened system, the gain compression is described for all pulse repetition rates. No temporal distortion of the pulse shape is observed under any conditions.

Two TW optical amplifiers, made by deposition of SiO_x anti-reflection coatings ($R < 3\times10^{-4}$) on the two facets of semiconductor diode laser chips, were used in this study. The laser chips used had channeled substrate buried heterostructure (CSBH) and semi-insulating etched mesa buried heterostructure (SI-EMBH) structures, with peak chip gains of 21 dB and 17 dB at wavelengths of 1.25 and 1.30 μm, respectively, and were 250 μm long. Two different sources of ultrashort optical pulses, an InGaAsP film laser [3], which was optically pumped by 0.5 ps pulses from a modelocked cw dye laser, and a gain-switched InGaAsP diode laser, were used to study the CSBH and SI-EMBH amplifiers, respectively. The film laser had repetition rate of 200 kHz, duration of 3 -10 ps, pulse energy input to the amplifier up to 200 fJ, and for these experiments was used over the wavelength range from 1.22 to 1.27 μm. Pulse shapes were measured with 0.5 ps resolution, before and after amplification, using upconversion sampling with a time-delayed replica of the 0.5 ps dye laser pulse, by sum frequency generation in a $LiIO_3$ crystal. The gain-switched diode laser emitted a train of pulses of ~25 ps duration at 1.31 μm wavelength, up to 100 fJ energy input to the amplifier, and at repetition rates up to 6 GHz.

Measured gain for the CSBH amplifier (which is 9 dB lower than chip gain, due to coupling losses) for 9 ps pulses at 1.257 μm is shown as a function of input energy in Fig. 1. To analyze the results, we model the amplifier as an ensemble of strict two-level systems [4]. The pulse duration is assumed short compared to the carrier lifetime τ and long compared to all carrier intraband relaxation and thermalization times which make the gain effectively homogeneous. In this case, the amplifier gain in dB, $G = 10\log(E_{out}/E_{in})$, can be written [5]

$$G = G_0 - 4.34E_{out}/E_{SAT} \quad , \tag{1}$$

where

$$E_{SAT} = h\nu ALN/0.23G_0 \quad . \tag{2}$$

In the above expressions, E_{out}, E_{in}, and E_{SAT} are the output, input, and saturation energies, respectively, G_0 is the small signal gain, $h\nu$ is the photon energy, A and L are the cross-sectional

* Permanent address: Electromagnetics Institute, Technical University of Denmark, DK-2800, Lyngby, Denmark.

Springer Series in Chemical Physics, Vol. 48: **Ultrafast Phenomena VI**
Editors: T. Yajima · K. Yoshihara · C.B. Harris · S. Shionoya

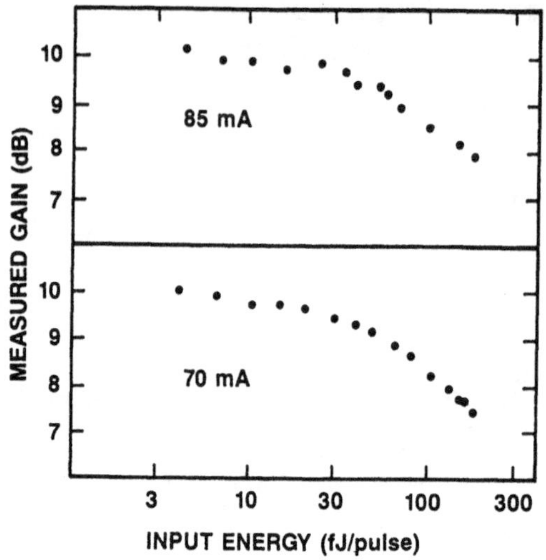

Figure 1: Measured gain vs input pulse energy for the CSBH amplifier, using film laser pulses at 200 kHz rate.

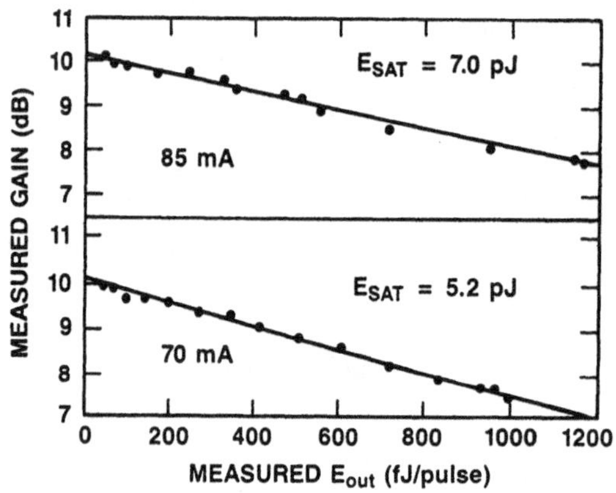

Figure 2: Data of Fig. 1 plotted vs E_{out}.

area and length of the active region in the amplifier, and N is the carrier density in the active region. Eq. (1) is valid when $E_{out}/E_{SAT} < 1$. The data of Fig. 1 is replotted in the form suggested by (1) in Fig. 2. The linearity of the plot supports the two-level system model. Values of E_{SAT} are derived from the slope of the plots, and agree quantitatively with estimates made using Eq. (2) and device parameters. Equation (2) shows the variation of E_{SAT} with bias (N) and wavelength (through G_0).

For pulses as short as 3 ps no temporal distortion was observed. Figure 3 shows the superimposed amplified pulse shapes for three conditions. With 52 fJ/pulse input energy, the amplifier gain is compressed by ~2.5 dB, while at 2.6 fJ/pulse input energy, the gain has its small signal value. Even under these three disparate conditions, the pulse shapes are identical to within experimental uncertainty. The lack of pulse distortion is explained by small group velocity dispersion in the amplifier [2,6] and sufficiently slow transitions in the pulse shape so that there is no distortion due to gain compression. [7]

When the period between input optical pulses, T, becomes comparable to or smaller than the gain recovery time (which is also the carrier lifetime τ) the overall gain of the amplifier will be reduced, because the gain will not recover to its original small signal value between pulses. This effect can lead to pattern-dependent gain and inter-symbol interference for a randomly modulated stream of pulses. Using a simple rate equation analysis it can be shown that the gain recovers with

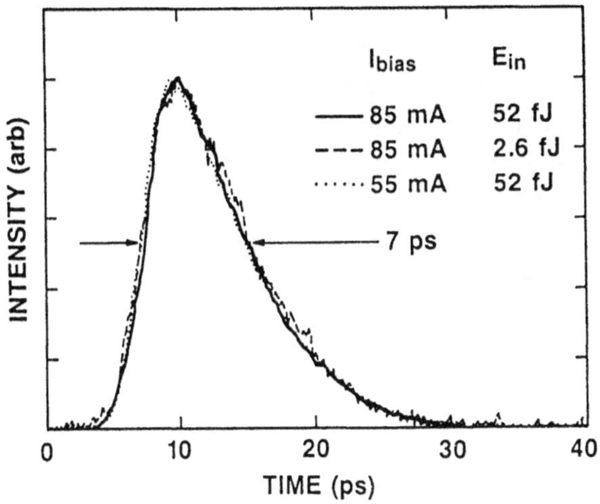

Figure 3: Pulse shapes after amplification in CSBH amplifier of a 7 ps film laser pulse at 1.249 μm, for conditions as indicated.

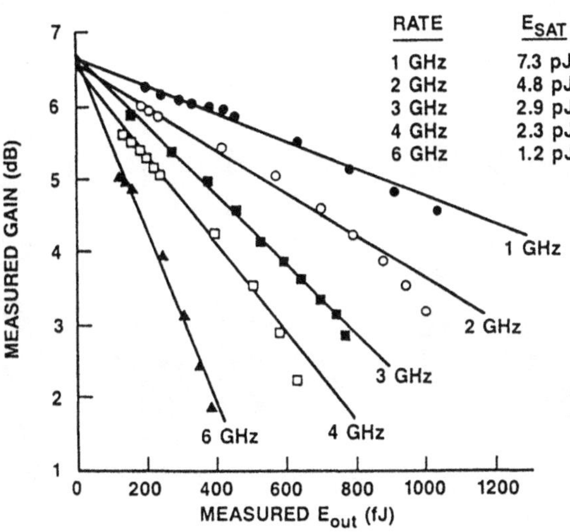

Figure 4: Measured gain of SI-EMBH amplifier using gain-switched laser diode pulses at 1.31 μm, for multigigahertz rates.

an exponential time constant τ between pulses. The repetition rate dependence of the gain can be derived from the two-level system model, and is

$$G = G_0 - 4.34E_{out}/E_{SAT}^{(eff)} \quad , \tag{3}$$

where

$$E_{SAT}^{(eff)} = (1 - e^{-T/\tau})E_{SAT} \quad . \tag{4}$$

Gain was measured as a function of pulse energy for the SI-EMBH amplifier at repetition rates from 1 to 6 GHz. Figure 4 shows that the effective saturation energy does decrease with repetition rate, as predicted by (4). The values of $E_{SAT}^{(eff)}$ are fit by taking E_{SAT} = 6 pJ and τ = 400 ps. The gain recovery time for this amplifier has been measured directly in a pump-probe experiment, and has a value of 360 ± 20 ps. The agreement between data and the predictions of the two-level system model further supports the model.

1. R. S. Tucker, G. Eisenstein, S. K. Korotky, L. L. Buhl, J. J. Veselka, G. Raybon, B. L. Kasper, and R. C. Alferness, Electron. Lett., **21**, 1270 (1987).

2. M. P. Kesler and E. P. Ippen, Appl. Phys. Lett., **51**, 1765 (1987).

3. J. M. Wiesenfeld and J. Stone, IEEE J. Quantum Electron., **QE-22**, 119 (1986).

4. N. J. Frigo, IEEE J. Quantum Electron., **QE-19**, 511 (1983).

5. J. M. Wiesenfeld, G. Eisenstein, R. S. Tucker, G. Raybon, and P. B. Hansen, Appl. Phys. Lett., **53**, xxx (1988).

6. J. P. Van der Ziel and R. A. Logan, IEEE J. Quantum Electron., **QE-19**, 164 (1983).

7. L. M. Frantz and J. S. Nodvik, J. Appl. Phys., **34**, 2346 (1963).

A Few Tens of Picoseconds Light Source of Continuous Spectrum with Cherenkov Radiation

S. Owaki[1], *T. Okada*[1], *Y. Kimura*[2], *S. Nakahara*[3], *and K. Sugihara*[3]

[1]The Institute of Scientific and Industrial Research, Osaka University, 8-1 Mihogaoka, Ibaraki, Osaka 567, Japan
[2]Dept. of Applied Physics, Konan University, 8-9-1 Okamoto, Higashinada, Kobe, Hyougo 658, Japan
[3]Dept. of Mechanical Engineering, Kansai University, 3-3-35 Yamatecho, Suita, Osaka 564, Japan

When a high energy electron passes through dielectric materials at a velocity exceeding that of light in the material, Cherenkov radiation is emitted into an annular cone about the forward direction of the electron. If an electron of energy of a few tens of MeV and a dielectric material of high refractive index are used for Cherenkov radiation, the number of emitted photons is only a few hundred within a spectral region from the absorption edge of the material to visible wavelengths[1]. Cherenkov radiation, however, produced with a lot of electrons bunched in a short time may be utilized for an intense pulsed light source with a continuous spectrum [2,3].

The electron linear accelerator (linac) of Osaka University can bunch electrons in a single wave of accelerating RF with velocity modulation and accelerate them to a maximum energy of 38 MeV with a repetition rate of up to 120 Hz [4]. The width of the single bunch pulse is estimated to be a few tens of picoseconds and about 10^{11} electrons can be contained in a single bunch pulse (maximum charge per pulse is about 50 nC). When the electron pulses from the linac are used for the excitation of Cherenkov radiation, considerably intense and short light pulses can be obtained because the time structure of the Cherenkov radiation is essentially the same as that of the electrons. Electron pulses of 28 MeV and 15 nC were used in these experiments.

The Cherenkov angle $\boldsymbol{\theta}$, a half vertical angle of the annular cone, and the number of photons **N** produced by an electron within a spectral region defined by wavelengths λ_1 and λ_2 are expressed by the following two equations[1]:

$$\cos\theta = 1/n\beta \qquad (1)$$

and

$$N = (2\pi e^2/\hbar c) \times (1/\lambda_2 - 1/\lambda_1)(1 - 1/n^2\beta^2)L \qquad (2),$$

where **n** is the refractive index of the dielectric material, $\boldsymbol{\beta}$ is the velocity of the electron relative to that of light in a vacuum, $2\pi e^2/\hbar c$ is the fine structure constant and **L** is the path length that the electron travels through the material. The most desirable characteristic as a light source is that the light emission is intense and directive. It is, however, obviously impossible from eqs.(1) and (2) to obtain both large **N** and small $\boldsymbol{\theta}$ simultaneously. As the Cherenkov light is utilized somewhere away from the radiator, because of high energy radiation protection, the directivity of the radiation rather than the intensity may be important for its transportation. The variables to determine $\boldsymbol{\theta}$ and **N** are only **n** and **L** while the electron energy of a linac, the value of $\boldsymbol{\beta}$, cannot change much

Springer Series in Chemical Physics, Vol. 48: **Ultrafast Phenomena VI**
Editors: T. Yajima · K. Yoshihara · C.B. Harris · S. Shionoya

because it is a linac characteristic. Then, selection of a dielectric material as a Cherenkov radiator could be done by experiment.

Three dielectric materials were examined as Cherenkov radiators and their optical features were compared. Among high purity fused silica (Suprasil) whose refractive index is the highest, xenon gas of pressure of a few atm, and air, the air in free space was found to be most convenient and easily available as a Cherenkov radiator.

An optical system was installed to take the Cherenkov radiation from the air near the electron beam window as shown in Fig.1. The light peak power with a single bunch electron pulse of 15 nC charge was estimated by a Si-photodiode to be about 10 kW in the spectral region from 200 to 600 nm.

Spectral distributions of the Cherenkov radiation from the air radiator were investigated with the method of pulse height integration. Experimental results are shown in Fig.2. The spectrum has a peak at 200 nm and a continuous band whose intensity decreases towards longer wavelengths. The trends of both experimental and calculated spectra (solid lines) are similar, though there is some discrepancy of the intensity. The dashed lines show about 1/3 of the calculated intensity in the UV to visible and 1/6 in the visible to IR region. The discrepancy arises from both overestimation of the number of electrons effectively used for the Cherenkov radiation and light collection loss.

The precise pulse shapes of the Cherenkov radiation were observed by a streak camera system with a time resolution of 10 ps described in a pre-

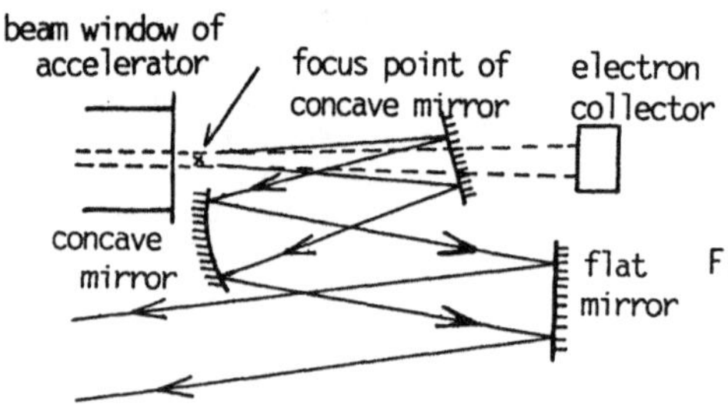

Fig.1 Generation of Cherenkov radiation from air and the optical system.

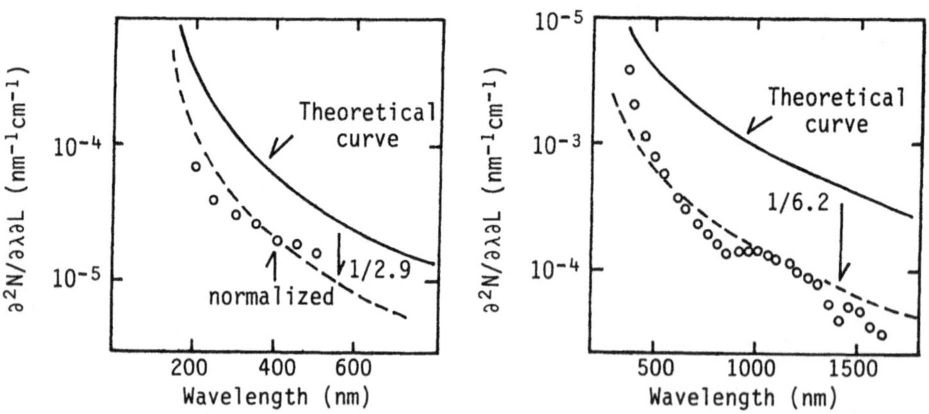

Fig.2 Spectral distributions of Cherenkov radiation from air with 30 MeV electron.

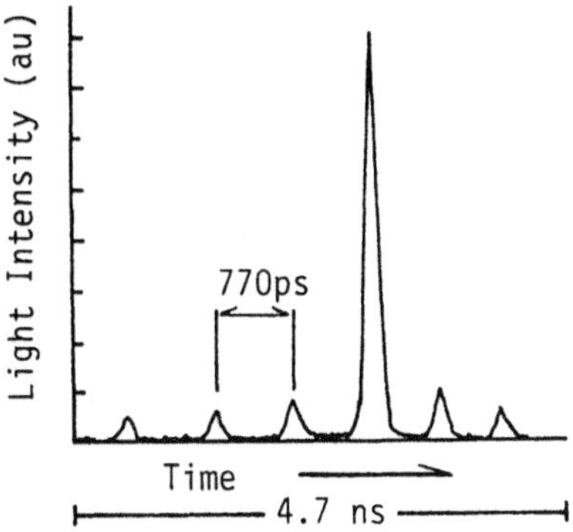

Fig.3 Light pulse shape of Cherenkov radiation from air with a single bunch electron pulse.

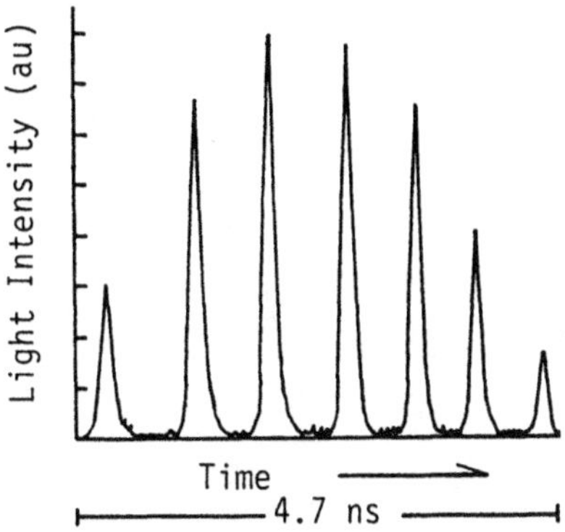

Fig.4 Light pulse shape of Cherenkov radiation from air with a 5 ns electron pulse.

vious paper [5]. The streak patterns provided by a computer are shown in Figs. 3 and 4. Recent data of the minimum pulse width of the Cherenkov radiation with the single bunch electrons were about 40 ps. This value increases 50 % with the single bunched charge increase after improvements of the linac. A light pulse shape of the Cherenkov radiation with an unbunched electron pulse of nominal 5ns is shown in Fig. 4.

In order to estimate the practical directivity of the Cherenkov radiation, cross-sectional intensity distributions of the electron beam and the Cherenkov radiation at a few points along the both beam axis were observed with a fiber optics technique. Divergences of the electron beam and the Cherenkov radiation, 2θ, are evaluated to be about 1° and 2° respectively. The calculated value of θ from eq.(1) in the case of no divergence of the electron beam is about 1° in the visible spectral region.

These experimental results show that practical use of Cherenkov radiation is possible for some kinds of physical, chemical and biological research. In addition, advantages of this light source are as follows:

1) The pulse width can be easily changed from a few tens of pico- to several nanoseconds with operation mode of the linac.
2) Pulse repetition rate is variable from manual operation to 120 Hz.
3) The region of continuous spectral distribution of the Cherenkov radiation can be selected with an appropriate dielectric material and the energy of the electron.

References

1) J.V.Jelley : In Cherenkov Radiation and Its Application (Pergamon Press, London 1958) p.15
2) M.A.Piestrup, R.H.Pantell, H.E.Puthoff and G.B.Rothbart : J. Appl.Phys. 44,5160(1973)
3) M.A.Piestrup, R.A.Powell, G.B.Rothbart, C. Chen and R.H.Pantell :Appl. Phys.Lett. 28,92(1976)
4) H. Sakurai, M. Kawanishi, K. Hayashi, T. Okada, K. Tsumori, S.Takeda, N. Kimura and T. Yamamoto : Mem. Inst. Sci. Ind. Res., Osaka Univ. 39, 21(1982)
5) S. Owaki : Jpn. J. Appl. Phys. 22,176(1983)

Part II

Generation of Short-Wavelength Ultrashort Light Pulses

Generation of High Power UV Femtosecond Pulses

S. Szatmári and F.P. Schäfer*

Max-Planck-Institut für Biophysikalische Chemie,
Abteilung Laserphysik, Postfach 2841, D-3400 Göttingen,
Fed. Rep. of Germany
*Permanent address: Research Group on Laser Physics of
the Hungarian Academy of Sciences, JATE University, Dóm tér 9,
H-6720 Szeged, Hungary

The bandwidth of most excimer lasers allows the amplification of subpicosecond pulses. However, the generation of subpicosecond seed pulses at the desired wavelength and their synchronization to the excimer amplifier needs a highly complex and expensive apparatus, where several different lasers and electronics have to work together properly. Using XeCl as an amplifying medium [1,2], the shortest pulse duration reported is 160 fs [2], when amplified and frequency-doubled CPM-laser pulses were used as seed pulses. At 248 nm ~450 fs pulses were obtained [3,4], when frequency-converted pulses from a Nd-YAG laser-pumped mode-locked dye laser [3], or from a subpicosecond XeCl laser [4] were amplified in the KrF amplifier. We have reported earlier an alternative method which is based on the use of a hybrid excimer-dye laser arrangement. In that setup a twin tube excimer laser (Lambda Physik EMG 150) is used as a pump source for a subpicosecond dye laser setup and as an amplifier [5,6]. With the use of this technique essentially similar results were obtained for both excimers, if the effect of the additional pulse compression reported there is not taken into account. The pulse duration directly at the output of the excimer amplifier was measured as 220 fs [5] and 370 fs [6] for XeCl and KrF, respectively. In the case of KrF further temporal compression of the amplified pulses resulted in 80 fs pulse duration [6]. A similar pulse compression experiment for XeCl led only to a slight decrease of the pulse duration from 220 fs to 170 fs [7]. Recent development of the excimer laser-pumped cascade ps dye laser setup [8] and the introduction of a new, simple distributed feedback dye laser (DFDL) [9] made it possible to simplify the construction of our hybrid excimer/dye laser system, allowing a much stabler operation, while preserving or improving the original output characteristics. Since KrF is found to show much better performance as a subpicosecond amplifier [10] than XeCl and pulse compression has been far more successful at this excimer wavelength [6], the new simplified setup reported here was first tested at the KrF wavelength.

The experimental arrangement is shown in Fig.1. The EMG 150 Lambda Physik excimer laser is used as a pump laser for pumping a special subpicosecond dye laser-amplifier arrangement, and as an amplifier for the amplification of the frequency-doubled output pulses of the dye laser setup. The oscillator channel of the EMG 150 laser is filled with the

Springer Series in Chemical Physics, Vol. 48: **Ultrafast Phenomena VI**
Editors: T. Yajima · K. Yoshihara · C.B. Harris · S. Shionoya

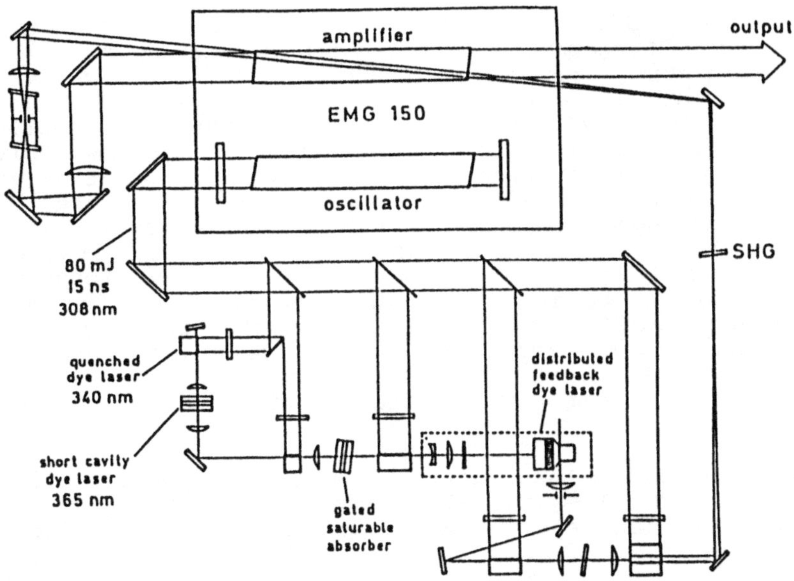

Fig.1. Experimental arrangement

standard XeCl fill, delivering 80 mJ, 15 ns pump pulses at 308 nm. The pump energy is distributed among the various dye cells as indicated in Fig.1. The pump beam coupled out by the first two quartz plates is used for pumping a newly developed, simple dye laser setup [8]. This makes use of two cascade dye lasers, two amplifier stages, and a gated saturable absorber (GSA) in between the amplifiers. The purpose of the dye lasers is to sharpen the leading edge, while the GSA plus the saturated amplifiers cut the trailing part of the pulses. The first dye laser operates with 5×10^{-3} M paraterphenyl in cyclohexane as an active medium in a 5 mm long cavity, yielding a pulse at 340 nm, with ~100 ps risetime. The second dye laser has a cavity length of 0.5 mm and contains a solution of 1×10^{-2} M butyl-PBD in methanol. Its output has ~10 ps risetime, and the 1 nm broad emission is centered at 365 nm. In the following amplifiers 4×10^{-4} M solution of butyl-PBD in ethanol is used. The gated saturable absorber consists of a solution of BBOT and a tilted cavity formed by two identical dichroic mirrors having approximately 10 % reflectivity at 365 nm and 100 % reflectivity at 430 nm, acting at the same time as dye cell windows. The combined pulse forming effect of the oscillators, saturated amplifiers and the GSA results in an output pulse duration of ~8 ps at 365 nm, with 1 mm cavity length, with a solution of 1×10^{-3} BBOT in ethanol is applied in the GSA. These pulses have an energy of typically ~4 µJ and are then used for pumping the DFDL master oscillator.

The DFDL used here is a tunable, achromatic arrangement first described in [9], utilizing a transmission grating, a microscope objective and a special dye cell with the active medium. In this DFDL the interference fringes, which are necessary for DFDL operation, are created by imaging a coarse transmission grating onto the inner surface of the dye cell by the use of a microscope objective (Fig.2). The pump beam is first expanded by a cylindrical telescope to permit a pencil-like illumination of the grating and the active medium. Only the two diffracted first orders are used for the formation of the interference pattern, while the zeroth order is blocked by a stop. This arrangement ensures the creation of a small size, high visibility interference pattern even with a pump beam of low spatial and temporal coherence. That is why a broadband dye laser pulse can be directly used for pumping, without any spectral selection. The other advantage of the arrangement is its easy tunability by translation of the transmission grating and proper choice of the refractive index of the active medium. In this way the whole visible spectrum can be covered (Fig.3).

The output pulse is then amplified in a two stage amplifier of standard design. The typical energy of the DFDL pulse is ~1 µJ after the first and ~20 µJ after the second amplifier. When the wavelength of the DFDL is set to be twice the wavelength of the excimer amplifier, the amplified and frequency doubled DFDL pulses can be used as seed pulses for amplification in the excimer amplifier. Frequency doubling is done just before the excimer amplifier by a thin BBO crystal.

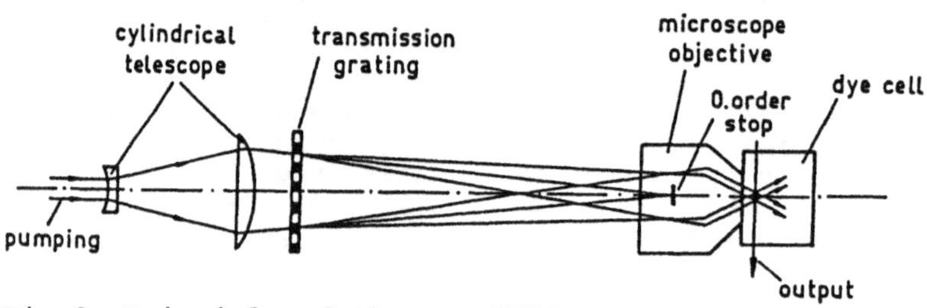

Fig.2. Principle of the new DFDL

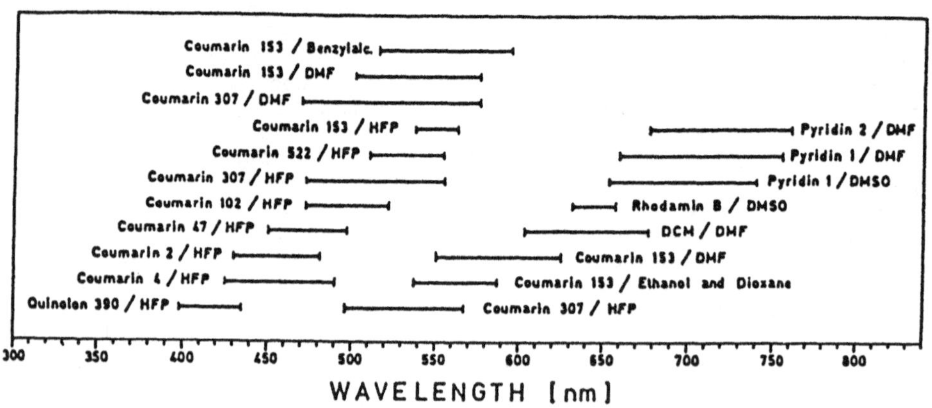

Fig.3. Tuning range of the new DFDL

The slightly divergent ultraviolet pulse has an energy of 2-3 µJ. This pulse is then amplified in a first pass through the second channel of the EMG 150, which is filled with KrF. After the first pass, the beam is magnified by a beam expanding telescope and spatially filtered by an evacuated pinhole. Then it is sent through the amplifier in a second pass. By this double-pass amplification scheme the pulse energy is boosted up to 13 mJ, with no more than 6 % amplified spontaneous emission (ASE) background. The cross section of the output beam is 11 x 25 mm^2.

After amplification, the uv pulses are sent through a double prism pulse compressor consisting of two 60° quartz prisms of 20 mm baselength and two dielectric mirrors to use the two prisms in a double pass pulse compressor arrangement. For the measurement of the pulse duration the general autocorrelation technique based on two photon ionizations of NO gas is used. The autocorrelator is shown schematically in Fig.4. This makes use of a 5 mm thick CaF window as a beam-splitter and two retroreflectors. In this design the two partial beams see exactly the same conditions in the two arms of the autocorrelator. By translation of one retroreflector in the direction perpendicular (and parallel) to the beams, complete spatial (and temporal) overlap of the two partial beams can be adjusted at the signal output. Then the ratio of the ion current of the signal and the reference channel is recorded as a function of the delay of the movable arm of the autocorrelator. The pulse duration before pulse compression is measured to be ~500 fs with significant chirp. After compression we have obtained a set of reproducible autocorrelation functions indicating a pulse width around 60 fs. The shortest autocorrelation function is shown in Fig.5. The autocorrelation width is 70 fs, corresponding to 45 fs pulse duration, with the generally used $sech^2$ assumption for the pulse shape.

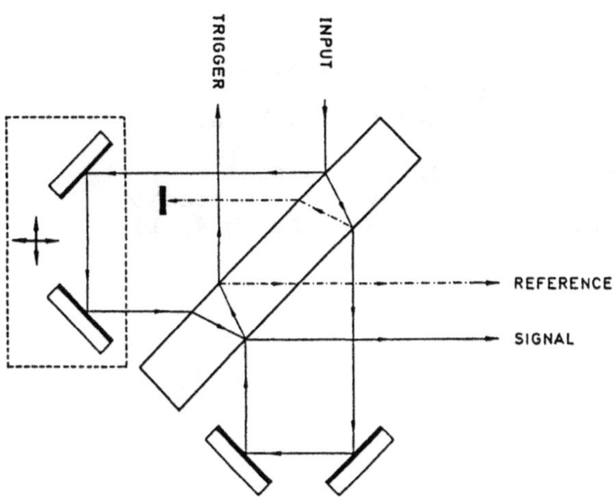

Fig.4. Schematic of the autocorrelator

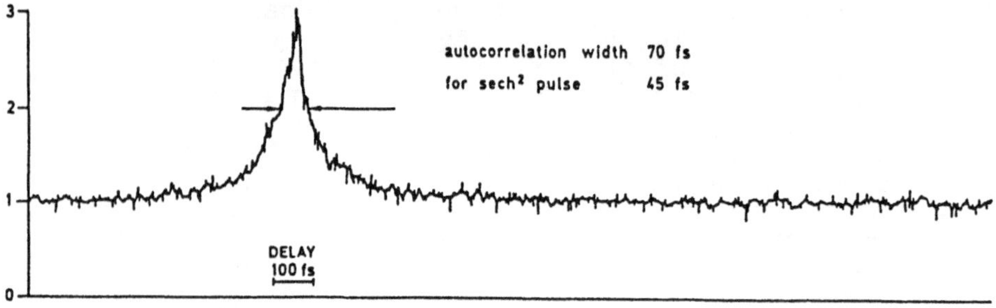

Fig.5. Autocorrelation curve of the shortest compressed uv pulses at 248 nm

In conclusion we have developed a simple hybrid excimer-dye laser capable of generating subpicosecond pulses at most of the excimer wavelengths. In this paper the generation of subpicosecond pulses with 13 mJ energy is reported at 248 nm. Preliminary investigations were made for the optimum pulse compression, which resulted in ~60 fs compressed pulse duration with very weak temporal wings. The shortest pulse duration is 45 fs.

References

1. J.H. Glownia, G. Arjavalingam. P.P. Sorokin, J.E. Rothenberg: Opt. Lett. **11**, 79 (1986)
2. J.H. Glownia, J. Misewich, P.P. Sorokin: CLEO '87, Digest of Technical Papers, 366 (1987)
3. A.P. Schwarzenbach, T.S. Luk, I.A. McIntyre, U. Johann, A. McPherson, K. Boyer, C.K. Rhodes: Opt. Lett. **11**, 499 (1986)
4. J.H. Glownia, J. Misewich, P.P. Sorokin: J. Opt. Soc. Am. B **3**, 1573 (1986)
5. S. Szatmári, B. Rácz, F.P. Schäfer: Opt. Commun. **62**, 271 (1987)
6. S. Szatmári, F.P. Schäfer, E. Müller-Horsche, W. Mückenheim: Opt. Commun. **63**, 305 (1987)
7. Q. Zhao, F.P. Schäfer, S. Szatmári: Appl. Phys. B **46**, 139 (1988)
8. S. Szatmári: Opt. Quant. Electron. (in press)
9. S. Szatmári, F.P. Schäfer: Appl. Phys. B (in press)
10. S. Szatmári, F.P. Schäfer: J. Opt. Soc. Am. B **4**, 1943 (1987)

Subpicosecond UV Pulse Generation for Multiterawatt XeCl and KrF Lasers

M. Watanabe, A. Endoh, N. Sarukura, and S. Watanabe

Institute for Solid State Physics, University of Tokyo,
Roppongi 7-22-1, Minato-ku, Tokyo 106, Japan

A multiterawatt (TW) excimer laser is an attractive light source for multi-photon processes and XUV lasers.[1] Over the past few years, many efforts have been devoted to reduce the pulse width from picosecond (ps) to subpicosecond (sub-ps) region.[2,3,4] The development of such a system is now in rapid progress in our Institute in a ps[5] and sub-ps region. The schematic of the entire system is shown in Fig.1. The amplification of sub-ps pulses has been performed in XeCl and KrF. A peak power of 1TW was obtained at XeCl in 310fs with an amplified spontaneous emission (ASE) less than 2% in energy within the discharge stage.[6] A 4-TW peak power was obtained in 390fs with an ASE of 1.8% at KrF by using the entire system. In this report, sub-ps UV pulse generation for the multiterawatt XeCl and KrF system is described.

The schematic diagram to generate sub-ps pulses is shown in Fig.2. In XeCl, the synchronously pumped dye laser generated sub-ps pulses at 616nm by using Rhodamine 590. As a pump source, a cw mode locked YAG laser (Spectra Physics series 3000) was employed. The fiber-grating compressor (Spectra Physics 3690) was inserted before frequency doubling. Autocorrelation measurement showed the pulse width at 616nm to be 320fs with some temporal wings present. A single pulse among the pulse train was amplified in a four-stage dye amplifier chain, which was pumped by a 5.5-ns, 75-mJ XeCl laser. Kiton Red 620 and Sulforhodamine 640 were used alternately along the stages of the amplifier chain. The growth of ASE was suppressed because the

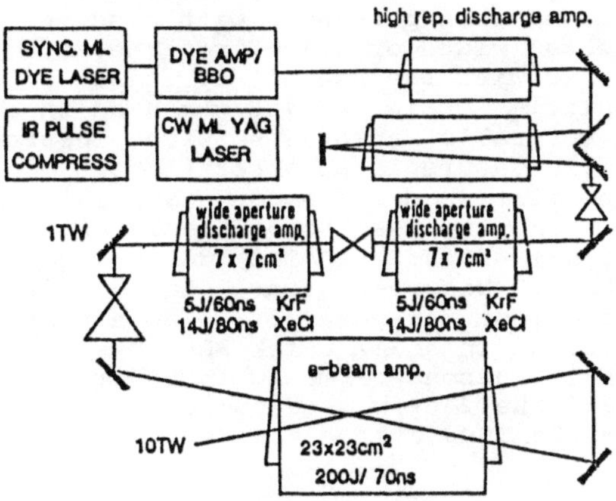

Fig.1 Multi-terawatt excimer laser system.

Springer Series in Chemical Physics, Vol.48: **Ultrafast Phenomena VI**
Editors: T. Yajima · K. Yoshihara · C.B. Harris · S. Shionoya

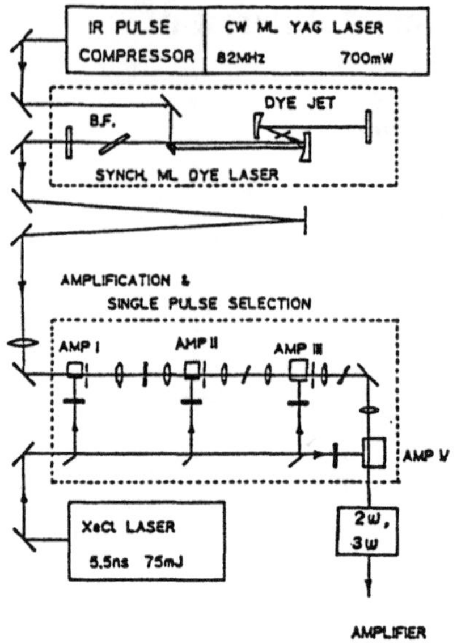

Fig.2 Schematic diagram to generate sub-ps pulses.

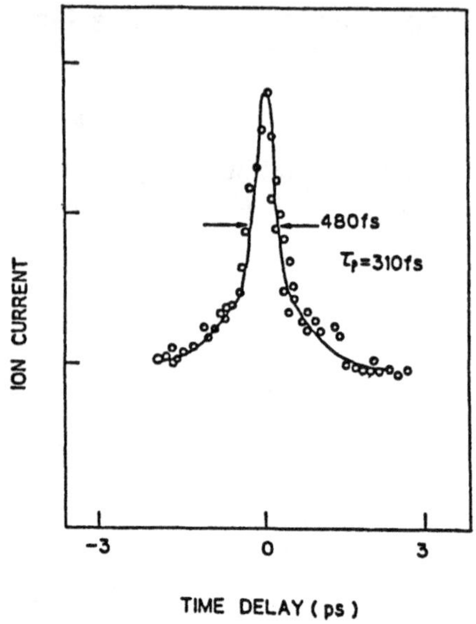

Fig.3 Autocorrelation trace of amplified 308-nm pulses.

spectral gain peak of Kiton Red 620 is closely positioned to the absorption peak of Sulforhodamine 640. Malachite Green saturable absorber dye jet was inserted between the second and the third stages. The output energy was 130μJ with 3% ASE. This was frequency doubled in a 1mm-thick BBO crystal with an efficiency of 18%. The UV pulses were amplified by two high repetition rate XeCl preamplifiers, which have the same aperture size of 2cm x 1cm, and active lengths of 30cm and 60cm, respectively. The autocorrelation trace by using two-photon ionization in triethylamine vapor showed the pulse width to be 310fs (Fig.3). A typical output energy was 7mJ with a negligible small content of an ASE background.

In KrF, sub-ps pulses were also obtained by the same schematic of Fig.2 with replacement of dyes and some modifications. In this experiment, 745-nm pulses were generated with the pulse width of 290fs and the spectral width of 2.1nm from the hybrid synchronously pumped dye laser with the use of Pyridine 2/DDI combination.[7)] In the dye amplifier chain, LD-700 was used in all stages. To reduce ASE, saturable absorbers were inserted between every stage. The solid saturable absorber (Schott glass RG850) was placed between the first and the second stages, and HITC saturable absorbrer dye jets were used among the later three stages. The output energy was 480μJ. Explicit difference in pulse width and spectrum was not observed after the amplifier chain. But the temporal wings almost disappeared. The amplified 745-nm pulse was frequency doubled in a 1mm-thick KDP crystal and then sum-frequency mixed with its second harmonic in a 1mm-thick BBO crystal. The energy of the 248-nm seed pulse was a few microjoules and then amplified to 10mJ by the preamplifiers.

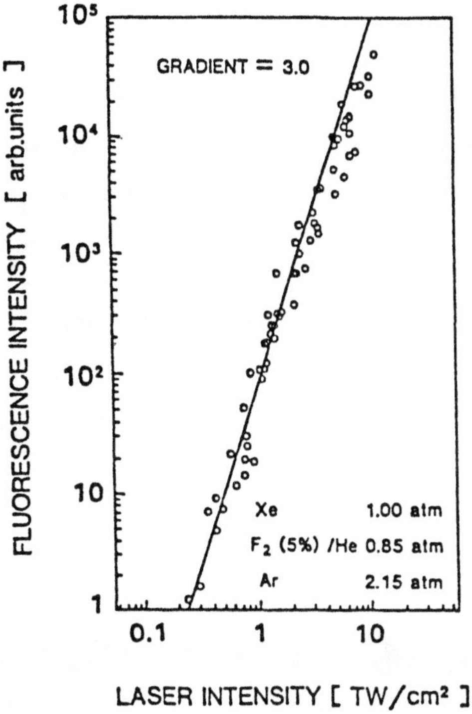

Fig.4 Cubic dependence of the XeF C-A fluorescence on laser intensity.

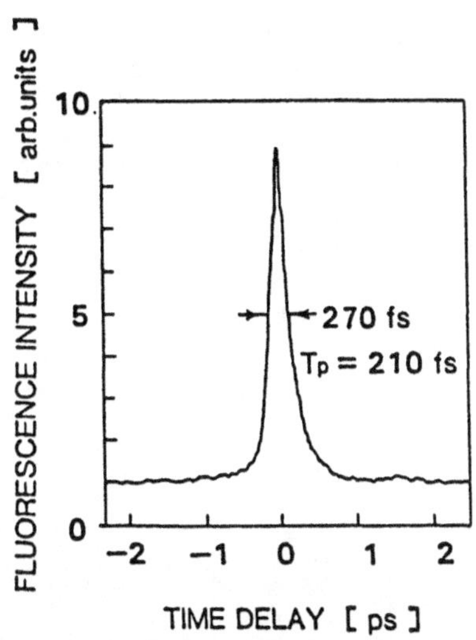

Fig.5 Autocorrelation trace of a single amplified 248-nm pulse.

The fluorescence (172nm) of Xe_2 was used before for the single shot pulse width measurement in KrF.[8)] We devised a new method to measure the sub-ps KrF pulses by using a C-A transition (480nm) of XeF produced by three-photon ionization of Xe.[9)] Figure 4 shows cubic dependence of the fluorescence on laser intensity for subpicosecond KrF pulses. The typical arrangement with the triangular configuration was employed. The pulses were focused by a 60-cm focal-length lens and collided precisely at the focal point in the gas cell. The fluorescence image was enlarged by a factor of 5 - 10 with a camera lens, and detected by the SIT camera. The pulse width after the preamplifiers was measured to be 210fs by using this method. The single shot autocorrelation trace is shown in Fig.5. The observed contrast ratio was close to the theoretical value of 10:1. The pulse was amplified to 1.5J by the entire system including the electron-beam pumped amplifier. The pulse width of the final output was measured to be 390fs. The observed increase of the pulse width is mostly explained by the linear dispersion of the optical elements. This will be compensated by using an appropriate element with negative dispersion such as a prism pair in the early stage.

References

1) C.K.Rhodes, Science 229, 1345 (1985)
2) J.H.Glownia, J.Misewich and P.P.Sorokin, J.Opt.Soc.Am.B 4, 1061 (1987)
3) A.P.Schwarzenbach, T.S.Luk, I.A.McIntyre, A.McPherson, K.Boyer and C.K.Rhodes, Opt.Lett. 11, 499 (1986)
4) S.Szatmári, F.P.Schäfer, E.Müller-Horsche and W.Mückenheim, Opt.Commun. 63, 305 (1987)
5) A.Endoh, M.Watanabe and S.Watanabe, Opt.Lett. 12, 906 (1987)
6) S.Watanabe, A.Endoh, M.Watanabe and N.Sarukura, to be published in Opt.Lett.
7) M.D.Dawson,T.F.Boggess and A.L.Smirl, Opt.Lett. 12, 254 (1987)
8) M.H.R.Hutchinson, I.A.McIntyre, G.N.Gibson and C.K.Rhodes, Opt.Lett. 12, 102 (1987)
9) N.Sarukura, M.Watanbe, A.Endoh and S.Watanabe, submitted to Opt. Lett.

Subpicosecond, High-Brightness Excimer Laser Systems

A.J. Taylor, R.B. Gibson, J.P. Roberts, C.S. Lester, T.R. Gosnell, S.E. Harper, and C.R. Tallman

Los Alamos National Laboratory, Group CLS-5,
Los Alamos, NM 87545, USA

Subpicosecond, high-brightness excimer laser systems are being used to explore the interaction of intense coherent ultraviolet radiation with matter. Applications of current systems include generation of picosecond x-ray pulses, investigation of possible x-ray laser pumping schemes, studies of multiphoton phenomena in atomic species, and time-resolved photochemistry. These systems [1,2], based on the amplification of subpicosecond pulses in small aperture (~1 cm^2) XeCl or KrF amplifiers, deliver focal spot intensities of ~10^{17} W/cm^2. Scaling to higher intensities, however, will require an additional large aperture amplifier which preserves near-diffraction-limited beam quality and subpicosecond pulse duration [3]. We describe here both a small aperture KrF system which routinely provides intensities >10^{17} W/cm^2 to several experiments, and a large aperture XeCl system designed to deliver ~1 J subpicosecond pulses and yield intensities on target in excess of 10^{19} W/cm^2. We also discuss the effects of two-photon absorption on large-aperture, high-brightness excimer lasers.

The small aperture KrF system consists of a "front-end" which generates 248-nm seed pulses, followed by two KrF amplifiers. The seed pulses are initially generated at 648 nm with a mode-locked dye laser that uses DCM as the gain dye and DTDCI as the absorber dye. The dye laser is synchronously pumped with the frequency-doubled output of a cw mode-locked Nd:YAG laser. The pulses are then amplified at a 3 Hz repetition rate and frequency-doubled in a 2-mm-thick BBO crystal. The resulting pulses at 324 nm are finally sum-frequency mixed with amplified 1064-nm pulses in a second 2-mm BBO crystal to produce 5- to 10-μJ subpicosecond seed pulses at 248 nm. These pulses are then amplified by two Lambda Physik EMG 200 Series KrF amplifiers, separated by a vacuum spatial filter to suppress ASE and improve beam quality. The output beam diameter is 17 mm and the final output energy at 248 nm is 25 ± 3 mJ with <5% ASE. The pulsewidth, measured using two-photon ionization in NO, is 700 fs. The focused spot size achievable with this system has been determined indirectly by measuring the confocal parameter of a beam focused by f/3 optics. The inferred focal spot diameter is 3.6 μm (twice the diffraction limit), which implies an intensity at the focal plane of 3.5 x 10^{17} W/cm^2. For all experiments parabolic mirrors are used as the focusing optics to preserve pulsewidth, minimize aberrations, and avoid nonlinear absorption and refraction.

This system is routinely operated as a source for several physics experiments [4]. An x-ray spectroscopy experiment has shown that multiphoton processes in a solid aluminum target produce an aluminum ion plasma exhibiting line radiation at energies exceeding 2 keV. Another experiment has shown that highly charged ion states can be produced by multiphoton ionization. In xenon, for example, absorption of 213 248-nm photons yields a Xe^{+11} ion.

Springer Series in Chemical Physics, Vol. 48: **Ultrafast Phenomena VI**
Editors: T. Yajima · K. Yoshihara · C.B. Harris · S. Shionoya

We are currently building a second high-brightness system, based on amplification in XeCl, that uses a large aperture (100 cm^2) final amplifier designed to deliver pulse energies approaching 1 J at a maximum repetition rate of 1 Hz. The seed pulse generator scheme is sketched in Fig. 1. Pulses of 175 fs duration at 616 nm are initially generated in a linear-cavity, dispersion-compensated dye laser (Rhodamine 6G/DODCI) that is synchronously pumped by a cw mode-locked Nd:YAG laser. The Nd:YAG laser also provides 1.06-μm seed pulses to a regenerative amplifier, whose frequency-doubled output longitudinally pumps a three-stage dye amplifier. This synchronous amplification scheme has the advantages of low amplified spontaneous emission, good beam quality, and the elimination of a separate pump laser for the dye amplifier. We observe no temporal broadening of our 175 fs pulses through this amplifier. The amplified 616-nm pulses are then frequency doubled to 308 nm in a BBO crystal. Pre-amplification of these 30-μJ uv pulses to the 3-mJ level is accomplished with a single small aperture commercial XeCl discharge amplifier. The beam is then expanded in a vacuum spatial filter before entering the final amplifier. This 10 x 10 cm^2 aperture device consists of two independently pumped discharge gain regions which share a common x-ray preionizer. The small signal gain, g_0l, in each discharge region is five. The discharges are pumped by low-jitter, thyratron-switched pulse modulators with three-stage magnetic pulse compression. The resultant jitter is less than 5 ns over a ~50 ns gain time. In order to maintain near-diffraction-limited beam quality at a sustained 1-Hz repetition rate, a transverse gas flow system is used. The wave front distortion at 632.8 nm is less than λ/20 over 80% of the aperture and the hot gas clears within 30 ms after a shot. Barring additional distortion due to nonlinear refraction in the output window, focal-spot intensities >10^{19} W/cm^2 should be obtained with this system.

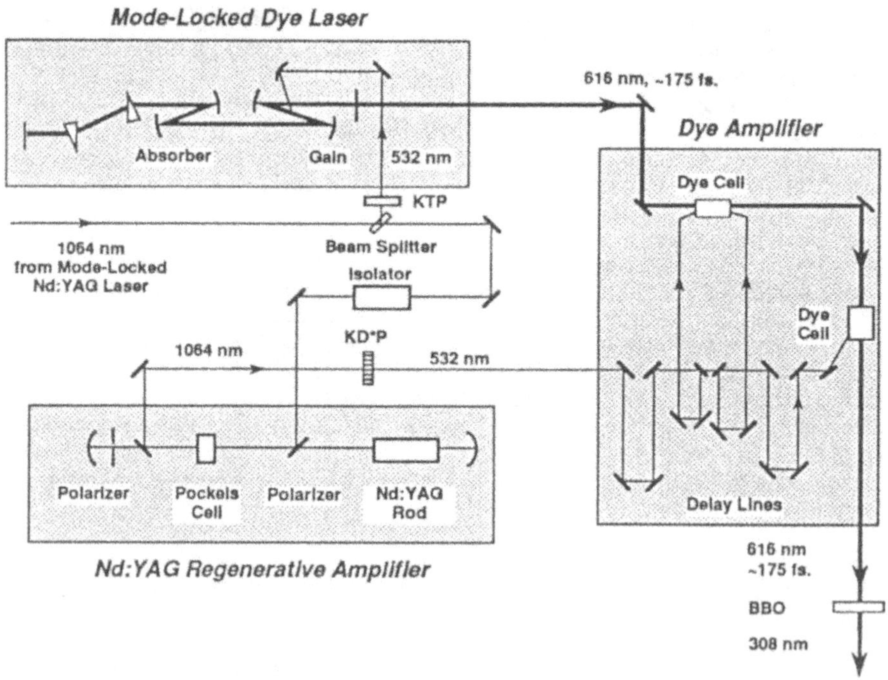

Fig. 1. Subpicosecond, 308-nm seed pulse generator

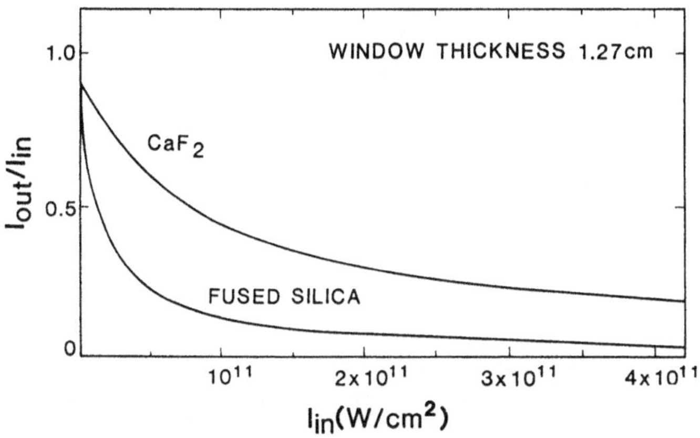

Fig. 2. I_{out}/I_{in} versus I_{in} for 0.5-inch-thick CaF_2 and fused silica samples

An understanding of nonlinear optical phenomena such as two-photon absorption and nonlinear refraction in ultraviolet window materials at excimer wavelengths is essential for the design of large aperture amplifiers. We have measured the two-photon absorption coefficient β at 248 nm for those materials (fused silica and CaF_2) which can be obtained in large apertures. For fused silica $\beta = 4.5 \pm 2.2 \times 10^{-11}$ cm/W, while for CaF_2, $\beta = 8.3 \pm 4.1 \times 10^{-12}$ cm/W. To evaluate the implications of these values of β for KrF laser systems, we plot, in Fig. 2, I_{out}/I_{in} versus I_{in} for 1.27-cm-thick windows of fused silica and CaF_2. Both two-photon absorption and its associated pulsewidth broadening are included in the calculation of I_{out}. At $I_{in} = 3 \times 10^{10}$ W/cm^2, a typical intensity for current systems, $I_{out}/I_{in} = 0.33$ for fused silica and $I_{out}/I_{in} = 0.66$ for CaF_2. Therefore, for large aperture, KrF-based, high-brightness lasers, the material for the output window, as well as for any subsequent windows should be carefully chosen and the total thickness after the gain medium minimized. In contrast, XeCl-based high-brightness lasers, where the photon energy is 4 eV, become attractive, since two-photon absorption in CaF_2 is not possible.

References

1. J. H. Glownia, J. Misewich, and P. P. Sorokin, J. Opt. Soc. Am. B 4, 1061 (1987).
2. A. P. Schwarzenbach, T. S. Luk, I. A. McIntyre, U. Johann, A. McPherson, K. Boyer, and C. K. Rhodes, Opt. Lett, **11**, 499 (1986).
3. S. Watanabe, A. Endoh, M. Watanabe, and N. Surakura, Opt. Lett. **13**, 580 (1988).
4. J. A. Cobble, G. A. Kyrala, A. J. Taylor, A. A. Hauer, P. H. Y. Lee, D. E. Casperson, L. A. Jones, and G. T. Schappert, IQEC '88, paper TuD-7.

Sub-100-fs Pulse Generation from Continuously Pumped Coumarin Dye Lasers in the Green-Blue

P.M.W. French and J.R. Taylor

Femtosecond Optics Group, Physics Department, Imperial College, London SW7 2BZ, UK

Until recently the generation of tunable subpicosecond optical pulses in the spectral region below ~550 nm has been practically precluded by the lack of suitable pump lasers. The development of u.v.enhanced argon ion lasers capable of giving up to ~6W all lines u.v. average power has, however, extended the potential tuning range of passively mode locked c.w. dye lasers down to ~410 nm. This new spectral region is of particular interest when frequency doubled to yield femtosecond optical pulses around 248 nm for amplification in high power Krypton Fluoride laser systems and from ~205 nm to 400 nm for time-resolved spectroscopy of organic molecules. In this work we report on the direct generation of femtosecond pulses from 488 nm to 554 nm using passively mode locked c.w. Coumarin dye lasers. Table 1. shows the new combinations of the gain and absorber dyes demonstrated and gives the tuning ranges and minimum pulse-widths achieved (assuming a $sech^2$ pulse profile).

Table 1. New passively mode locked c.w. dye lasers

GAIN/ABSORBER	SPECTRAL RANGE		PULSE DURATIONS	
Coumarin 102/DOCI	487-510	nm	$t_p > 93$	fs
Coumarin 6H / DOCI	492-507	nm	$t_p > 110$	fs
Coumarin 102 / DQTI	492-512	nm	$t_p > 140$	fs
Coumarin 102 / D9MOCI	488-511	nm	$t_p > 140$	fs
Coumarin 102 / DPQI	494-511	nm	$t_p > 150$	fs
Coumarin 1 / DOCI	501-508	nm	$t_p > 130$	fs
Coumarin 6 / DI	518-554	nm	$t_p > 95$	fs

Figure 1. shows the absorption profiles of the saturable absorbers and it should be noted that for a given wavelength it is possible to generate femtosecond pulses with a wide range of absorption cross-sections (and hence "S" parameters [1]). The cavity configuration is shown in Figure 2. A colliding-pulse ring laser with the standard prism sequence was employed, after the manner of Valdmanis et al. [2]. The mirrors were all single-stack dielelectric coatings of 100% reflectivity at normal incidence except mirror M6 which was of 99% reflectivity . Control of the laser wavelength and bandwidth was achieved using the aperture A located at the midpoint of the prism sequence where there is a spatial distribution of the laser spectrum. Translation of one of the prisms along its axis of symmetry was used to vary the intracavity group velocity dispersion. The active folding mirrors (M2 , M3) were of 50 mm radius of curvature and the passive folding mirrors (M4 , M5) of 25 mm. Mirror M1, of 40mm radius of curvature, focussed the pump radiation into a dye jet of ~400um thickness. This unusually thick jet was necessary to achieve a reasonable absorption of the pump beam since the absorption cross-sections of the Coumarin dyes are low compared to that of Rhodamine 6G. A standard Coherent Radiation nozzle was used for the absorber jet stream. The

Springer Series in Chemical Physics, Vol. 48: **Ultrafast Phenomena VI**
Editors: T. Yajima • K. Yoshihara • C.B. Harris • S. Shionoya

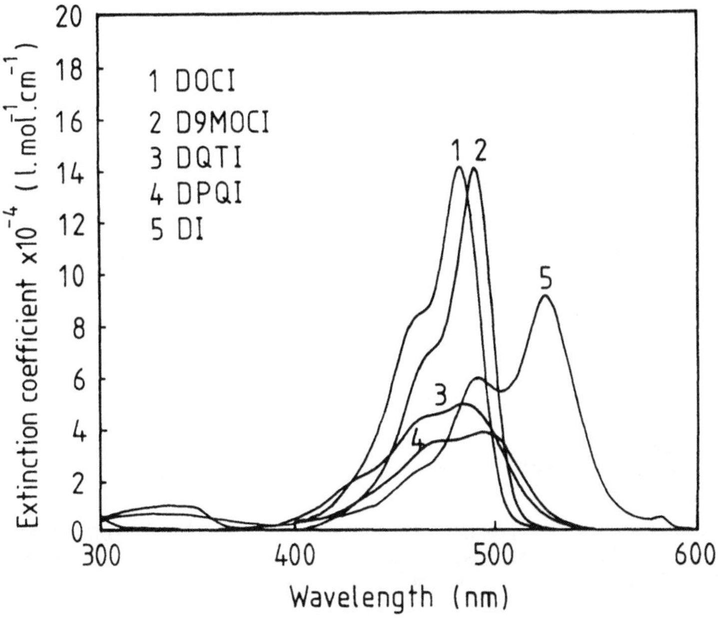

Figure 1. Absorption profiles of saturable absorbers used

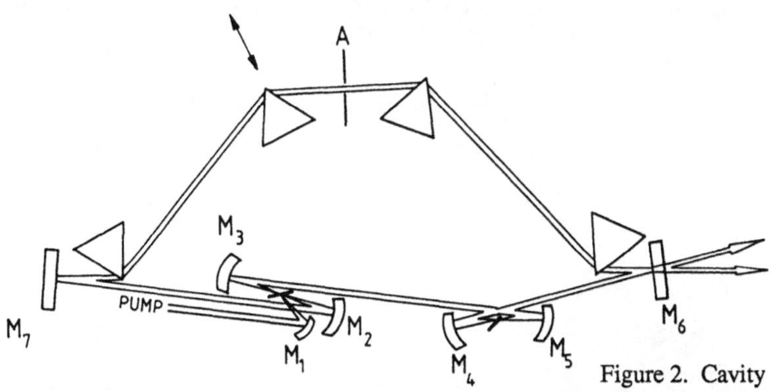

Figure 2. Cavity configuration

pump source was a Coherent 100 UV-E argon ion laser which provided the 488 nm pump radiation for the Coumarin 6 laser and the all lines u.v. (up to ~6W) for the other systems.

All the other systems reported here exhibited the high stability characteristic of all passively mode locked dye lasers which are, when available [3], generally preferable to synchronously pumped schemes because the difficulties and expense of active mode locking and cavity length stabilization are avoided. In addition, passively mode locked laser systems generate qualitatively better pulses as evinced by the autocorrelation traces not exhibiting the exponential wings indicative of pulse profile fluctations.

Figure 3. shows an autocorrelation trace of pulses obtained at 497 nm from the Coumarin 102 / DOCI system which were within 20% of their transform-limited duration - as were the durations of

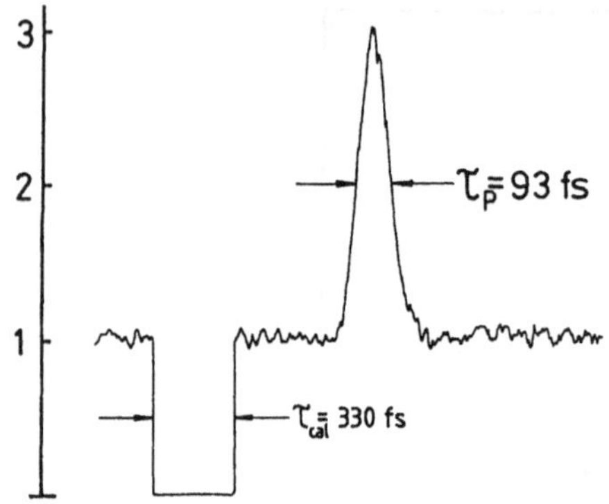

Figure 3. Autocorrelation trace of pulses at 497 nm

the pulses from the other optimised systems. Average output powers were typically of the order of 5 mW per beam for pump powers around 4W. It is expected that similar lasers will be developed to operate at yet shorter wavelengths.

The authors gratefully acknowledge the assistance of M.M.Opalinska and J.A.R.Williams and the support of the Royal Society and the Science and Engineering Research Council. This work was carried out in association with the Blackett Laboratory Laser Consortium.

References

1) G.H.C.New, IEEE J. Quantum Electron. QE-10, 115 (1974)
2) J.A.Valdmanis, R.L.Fork and J.P.Gordon, Opt. Lett. 10 , 559 (1985)
3) P.M.W.French, J.A.R.Williams and J.R.Taylor, Rev. Phys. Appl. 22 , 1651 (1987)

Generation of Intense Tunable Femtosecond Pulses in the Deep Blue Spectral Region

D.K. Negus, B.C. Couillaud, and R. Brady

Coherent, Inc., Laser Products Division, 3210 Porter Drive, Palo Alto, CA94303, USA

The direct generation of tunable ultrafast pulses in the blue region of the optical spectrum has previously been limited to a few attempts at synchronously pumping a three mirror dye laser cavity by a modelocked ion laser operating in the near UV [1]. The relatively low power of the Argon laser in this region requires that one simultaneously modelocks the three closely spaced lines at 333,351, and 364 nm. The result is an incompletely modelocked pulse sequence with a 300 ps halfwidth and low output power, which cannot be effectively used as a dye laser pump source. Frequency doubling of near IR dye laser output has the disadvantage of being limited to UV powers in the several milliwatt range.

Direct generation of femtosecond pulses in more complex dye laser structures, such as two jet hybridly modelocked dye lasers or more recently developed dispersion compensated variants of them [2], requires a pump source delivering several hundreds of milliwatts average power in a modelocked train of pulses stable in shape, phase and amplitude.

Recent advances in high power fundamental mode Nd^{+3}:YAG lasers capable of generating over 22 watts of modelocked average output at 1064 nm [3] coupled with new nonlinear optical material developments [4] have made the third harmonic of the modelocked YAG a viable alternative for synchronous pumping of femtosecond dye lasers. We report the first demonstration of high average third harmonic (355 nm) powers of nearly 2 watts, from a cw modelocked YAG laser (76 MHz) and the successful utilization of this new capability as a pump source for a dispersion compensated two jet Stilbene 3 dye laser.

Third harmonic generation was achieved by sum frequency mixing the second harmonic of the YAG at 532 nm with the fundamental at 1064nm in a Type I angle-tuned (critically) phase-matched βBaB_2O_4 crystal. In this mixing process the third harmonic power, P_{355}, is proportional to the product of the power in the fundamental P_{1064}, and the second harmonic P_{532}, that is,

$$P_{355} = \eta P_{1064} P_{532} \quad . \tag{1}$$

Figure 1 shows the third harmonic output power (without accounting for Fresnel losses) plotted versus the product of the fundamental and second harmonic powers for an uncoated 6.9 mm length βBaB_2O_4 crystal. This data was taken at a constant YAG fundamental output power of 23 watts by varying the fraction of 1064 converted to 532. The pulsewidth of the fundamental was 80ps. A least squares fit to this data yields a slope of $\eta = 0.021 W^{-1}$. The total power density in the βBaB_2O_4 crystal was $\approx$ 250 $MW\text{-}cm^{-2}$.

The large Poynting vector walk-off of βBaB_2O_4 (51 mrad at 1064nm) results in a highly distorted third harmonic beam. Thorough analysis of the beam propagation after the crystal led to the design of a compensating optic which transformed the output into a low divergence beam with an aspect ratio close

Springer Series in Chemical Physics, Vol. 48: **Ultrafast Phenomena VI**
Editors: T. Yajima · K. Yoshihara · C.B. Harris · S. Shionoya

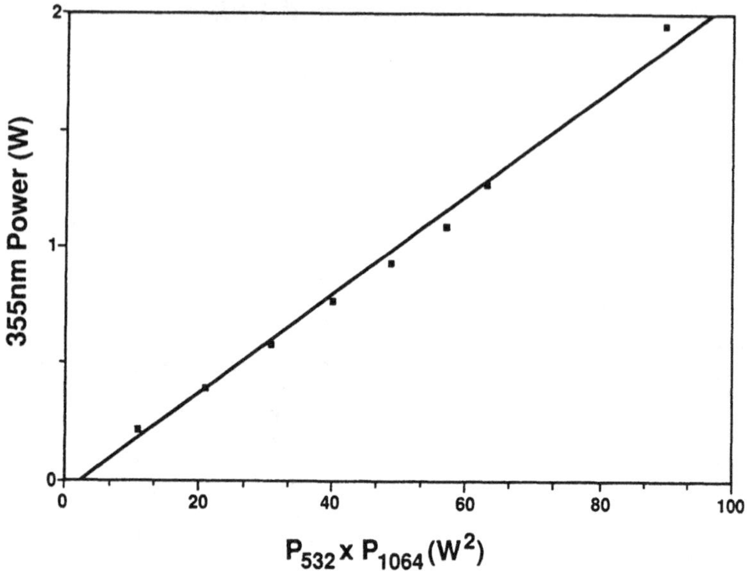

Figure 1: CW modelocked third harmonic output power versus the product of the fundamental and second harmonic power at constant YAG 1064 output of 23 watts.

to 1.0. A qualitative measure of the magnitude of the low order (gaussian-like) component in this beam is given by the overall conversion efficiency obtained when the tripled output is used to synchronously pump a dye laser. Conversion efficiencies as high as 35% have been observed at 435 nm in a single jet fundamental mode Stilbene 3 dye laser pumped with 1.0 W at 355 nm.

Hybrid modelocking in a 2 jet resonator configuration required identification of a material exhibiting saturable absorption in the spectral region 420-470 nm. Several dyes were investigated for this purpose and Coumarin 535 was found to have the most desirable properties. The result was an output tunable from 420-470 nm with output powers of 40-60 mW for a pump power of 600mW at 355 nm. Synchronous pumping of this uncompensated 2 jet laser however, resulted in pulses with a large residual chirp. For this laser configuration, using a noncolinear autocorrelation scheme, it was observed that due to spectral filtering of the chirped pulse by the autocorrelator SHG crystal, the pulsewidth varied from 0.5 to 1.2 ps depending upon the incidence angle for the beams on the βBaB_2O_4 crystal used for autocorrelation. This chirp is substantially larger for this hybrid 2 jet laser operating in the 420-470nm region than for the visible and near infrared lasers commonly used, possibly due to inherently higher material dispersions for intracavity optical components in this wavelength region.

To compensate this residual chirp a 2 jet resonator containing 4 brewster angle fused silica prisms was constructed [5]. The interprism separation as defined in reference [5] was 32 cm. Figure 2 shows the variation of output power and pulsewidth at 452nm versus the pathlength of intracavity glass for the dispersion compensated Stilbene3/C535 laser pumped with 1.8 W at 355nm.

The variation of the autocorrelation trace as a function of intracavity glass pathlength is shown in Fig.3. Each autocorrelation trace represents the average of 5 "real time" traces collected over a period of 1 second and is therefore

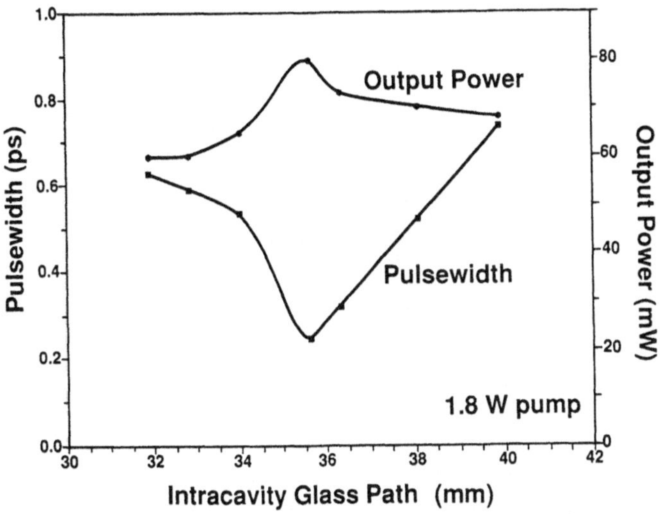

Figure 2: Variation of output power and pulsewidth (assuming $sech^2$) versus intracavity glass pathlength for dispersion compensated 2 jet hybrid modelocked Stilbene 3/C535 dye laser pumped with 1.8 W at 355nm.

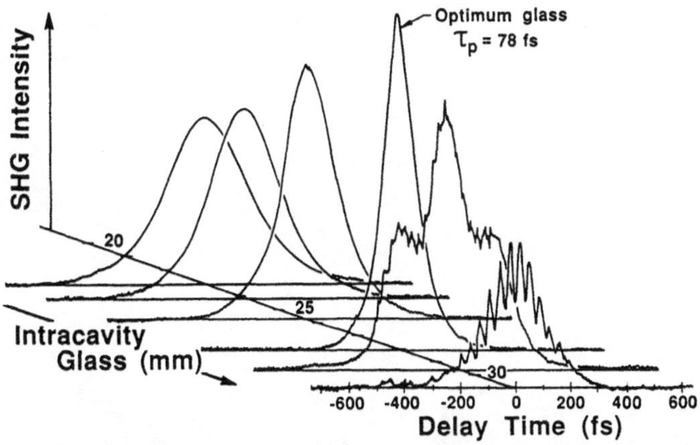

Figure 3: Variation of autocorrelation trace with intracavity glass pathlength for dispersion compensated 2 jet hybrid modelocked Stilbene 3/C535 dye laser operating at 452 nm.

the sum of approximately 1.9×10^4 pulses. The autocorrelation trace corresponding to the position of optimum glass pathlength (35.6mm) is observed to have a width of 355fs. Current theoretical and experimental understanding of the modelocking process in this laser system indicates that a hyperbolic secant squared pulse shape is expected which would imply a pulse FWHM = 220fs.

The use of the frequency tripled output of modelocked YAG lasers as a pump source for synchronous dye lasers provides a convenient means of directly generating tunable ultrafast pulses in the blue end of the optical spectrum. Ultrafast time domain studies using dyes such as Stilbene 1, Stilbene 3, and Coumarin 102 become feasible for the first time.

REFERENCES

1. J. N. Eckstein, A.J. Ferguson and T. W. Hänsch C. A. Minard and C. K. Chan, Opt. Com. 27, 466 (1978)
2. M.W. Dawson, T.F. Boggess, D. W. Garvey, and A. L. Smirl, Opt. Com. 60, 79 (1986)
3. Coherent Antares 76s
4. Chen Changtian, Wu Bochang, Jiang Aidong and You Guiming, Scientia Sinica B82, 235 (1985)
5. R.L. Fork, O.E. Martinez, and J.P. Gordon, Optics Lett. 9, 150 (1984)

New Developments in Ultraviolet and High Intensity Femtosecond Sources

M.C. Downer, G. Focht, T.R. Zhang, W.M. Wood, D.H. Reitze, and G.W. Burdick

Physics Department, University of Texas at Austin,
Austin, TX78712, USA

1. Intracavity Frequency Doubling: A New Class of Ultraviolet Femtosecond Sources

Despite extensive development of new femtosecond dye lasers operating from green to near infrared [1], femtosecond source lasers in the blue and ultraviolet have been lacking. Using the simple, non-perturbative technique of intracavity frequency doubling, we have developed an efficient, high repetition rate source of synchronized ultraviolet and red femtosecond pulses [2]. This new UV source is useful in photoexcitation of molecular absorption bands, photoionization, photoemission, and tight focusing applications such as electro-optic sampling of high speed integrated circuits.

Figure 1a shows the intracavity subresonator which we have added to a colliding pulse mode-locked (CPM) laser to focus intracavity red pulses into a thin KDP crystal cut for Type I phase matched second harmonic generation at Brewster angle incidence. Linear dispersion of the crystal was compensated by an intracavity prism configuration. Red pulse durations of 50-70 fs were preserved, without bandwidth limitation or power reduction, as shown in Figs. 1b and 1c. Using a 1 mm thick KDP crystal an ultraviolet beam of ~ 1 mW average power (~ 10 pJ /pulse) was extracted through the dichroic mirror as shown in Fig. 1a. Average UV powers as high as 20 mW have been obtained with the more efficient BBO doubling crystal [3]. UV pulse durations range from 40 fs to 170 fs depending on crystal thickness.

In contrast to our observations, previous attempts at intracavity frequency doubling of actively and synchronously mode-locked lasers [4] resulted in mode-locking instability, pulse broadening, and bandwidth limitation, even though these lasers operated in the picosecond, rather than the femtosecond, time domain. *Passively* mode-locked lasers, on the other hand, possess a built-in stabilization mechanism which avoids these problems, as depicted in Fig. 2a. Temporal broadening occurs in the crystal because power dependent loss selectively attenuates

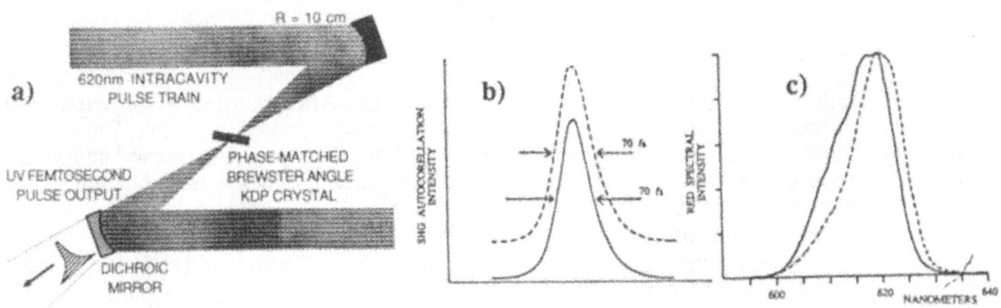

Fig. 1 a) Schematic of the subresonator for intracavity frequency doubling; b) autocorrelation traces and c) spectral intensity profiles of the fundamental red pulses without the intracavity doubling crystal (dashed curves) and with the intracavity crystal tuned for maximum output power (solid curves)

Springer Series in Chemical Physics, Vol. 48: **Ultrafast Phenomena VI**
Editors: T. Yajima · K. Yoshihara · C.B. Harris · S. Shionoya

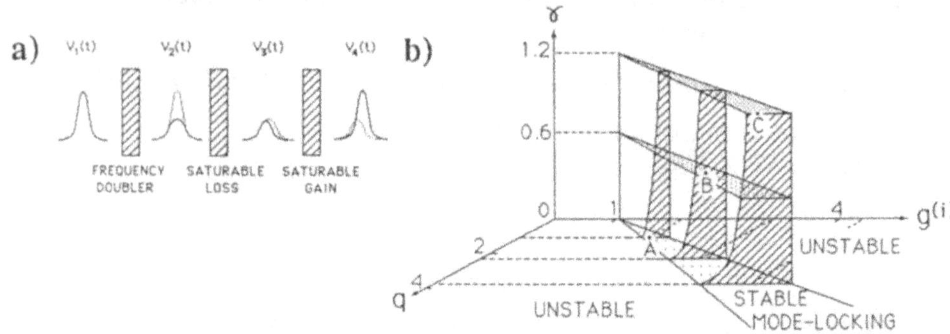

Fig. 2 a) Changes in intracavity pulse envelope which occur upon passage through the major components of a passively mode-locked laser with intracavity frequency doubler; b) graphical representation of the stable mode-locking regime as a function of saturable loss q, saturable gain g(i), and a dimensionless second harmonic generation parameter γ, based on Haus' passive mode-locking theory. The points A, B, and C represent stable operating conditions at 0, 1.2%, and 2.4% conversion efficiency which correspond to equal intracavity pulse duration (70 fs) and energy (5 nJ)

the peak of the pulse. Passage through a saturable absorber, on the other hand, selectively attenuates and therefore sharpens the leading edge of the pulse. Analogously gain saturation sharpens the trailing edge. Adjustment of absorber and gain saturation levels can therefore precisely compensate the pulse broadening caused by the doubling crystal. Intracavity doubled *passively* mode-locked lasers therefore constitute a whole new class of blue and ultraviolet femtosecond sources.

We have quantified these general concepts by modifying Haus' theory of the passively mode-locked laser to include an intracavity frequency doubler [5]. Fig. 2b graphically illustrates our model of the stable operating regime as a function of saturable loss (q), saturable gain (g(i)), and second harmonic conversion efficiency (γ). Note that stable operation can always be recovered by adjusting gain or loss (i.e. adjusting pump power or intracavity focus into saturable absorber). Similar calculations from our model [5] demonstrate the preservation of fundamental pulse duration and energy within the stable regime for conversion efficiencies ranging from 0 to 5%, in accord with experimental observations [2,3].

2. Tight Focusing of Millijoule Femtosecond Pulses: Light Intensity > 10^{16} W/cm^2

Amplification of femtosecond pulses to millijoule and higher energies has opened up the study of the interaction of matter with radiation fields of unprecedented intensity. The output of high power femtosecond amplifiers, however, often suffers from poor transverse beam quality and limited focusability caused by severe phase front distortion during the amplification process. We have constructed a high power dye amplifier pumped by a Q-switched Nd: YAG laser which achieves near diffraction-limited focusing of 100 fs millijoule pulses without any spatial filtering [6]. The peak intensity at our tightest focus (~ 2 μm radius) exceeds 10^{16} W/cm^2, where breakdown of air and other gases at pressures of 1-10 atm is easily observed.

Critical to achieving this result was a final stage gain cell of conical axicon geometry [7], as shown in Fig. 3a. The first three stages (not shown) use standard transversely pumped gain cells to pre-amplify femtosecond pulses from a CPM to ~ 0.1 mJ at 10 Hz repetition rate. These pulses are then expanded to 1 cm diameter and injected into the gain region of the axicon, an 8 cm long cylindrical flow tube containing Rh640 centered on the axis of a solid glass cone. Incident to the base of the cone and expanded to match its 8 cm radius, counter-propagating

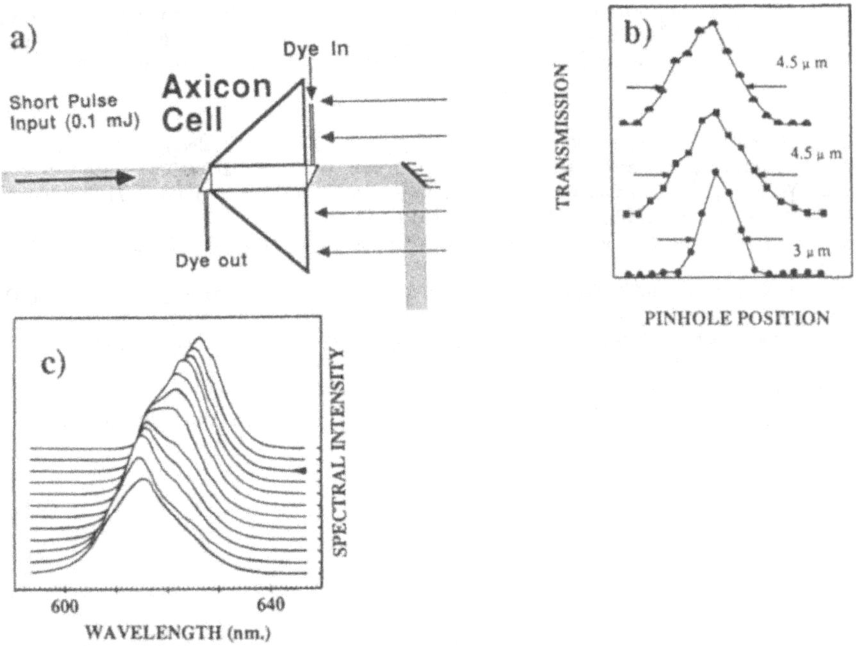

Fig. 3 a) Final stage geometry for the femtosecond axicon dye cell amplifier; b) beam waist profile of unamplified CPM output (bottom curve), pre-amplified output prior to axicon cell (middle curve), and fully amplified output from axicon cell, all at focus of a 7x microscope objective; c) Spectra of 100 fs millijoule pulses transmitted through air breakdown spark, with focal intensity increasing toward the front in increments of 0.1 optical density. Arrow denotes breakdown threshold; successive curves have been displaced from each other in an exact vertical direction to emphasize blue shift as intensity increases

220 mJ , 532 nm pump pulses internally reflect from the sides and enter the gain medium at radial incidence, thereby maximizing pump absorption efficiency, Gaussian beam quality, and focusability [6,7]. Figure 3b shows measurements of transmission of the beam focused at f/2 through a 3 μm pinhole translated in 1 μm steps across the beam waist, and demonstrates the extremely tight focusability of millijoule femtosecond pulses amplified in such a cell. While a slight degradation in focusability occurs in the pre-amplification stages, the axicon cell preserves focusability perfectly.

Figure 3c shows the pronounced blue shift in the spectrum of 100 fs pulses tightly focused in air. This blue shift is caused by a rapid negative refractive index change as a plasma is created during the pulse. Below breakdown threshold (back spectrum in Fig. 3c) the spectrum is centered at 626 nm and is indistinguishable from that of the unfocused pulse. Above threshold the blue shift increases with focal intensity until saturating at a center wavelength of 615 nm at a focal intensity of 2×10^{15} W/cm^2 (third spectrum from front) and above. The blue shift saturates at 11 nm because air within the focal volume has been completely ionized. At higher gas pressures and with different gases we have observed blue shifts as large as 100 nm, causing a visible change in beam color. Self-focussing and continuum generation are suppressed because of the extremely short interaction length (~100 μm) with tight focusing. Such blue shifts can complement other diagnostics (e.g. photoelectron spectroscopy) of the ionization process in the strong field, above threshold, collisionless regime. They also provide a simple means of tuning powerful femtosecond pulses with virtually no loss in energy.

This research was supported by the Joint Services Electronics Program (Contract F49620-86-C-0045), the Robert A. Welch Foundation (Grant F-1038), an NSF Presidential Young Investigator Award (Grant DMR-8858388), and an IBM Faculty Development Award.

1. P.M.W. French and J.R. Taylor, *Ultrafast Phenomena V* (Springer-Verlag, Heidelberg 1986), p.11; M.D. Dawson, T.F. Boggess, D.W. Garvey, and A. L. Smirl, IEEE J. Quant. El. **22**, 290 (1987); J. Dobler, H.H. Schultz, and W. Zinth, Opt. Commun. **61**, 407 (1986); W.H. Knox, J. Opt. Soc. Am. B4, 1771 (1987).
2. Glenn Focht and M.C. Downer, IEEE J. Quant. El. **24**, 431 (1988).
3. D.C. Edelstein, E.S. Wachman, L.K. Cheng, W.R. Bosenberg, and C.L. Tang, Appl. Phys. Lett. **52**, 2211 (1988).
4. M. Yamashita, K. Yamada, and T. Sato, IEEE J. Quant. El. **18**, 95 (1982); M. Yamashita, W. Sibbett, D. Welford, and D.J. Bradley, J. Appl. Phys. **51**, 3559 (1980).
5. T.R. Zhang, Glenn Focht, P.E. Williams, and M.C. Downer, IEEE J. Quant. El., in press (1988); H.A. Haus, IEEE J. Quant. El. **11**, 736 (1975).
6. W.M. Wood, Glenn Focht, and M.C. Downer, Opt. Lett., in press (1988).
7. F.P. Schäfer, Appl. Phys. B39, 1 (1986); G. Kuhnle, G. Marowsky, G.A. Reider, Appl. Opt., in press.

Part III

Propagation, Control and Measurement of Ultrashort Light Pulses

Dispersive Pulse Shaping and Soliton-like Pulses in a Passively Mode-Locked Dye Ring Laser

D. Kühlke, T. Bonkhofer, U. Herpers, and D. von der Linde

Universität-GHS-Essen, Institut für Laser und Plasmaphysik, D-4300 Essen 1, Fed. Rep. of Germany

Recent investigations have revealed that in passively mode-locked dye lasers two mechanisms are responsible for the formation of femtosecond pulses. These are saturation of the absorption and gain [1] and the combined action of self-phase modulation and group velocity dispersion (dispersive pulse shaping) [2]. However, the actual contribution of each of the two mechanisms to the pulse shaping has not been investigated in detail.

In this paper we report on measurements on a colliding-pulse mode-locked dye laser which enabled us to estimate the influence of either mechanism to the relative pulse shortening per round trip in the steady state regime. We found that the shortest pulses occur when the dispersive pulse shaping dominates. The variation of the absorber concentration and the pump power revealed that for low concentrations and low pump powers pulse shaping is mainly governed by saturation effects whereas for high concentrations and high pump powers dispersive effects dominate.

We used a standard rhodamine 6G/DODCI CPM dye laser with a four prism sequence to control the group velocity dispersion [2]. We measured energy, duration and spectra of the pulses, beam diameters in the dye jets, dye concentrations and the optical path through the four prism sequence.

To evaluate the influence of the two pulse shaping mechanisms we considered a stationary round trip model. Pulse broadening per round trip due to the bandwidth limitation of gain and absorption is compensated by pulse shortening per round trip. We calculated the decrease of the pulse duration per round trip due to saturation effects from a rate equation model for the dynamics of a two-level system [1]. To evaluate the influence of dispersive effects we assumed that the phase modulation of the pulses is compensated by the group velocity dispersion. The actual dispersive pulse shortening is then calculated from the group velocity dispersion of the four prism sequence which is neccessary to obtain shortest pulses.

Fig. 1a shows the estimated relative change of the pulse duration caused by saturation (open squares) and by dispersive effects (full squares) as a function of the absorber position relative to the beam waist. The arrows indicate the positions where the shortest pulses (55 fs) occur. The data indicate that at these positions the dispersive pulse shaping is dominant and that the contributions due to saturation effects are practically neglegible. The pulse duration is 175 fs when the absorber is placed at the beam waist. Here the relative change of the pulse duration is distinctly less and saturation pulse shaping is the dominant process.

We also performed measurements for different absorber dye concentrations. For each concentration the absorber position and pump power was adjusted to obtain shortest pulses. Fig. 1b shows the estimated relative decrease of the

Springer Series in Chemical Physics, Vol. 48: **Ultrafast Phenomena VI**
Editors: T. Yajima • K. Yoshihara • C.B. Harris • S. Shionoya

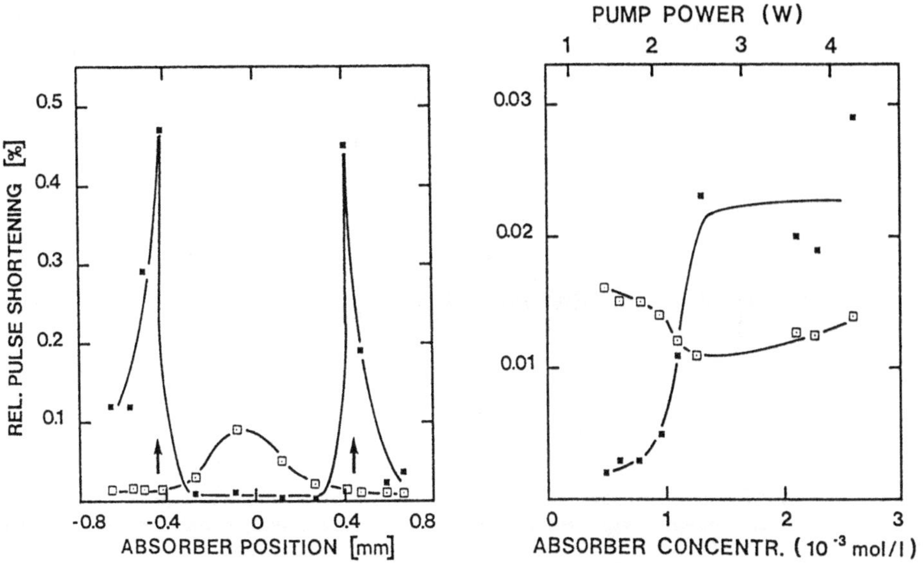

Fig. 1a   Fig. 1b

Relative shortening of the pulse duration per round trip as a function of the absorber distance from the beam waist (a), and of the absorber concentration and the pump power (b) due to saturation effects (open squares) and dispersive effects (closed squares). The arrows in 1a indicate the absorber positions of shortest pulses.

pulse duration per round trip as function of the absorber concentration and the pump power. The plot reveals that at low absorber concentrations (<1 mMol/l) and small pump powers (< 2W) the saturation pulse shaping is the dominant process. The pulse duration is about 200 fs. When the pump power is increased to 2.5 W the pulse duration decreases to 125 fs and dispersive pulse shaping becomes the leading mechanism. The reason is that the rise of the pump power increases the intracavity power and, consequently, the self-phase modulation produced in the dye solvents. On the other hand, to obtain shortest pulses the absorber jet is moved away from the beam waist with increasing pump power. The resulting increase of the beam area causes a decrease of the absorber saturation. The longer pulses and the less pulse shortening compared with Fig. 1a are probably caused by a non-optimal cavity adjustment.

Dispersive pulse shaping can lead to pulses resembling higher order solitons [3]. N = 3 solitons show periodic changes of the pulse shape, but the pulse energy is constant [4]. The autocorrelation averaged over the soliton period consists of a central maximum and two symmetric secondary peaks. To detect the periodic changes we took advantage of the fact that upon second harmonic generation changes of the pulse shape of the fundamental are transformed into variations of the pulse energy of the second harmonic. The modulation frequency and the peak intensity variation can then be evaluated from the power spectrum. We have observed soliton-like pulses in our CPM laser when the group velocity of the cavity was adjusted to a slightly greater value than that required for shortest pulses. Figure 2 shows a 2 MHz section of the power spectrum of the second harmonic pulse train. There is the sharp central component corresponding to the cavity round trip frequency and sideband frequencies of 480 kHz which does not appear in the power spectrum of

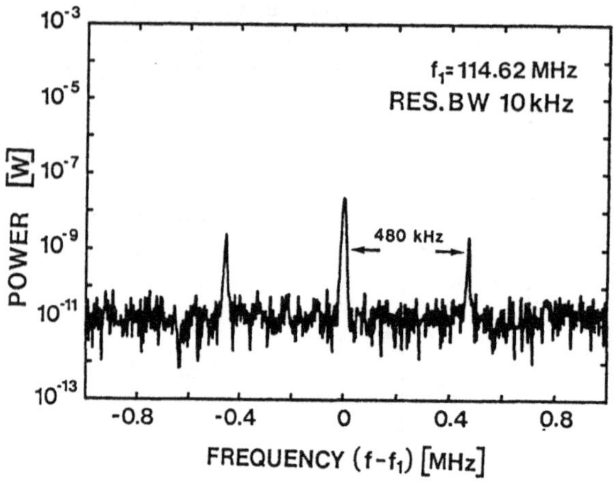

Fig. 2 Section of the power spectrum of the second harmonic of a N = 3 soliton-like pulse train centered around 114.6 MHz. The frequency resolution is 10 kHz.

the fundamental pulse train. This frequency is attributed to half a soliton period. From the power ratio of the side and the central peak we estimated the variation of the peak intensity of the fundamental pulse train to be 64%. This agrees well with the modulation depth of 71% that is evident from the numerically calculated evolution of a N = 3 soliton [4].

The pulse peak intensity estimated from our laser parameters and the observed soliton frequency also agree quite well with the values theoretically expected for a N = 3 soliton [4]. Thus we think that these phenomena can be attributed to soliton-like behavior.

Finally we note that in some cases autocorrelations with three humps resembling N = 3 soliton behavior were seen but the side bands in the second harmonic spectra were absent. It appears that in these cases the structure of the autocorrelation was due to different pulsing modes of the laser, possibly a temporally stable double pulse structure.

1. H. A. Haus, IEEE J. Quantum Electron. QE-11, 736 (1975)
2. J. A. Valdmanis, R. L. Fork, J. P. Gordon, Opt. Lett. 10, 131 (1985)
3. F. Salin, P. Grangier, G. Roger, A. Brun, Phys. Rev. Lett. 56, 1132, (1986) and Phys. Rev. Lett. 60, 569 (1988)
4. L. F. Mollenhauer, R. H. Stolen, J. P. Gordon, Phys Rev. Lett. 45, 1095 (1980))

Control and Characterization of Soliton-like Pulses in a Femtosecond Dye Laser

W.L. Nighan, Jr., T. Gong, and P.M. Fauchet

Princeton Laboratory for Ultrafast Spectroscopy,
Department of Electrical Engineering, Princeton University,
Princeton, NJ 08514, USA

The balance of positive self-phase modulation (SPM) and negative group velocity dispersion (GVD) that enhances ultrashort pulse formation in a dispersion compensated, colliding pulse modelocked (CPM) dye laser [1,2] has been compared to the shaping mechanism for the N=1 soliton in an optical fiber [3,4]. Periodic pulseshaping reminiscent of higher order solitons has also been observed [5-7]. It is however unclear how far the analogy with solitons in an optical fiber can be extended. We report here results of our investigation of periodic pulseshaping in a 7 mirror/4 prism CPM laser that delivers pulses as short as 35fs at 100 MHz. The stability achieved allows for reproducibility and full time-resolved characterization.

We control the order, period and duration of the soliton-like pulses (henceforth abbreviated as solitons) by adjusting the group velocity dispersion (GVD) and self-phase modulation (SPM) within the laser cavity. Starting from the shortest pulse (N=1) configuration, higher order solitons are achieved by (1) increasing the amount of SPM by translating the absorber jet much closer to the intracavity focus or using a thicker (+ 10 μm) region of the jet, (2) increasing the magnitude of the negative GVD by translating a prism to reduce the amount (- 150 μm) of glass traversed, (3) optimizing the alignment of the laser cavity. The clockwise and counterclockwise beams are typically asymmetrized by step (3), indicating that a precise balance of gain and loss is required to support the higher order solitons.

The single sweep real-time intensity autocorrelations of Fig. 1 are an illustration of this control. Solitons of order 3, 2 and 1 are seen as the absorber jet is translated away from the focus. Modulation at the soliton frequency is clearly evident in these single sweeps. Full and accurate characterization of the solitons is possible by sampling intensity autocorrelations, interferometric autocorrelations and spectra during the soliton period. This is achieved by using the deep ($\sim$ 50%) modulation in a narrow spectral bandwidth near the central wavelength to generate a short gate pulse train for a boxcar sampler and for an optical multichannel analyzer (OMA). Both intensity and interferometric autocorrelations taken at points in the 2.5 μsec period of the N=2 soliton of Fig. 1(b) are shown in Fig. 2(b) and Fig. 2(c). Figure 2(a) shows spectra sampled from the period of a similar N=2 soliton. The autocorrelation traces show that the pulse is always composed of two coherent peaks with a separation that increases up to nearly one picosecond. The signal enhancement achieved by interferometric autocorrelation allows observation of the weaker peak even where it is no longer seen with intensity autocorrelation. The spectra indicate the presence of two major components whose peak positions and relative amplitude vary periodically.

The evolution characterized in Fig. 2 is reminiscent of the results of SALIN et al. [6], who interpreted their sampled intensity autocorrelations and spectra as evi-

Springer Series in Chemical Physics, Vol. 48: **Ultrafast Phenomena VI**
Editors: T. Yajima · K. Yoshihara · C.B. Harris · S. Shionoya

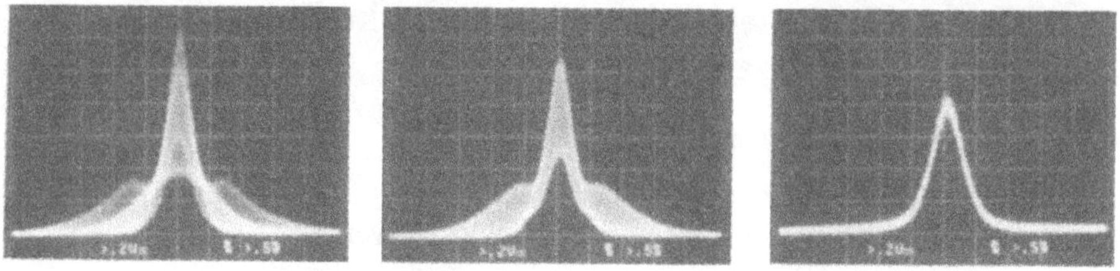

Fig.1. Single sweep real-time autocorrelation traces for different jet positions.
(a) N=3: Jet at focus (b) N=2: Jet 50 μm from focus (c) N=1: Jet 250 μm from focus

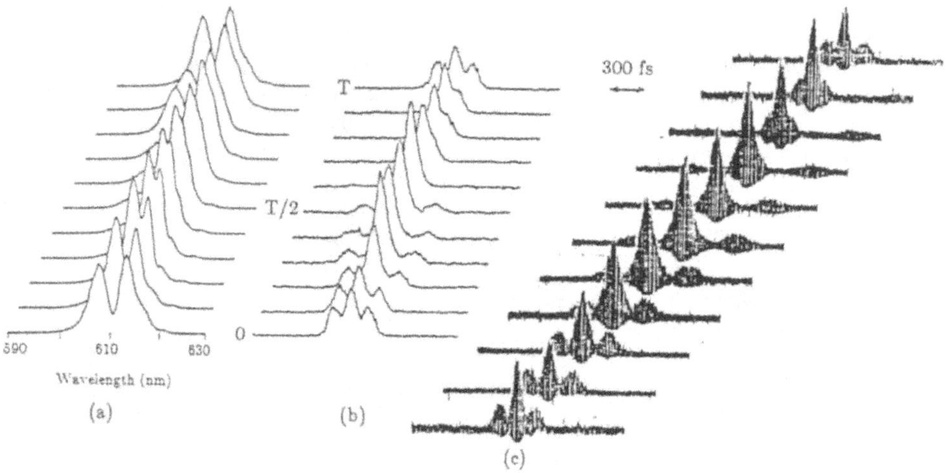

Fig.2. Time-resolved evolution of the N=2 soliton of Fig.1(b).
(a) Spectra (b) Intensity autocorrelations (c) Interferometric autocorrelations

dence for an asymmetric N=2 soliton. Our output pulse energy is modulated by $\sim 10\%$ with a period of 2.5 μsec, which corresponds to the soliton period. The two peaks of this soliton vary in separation (Fig. 3) but are coherent with one another (unlike the results of Ref. 7 which, however, were obtained under different conditions) and never merge into a single peak. These characteristics may correspond to a periodic motion of the poles and residues used to describe the N=2 soliton [5,8]. Fig. 4 shows the linearity of the soliton dispersion relation as defined in [5]. The transition from N=2 to N=3 as described in Fig. 1 is analogous to increasing the input intensity in a fiber. However, the period of the evolution abruptly doubles upon this transition, unlike what happens in a fiber. The spectral components do, however, exhibit N-1 frequencies of modulation. The sampled autocorrelations for the N=3 case show that the pulse is narrowed twice in its period, as predicted for asymmetric solitons of this order in optical fibers. The abrupt change in period upon the N=2 to N=3 transition may result from the alteration in saturation effects that accompanies the translation of the jet to a point of higher intracavity intensity. The average output pulse energies remain nearly constant through the transition.

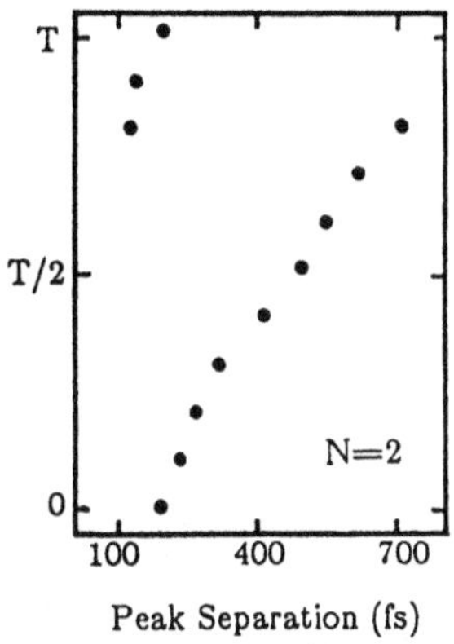

Fig.3. Peak separation during N=2 soliton period.

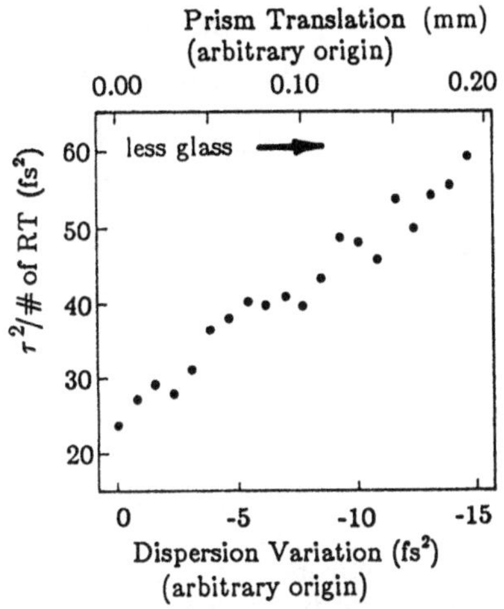

Fig.4. N=2 soliton dispersion relation. τ = pulsewidth of soliton at single-peaked point of period. # of RT= number of cavity roundtrips corresponding to a single soliton period.

We have demonstrated control of soliton pulseshaping in a femtosecond CPM laser. Accurate characterization is possible since the pulseshapes are stable and reproducible. Similarities and differences with solitons in an optical fiber have been pointed out. The principles are general and could be applied to other short pulse lasers in different wavelength regimes.

This work is supported by NSF (ECS-8606531), the NSF Presidential Young Investigator Program (ECS-8657263) and ARO (DAAL03-87-K-0145).

References

1. R.L. Fork, B.I. Greene and C.V. Shank, Appl. Phys. Lett. **41**, 671 (1981).
2. J.A. Valdmanis, R.L. Fork and J.P. Gordon, Opt. Lett. **10**, 131 (1985).
3. J.C. Diels, W. Dietel, J.J. Fontaine, W. Rudolph and B. Wilhelmi, J. Opt. Soc. Am. **B2**, 680 (1985).
4. O.E. Martinez, R.L. Fork and J.P. Gordon, J. Opt. Soc. Am. **B2**, 753 (1985).
5. F. Salin, P. Grangier, G. Roger and A. Brun, Phys. Rev. Lett. **56**, 1132 (1986).
6. F. Salin, P. Grangier, G. Roger and A. Brun, Phys. Rev. Lett. **60**, 569 (1988).
7. F.W. Wise, I.A. Walmsley and C.L. Tang, Opt. Lett. **13**, 129 (1988).
8. H.A. Haus and M.N. Islam, IEEE J. Quantum Electron. **QE-21**, 1172 (1985).

Solitons in the Region of the Minimum Group Dispersion Wavelength of a Single Mode Optical Fibre

A.S. Gouveia-Neto, M.E. Faldon, and J.R. Taylor

Femtosecond Optics Group, Physics Department, Imperial College, Prince Consort Road, London SW7 2BZ, UK

The potential application of solitons [1] in future all-optical, high bit rate, communication systems has been recognised by many authors. The power to establish a fundamental soliton is inversely proportional to the square of the pulse duration and directly proportional to the fibre dispersion [2]. For pure silica fibres, the zero of the second order dispersion occurs at 1.27μm. However, due to the contribution of the waveguide dispersion, the overall zero component of the second order dispersion is shifted to longer wavelengths. For pulses launched in this wavelength range, the effects of third order dispersion are predominant in determining the required power levels to establish solitons, while the non-linear Schrodinger Equation (NLSE) involving second order dispersion, which has been successful in modelling soliton propagation, is no longer valid. It has recently been shown by Wai et al.[3] that solitons or more precisely solitary waves do evolve from pulses launched at or in the region of the minimum dispersion where the effects of third order dispersion are considered in the NLSE. In addition , the power levels required to establish the solitons are substantially reduced from those required to generate similar solitons further into the anomalously dispersive regime.

We have examined experimentally the spectral and temporal evolution of pulses launched with their spectral bandwidth spanning the minimum dispersion wavelength. In good qualitative agreement with theoretical prediction [3] it was observed that as the soliton was formed, the nonlinearity gave rise to a clear spectral fragmentation with the associated generation of solitary and dispersive waves.

Figure 1 shows a schematic of the experimental arrangement. The source of pulses was derived from a c.w. mode locked Nd:YAG laser at 1.32μm. Using a standard fibre-grating pair compressor arrangement the 100ps pulses were reduced in duration to ~1.5ps , with an average power of 280mW. In order to derive a source of pulses which was tunable in the region of the minimum dispersion wavelength of several fibre species, use was made of stimulated Raman conversion of multisoliton pulses [4] or the soliton self frequency shift, as it is alternatively known [5]. By simply adjusting the power of the 1.5ps compressed pulses at 1.32μm launched into the 140m length of fibre F_1 with a dispersion minimum wavelength λ_0= 1.27 μm the amount of self frequency shift was controlled to permit the generation of a wavelength component which spanned the minimum dispersion wavelength of fibre F_2.

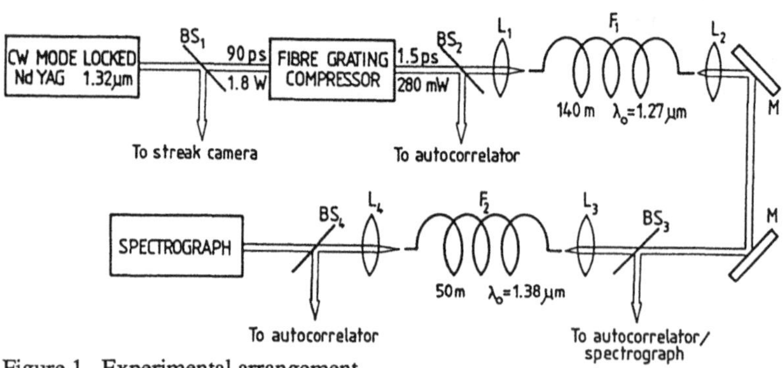

Figure 1. Experimental arrangement

Springer Series in Chemical Physics, Vol. 48: **Ultrafast Phenomena VI**
Editors: T. Yajima · K. Yoshihara · C.B. Harris · S. Shionoya

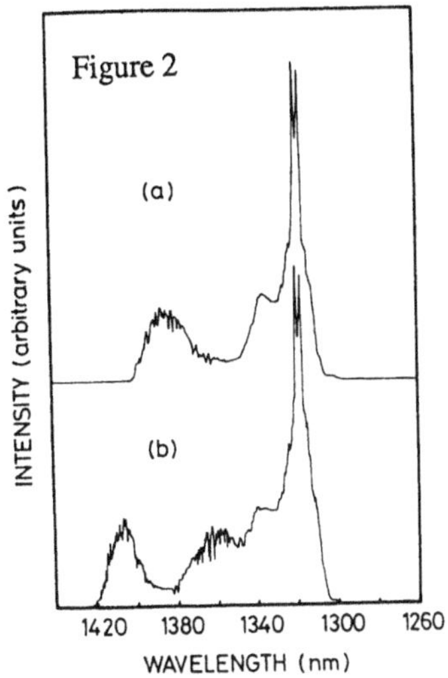

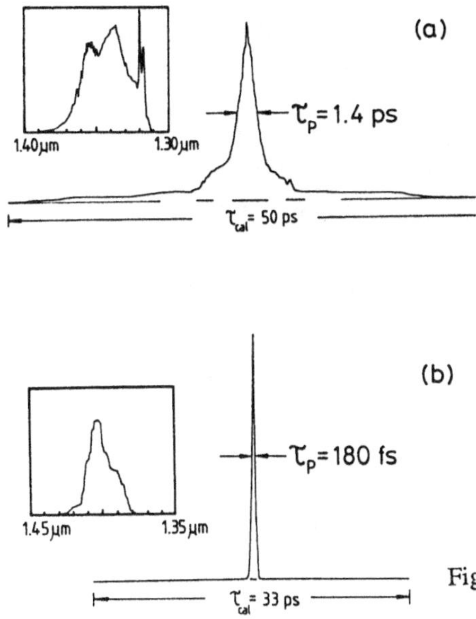

Figure 3

Figure 2. (a) Input and (b) output spectra from a fibre with a minimum dispersion wavelength at 1.38μm.

Figure 3. Autocorrelation traces of the components in the (a) normally and (b) anomalously dispersive regimes. The corresponding spectra are shown.

Figure 2(a) shows the spectrum of the pulses which were launched into F2. It would have been preferable to filter out the self shifted spectrum around 1.38μm and launch it alone into F2, however, such a filter was initially unavailable and the remaining portion of the spectrum did tend to mask some of the spectral shifting effects. Figure 2(b) is the corresponding output spectrum from F2 at a launched total power of 25mW, of which approximately 6mW was in the component at 1.38 μm. This clearly shows a spectral splitting around 1.38 μm, with the spectrum of the soliton centering around 1.41 μm, while in the normally dispersive regime the peak shifted to around 1.36μm.

The variation of the wavelength shift with the amplitude of the pump pulse revealed the effect in the normally dispersive regime to be approximately 1.3 times that in the anomalous region. Theory has predicted that this ratio should be 1.7 [3]. However, it is possible that soliton self frequency shift effects would increase the corresponding wavelength shifts and so decrease this ratio.

Because of the clear spectral separation of spectra in the normal and anomalously dispersive regimes, spectral filtering permitted the selection of the appropriate component. Figure 3 shows background free autocorrelation traces together with inserts of the corresponding spectral components in the two regimes. From the measured 330W peak power of the pulses in the anomalous regime it was calculated that the pulse was propagating as a single soliton in the fibre. In the normally dispersive region, temporal broadening due to dispersion was apparent (see fig.3(b)).

In conclusion we have shown reasonable agreement with theoretical prediction that pulses launched into a single mode fibre which span the dispersion mimimum wavelength can generate solitary waves. These exhibit a frequency down shift and their required power levels are substantially lower than those necessary for generation further into the anomalously dispersive region. Such a technique may be applicable to the use of semiconductor lasers as potential sources of solitons.

The financial support of British Telecom and the SERC is gratefully acknowledged.

References

1. A. Hasegawa and F. Tappert : Appl. Phys. Lett . 23 ,142 (1973)
2. L.F. Mollenauer, R.H. Stolen and J.P. Gordon : Phys. Rev. Lett. 45,1095 (1980)
3. P.K.A. Wai, C.R. Menyuk , H.H. Chen and Y.C. Lee : Opt. Lett. 12, 628 (1987)
4. E.M. Dianov , A.Ya. Karasik , P.V. Mamyshev , A.M. Prokhorov , V.N. Serkin , M.F. Stelmakh and A.A. Fomichev : JETP Lett. 41, 294 (1985)
5. F.M. Mitschke and L.F. Mollenauer : Opt. Lett. 11, 659 (1986)

Femtosecond Pulse Tailoring for Dark Soliton Propagation Studies

A.M. Weiner[1], *J.P. Heritage*[1], *R.J. Hawkins*[2], *R.N. Thurston*[1], *E.M. Kirschner*[1], *D.E. Leaird*[1], *and W.J. Tomlinson*[1]

[1]Bell Communications Research, 331 Newman Springs Road, Red Bank, NJ 07701, USA
[2]Lawrence Livermore National Laboratory, Livermore, CA 94550, USA

Intense ultrashort pulses propagate in optical fibers under the influence of group velocity dispersion (GVD) and self-phase-modulation. Balancing these processes leads to optical solitons which propagate without distortion down long lengths of fiber. Bright soliton propagation in the negative GVD portion of the spectrum has been investigated in detail. Dark solitons, which are predicted for positive GVD [1], have been studied much less extensively [2,3], in part due to the difficulty of generating the required dark input pulses. We have developed a technique for synthesizing arbitrarily shaped femtosecond pulses [4], and we apply this technique to generate dark pulses for nonlinear propagation studies. In this way we succeed for the first time in clearly demonstrating fundamental dark soliton propagation.

Dark soliton solutions to the nonlinear Schrodinger equation (NLSE) consist of a rapid dip in the intensity of a CW background. The intensity of the fundamental dark soliton drops to zero at $t=0$; we call this a "black" soliton. Other dark solitons exist with lesser hole depths; we term these "gray" solitons. The black soliton is an antisymmetric function of time, with an abrupt π phase shift at $t=0$. Gray solitons exhibit smaller and more gradual phase modulations at their center [1].

With our pulse shaping technique it was most efficient to generate 100-200 fsec dark pulses on background pulses 1-4 psec in duration. For this reason we performed extensive computer simulations to test whether dark pulses could exhibit stable soliton propagation with finite duration background pulses. Figure 1 shows the results of a simulation for an odd-symmetry dark pulse on a Gaussian background for a peak power equal to the soliton power. Although the background acquires a chirp and eventually broadens, the dark pulse propagates almost without distortion. Simulation results for an even-symmetry dark pulse are shown in Fig. 2. The input is not a soliton solution to the NLSE, and the dark pulse breaks up into a pair of gray solitons [5].

The pulse shaping apparatus has been described previously. Briefly, 75-fsec, 620-nm pulses from a CPM dye laser and a copper vapor laser pumped dye amplifer are shaped within a temporally nondispersive lens and grating apparatus, consisting of a pair of diffraction gratings placed at the focal planes of a unit magnification, confocal lens pair [4,6]. Spatially patterned amplitude and phase masks are positioned midway between the gratings where the optical frequency components are spatially dispersed. The shape of the output pulse is the Fourier transform of the pattern transferred by the masks onto the spectrum. In the present study, intense shaped pulses are launched into a 1.4-meter length of single-mode, polarization preserv-

Springer Series in Chemical Physics, Vol. 48: **Ultrafast Phenomena VI**
Editors: T. Yajima · K. Yoshihara · C.B. Harris · S. Shionoya

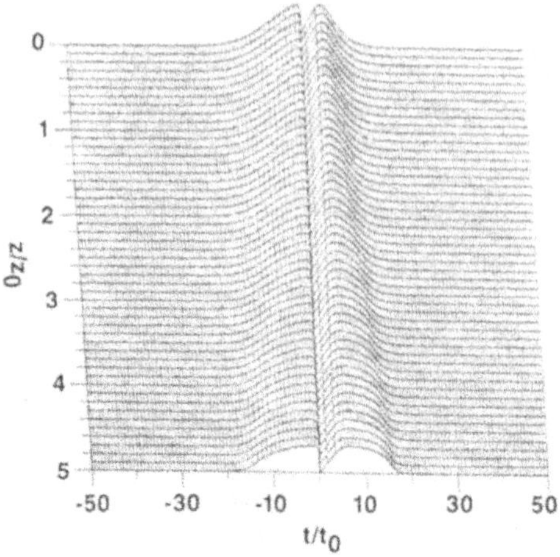

Fig. 1 Propagation simulation for an odd-symmetry dark pulse on a Gaussian background, with peak power equal to the soliton power.

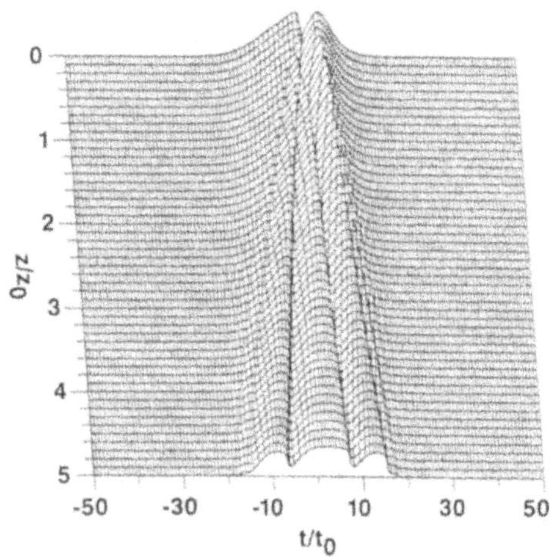

Fig. 2 Propagation simulation for an even-symmetry dark pulse on a Gaussian background, with the same peak power as in Fig. 1.

ing fiber (Newport Corp. F-SPV). Both symmetric and antisymmetric dark pulses are investigated. The intensity profile of the pulses emerging from the fiber are measured by cross-correlation with 75-fsec pulses directly from the amplifier.

An example of our data is shown in Fig. 3. The input consists of a 185-fsec, antisymmetric dark pulse on a 1.76 psec Gaussian background (Fig. 3a). Figures 3b-3e depict the waveforms emerging from the fiber at various power levels. At low power SPM is negligible, and GVD broadens the central hole to more than 600 fsec. As the power is increased, the hole narrows. At 300 W the hole is 195 fsec wide, close to the width of the input hole. The data are in quantitative agreement with numerical solutions to the NLSE, which are shown as solid lines. At high power a pair of gray solitons is predicted in addition to the central black soliton; one gray soliton can be discerned in the data (Fig. 3e). Our measurements constitute the first clear observation of the fundamental dark soliton in fibers.

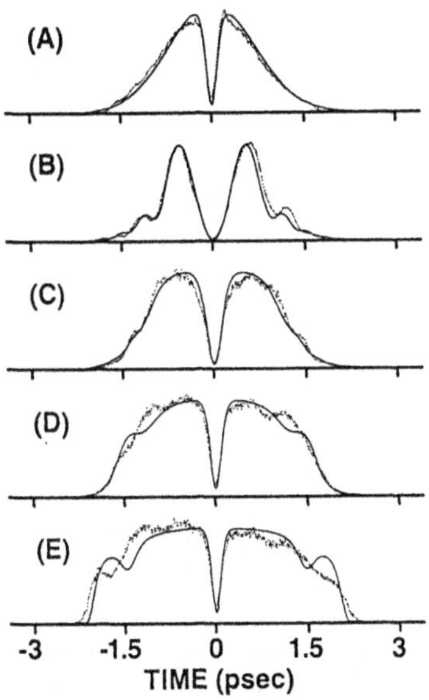

Fig. 3 Measured (dots) and calculated (solid lines) cross-correlations (A) for odd-symmetry input pulses and (B-E) for pulses exiting the fiber. (B) P = 1.5 W; (C) 50 W; (D) 150 W; (E) 300 W.

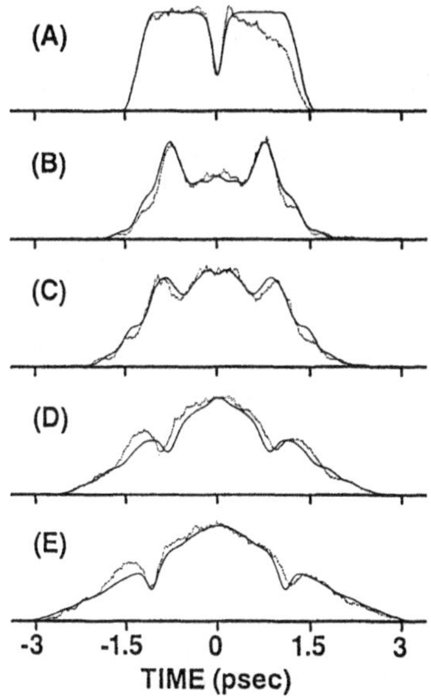

Fig. 4 Measured (dots) and calculated (solid lines) cross-correlations (A) for even-symmetry input pulses and (B-E) for pulses exiting the fiber. (B) P = 2.5 W; (C) 50 W; (D) 150 W; (E) 285 W.

We also performed experiments using even-symmetry dark pulses; data for an even dark pulse on a square-like background are shown in Fig. 4. Similar data were reported by KROKEL et al for effectively infinite duration background pulses [2]. At low power the dark pulse is completely obscured by dispersion. At higher power the data shows two shallow holes superimposed on a smooth background. According to the numerical simulations, these holes may be low contrast gray solitons. At both low and high powers, the data are in quantitative agreement with numerical solutions to the NLSE (solid lines).

In summary, our data provides compelling evidence for dark soliton propagation in optical fibers and confirms the crucial importance of the dark pulse phase profile, as predicted by theory.

References

1. A. Hasegawa and F. Tappert, Appl. Phys. Lett. 23, 171 (1973)
2. D. Krokel, N.J. Halas, G. Giuliani and D. Grischkowsky, Phys. Rev. Lett. 60, 29 (1988)
3. P. Emplit, J.P. Hamaide, F. Reynaud, C. Froehly, and A. Barthelemy, Opt. Commun. 62, 374 (1987)
4. A.M. Weiner, J.P. Heritage and J.A. Salehi, Opt. Lett. 13, 300 (1988); A.M. Weiner, J.P. Heritage, and E.M. Kirschner, J. Opt. Soc. Am. B5 (1988), in press
5. K.J. Blow and N.J. Doran, Phys. Lett. 107A, 55 (1985)
6. C. Froehly, B. Colombeau, and M. Vampouille: in Progress in Optics XX, ed. by E. Wolf (North-Holland, Amsterdam 1983)

White Light Interferometric Measurements of Femtosecond Group Delay in Optical Components with 1-fs Precision

W.H. Knox [1], *N.M. Pearson* [1], *K.D. Li* [1], *and C.A. Hirlimann* [2]

[1] AT&T Bell Laboratories, Holmdel, NJ07733, USA
[2] University P. et M. Curie, Paris, France

In femtosecond spectroscopy one of the most important problems has been the lack of direct information about the frequency dependence of the group delay for common optical components such as mirrors, filters, polarizers. For several years since the invention of the colliding pulse modelocked laser (CPM) [1] a well-known recurring problem has been the occasional "bad mirror", which prevents formation of high quality pulses of 50 fs or less. Using fiber-grating pulse compression, pulses of 8 fs duration were produced [2], however, this result was strictly limited by the cubic phase error in the gratings [3]. By including a four-prism sequence, the cubic phase error was effectively cancelled and pulses of 6 fs duration were obtained [4]. In this regime, many common optical components are useless because of the variation of the group delay for light over the pulse bandwidth. Novel phase correctors will be required for error cancellation, and a direct and simple measurement technique is needed for further progress in generation and measurement. We present the first direct measurements of the frequency dependence of the group delay for a number of optical components which are commonly used in femtosecond spectroscopy [5].

We use a Michelson interferometer with a white light "point source" (Oriel #6342). A precision translator moves one mirror in 45 nm steps and a photomultiplier records the intensity transmitted through a tunable interference filter in the 400-750 nm range. We first obtain white light fringes and carefully compensate the path lengths [6]. We use a simple algorithm to find the time center of the fringe packet as we change the wavelength of observation. We find that the time center of the fringe packet is constant to within 1 fs over the range of 400-750 nm. Thus, any optical device which is placed into one arm will cause a time displacement of the fringe packets if the group delay varies by more than 1 fs over this range. The time displacement is exactly the relative group delay at that frequency, as long as the group delay does not vary too rapidly over the filter bandwidth. We note that this is not the same as FTIR instruments, because devices are not placed inside the interferometer in an FTIR instrument. In the present experiment, our filter bandwidth is 10 nm. In the most recent embodiment, grating tuning is used which permits adjustable bandwidth. Figure 1 shows the results for laser mirrors from an operational CPM laser running at 50 fs pulsewidth. The 3% output coupler is quite flat within its design range, but exhibits oscillations in the blue. The corner mirror at 45 degrees is also quite flat over the bandwidth of a pulse even as short as 27 fs, however, we find that the group delay is even flatter in "S" polarization. It is now well-known that a single-stack design centered at 610 nm is the best for intracavity applications. In "S" polarization, these mirrors should be suitable for pulses of less than 10 fs duration as well. Figure 2 shows results for a common broadband dielectric mirror. These mirrors are known to distort 16 fs pulses [7]. The time steps due to the three reflection stacks are clearly observed. Figure 3 shows results for a pair of flint Brewster prisms which are now commonly used for dispersion

Springer Series in Chemical Physics, Vol. 48: **Ultrafast Phenomena VI**
Editors: T. Yajima · K. Yoshihara · C.B. Harris · S. Shionoya

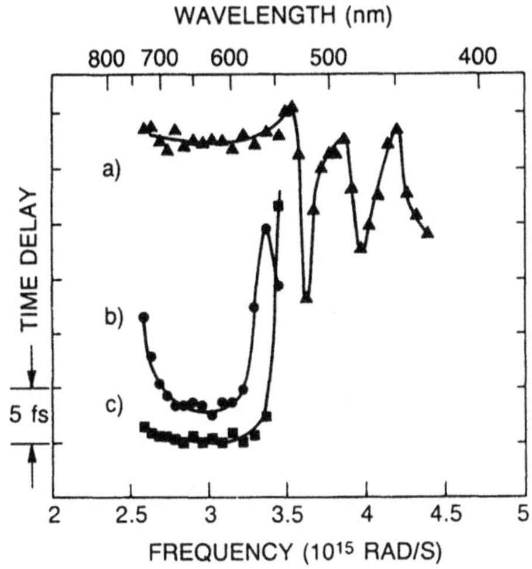

Figure 1. Results for CPM mirrors : (a) 3% output coupler, (b) corner mirror in "P" and (c) in "S" polarization.

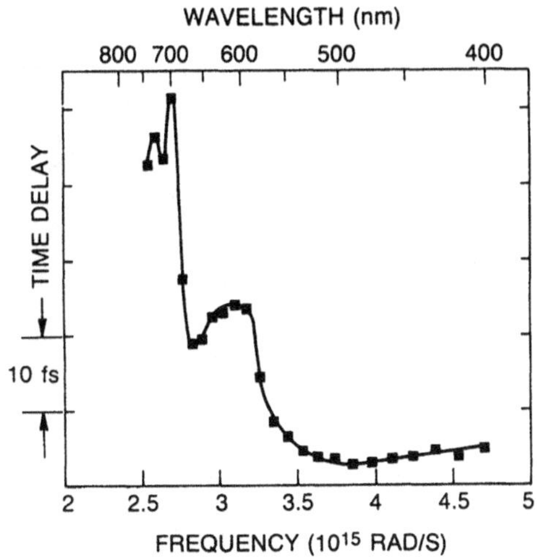

Figure 2. Broadband dielectric mirror NRC BD.1 at normal incidence.

correction [8]. The tuning through positive GVD to negative GVD is clearly observed in the range of 625 nm. Strongly absorbing elements also introduce errors. Figure 4 shows results for two common laser dyes. A reference cell has been included in the second arm of the interferometer with solvent alone, so we measure the dispersion due to the dye molecules in solution only. Figure 5 shows results for a Gires-Tournois interferometer which has its first resonance around 625 nm. We made two devices [9] consisting of a single high index layer on different substrates: single stack centered at

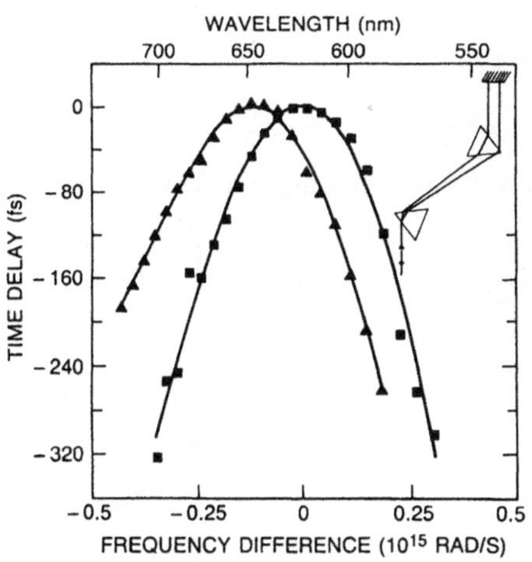

Figure 3. Flint Brewster prisms for two glass path length settings.

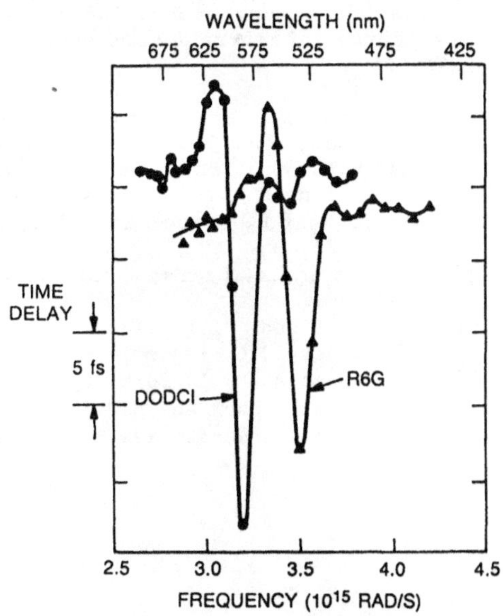

Figure 4. Laser dyes R6G and DODCI in methanol at optical density 2.0

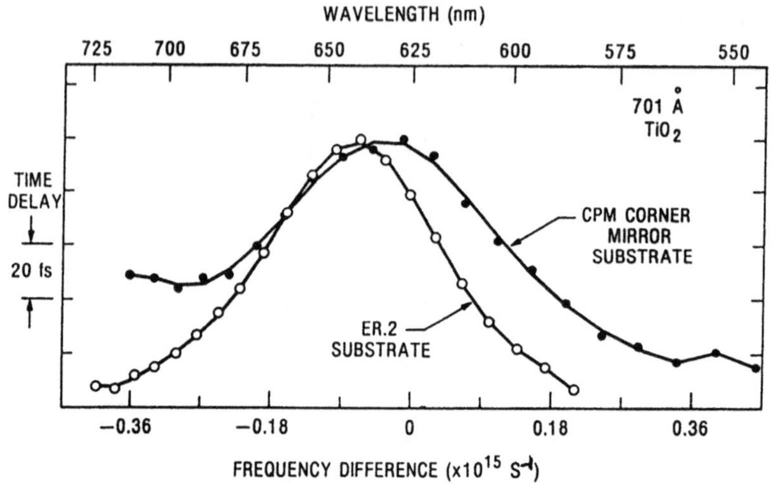

Figure 5. Resonant Gires-Tournois interferometers consisting of 70.1 nm TiO_2 film on CPM corner mirror substrate and NRC ER.2 substrate. Twelve bounces are used to obtain these results. The center wavelength can be angle-tuned.

610 nm and the metal-dielectric stack NRC ER.2 mirror. The choice of substrates is very important. Both devices provide negative cubic phase which can be used to cancel grating errors. These devices become part of a collection of phase correctors which can be added in series to obtain various shapes for specific applications.

In summary, we have made direct measurements of the frequency dependence of the group delay of optical elements which are commonly used in femtosecond spectroscopy. These techniques can be extended to waveplates, polarizers, interference and absorbing glass filters and combinations of elements.

[1] R.L. Fork, B.I. Greene and C.V. Shank. Appl. Phys. Lett. 38, 671 (1981).

[2] W.H. Knox, R.L. Fork, M.C. Downer, R.H. Stolen, C.V. Shank and J.A. Valdmanis, Appl. Phys. Lett. 46, 1120 (1985).

[3] W.J. Tomlinson and W.H. Knox, JOSA B4, 1404 (1987).

[4] R.L. Fork, C.H. Brito-Cruz, P.C. Becker and C.V. Shank, Opt. Lett. 12, 483 (1987).

[5] W.H. Knox, N.M. Pearson, K.D. Li and C.A. Hirlimann, Opt. Lett. 13, 574 (1988).

[6] A.A. Michelson, "Light Waves and Their Uses" (U. Chicago Press, 1902).

[7] A.M. Weiner, J.G. Fujimoto and E.P. Ippen, Opt. Lett. 10, 71 (1985).

[8] J.P. Gordon and R.L. Fork, Opt. Lett. 9, 153 (1984).

[9] K.D. Li, W.H. Knox and N.M. Pearson, to be submitted to Opt. Lett.: K.D. Li, Bell Laboratories Summer Research Project 1986.

Caution! IR Pulses May Be Distorted After Propagation in Atmospheric Air

A. Seilmeier, M. Wörner, and W. Kaiser

Physik Department E11, Technische Universität München, D-8000 München, Fed. Rep. of Germany

The investigation of pulse propagation effects in the infrared is of growing interest on account of recent advances in the generation of tunable picosecond light pulses of wavelengths between 5μm and 12μm /1/. Of special importance is pulse deformation in atmospheric air, which is the favorite medium for time resolved spectroscopy.

There exist numerous strong absorption lines in the infrared spectrum of atmospheric air corresponding to rovibrational transitions of H_2O and CO_2. The transmission of energy of picosecond infrared pulses is not substantially affected by these absorption lines on account of their very small line widths of a few 10^{-1} cm^{-1} (spectral width of the pulses $\Delta\tilde{\nu}$=10cm^{-1}). The pulse shape, however, may be changed considerably after passage of the pulse over the short distance of tens of centimeters in air due to coherent interaction with resonant absorption lines.

In this paper we present experimental data on coherent pulse distortion which - to our knowledge - have never been seen previously in laboratory air. Infrared pulses (t_p = 2ps) tunable between 4 and 12 μm are produced via difference frequency generation in $AgGaS_2$. After passage of a distance of 80 cm in air they are upconverted with a visible light pulse. The cross-correlation curve gives direct information on the pulse shape of the infrared pulse.

In Fig. 1a a part of the infrared absorption spectrum of air under normal conditions is depicted. There exist three wavelength regions with strong absorption lines. At λ~6μm and at λ~2.8μm the ν_2 bending mode and the OH stretching mode of H_2O are located, respectively. Around λ=4.3μm the CO_2 stretching mode is observed. In this paper we focus on the rovibrational lines between 5.3μm and 7.7μm. Part of the infrared spectrum is redrawn on an expanded scale in Fig. 1b showing the very complex rotational structure of the vibrational absorption band. The absolute absorbance of the lines depends on the humidity of the laboratory air. Fig. 1c shows one single rovibrational line at 1576.2 cm^{-1} measured with a spectral resolution of 0.04 cm^{-1}. The spectral width of 0.23 cm^{-1} corresponds to a dephasing time T_2=50 ps.

In the following it is shown that the pulse shape is significantly distorted by interaction with rovibrational lines. Cross correlation curves are presented for two infrared frequencies in Fig.2. The first example is an IR pulse at 2017 cm^{-1} where air is highly transparent (see Fig. 1a). In Fig.2a one sees the corresponding sum-frequency signal with steep wings rising and falling over four orders of magnitude. Passage of the infrared pulse through air preserves the Gaussian pulse shape.

Springer Series in Chemical Physics, Vol. 48: **Ultrafast Phenomena VI**
Editors: T. Yajima • K. Yoshihara • C. B. Harris • S. Shionoya

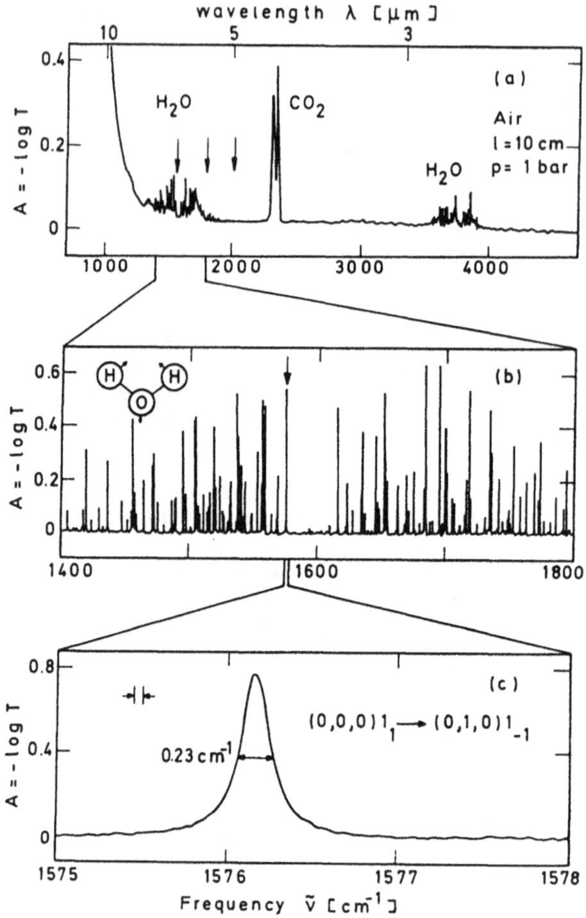

Fig. 1 Infrared absorption of air under normal conditions
a) Survey of the absorption between 2 and 8 μm, b) rovibrational lines of the ν_2 bending mode of H_2O, c) single rovibrational line at 1576.2 cm^{-1}.

Quite different is the situation for an IR pulse at $\tilde{\nu}$ = 1576 cm^{-1} which interacts strongly with a rovibrational transition of water at $\tilde{\nu}$ = 1576.2 cm^{-1} (see Fig. 1c) . The cross-correlation curve depicted in Fig.2b shows a pronounced distortion. One sees a long tail with a relative signal value of 10^{-3} at a delay time as long as 50 ps.

The observations may be explained as follows. The electric field of the infrared pulse generates a coherent polarization of the dipole moment of the rovibrational line which subsequently reemits an electric field with opposite sign. The interplay of the macroscopic polarization and the electric field generates the tail of the infrared pulse. In Fig. 2b the absorption length is large (αl=13). In this situation the signal decays due to destructive interference of the electric field components and due to dephasing with T_2=50 ps /2/.

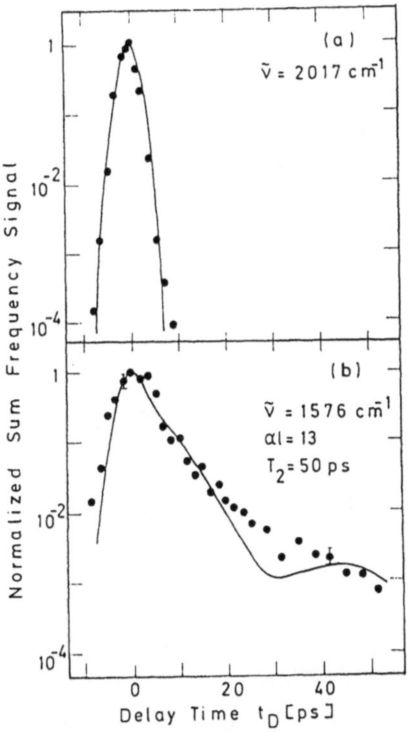

Fig. 2 Cross correlation of an infrared pulse with a visible pulse
a) at $\tilde{\nu}_{IR}$ = 2017 cm^{-1} in a highly transparent region,
b) at $\tilde{\nu}_{IR}$ = 1576 cm^{-1}, where a rovibrational line is located.

The solid line in Fig. 2b is the result of a model calculation based on a two level system. The coherent pulse propagation is treated within the framework of small area pulse propagation ($\theta=(\mu/\hbar)\int Edt=0.1$) /2/. The deviation of the calculated curve from the experimental data around t_D=30ps is observed when only the line at $\tilde{\nu}$ = 1576.2 cm^{-1} is taken into account. Including other weaker neighboring rovibrational lines in the calculations eliminates the minimum in the calculated curve at t_D = 30 ps.

The extended wing due to coherent interaction is strongly reduced by performing the experiment in a dry nitrogen atmosphere where the H_2O concentration and consequently αl is lower. A reduction of the tail of the pulse by two orders of magnitude is easily obtained.

Similar pulse distortion is observed at other frequency positions within the rovibrational manifolds of H_2O and CO_2. Corresponding experimental data are discussed elsewhere.

We point to the importance of the pulse distortion for time resolved infrared spectroscopy. At frequencies, where air exhibits absorption lines, caution has to be exercised to interpret the observed decay times. When a transmitted signal is measured over several orders of magnitude for a detailed study of a relaxation process, one has to be aware of the pulse distortion in laboratory air. In certain wavelength regions accurate data require the elimination of H_2O and CO_2 molecules in the optical path of the IR pulses.

References

1. T. Elsaesser, H. Lobentanzer, A. Seilmeier: Optics Commun. 52, 355 (1985)
2. H. J. Hartmann, A. Laubereau J. Chem. Phys. 80, 4663 (1984)

Pulse Front Distortion in Lens Systems

S. Szatmári and G. Kühnle*

Max-Planck-Institut für Biophysikalische Chemie,
Abteilung Laserphysik, Postfach 2841, D-3400 Göttingen,
Fed. Rep. of Germany

*Permanent address: Research Group on Laser Physics
of the Hungarian Academy of Sciences, JATE University,
Dóm tér 9, H-6720 Szeged, Hungary

It is known from early publications [1] that dispersion - the difference between the phase and group transit times - in prisms introduces a continuously increasing temporal delay across the beam. The time difference between the phase and pulse fronts (ΔT) is given as a function of the length (ΔL) and of the dispersion ($dn/d\lambda$) of the material as

$$\Delta T = -\frac{\lambda}{c}\,\Delta L\,\frac{dn}{d\lambda}\,. \tag{1}$$

If ΔL is different across the beam, the shape of the pulse front will differ from that of the wave front. The simplest example is a lens which introduces different delays for the various parts of the beam [2]. Recent developments of large aperture, femtosecond uv excimer lasers [3-5] necessitate the study of the above problem, since pulse front distortion caused already by a single thin lens of these short wavelengths is comparable to the pulse duration.

The pulse front delay introduced by a single planoconvex lens can be determined by calculating the spatial dependence of ΔL, and substituting into (1), as

$$\Delta T = -\frac{\lambda}{c}\,\frac{h^2}{2}\,\frac{1}{r}\,\frac{dn}{d\lambda},$$

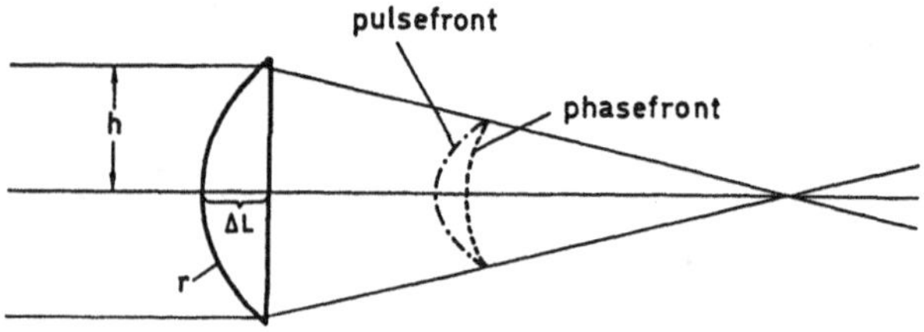

Fig.1. Schematic of a planoconvex lens for the calculation of the pulse front delay

Springer Series in Chemical Physics, Vol. 48: **Ultrafast Phenomena VI**
Editors: T. Yajima · K. Yoshihara · C.B. Harris · S. Shionoya

where r is the radius of curvature of the lens and h is the distance from the optical axis (Fig.1). For a lens system the final distortion is the sum of the distortion of each of the lenses:

$$\Delta T = -\frac{\lambda}{c}\frac{h^2}{2}\sum_i \frac{1}{r_i}\frac{dn_i}{d\lambda}.$$

For a perfect lens system ΔT has to be zero, then one gets the general formula for achromats. This means that by the use of achromats the spatially dependent pulse front delay can be avoided. However, in the ultraviolet the transparent optical materials have similar dispersion, which makes the construction of uv, high power achromats quite complicated.

One way to avoid distortion is to use reflective optics. The other is to allow pulse front distortions and to compensate the distortion later, by a correction device. Possibly the simplest realization of such a device is shown in Fig.6 consisting of a lens and a mirror. The pulse front compensation takes place in the concave lens, where the outer rays are delayed with respect to the middle one.

For the measurement of the pulse front distortion the newly developed uv laser system was used (described in Ref. [5]), delivering ~15 mJ, ~500 fs pulses at 248 nm. These pulses were compressed to ~60-80 fs and measured by a special auto-correlator (Fig.2). In this autocorrelator the position of the first partial beam is fixed at the aperture, the other can be shifted by shifting the retroreflector in the direction perpendicular to the beam. Therefore different parts of the

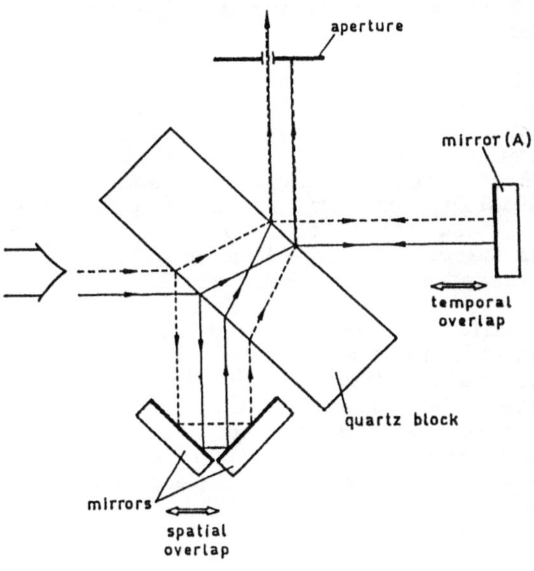

Fig.2. Schematic of the autocorrelator

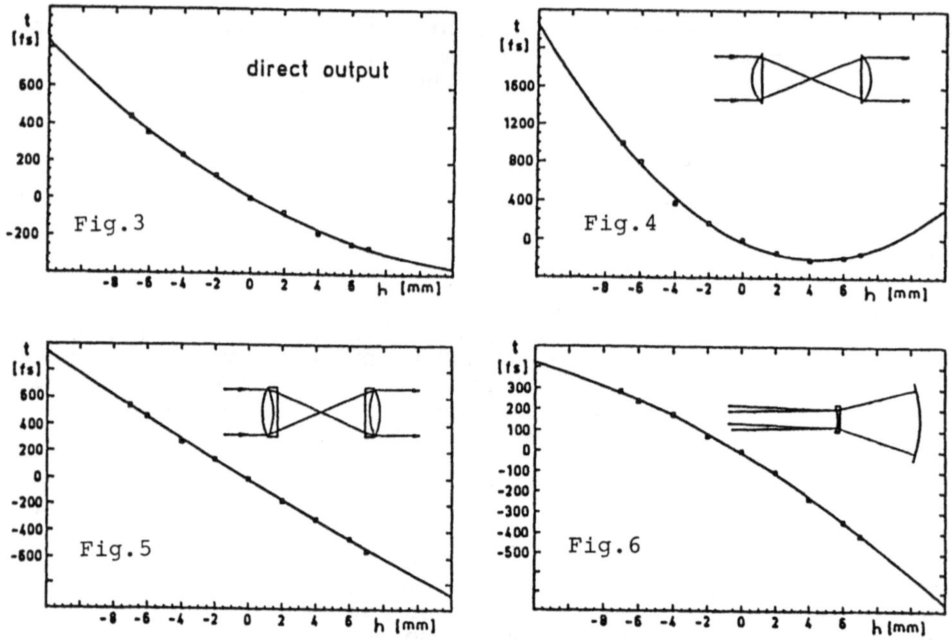

Fig.3.-6. Measured pulse front of the direct output, after a Kepler telescope, after an achromatic telescope, and after a compensating device

input beam can be combined behind the aperture, where the autocorrelator measurement is done. When the position of mirror (A) corresponding to complete temporal overlap is recorded as a function of the shift of the second partial beam the curvature of the pulse front can be determined. This is shown in Figs.3-6. Figure 3 shows the measured pulse front of the direct output of our high aperture fs excimer laser system. The pulse front is already distorted by the lenses used in the system. For Fig.4, the beam was passed through a Kepler telescope formed by two f = 104 mm cylindrical lenses. In this case, the pulse front is clearly more distorted by comparison with Fig.3. Figures 5 and 6 give examples for the use of pulse distortion free optics (achromats) and compensating devices, each indicating slight, or more pronounced overcompensation. We also calculated the curvature of the pulse front for these four cases and found good quantitative agreement between the calculated and measured values.

In conclusion we have studied the pulse front distortion of refractive optics. We derived a simple formula describing the spatial dependence of the pulse front when the pulse has passed through a lens system. We showed that no pulse front distortion occurs in achromats. A pulse front compensating device is also suggested introducing a distortion of opposite sign. A special autocorrelator was developed for the measurement of the shape of the pulse front. With this device we found good agreement between the experimental and the predicted pulse front distortions for simple lenses (telescopes), achromats, and also our compensating device.

1. M. Topp and G.C. Orner: Opt. Commun. **13**, 276 (1975)
2. H. Staerk, J. Ihlemann, A. Helmbold: Laser und Opto-elektronik **20** (1988) (in press)
3. A.P. Schwarzenbach, T.S. Luk, I.A. McIntyre, U. Johann, A. McPherson, K. Boyer, C.K. Rhodes: Opt. Lett. **11**, 499 (1986)
4. S. Szatmári, F.P. Schäfer, E. Müller-Horsche, W. Mückenheim: Opt. Commun. **63**, 305 (1987)
5. S. Szatmári, F.P. Schäfer: Opt. Commun. (1988) (in press)

Blue Shifting of Intense Femtosecond Pulses in Gas Breakdown and Solid State Plasmas

M.C. Downer, G. Focht, D.H. Reitze, W.M. Wood, and T.R. Zhang

Physics Department, University of Texas at Austin, Austin, TX78712, USA

We have shown that millijoule femtosecond pulses can be focused to spot radii as small as 2 microns when amplified in a "conical axicon" cell, thus creating peak intensities of 10 petawatts (PW)/cm^2 [1]. We now use this source to generate dense gas breakdown plasmas in the strong field [2], collisionless limit while using the plasma response on the light field in the form of blue shifts and defocusing as diagnostics to characterize the femtosecond ionization process [3]. Such diagnostics are compatible with a wide range of gas pressures.

Figure 1a shows how the blue shift of 90 fs pulses increases with light intensity after focusing at f/2 through a breakdown spark in 1 atm. neon. Numerous observations support the dominance of plasma generation, and the weakness of $\chi^{(3)}$ effects, in producing this shift: the similarity of the blue shifts in different gases (neon, air [1], helium, argon at 1 atm) having third order hyperpolarizabilities which vary by more than a factor of 20 [4]; correlation of these blue shifts with appearance of a breakdown spark; correlation with beam defocusing caused by the plasma's lower refractive index; the consistent absence of any red-shifted spectral components (in contrast to spectral *broadening* observed by others [5,6] in the absence of gas breakdown); the absence of supercontinuum generation from intense femtosecond excitation of neon up to 40 atm. in previous studies [6]; the absence of self-focusing or filamentation under breakdown conditions. The blue shift is caused by the ultrafast negative refractive index change when a plasma is created during the pulse. The temporal asymmetry of the index change is strong with femtosecond pulses, since electrons neither recombine nor escape from the focal region during the pulse, thus eliminating a compensating red shift, seen with longer pulses, when the index recovers [3]. According to a Drude model the blue shift can be approximated as [3]

$$\Delta\lambda = (\lambda^3 e^2 / 2\pi mc^3)\, d/dt \int^L N(x)\, dx \qquad (1)$$

where N(x) is the plasma density along the propagation direction x, and the interaction length L is approximately the confocal beam parameter. The blue shift in neon increases steadily up to our highest intensity (10 PW/cm^2), rather than saturating because of complete single ionization within the focal volume at 2 PW/cm^2, as observed in air [1]. This difference may result from the lower ionization potential of nitrogen and oxygen compared to neon. Figure 1b shows a blue shift caused by 1 μJ 90 fs pulses generating a dense (~10^{20} cm^{-3}) electron-hole plasma in sapphire by three-photon absorption when focused at f/10 into a 1 mm sample. In contrast to the gases, there is an accompanying red-shifted feature. Nevertheless the resulting spectra are accurately fit by combining plasma effects described by Eq.1 with self-phase modulation resulting from the third order hyperpolarizability, as shown by the dots in Fig. 1b.

Figure 2 shows other characteristics of the blue-shifted pulses. Figure 2a shows the defocusing which accompanies blue shifting, manifested as an increased beam divergence angle. This effect is strongest in sapphire because of the high plasma density and long interaction length used. In other solids this effect has been used for optical limiting [7]. In gases at 1 atm with f/2 focus, plasma defocusing causes a 30% increase in divergence angle which can be correlated quantitatively with the blue shift. The blue shifted beam thus maintains sufficiently good beam quality to be useful spectroscopically. The autocorrelation traces in Fig. 2b demonstrate that the glass f/2 focusing optics cause 10% temporal broadening of low intensity pulses (no plasma

Springer Series in Chemical Physics, Vol. 48: **Ultrafast Phenomena VI**
Editors: T. Yajima • K. Yoshihara • C.B. Harris • S. Shionoya

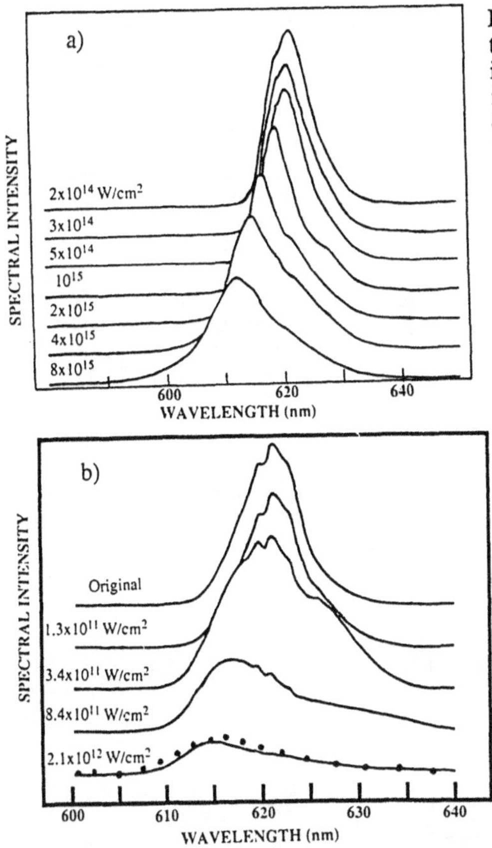

Fig. 1 a) Spectra of 90 fs pulses transmitted through neon breakdown spark, with focal intensity increasing toward the front in increments of 0.2 optical density. Successive curves have been displaced from each other in an exact vertical direction to emphasize blue shift as intensity increases. b) Spectra of 90 fs pulses after generating a dense electron-hole plasma in sapphire by a three-photon absorption process. Note blue shift accompanied by a weaker red-shifted feature. Dots represent a theoretical fit

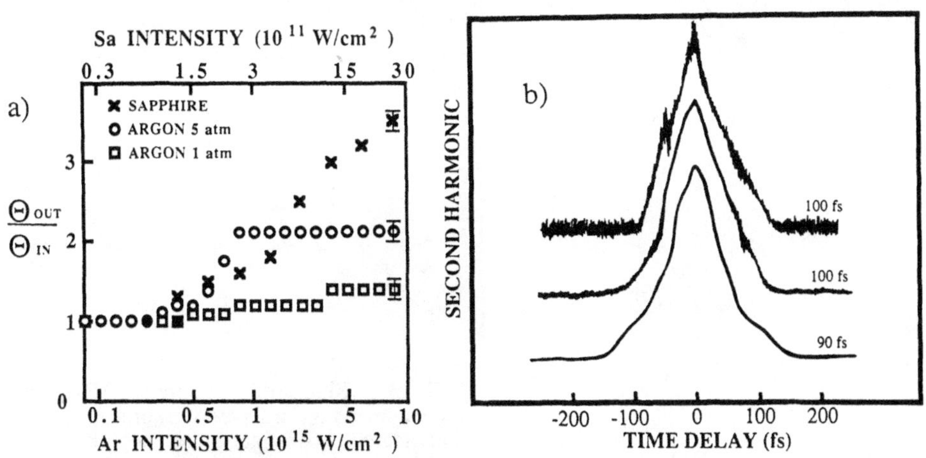

Fig. 2 a) Defocusing vs. peak focal intensity for sapphire, 1 atm argon, and 5 atm argon. The filled square and circle denote the first appearance of a breakdown spark. b) Autocorrelation traces of unfocused millijoule pulses (lower trace), pulses after focusing at f/2 in air by 7x microscope objective and recollimation, but attenuated to avoid breakdown spark (middle trace), and full intensity pulses after focusing at f/2 through intense breakdown spark and recollimation (upper trace). Cited pulse durations assume sech^2 pulse shape formed), and that at high intensity the ionization process causes no further measurable broadening. As gas pressure increases above 1 atm we observe only a slight further increase in blue shift, as shown by the blue shifted neon spectra in Fig. 3a. The saturation of the blue shift at 2 to 3 atm probably results from increased defocusing caused by the higher plasma density in the interaction region. Consequently focal spot size is increased and peak intensity reduced.

The behavior of argon at 5 atm contrasts strongly with other gases we have studied. Just above threshold the blue shift behaves as in other gases, and is well explained by Eq. 1. At an intensity near 10 PW/cm^2, however, we observe a sudden spectral broadening, with components blue shifted well beyond those observed in other gases, as shown in Fig. 3b. The transmitted light contains yellow, green, and even blue components visible to the naked eye, although still no components which are red-shifted from the original pulse spectrum. Onset of double ionization, for which the threshold is lower in the heavy rare gases, may cause an increase in the rate of change of plasma density dN/dt, and may thus partly explain the sudden spectral change. However, the absence of increased defocusing (see data for 5 atm argon in Fig. 2b), as expected from increased plasma density, together with the pressure dependence of the effect, suggest the role of self phase modulation in the argon plasma.

Highly focusable femtosecond pulses from an axicon amplifier provide a powerful new tool in ultrahigh intensity spectroscopy. The blue shifts observed here may be related to red shifted

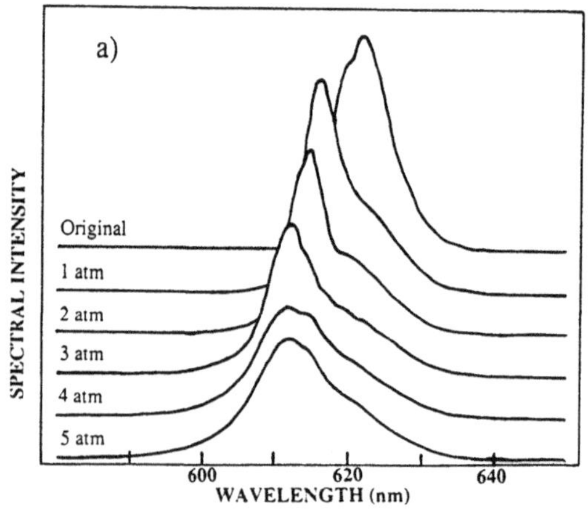

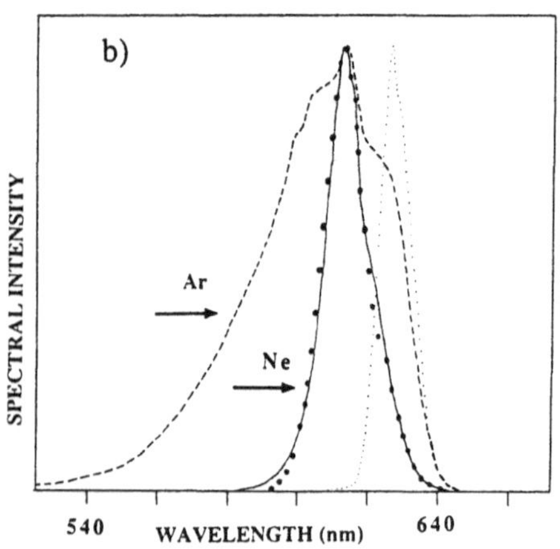

Fig. 3 a) Blue shifted spectra at 10 PW/cm^2 focal intensity for 5 pressures of neon, compared to original pulse spectrum.
b) Original pulse spectrum (dotted curve) and blue shifted spectra following transmission through plasmas generated in 5 atm neon (solid curve) and argon (dashed curve). Large dots are a theoretical fit based on Drude model (see Eq. 1 of text)

photoelectron energies observed in recent above threshold ionization (ATI) experiments [8]. Dynamic collisionless absorption in plasmas, important when electron acceleration in strong light fields becomes relativistic, may be observable with such pulses [9]. Finally this source should be useful in generating picosecond soft X-ray pulses at noble metal surfaces [10].

This research was supported by the Robert A. Welch Foundation (Grant F-1038), the Joint Services Electronics Program (Contract F49620-86-C-0045), an NSF Presidential Young Investigator Award (Grant DMR-8858388), and an IBM Faculty Development Award.

1. M.C. Downer et. al., "New Developments in Ultraviolet and High Intensity Femtosecond Sources" in this volume; Wm. M. Wood, Glenn Focht, and M.C. Downer, Opt. Lett., in press (1988).
2. L.V. Keldysh, Sov. Phys. JETP 20, 1307 (1965).
3. E. Yablonovitch, Phys. Rev. Lett. 32, 1101 (1974); Phys. Rev. A 10, 1888 (1974); Phys. Rev. Lett. 60, 795 (1988); P.B. Corkum, NATO ASI Series-Physics B 171, 157 (1988).
4. H.J. Lehmeier, W. Leupacher and A. Penzkofer, Opt. Commun, 56, 67 (1985).
5. D. Kuhlke, U. Herpers, and D.von der Linde, Opt. Commun. 63, 275 (1987); J.H Glownia, J. Misewich, and P.P. Sorokin, J.Opt. Soc. Am. B3, 1573 (1986).
6. P.B. Corkum, C. Rolland, and T. Srinivasan-Rao, Phys. Rev. Lett. 57, 2268 (1986).
7. T.F. Boggess, Jr., A.L.Smirl, S.C. Moss, I.W. Boyd, and E. W. van Stryland, IEEE J. Quant. El. 21, 488 (1985); S. Guha, E.W. Van Stryland, and M.J.Soileau, Opt. Lett. 10, 285 (1985).
8. E. Yablonovitch, Phys. Rev. Lett. 60, 795 (1988); R.R. Freeman, P.H. Bucksbaum, H. Milchberg, S. Darack, D. Schumacher, and M.E. Geusic, Phys. Rev. Lett. 59, 1092 (1987).
9. R. Dragila and H. Hora, Phys. Fluids 25, 1057 (1982).
10. O.R. Wood II, W.T. Silfvast, H.W.K. Tom, W.H. Knox, R.L. Fork, C.H. Brito-Cruz, M.C. Downer and P.J. Maloney, Appl. Phys. Lett., to be published.

Time Domain Phase Conjugation and Twin Fields of Picosecond Light Continuum

A. Jankauskas, A. Piskarskas, and A. Stabinis

Vilnius University Laser Research Centre, 232054 Vilnius, Sauletekio ave 9, Lithuanian SSR, USSR

Results of squeezed states observation in powerful spectral quasicontinuum emitted by superluminescent optical parametric oscillator (OPO) are presented. We show that squeezing in OPO at high-intensity field is caused by phase conjugation.

1. An alternative system capable to form squeezed states at high-intensity field may be degenerate OPO /1-4/. Squeezed states formation in degenerate OPO radiation at high gain has a classical analogy, because in OPO due to phase selectivity only one quadrature component of the input signal is highly amplified /5-6/. Twin wave packets may be formed also in nondegenerate OPO /7,8/.

2. The physical mechanism of squeezed states generation in OPO at high gain is a phase conjugation. Conditions of phase conjugated waves generation in OPO are investigated /9/. Under such conditions the complex amplitudes A_1, A_2 of waves at frequencies ω_1 and ω_2 ($\omega_1+\omega_2=\omega_3$, where ω_3- pump frequency) at high amplification are in relation $A_2(t)=\sqrt{\sigma_2/\sigma_1}A_1^*(t)$, where σ_1, σ_2 - nonlinear wave interaction coefficient and t - time. Conjugation in OPO may be tested, for example, by investigation of sum frequency spectrum. The conjugated waves generated by OPO we assume as noisy pulses of correlation time τ_k. Let us define correlation function B_s of the signal at sum frequency ($\omega_s=\omega_1+\omega_2$) with the amplitude $A_s \sim A_1A_1^*$ as $B_s(t,T,T_0) \sim \langle A_s(t) A_s^*(t+T+T_0)\rangle$, where T_0 - time delay between phase conjugated waves. For the spectral intensity at sum frequency we have

$$G_s(\Omega,T_0) \sim \exp\left(-\frac{\tau_0^2}{\tau_k^2}-\frac{\Omega^2\tau_0^2}{4}\right) + \frac{\tau_k}{\tau_0}\exp\left(-\frac{T_0^2}{\tau_0^2}-\frac{\Omega^2\tau_k^2}{4}\right). \qquad (1)$$

Thus, the spectrum is defined by the ratio T_0/τ_k. Fig. 1 shows the spectral intensity $G_s(\Omega)$ for three values T_0/τ_k at $\tau_k/\tau_0 = 0.1$. When $T_0 = 0$(a) a narrow peak related to noisy pulse envelope appears. The energy in the spectrum is equally divided between the peak and the background. The background is the result of amplitude fluctuations of the noisy pulse. At $T_0=\tau_k$(b) waves are partially conjugated and the peak goes down. If $T_0=2\tau_k$(c) the peak in spectrum is not evident. The narrow peak shows the presence of conjugated fields at frequencies ω_1 and ω_2 in OPO radiation (twin fields). For the first time the narrow peak in sum frequency spectrum was observed when nondegenerate α-HIO_3 OPO experimentally have been analysed [10].

3. In degenerate OPO $\omega_1=\omega_2$, $A_2=A_1^*$ and radiation field at high gain may be considered as a superposition of phase conjugated waves with an amplitude $A_1+A_1^*$. Sum frequency wave amplitude has thus the form

$$A_\Sigma \sim (A_1 + A_1^*)^2 = A_1^2 + 2A_1A_1^* + A_1^{*2} . \qquad (2)$$

Springer Series in Chemical Physics, Vol. 48: **Ultrafast Phenomena VI**
Editors: T. Yajima · K. Yoshihara · C.B. Harris · S. Shionoya

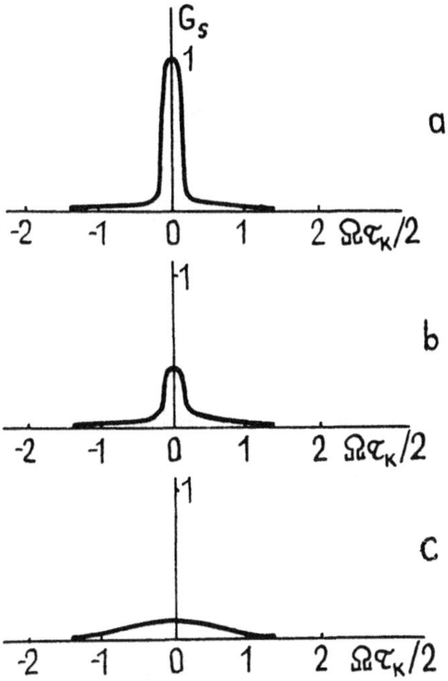

Fig. 1

If the signal shows fast phase modulation, spectrum $G_s(\Omega)$ consists also of two parts: the narrow peak caused by the second term $2A_1A_1^*$ and the wide background caused by the first and the third terms (A_1^2 and A_1^{*2}). At $\tau_0 \gg \tau_k$ the spectral intensity of the narrow peak is higher than the spectral intensity of the wide background. Appearance of narrow peak in second harmonic spectrum testifies an inequality of the quadrature components in the field of degenerate OPO radiation and the squeezed states formation. On the other hand, it means that second order coherence in degenerate OPO radiation takes place. This kind of peak was observed in /7/ and in radiowaves region in /11/.

4. Recently we investigated the spectral quasicontinuum of superluminescent OPO. The two CDA crystals (l=4 cm, 1-st phase-matching type) in series were pumped by a single 4-5 ps pulse of a frequecy doubled phosphate glass laser. A 90° phase matching angle allows us to obtain a large spectral bandwidth of OPO radiation. At degeneracy it exceeds 1000 cm^{-1}, which corresponds to a correlation time of about 15 fs. The second harmonic was generated in KDP crystal (l=2 mm). The feature of the second harmonic spectrum (Fig. 2) is the narrow peak (a), as already mentioned above. It confirms that OPO generates phase conjugated waves and one of the quadrature components is suppressed. We are led to consider this fact as a manifestation of squeezed states in parametric quasicontinuum. Poor contrast of peak and background shows rather low conjugation quality in quasicontinuum. Because of the fact that the correlation time of chaotic (noisy) phase modulation is very short, a glass plate of 4 cm thickness in front of KDP crystal leads to the peak broadening of about 10 times (b). Two times thicker glass plate leads to the peak disappearance (c).

5. The squeezed states of OPO radiation considered above are caused by time-domain phase conjugation. On the other hand due to phase conjugation

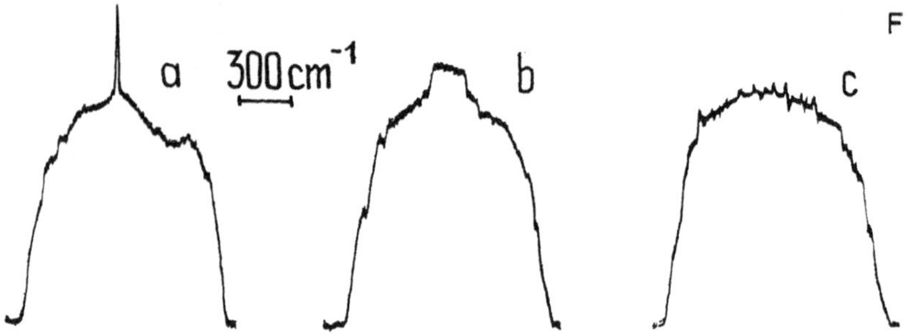

Fig. 2

in space domain (wave front reversal) squeezing of spatial quadrature components is possible. By nonlinear mixing of OPO waves, radiation at sum frequency with narrow spatial spectrum (highly directed beam) is observed /13/.

References

1. D.F. Smirnov, A.S. Troshin: Usp. Fiz. Nauk 153, 233 (1987)
2. L. Wu., J.J. Kimble, J.L. Hall et al.: Phys. Rev. Lett. 57, 2520 (1986)
3. L. Wu, M. Xiao, H.J. Kimble: J. Opt. Soc. Am. B4, 1465 (1987)
4. R.E. Slusher, P. Grangier, A. La Porta et al.: Phys. Rev. Lett. 59, 2566 (1987)
5. S.A. Akhmanov, T.E. Diakov, A.S. Chirkin: Vvedenije v statisticheskuju radiofiziku i optiku. M.: Nauka (1981)
6. S.A. Akhmanov: Usp. Fiz. Nauk 149, 361 (1986)
7. I. Abram, R.K. Raj, J.L. Oudar et al.: Phys. Rev. Lett. 57, 2516 (1986)
8. S. Reynaud: Europhys. Lett. 4, 427 (1987)
9. A. Piskarskas, A. Stabinis, A. Jankauskas: Sov. Phys. Usp. 29, 869 (1986)
10. A. Varanavicius, R. Grigonis, A. Piskarskas et al.: Pisma v ZhTF 6, 1447 (1980)
11. V.F. Marchenko, Y.M. Petrin, I.T. Trofimenko: Radiotechnika i elektronika 30, 1653 (1985)
12. A.M. Bonch-Bruevich, S.G. Przibelskij, N.A. Chigir: ZhETF 80, 565 (1981)
13. A. Varanavicius, R. Grigonis, A. Piskarskas et al.: ZhTF 51, 658 (1981)

Picosecond to Femtosecond Optical Synthesizers

T. Kobayashi, A. Morimoto, M. Doi, Bong Young Lee, and T. Sueta

Engineering Science, Osaka University, Toyonaka,
Osaka 560, Japan

The development of an "ultrafast optical synthesizer" which makes it possible to form arbitrarily shaped optical pulses or optical signals as well as transform-limited pulses in the picosecond to femtosecond range will extend the application area of optical electronics.

The typical method of synthesizing ultrafast optical signals is to produce widely spread optical frequency components and to control their amplitudes and phases as desired. In this point of view, first a method using an electrooptic deflector which produces frequency components distributing spatially according to their frequencies was used for synthesizing in the several-ten picosecond range [1]. In a similar way, frequency masking within a grating pair compressor was used for pulse shaping in the picosecond to femtosecond range [2]. These methods are not convenient for practical applications because of the difficulty in controlling precisely the frequency components and the large size of the system.

Here we propose various kinds of new optical synthesizers which have the possibility of generating arbitrarily shaped signals in the picosecond or possibly femtosecond range. These seem to be more convenient for practical use [3].

1. Basic consideration

Figure 1 shows the basic steps of the proposed synthesizing process. The first step is to provide frequency components distributed in a wide range. The second is to separate (demultiplex) each frequency component. The third is to control the amplitude and phase/group delay of each frequency component. The final step is to compose (multiplex) the frequency components and to form desired pulse shape. A grating, a Fabry-Perot filter, or a Michelson interferometer is employed as a demultiplexer and a multiplexer. If the power efficiency is out of consideration, a beam-splitter array or a single slit may be used as the latter.

Realistic examples for the practical construction of the synthesizer are

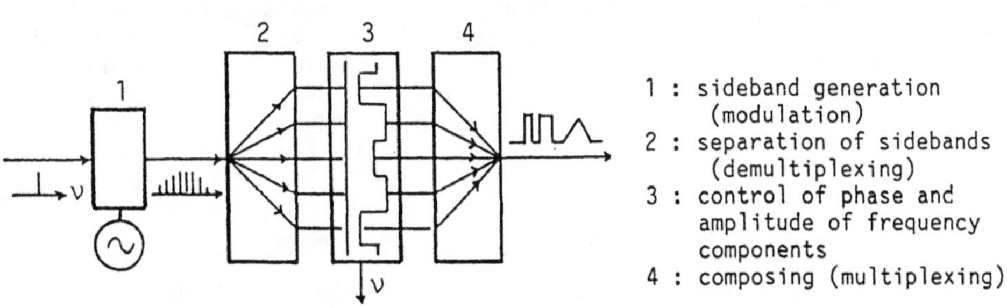

Fig.1 Basic procedures of the proposed high-speed optical synthesizer.

Springer Series in Chemical Physics, Vol. 48: **Ultrafast Phenomena VI**
Editors: T. Yajima · K. Yoshihara · C.B. Harris · S. Shionoya

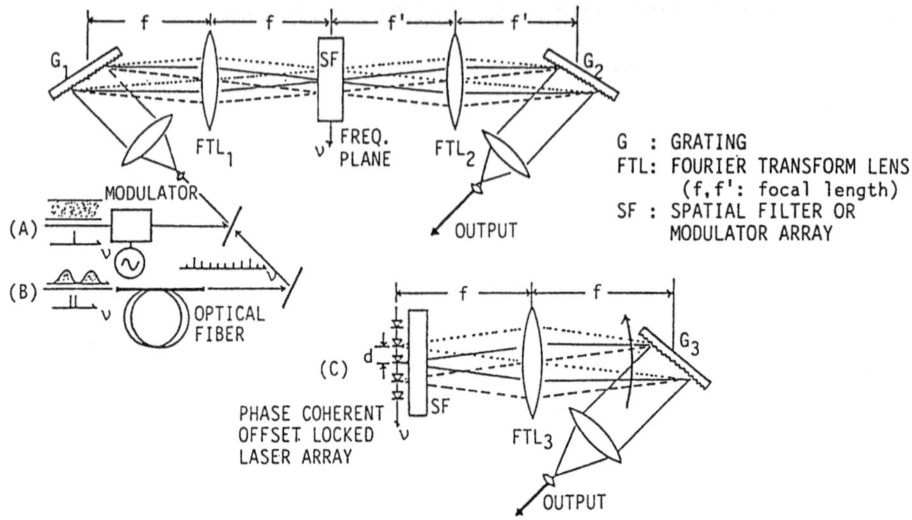

Fig.2 Realistic examples for the optical synthesizer in the ps-fs range.

shown in Fig.2. These systems are essentially equivalent to the synthesizer using the electrooptic deflector [1] and analogous to a phased array antenna or a frequency-domain multiplexer in the optical communication system. In Fig.2(C), a frequency- and phase-locked laser array is used. In this case the phase coherent frequency offset locking is made, i. e. element lasers are placed with the same distance and with the constant frequency intervals.

For the case of the periodic input with discrete line spectra, every frequency component can be separated spatially without any overlap and its amplitude and phase can be controlled individually by a spatial filter, or a modulator array (c.f. for single shot pulse input with continuous frequency spectra, neighboring frequency components overlap each other because of the finite frequency or spatial resolution of the system and individual control of the frequency components is impossible). Then the synthesized output is obtained after multiplexing. We can compose/synthesize arbitrary shape by phase control only within 2π (corresponds to one wavelength) without control of group delay of the frequency components. The shortest pulsewidth or the shortest temporal structure of the synthesized output is limited by the width of frequency-component spread.

The electrooptic method (A) has an advantage in controllability. The method (B) using self-phase modulation caused by input with two frequency components can provide very high repetition rate as THz and ultrashort pulsewidth as a few fs. This method can apply not only to zero dispersion region but also to normal or anomalous dispersion region [4] of the optical fiber. Soliton (or dark soliton)-like signals produce the wide sidebands with their propagation. The method (C) provides a portable synthesizer with an LD array, and may be useful for practical applications in the future.

2. Results of preliminary experiment and other discussions.

We have done preliminary pulse synthesis experiments using method (A). We used an Ar laser as a light source. Optical sidebands were produced by a 9.35GHz $LiTaO_3$ phase modulator [5]. Fig.3 shows spatially separated FM

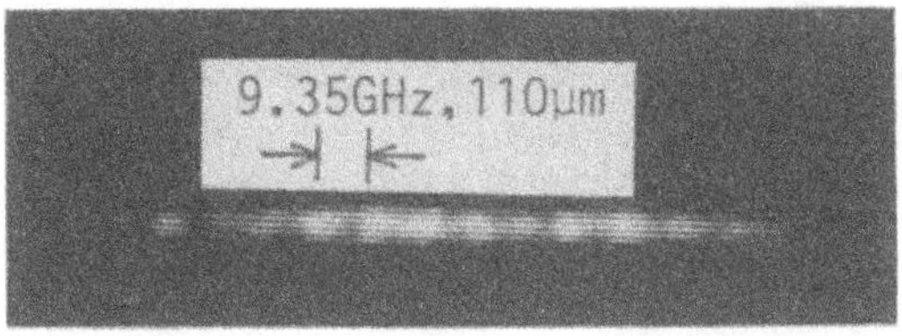

Fig.3 Spatially separated optical sidebands produced by EO modulation.

sidebands after demultiplexing by a grating. It is seen from the figure that the spatial separation between adjacent sidebands was 110µm and every frequency component is clearly separated without any overlap (the frequency intervals are 9.35GHz). Fig.4 shows the examples of the streak traces of synthesized signals obtained (a) by eliminating central five sidebands and (b) by giving parabolic phase shift to every sideband. Both results agree well with theoretical ones. Fig.5 shows the examples of calculated binary signals formed by controlling the sidebands using a 15 segment modulator array. These word patterns are very clear and arbitrarily changeable. Applying the 640GHz (FWHM) FM sideband recently reported [5] to this method, generation of 32bit words with 350Gbit/s rate will be possible.

We also tried an experiment of simple synthesis by substituting a Fabry-Perot frequency filter for the last three steps in Fig.1. As a result, we succeeded in generating a pulse train with 9-112 gigapulse per second [6]. We are also studying the method (B) and (C).

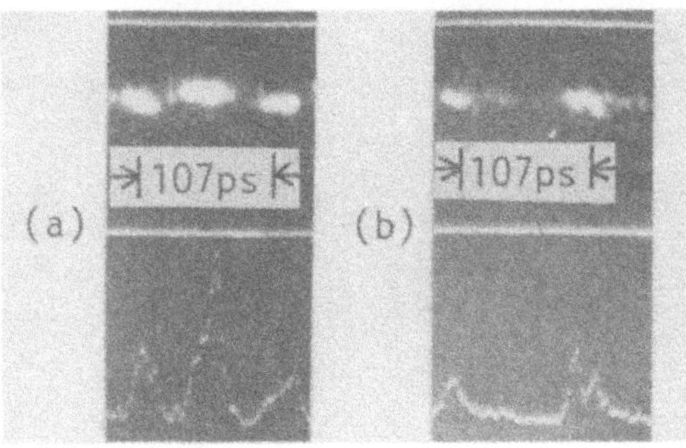

Fig.4 Streak traces of synthesized output.
Obtained (a) by eliminating central 5 sidebands and (b) by putting parabolic phase shift.

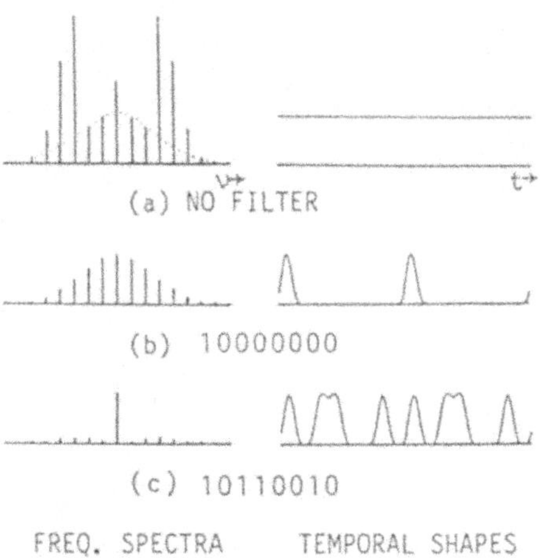

Fig.5 Typical examples of synthesizing word pattern (calculation).

References.

1. T. Kobayashi, H. Ideno, and T. Sueta: IEEE J. Quant. Electron. QE-16, 132 (1980) / T. Kobayashi: Japan Patent, No.1268338 (1977).
2. J. P. Heritage, A. M. Weiner, and R. N. Thurston: In Ultrafast Phenomena V, ed. by G. R. Fleming and A. E. Siegman (Springer Verlag, Berlin, 1986) p.34.
3. T. Kobayashi, Y. Fukushima, H. Yao, K. Amano, A. Morimoto, and T. Sueta: Proc. Meeting Jpn Soc. Appl. Phys. 27p-F-12 (1986) / 30p-ZG-14 (1987).
4. K. Tai, A. Tomita, J. L. Jewell, and A. Hasegawa: Appl. Phys. Lett. 49, 236 (1986).
5. K. Amano, T. Kobayashi, H. Yao, A. Morimoto, and T. Sueta: IEEE/OSA J. Lightwave Technology, LT-5, 1454 (1987).
6. A. Morimoto, H. Yao, T. Kobayashi, and T. Sueta: IQEC '1988, ThH-6, to be presented.

Synthesis and Applications of Arbitrarily Shaped Optical Pulses in Coherent Spectroscopy and Nonlinear Pulse Propagation

M. Haner[1] *and W.S. Warren*[2]

[1]Department of Physics, University of California, Berkeley, CA 94720, USA
[2]Department of Chemistry, Princeton University, Princeton, NJ 08544, USA

1. Introduction

The capability to artificially tailor the phase and amplitude modulation of a laser pulse has proven to be valuable in many experiments. Precise control over the radiation field improves frequency selectivity in molecular spectroscopy and can compensate for such experimental complications as optical density effects or inhomogeneities[1]. Theoretical work has shown that phase and amplitude control can enhance pumping of multiphoton and overtone transitions or improve localized excitation[2]. In addition, pulse shaping is an important aspect in a variety of nonlinear experiments, for example in optical soliton generation[3,4]. At the last ultrafast conference, we reported two developments in coherent laser spectroscopy: synthesis of arbitrarily shaped laser pulses with ≈100ps temporal resolution using an integrated electrooptic device and generation of phase coherent optical pulse sequences with accurate interpulse phase control[5]. With continued improvements in both the electrooptic modulator and high speed programmable GaAs drive circuitry, we have extended the resolution to ≤15ps on pulses of 100-600ps duration and demonstrated amplification to 10-20kW peak powers using a copper vapor laser/dye amplifier[6]. In addition to direct generation methods, the integrated electrooptic device can function as a time domain filter and can be incorporated into a fiber grating pulse compressor[7]. With this technique, we have demonstrated pulse synthesis with approximately 90fs resolution on pulses of 0.5-1.5ps in duration. More complex experimental schemes have also been demonstrated for synthesizing phase and amplitude modulated optical pulses with ≤10ps resolution[3]. We present here new developments in our pulse shaping system, the capability to generate higher peak powers (≈5-10MW) and to actively compensate for pulse distortion in a pulsed dye amplifier. To achieve this, we have incorporated a chirped pulse amplifier (CPA) into our experimental set-up providing amplification of $\approx 10^3$-10^4[8]. This technique allows for the shaped pulse to be amplified at lower peak powers and reduces pulse distortion in the amplifier. Residual artifacts can be compensated for by adjusting the temporal modulation on the chirped pulse before amplification using our EO-modulator apparatus.

2. Experimental Methods and Results

Synthesis of subpicosecond optical pulses is accomplished by integrating a high speed voltage programmable EO-intensity modulator into a fiber-grating pulse compressor. The integrated EO-modulator functions as a time domain filter, so that a given pulse may be discretely modulated to replicate an arbitrary temporal envelope within ≈15ps resolution. To determine the correct input pulse shape $\{E(t')\}$ to the grating compressor necessary to produce a desired target pulse shape $\{E(t)\}$, we calculate a transfer function $G(\omega)$ which describes the gaussian beam diffraction and quadradic dispersion of the grating pair in a precise geometry[9]. Application of the numerical transfer function to the target shape, followed by Fourier transformation to the time domain, provides the discrete modulation waveform programmed into the drive circuitry for the EO intensity modulator.

Springer Series in Chemical Physics, Vol. 48: **Ultrafast Phenomena VI**
Editors: T. Yajima • K. Yoshihara • C.B. Harris • S. Shionoya

The electrooptic time domain filter consists of an EO-waveguide intensity modulator integrated with a high speed GaAs waveform generator[6,7]. The -3dB and -6dB electrical bandwidths of modulator were found to be 14GHz and 19GHz respectively, with an overall impedance of 39.5Ω and a transmission loss of -1.5dB/cm. The GaAs FET drive circuit has an effective bandwidth of 15.5GHz providing ≈15ps risetime and 33ps FWHM resolution[6]. Data generated from the pulse autocorrelation and spectral bandwidth are compiled, then a least squares fit of the crosscorrelation data with the target pulse is computed allowing any adjustments to be programmed into the GaAs drive circuitry. The compressed pulse can be optimized by direct voltage programming of the GaAs drive circuit.

Figure 1 illustrates a synchronously pumped, cavity dumped, dye laser operating at λ=0.65μm producing pulses of 5.4ps. The cavity dumper is externally synchronized to operate at 4kHz, providing 3-4kW of peak power in the pulse train. A silica fiber with a dispersion figure of 0.107 $m^{-1}ps^{2}$ and a normalized pulse amplitude are used to calculate an optimum fiber length and an overall compression factor of 10-20 for our pulse synthesis. Individual pulses emerging from the fiber are temporally broadened to 110-120ps while the spectral bandwidth is increased to 25-30cm^{-1}. The pulse train is attenuated to <100μW before coupling into the EO-modulator. The chirped pulses are temporally restructured by applying a modulation dictated by the desired pulse shape and the grating transfer function. The reshaped chirped pulses coupled out of the modulator (<10nJ per pulse) are amplified using a two-stage, multiple pass, copper vapor laser (CVL)/dye amplifier. The output of a 40W CVL is spatially filtered, collimated, then used to pump two stages of the amplifier. In each amplifier stage, the chirped pulse traverses the gain medium about 4-8 times, yielding an end stage pulse energy of 5-6μJ with no significant pulse broadening. Compression is achieved in a double-pass grating compressor with 1600l/mm holographic gratings, the energy yield in the target pulses is typically 3-4μJ at 4kHz repetition rate. The high peak power (5-7MW) in the output pulses can produce a large fractional population inversion or induce nonlinear excitation in a variety of spectroscopic applications[10].

An important advantage to the CPA scheme is that the target pulse resolution can be increased by further stretching the temporal width of the chirped pulse. For example, if the chirped pulse coupled into the EO-modulator is lengthened to ≈600ps the corresponding resolution in the target pulse is increased to roughly 40fs. An alternative approach would be to replace the picosecond pulse and fiber in figure 1 with a colliding pulse modelocked (CPM) laser (τp ≤100fs) and an antiparallel grating pair configured to produce positive group velocity dispersion. The stretched pulse can be restructured, amplified, then recompressed to generate an arbitrarily shaped pulse. A 1000:1 expansion followed by amplification and subsequent recompression has been demonstrated with pulses ≤100fs[11].

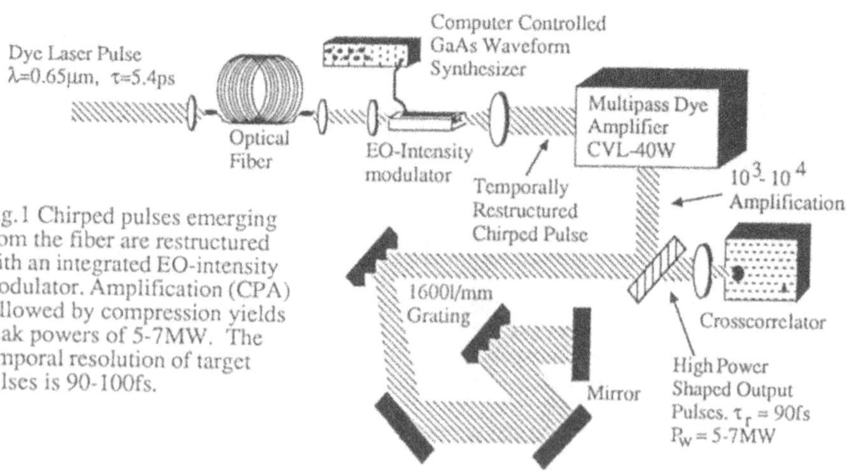

Fig.1 Chirped pulses emerging from the fiber are restructured with an integrated EO-intensity modulator. Amplification (CPA) followed by compression yields peak powers of 5-7MW. The temporal resolution of target pulses is 90-100fs.

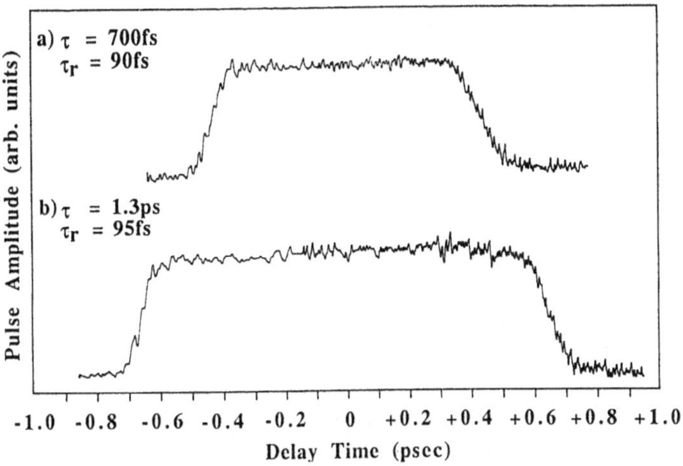

Fig. 2 Crosscorrelation traces of two optical square pulses of approximately 700fs(a) and 1.3ps(b) duration, both of the pulses have comparable risetimes of ≈90fs. The data have been deconvolved of the 180fs $sech^2$ reference pulse artifact.

Figure 2 illustrates two rectangular optical pulses synthesized with our experimental set-up, the temporal resolution is 90-100fs on pulses of 0.5-1.5ps in length. We have used variable length rectangular laser pulses to probe carrier relaxation in GaAs photoconductive switches. The rectangular pulses preserve risetime resolution needed to make accurate electrical measurements, but allow the electron-hole plasma to achieve a steady state in the lattice. Optical square pulses can also be useful in experimental measurements of nonlinear and dielectric responses in materials.

This work was supported by the National Science Foundation under grant CHE-8719545, and by DARPA under contract N60921-87-C-0023 monitored by NSWC. MH would like to thank the Miller Institute for Basic Research in Science at the University of California.

3. References

1. W.S. Warren and M.S. Silver, Adv. Mag. Res. (in press); C.P. Lin, J. Bates, J. Mayer, and W.S. Warren, J. Chem. Phys., **86**, 3750(1987).
2. W.S. Warren and A.H. Zewail, J. Chem. Phys., **78**, 3583 (1983); J.C. Diels and S. Besnainou, J. Chem. Phys., **85**, 6347 (1986); H. Rabitz, Atomic and Molecular Processes with Short Intense Laser Pulses (A.D.Bandrauk, ed., Plenum, NY, 1988).
3. M. Haner and W.S. Warren, CLEO-88, PD20-1, Anaheim, CA(1988); Phys. Rev. Lett. (in press.); M.Haner and W.S. Warren, Phys. Rev. A, to be submitted.
4. A.M. Weiner, J.P.Heritage, R.J. Hawkins, R.N.Thurston, E.M. Kirschner, D.E. Leaird, and W.J.Tomlinson, CLEO-88, PD24-1, Anaheim, CA (1988).
5. M. Haner, F. Spano, and W.S. Warren, **Ultrafast Phenomenon V** (G. Fleming and A. Siegman, eds.; Springer Verlag, 1986); F. Spano, M. Haner, and W.S. Warren, Chem. Phys. Lett., **135**, 97(1987).
6. M. Haner and W.S. Warren, Applied Optics, **26**, 3687(1987).
7. M. Haner and W.S. Warren, Appl. Phys. Lett., **52**, 1458 (1988); M. Haner and W.S. Warren, Opt. Lett., **12**, 398(1987).
8. P. Maine, D. Strickland, P. Bado, M. Pessot, and G. Mourou, IEEE J. Quantum Electron., **QE-24**, 398 (1988).
9. E.B. Treacey, IEEE J. Quantum Electron., **QE-5**, 454(1969).
10. W.S. Warren and M. Haner, NATO workshop on Interactions of Short Intense Laser Pulses with Atoms and Molecules, Sherbrooke, Quebec, July, 1987.
11. M. Pessot, P. Maine, and G. Mourou, Opt. Commun., **62**, 419 (1987).

A New Method for Measuring Ultrashort Optical Pulses

K. Naganuma, K. Mogi, and H. Yamada

NTT Opto-electronics Laboratories, 3-1, Morinosoto Wakamiya,
Atsugi-shi, Kanagawa, 243-01, Japan

1. Introduction

Assumption-free pulse reconstruction including chirp from SHG signals is necessary for detailed light pulse characterization. The methods previously reported are limited to specific application areas: the heterodyne method [1] requires much shorter sampling pulses, while the femto-nitpicker [2] serves only in the femtosecond domain where pulse can be greatly broadened by a glass of reasonable size. In this paper, a new measuring method is proposed and an experimental demonstration reported.

2. Conditions for Pulse Reconstruction and a Practical Algorithm

The light spectrum (or its Fourier-transformed equivalent G_2: interferogram), SH light spectrum (or F_2: SH interferogram), and intensity autocorrelation, G_2, are sufficient data for assumption-free reconstruction of time-limited (i.e. non-zero only in a finite time domain) pulse, with time inversion the only remaining ambiguity. This proposition is proved by showing that the Taylor coefficients of the pulse electric field at each domain edge are successively determined from the Taylor coefficients of the given data.

We present a practical iterative algorithm for the pulse reconstruction, as shown in Fig. 1. The algorithm retrieves the phase of the Fourier amplitude $\tilde{E}$ using the fundamental Fourier modulus $|\tilde{E}|$, SH Fourier modulus $|\tilde{u}|$ and $|\tilde{I}|$ (obtained from $|\tilde{I}|^2 = \mathcal{F}[G_2]$). Fig. 2 shows one of the simulation results of the algorithm. For an assumed pulse intensity and phase (solid line), the above three Fourier moduli are calculated and input to the iteration loop. The reconstructed pulse intensity and phase (shown by dots) excellently coincide with the original, which shows that the algorithm really reconstructs the rather complicated pulseform.

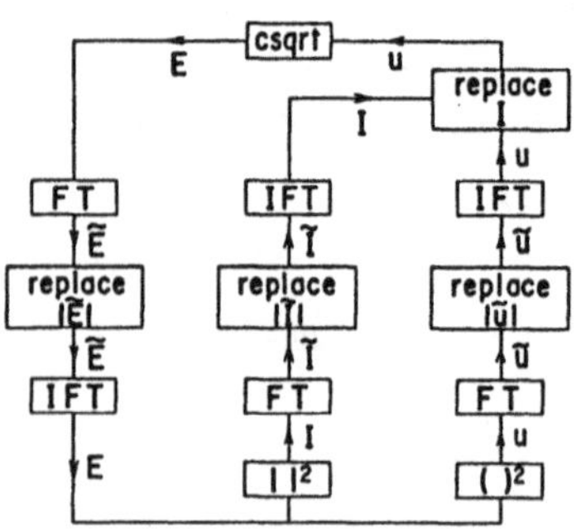

Fig.1 Iteration scheme (FT: Fourier-transform, IFT: inverse FT). In Fourier space, the modulus of the trial function is replaced by given data while its phase is retained. The difference between the trial function modulus and the given one is used to judge convergence.

Springer Series in Chemical Physics, Vol. 48: **Ultrafast Phenomena VI**
Editors: T. Yajima · K. Yoshihara · C.B. Harris · S. Shionoya

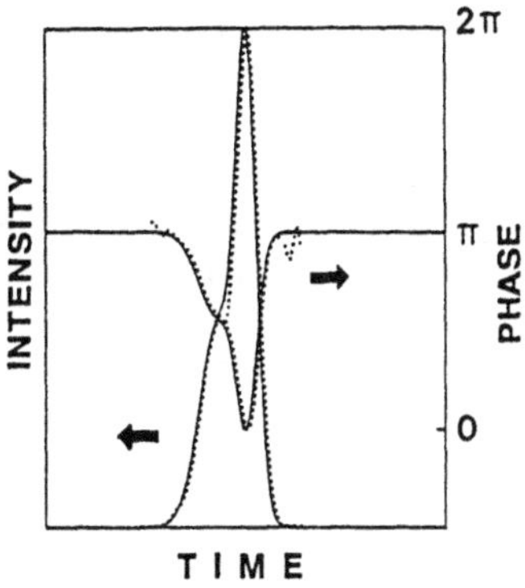

Fig.2 Iterative algorithm test result. Asymmetric pulseform with side-lobe and the phase shift due to fast SPM are assumed. The reconstructed pulse (dots) coincides with the original (solid line).

3. Experimental Results

The measurement setup is shown in Fig. 3. G_1 and the fringe-resolved SHG correlation (FRSHG) are simultaneously measured using a Michelson interferometer. The trigger generator senses the path difference by observing a He-Ne laser interference signal, and generates a trigger signal every time the path difference changes by $\lambda_{He-Ne}/2$.

$|\tilde{E}|$ is obtained from G_1 by a Fourier transform, while $|\tilde{I}|$ and $|\tilde{u}|$ are derived from FRSHG. Such a SH spectrum $|\tilde{u}|^2$ measurement through FRSHG signal is superior to measurements with a high resolution spectrometer, especially for the very weak SH light generated by LD pulses. The iteration is then executed by computer. We call this procedure IRIS (Iterative pulseform Reconstruction from Interferometric Signals).

To demonstrate the method, pulses generated by a gain-switched 1.3 μm DFBLD were measured. Measured correlations (Fig. 4a) and the reconstructed pulse (Fig. 4b) are shown for the case of applying a 1.25GHz sinusoidal rf current of 1.6 I_{th} amplitude without dc bias. A nearly linear chirp is clearly observed. In Fig. 4b, the time direction

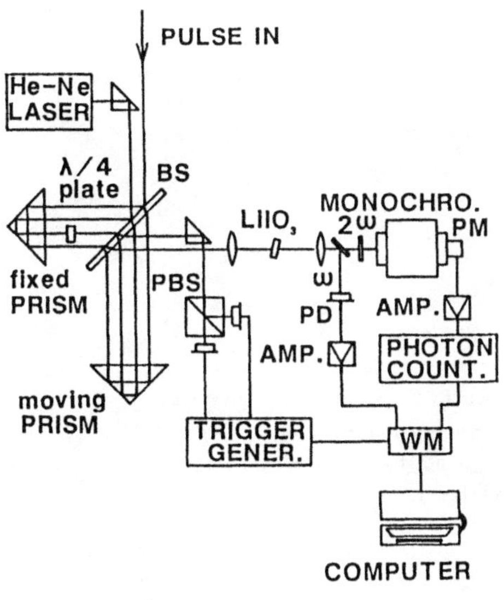

Fig.3 Pulse measurement apparatus (BS: beam splitter, PBS: polarization beam splitter, PM: photomultiplier, PD: photodiode, WM: wave-memory). A photon counter is installed to detect the weak SH light in the case of the LD pulse.

is chosen so that the chirp becomes negative, considering the chirp direction due to fast carrier density change. For the first time, LD chirp is measured in detail.

Recently, LD pulse compression by an optical fiber [3] was reported, where the negative chirp was compensated by the positive fiber dispersion. Precise measured chirp information is useful to optimize the dispersive media parameters on compression.

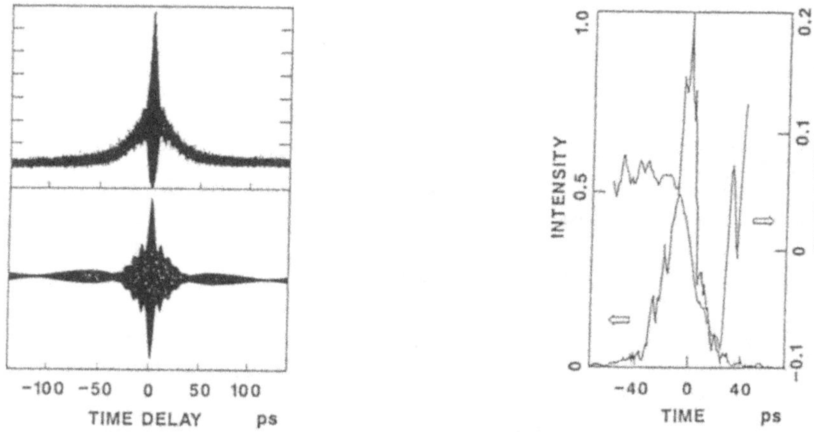

Fig.4 An example of gain-switched DFBLD measurement results.
(a) Measured interferometric signal traces each consisting of 256k points with 1.0554fs delay time interval; upper trace: FRSHG, lower trace: interferogram.

(b) Pulseform reconstructed from the above interferometric signals.

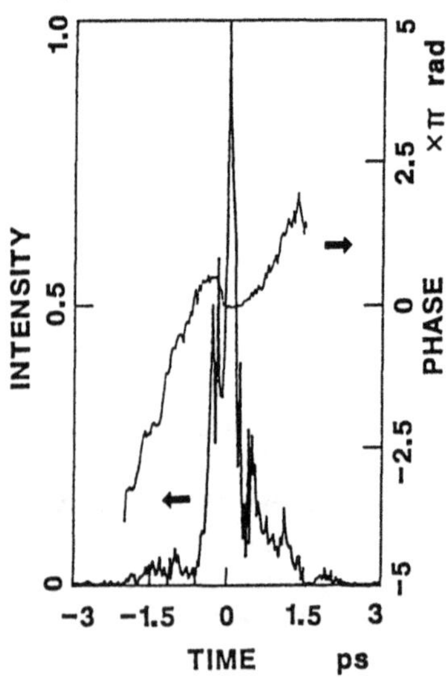

Fig.5 An example of subpicosecond dye laser pulse measurements. Pulses were generated by a passively mode-locked dye laser.

IRIS also performs successfully with subpicosecond dye laser pulses. A reconstructed mode-locked dye laser pulse is shown in Fig. 5. To suit the visible light, interferometric signals are measured with $\lambda_{He-Ne}/8$ path difference pitch enough to satisfy the sampling theorem. In conclusion, this method is highly practical and can be used many areas of application.

References

1. Rothenberg and Grischkowsky: Opt. Lett. 12, 99 (1987)
2. Diels, Fontaine and Rudolph: Revue Phys. Appl. 22, 1605 (1987)
3. Takada, Sugie and Saruwatari: IEEE J. Lightwave Technol. 5, 1525 (1987)

A Phase Sensitive Single Pulse Autocorrelator for Ultrashort Laser Pulses

G. Szabó[1], *Z. Bor*[2], *and A. Müller*[3]

[1] JATE University, Department of Experimental Physics, Dóm tér 9, H-6720 Szeged, Hungary

[2]Research Group on Laser Physics of the Hungarian Academy of Sciences, JATE University, Dóm tér 9, H-6720 Szeged, Hungary

[3]Max-Planck-Institut für Biophysikalische Chemie, Abteilung Laserphysik, Postfach 2841, D-3400 Göttingen, Fed. Rep. of Germany

Single shot autocorrelation methods [1,2,3,4,5] yield, so far, no useful information about the phase structure of the pulses. We present here, for the first time, a phase sensitive single shot technique suitable to study picosecond and femtosecond pulses.

The pulse to be analyzed is horizontally expanded by a cylindrical telescope before it enters the modified Michelson interferometer (Fig. 1). A tilt of the pulse fronts in the horizontal plane is introduced by two prisms located in front of the two interferometer mirrors [6]. The beams coming from the interferometer are set slightly non-parallel with an angle of only a few milliradians. They are focussed onto the second harmonic crystal by a cylindrical lens. Its focal plane is imaged onto the input plane of the detector where the horizontal intensity distribution of the second harmonic (SH) is detected by an image intensifier coupled to an optical multichannel analyzer.

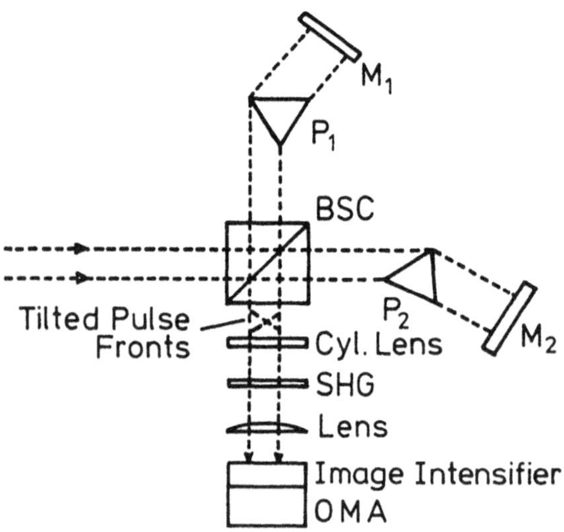

Fig. 1

Springer Series in Chemical Physics, Vol. 48: **Ultrafast Phenomena VI**
Editors: T. Yajima • K. Yoshihara • C.B. Harris • S. Shionoya

The distribution of SH as a function of the horizontal spatial coordinate constitutes a special kind of **phase sensitive autocorrelation function** (PSA). Its envelope is defined by the pulse duration and the tilt of the pulse fronts; the period of the modulation depends on the small angle between the two interferometer arms, and the modulation depth on the phase structure of the pulse. The key point of the method presented is that both the delay axis and the fringe period may be arbitrarily chosen and can, thus, be matched to the spatial resolution of the detector.

Assuming a Gaussian pulse shape and type I phase matching and allowing for linear chirp the expected signals have been modelled, and the time integrated SH intensity in the focal plane of the cylindrical lens has been calculated numerically.

Figures 2a and b each show an experimentally obtained **single shot PSA** (solid line) together with the result of the numerical integration of (1) (dashed line). The experimental trace of Fig. 2a corresponds to a 3.5 ps long pulse which is fully coherent. In contrast, fitting the experimental curve of Fig. 2b one has to assume chirp. The best agreement (dashed line on Fig. 2b) was achieved by a pulse duration of 3.5 ps and a three times 2π phase shift during the pulse duration. The sign of the chirp cannot be determined from the PSA. As Fig. 2 shows, the method presented here is well suited to trace the chirp in single shots. Testing the ultimate limit of the temporal resolution was not yet possible (the shortest measured pulse duration was 250 fs) but we estimated it to be 50 fs. This limit is caused by the group velocity dispersion in the delay prisms. By replacing the prisms with gratings for pulse front tilt, further improvement of the temporal resolution may be possible.

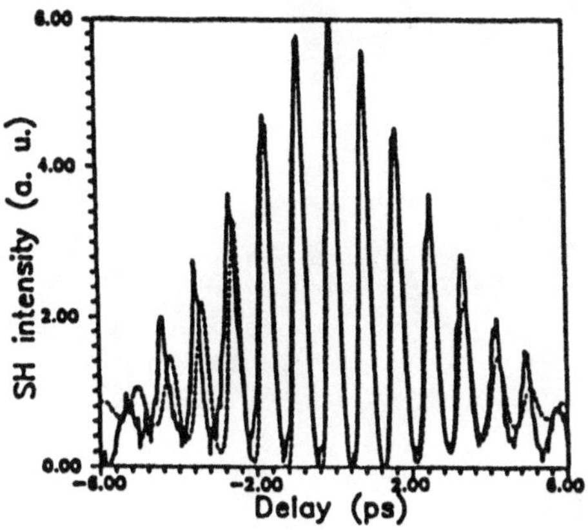

Fig.2a

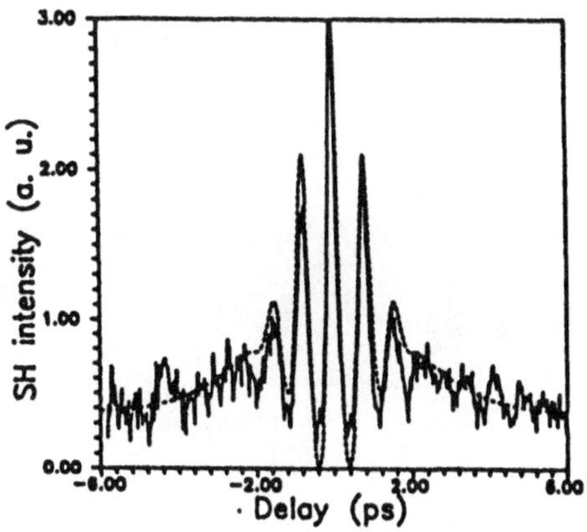

Fig.2b

References

1. J.A. Giordmaine, P.M. Rentzepis, S.L. Shapiro, K.W. Wecht: Appl. Phys. Lett. 11, 216 (1967)
2. J. Janszky, G. Corradi, R.N. Gyuzalian: Opt. Commun. 23, 293 (1977)
3. C. Kolmeder, W. Zinth, W. Kaiser: Opt. Commun. 30, 453 (1979)
4. G. Szabó, A. Müller, Zs. Bor: Appl. Phys. B31, 1 (1983)
5. S.M. Saltiel, K.A. Stankov, P.D. Yankov, L.I. Telegin: Appl. Phys. B40, 25 (1986)
6. Zs. Bor, B. Rácz: Opt. Commun. 54, 165 (1985)

Single Shot Measurement of Optical Phase Modulation with Subpicosecond Resolution by Temporal Holography

F. Reynaud[1], *F. Salin*[2], *A. Barthélémy*[1], *A. Brun*[2], *and C. Froehly*[1]

[1]Institut de Recherche en Communications Microondes et Optiques, Equipe Optique, Faculté des Sciences, F-87060 Limoges Cedex, France

[2]Institut d'Optique Théorique et Appliquée, CNRS, U.A. 14, Centre Universitaire d'Orsay, BP 43, F-91406 Orsay Cedex, France

Amplitude and phase distortions induced by optical nonlinearity are undetectable by the usual means of picosecond and subpicosecond analysis (streak camera, second harmonic generation...). Recently Rothenberg and Grischkowsky [1] introduced a method to measure the phase variation of picosecond pulses that were chirped by self-phase modulation in a single mode fiber. Another possible way is based on a less recent method of spectral interferometry also known as "temporal holography" [2]. Until now this measurement technique has only been applied to the measurement of impulse response and transfer function of optical linear filters. We present here the complete determination of phase modulation induced by nonlinear effects in a single mode fiber on a picosecond pulse using this method [3].

A pulse is sent into a Mach-Zehnder interferometer. One arm is free space and serves for the production of a reference wave. The optical system under test is placed in the other arm. After recombination the two beams are injected into a spectroscope. As the phase shift introduced by the optical system varies with the frequency, the recorded spectra is modulated by fringes. If there is only a difference Δt in the delay introduced by each arm (i.e. a different slope for the spectral phases $\Delta t = d\phi/d\omega$), one observes a periodic fringe pattern with a period $\Delta\upsilon = 1/\Delta t$. If there is also a difference in group time dispersion (i.e. a curvature of the spectral phase distributions $dt/d\omega = d^2\phi/d\omega^2$), the fringe spacing will vary linearly from one edge of the spectrum to the other.

We have demonstrated this technique by recording the self-phase modulation of a 800 fs pulse after passage through a single mode optical fiber. When a short and intense pulse is sent into an optical fiber it experiences a temporal phase shift that depends on the instantaneous intensity $\theta(t) = 2\pi n_2 I(t)\, z/\lambda$, where n_2 is the nonlinear index of the fiber, z its length and λ the pulse wavelength. This results in a broadening of the pulse spectrum. In order to ensure that the pulse exiting the fiber has a spectral width smaller than that of the reference pulse, we have added at the entrance of the interferometer test-arm a filtering sta-

Springer Series in Chemical Physics, Vol. 48: **Ultrafast Phenomena VI**
Editors: T. Yajima · K. Yoshihara · C.B. Harris · S. Shionoya

ge that transforms the input pulse into a longer transform-limited pulse, with a corresponding narrower spectrum. This lengthened pulse is then sent into the fiber and propagates in a self-phase modulation regime leading to a broadening of its spectrum.

The complete set-up is illustrated in Figure 1. The source employed is a CPM dye laser whose emission at 615 nm is amplified by passage through a four-stage dye amplifier giving near-transform-limited 60 femtosecond pulses in the GW range at a 10 Hz repetition rate. The laser beam is separated into two parts by the beam splitter 1, at the input of the interferometer. The reference arm is free space and its length, about 18 m, is adjusted by the displacement of a cube corner. In the test arm, the input pulse is first lengthened to about 800 fs by the prism dispersive line and then sent into a single mode optical fiber in which is suffers the combined effects of self-phase modulation and normal group velocity dispersion. The fiber length is 5.75 m and its measured chromatic dispersion is 440 ps/km.nm at 600 nm. The fiber output is recombined with the reference pulse through the beam splitter 2 and sent into a grating spectrograph. The width and the exact shape of the detected spectra depend on the optical power coupled in the fiber and fluctuate from shot to shot. In order to obtain reliable results, all measurements are done on single shot recordings. A typical spectrum is shown in Figure 2 together with the reference spectrum. The mean fringe spacing can be adjusted by slightly changing the reference arm length. Starting from these recordings, we deduce the value of the spectral phase of the pulse, after its nonlinear propagation across the fiber, for a discrete set of wavelengths (see Figure 3). The linear part (which has no meaning since it only depends on the temporal delay between the two pulses) has been subtracted. These experimental points fit well a parabolic varia-

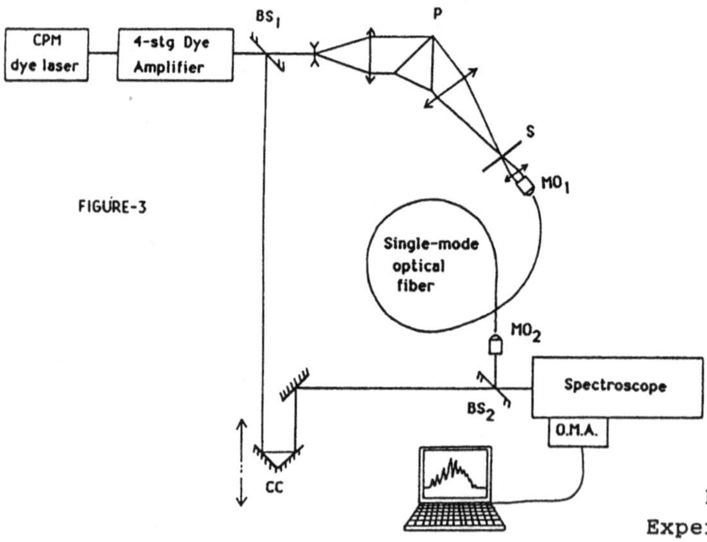

Figure 1 : Experimental set-up

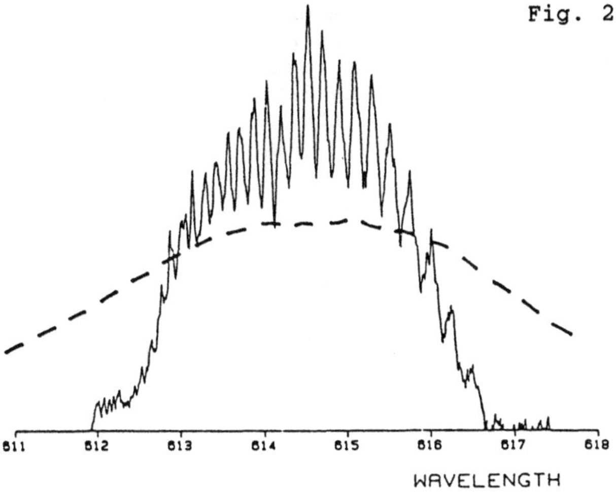

Fig. 2 : Experimental modulated spectrum.

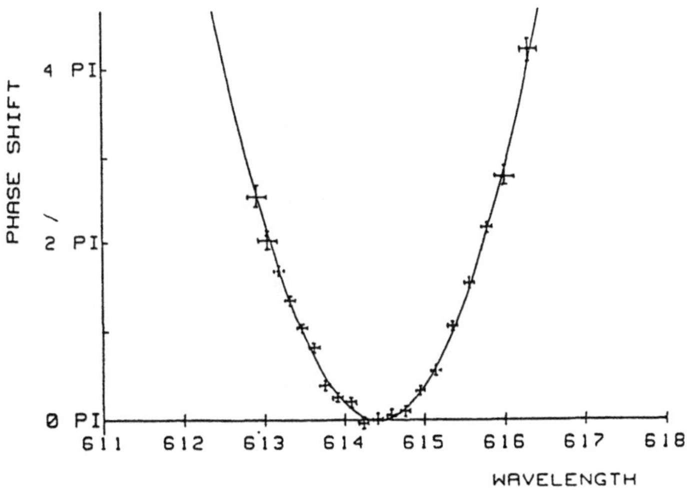

Fig. 3 : Relative phase of the chirped pulse.

tion, also shown in Figure 3 (solid line), indicating the presence of a nearly linear frequency modulation onthe test pulse, in agreement with theoretical predictions of the nonlinear Schrödinger equation.

In conclusion, we have proposed a powerful new tool for the <u>coherent</u> analysis of distortions introduced by optical nonlinearities on short pulses (in the subpicosecond range). The technique described is the first to our knowledge that is able to analyse a single event. This is an important point since the reproductibility of femtosecond pulses obtained by pulse compression and/or amplified in dye amplifiers is usually not good. The only competing method in this field is restricted to phase-shifts larger than 2π and needs repetitive events to achieve a measurement [1]. Nevertheless we have to note that the interferometric technique presented here needs near-transform-limited reference pulses or at least pulses whose phase distortions are much smaller than the phase modulation of the pulse under test. This research has received the financial support of the European Community under grant ST2J00009 2F, and of the DRET (division optique).

References :

[1] J.E. Rothenberg and D. Grischkowsky
Opt. Lett. 12, 99 (1987).

[2] C. Froehly, A. Lacourt and J.C. Viennot
J. of Optics 4, 183 (1973).

[3] F. Reynaud, F. Salin and A. Barthélémy
Submitted to Opt. Lett.

A Method of Evaluation of Ultrashort Light Pulses Based on Self-Phase Modulation

H. Yoshiyama[1], *Y. Shio*[1], *A. Imaizumi*[1], *H. Motoyama*[1], *M. Nakajima*[1], *S. Tanaka*[1], *H. Kobayashi*[1], *A. Watanabe*[2], *and H. Saito*[2]

[1]Department of Electronics, Tottori University, Koyama, Tottori 680, Japan
[2]Okayama University of Science, Ridai, Okayama 700, Japan

1. Introduction

Phase characteristics of optical light pulses, so-called chirp, play an essential role in femtosecond light pulses. One of the vexing things is that we can not determine the pulse intensity shape I(t) and the phase shift $\phi(t)$ independently.

Recently, a new method has been developed to measure ultrashort light pulses [1]. We, also, have developed a similar but different evaluation method for the chirp. In this paper, we present our evaluation method.

2. Method of evaluation of ultrashort light pulses

The optical electric field E(t) of the steady state is given by $E(t)=\sqrt{I(t)}\exp[i\phi(t)-i\omega_0 t]$, where ω_0 is the central frequency of the electric field. When the phase characteristics are determined by self-phase modulation (SPM), coming from the optical Kerr effect and/or saturable loss effect of the absorber, $\phi(t)$ is described as follows:

$$\phi(t) = p\int_{-\infty}^{t}\exp[(t'-t)/\tau_r]\cdot I(t')\cdot\exp[(J(t')-J(t))/S]dt' \,, \quad (1)$$

$$p = \frac{\omega_0 L_e}{c}\left(n_2/\tau_r - n_0/S\right) \,, \qquad J(t) = \int_{-\infty}^{t} I(t')dt' \,, \quad (2)$$

where n_0 and n_2 are linear and nonlinear refractive indexes of the absorber, τ_r the recovery time of the optical Kerr effect, S the saturation fluence and L_e the effective length of the absorber. In eq.(1), for $p>0$, the optical Kerr effect is predominant [2], and for $p<0$, the saturable loss effect is predominant [3].

Figure 1 shows an iterative algorithm to reconstruct I(t) and $\phi(t)$ from the experimental data of spectrum $I(\omega)$ and second-harmonic (SH) autocorrelation $G_2(\tau)$. The algorithm can determine the phase $\phi(\omega)$ and the phase $\psi(\omega)$. The phase $\psi(\omega)$ corresponds to the Fourier amplitude $V(\omega)$

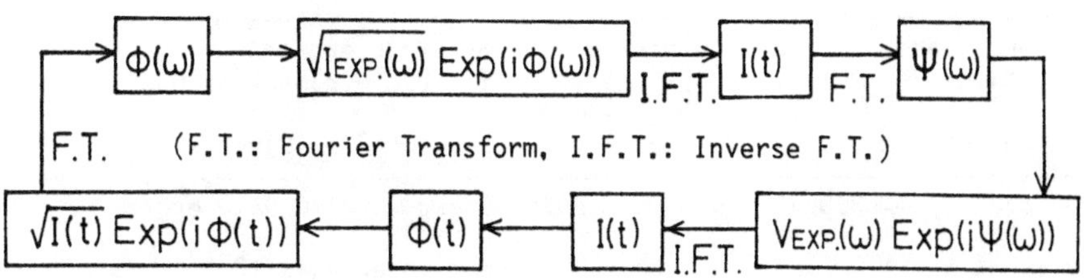

Fig. 1 Flow chart of the calculation. To obtain $\phi(\omega)$, eq.(1) is used.

Springer Series in Chemical Physics, Vol. 48: **Ultrafast Phenomena VI**
Editors: T. Yajima · K. Yoshihara · C.B. Harris · S. Shionoya

of the pulse intensity I(t), where $V(\omega)^2$ is Fourier modulus of $G_2(\tau)$. For the iterative calculations, the experimental values of $I_{exp}(\omega)$ and $V_{exp}(\omega)$ are used for $I(\omega)$ and $V(\omega)$. Only when p, S and τ_r take the experimentally reasonable values, the iterative calculation converges well.

3. Evaluation of experimental results

We measured the experimental data of $I(\omega)$ and $G_2(\tau)$ at the same time. Our CPM ring dye laser consists of two dye jet streams: one is gain medium (Rh-6G) and the other is saturable absorber (DODCI).

Figure 2 shows the experimental values of (a)$I(\lambda)$ (λ: wavelength) and (b)$G_2(\tau)$, together with the calculated ones. The agreement between our calculation and the experimental is very good. The parameters used to reconstruct the pulses are listed in Table 1. The positive value of p implies that the optical Kerr effect is predominant rather than the saturable loss effect. It causes a broad wing to the spectrum at long-wavelength side [2], as shown in Fig.2(a).

Figure 3 shows the reconstructed pulse together with the frequency shift $\Delta\omega(t)$ ($=-d\phi(t)/dt$), being calculated using the values shown in Table 1. The positive chirp resulting from the optical Kerr effect of fast

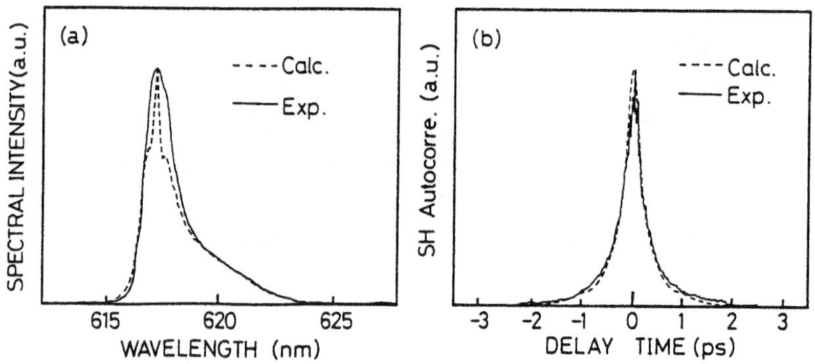

Fig. 2 (a)Spectral intensity and (b)conventional SH autocorrelation. Solid lines show the experimental results, while broken lines show the calculated ones using the reconstructed pulse.

Table 1 Parameters used for iterative calculation

Parameters of absorber (DODCI)	Estimation from calculation	Estimation from laser condition
Saturation fluence	$S=6\times10^4 erg/cm^2$	$S=6.4\times10^4 erg/cm^2$
Recovery time	$\tau_r=0.3ps$	$\tau_r=(0.2-2)ps$ (for CS_2)
Coefficient of SPM	$pI(0)=22ps^{-1}$	$pI(0)=(-3-10)ps^{-1}$
Laser frequency $\omega_0=3.05\times10^{15}s^{-1}$	Pulse energy $J(\infty)=0.9\times10^4 erg/cm^2$	
Linear index $n_0=(1-3)\times10^{-5}$ Effective thickness $L_e=(1-10)cm$	Nonlinear index $n_2=3\times10^{-16}cm^2/W$	

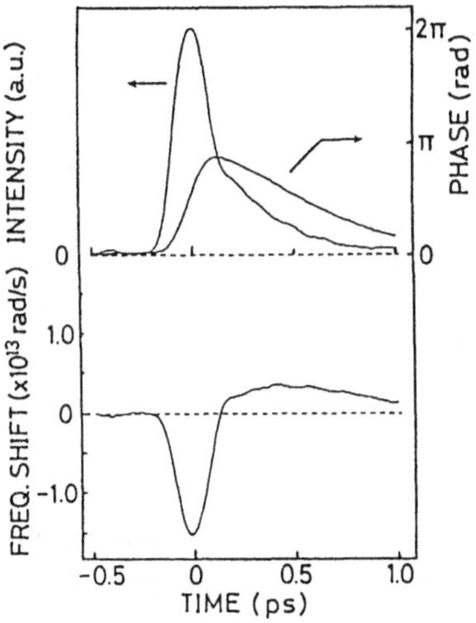

Fig. 3 Pulse intensity, phase shift and frequency shift of the reconstructed pulse. The parameters used are listed in Table 1.

recovery time can be observed around the peak of pulse intensity. It should be noted that the pulse width (t_p=0.2ps) is comparable to the recovery time (τ_r=0.3ps).

4. Discussion

In order to consider the effects of group-velocity dispersion (GVD) on SPM pulses, the recovery time of the SPM should be considered, when the pulse width is comparable to the recovery time. We, consequently, have introduced the recovery time into the theory developed by MARTINEZ et al. /4/.

Using our modified calculation, we calculate the GVD (g) dependence of relative pulse width $\xi=t_p(g)/t_p(0)$. For the fast recovery time, i.e. for $\tau_r/t_p(0)\ll 1$, a large reduction of ξ can be observed at a large negative value of g, while for the slow recovery time, i.e. for $\tau_r/t_p(0)\gg 1$, the reduction of ξ can not be observed at any value of g.

Next, using the reconstructed pulse, we calculate the GVD dependence of the pulse width. The result shows that the pulse compression is about 32%, when GVD is given to be negative outside the laser oscillator. This result comes from the large optical Kerr effect with a fast recovery time.

References

1. K. Naganuma, J. Noda: Extended abstracts (The 34th Spring Meeting, 1987); The Japan Society of Applied Physics and Related Societies, p.684, 30p-ZG-5 (in Japanese)
2. Y. Ishida, K. Naganuma, T. Yajima: IEEE J. QE-21, 69 (1985)
3. W.B. Jiang, D.C. Sun, F.M. Li: Opt. Commun. 64, 449 (1987)
4. O.E. Martinez, R.L. Fork, J.P. Gordon: Opt. Lett. 9, 156 (1984)

Measurement of Femtosecond Pulsewidths Using Interference Autocorrelation

Gong Zhenglie[1], *Huang Zhengyi*[1], *Xu Ziguang*[1], *Qian Shurong*[2], *Xiang Wanghua*[2], *and Wang Qingyue*[2]

[1]Tianjin Institute of Science and Technology, Tianjin, People's Republic of China
[2]Tianjin University, Tianjin, People's Republic of China

1. Abstract

A technique for the measurement of femtosecond pulsewidths is presented that uses the equality-width of the envelope of the interference correlation curve and computer processing of the curve.

2. Introduction

In the measuring technique for ultrashort pulses with type I phase matching, all the correlation information of the pulses is expressed (if the delay velocity of two autocorrelation pulses is slow enough) by the equation

$$\begin{aligned}\langle I(\tau)\rangle = \int_{-\infty}^{+\infty} \{ & E^4(t) + E^4(t+\tau) + 4E^2(t)E^2(t+\tau) \\ & + 2E^2(t)E^2(t+\tau)\cos 2\omega\tau + 4E(t)E(t+\tau) \\ & \times [E^2(t) + E^2(t+\tau)] \cos \omega\tau \}\, dt \quad . \end{aligned} \tag{1}$$

The curve described by (1) is called the interference correlation measurement (IFCM) curve. The ratio of the peak to background is 8:1, and so it is also called an 8:1 curve. When the delay velocity is fast enough, just the first three terms in (1) are recorded and the ratio of peak to background is 3:1; the curve is called a 3:1 curve, and is written as $\langle I_1(\tau)\rangle$. This is the intensity correlation measurement (ISCM) curve which is usually used to measure pulsewidths.

It is of interest to raise the precision of measurement of femtosecond pulsewidths. The methods of equality-width and of microcomputer processing are used to attempt this in this paper.

3. Equality-Width

The first method involves finding the equality-width of the up-envelope of the 8:1 curve that is equivalent to the FWHM of the 3:1 curve for the same pulses. The former's size on the recording sheet is longer than the latter's.

Springer Series in Chemical Physics, Vol. 48: **Ultrafast Phenomena VI**
Editors: T. Yajima · K. Yoshihara · C.B. Harris · S. Shionoya

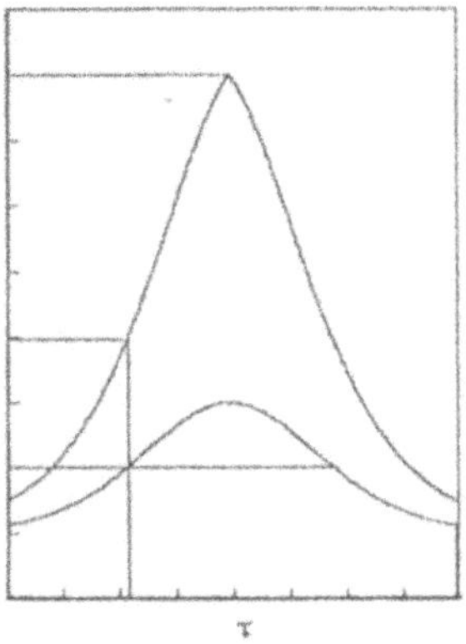

Fig.1 Position of equality width for $sech^2x$

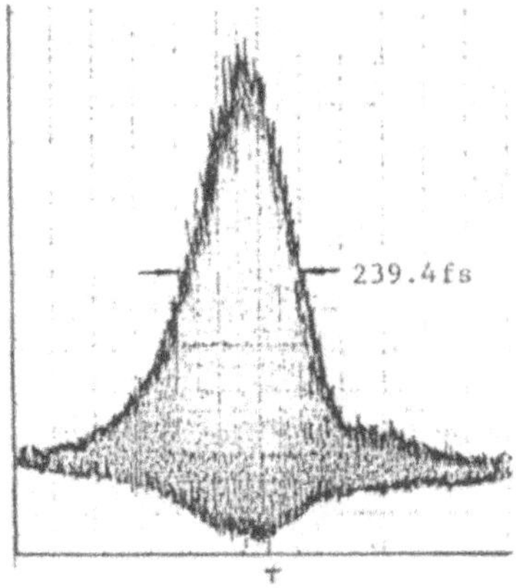

Fig.2 Experiment curve for IFCM

The position of the equality-width is different for different pulse shapes. For pulses with a peak height of 8, the height positions of the equality-width are 3.984, 4.875, 4.928, 3.533 and 3.042 for shapes of $\mathrm{sech}^2 t/T$, $\exp(-2t^2/T^2)$, $\exp(-2|t|/T)$, $(1+t^2/T^2)^{-1}$ and $\mathrm{sinc}^2(t/T)$, respectively. It has been calculated that the pulse shape for a CPM dye laser is of hypersecant type [1], so the position of the equality-width is given by the value of 4 (Fig. 1). The pulses from a CPM dye laser with a six-mirror ring cavity were measured by the optical correlator, which consists of prisms [2]. The equality-width was 239.4 fs (Fig. 2), and the corresponding pulsewidth was 154.5 fs. In order to verify this result, the FWHM of the 3:1 curve was measured; the pulse width was 158.7 fs. The difference between the two results is 4.2 fs, which is within the error range.

Both theoretical analysis and experimental results show that there is a fast velocity limit V_8 for the 8:1 curve and a slow velocity limit V_3 for the 3:1 curve, which are $V_8 = \lambda/8t_r$ and $V_3 = \lambda/t_r$, respectively, where t_r is the response time of the recorder and λ is the central wavelength of the CMP laser. Since $V_3/V_8 = 8$, at the same speed of recording sheet, the size of the equality-width on the sheet is at least 8 times as long as the FWHM of the 3:1 curve, although their pulse durations are the same. That is, the resolution time of IFCM is at least seven times higher than that of ISCM. In fact, the delay velocity can be slowed down greatly. Therefore, it is not difficult to raise the resolution time by one order of magnitude.

4. Computer Processing

The mathematical model of interference correlation signals sampled and processed by computer is presented. The basic idea is to simulate the physical process of ISCM by computer and to combine IFCM and ISCM. We have

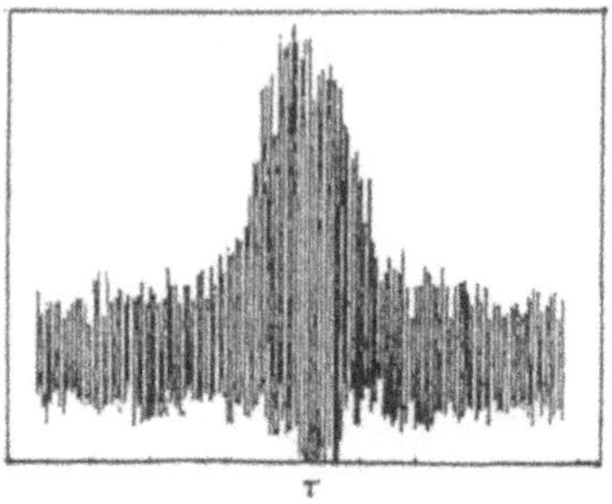

Fig.3 Information of IFCM sampled by computer

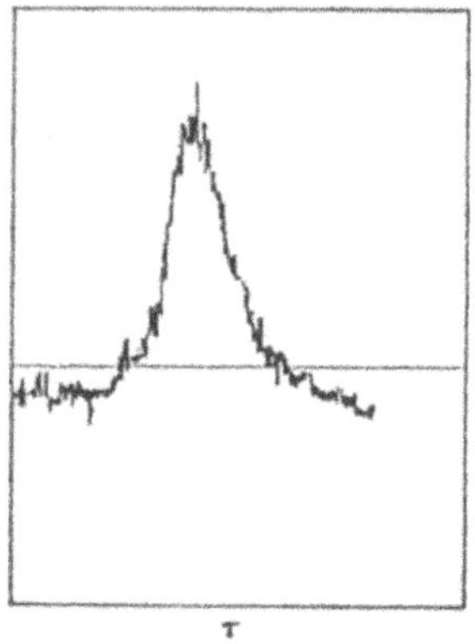

Fig.4 Result processed by computer

proved that the area under the interference correlation curve is equal to the area under the second-order intensity correlation curve if they are recorded at equal delay velocities (in fact, the latter cannot be obtained at this velocity experimentally), and the former is a periodic oscillating function. We have

$$\langle I_1(\tau)\rangle = (1/p) \int_{\tau}^{\tau+p} \langle I(\tau)\rangle d\tau \quad . \tag{2}$$

Here, the left hand side of the equation is the general intensity correlation function and the right hand side is the average of the interference autocorrelation function over one period. Its accuracy is very high because the computer can sample data very densely. It is important that the mathematical model of computer processing can reproduce the physical course followed by ISCM mor accurately than the chart recorder.

A plot of interference correlation information sampled by computer and the general intensity correlation curve obtained by processing the information according to the above mathematical model are shown in Figs. 3 and 4, in which 12 288 data were sampled by computer in 35 s.

The position of the equality-width on the envelope should be revised for a pulse with chirp, but it is not necessary to consider chirp for computer processing.

5. Reference

1. J. Herrmann et al.: Appl. Phys. B **26**, 197–202 (1981)
2. Gong Zhenglie et al.: Acta Opt. Sinica **7**, 935–938 (1987)

This work is supported by the National Natural Science Foundation of China.

Recent Advances Towards a 100fs-Resolution Streak Camera

A. Finch[1], *Y. Liu*[2], *H. Niu*[2], *W. Sibbett*[1], *W.E. Sleat*[1], *D.R. Walker*[1], *Q.L. Yang*[2], *and H. Zhang*[2]

[1]J.F. Allen Physics Research Laboratories,
Department of Physics and Astronomy, University of St. Andrews,
North Haugh, St. Andrews, Fife KY169SS, UK
[2]Xian Institute of Optics and Precision Mechanics,
Xian, People's Republic of China

1. Introduction

In view of the generation and application of laser pulses in the femtosecond regime it has become necessary to further improve the performance of linear-response streak cameras. The approach that we have taken involves several refinements to a previous design of electron-optical streak tube [1]. These include (i) a photocathode having reduced photoelectron energy distribution, (ii) optimised electron-optical lens configuration, (iii) ultrahigh-sensitivity travelling-wave deflection arrangement and (iv) high quantum-efficiency phosphor screen.

The purpose of this paper is to detail the pertinent design aspects of the tube and present experimental data to indicate that the camera, in its present state of development, has a limiting temporal resolution of approximately 300fs. Design modifications aimed to further improve this resolution to around 100fs will also be briefly discussed.

2. Tube Design and Construction

The new design of tube [2] is represented schematically in Fig. 1 and is based on the electrode configuration of the Photochron IV which has been demonstrated to have a single-shot temporal resolution of 800fs and a synchroscan resolution ~900fs [3,4]. Notable improvements in the new tube design relate to the structure of the photocathode and deflection system. The photocathode is especially thin to minimise any electron emission time distribution and its composition, with increased bandgap, gives rise to a calculated initial electron energy spread of ~100meV at 620nm.

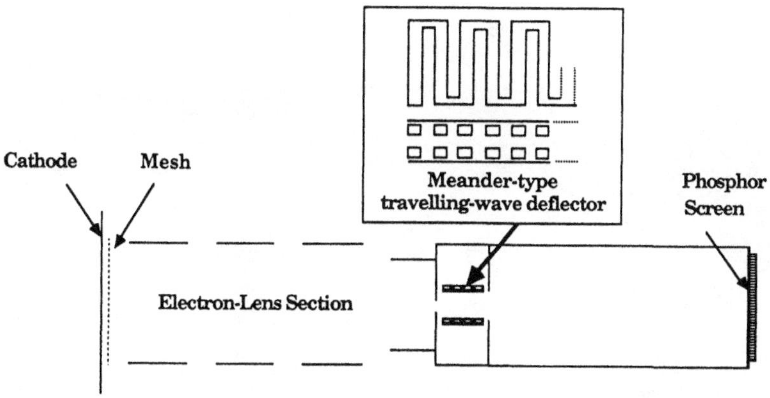

Fig. 1 Schematic of new streak tube

Springer Series in Chemical Physics, Vol. 48: **Ultrafast Phenomena VI**
Editors: T. Yajima · K. Yoshihara · C.B. Harris · S. Shionoya

The redesign of the deflection system has proved very successful. By adopting a meander-type travelling-wave configuration which crosses the electron path a total of 23 times, the deflection sensitivity and hence achievable streak speeds have been significantly enhanced. The meander line, constructed from $1mm^2$ copper bar has a total length of 4cm, with a 5mm spacing between the two deflector lines. The lines were impedance matched to the 50Ω drive cable via a tapered transmission line inside the image tube and a 50Ω termination was externally connected to the deflector system. For this arrangement the bandwidth of the deflector network exceeded 2.5GHz, yielding an experimentally measured static and dynamic deflection sensitivity of 8.5cm/kV, with streak speeds of $2x10^{10}$cm/s.

The operating voltages of the tube photocathode and mesh were −8kV, −3kV respectively relative to the grounded anode. This gave, for a mesh to cathode spacing of 1.2mm, an electric field in the vicinity of the photocathode of ~40kV/cm and an electron-optical magnification of 2. A static limiting spatial resolution of 50 lp/mm was observed at the output phosphor screen. The enhanced quantum efficiency of the phosphor implied a reduced photoelectron current requirement and therefore minimised possible space charge effects.

3. Dynamic Operation

The experimental arrangement that has been employed in the assessment of the single-shot streak performance of the camera is shown in Fig. 2. The image from the streak tube output screen was fibre-optically coupled to a microchannel plate intensifier which provided a luminous gain of 10,000 and a spatial resolution of ~50 lp/mm at its photocathode. This was compatible with the dynamic performance of the streak tube. Ultrashort pulses from a colliding-pulse, passively mode-locked ring dye laser [5] were employed directly as test optical signals and the streak images were recorded using an optical multichannel analyser.

A double avalanche transistor chain generated a differential voltage ramp of 3kV in 1.5ns across the deflector. An initial problem encountered with this circuit was its slow recovery time (1/e point ~800μs) which tended to generate an intense "fly-back" on the streak camera output. This masked any streaked image, even when the input light signal was mechanically chopped. By applying an appropriate electronic gating signal to the photocathode it was possible to turn off the camera immediately after streaking. This method effectively provided single pulse selection and, as a result, no external Pöckels cell arrangement was required.

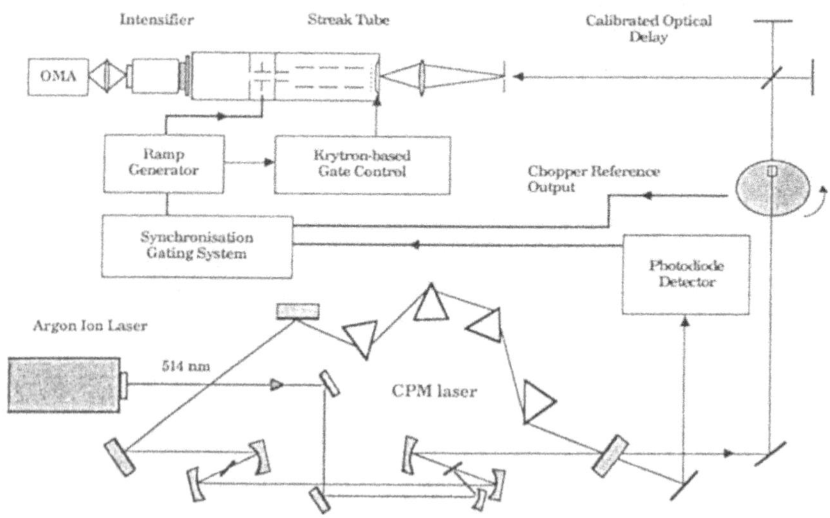

Fig. 2 Experimental arrangement

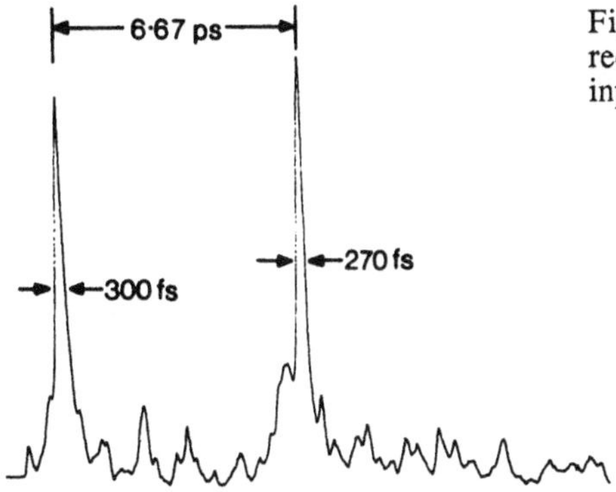

Fig. 3 Intensity profiles of recorded streak images for an input pulse separation of 6.67 ps.

The CPM laser when operated deliberately with excess negative group velocity dispersion in the cavity produced pulses of ~150fs duration with a central wavelength of ~630nm and energies that were suitable for recording purposes. Typically, the pulse energies were ~20pJ per pulse at the input slit. In preliminary studies the measured temporal resolutions were around 500–600fs and were found to be limited by a dynamic defocusing effect from the deflector network. This was found to result from slight asynchronism in the switching characteristics of each transistor chain which gave rise to non-symmetric ramp generation. By optimising focus voltages during streaking this effect could be eliminated. Under these optimal operating conditions the temporal resolution was substantially improved to within the range of 300-400fs and the streak result reproduced in Fig. 3 shows intensity profiles which indicate a temporal resolution around 300fs. [The background noise is due to light transmitted through the photocathode which illuminated the back of the phosphor.]

4. Conclusions

The measured dynamic performance of this femtosecond streak tube compares well with the theoretical prediction of ~250fs for the applied operating voltages. With improved tube designs it is hoped to operate at higher cathode voltages and generate higher electric fields in the vicinity of the photocathode by transient voltage pulsation of the mesh. The significant advantages of travelling-wave deflection are obvious; high streak scan speeds have been achieved with comparatively small ramp voltages without the need to resort to laser pulse amplification and Auston-type switching elements. The single-shot/gating system reported here proved highly convenient giving repeatable streak results with a jitter of less than 50ps. It is our intention to use this camera system to directly study "solitonic" effects [6] in the CPM laser.

Acknowledgements

The support of this collaborative programme by The Royal Society and The Chinese Academy of Sciences is gratefully acknowledged. Acknowledgement is also made for personal support as a Chinese Royal Fellow (Y.L) and for visits (H.N, Q.L.Y) by The Royal Society and The Chinese Academy of Sciences. Two of us (A.F and D.W) express gratitude for CASE studentship support by the Science and Engineering Research Council.

References

1. W. Sibbett, H. Niu and M.R. Baggs: Rev. Sci. Instrum., 53, 758 (1982)
2. Y. Liu, H. Niu, Q.L. Yang and H. Zhang: Proc. 5th National Conf. on High Speed Photography and Photonics, China (1985)
3. M.R. Baggs, R.T. Eagles, W. Margulis, W. Sibbett and W.E. Sleat: Adv.Electron. and Electron Phys., 64B, 617 (1985)
4. A. Finch, W.E. Sleat and W. Sibbett: Rev. Sci. Instrum. (in press)
5. A. Finch, G.F. Chen, W.E. Sleat and W. Sibbett: J. Mod. Opt., 35, 345 (1988)
6. F. Salin, P. Grangier, G. Roger and A. Brun: Phys. Rev. Lett., 56, 1132 (1986)

Part IV

Opto-Electronics and Communications

Advanced Optical Communications Technologies Utilizing Ultrashort Optical Pulses

M. Saruwatari, K. Nakagawa, S. Kawanishi, and A. Takada

NTT Transmission Systems Laboratories, Take 1-2356, Yokosuka-shi, Kanagawa 238, Japan

1 Introduction

Optical fiber communications systems are expected to accommodate the overwhelming information increase that the coming digital information networks should require. Recently, much effort is being applied toward improving lightwave communication technologies that will lead to high-speed transmission, long repeater spacing, and efficient use of lightwave frequencies. However, present state-of-the-art optical technology does not fully utilize the ability of lightwaves as an ultrahigh frequency carrier. There is much room for improvement of information transmission capacity, optical receiver sensitivity, optical signal processing speed, and so on. In this paper, we report several novel techniques applicable to future ultrahigh-speed optical communications that utilize ultrashort optical pulses. They include ultrashort optical pulse generation/compression using laser diodes, all-optical multi/demultiplexing, chirped pulse transmission and waveform detection techniques.

2 Optical Pulse Generation/Compression Technique

To develop high-speed optical communications systems, ultrashort optical pulses with high repetition-rate must be generated. At present, subpicosecond optical pulses have been obtained with collision-pulse mode-locked (CPM) dye lasers [1] or mode-locked solid-state laser systems. Both are employed with the sophisticated pulse compression technique which utilizes the self-phase modulation in single-mode (SM) fibers and the chirping compensation with grating-pairs. However, these systems are not suitable because of their low repetition rate of around 100 MHz and their lack of compactness.

Laser diodes (LD) are most promising as the optical source owing to their fast response, compactness, easy handling, etc. Incidentally, LDs have an advantage in that their oscillation wavelengths coincide with a minimum-loss and zero-dispersion region of conventional optical fibers. To date, ultrashort high repetition rate optical pulses have been generated from LDs by Q-switching, active/passive mode-locking incorporating an external cavity, and gain-switching using short electrical/optical pulse pumping or strong RF modulation. Among them, the gain-switching technique [2-6] is convenient as a high-speed and short optical signal generator because of its simplicity, stability and controllability of repetition rate.

Figure 1 illustrates the experimental setup of a pulse generation system using a gain-switched distributed feedback (DFB) LD. The 1.3µm wavelength DFB-LD is driven by a strong sinusoidal RF signal superimposed on a DC bias current. It is clarified that generated optical pulses exhibit red-shift frequency chirping due to carrier-induced refractive index change associated with intense short pulse generation. Therefore, by merely traveling through SM fibers with normal dispersion at 1.3µm, the red-shift chirping of the original pulses can be compensated for, and the pulses are compressed to the nearly transform-limited

Springer Series in Chemical Physics, Vol. 48: **Ultrafast Phenomena VI**
Editors: T. Yajima · K. Yoshihara · C.B. Harris · S. Shionoya

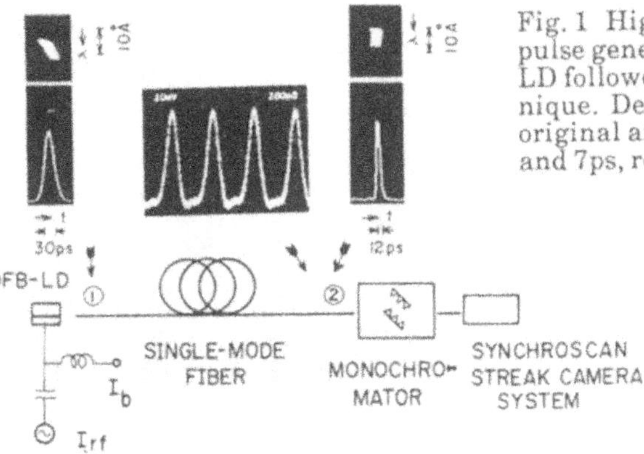

Fig. 1 Highly repetitive short optical pulse generation from a gain-switched LD followed by fiber compression technique. Deconvoluted pulse widths of original and compressed pulses are 26ps and 7ps, respectively.

pulses [5]. Pulse waveforms and time-resolved spectrum characteristics measured with a synchroscan streak camera are also shown in Fig. 1. Left-hand and right-hand photographs correspond to those before and after the chirping compensation using dispersion-shifted SM fibers. The 26 ps original optical pulses have been successfully compressed to 7 ps pulses with a 4.4 GHz repetition rate [5]. In this manner, 12 GHz[6] ultrashort optical pulses at 1.3 or 1.55 µm have been generated.

For the application to optical communications, repetitive pulses must be modulated directly or with an external modulator. Recently, randomly modulated optical signals are also being generated with gain-switched LDs by adopting an appropriate undershoot in the electrical signal. At present, the maximum bit rate is limited to a relatively low value, compared with the attainable pulse duration. This is because not only LDs but also electrical circuits including an LD driver and a signal generator have limited bandwidths. To solve this limitation, another approach such as optical multiplexing techniques are expected to be satisfactory.

3 All-Optical Multi/Demultiplexing Techniques

One way of realizing ultrahigh-bit rate optical communication is to multiplex several lower bit rate signals to one higher rate signal, and to demultiplex the transmitted higher bit rate signal when needed, all-optically.

The experimental setup of a multiplexing circuit is shown in Fig. 2. Optical pulses with 30 ps duration are generated from a gain-switched 1.3 µm DFB-LD at 3.125GHz repetition rate. After being randomly modulated with the external modulator, the 30 ps pulse stream is compressed to 8.6 ps with the SM fiber. The modulated and compressed pulse signals are then led to the 5-stage multiplexer composed of 6 pieces of 2x2 fiber couplers and 5 pieces of fiber delay lines. This multiplexes the original bit rate by 2^5 (=32) times. The photograph in Fig. 2 demonstrates the 32-time multiplexed signals, that is, a 100 Gb/s pulse pattern observed with the streak camera [7]. The mark"1" and space"0" bits are clearly distinguished. By selecting an appropriate delay time, the pseudo-random optical pattern can also be generated with this configuration. This technique is applicable to the evaluation of high speed optical transmission and optical logic circuits.

All-optical time-division demultiplexing of an optical pulse train has also been demonstrated using the Kerr effect in polarization-maintaining SM fibers as shown in Fig. 3. Control pulses and signal pulses are generated from the 1.06 µm mode-locked YAG laser at 82MHz and the 1.3 µm gain-switched DFB-LD at 2GHz. Because only the signal pulses superimposed with the control pulses change their

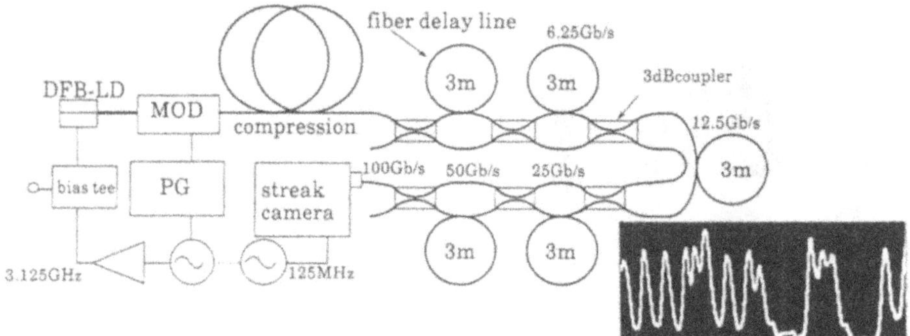

Fig.2 Experimental setup of a multiplexing circuit. 100Gb/s modulated pulses from the last fiber coupler are also shown.

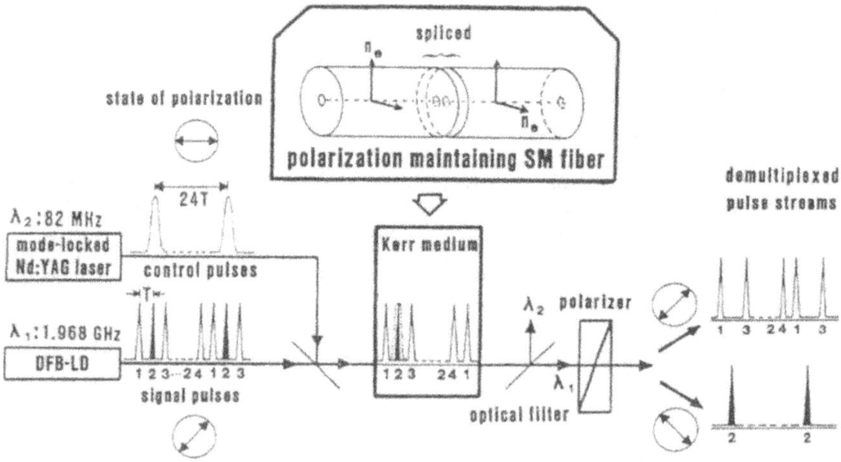

Fig. 3 Ultra-fast all-optical time-domain demultiplexer utilizing optical Kerr effect in polarization-maintaining single-mode fibers.

polarization direction by 90 degrees, the signal pulse stream can be demultiplexed with the polarization splitter into two pulse streams according to the control pulse stream. To stabilize the polarization fluctuations of the signal pulses caused by temperature/pressure-induced birefringence change, two birefringent fibers with the same length are spliced so that their fast axes cross at right angles [8]. This results in overall birefringence compensation without sacrificing the polarization maintaining property of the fibers. With the control pulses of 34 W at peak, the 2-GHz pulse stream is completely demultiplexed for the 20 m spliced fibers [8-9].

The present Kerr-switching duration of about 100ps is determined by the control pulse width of 80 ps and by the walk-off of 100ps between signal and control pulses. The latter is caused by the chromatic/polarization dispersions, and also limits the effective interaction fiber length, which is inversely proportional to the required control power [9]. When much shorter control pulses and no-walk-off conditions are adopted, less than 1 ps operation can be expected with much lower control power. A preliminary experiment using a 1.3 μm YAG laser and 150m fibers has shown that the required peak control power is decreased to 3.4 W [9].

4 Optical Pulse Transmission Through Fibers

Ultrashort optical pulses have very wide spectra inversely proportional to their pulse widths, and relatively high peak intensities. As a result, when they are transmitted through long fibers, they are apt to broaden due to chromatic dispersion. They also may be affected by various nonlinear optical effects in fibers, e.g., self-phase modulation (SPM), Brillouin/Raman scatterings/amplifications. In connection with the nonlinear effects, optical soliton transmission utilizing both SPM and anomalous dispersion in fibers has also been studied for application to long distance transmission [10]. Leaving these nonlinear phenomena to another paper [11], this paper touches on chirped optical pulse transmission through fibers.

Optical pulses generated from LDs exhibit the frequency chirping due to carrier-induced refractive index change as cited previously. It is clarified that the chirped optical signals from directly modulated LDs can be transmitted through long fibers without receiver sensitivity degradation [12-13][6] due to no waveform broadening after transmission. Because the gain-switched LD pulses have nearly linear red-shift chirping, the pulse compression effect due to normal dispersion of fibers can be fully utilized for long distance transmission. Figure 4 shows the relationship between the measured pulse width versus the normal dispersion value (ps/nm) when using the gain-switched DFB-LD and the high-dispersion (-10 ps/nm/km) fibers [6]. This indicates that by using conventional fibers with -1 ps/nm/km dispersion, 20 ps chirped pulses corresponding to 50Gb/s signal can be sent through about 100km distance without waveform broadening.

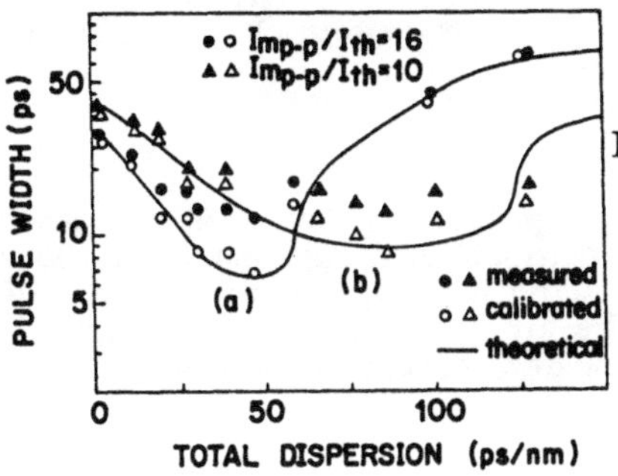

Fig.4 Relationship between the measured pulse width versus the normal dispersion value. Theoretical values are also indicated by solid lines. $I_{m_{p\text{-}p}}$ is the peak-to-peak amplitude of drive current applied to LD.

5 High-Speed Optical Detection Technique

Design and evaluation techniques for wideband optical receivers are important for developing future ultrahigh-speed optical transmission systems. A heterodyne detection method [14-15] using two DFB-LDs [16] as the beat signal source is a powerful technique for measuring the frequency response of wideband optical receivers because this technique has, in principle, no bandwidth limitation [15-16]. This technique aids the optimum design of ultrahigh-speed optical receivers. The developed receiver configuration with a 35 μmφ InGaAs-PIN photodiode (0.07 pF capacitance and 82% quantum efficiency) followed by two stage HEMT amplifiers [17] is shown in Fig. 5. By optimizing the gate bias condition, wideband optical receivers with 15GHz bandwidth (3 dB down) and nearly flat frequency response are realized as shown in Fig. 5. The noise figures of 3.6 dB and 2.8 dB (average value of 2-18GHz) are measured at room temperature and low (158K) temperature operations, respectively. This corresponds to the receiver sensitivity of -18.6 dBm in the cases of 20 Gb/s transmissions at room temperature.

Optical waveform measurements with high resolution are also required for ultrahigh-speed optical systems. An optical sampling technique [18], using

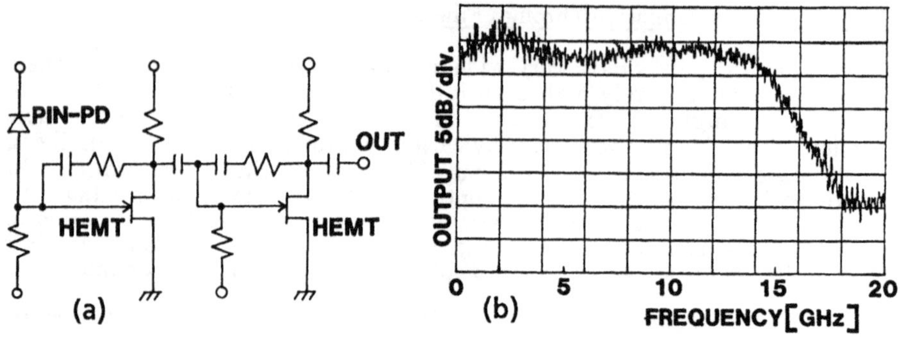

Fig. 5 Frequency response of a PIN-PD + HEMT optical receiver. (a) Circuit configuration, (b) Measured frequency response by optical heterodyne detection.

ultrashort optical pulses as the sampling pulses, offers an attractive method that avoids problems associated with high-speed detectors/electronics. This sampling method utilizes the femtosecond response of optical sum-frequency generation (SFG) in nonlinear optical material. The cross-correlation signal between the measured and the sampling pulses is obtained through the SFG process. By adopting a highly efficient $LiNbO_3$ waveguide, where the SFG light is emitted as the 'Cherenkov radiation' mode, and using compressed gain-switched DFB-LD pulses as the sampling pulses, the sampled optical waveform of 25 Gb/s word pattern is measured at time-resolution less than 10 ps [19].

6 Conclusion

Novel techniques utilizing ultrashort optical pulses are reviewed as future applications of high-speed optical communications. Although some of these techniques are not sufficiently mature to be actually used, they are expected to enhance new conceptual techniques such as all-optical processing, photonic switching, optical computers, and others. These new techniques are indispensable to the coming information society.

References

1. C.H.Brito-Cruz, R.L.Fork, C.V.Shank: CLEO'87, MD-1(1987)
2. H.Ito, H.Yokoyama, S.Murata and H.Inaba: Elect. Lett. 15, 738 (1979)
3. T.Kobayashi, A.Yoshikawa, A.Morimoto, Y.Aoki and T.Sueta: IQEC'80, W-1
4. P.Liu, C.Lin, I.Kaminow and I.Hsieh: IEEE J. Quant. Elect. QE-17, 671 (1981)
5. A.Takada, T.Sugie and M.Saruwatari: Elect. Lett. 22, 1347 (1986)
6. A.Takada, T.Sugie and M.Saruwatari: IEEE J. Light. Tech. LT-5, 1525 (1987)
7. A.Takada and M.Saruwatari: to be submitted
8. M.Saruwatari, T.Morioka and A.Takada: CLEO'87, ThO-2 (1987)
9. T.Morioka and M.Saruwatari: to be published in IEEE J.Select.Areas Commun.
10. L.F.Mollenauer and K.Smith: CLEO'88, TuP-5 (1988)
11. L.F.Mollenauer, J.P.Gordon and M.N.Islam: IEEE J. Quant. Elect. QE-22, 157 (1986)
12. K.Iwashita, K.Nakagawa, R.Nakano and R.Suzuki: Elect. Lett. 18, 873 (1982)
13. C.Lin and A.Tomita: Elect. Lett. 19, 837 (1983)
14. L.Piccari and P.Spano: Elect. Lett. 18, 116 (1982)
15. E.D.Hinkley, T.C.Harman and C.Freed: Appl. Phys. Lett. 13, 49 (1968)
16. S.Kawanishi and M.Saruwatari: Elect. Lett. 22, 337 (1986)
17. S.Kawanishi and M.Saruwatari: to be submitted
18. D.L.Franzen, Y.Yamabayashi and T.Kanada: Elect. Lett. 23, 289 (1987)
19. Y.Yamabayashi, A.Takada and M.Saruwatari: ICWO'88, WD-5 (1988)

Femtosecond All-Optical Switching in Nonlinear Directional Couplers

A.M. Weiner, Y. Silberberg, S.R. Friberg, B.G. Sfez**, and P.W. Smith*

Bell Communications Research, 331 Newman Springs Road,
Red Bank, NJ 07701, USA
*Present address: NTT Basic Research Laboratories, Tokyo, Japan
**Present address: Ecole Nationale Superieure des Télécommunications, Paris, France

All-optical switching devices are of considerable interest for future ultrahigh speed transmission systems because of their potential for subpicosecond response and the prospect of eliminating costly optical-electronic conversions. In the last few years, a number of switching devices have been demonstrated using glass optical fibers, including the birefringent fiber polarization switch [1], the optical fiber Kerr gate [2], and the dual-core fiber nonlinear directional coupler [3,4]. In this paper we report substantially complete switching of 100 fsec optical pulses by a glass fiber nonlinear coupler [5]. To our knowledge this is the fastest switching time ever measured in a guided-wave all-optical device.

Figure 1 shows a schematic diagram of a nonlinear coupler. At low intensity the device behaves as a conventional directional coupler; light introduced into waveguide (1) transfers into waveguide (2) in the coupling length L_c. Higher intensity light induces changes in the refractive index, detunes the coupler, and hence remains primarily in the input guide. The transition from low power to high power behavior occurs at the critical power P_c, at which the power splits evenly between the two waveguides. Theoretical switching characteristics for a nonlinear coupler of length L_c are shown in Fig. 2 for CW input light. Low power signals introduced into guide (1) emerge from guide (2), whereas high power signals ($P \gg P_c$) emerge from guide (1) [4].

For an input signal in the form of a pulse, the pulse breaks up according to its instantaneous intensity [6]. Figure 3 shows calculated temporal profiles of the output pulses, ignoring group velocity dispersion effects, for $\mathrm{sech}^2(t)$ input pulses with peak power $2P_c$. The low intensity wings of the input pulse emerge from guide (2), whereas the intense central portion of the pulse remains in guide (1). In an experiment in which pulse break-up is not temporally resolved, the measured switching curves are the CW curves integrated over the pulse intensity profile. The dashed line in Fig. 2 shows the result for a $\mathrm{sech}^2(t)$ pulse shape.

Our nonlinear coupler consists of a 5-mm length of fused quartz dual-core optical fiber, shown in Fig. 4. The fiber contains two 2.8-μm cores,

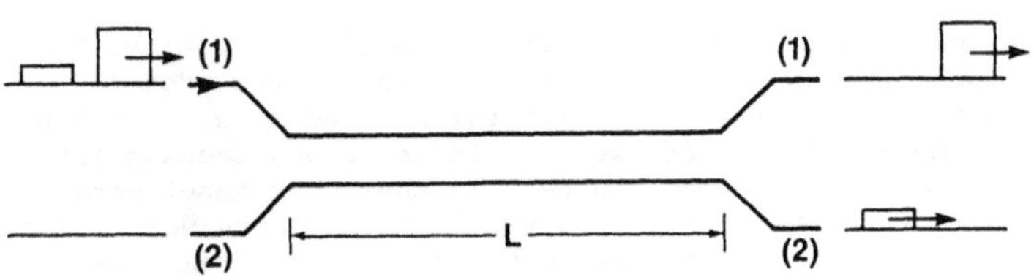

Fig. 1 Schematic of a nonlinear coupler.

Springer Series in Chemical Physics, Vol. 48: **Ultrafast Phenomena VI**
Editors: T. Yajima · K. Yoshihara · C.B. Harris · S. Shionoya

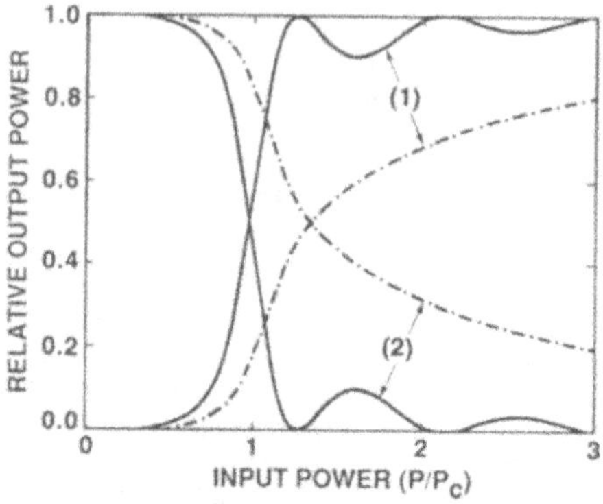

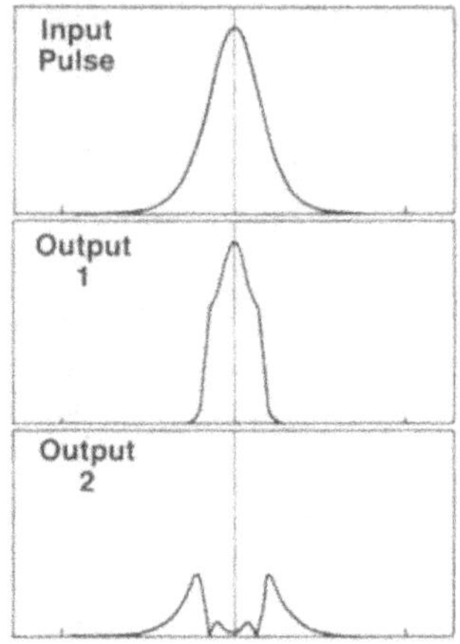

Fig. 2 Relative output power from guides (1) and (2), calculated as a function of input power (normalized to P_c). Solid lines: CW response. Dash-dot lines: Response to $sech^2(t)$ input pulses.

Fig. 3 Temporal profile of the output pulses from guides (1) and (2), computed for $sech^2(t)$ pulses with peak power $2P_c$.

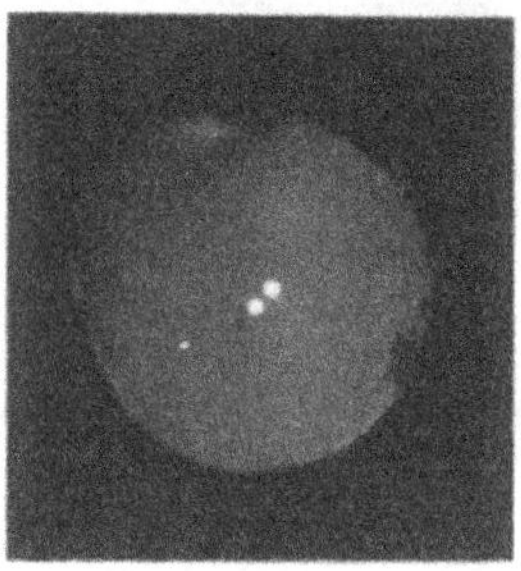

Fig. 4 Photograph of the dual-core fiber nonlinear coupler.

with an index difference of 0.003 and a core separation of 8.4 μm. Each fiber core is single-mode for wavelengths longer than 500 nm. We characterized the coupler by performing a series of measurements in which we focused a low power white light source into a single fiber core and measured the relative output power from each of the two cores as a function of wavelength. From the data we estimate a coupling length of 4.7 mm at 620 nm. We also verified that the coupling length was independent of polarization and that the coupler maintained the polarization state of the input light.

Switching experiments were performed using pulses from a CPM dye laser and a copper vapor laser pumped dye amplifier [7]. The system produced 620 nm, 100 fsec pulses with energies of several hundred nJ at an 8.6 kHz repetition rate. Amplified pulses were attenuated and focused into a single fiber core, which we denote guide (1); the other input core was carefully blocked by a razor blade. The output from each core was focused onto a separate power meter, and the average power emerging from each core was measured as a function of the input power.

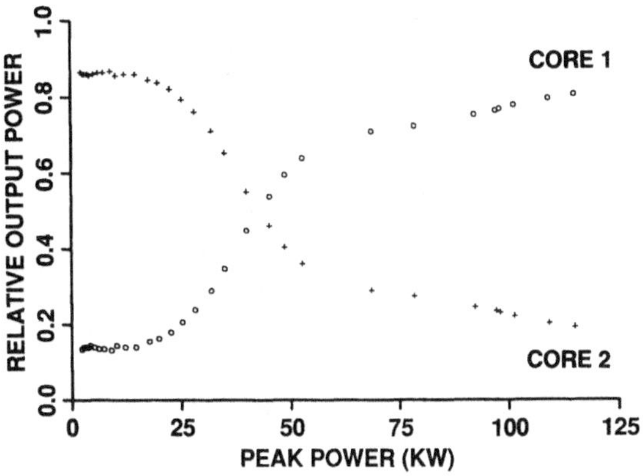

Fig. 5 Relative output power from guides (1) and (2) for the 5 mm, dual-core fiber nonlinear coupler. The data are the response for 100 fsec input pulses.

The fraction of the output power emerging from each of the cores is plotted in Fig. 5 as a function of total power. At low power 86% of the light emerged from core 2; at the highest power 81% emerged from core 1. Similar data were obtained for several pieces of dual core fiber, for various laser pulse widths in the range 100-200 fsec, and for various polarization angles. The response is in good agreement with the dashed line in Fig. 2. From the data we estimate $P_c \simeq 32$ kW, within a factor of two of the 60 kW value obtained from the formula $P_c = A\lambda/n_2L_c$, using the known nonlinear coefficient for silica $n_2 = 3.2 \times 10^{-16}$ cm^2/W and using 15 μm^2 as a rough estimate of the effective mode area.

In order to demonstrate break-up of the input pulses, we performed autocorrelation measurements of the pulses emerging from the two fiber cores. Figure 6 shows autocorrelation traces of the pulses emerging from the two cores, for 200 fsec input pulses at a peak power of $\simeq 2P_c$. Although the trace corresponding to guide 1 is similar to that of the input pulse, the trace corresponding to core 2 is triply peaked, indicating a pulse doublet emerging from this core. Each of the three autocorrelation peaks is significantly narrower than the input autocorrelation. These data confirm our picture of pulse fragmentation and demonstrate femtosecond switch-on as well as switch-off times for our nonlinear coupler.

In addition to the nonlinear reshaping evident in Fig. 6, output pulses from the nonlinear coupler are temporally and spectrally broadened by group velocity dispersion (GVD) and self-phase-modulation (SPM). Figure 7 shows spectra of pulses emerging from guide (1) for various power levels. At low powers the output spectra from both guides are identical to the input spectrum, with a 10 nm width. The spectrum from guide (1) broadens to $\simeq 20$ nm at P_c and to $\simeq 40$ nm at $2P_c$. These spectrally broadened pulses should be especially suitable for high quality pulse compression since the low intensity wings are suppressed by the nonlinear reshaping mechanisms. The spectra corresponding to guide (2) (not shown) broaden to a similar extent but develop features more complicated than those seen in Fig. 7. Spectral broadening occurs for any fiber switching device based on nonlinear phase shifting and may necessitate pulse regenerators in systems with cascaded fiber switching elements.

The speed of the nonlinear coupler switch is limited by GVD. The temporal broadening of pulses emerging from guide (1) can easily be estimated if we neglect nonlinear reshaping and refer to the literature on pulse propagation under the combined influence of SPM and GVD. We start by

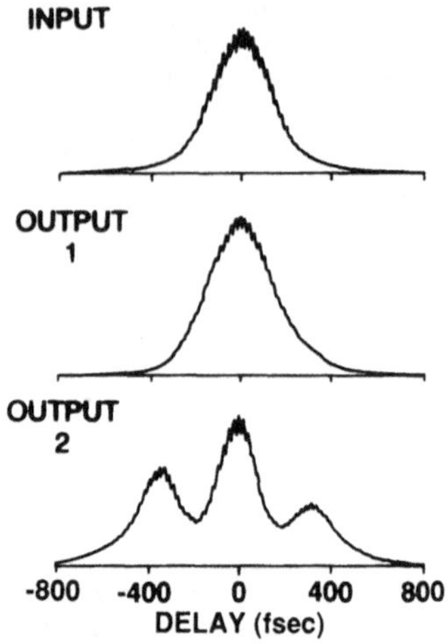

Fig. 6 Autocorrelation traces of pulses emerging from the coupler, for 200 fsec input pulses at a peak power of 2.2 P_c.

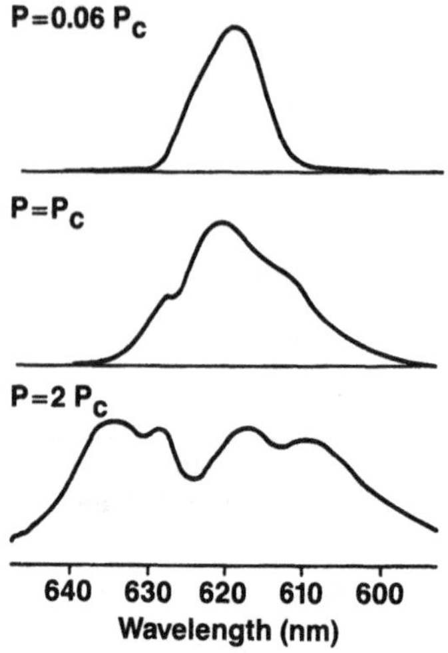

Fig. 7 Spectra of pulses emerging from guide (1), for 100 fsec input pulses.

expressing the experimental parameters in dimensionless 'soliton' units [8]. The normalized field amplitude A is then given by

$$A = 2\ [(z_o/L_c)(P/P_c)]^{1/2} \quad ,$$

where z_o is the dimensionless soliton period. The amount of broadening increases with the peak input power. For $P = P_c$, $\lambda = 620$ nm, and 100 fsec input pulses, we obtain $A = 8.5$, $L_c/z_o = 0.055$, and approximately 60% temporal broadening. For $P = P_c$ and 200 fsec input pulses, we predict $A = 17$, $L_c/z_o = 0.014$, and 25% temporal broadening; this estimate is consistent with the autocorrelation data for 200 fsec input pulses since the expected broadening due to GVD closely compensates the anticipated temporal narrowing due to pulse fragmentation. At a wavelenth of 1.3 μm, the GVD is zero; and other factors, such as the wavelength dependence of the coupling length, will limit the maximum speed to several tens of femtoseconds.

Further work will be aimed at reducing the switching power and minimizing incomplete energy exchange due to pulse fragmentation. The critical power could be reduced by fabricating longer couplers from more highly nonlinear material. For example, a 50 cm device made from SF-59 glass would switch at $\approx$ 10 W. Pulse fragmentation could be circumvented by performing experiments with ultrashort square pulses. Square pulses have been generated by masking of spatially dispersed frequency components within a temporally nondispersive grating apparatus [9]; Fig. 8 shows a representative cross-correlation measurement of a 1 psec square pulse with a 100 fsec rise time. Measurements performed using such ultrashort square pulses would permit demonstration of very sharp switching transitions predicted for devices several L_c in length [4].

In summary, we have demonstrated all-optical switching of 100 fsec pulses by a dual core fiber nonlinear coupler. This is the fastest switching time measured for any guided-wave switching device.

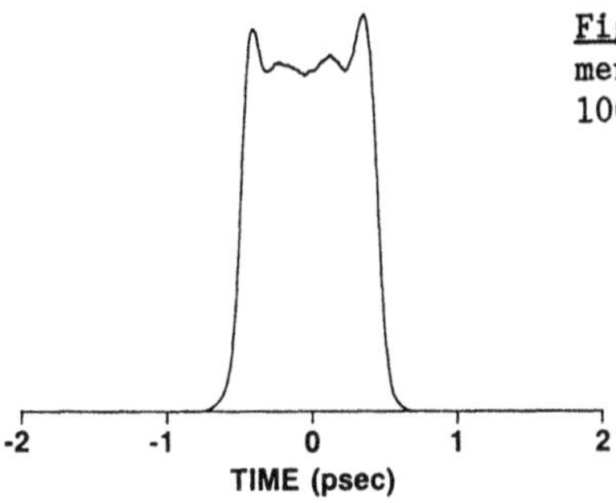

Fig. 8 Cross-correlation measurement of a 1-psec square pulse with a 100-fsec rise time [9].

We gratefully acknowledge D.E. Leaird for expert technical assistance and E. Snitzer of Polaroid for providing the dual core fiber.

References

1. S. Trillo, S. Wabnitz, R.H. Stolen, G. Assanto, C.T. Seaton, and G.I. Stegeman, Appl. Phys. Lett. 49, 1224 (1986).
2. K. Kitayama, Y. Kimura, and S. Seikai, Appl. Phys. Lett. 46, 317 (1985); N.J. Halas, D. Krokel, and D. Grischkowsky, Appl. Phys. Lett. 50, 886 (1987).
3. D.D. Gusovskii, E.M. Dianov, A.A. Maier, V.B. Neustruev, E.I. Shklovskii, and I.A. Shcherbakov, Sov. J. Quantum Electron. 15, 1523 (1985); S.R. Friberg, Y. Silberberg, M.K. Oliver, M.J. Andrejco, M.A. Saifi, and P.W. Smith, Appl. Phys. Lett. 52, 1135 (1987).
4. S.M. Jensen, IEEE J. Quantum Electron. QE-18, 1580 (1982).
5. S.R. Friberg, A.M. Weiner, Y. Silberberg, B.G. Sfez, and P.W. Smith, Opt. Lett. 13 (1988), in press.
6. K. Kitayama and S. Wang, Appl. Phys. Lett. 43, 17 (1983).
7. W.H. Knox, M.C. Downer, R.L. Fork, and C.V. Shank, Opt. Lett. 9, 552 (1984).
8. W.J. Tomlinson, R.H. Stolen, and C.V. Shank, J. Opt. Soc. Am. B1, 139 (1984).
9. A.M. Weiner, J.P. Heritage and R.N. Thurston, Opt. Lett. 11, 153 (1986); A.M. Weiner, J.P. Heritage and E.M. Kirschner, J. Opt. Soc. Am. B5, (1988), in press.

Picosecond Switching of Surface-Emitting Laser Diodes

K. Kojima and K. Kyuma

Central Research Laboratory, Mitsubishi Electric Corporation,
1-1 Tsukaguchi-honmachi 8-chome, Amagasaki, Hyogo 661, Japan

1. Introduction

There has been increasing interest in optical bistable devices for the applications of optical computing and optical communication. There are various kinds of optical bistable devices, such as bistable laser diodes with saturable absorbers[1-3], laser diode amplifiers[4], optical thyristors[5], Self Electro-optic Effect Devices (SEED's)[6], and nonlinear etalons[7]. Among these devices, bistable laser diodes with saturable absorbers have the advantages of high output power, high on/off ratio, relatively low switching energy, and insensitivity to input light wavelength. However, they are not applicable to two-dimensional arrays in principle, since they use cleaved facets. Moreover, their switching speed is limited to about 200ps, and further improvement is desired.

We have recently reported surface-emitting multiquantum well (MQW) distributed Bragg reflector (DBR) lasers using second-order gratings[8]. They have the advantages of single longitudinal mode operation and very narrow beam divergence (less than 0.2°). When the lasing wavelength was detuned to shorter wavelength, a bistable operation was also obtained due to the increased absorption in the unpumped DBR section[9]. Very fast switching operation was observed due to the increased differential gain (the gain coefficient differentiated by the carrier density) of the MQW structure. In the first half of this paper, we describe the bistable operation and fast switching characteristics of the surface-emitting lasers.

Beam switching or deflecting devices are also expected in various applications such as optical computing. Since grating-coupled surface-emitting lasers have narrow beam divergence, high performance beam control operation is expected by changing the lasing wavelength. In the latter half of this paper, we describe the beam switching operation of surface-emitting MQW-DBR lasers due to the lasing mode control.

2. Bistable switching operation

Figure 1 shows the schematic of the device structure. A MQW structure was used both in the gain section and the DBR section. Since the grating is second-order, the DBR section also acts as a grating-coupler as well as a saturable absorber. To increase the absorption, the lasing wavelength was detuned so that it is 8nm shorter than the gain peak. The length of the gain section and the DBR section are 150μm and 300μm, respectively. Figure 2 shows the current vs power characteristic. The threshold current was 38mA and the output power was 2mW. The on/off ratio was greater than 50:1.

The turn-on characteristic with a current trigger measured by a streak camera is shown in Fig.3. The rise time was measured to be 12ps when the current height was eight times the threshold current, although the rise time was almost independent of the applied current height. The delay time

Springer Series in Chemical Physics, Vol. 48: **Ultrafast Phenomena VI**
Editors: T. Yajima · K. Yoshihara · C.B. Harris · S. Shionoya

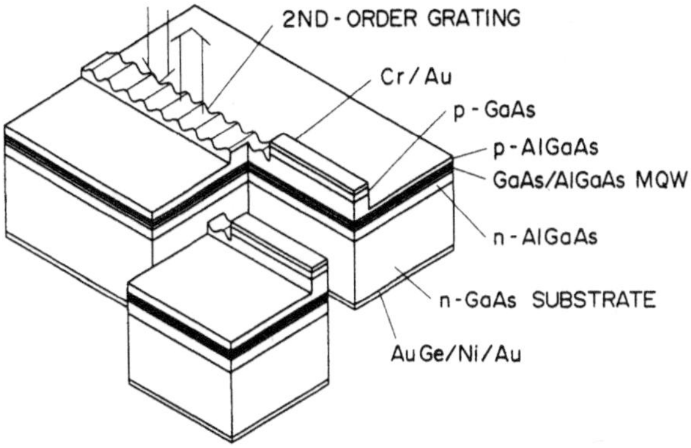

Fig.1 Schematic of the surface-emitting MQW-DBR laser

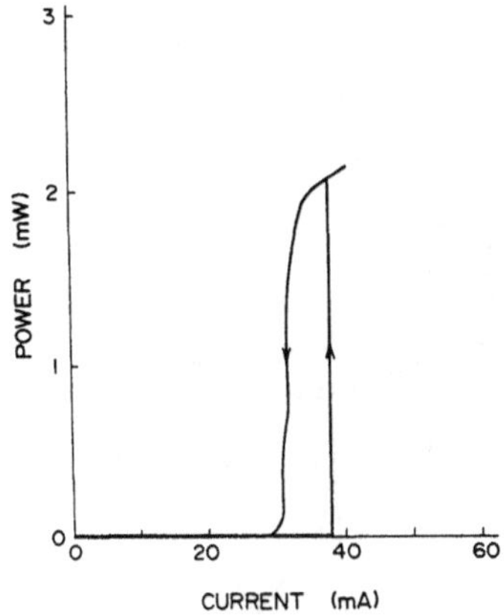

Fig.2 Current vs power characteristic

after the current pulse was applied was about one nanosecond when no bias current was applied, and it was reduced to less than 100ps with increasing the bias current. The fall time was measured to be 60ps when the current was switched-off from eight times the threshold.

The measured fast response characteristics can be explained as follows. The delay time mainly depends on the speed of the bleaching of the saturable absorber, so increasing the bias current, pulse current height and the differential gain is important. On the other hand, the rise time is determined by the increase of the population inversion by the saturable absorber, and it can be reduced by increasing the hysteresis width as well as increasing the differential gain.

One great advantage of this device is that it can be triggered by optical pulses injected from the surface, since there is no electrode above the saturable absorber. The turn-on characteristics were measured by biasing the bistable laser slightly below threshold and injecting light pulses into the saturable absorber with a cylindrical lens attached to the device.

A laser diode (wavelength:830nm) was first used as a trigger light source. The minimum power and energy required for switching were 1mW and

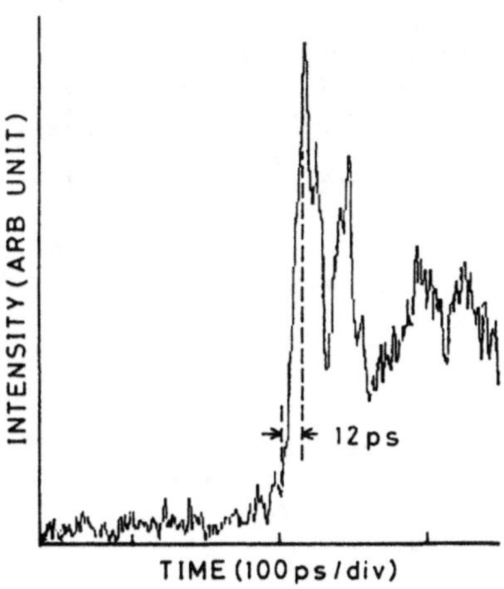

Fig.3 Turn-on characteristic with a current trigger

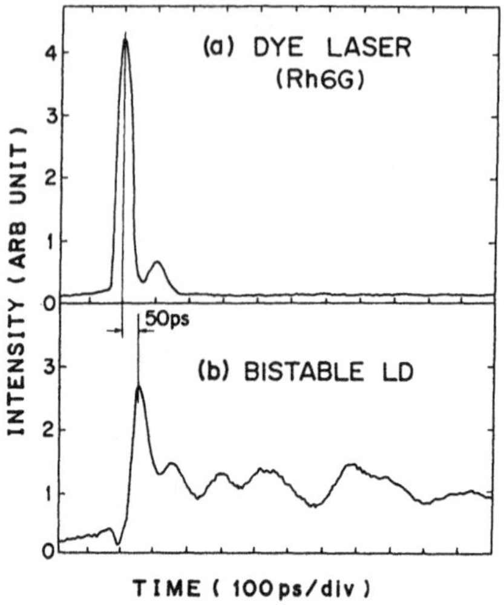

Fig.4 Response of the bistable laser with a dye laser pulse

5pJ, respectively. The turn-on delay time and the rise time were 110ps and 250ps, respectively, when the input power was 20mW.

To measure the ultimate switching speed with short pulses, a dye laser with R6G was used as a trigger source. The lasing wavelength was 610nm and the pulse width was 800fs, however, the wavelength difference from the laser diode is not significant. Figure 4 shows the input and the output light when the input pulse energy was 2nJ. The total switching time (delay time + rise time) was measured to be 50ps. The switching time was almost saturated in this input energy range, and it increased to about 100ps when the input pulse energy was 100pJ.

In bistable lasers with saturable absorbers, the lasing usually occurs about 10nm longer than the gain peak due to the lower absorption coefficient of the saturable absorber at longer wavelength. However, the differential gain is much smaller at this wavelength. In the current device, the lasing wavelength was detuned to 8nm shorter than the gain peak by so designing the grating period. By a theoretical calculation the differential gain of the detuned MQW structure is more than five times larger than that without wavelength selection mechanism (Fabry-Perot bistable lasers).

Dynamic characteristics were also numerically calculated using rate equations[3,10]. In the calculation, we assumed that an optical pulse with an infinitesimal width bleaches the saturable absorber with a 100% efficiency. The result shows that the switching times of 40ps and 20ps are expected with the input optical pulse energy of 10pJ and 20pJ, respectively, with an undoped MQW structure. Compared with the experimental results, if we can increase the coupling efficiency of the input light into the saturable absorber, higher switching speed with less optical energy will be realized. The calculation also shows that the switching speed becomes about twice when a p-doped MQW structure with an acceptor concentration of $1X10^{19}$ cm^{-3} is used as an active layer, since the differential gain is further increased.

3. Beam switching operation

A beam switching operation was also observed in a surface-emitting MQW-DBR laser with an additional control section. The device structure is shown in Fig.5. The length of the gain section, DBR section, and the control section are 270μm, 300μm, and 300μm, respectively.

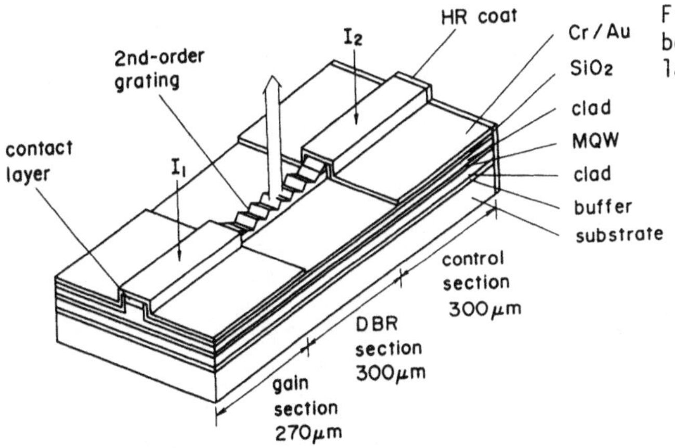

Fig.5 Schematic of the beam-switching MQW-DBR laser

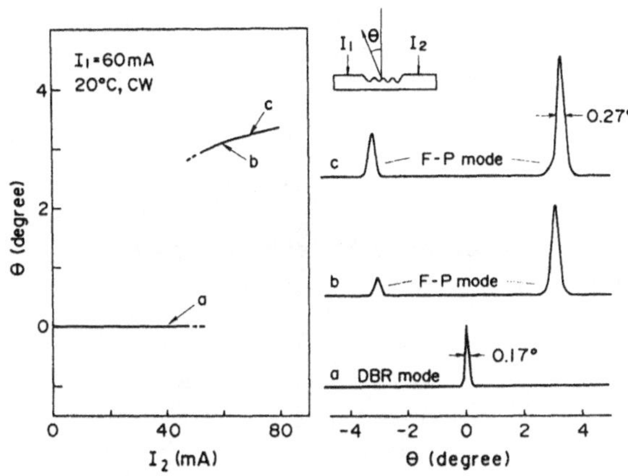

Fig.6 Beam emission angle vs control current I_2 and far-field patterns

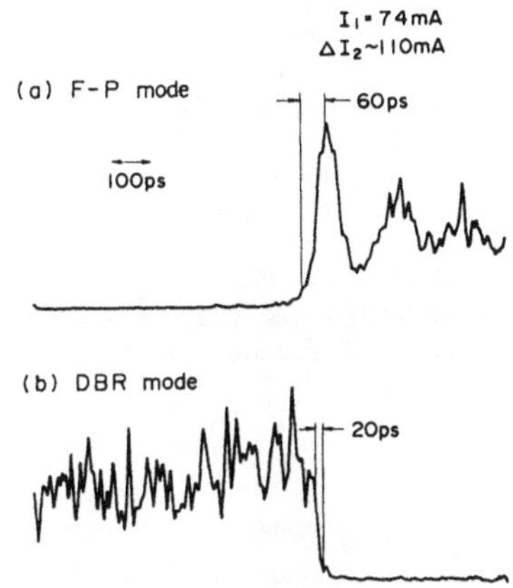

Fig.7 Time-resolved intensity of the F-P and the DBR modes

The operation principle is as follows. When only the gain section is pumped by an injection current of I_1, the laser oscillates in a DBR mode and the output beam is normal to the surface. With increasing the control section current I_2, the light reflected from the control section facet becomes larger, and the laser oscillates in a Fabry-Perot (F-P) mode. Figure 6 shows the beam emission angle vs control section current I_2. The beam was deflected by 3°by the wavelength difference of 13nm.

The time-resolved intensity of each beam is shown in Fig.7 when a current pulse was applied to the gain section with no bias current. The F-P mode beam intensity increases almost corresponding to the carrier density increase in the control section, while the DBR mode beam intensity decreases since the carriers in the gain section are consumed for the stimulated emission of the F-P mode. The F-P mode beam switched-on in 60ps, while the DBR mode beam switched-off in 20ps. By shortening the cavity length, higher switching speed will be possible.

4. Conclusion

The bistable operation and beam switching operation were measured in surface-emitting MQW-DBR lasers. The bistable operation can be controlled by an external trigger light injected from the surface, and extension to two-dimensional arrays will be possible. The optical switching time was 50ps. The beam switching operation was realized by using the lasing wavelength change with changing the control current. The beam deflection angle was 3°, and the switching time was 60ps. These devices are expected to play an important role in future optical computing systems, and further research work is expected.

References

1. G.J. Lasher: Solid-State Electron. 7,707 (1964)
2. H. Kawaguchi and G. Iwane: Electron. Lett. 17, 167 (1981)
3. H.-F. Liu, Y. Hashimoto, and T. Kamiya: IEEE J. Quantum Electron. QE-24, 43 (1988)
4. W.F. Sharfin and M. Dagenais: Appl. Phys. Lett. 48, 321(1986)
5. Y. Tashiro, K. Kasahara, N. Hamao, M. Sugimoto, and T. Yanase: Jpn. J. Appl. Phys. 26, L1014 (1987)
6. N. Peyghambarian and H.M. Gibbs: Optical Engineering 24, 68 (1985)
7. D.A.B. Miller, D.S. Chemla, T.C. Damen, T.H. Wood, C.A. Burrus, Jr., A.C. Gossard, and W. Wiegmann: IEEE J. Quantum Electron. QE-21, 1462 (1985)
8. K. Kojima, S. Noda, K. Mitsunaga, K. Kyuma, and K. Hamanaka, Appl. Phys. Lett. 50, 1705 (1988)
9. K. Kojima, K. Kyuma, S. Noda, J. Ohta, and K. Hamanaka: Appl. Phys. Lett. 52, 942 (1988)
10. H. Shoji, Y. Arakawa, and Y. Fujii: Proceeding of Tech. Group Meet. IECE of Japan, paper OQE87-54 (1987)

Ultrafast Optical Switching Based on Stimulated Emission in GaAs/AlGaAs Multiple Quantum Wells

J.L. Oudar [1], *C. Tanguy* [1,2], *J.P. Chambaret* [2], and *D. Hulin* [2]

[1]Centre National d'Etudes des Télécommunications, 196, Avenue Henri Ravera, F-92220 Bagneux, France

[2]Laboratoire d'Optique Appliquée, Ecole Polytechnique-ENSTA, F-91120 Palaiseau, France

All-optical gates are intrinsically very fast when they involve only virtual light-matter interactions [1]. However this usually requires very high (GW/cm²) intensities, which restricts their use to relatively modest repetition rates, due to the unavailability of adequate lasers. On the other hand optical gates and bistable devices based on absorption saturation or carrier-induced nonlinear refraction in semiconductors have much lower intensity requirements. As is well known, the switch-off time of such devices is limited by the carrier lifetime, which led several groups to deliberately increase the nonradiative recombination rate in their devices. The drawback of this approach is that a lot of heat is generated through these nonradiative transitions, which puts another restriction on the useful repetition rate of these optical gates. It would be interesting to develop a practical means to drain the stored energy very quickly out of the device material. Stimulated emission induced by ultrashort light pulses seems to be an adequate answer to this problem, provided that it can be achieved efficiently with reasonable pulse energies.

Very recently it has been found that low temperature Multiple Quantum Well (MQW) samples could exhibit a surprisingly fast recovery of absorption (within 10 ps at 15 K) after excitation by intense ultrashort light pulses [2]. This has been attributed to the effect of the high stimulated recombination rate caused by amplified spontaneous luminescence guided along the MQW structure, as confirmed by streak camera observations of the edge-emitted luminescence [2], and measurements of time-resolved luminescence spectra with subpicosecond resolution [3].

Our experimental results on the spot size, intensity, and temperature dependences of this effect point to a strong enhancement of stimulated recombination at low (intermediate) temperatures. At a given temperature in the range 100-160 K, a faster recovery is observed for larger spot diameters but similar power densities (about 30 μJ/cm²). This is expected since the amplification of guided luminescence increases rapidly with the spot diameter d. The temperature variation of the intrinsic gain coefficient, estimated from the recovery time data, is particularly strong around 140 K. This is attributed to the excitonic effects which cause a large gain enhancement due to stronger electron-hole (e-h) correlations at low temperatures [4,5].

As a result it is relatively easy to achieve a substantial gain in such samples, even at modest e-h pair density. Such a large gain, which leads to lifetime self-shortening in the case of a large excited region, can be maintained for a longer time if the excited region has smaller dimensions. Thus it is possible to control the carrier lifetime by an additional beam at an

Springer Series in Chemical Physics, Vol. 48: **Ultrafast Phenomena VI**
Editors: T. Yajima · K. Yoshihara · C.B. Harris · S. Shionoya

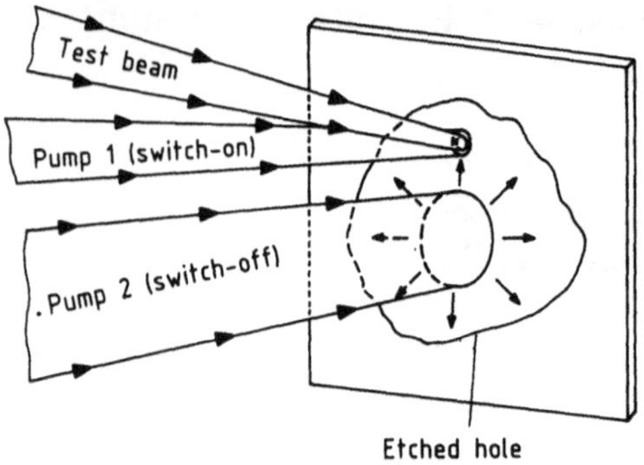

Fig. 1: Schematics of the three-pulse experiment

appropriate wavelength within the gain spectral region of the material. In order to demonstrate such an external control of the absorption recovery, a three-pulse experiment was performed, as schematized on Fig.1. This experiment involves essentially two control pulses, called pump 1 for switch-on and pump 2 for switch-off respectively, while a third test pulse is used to read the transmission state of the MQW optical gate. These pulses are spectrally selected portions of a white light continuum generated by a high intensity, 100 fs pulse at 620 nm.

The sample consists of 60 GaAs wells of 75Å width separated by $Al_{.3}Ga_{.7}As$ barriers of 85Å, sandwiched between two $Al_{.3}Ga_{.7}As$ layers of 1 μm thickness. The switch-on pulse was focused on a spot size of approximately 100 μm diameter, and its energy was in the 10 nJ range in the wavelength interval 715-810 nm. The sample temperature was adjusted to 210 K to avoid the lifetime self-shortening observed at lower temperatures. A substantial amount of absorption saturation then persists in the interband transition region for a time duration longer than 100 ps. The switch-off light is the guided amplified luminescence produced by the pump 2 pulse. This pulse is of similar power density as pump 1, but is spatially separated from it, and extends over a larger diameter (about 1 mm) and spectral width (645-810 nm). A strong stimulated emission results from the relatively large size of photoexcited region 2, and the guided amplified luminescence can reach the photoexcited region 1, producing a sudden decrease of carrier density. The experimental data of Fig. 2 show that recombination in region 1 can effectively be controlled by the stimulated light produced in region 2. The switching contrast achieved here is about a factor of 2. It could be increased by reducing the size of region 1, which would maintain a higher stabilized carrier density before switch-off. The delay of about 15 ps between pump 2 and the actual switch-off is partly due to the dynamics of stimulated emission in region 2, and partly to the light propagation time along the MQW. Finally we note that a careful waveguiding design should significantly reduce the power requirements.

In summary we have demonstrated that the recovery time of optical gates based on absorption saturation can be speeded up through the use of stimulated emission. The large gain achieved in multiple quantum wells at moderate excitation densities allows a controlled enhancement of the carrier recombination rate and provides a practical means to quickly remove the stored energy out of the device material. This principle can be extended to controlling the switch-off of other optical gates and bistable (nonlinear Fabry-Pérot) devices based upon carrier-induced nonlinear refraction.

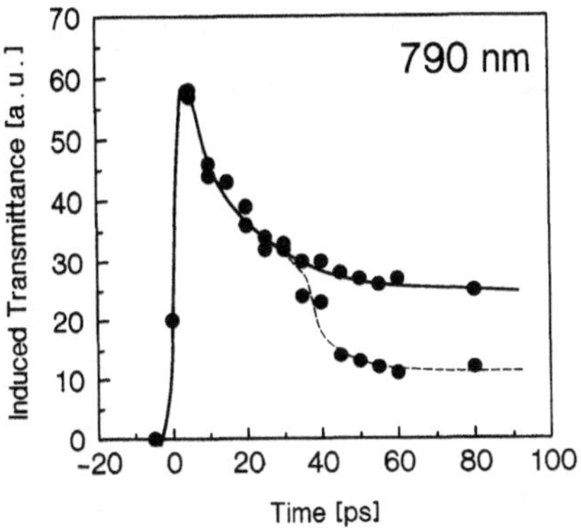

Fig. 2: External control of absorption saturation in the optical gate. Full curve: pump 1 only. Dashed curve: pump 2 arrives 20 ps after pump 1. The test pulse, at 790 nm, is in the absorption region just above the MQW light-hole exciton energy. The sample temperature is 210 K.

References

1. D. Hulin, A. Mysyrovicz, A. Antonetti, A. Migus, W.T. Masselink, H. Morkoc H.M. Gibbs and N. Peyghambarian, Appl. Phys. Lett. 49, 749 (1986).
2. J. Dubard, J.L. Oudar, F. Alexandre, D. Hulin and A. Orszag, Appl. Phys. Lett. 50, 821 (1987); ibid. 50, 1696 (1987).
3. D. Hulin, M. Joffre, A. Migus, J.L. Oudar, J. Dubard and F. Alexandre, J. de Phys. (Paris), C5, 267 (1987).
4. J. L. Oudar and J. A Levenson, XVIth Int. Quantum Electron. Conf., paper ThB7 (July 88).
5. S. Schmitt-Rink, C. Ell and H. Haug, Phys. Rev. B33, 1183 (1986).

Picosecond Characterization of InGaAs/InAlAs Resonant Tunneling Barrier Diode by Electro-Optic Sampling

A. Tackeuchi, T. Inata, S. Muto, and E. Miyauchi

Fujitsu Laboratories, Ltd., 10-1 Morinosato-Wakamiya,
Atsugi, Kanagawa 243-01, Japan

Recently, it has been shown that an InGaAs/InAlAs resonant tunneling barrier (RTB) diode lattice matched to an InP substrate has better static characteristics [1] than conventional GaAs/AlGaAs RTB diodes. This is essentially because the InAlAs barrier layer is a direct-gap semiconductor and has a lower electron effective mass than the AlGaAs barrier layer. This paper describes the first dynamic characterization to our knowledge of the InGaAs/InAlAs RTB diode using the electro-optic sampling (EOS) technique [2].

When a load resistance is placed in series with a RTB diode, the circuit has two stable conditions. This bistability lends the RTB to memory device applications. The important question is how fast the RTB can switch from one stable point to the other. We studied the switching time using EOS. EOS was already applied to a GaAs/AlAs RTB diode by G. Mourou et al.[3], who showed that a RTB diode can respond in the picosecond region, close to EOS time resolution. Here, we used RTB structures with switching times that are supposed to be much longer than the time resolution to enable signal analysis. The simple equivalent circuit model analysis was performed to separate the intrinsic response time of the RTB diode from the observed signal.

The electro-optic sampling head consisted of a photoconductive switch, the InGaAs/InAlAs RTB diode, and 20 micron coplanar transmission lines. These were monolithically formed on a semi-insulating (100) InP substrate. A $LiTaO_3$ crystal, coated with a highly reflective dielectric mirror, is pasted on the coplanar lines to detect electric field changes near the coplanar lines. Optical pulses of 150 fs duration, generated by a colliding pulse mode-locked laser, were used to excite the photoconductive switch and detect the electrical signal.

Two different RTB structures were studied. One with two 5.3 nm InAlAs barriers and a 4.4 nm InGaAs quantum well (7 X 7 μm^2), and the other with two 6.5 nm barriers and a 4.4 nm quantum well (10 X 10 μm^2). Fig.1 shows the I-V characteristics of the RTB diode with the 5.3 nm InAlAs barrier layers. During these measurements, the DC bias was kept just outside the negative differential resistance (NDR) region. When a step-like electrical pulse (typical rise time: 1.9 ps) produced by the photoconductive switch was applied to the RTB diode, the diode was switched into the middle of the NDR region. Fig.2 shows the electro-optic sampling signal corresponding to the current drop

Springer Series in Chemical Physics, Vol. 48: **Ultrafast Phenomena VI**
Editors: T. Yajima · K. Yoshihara · C.B. Harris · S. Shionoya

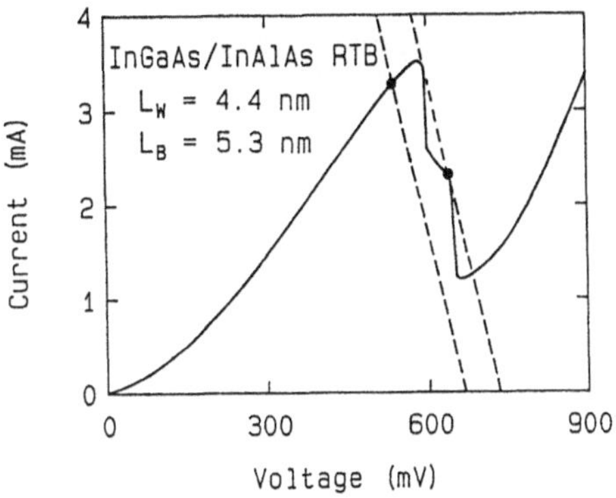

Fig.1 I-V characteristics of the RTB diode

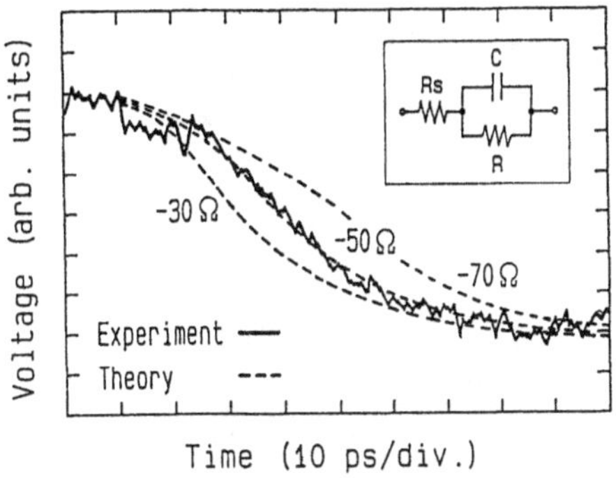

Fig.2 Response signal of the RTB diode with 5.3 nm barriers

through the RTB diode during switching. A current drop with a fall time of 50 ps was observed. A fall time of 210 ps was observed for the RTB diode with the 6.5 nm barriers.

The observed switching time is strongly affected by a load resistance. To separate the extrinsic effect due to the load resistance from the observed switching time, we simulated the switching behavior using the simplest equivalent circuit model of the RTB diode. Fig.2 includes the RTB diode model, which consists of the negative differential resistance, R, in parallel with a capacitance, C, and a series resistance, Rs. For this model, the observed switching time, τ_{obs}, is approximated by

$$1/\tau_{obs} = 1/|RC| - 1/(2Z+Rs)C. \qquad (1)$$

Here, the intrinsic response time, |RC|, is the minimum response time realized in the high-impedance load limit. The series resistance, Rs=15.5 ohms, was obtained from I-V characteristics of the control diode, which has the same structure as the RTB diode except that RTB section is replaced by an n-InGaAs layer. The capacitance, C=0.2 pF (7 X 7 μm^2), and C=0.4 pF (10 X 10 μm^2) was calculated in consideration of the charge distribution in and around the RTB. The characteristic impedance, Z, of the coplanar lines was designed to be 46 ohms. The I-V characteristics of the RTB diode were simplified as the middle of the NDR region was infinite. The negative resistance, R, between the current peak and the middle of the NDR region was used as a fitting parameter. The dashed curves in Fig.2 were obtained with negative differential resistance, R, of -50 ohms for the RTB diode with the 5.3 nm barriers. A similar analysis applied to the diode with the 6.5 nm barriers gave a negative differential resistance of -170 ohms.

The intrinsic response time, |RC|, obtained in this way for the 5.3 nm barrier diode and the 6.5 nm barrier diode were 10 ps and 68 ps, respectively. Note that, for both cases, the intrinsic response times are considerably shorter than the observed response times of 50 ps and 210 ps. Fig.3 shows the

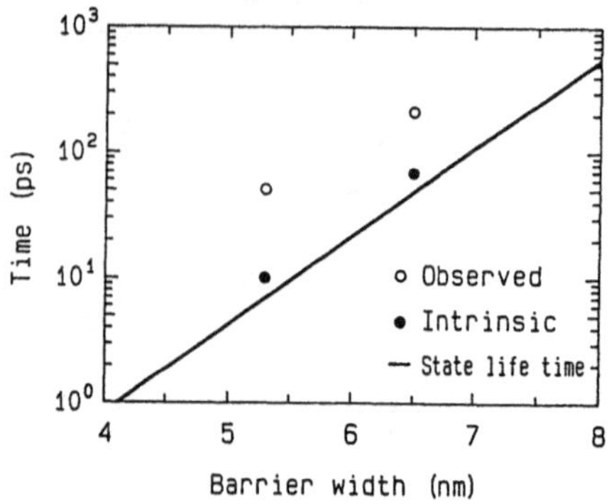

Fig.3 Response times of the RTB diodes

relation between the response time and the state life time, τ(= $h/\Delta E_{FWHM}$, where ΔE_{FWHM} is the full energy width at half maximum of tunneling probability). The intrinsic response times are close to the calculated state life times of 7 ps and 43 ps. Generally, the intrinsic response time seems to be close to the state life time for the symmetric RTB diode. Therefore we can assume that the intrinsic response time of the RTB diode can be reduced to less than 1 ps with a reduced barrier thickness of 4 nm.

In summary, the switching of InGaAs/InAlAs RTB diodes was observed by electro-optic sampling for the first time to our knowledge. The intrinsic response time of the RTB diode was obtained by using a simple equivalent circuit analysis. The intrinsic response time is found to be approximately equal to the state life time.

We thank M. Yamashita of the Electrotechnical Laboratory for advice on colliding pulse mode-locked lasers. We also thank S. Hiyamizu (now at Osaka University), T. Yamaoka, O. Ohtsuki, and K. Dazai for their encouragement and S. Sasa, Y. Sugiyama, Y. Nakata, T. Fujii, T. Mori, H. Ohnishi, and T. Futatsugi for their helpful advice.

References

1. Y. Sugiyama, T. Inata, S. Muto, Y. Nakata, and S. Hiyamizu: Appl. Phys. Lett. 52, 314 (1988)
2. J.A. Valdmanis, G.A. Mourou, and C.W. Gabel: Appl. Phys. Lett. 41, 211 (1982)
3. G. Mourou, K. Meyer, J. Whitaker, M. Pessot, R. Grongin, and C. Caruso: Proc. 2nd OSA-IEEE (LEOS), Incline Village, Nevada, 1987 (Springer-Verlag, Berlin, 1987) p. 40

Tunneling-Time Measurements of a Resonant Tunneling Diode

J.F. Whitaker [1], T.B. Norris [1], G. Mourou [1], T.C.L.G. Sollner [2], W.D. Goodhue [2], X.J. Song [3], and L.F. Eastman [3]

[1]Laboratory for Laser Energetics, University of Rochester, 250 East River Road, Rochester, NY 14623, USA
[2]Lincoln Laboratory, Massachusetts Institute of Technology, Lexington, MA 02173, USA
[3]Department of Electrical Engineering, Cornell University, Ithaca, NY 14853, USA

Since the introduction by TSU and ESAKI [1] of the notion of tunneling in a finite superlattice, the study of transport in single- and double-barrier heterostructures has raised considerable interest. We have used two different methods to make measurements on tunneling devices in order to demonstrate bistable operation and to provide insight concerning tunneling mechanisms. Using the electro-optic sampling technique, a switching time of 1.9 ps has been resolved for a resonant tunneling diode (RTD) operating in its bistable mode. In addition, time-resolved photoluminescence (PL) has been used to determine the field dependance of the tunneling rate of several single-barrier heterostructures.

The RTD used in the switching time measurements was grown by molecular beam epitaxy to have two barriers of 1.5-nm-thick AlAs separated by a GaAs well 4.5-nm thick. The outer regions of GaAs contained Si doping to an electron density 2×10^{17} cm^{-3}. This resulted in a peak current density of about 4×10^4 A/cm^2. The RTD was etched to a 4-μm diameter mesa on an n^+-GaAs substrate, and the capacitance near the current peak was ~20 fF. The series resistance was 15 Ω [2].

A schematic diagram of the circuit used to test the time response of the RTD is shown in Fig. 1. Coplanar electrodes of gold were patterned on a GaAs photoconductive switch and lithium tantalate sampling crystal. The RTD substrate was mounted so that the mesa faced the switching element as shown; a small wire, etched to a 1-μm diameter at its tip, was used to make contact to the mesa. Synchronized optical pulses from a dye laser of about 80-fs duration were used to activate the photoconductive switch at the input to the device and to measure the change in index of refraction of the electro-optic sampling crystal induced by the electric fields at the output of the device. When the switch was closed, additional current

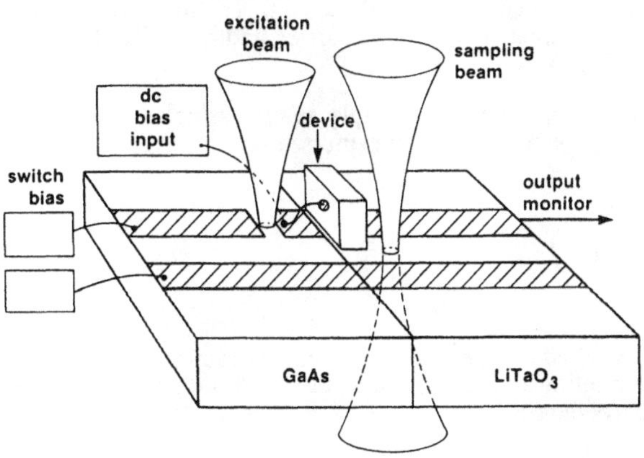

Fig. 1 Schematic diagram of test geometry

Springer Series in Chemical Physics, Vol. 48: **Ultrafast Phenomena VI**
Editors: T. Yajima · K. Yoshihara · C.B. Harris · S. Shionoya

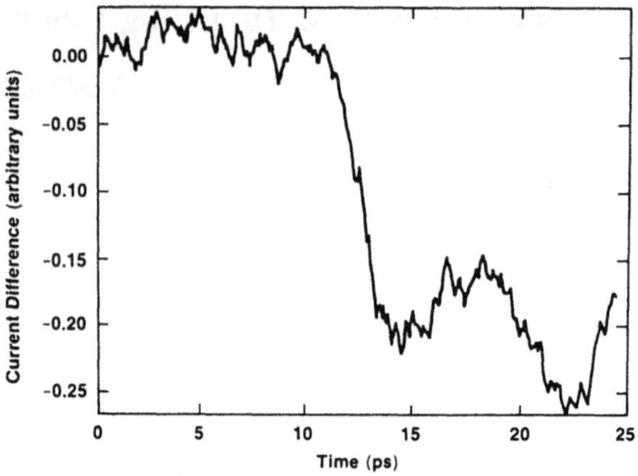

Fig. 2 Bistable-mode operation of RTD showing 1.9 ps rise time of switching

flowed through the RTD, a potential difference developed across the diode, and its operating point changed. Since the RTD has an I-V characteristic dominated by a negative differential resistance, an increase in voltage can lead to a decrease in current. (For additional details, see [3].) Due to the speed with which electrons may tunnel through this double-barrier structure, the switching time may also be very brief. The rise time of this switching has been experimentally resolved as 1.9 ps (Fig. 2).

We have also applied the technique of time-resolved photoluminescence (PL) to study the time dependence of tunneling. This technique has been used by TSUCHIYA et al. [4], to study the tunneling escape rate of electrons from a quantum well (QW) in a double-barrier structure versus the barrier width. In this study, we have been particularly interested in the dependence of the tunneling escape rate through a single barrier on an applied perpendicular electric field.

The structure of the samples under study was a single 30 Å GaAs QW, bounded on one side by a thick (0.2 µm) $Al_xGa_{1-x}As$ barrier, and on the other side by a thin $Al_xGa_{1-x}As$ barrier and 0.2-µm thick GaAs region. This undoped structure is situated in the intrinsic region of a p-i-n diode. The diode mesa diameter is 600 µm, and ohmic contacts were made on both n^+ buffer and p^+ cap, except a 200-µm dia hole in the cap was left for optical access. The diodes were operated in reverse bias and were held in a cryostat at a temperature of about 6 K. The electric field in the QW region was estimated from C-V measurements. Optical excitation was provided by a synchronously pumped dye laser, which was tuned so that only electron-heavy hole pairs were generated. The PL was imaged through a grating monochromator onto the entrance slit of a synchroscan streak camera. The time-resolved PL data is shown for three samples with Al conc. x = 0.3 and barrier width b = 85, 111, and 121 Å as a function of applied field in Fig. 3.

The field dependence of the tunneling follows a simple semiclassical model. The tunneling time of a bound electron through a barrier may be expressed as

$$\tau_T = c \exp\left(\frac{2}{\hbar}\int_0^b \sqrt{2m\,(V-E-Fz)}\ dz\right) , \tag{1}$$

where b is the barrier thickness, F the applied field, and c a constant determining the tunneling time at zero field. The calculated tunneling time is shown as the solid lines in Fig. 3; this simple model evidently properly displays the field dependence of the tunneling rate.

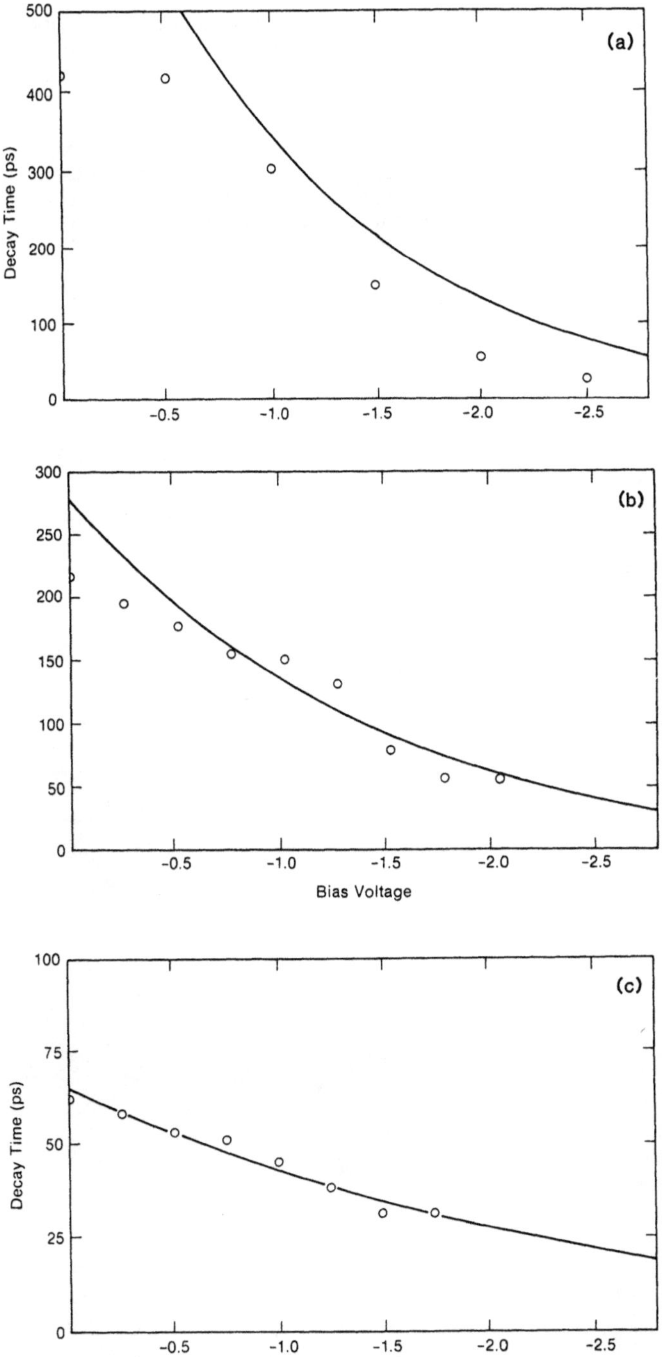

Fig. 3 PL decay rate versus estimated applied electric field, for different barrier widths (a) b = 121 Å, (b) b = 111 Å, (c) b = 85 Å. The solid lines are theoretical predictions from Eq. (1) in the text

This measurement, along with that of the switching time of a double-barrier diode, contributes necessary information to the understanding of the transport mechanism in heterostructures. The switching-time measurement also demonstrates that appropriately designed devices may be useful in electronic, high-speed logic applications of the future.

This work was supported by the Laser Fusion Feasibility Project at the Laboratory for Laser Energetics which has the following sponsors: Empire State Electric Energy Research Corporation, New York State Energy Research and Development Authority, Ontario Hydro, and the University of Rochester. Additional support was provided by the United States Air Force Office of Scientific Research under contract F49620-87-C-0016 to the Ultrafast Optical Electronics Center at the Laboratory for Laser Energetics of the University of Rochester. Such support does not imply endorsement of the content by any of the above parties. The work at Lincoln Laboratory was supported by the U. S. Army Research Office, NASA and the U. S. Air Force.

1. R. Tsu and L. Esaki, Appl. Phys. Lett. 22, 562 (1973).
2. W. D. Goodhue, T. C. L. G. Sollner, H. Q. Le, E. R. Brown, and B. A. Vojak, Appl. Phys. Lett. 49, 1086 (1986).
3. J. F. Whitaker, G. A. Mourou, T. C. L. G. Sollner, and W. D. Goodhue, to appear in Appl. Phys. Lett. August 1, 1988.
4. M. Tsuchiya, T. Matsusue, and H. Sakaki, Phys. Rev. Lett. 59, 2356 (1987).

Electro-Optic Sampling of a Flip-Chip with a Distributed Feedback Laser Diode

K. Joshin, K. Kamite, T. Mimura, and M. Abe

Fujitsu Laboratories, Ltd., Atsugi, 10-1 Morinosato Wakamiya, Atsugi 243-01, Japan

1. Introduction

High-speed semiconductor device technology is progressing rapidly, especially with respect to compound semiconductor devices such as HEMTs. The signal delays of these devices have reached less than ten picoseconds in a single gate, and are in the sub-nanosecond range at the LSI circuit level. As device switching speeds increase, signal delays within bonding wires and packages become more important to reduce the total device delay. The flip-chip technique is one possible way to eliminate these excess delays. However, it is difficult for conventional coaxial probes and electron beam probes to make contact with flip-chips and to measure fast switching speeds. We used the E/O sampling technique with a distributed feedback laser diode to measure the signal delay of metal lines on a GaAs flip-chip.

2. Measurement setup

The E/O sampling procedure used was almost the same as that described in reference /1/. We used an InGaAsP DFB laser diode with a 1.3 μm wavelength radiation, which was gain-switched and driven through a comb generator. The optical pulse width was between 20 and 25 ps. The reflected light from a GaAs chip was received by an InGaAs pin photodiode through one-half and one-quarter wave plates and a polarizing beam splitter. The sampled waveforms were displayed on an oscilloscope triggered by a mixer at a rate of offset frequency between the laser drive frequency and the input frequency to a GaAs chip.

3. Results

Figure 1(a) shows a GaAs flip-chip patterned with metal lines and mounted on a printed circuit board. The metal lines on the chip and the board were connected through 200 μm-diameter and 70 μm-thick metal bumps. Sinusoidal signals traveled alternately along the lines on the chip and the printed circuit board. The chip was 10 mm square. Figure 1(b) shows the sampled waveforms at distances of 0, 3, 7, and 11 mm from an input port of the chip. The input signal frequency was 990 MHz and the offset frequency was 20 KHz. The delay was estimated to be 32 ps/mm, a value greater than the GaAs stripline delay. This seems to be due to parasitic capacitance and inductance of the bump contacts.

Springer Series in Chemical Physics, Vol. 48: **Ultrafast Phenomena VI**
Editors: T. Yajima • K. Yoshihara • C.B. Harris • S. Shionoya

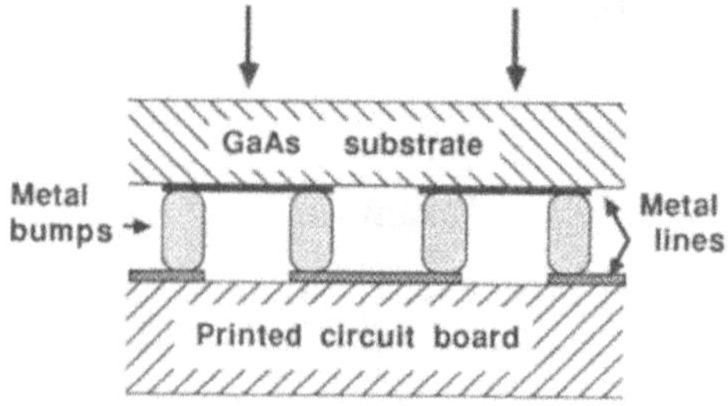

(a)

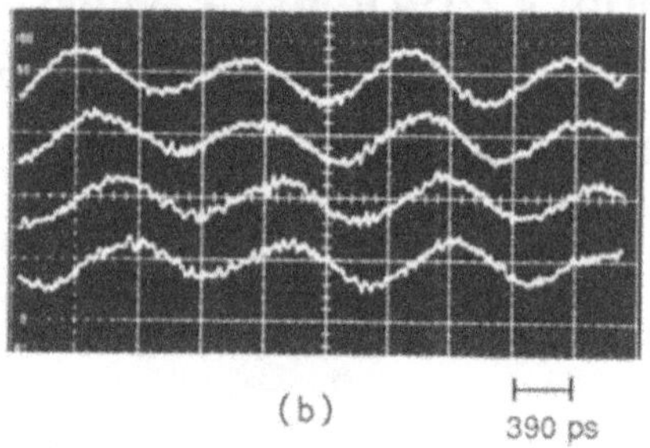

(b)

Fig. 1 (a) Cross section of a GaAs flip-chip mounted on a printed circuit board. (b) Delay of 990 MHz sinusoidal signal in the flip-chip. Waveforms were sampled at the input (top) and at distances of 3, 7, and 11 mm from the input port of the chip.

To check the accuracy of our measurement system, we measured the delay of a 1 μm gate-length HEMT inverter chain by backside probing to 80 μm square pads. Figure 2 shows the logic diagram and the signal waveforms at output ports A and B, which were separated by 32 inverters. The input and offset frequency were the same as above. We compared the delays estimated by the E/O sampling of this inverter sequence and by the electric frequency measurement of a ring-oscillator on the same wafer. The delay of a HEMT inverter was 54 ps in the former and 53 ps in the ring-oscillator. The agreement between the E/O sampling and the electric measurement was quite satisfactory.

To check the harmonic mixing capability of our E/O sampling system, we also measured two frequency-different signals in a frequency divider. The frequency divider was made of a HEMT flip-flop by connecting the $\overline{Q}$ port with the D port in Fig. 3(a). The flip-flop was driven by 0.8 V peak-to-peak sinusoidal signals at a rate of 2 GHz + 10 KHz. The laser diode drive frequency was 1 GHz. The oscilloscope was triggered at a rate

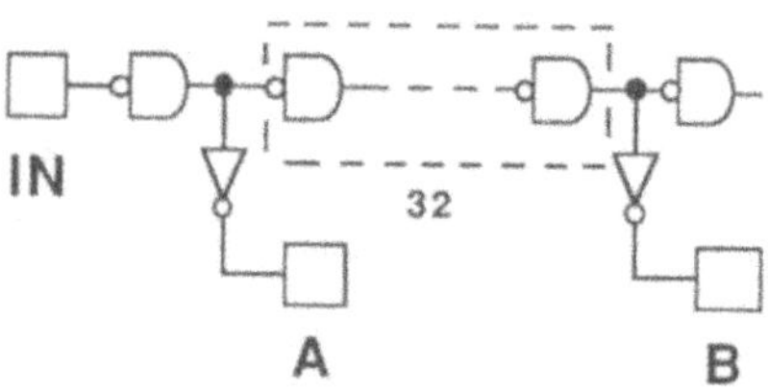

(a)

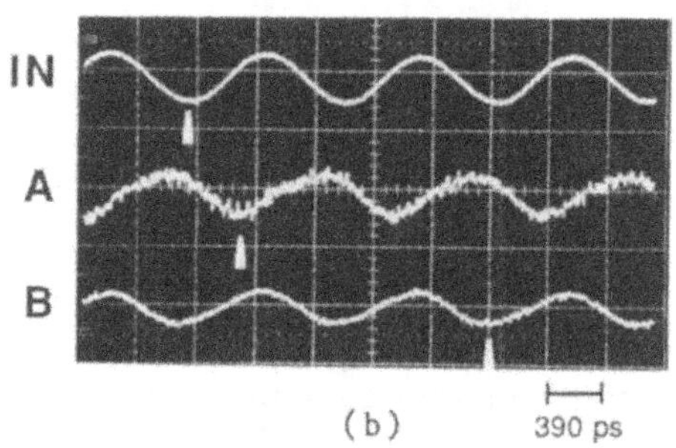

(b)

Fig. 2 (a) HEMT inverter chain circuit diagram. (b) Sampled waveforms observed at the input (top) and at outputs A (middle) and B (bottom)

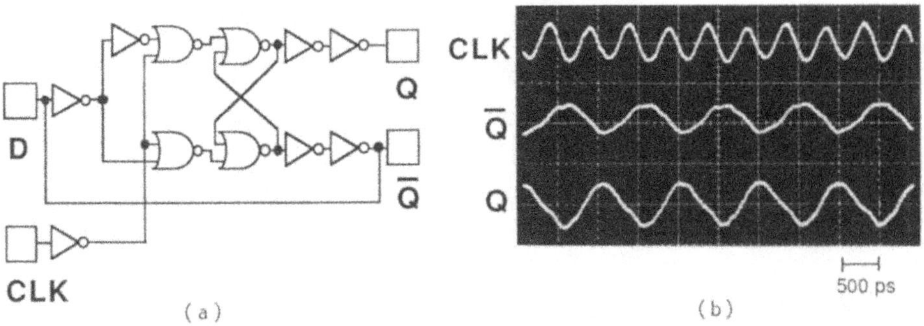

Fig. 3 (a) HEMT flip-chip circuit diagram. (b) Sampled waveforms observed at the CLK input (top) and at outputs $\overline{Q}$ (middle) and Q (bottom)

of 5 KHz by another phase-locked synthesizer instead of the mixer. Input and 1/2 frequency-divided signals could be sampled at the CLK input pad, $\overline{Q}$ and Q output pads and are shown in Fig. 3(b). In our system, the harmonic sampling capability extends to the 10 GHz frequency range for 0.8 V peak-to-peak sinusoidal signals on a GaAs stripline.

4. Summary

We measured the signal delay of a GaAs flip-chip, a HEMT inverter chain, and a HEMT flip-flop and demonstrated the effectiveness of E/O sampling with a DFB laser diode for these devices.

Acknowledgment

The authors wish to thank H. Isikawa, T. Yamada, N. Kobayashi, and S. Notomi for their support of this work.

Reference

1. J. M. Wiesenfeld, R. S. Tucker, A. Antreasyan, C. A. Burrus, A. J. Taylor: In Picosecond Electronics and Optoelectronics 2, ed. by F. J. Leonberger, C. H. Lee, F. Capasso, and H. Morkoc, Springer Ser. Electronics and photonics vol. 24, 25 (1987)

A New Scheme of Resolution Improved Electrooptic Sampling

T. Kamiya, R. Takahashi, H. Kamiyama, H.F. Liu, and I. Tanaka

Department of Electronic Engineering, University of Tokyo,
7-3-1, Bunkyo-ku, Hongo, Tokyo 113, Japan

1. Introduction

Based on the pioneering works, wider application of electrooptic sampling in high speed electronics is anticipated. Usually a mode-locked YAG laser or a dye laser has been employed as the pulse source. Because of their large physical dimension, high power consumption, and lack of long term stability, replacement with semiconductor lasers would be particularly advantageous for industrial applications if both power and resolution requirements are met by the latter: Optically sampled oscilloscopes with desk-top size would then be realized. One problem encountered in using semiconductor lasers may be the limitation in time resolution because only somewhat wider pulse width is available.

We propose here a new scheme of resolution improved sampling, by introducing a kind of deconvolution procedure to the acquired convolution data. With iterated deconvolution procedures, it is predicted that time resolution much smaller than the probe pulse width itself should be achieved.

2. Principle of Resolution Improved Sampling

Let an unknown waveform be denoted by h(t), probe pulse waveform by g(t), then the sampled data at delay time τ is given by the convolution integral

$$f(\tau) = \int_{-\infty}^{\infty} h(t) g(\tau - t) dt \equiv \int_{\infty}^{\infty} \bar{h}(x) \bar{g}(x_d - x) dx \qquad (1)$$

where $x \equiv 2\sqrt{2\log^2}\, t/t_o$ is the normalized time variable, with t_o as FWHM value of probe pulse. The resolution of the data is restricted by the pulse width t_o. To restore the original waveform, the application of deconvolution procedure should be effective: By Fourier-transforming eq.(1), we would get the Fourier spectrum of the unknown function $H(\omega)=F(\omega)/G(\omega)$, where F and G are the Fourier transforms of the convolution and the probe pulse, respectively. We have only to perform inverse Fourier transform of H to recover the original waveform in time domain. This simplest procedure, however, is interrupted with the addition of white noise spectrum to F, because decay of G at high frequency tail (corresponding to finite pulse width) would amplify the noise spectrum unlimitedly. Therefore truncation of Fourier spectrum at certain cutoff frequency and extrapolation of spectrum beyond this cutoff frequency are necessary. There have been many analysis procedures proposed in such fields as Fourier spectroscopy or image processing, and our problem is essentially a time domain analog of these spatial Fourier transform procedures.

General procedure can be summarized as follows: the unknown function $\bar{h}(x)$ is to be expanded in a functional space with the set of eigenfunctions $\phi_n(x)$

$$\bar{h}(x) = \sum_n C_n \phi_n(x) \qquad (2)$$

Springer Series in Chemical Physics, Vol. 48: **Ultrafast Phenomena VI**
Editors: T. Yajima · K. Yoshihara · C.B. Harris · S. Shionoya

where C_n is the expansion coefficient of $\bar{h}(x)$. Then the convolution integral is

$$\bar{f}(x_d) = \sum_n C_n \int_{-\infty}^{\infty} \phi_n(x)\bar{g}(x_d - x)dx \ . \tag{3}$$

Using experimentally obtained set of convolutions $f(x_{di}) = y_i$, minimization of mean square error

$$Q = \sum_{i=-N}^{N}\{y_i - f(x_{di})\}^2 \qquad (\partial Q/\partial C_k) = 0 \tag{4}$$

is performed. From the conditions $(\partial Q/\partial C_k) = 0$, we have the simultaneous linear equations for C_n

$$\langle y_i \int_{-\infty}^{\infty} \phi_k(x)\bar{g}(x_{di}-x)dx\rangle = \sum_n C_n \langle \int\phi_n(t)\bar{g}(x_{di}-x)dx \cdot \int\phi_k(t)\bar{g}(x_{di}-x)dx\rangle . \tag{5}$$

Here $\langle \ldots \rangle$ denotes the average over sampled points. Use of standard Fourier integral, namely choice of sinusoidal functions as basis functions, is not always the most adequate, because of the slow convergence in the Fourier expansion of step function. In the next section, we will show that the choice of Hermite-Gaussian functions as basis is suitable for representing pulse waveforms with internal structures.

3. Formulation of Deconvolution Based on Hermite Gaussian Expansion

An optical pulse is approximated by a Gaussian function:

$$\bar{g}(x) = \exp(-x^2/2) \ . \tag{6}$$

We expand the unknown electrical waveform with a set of Hermite-Gaussian functions:

$$\phi_n(x) = H_n(x/\sigma) \exp(-x^2/2\sigma^2), \quad \sigma = t_1 / t_0 \tag{7}$$

where t_1 is the roughly estimated FWHM of the unknown pulse.
Then the simultaneous equations for expansion coefficient for a particular case of $\sigma = 1$ yields

$$\langle y_i \, x_{di}^k \, \pi^{1/4}/\sqrt{2^k k!} \, \exp[-x_{di}^2/4]\rangle$$

$$= \sum_n C_n \langle \pi^{1/2}/\sqrt{2^{n+k} \, n! \, k!} \;\; x_{di}^{n+k} \exp[-x_{di}^2/2]\rangle \tag{8}$$

where truncation order is determined by the condition that the magnitude of the integral of the left hand side computed by using experimental data becomes equal to noise level (S/N = 1).

To improve the accuracy further by extrapolations, we employed an iterative procedure: Starting from the approximate unknown function expressed by the truncated series expansion we compute the extrapolated waveforms outside of measured data. Using them together with a set of experimental data we have a set of matrix equation of eq.(8) with a greater matrix dimension by one. This procedure is repeated several times to yield a converged approximate waveform.

One typical waveform of asymmetric pulse with internal structure was tested:

$$\bar{h}(x) = 2\cos^2\{2\pi(x + 3/2)/3\} + \cos^2\{\pi x/2\}$$

$$+ (3/5)\cos^2\{2\pi(x - 3/2)/3\} . \tag{9}$$

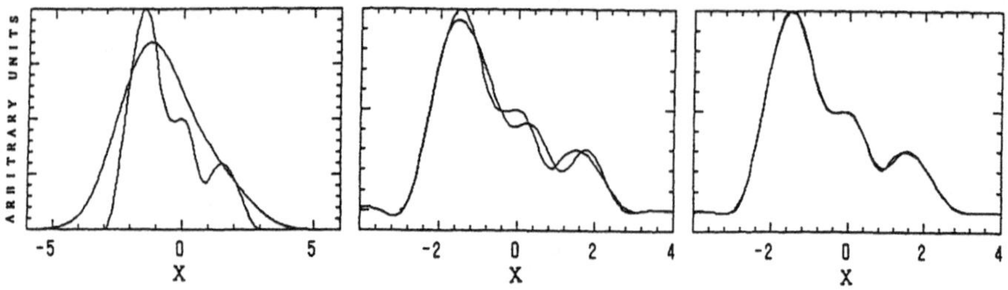

(a) original waveform and its convolution with $\exp(-x^2/2)$.

(b) 14th order signal recovery with 1% noise level.

(c) 9 time iteration up to 14th order approximation.

Fig. 1 Signal recovery for an asymmetric pulse with internal structure.

The noise level of 1%, as large as the maximum amplitude of experimental data, was assumed. The delay time unit is assumed to be 5% of probe pulse width. A total of 401 convolved data were used for the waveform restoration computation. Figure 1(a) is the comparison made between original and convolved waveforms. Internal structure is totally smeared out by using a probe pulse whose pulse width is larger than the internal structure of the measured pulse. Fig. 1(b) shows the restored waveform in comparison with the original one, where the expansion coefficients up to 14th order was calculated. Without extrapolation procedure we have succeeded in recovering the internal structure although some quantitative difference is remaining. Figure 1(c) is the restored signal by the iterative calculation starting from 6th order up to 14th order. Quantitative agreement is satisfactory.

4. Conclusion

In conclusion, it was proposed that by a kind of deconvolution procedure, the electro-optic sampling measurement should gain higher time resolution. Specifically, even with the presence of 1% noise level to the convolved data points, improvement of nearly one order better than the probing pulse width is expected as far as the pulse shape can be approximated by Gaussian function. Therefore, for the purpose of increasing resolution, precise control of pulse waveform and jitter reduction are as effective as shortening of pulse width.

Metal-Semiconductor-Metal Photodiode on GaInAs Exhibiting Very Fast Response

O. Wada, H. Nobuhara, H. Hamaguchi, T. Mikawa, A. Tackeuchi, and T. Fujii

Fujitsu Laboratories, Ltd., 10-1, Morinosato-Wakamiya,
Atsugi, 243-01, Japan

1. Introduction

Metal-semiconductor-metal(MSM) photodiodes(PDs) formed on GaAs [1, 2] have so far exhibited marked advantages of simple, lateral structure and excellent performance including ultra fast response speed [3] for short-wavelength (~0.8 µm) applications. On the other hand, the fabrication of MSM-PDs on GaInAs for the application in the long-wavelength region (1.3 µm-1.55 µm) has been retarded by excessive dark current caused by low Schottky barrier height (~0.2 eV) in the metal-GaInAs(InP) system. We describe in this paper the first fabrication of a MSM-PD on GaInAs, in which a large barrier height is realized by a new structure consisting of a large-bandgap AlInAs layer followed by an AlInAs/GaInAs graded superlattice. This MSM-PD has exhibited low dark current, low capacitance characteristics, and the speed measurement carried out by electro-optic sampling technique has demonstrated its very fast response with a FWHM of 14.7 ps.

2. Diode Structure and Fabrication

The structure of the present GaInAs MSM-PD is shown in Fig. 1. The wafer used is a nominally undoped epitaxial wafer grown by pulsed molecular-beam epitaxy [4] on a semi-insulating InP substrate. The GaInAs photoabsorption layer was 1.5 µm. To reduce the dark current, an undoped AlInAs barrier height enhancement layer (70 nm thick) was employed. Also, to minimize the effect of carrier trapping which may be caused by a large band discontinuity at the AlInAs/GaInAs interface, a graded superlattice (200 nm thick) was incorporated between the AlInAs and GaInAs layers. This superlattice was composed of pairs of ultra thin (one to nine atomic layers) AlInAs and GaInAs layers where the width ratio varies so that the effective bandgap is graded linearly in depth as shown by dotted lines in Fig. 1. The same structure was

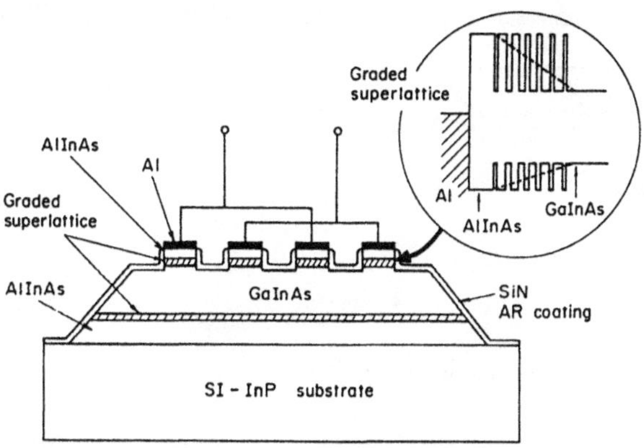

Fig. 1 Cross-section of GaInAs MSM-PD.

Springer Series in Chemical Physics, Vol.48: **Ultrafast Phenomena VI**
Editors: T. Yajima · K. Yoshihara · C.B. Harris · S. Shionoya

used at the bottom interface for preventing generation of two-dimensional carrier gas there. Al contacts had interdigitated patterns (1.1 µm line and 1.4 µm space, approximately 20 µm x 20 µm area). Shallow mesa was formed at the periphery of contacts for homogenizing the field in the GaInAs layer. The photosensitive area was coated by a SiN antireflection film.

3. Diode characteristics

The dark current was found to be dependent steeply on the bias voltage but it was still below 3 µA at 20 V and as low as 100 nA at 10 V. This has confirmed the effect of barrier height enhancement by the introduction of an AlInAs layer. The capacitance was found to decrease as the bias voltage increases and then saturate at approximately 40 fF above 5 V. Since the observed capacitance included an appreciable fraction of stray component caused by the contact structure for signal extraction, the intrinsic capacitance of the MSM-PD [1] is expected to be much lower than this. Photocurrent characteristic was measured using a 1.3 µm wavelength CW laser. The sensitivity steeply increased in the bias voltage region where the GaInAs layer is depleted and then became more insensitive to the voltage beyond 2 V. The quantum efficiency determined at 10 V is typically 50% which corresponds to an internal quantum efficiency more than 80% when contact shadowing is taken into account.

Response speed characterization was performed using standard electro-optic sampling technique [5] with a CPM dye laser pulse (λ = 620 nm, 100 Mpps) having a FWHM of 260 fs. Device mounting and sampling setup used are illustrated in Fig. 2. The result obtained at the bias voltage of 10 V is shown in Fig. 3. The response speed is very fast; the rise time and FWHM are determined to be 8.5 ps and 14.7 ps, respectively. The weak ringing in the tail is presumed to be due to mismatch caused by the bonding wire. The pulse width was found to be independent of the bias voltage(2.5 V to 20 V) and the excitation power(90 µW to 900 µW average) variations, suggesting that there are no speed-limiting effects associated with the graded-superlattice structure. The observed FWHM is consistent with the carrier transit time in this diode (1.4 µm gap). This corresponds to a bandwidth of 23 GHz and is, to the best of our knowledge, the fastest speed ever reported for lateral structure photodiodes on GaInAs.

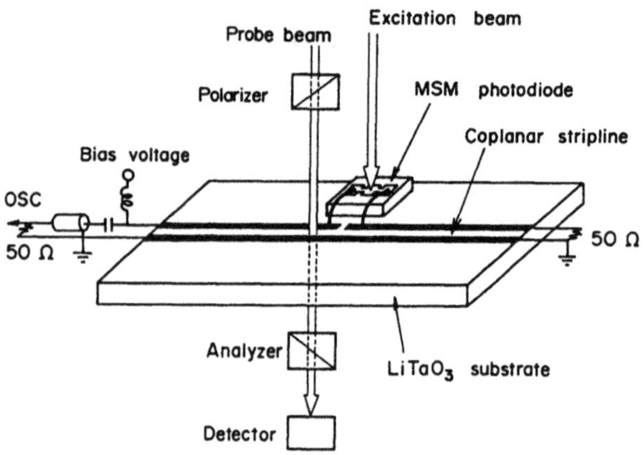

Fig. 2 Setup of electro-optic sampling measurement.

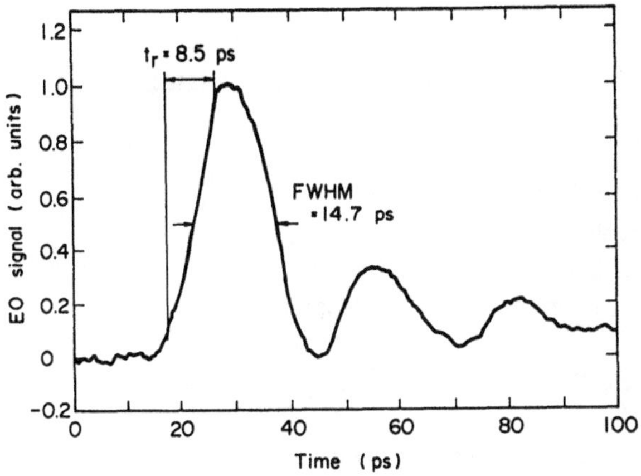

Fig. 3 Temporal response of MSM-PD measured by electro-optic sampling technique at the bias voltage of 10 V.

4. Conclusion

A lateral structure AlInAs/GaInAs MSM-PD has been fabricated. This photodiode has exhibited low dark current, low capacitance and very fast response characteristics, and will be significant for the application to not only discrete photodetectors but also optoelectronic integrated receivers for optical communication, signal processing and measurement at ultra high speed.

References

1. M. Ito and O. Wada, IEEE J. Quantum Electron. QE-22, 1073 (1986).
2. O. Wada, H. Hamaguchi, M. Makiuchi, T. Kumai, M. Ito, K. Nakai, T. Horimatsu and T. Sakurai, IEEE J. Lightwave Technol. LT-4, 1694 (1986).
3. B. J. Van Zeghbroeck, Ch. Harder, J-M. Halbout, H. Jackel, H. Meiner, W. Patrik, P. Vettiger and P. Wolf, Tech. Dig. IEDM 1987, Washington, D.C., p. 279.
4. T. Fujii, Y. Nakata, Y. Sugiyama and S. Hiyamizu, Jpn. J. Appl. Phys. 25, L254 (1986).
5. J. A. Valdmanis, G. Mourou and C. W. Gabel, Appl. Phys. Lett., 41, 211 (1982).

Picosecond HEMT Photodetector: Improvement of Response Speed at Low Temperatures

T. Umeda[1], *Y. Cho*[1], *H. Tanaka*[2], *and Nion Sock Chang*[2]

[1]The Institute of Scientific and Industrial Research, Osaka University, 8-1 Mihogaoka, Ibaraki, Osaka 567, Japan
[2]Osaka Electro-Communication University, Department of Communication Engineering, 18-8 Hatsumati Neyagawa, Osaka 572, Japan

A short and narrow gate GaAs/AlGaAs HEMT was tested as a picosecond photodetector. Its photoresponse was measured by the autocorrelation technique and found to be as fast as 18 ps FWHM. At a low temperature of 126 K, the response speed was improved to 13 ps. This improvement is considered to be evidence for the effective contribution of the 2-dimensional electron gas to the high-speed response of this photodetector.

Introduction

Since the advent of the HEMT (high electron mobility transistor) structure in GaAs MESFET (metal-semiconductor field effect transistor) devices, its application to a high-speed photodetector has been attempted by several authors[1-3] with the expectation of possible utilization of a 2-dimensional high-mobility channel (so-called 2-D electron gas) along its included hetero-interface. Chen et al. obtained an impulse response speed of 27 ps (FWHM, full width at half-maximum) with a MODFET (modulation-doped field-effect transistor, same as HEMT) of rather large device dimensions[1]. We also reported an impulse response of 22 ps (FWHM) on a HEMT[3]. Meanwhile, our investigation[4] of the influence of gatewidth upon the photoresponse speed of GaAs MESFET's indicated that gatewidths more than several tens of micrometers may, due to their included distributed capacitance proportional to the gatewidth, mask other causes of response speed limitation.

The actual contribution of the high mobility of the 2-D electron gas to those high speed responses obtained by HEMT photodetectors as mentioned above, besides the contribution from the narrow gatewidth used in those experiments, must be checked. The most direct check of this problem is the temperature dependence of the response speed of the HEMT photodetector, since the main factor for determining the electron mobility is the phonon scattering for the 2-D electrons, while it is the impurity scattering for electrons in a bulk material. We report here a response speed improvement of a HEMT photodetector by cooling.

Experimental Setup

A HEMT photodetector measured here has a GaAs-AlGaAs HEMT structure (aluminum content of AlGaAs is 0.28), which is the same as that used in our previous report[2,3] with a short-length and narrow-width gate. The gatewidth and gate length are 10 μm and 2 μm, respectively.

Autocorrelation measurement requires two identical devices to be measured. In Fig.1, $HEMT_1$ and $HEMT_2$ were patterned separated by a short distance on a single common substrate. The cascaded total drain bias voltage over two HEMTs was V_{DS}=4 V. Gates connected in parallel were

Springer Series in Chemical Physics, Vol. 48: **Ultrafast Phenomena VI**
Editors: T. Yajima · K. Yoshihara · C.B. Harris · S. Shionoya

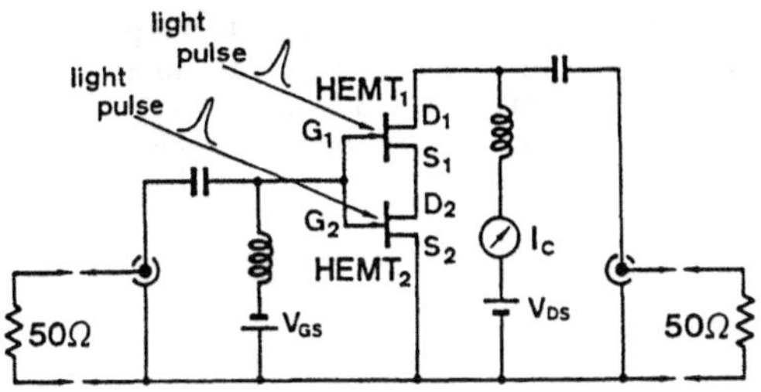

Fig.1 Test circuit configuration for the GaAs HEMT photodetector.

biased with a common bias supply of a voltage V_{GS}=-2 V which was sufficiently deep to keep those two HEMTs in pinch-off in the dark condition. A test optical pulse train was generated from a synchronously Ar-laser-pumped Rhodamine-6G dye mode-locked laser (λ=600 nm). The optical pulsewidth used was around 10 ps. Optical pulse trains thus prepared were focused accurately on each HEMT over respective source-gate-drain spaces through a single 20x microscope objective with the help of a microscope illumination-field TV monitor system[5].

Autocorrelation Measurement and Discussion

The average current I_c flowing through cascade-connected $HEMT_1$ and $HEMT_2$ under the trained light pulse illumination is approximately proportional to an autocorrelation function, that is, $I_c(\tau) \propto \int f(t)f(t-\tau)dt$, where τ is a mutual delay between two light pulse trains, and f(t) is a response current waveform, which is to be measured, of each individual HEMT photodetector.

The observed autocorrelation plot ($\propto I_c(\tau)$) is shown in Fig.2. The plot shows an excellent fit with a Gaussian (broken curve) and its correlation width (FWHM) is read to be 25 ps. Assuming a Gaussian autocorrelation function, the time response f(t) of the measured HEMT photodetector is concluded to be $25/\sqrt{2}$ [6]= 18 ps (also in FWHM).

The measurement described above was done at room temperature (~20 °C). To see the temperature dependence of the device, the same HEMT photodetector was cooled down to 126 K, where the phonon scattering effect on the mobility is considered to be considerably suppressed. Cooling was attained by jetting cooled helium gas using liquid nitrogen over the chip of the device through a jet nozzle. The cooled helium gas served as a coolant gas as well as a purging gas for air. The observed autocorrelation plot and its fitting curve are shown also in Fig.2. The obtained autocorrelation width (FWHM) was 19 ps, and assuming a Gaussian from the fitting curve (solid curve), as in the previous measurement, the photoresponse of the device is concluded to be $19/\sqrt{2}$ = 13 ps (also FWHM).

The observed improvement in the response speed of the HEMT photodetector produced by cooling is considered to be evidence proving the effective contribution of the 2-D electron gas to the high-speed response of this device, since the conditions under which these experiments were done at room temperature and at a low temperature were unchanged except for the temperature. Also, the high speed response of the GaAs HEMT photodetector even at room temperature can be explained using a model we proposed earlier[3,4], called the "depletion region direct-deformation model" for the mechanism of the photoresponse of MESFETs.

Although the response speed of 13 ps observed at a low temperature of 126 K is the fastest obtained heretofore by a GaAs MESFET to our knowledge, the light pulses used had a width of around 10 ps, which was not short enough to measure the response of around 10 ps. The intrinsic response speed of this GaAs MESFET must be much faster.

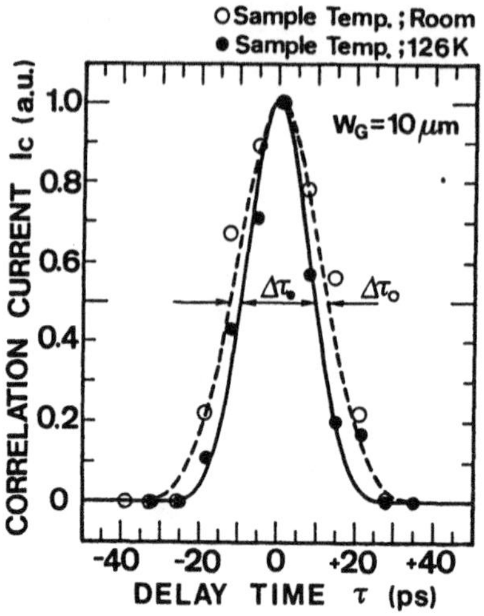

Fig.2 Obtained autocorrelation plot of impulse responses of the HEMT photodetector(● and solid curve, 126 K; O and broken curve, room temp.).

Conclusion

The response speed of a short and narrow gate GaAs/AlGaAs HEMT photodetector was measured by the autocorrelation technique and found to be as fast as 18 ps FWHM at room temperature and 13 ps at a temperature of 126 K. This improvement is considered to be evidence indicating the effective contribution of the 2-D electron gas to the high-speed response of this photodetector. The mechanism of this high speed response of the HEMT photodetector was discussed using the model of photoresponse of MESFETs we proposed earlier, and it was pointed out that the high electron mobility of the 2-D electron gas and the built-in field associated with the hetero-structure, which can be designed and formed by an adequate combination of alloy semiconductor materials, play a main role. The intrinsic response speed of the HEMT photodetector tested is believed to be much faster but could not be measured because of the rather broad light pulses used.

The authors wish to thank Dr. A. Shibatomi and Dr. K. Kondo of Fujitsu Laboratories for their discussions and great help in supplying HEMT samples throughout this work.

References

1) CHEN, C.Y., CHO, A.Y., BETHEA, C.G., GARBINSKI, P.A., PANG, Y.M., LEVINE, B.E., and OGAWA, K.: "Ultrahigh speed modulation-doped heterostructure field-effect photodetectors", Appl. Phys. Lett., 1983, 42, pp.1040-1042.
2) UMEDA, T., and CHO, Y.: "Narrow gate width GaAs MESFET high-speed photodetector", Trans. IECE Japan, Part C, 1985, J68-C, pp.1132-1134 [in Japanese].
3) UMEDA, T., CHO, Y., and SHIBATOMI, A.: "Picosecond HEMT photodetector", Jpn. J. Appl. Phys., 1986, 25, pp.L801-L803.
4) UMEDA, T., and CHO, Y.: "High-speed photodetectors using the GaAs MESFET", Trans. IECE Japan, Part C, 1985, J68-C, pp.263-269 [in Japanese].
5) UMEDA, T., and CHO, Y.: "Effect of incident light illumination shape on responsivity of GaAs MESFET photodetector", Jpn. J. Appl. Phys., 1985, 24, pp.L367-L369.
6) VON der LINDE, D.: "Experimental study of single picosecond light pulses", IEEE J. Quantum Electron., 1972, QE-8, pp.328-338.

Picosecond Transient Propagation Studies on Thin-Film Y-Ba-Cu-O Transmission Lines

J.M. Chwalek[1], *D.R. Dykaar*[1], *J.F. Whitaker*[1], *T.Y. Hsiang*[1], *G. Mourou*[1], *D.K. Lathrop*[2], *S.E. Russek*[2], *and R.A. Buhrman*[2]

[1]Laboratory for Laser Energetics and Department of Electrical Engineering, University of Rochester, 250 East River Road, Rochester, NY 14623, USA

[2]School of Applied and Engineering Physics, Cornell University, Ithaca, NY 14853, USA

By studying the propagation of ultrafast transients on superconducting transmission lines one cannot only directly assess the practical limitations of such superconducting structures, but also gain insight into the nature of the energy gap, quasiparticle excitation effects, and other loss mechanisms. In the new granular superconductor $YBa_2Cu_3O_{7-x}$ (YBCO), losses due to flux flow and grain boundary effects are possible [1]. Workers at IBM recently reported results for aluminum transmission lines over a YBCO groundplane, showing no absorption for pulses propagating on the normal aluminum lines [2]. This paper reports on preliminary studies of picosecond electrical transients propagated on thin-film, coplanar transmission lines made out of the YBCO material.

The cryogenic measurement technique is electro-optic based and is a variation of a technique reported elsewhere [3]. The experimental setup is shown in Fig. 1(a). A GaAs photoconductive switch provided the initial step transient for the transmission line under study. The sampling was performed in a reflection mode using two high-reflection coated $LiTaO_3$ crystals placed at either end of the transmission line. The edge of the crystals facing toward one another defined the input and output sampling points as shown in Fig. 1(a).

The transmission lines were 5.4 mm long and consisted of two 30-μm wide lines separated by 30 μm in the first sample, and 15 μm linewidths and spacing in the second sample. A 360 nm thick (1.1 μm in sample 2) layer of YBCO was deposited by a high pressure reactive evaporation process described in Ref. 4 on a yttria-stabilized ZrO_2 substrate. The lines were patterned by a standard photolithographic process and argon ion beam etching. The electrical properties of the films used were very similar to that reported in Ref. 4. The critical temperature for sample 1 was 60K and 78K for sample 2.

The measurements on sample 1 started with it immersed in superfluid helium. For a given switch signal, the input and output transients over a propagation distance of 5 mm was measured. The result, shown in Fig. 1(b), clearly shows that a YBCO transmission line can support distortion-free propagation of picosecond electrical pulses over a substantial propagation distance. The signals had rf current densities in excess of 10^5 A/cm^2. The limiting factor was the ability to produce larger amplitude input signals. Next, a current density of about 4 x 10^4 A/cm^2 was maintained as the temperature was increased. At this current density, the rise time showed no degradation until a temperature of 17–18K. This defined a critical temperature for distortion-free propagation for that current density.

In a subsequent experiment (using sample 2), the output transient was measured as a function of temperature for a fixed input signal. As shown in Fig. 2(a), there is substantial dispersion measured. The degradation in signal rise time is independent of temperature, implying the losses are not due to the YBCO film but to other loss mechanisms such as substrate loss. For current densities of approximately 10^4 A/cm^2, degradation of the output waveform did not occur until temperatures >65 were reached. This sample (sample 2) unlike the first, was covered by a residue which may have contributed to the high losses measured in the experiment. As can be seen from Fig. 2(a), an increasing phase delay is measured as

Springer Series in Chemical Physics, Vol. 48: **Ultrafast Phenomena VI**
Editors: T. Yajima · K. Yoshihara · C.B. Harris · S. Shionoya

1(a)

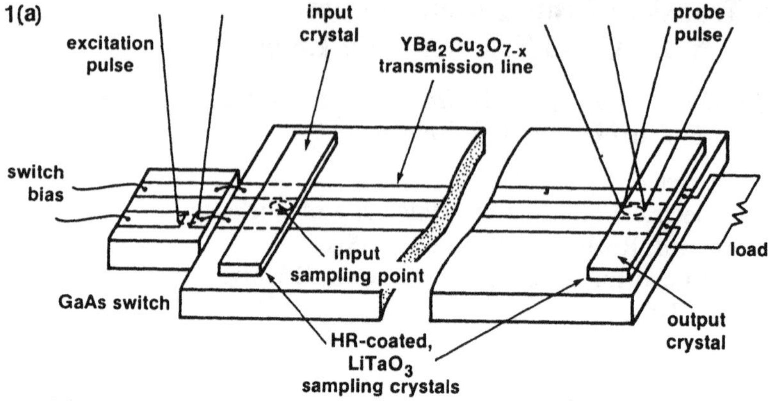

1(b)

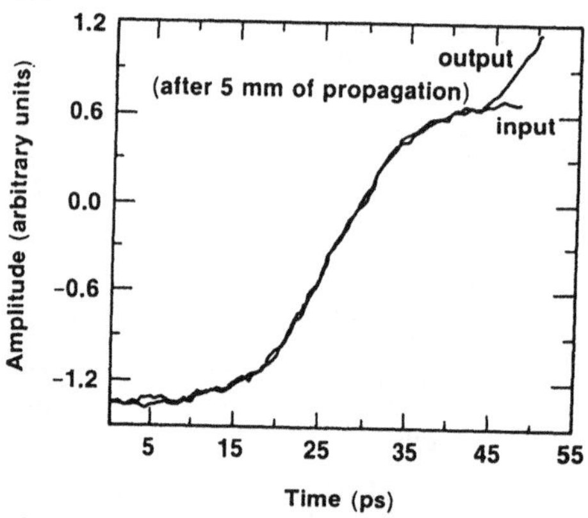

Fig. 1(a) Experimental configuration of the samples; 1(b) Experimentally measured picosecond electrical transient propagation on the coplanar YBCO transmission line (sample 1: 30 μm lines). The deviation at the very end of the output transient is due to a reflection at the end of the transmission line

temperature is increased. From the Mattis-Bardeen theory, a decrease in phase velocity is expected as the temperature approaches the critical temperature. The waveforms were simulated using a previously reported numerical simulation program, which has been used successfully to model propagation on ordinary superconducting transmission lines [5]. The percent change in phase delay, as calculated relative to the initial delay at a temperature of 2K, is plotted against temperature and is shown in Fig. 2(b). The percent change was found to be very weakly dependent on the substrate parameters and very strongly dependent on the YBCO parameters [6]. While different methods yield a range of values for the energy gap of YBCO, the corresponding gap frequency is so much greater than the frequency content of our signal that there is negligible effect on the simulated results due to a variance in the energy gap. Good agreement is obtained, however, there is a slight underestimation of the delay in the simulations at temperatures near T_c. This is likely due to the effect of finite current densities present in the experiment not accounted for in the simulation.

In conclusion, we have demonstrated the propagation of undistorted signals on thin-film YBCO transmission lines at current densities in excess of 10^5 A/cm^2. The results were successfully simulated using Mattis-Bardeen complex conductivities. Substrate losses were found to be a limiting factor for the system performance in the high-frequency regime.

2 (a)

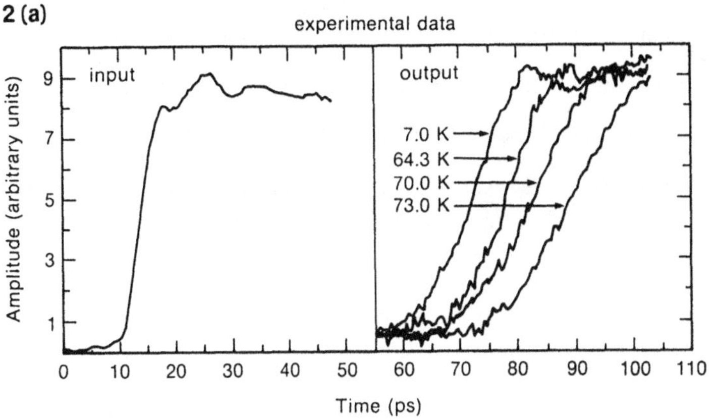

2 (b)

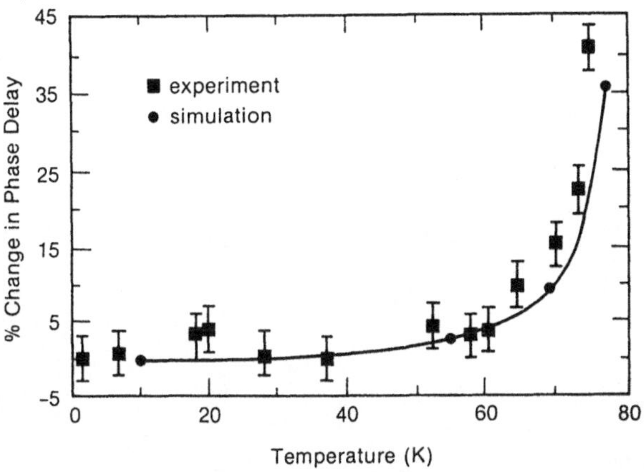

Fig. 2(a) Measured waveforms from sample 2. The left-hand waveform is the input transient and the sequence of waveforms on the right show the transient response at the output sampling point as a function of temperature. Note that the output amplitudes have been normalized to the input amplitude; 2(b) Relative phase delays for the simulated (using Mattis-Bardeen complex conductivities) and measured waveforms

The experiments conducted at the University of Rochester were supported by the United States Air Force Office of Scientific Research under contract F49620-87-C-0016 to the Ultrafast Optical Electronics Center at the Laboratory for Laser Energetics of the University of Rochester, the Army Research Office under contract DAAL03-86-K0152, and the National Science Foundation through grant DMR-85-06689. Additional support was supported by the sponsors of the Laser Fusion Feasibility Project at the Laboratory for Laser Energetics which has the following sponsors: Empire State Electric Energy Research Corporation, New York State Energy Research and Development Authority, Ontario Hydro, and the University of Rochester. The work at Cornell University was supported by the Office of Naval Research and by the National Science Foundation through support of the National Nanofabrication Facility at Cornell.

1. M. K. Wu, J. R. Ashburn, C. J. Torng, P. H. Hor, R. L. Meng, L. Gao, Z. J. Huang, Y. Q. Wang, and C. W. Chu, Phys Rev. Lett. 58, 908 (1987).
2. D. Grischkowski and I. N. Duling III, OSA meeting, Rochester, NY, 1987.

3. D. R. Dykaar, R. Sobolewski, J. M. Chwalek, T. Y. Hsiang, and G. A. Mourou, in Advances in Cryogenic Engineering, ed. R. W. Fast (Plenum, New York, 1988), Vol. 33, pp. 1097.
4. D. K. Lathrop, S. E. Russek, and R. A. Buhrman, Appl. Phys. Lett. 51, 1554 (1987) and references therein.
5. T. Y. Hsiang, J. F. Whitaker, R. Sobolewski, D. R. Dykaar, and G. A. Mourou, Appl. Phys. Lett. 51, 1551 (1987).
6. D. R. Dykaar, R. Sobolewski, J. M. Chwalek, J. F. Whitaker, T. Y. Hsiang, G. A. Mourou, D. K. Lathrop, S. E. Russek, and R. A. Buhrman, Appl. Phys. Lett. 52, 1444 (1988).

Greater than 100 GHz Traveling Wave Modulator

J. Nees, S. Williamson, and G. Mourou

Laboratory for Laser Energetics, University of Rochester,
250 East River Road, Rochester, NY 14623, USA

A GaAs traveling wave modulator with a bandwidth greater than 100 GHz has been demonstrated. The dramatic bandwidth enhancement is due to the elimination of the velocity mismatch by the addition of a GaAs superstrate in contact with the optical waveguide.

Signal processing, optical communication, and computer technology require a broadband modulator with low operating power. The optical waveguide, due to the confinement of the optical beam over a long distance, makes optical modulation with long interaction length possible [1–6]. The driving power for a traveling wave modulator, of impedence Z_0, is given by the expression

$$P = \left[\frac{d}{\ell}\left(\frac{\lambda}{n^3 r}\right)\right]^2 / Z_o \tag{1}$$

where n is the index of refraction, λ the wavelength in free space, r is the effective electro-optic coefficient, d is an effective electrode separation, and ℓ is the waveguide length.

The modulator bandwidth is determined on one hand by the dispersive properties of the transmission line due to modal dispersion [7], and skin effect losses [8], and, on the other hand, by the phase velocity difference [9] between the optical and electrical waves. It is the velocity mismatch which is presently limiting the time response of the traveling wave modulator to 20 GHz [10].

To minimize modal dispersion, it is necessary to use transmission lines with small cross sectional dimensions. Due to the air-substrate interface, quasi-TEM propagation occurs only when the cross-sectional dimensions of the transmission lines are small compared to the wavelength of the driving signal. For coplanar lines, the effective dielectric constant is frequency dependent as shown in Fig. 1. The cut-off frequency, ν_c, of the surface wave TE_1 mode [10] follows the form

$$\nu_c \sim \frac{c}{4d\sqrt{\varepsilon_{sub} - \varepsilon_{sup}}} \tag{2}$$

where c is the speed of light and the dielectric constant of the substrate, ε_{sub}, is greater than the dielectric constant of the superstrate, ε_{sup}.

Contrary to modal dispersion, skin effect losses demand electrode dimensions to be as large as possible. Consequently, for high frequency operation, modal dispersion and skin effect losses are irreconcilable. We will see later, that in order to avoid modal dispersion and skin effect simultaneously, it is necessary to use superconducting electrodes. Figures 2(a) and 2(b) illustrate the effects of modal dispersion and skin effect losses on the short electrical pulse propagation. The two figures are the result of computer simulation. Figure 2(a) shows the broadening of a 1 ps rise time step function after 10 mm propagation on a gold coplanar

Springer Series in Chemical Physics, Vol. 48: **Ultrafast Phenomena VI**
Editors: T. Yajima · K. Yoshihara · C.B. Harris · S. Shionoya

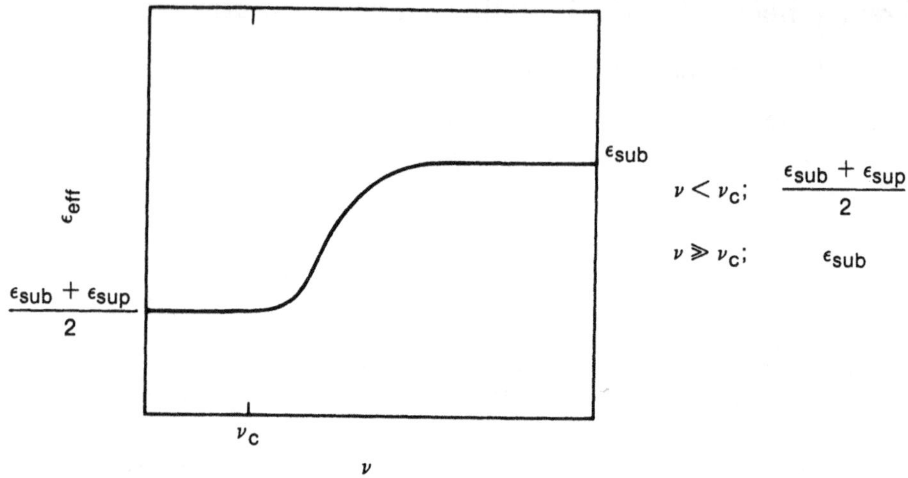

Fig. 1 Electrical dispersion due to dielectric mismatch

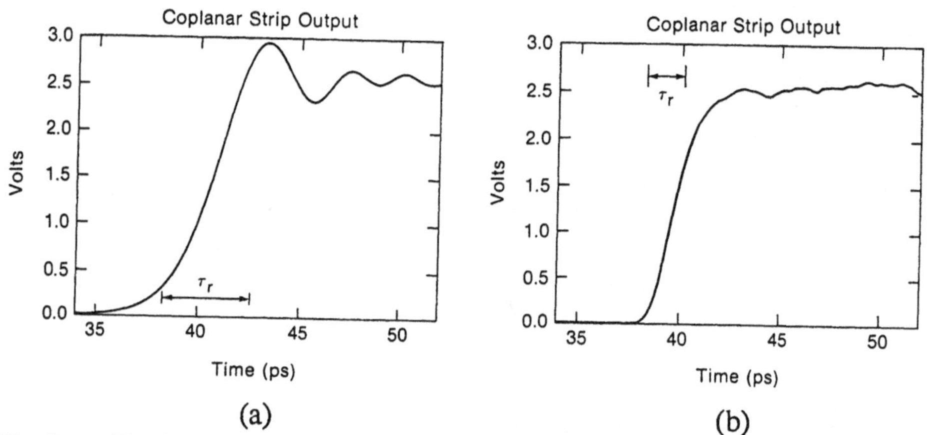

Fig. 2 Simulated propagation of signals with and without dielectric mismatch

transmission line evaporated onto a GaAs substrate. The width and electrode separation are both 20 μm. It can be seen after 10 mm, the rise time has degraded from an initial value of 1 ps to 5 ps. In Fig. 2(b), the modal dispersion has been turned off and the observed decrease in rise time and signal amplitude are due solely to the skin effect attenuation.

As mentioned earlier, the most important mechanism limiting the speed of the electro-optic modulator is the velocity mismatch between the optical and electrical signals. This effect limits the rise time τ_r of the modulator to a value

$$\tau_r = \Delta n \cdot \frac{\ell}{c} \tag{3}$$

where $\Delta n = \sqrt{\varepsilon_{eff}} - n$ and n is the substrate index of refraction.

By placing a GaAs superstrate in contact with a GaAs/GaAlAs waveguide, the condition $\sqrt{\varepsilon_{eff}} = n$ can be satisfied. This enables us, for the first time, to suppress the effect of the velocity mismatch in longer, more efficient modulators.

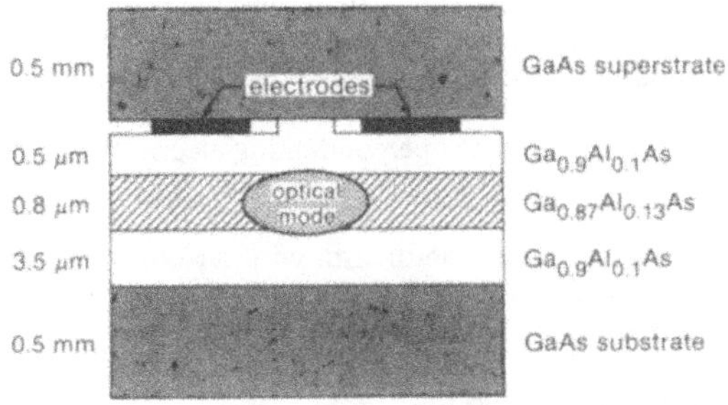

Fig. 3 >100 GHz modulator

A diagram depicting this modulator is shown in Fig. 3. A waveguide was prepared on a GaAs substrate. Vertical confinement was provided by a series of MBE grown GaAlAs layers and horizontal confinement by a ridge etched in the upper layer. The coplanar striplines were 100 μm wide and separated by 20 μm. The film is gold of 5000 Å thickness. A 10 μm photoconductive switch interrupted one line; 0.5 mm from the modulator's input end. The resistance of the metalization over the 5 mm modulator length was 7.5 Ω. The GaAs superstrate was then placed in contact with the coplanar transmission lines.

After passing through a quarter-wave plate about 10% of the compressed IR light from a Nd:YAG laser was coupled into a 4 μm core fiber 40 cm in length. This fiber was then butt coupled to the input of the modulator waveguide. A 20X microscope objective was used at the output to image the modulated light through a polarizer and onto a reverse biased PIN photodiode. The remaining 90% of the compressed IR beam was converted to 532 nm at 2% efficiency in KTP to form the switch pump pulse. The green light was first chopped at the lock-in reference frequency of 4 MHz by an acousto-optic modulator and passed through an optical delay line to allow adjustment of pump/probe timing. Finally, the pump pulse was focused onto the photoconductive switch. A step function electrical pulse was generated and the rise time measured by the short probe pulses which correspond directly to the response time of the modulator as it would be used with cw optical input. The result of modulating with a subpicosecond step function and a 1 ps probe pulse is shown in Fig. 4. The 10%–90% rise time is found to be 3.2 ps. If a Gaussian rising edge is assumed, this rise time corresponds to 110 GHz for 3dB roll-off.

A measurement of the half-wave voltage of this modulator at 4 MHz yielded a value of 288 V. This is nearly a factor of 4 above the expected value of 75 V. The loss in sensitivity was probably due to depolarization of the modulated beam in the input fiber.

As more devices begin to function at frequencies above 100 GHz and as the demand for test equipment increases, the need for an ultrafast modulator will continue to grow. We have

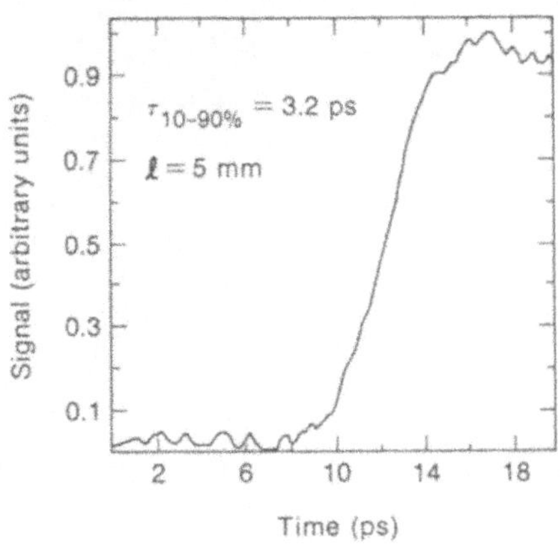

Fig. 4 >100 GHz modulator response

demonstrated a GaAs modulator with velocities nearly matched. This is simply achieved by putting a GaAs superstrate in contact of a GaAs waveguide. Further improvement can be achieved by varying the aluminum mole fraction in a GaAlAs superstrate. This approach should make the construction of 500 GHz possible with lower half-wave voltage. At these frequencies, the skin effect losses will become dominant and superconducting electrodes will have to be used. The superconducting electrodes will have to possess a large energy gap in excess of 0.5 THz. Superconductors such as Nb with a $T_c = 9°K$ and an energy gap corresponding to 700 GHz could be used. High T_c superconductors with an energy gap corresponding to 10 THz could even be a better choice. Recent experiments have shown that picosecond electrical pulses could propagate over distances of 1 cm on high T_c superconducting lines without detectable absorption or dispersion. Furthermore, the dielectric constants of some high T_c substrates nearly match that of GaAs.

The authors wish to thank Chris Chorey and Kul Bhashin of NASA Lewis in Cleveland for supplying GaAs/GaAlAs waveguides. We also thank Brian Olmsted at the University of Rochester and Doug Shire at Cornell University for generating the electrode mask and for depositing electrodes. This work was supported by the Air Force Office of Scientific Research, University Research Initiative under contract number F49620-87-C-0016 and the Sponsors of the Laser Fusion Feasibility Project at the Laboratory for Laser Energetics.

1. M. Izutsu, Y. Yamane, and T. Sueta, IEEE J. Quant. Electron. QE-13, 287 (1977).
2. K. Kubota, J. Noda, and O. Mikami, IEEE J. Quant. Electron. QE-16, 754 (1980).
3. F. J. Leonberger, Opt. Lett. 5, 312 (1980).
4. C. M. Gee, G. D. Thurmond, and H. N. Yen, Appl. Phys. Lett. 43, 998 (1983).
5. P. Buchman, H. Kaufmann, H. Melchoir, and G. Guekos, Appl. Phys. Lett. 46, 462 (1985).
6. D. M. Materna, Case Western Reserve University, Department of Electrical Engineering and Applied Physics, M.S. Thesis (1986).
7. J. F. Whitaker, T. B. Norris, G. Mourou and T. Y. Hsiang, IEEE Trans. Microwave Theory Tech MTT-35, 41 (1987).
8. J. F. Whitaker, R. Sobolewski, D. R. Dykaar, T. Y Hsiang, and G. A. Mourou, IEEE Trans. Microwave Theory Tech. MTT-36, 277 (1988).
9. See for instance, I. P. Kaminov, An Introduction to Electro-Optic Devices (Academic Press, New York, 1974), p. 228.
10. See for instance, E. Yamashita, K. Atsuki, and T. Ueda, IEEE Trans. Microwave Theory Techn. MTT-27, 1036 (1979).

Part V

Applications to Solid-State Physics

Ultrafast Scattering and Energy Relaxation of Optically Excited Carriers in GaAs and AlGaAs

W.Z. Lin, R.W. Schoenlein, M.J. LaGasse, B. Zysset, E.P. Ippen, and J.G. Fujimoto

Department of Electrical Engineering and Computer Science,
Research Laboratory of Electronics,
Massachusetts Institute of Technology, Cambridge, MA 02139, USA

Advances in femtosecond pulse lasers, amplifiers and measurement techniques now make it possible to study the primary events in semiconductors that determine, and ultimately limit, the characteristics of ultrahigh speed electronic and opto-electronic devices[1]. Previous investigators have applied a variety of experimental techniques to study femotsecond carrier dynamics. The initial scattering processes of excited carriers in GaAs and AlGaAs have been investigated using absorption saturation[2]. Continuum probe measurements have been applied to observe hole burning and nonthermal carrier distributions[3]. Orientational relaxation of excited carriers has been investigated using femtosecond four wave mixing[4].

In this paper we describe femtosecond studies of induced absorption saturation in GaAs and AlGaAs that investigate initial carrier scattering, energy relaxation, and intervalley scattering processes. To obtain complementary information on carrier dynamics, measurements were performed on a variety of sample compositions using three different laser systems to produce different excitation and probe wavelengths. Transient absorption saturation measurements were performed using identical pump and probe pulses from a CPM oscillator to investigate the initial scattering of carriers out of their optically excited states[5]. Pump and continuum probe measurements were performed with a copper vapor laser amplifier to observe the evolution and energy relaxation of the excited carrier distributions[6]. Finally, a tunable, synchronously-pumped femtosecond dye laser was used to study intervalley scattering[7].

The samples used in our studies were 0.5 μm layers of GaAs or $Al_xGa_{1-x}As$ grown by liquid phase epitaxy and clad with transparent layers of $Al_{.85}Ga_{.15}As$. All measurements were performed at room temperature with orthogonally polarized pump and probe pulses to reduce coherent coupling. Measurements performed using both parallel and orthogonal polarizations as well as studies by previous investigators[3,4] indicate that orientational relaxation occurs too rapidly to contribute significantly to our measurements. Figure 1 shows the transient absorption saturation in GaAs observed using pump and probe pulses from the CPM oscillator (photon energy $\sim$ 2 eV). The temporal response has two components, an initial ultrafast transient that occurs on the timescale of the pulse and a longer, picosecond decay. The femotsecond response may be attributed to the initial scattering of the carriers out of their optically excited states. The picosecond response results from cooling of the hot carrier distribution through electron phonon interactions. For excitation densities in the range 10^{17} - $10^{18}/cm^3$ the initial relaxation time is shorter than our pulse duration of 35 fs. Detailed analysis of the data is complicated by the fact that carriers are excited from the heavy and light hole bands high into the conduction band where intervalley scattering to both the X and L valleys can compete with intraband redistribution. Transitions from the split-off valence band to lower conduction band energies also occur.

Springer Series in Chemical Physics, Vol. 48: **Ultrafast Phenomena VI**
Editors: T. Yajima • K. Yoshihara • C.B. Harris • S. Shionoya

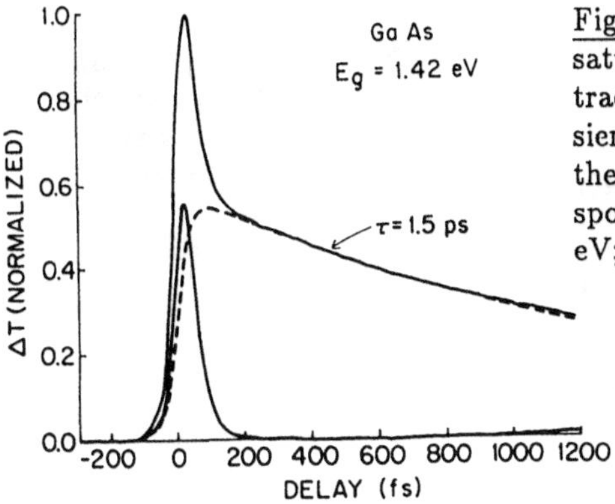

Fig. 1. Transient absorption saturation in GaAs. The lower trace is the rapid initial transient extracted by subtracting the longer (dashed line) response. $E_{pump} = E_{probe} = 1.97$ eV; $N = 10^{18}/cm^3$.

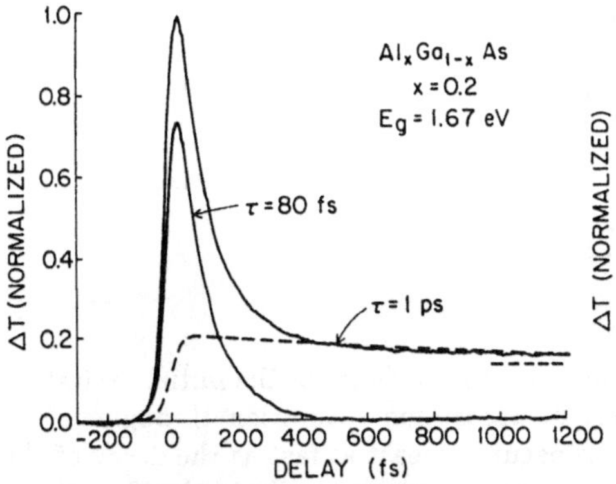

Fig. 2. Transient absorption saturation data for $Al_{.2}Ga_{.8}As$. Dashed lines indicate the long-lived background level and the picosecond exponential decay.

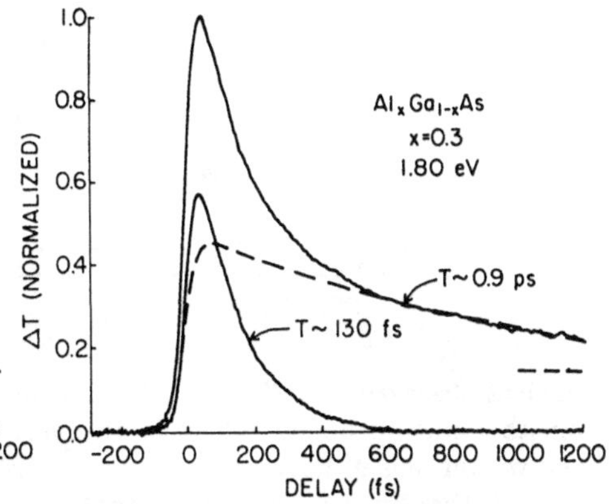

Fig. 3. Experimental data for $Al_{.3}Ga_{.8}As$. Rapid initial transient extracted as in Figs. 1 and 2.

Since the laser source was not tunable, studies were performed by using $Al_xGa_{1-x}As$ samples with differing composition. Experimental results for $x = 0.2$ and $x = 0.3$ are shown in Figures 2 and 3. The slow behavior in each trace has been fit by a combination of a 1 ps decay and a long lived background due to band filling by the cooled carriers. The initial relaxation time has increased dramatically to 80 fs and 130 fs respectively. This is a result of several factors. Since the bandgap increases with Al concentration, initially excited distributions are cooler. Energies of the X and L valleys also increase and the intervalley scattering processes become less likely. At the same time transitions from the split-off valence band are no longer allowed. For $x = 0.1$ all processes are still allowed and the initial relaxation occurs on a comparable time scale to that in GaAs.

In order to investigate the evolution of the carrier distributions and energy relaxation to the lattice we performed experiments using amplified pulses to produce a femtosecond continuum probe. Figure 4 shows a set of traces obtained for GaAs. Data obtained at probe photon energies near the 1.99 eV pump energy show rapid scattering of the carriers out of their initial states consistent with that in Figure 1. The fast

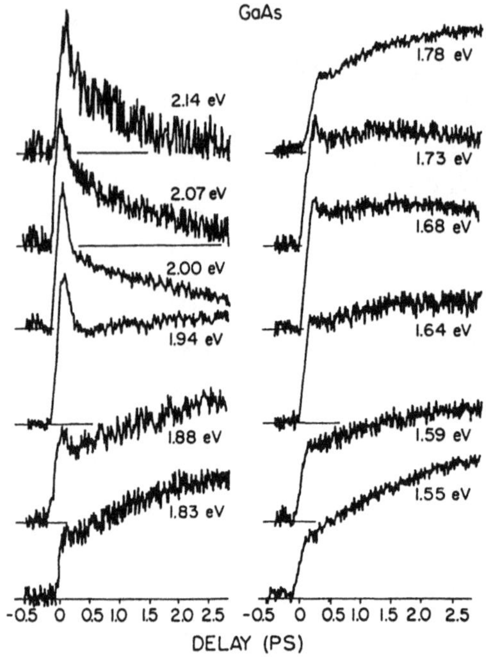

Fig. 4. Absorption saturation data obtained by selecting different values of E_{probe} from the continuum. Actual peak amplitudes range from 1% to 13%. $E_{pump} = 1.99$ eV.

transient observed at 1.73 and 1.68 eV is independent evidence of the initial excitation from the split-off band. There is an onset of absorption saturation throughout the band within tens of femtoseconds. Since this occurs at least as fast as the decay of the localized distributions, it suggests that hole scattering may contribute significantly to the observed absorption saturation data. The picosecond time behavior is consistent with cooling of the carrier distribution and a return of carriers from the satellite valleys as described by previous researchers[8]. The rising wing at intermediate photon energies is the result the split-off transition which is sensitive to carriers at the bottom of the conducton band.

In order to eliminate effects of the split-off band transitions, measurements were performed with $x = 0.2$ and 0.3 samples. Results for $x = 0.2$ are shown in Figure 5. This sample has a bandgap of 1.67 eV and split-off transitions do not occur for our 1.99 eV pump photons. The excess energy of the carriers is sufficient to permit scattering to the L valley. For carriers near the top of the distribution X valley scattering is also permitted. The sharper transient observed for the higher energies is in agreement with this hypothesis. Rapid onset of absorption saturation is again observed throughout the band. The increasing absorption saturation on the picosecond timescale in the 2.07 eV trace is the result of the split-off transition which detects carriers at the bottom of the conduction band. Figure 6 shows results obtained with the $x = 0.3$ sample where the split-off band transitions and X valley scattering is not allowed. Initial relaxation observed near the pump energy is consistent with that in Figure 3; probing near the bottom of the band (and at 2.14 eV via the split-off band) shows the subsequent increasing absorption saturation attributable to cooling. The time scale of this process is more rapid than that observed for the $x = 0.2$ sample, indicating that the cooling rate in the $x = 0.2$ sample may be slowed by the return of carriers from the satellite

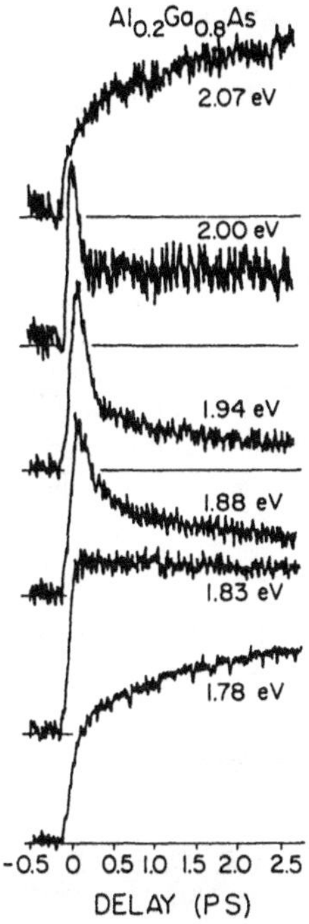

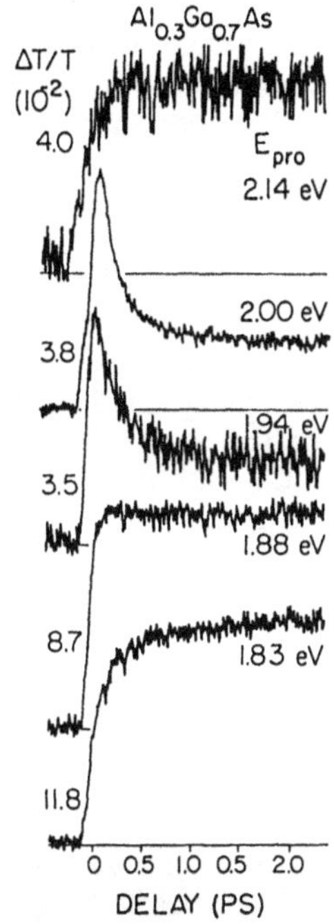

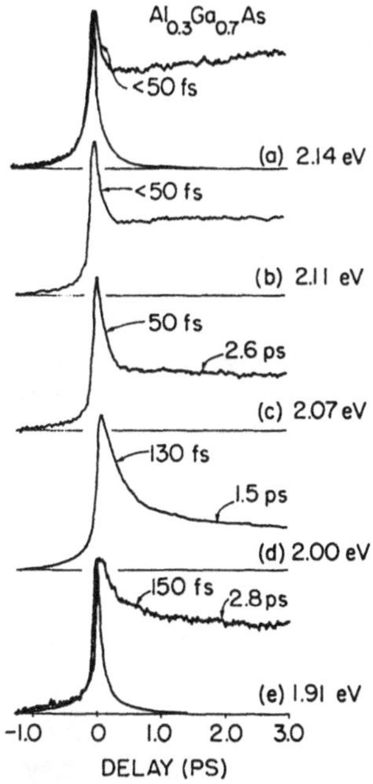

Fig. 5. (left) and Fig. 6. (center) Absorption saturation behavior observed at different continuum probe wavelengths for $x = 0.2$ and 0.3. $E_{pump} = 1.99$ eV.

Fig. 7. Dynamics observed using identical pump and probe pulses from a tunable source to study intervalley scattering.

valleys. Thus, if X valley scattering is allowed it appears to be substantially more effective in removing carriers from the Γ valley than L valley scattering.

Finally, to explicitly examine the role of intervalley scattering, we performed additional measurements with 70 fs pulses from a wavelength-tunable femtosecond system. Using the $x = 0.3$ sample with wavelengths ranging from 580 to 650 nm (2.14 to 1.91 eV) the excitation could be tuned through both Γ to X and Γ to L valley scattering transitions without the occurrence of transitions from the split-off band. Figure 7 shows representative results. For photon energies of 2.07 eV or greater the carriers have sufficient energy to scatter to both X and L valleys. The observed relaxation in this range was less than 50 fs and was limited by the laser pulse duration. As the excitation energy is decreased to 2.00 eV scattering to the X satellite valley can no longer occur and the relaxation time increases to 130 fs. At 1.91, eV, where Γ to L scattering is no longer allowed, the initial relaxation rate appears to slow further but detailed fitting is complicated by increased contributions from band filling.

The time constants measured by absorption saturation do not correspond directly to intervalley scattering times since the absorption changes are affected by all processes which remove carriers from their initial excited states. Nevertheless, our results indicate that Γ to X scattering is very rapid and that a significant number of carriers scatter to the X valleys when carrier energies are sufficient to permit this process. Scattering to the L valleys produces a weaker contribution to the initial carrier relaxation process. This finding is commensurate with our continuum probe measurements as well as with recent measurements utilizing temperature variation[9] and luminescence depolarization[10].

Acknowledgements

We gratefully acknowledge stimulating discussions with C. Stanton and K. Hess and thank R.A. Logan for providing the samples. This work was supported in part by the Air Force Office of Scientific Research Grant 85-0213, the Joint Services Electronics Program DAAL03-86-K002, and the National Science Foundation Presidential Young Investigators Program Grant ECS-8552701. W.Z. Lin was a visiting scholar from Zhongshan University, Guangzhou, People's Republic of China.

References

1. C.V. Shank, *Science*, **233**, 1276 (1986); S.A. Lyon, *J. Luminescence*, **35**, 121 (1986).
2. C.L. Tang and D.J. Erskine, *Phys. Rev. Lett.* **51**, 840 (1983); A.J. Taylor, D.J. Erskine, and C.L. Tang, *J. Opt. Soc. Am. B* **2**, 663 (1985).
3. J. L. Oudar, D. Hulin, A. Migus, A. Antonetti, and F. Alexandre, *Phys. Rev. Lett.* **55**, 2074 (1985); C. H. Brito-Cruz, R. L. Fork, and C. V. Shank, *IQEC'87 Technical Digest*, Baltimore, MD, 82 (1987).
4. J.L. Oudar, A.Migus, D. Hulin, G. Grillon, J. Etchepare, and A. Antonetti, *Phys. Rev. Lett.*, **53**, 384 (1984).
5. W.Z. Lin, J.G. Fujimoto, E.P. Ippen and R.A. Logan, *Appl. Phys. Lett.* **51**, 161 (1987).
6. R.W. Schoenlein, W.Z. Lin, E.P. Ippen and J.G. Fujimoto, *Appl. Phys. Lett.*, **51**, 1442 (1987); W.Z. Lin, R.W. Schoenlein, J.G. Fujimoto and E.P. Ippen, *J. Quant. Electron.* **24**, 267 (1988).
7. M.J. LaGasse, W.Z. Lin, B. Zysset and J.G. Fujimoto, *Proc. CLEO' 88*, Anaheim, CA paper FC2 (1988).
8. D.J. Erskine, A.J. Taylor, and C.L. Tang, *Appl. Phys. Lett.* **45**, 1209 (1984).
9. F.W. Wise, I.A. Walmsley and C.L. Tang, *Appl. Phys. Lett.*, **51** 605 (1987).
10. D.N. Mirlin, I. Ya. Karlik and V.F. Sapega, *Solid St. Commun.*, **65**, 171 (1988).

Time-Resolved Terahertz Conductivity of Photoinjected Hot Electrons in Gallium Arsenide

M.C. Nuss

AT&T Bell Laboratories, Crawfords Corner Road,
Room 4C-330, P.O. Box 400, Holmdel, NJ 07733, USA

Recently, we have directly measured the transient d.c.-mobility $\sigma(t)$ of hot electrons injected with 500 meV excess energy into the conduction band of GaAs using femtosecond electrical pulses synchronized to the optical injection pulse as a probe of the mobility [1]. These experiments have shown that transfer into the L-valleys is extremely fast and that 300 fs after the electron injection ~ 80% of all electrons reside in low mobility satellite valleys. In this article, we describe experiments that use the broad spectral bandwidth of the femtosecond electrical pulses to investigate the complete frequency response as well as the time response of the small-signal conductivity $\sigma(\omega,t)$ in the range from 0 to 2 THz as a function of time after injection of hot electrons. The THz frequency range covers both momentum and energy relaxation rates and deviations from a Drude-like conductivity give valuable information about nonequilibrium transport processes of hot electrons [2].

Figure 1 shows the experimental setup. A subpicosecond conical shock wave of electro-magnetic radiation is generated by optical rectification of a 100-fs optical pulse at 625 nm out of a CPM dye laser in a 1-mm electro-optic $LiTaO_3$-crystal [3]. The shock wave consists of a single cycle electrical pulse with a far-infrared spectrum extending from almost 0 to 4 THz and peaking at ~ 1 THz. This femtosecond electrical transient propagates towards a GaAs sample which is optically contacted to the side of the electro-optic crystal. The sample consists of an undoped 8800A grown GaAs layer capped with a 83A $Al_{0.3}Ga_{0.7}As$-layer grown by MBE on a semi-insulating substrate. A second optical pulse probes the incident and reflected waveforms via electro-optic sampling, measuring the amplitude and phase of the electrical field associated with the far-infrared transient.

Synchronized with the two other pulses, a third pulse injects electrons with an excess energy of ~500 meV into the GaAs conduction band. The wavefront of this optical excitation pulse is tilted using a transmission echelon to provide synchronization with the conical wavefront of the electrical pulse sweeping across the semiconductor sample.

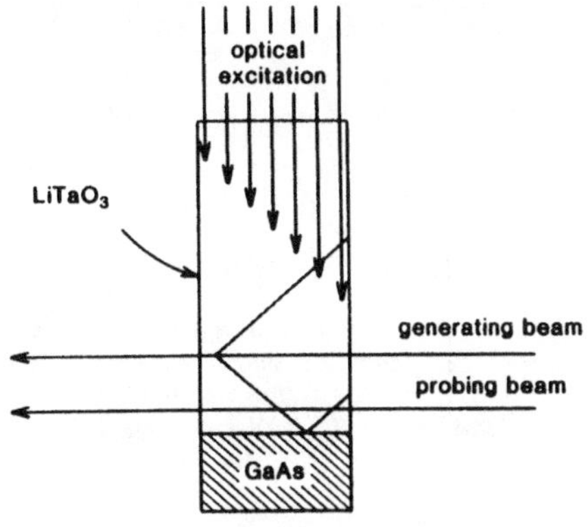

Fig.1: Experimental arrangement

Springer Series in Chemical Physics, Vol. 48: **Ultrafast Phenomena VI**
Editors: T. Yajima · K. Yoshihara · C.B. Harris · S. Shionoya

Unlike in our previous time-domain experiment [1], the generating pulse is scanned rather than the electro-optic sampling beam to map out the waveform of the incident and reflected electrical pulses. This way the temporal delay between the optical excitation and the probe beam remains constant, so that the electron gas always appears stationary during the scan of the electrical pulse. This allows us to Fourier-transform the entire waveform to determine the transient frequency response of the cooling nonequilibrium electron gas at any time after the injection of the hot electrons.

Solving the boundary condition at the interface between $LiTaO_3$ and GaAs yields

$$E_r = r_0 E_i - \frac{1}{Y} \cdot J_S \qquad (1)$$

where E_r, E_i are the reflected and incident electrical fields, Y is the sum of the admittances in $LiTaO_3$ and GaAs and J_S is the sheet current in the thin photoexcited layer at the interface. For the small electrical fields of the far-infrared transient (~ 50 V/cm), Ohms law applies:

$$J_S(\omega, T_{ex}) = \int_0^{T_{ex}} dt' \, \sigma_S(\omega, T_{ex} - t') \cdot E(\omega, t') \, , \qquad (2)$$

here, $\sigma_S(\omega)$ is the sheet conductivity and T'_{ex} denotes the delay of the optical excitation beam. $E(t) = E_i(t) + E_r(t)$ is the total electric field acting on the electrons at the interface. When the temporal variation of the electrical field is slower than the momentum scattering time, then Eq.2 simplifies to: $J_S(\omega, T_{ex}) = \sigma_S(\omega, T_{ex}) \cdot E(\omega, T_{ex})$. This relation is usually satisfied, since the momentum scattering times are between ~ 10-fs for hot electrons and 150-fs for electrons at the lattice temperature, being much shorter than the time over which the electrical pulse varies (0.5 - 1 ps). However, Eq.2 imposes a limit on the temporal resolution with which we can determine the frequency-dependent conductivity $\sigma_S(\omega)$ because each waveform has within a time constant on the order of the momentum scattering time a memory of the past. For the assumption of instantaneous response we can then solve Eq.1 to obtain the conductivity:

$$\sigma_S(\omega) = Y \cdot \frac{r_0 - r(\omega)}{1 + r(\omega)} \quad . \qquad (3)$$

The results for the real part of the sheet conductivity $\sigma_S(\omega)$ are shown in Fig.2 for delay times T_{ex} ranging from 300 fs to 14 ps after the injection of hot electrons with 500

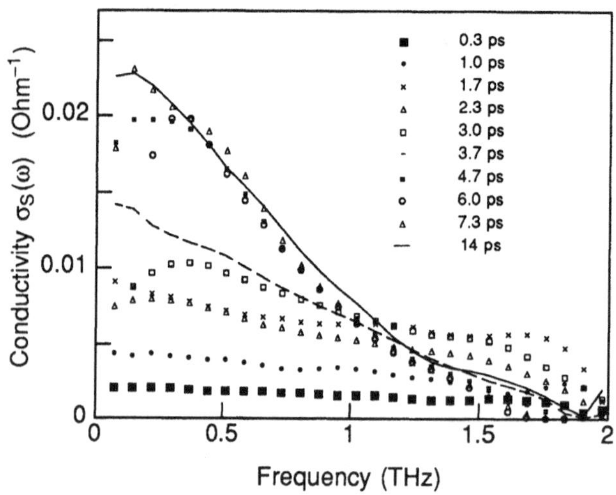

Figure 2: Conductivity $\sigma_S(\omega)$ for various time delays after photoinjection of hot electrons with 500 meV excess energy

meV excess energy. At very early times, the sheet conductivity is very low, supporting our earlier findings in the time domain that the electrons scatter within 300 fs very efficiently into the low-mobility L-valley. Starting at ~2 ps, a rising conductivity at frequencies slightly above 1 THz is observed. The appearance of this peak in the high-frequency mobility is similar to the a.c.-velocity-overshoot predicted by Monte-Carlo simulations for the conductivity of a hot-electron gas heated by an electric field [2]. The exact frequency dependence is different, however, and can probably be assigned to the non-thermal electron distributions resulting from optical excitation. In particular, the transient conductivity at ≥1 THz actually exceeds the conductivity of the relaxed electron distributions for times between 1.5 and 3 ps after the injection of hot electrons. After 4 ps, the conductivity at lower frequencies starts to increase and eventually becomes similar (but not identical) to the conductivity of a Drude-gas $\sigma_S^{drude} = n_S e^2 \tau / [m^* \cdot (1 + \omega^2 \tau^2)]$ with $\tau_m \approx 180$ fs and $n_S \sim 3\times10^{13}\mathrm{cm}^{-2}$ as best fit to the data after 14 ps.

An increase in injected carrier density to $n_S = 5\times10^{14}\mathrm{cm}^{-2}$ (not shown in this article) reduces the transient conductivity peak at >1 THz significantly. We believe that this is mostly due to the increased scattering with the holes that are injected optically at the same time as the electrons.

In conclusion, we have been able to measure the transient frequency-domain conductivity $\sigma(\omega)$ of electrons injected with a 100 fs optical pulse at 500 meV excess energy with ~ 300 fs time resolution as they cool down to the lattice temperature. Significant deviations from a Drude-like conductivity are observed. We have also shown that the conductivity at ≥ 1 THz of hot electrons injected optically can exceed the conductivity of the completely cooled distribution between 1.5 and 3 ps after electron injection. Increased scattering due to electron-hole scattering at higher excitation densities diminishes the overshoot effect drastically.

We would like to thank D. H. Auston for many stimulating discussions and suggestions and T. E. Harvey for his expert assistance with the experiment.

1. M. C. Nuss, D. H. Auston and F. Capasso, Phys. Rev. Lett. 58, 2355 (1987)
2. e.g. E. Constant in "Hot-Electron Transport in Semiconductors", ed. by L. Reggiani, Topics in Applied Physics, Vol.58, Springer-Verlag (1985)
3. D. H. Auston and M. C. Nuss, IEEE Journ. Quant. Electr. 24, 184 (1988)

Femtosecond Transients and Dynamic Stark Effect in Semiconductors

N. Peyghambarian[1], *B. Fluegel*[1], *S.W. Koch*[1], *J. Sokoloff*[1], *M. Lindberg*[1], *M. Joffre*[2], *D. Hulin*[2], *A. Migus*[2], *and A. Antonetti*[2]

[1]Optical Sciences Center, University of Arizona, Tucson, AZ 85721, USA

[2]Laboratoire d'Optique Appliquée, Ecole Polytechnique, ENSTA, F-91120 Palaiseau, France

1. Abstract

The optical Stark effect has been studied experimentally and theoretically in thin CdS platelets with femtosecond laser pulse excitation. For early times before the arrival of the pump pulse peak, we observe oscillatory behavior in the differential transmission spectra in the vicinity of the exciton resonance. The oscillations evolve into the dispersive feature which is characteristic for the optical Stark effect. In order to minimize the generation of real excitations the experiments are performed using low intensities and large laser-exciton detunings. In addition to the Stark effect of the exciton states, we also observe the Stark effect of the continuum states. The experimental results are analyzed theoretically solving the recently derived generalized semiconductor Bloch equations for femtosecond excitation conditions.

2. Summary

The optical Stark effect is a well-known phenomenon in atomic systems where an atomic transition shifts by application of an off-resonance light beam COHEN-TANNOUDJI et al.[1], GALITSKII et al.[2] and LIAO et al.[3]. The effect persists for the duration of the optical pulse and has been explained using the "dressed atom" picture.

In semiconductors, one deals with a combination of bound electron-hole-pair (exciton) resonances and unbound continuum states. The optical Stark effect of excitons in multiple quantum wells and in Cu_2O has been observed recently FRÖHLICH et al.[4], MYSYROWICZ et al.[5], VONLEHMEN et al.[6], TAI et al.[7] and FRÖHLICH et al.[8]. Since the effect occurs on the timescale of the excitation pulse and no real carriers are involved, the exciton Stark effect has been used to construct optical logic gates with sub-picosecond response times HULIN et al.[9].

When discussing the pure Stark effect in semiconductors, one has to keep in mind that, quite generally, there is always a competition between effects caused by real carriers and the electric field effects (Stark effect). The previous experiments have been conducted with pump pulses tuned relatively close to the exciton transition and, therefore, the unavoidable generation of real carriers may have influenced the observed behavior MYSYROWICZ et al.[5]. Even when the light frequency is detuned from the resonance such that one-photon carrier generation is very small, two-photon absorption may lead to a finite concentration of real excitations. These carriers modify the transmission spectra through energy renormalizations, as well as bleaching and broadening of resonances. In order to minimize these long-lived effects (real excitations decay on the timescale of the interband recombination time) and to approach the regime of the true optical Stark effect, we have employed large detunings and weak excitation pulse intensities.

The experimental arrangement is a usual pump-probe technique. The output of a colliding pulse mode-locked dye laser is amplified by a copper vapor laser and the output is divided into two parts. One part goes through an ethylene glycol jet and produces a broad-

Springer Series in Chemical Physics, Vol. 48: **Ultrafast Phenomena VI**
Editors: T. Yajima · K. Yoshihara · C.B. Harris · S. Shionoya

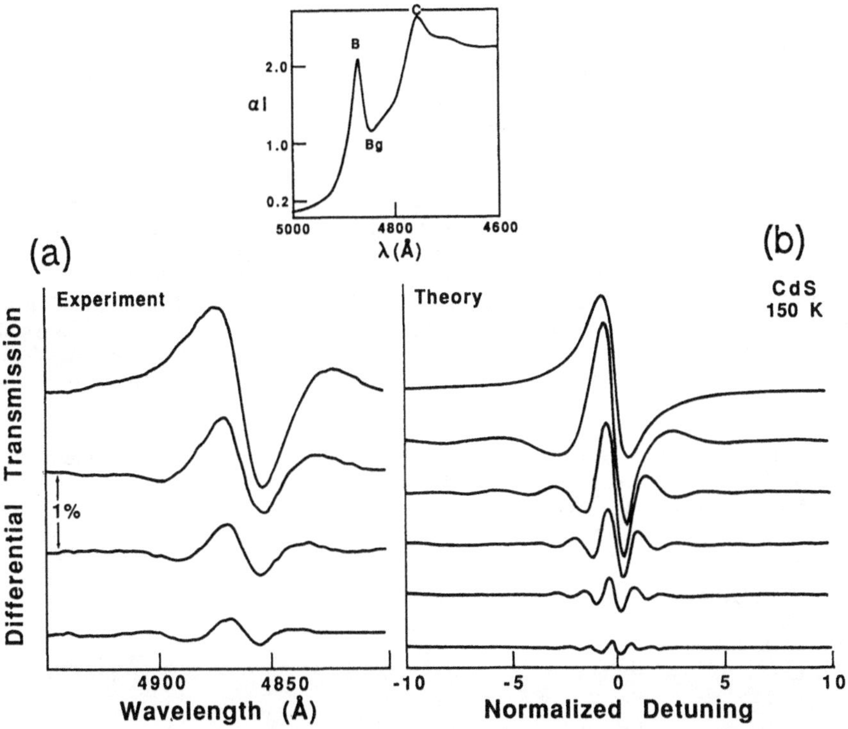

Fig. 1 a) The measured oscillatory behavior of DTS for negative time delays. The spectra are 50 fs apart. b) The calculated DTS for negative time delays. Inset: The linear absorption spectrum of CdS at 150 K for $E_p \parallel C$

band continuum to be used as a probe pulse. The other part is sent through a delay generator and is used as a pump pulse with 60 fs duration (FWHM of the autocorrelation trace). A spectrometer and an optical multichannel analyzer detect the probe transmission as a function of frequency in the absence and presence of the pump and at various time delays between the pump and probe pulses.

Figure 1a shows the measured differential transmission spectra, DTS = $(T-T_0)/(T_0)$, where T and T_0 are the probe transmission in the presence and absence of the pump pulse, near B-exciton in CdS. The linear absorption spectrum of CdS is shown in the inset of this figure. The sample thickness was a few tenths of a micrometer. In CdS, the valence band is split into one upper and two lower subbands. These bands are called A, B and C, respectively, and result in three excitonic levels. We used a probe polarization that was parallel to the crystal c-axis. The A-exciton is then dipole-forbidden and appears only as a shoulder on the low-energy side of the B-exciton, while the B and C-excitons appear fully in the absorption spectrum. The bandedge of the B-exciton is also labeled on the inset. In Fig. 1a the spectra which were taken in 50-fs intervals, are shown only for negative time delays when the peak of the pump pulse precedes the peak of the probe pulse. Spectral transmission oscillations are clearly observed FLUEGEL et al.[10], JOFFRE et al.[11], SOKOLOFF et al.[12] and KOCH et al.[13]. The oscillation period (the interfringe spacing) grows as the time delay, t_p, approaches zero.

In the neighborhood of zero time delay ($t_p = 0$), the oscillatory features disappear and one is left with a dispersive-looking transmission change which is generally interpreted as the optical Stark shift. As is emphasized in the theoretical analysis presented below, in general, one cannot clearly separate the Stark effect from the transmission changes caused by the total material-pulse interaction. Strictly speaking one has a pure "Stark shift" only for situations when the changes in the excitation pulse occur on timescales which are long in comparison with the medium coherence time. This situation is not realized under the present conditions of ≅ 60 fs pulses and ≅ ps exciton coherence time. Therefore, we refer to the complete scenario of the differential transmission changes for the case of non-resonant excitation as the "optical Stark effect". Figure 2 shows the measured DTS in the vicinity of the B-exciton and C-exciton. The Stark effect of the B-exciton and continuum states (bandedge and higher exciton states) of the B-band at T = 150 K is observed around zero time delay, as evidenced by the dispersive DTS feature in the vicinity of the excitonic peaks and a positive signal in the bandedge region. The effect of real excitations on the DTS is very small. This can be tested by comparing the measured spectrum at $t_p = 0$ with the DTS obtained from an artificial shift of the linear absorption. In Fig. 2, we have plotted, in addition to the measured DTS, the DTS obtained by artificially shifting the linear absorption. The good agreement between the two curves in Fig. 2 reveals that one may obtain at low intensities and for large detunings an almost pure Stark shift. In addition, the continuum states have also shifted to higher energies.

In order to theoretically analyze the observed transmission changes we use the generalized semiconductor Bloch equations as derived by LINDBERG and KOCH [14]. The coherent part of these equations can be written as

$$\frac{\partial}{\partial t} f_{e/h}(k) = i\left[\mu E(t) + \sum_{q\neq 0} V(q)P^*(q+k)\right]P(k) + c.c. \tag{1}$$

$$\left[\frac{\partial}{\partial t} - i\Delta\epsilon_{eff}(k)\right] P(k) = -i\left[\mu^* E(t)^* + \sum_{q\neq 0} V(q)P(k+q)\right][1 - f_h(k) - f_e(k)]\,, \tag{2}$$

where $f_{e/h}(k)$ is the probability to have an electron/hole with wavenumber k, E(t) is the applied field, V denotes the Coulomb potential, P(k) is the interband polarization of state k, μ is the interband dipole matrix element, and $\Delta\epsilon_{eff}$ denotes the renormalized single-particle energy, respectively. The term $\sum V(q)P(k+q)][f_h(k) + f_e(k)]$ in Eq. (2) is responsible for

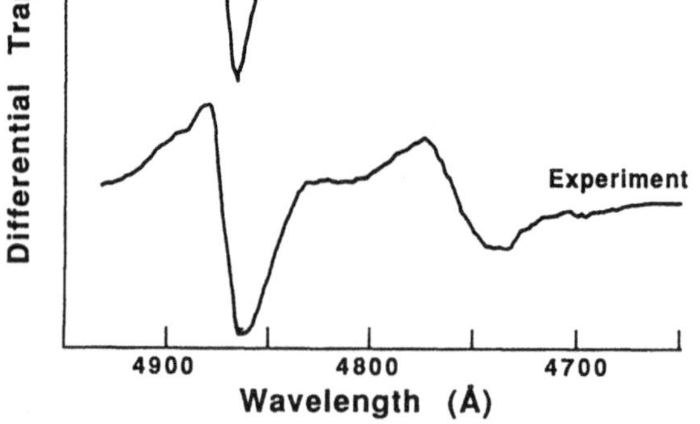

Fig. 2 The measured DTS at 150 K and its comparison with a pure shift of the absorption spectrum (see the text for description)

the so-called phase-space-filling nonlinearities. For the conditions in the discussed experiments, we assume that this term, and consequently also the exchange contribution to the bandgap renormalization are small and can be neglected. To evaluate Eqs. (1) and (2) under these conditions, it is advantageous to transform these equations to real space. We can then use the transformation $A_\lambda = \int dx\, \psi^*_\lambda(x)\, A(x)/\psi^*_\lambda(0)$ where A(x) stands for either $f_{e/h}(x)$ or P(x) and $\psi_\lambda(x)$ is the solution of the Wannier equation for the relative motion of the electron-hole pair. The index λ runs over the bound states (exciton states) as well as over the electron-hole continuum. We have shown in Ref. 14 that the transformed quantities P_λ and $f_{e/h,\lambda}$ obey coupled equations which have the same form as the optical Bloch equations for two-level systems. These equations can be solved for a given input field E(t) and the total material polarization is then obtained as $P_{tot} = \mu \sum \left| \psi_\lambda(0) \right|^2 P_\lambda$ + c.c. Knowing P_{tot} we compute the field transmitted through the sample and, hence, the differential transmission spectra using Maxwell's equations.

Using only a single exciton resonance, we obtain the spectra shown in Figure 1b for negative time delays. Comparison with Fig. 1a reveals the good agreement between theory and experiment. We see clearly that the transmission oscillations for negative time delays evolve into the dispersive-looking feature characteristic for the optical Stark shift. However, it is worthwhile to point out that a close inspection of the computed spectra shows that the amplitude of the transmission changes reaches its maximum value at negative time delays (around -50 fs for a 60 fs pulse). In Fig. 3 we show computed DTS for different time delays where we include not only the exciton resonances, but also the electron-hole continuum states. Around zero time-delay we obtain a differential transmission spectrum which closely resembles the experimental results shown in Fig. 2.

In summary, we have presented an experimental and theoretical analysis of the optical Stark effect of exciton and continuum states in semiconductors for different delay times between pump and probe pulses. In the experiments we have minimized the effects of real carriers by choosing a large detuning and low intensity of the excitation pulse. The discussed theory includes the dynamic variations of the applied field and is in this respect an extension of the results obtained under adiabatic elimination of the polarization dynamics SCHMITT-RINK and CHEMLA [17], SCHMITT-RINK et al.[18], COMBESCOT and COMBESCOT [19], ZIMMERMAN [20] and ELL et al.[21]. It can be shown that the importance of the dynamic coherent DTS modifications decreases for increasing temporal pulse widths. Accordingly, our theory justifies the adiabatic elimination of the polarization dynamics in this limit. For sufficiently long and strong pulses, where phase space filling and gap renormalization become important, the results of SCHMITT-RINK et al.[18], ZIMMERMAN [20] and ELL et al.[21] are an improvement of our results which neglect these effects.

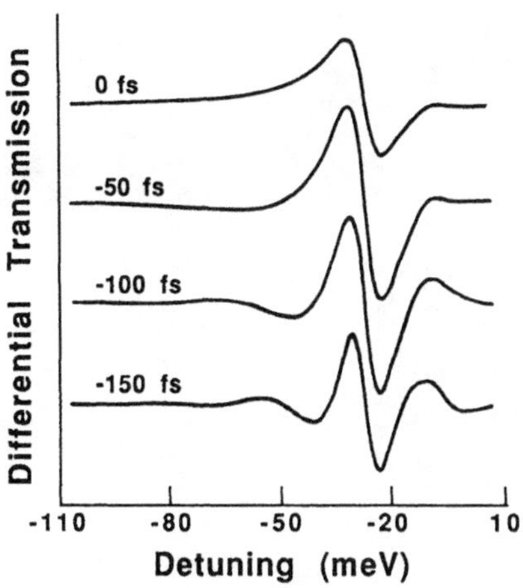

Fig. 3 Differential transmission spectra calculated for B-exciton in CdS at different time delays. Excitation pulse 60 fs, time delays (from bottom to top) -150 fs, -100 fs, -50 fs, 0 fs. The x-axis shows the detuning from the B-exciton bandedge (0) in units of meV. The y-axis is in arbitrary units, the curves are shifted with respect to another, but all curves in one plot are scaled the same way. The exciton linewidth (HWHM) is taken as 5 meV

3. Acknowledgements

The authors would like to acknowledge support from the Optical Circuitry Cooperative of the University of Arizona, the National Science Foundation (grant numbers 8610170, EET8620265 and travel grant), JSOP, NATO (travel grant numbers 86/0749 and 87/0736), ONR/SDIO (grant number N00014-86-K-0719), DARPA/RADC (grant number F30602-87-C-0009), and the John von Neumann Computer Center for the CPU time.

REFERENCES

1. C. Cohen-Tannoudji and S. Reynand, J. Phys. B. **10**, 345 (1977).
2. V. M. Galitskii, S. P. Goreslavskii, and V. F. Elesin, Sov. Phys. JETP **30**, 117 (1970).
3. P. F. Liao and J. E. Bjorkholm, Phys. Rev. Lett. **34**, 1 (1975).
4. D. Fröhlich, A. Nöhte, and K. Reimann, Phys. Rev. Lett. **55**, 1335 (1985).
5. A. Mysyrowicz et al., Phys. Rev. Lett. **56**, 2748 (1986).
6. A. VonLehmen et al., Opt. Lett. **11**, 609 (1986).
7. K. Tai, G. Hegarty, and W. T. Tsang, Appl. Phys. Lett. **51**, 152 (1987).
8. D. Fröhlich, R. Wille, W. Schlapp, and G. Weimann, Phys. Rev. Lett. **59**, 1748 (1987).
9. D. Hulin et al., Appl. Phys. Lett. **49**, 749 (1986).
10. B. Fluegel et al., Phys. Rev. Lett. **59**, 2588 (1987).
11. M. Joffre et al., Opt. Lett. **13**, 276 (1988).
12. J. P. Sokoloff et al., accepted for publication in Phys. Rev. B.
13. S. W. Koch, N. Peyghambarian, and M. Lindberg, J. Phys. C (Review), to be published
14. M. Lindberg and S.W. Koch, Phys. Rev. **B**, Aug. 15, 1988
15. M. Lindberg and S. W. Koch, Journ. Opt. Soc. Am. **B5**, 139 (1988).
16. M. Joffre et al., this issue.
17. S. Schmitt-Rink and D. S. Chemla, Phys. Rev. Lett. **57**, 2752 (1986).
18. S. Schmitt-Rink, D. S. Chemla, and H. Haug, Phys. Rev. B **37**, 941 (1988).
19. M. Combescot and R. Combescot, Phys. Rev. Lett. **61**, 117 (1988).
20. R. Zimmerman, Phys. Stat. Sol. **b146**, 371 (1988).
21. C. Ell, J. F. Müller, K. El Sayed, and H. Haug, private communication.

Bleaching of an Exciton Line Using Sub-T_2 Pulses: Artifact or Reality?

M. Joffre[1,2], *D. Hulin*[1,2], *J.P. Chambaret*[1], *A. Migus*[1], *A. Antonetti*[1], *and C. Benoit à la Guillaume*[2]

[1]Laboratoire d'Optique Appliquée, Ecole Polytechnique, ENSTA, F-91120 Palaiseau, France

[2]Groupe de Physique des Solides de l'Ecole Normale Supérieure, Université Paris VII, 5 place Jussieu, F-75005 Paris, France

Laser sources have now been pushed into the femtosecond domain raising therefore the problem of the temporal limits of time resolved absorption spectroscopy. In the case of absorption a spectrally broadband pulse probes the sample transmission before, during and after excitation by a pump pulse. A spectrometer and a slow detector placed after the sample measures the transmitted spectrum. In this case it is known that the temporal resolution Δt of the experiment may not be related to the spectral resolution $\Delta\nu$ as determined by the spectrometer. Therefore Δt can be much shorter than $1/\Delta\nu$. This looks very intriguing since it seems to get round the Heisenberg uncertainty relation. This problem is only worth mentioning when the spectrometer has something to resolve, i.e. in case of narrow enough absorption lines, that is in a medium with long coherence time T_2. In fact the uncertainty principle becomes important when T_2 gets longer than the light pulse duration, in which case new methods have to be devised for getting the true dynamics of the phenomena under investigation.

Figure 1 represents the spectral response of a gallium arsenide bulk sample excited by a pump pulse tuned far above the band gap. The basic physics here is the following: carriers created by the pump pulse screen the Coulomb interaction between electrons and holes forbidding the

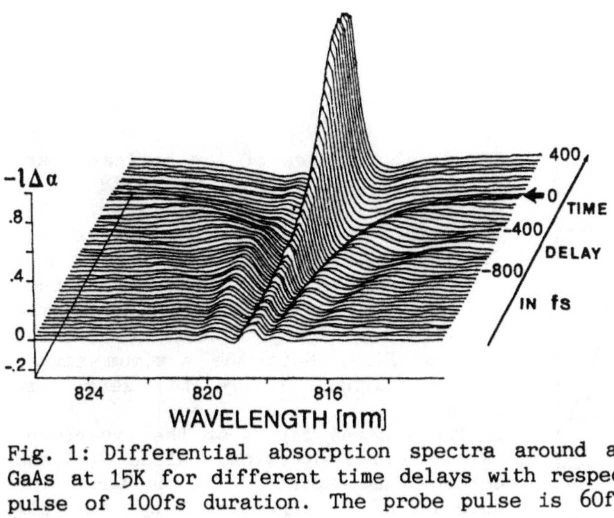

Fig. 1: Differential absorption spectra around an exciton line in bulk GaAs at 15K for different time delays with respect to a 670nm excitation pulse of 100fs duration. The probe pulse is 60fs long. Curves are 40fs apart.

creation of excitons and therefore making the excitonic resonance disappear. Two main striking features are readily apparent: a signal appears long before the zero (negative delays τ mean probe pulse arriving before the pump pulse) and manifests itself as strong spectral oscillations. This suggests that the interpretation of these data as time dependent absorption could be erroneous.

These paradoxal features result from a polarization decay time T_2 of the medium longer than both the time resolution Δt and the characteristic evolution time of the medium, here the ultrafast screening of Coulomb interaction[1]. The polarization of the medium after excitation by the ultrashort probe pulse reads

$$P(t) = P^{(1)}(t)(1 - G(t+\tau)) \tag{1}$$

where $P^{(1)}(t)$ is the linear polarization, in absence of the pump. G(t) describes the disappearance of the exciton resonance and varies abruptly from 0 to 1 around t=0. It is clear on (1) that the polarization induced by the probe pulse is affected by the pump pulse, even though there is no temporal overlap between the two pulses, explaining why there is a signal for negative time delays. The spectra observed in fig. 1 can be reproduced through a Fourier transformation of the differential polarization as deduced from (1).

Oscillatory features also appear in spectra corresponding to the exciton optical Stark shift[2,3]. This is included in the expression of the polarization:

$$P(t) = P^{(1)}(t)\exp\left(-i\int_0^t \frac{\Delta E(t'+\tau)}{h}dt'\right) \tag{2}$$

where ΔE is the exciton shift induced by the pump, theoretically known to follow the pump temporal profile. Fig. 2a represents both the amplitude of

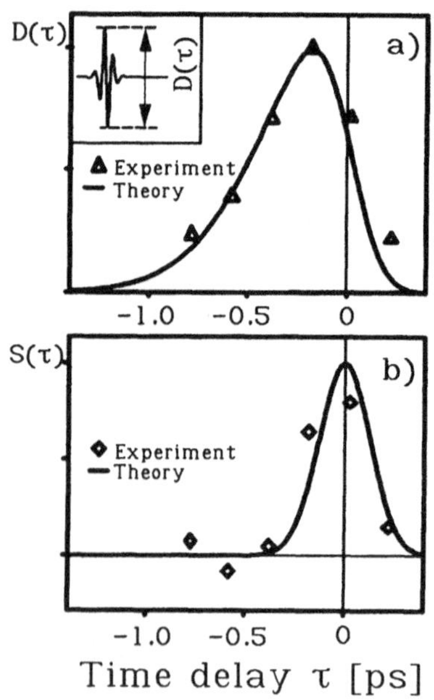

Fig 2: a) Dynamics of the optical Stark effect signal, called here $D(\tau)$ and defined as shown in the inset. The triangles are experimental while the full line is a theoretical fit. We have used a gaussian 300fs FWHM for the pump and a gaussian 530fs HWHM for the polarization decay associated to the exciton line. Note the maximum signal occurs slightly before the zero time delay

b) The squares represent the experimental recovered dynamics, called $S(\tau)$ and obtained as explained in the text. The pump pulse is represented by the full line the experimental differential signal and a theoretical fit based on (2). The signal appears also in this case long before the zero time delay. However the optical Stark shift being instantaneous the signal goes very fast back to zero for positive time delays, producing this very striking "reversed" dynamics.

From the two examples above it follows that monitoring a selected wavelength as a function of time delay τ does not yield the real dynamics of the medium. A method to get the true response is therefore needed.

As the problem comes from the wish of simultaneous temporal and spectral accuracy, it disappears when one gets rid of the spectral resolution. This can be done by integrating the differential spectra, so that the oscillations occuring for negative time delays average to zero. For the same purpose one can also open the slit of the spectrometer until the resolution is of the order of $1/\Delta t$, verifying therefore the uncertainty relation. This is the correct method to obtain the bleaching kinetics but cannot be used for the exciton shift since in this case the integrated spectrum is always zero.

The real optical Stark shift dynamics can be recovered by integrating the product of the differential spectra and of the detuning to the line center, as will be shown in a forthcoming publication[4]. Such an operation has been performed on experimental results and is plotted in fig. 2b. It shows that the recovered kinetics is indeed very close to the pump intensity temporal profile.

In summary we have studied phenomena faster than the coherence decay time of the investigated medium. We show that the observed variations of the absorption spectrum are slower than the actual physical effect but we give methods to recover the true dynamics.

REFERENCES

1. M. Joffre, D. Hulin, A. Migus, A. Antonetti, C. Benoit à la Guillaume, N. Peyghambarian, M. Lindberg and S.W. Koch, Opt. Lett. 13, 276 (1988)
2. B. Fluegel, N. Peyghambarian, G. Olbright, M. Lindberg, S.W. Koch, M. Joffre, D. Hulin, A. Migus and A. Antonetti, Phys. Rev. Lett. 59, 2588 (1987)
3. N. Peyghambarian et al., this issue
4. M. Joffre, D. Hulin, A. Migus and A. Antonetti, submitted to J. Mod. Opt.

Picosecond-Laser-Driven Shock Wave Dynamics in Semiconductors

X.Z. Lu, R. Garuthara, S. Lee, and R.R. Alfano

Institute for Ultrafast Spectroscopy and Lasers,
The City College of New York, New York, NY 10031, USA

1. Introduction

Picosecond-laser-driven shock waves offer a new technique to investigate material properties under very high pressure and pressure gradient with high time resolution, spatial locality and high repetition rate using a table top laser facility. Recently progress has been made by our group in the characterization of picosecond-laser-driven shock waves,[1] shock wave effects in water,[2] and semiconductors: GaSe,[3,4,5] CdSe,[6] and GaAs.[7] In this paper, we will present the main results of our investigations.

2. Experimental Technique

The experimental geometry is shown in Fig. 1. A pump-and-probe technique was used to observe shock effects on the photoluminescence (PL) from semiconductors.[4] The intense picosecond laser pulses of the pump beam were focused onto an Al foil. High pressure shock waves were generated by rapid localized heating and sudden plasma expansion in the Al foil against the glass cover.[8] The probe pulse was delayed by 35 ~ 50 ns relative to arrival of the pump pulse at the Al foil. Samples were polished down to 40 µm in thickness. The aluminum foil had a thickness of 20 µm. The pump beam was focused to a 450 µm diameter spot on the Al foil. The probe beam was focused to a 350 µm diameter spot on the sample. Since the pump beam damaged the Al foil, a different spot on the Al foil was employed after each shock. The PL was dispersed by a 1/4 m Jarrell-Ash spectrometer, detected by a silicon-intensified-target coupled to an optical multichannel analyzer OMA III, and stored in a PDP11/23+ computer. All spectra were recorded by single shots of the laser.

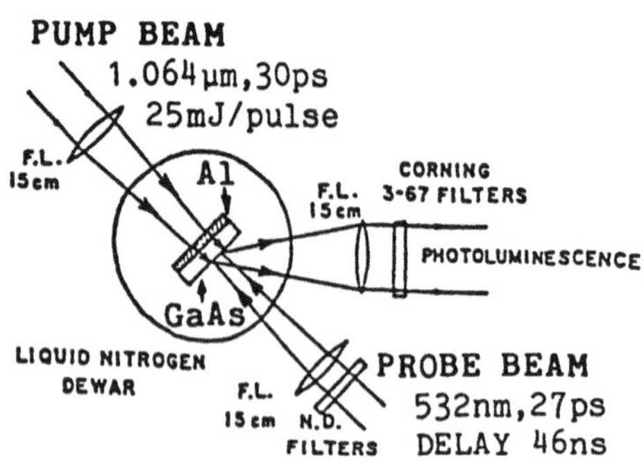

Fig. 1 The experimental geometry

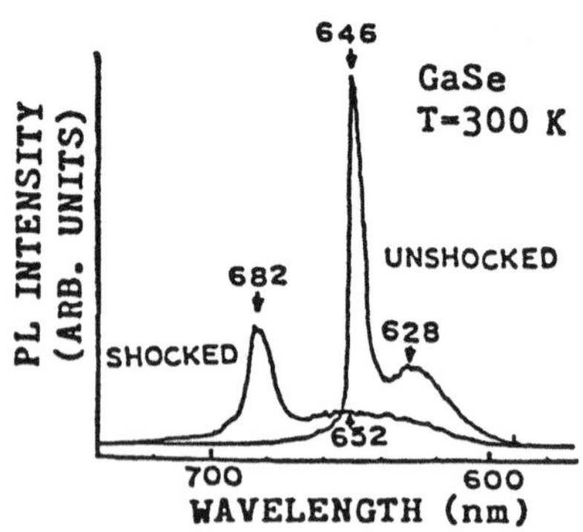

Fig. 2 Unshocked and shocked PL spectra from GaSe

Springer Series in Chemical Physics, Vol. 48: **Ultrafast Phenomena VI**
Editors: T. Yajima · K. Yoshihara · C.B. Harris · S. Shionoya

3. Results and Discussion

3.1 GaSe

Figure 2 shows shocked and unshocked PL spectra from GaSe for a probe intensity of 6.8×10^7 W/cm² at T = 300 K. A sharp stimulated emission line superimposes on the low energy tail of the broad spontaneous emission line. Both lines come from the exciton-exciton scattering process,[9] which can be expressed as $(E_K^{1s}, E_{K'}^{1s}) \rightarrow (h\nu, e\text{-}h)$. Under shock loading with a pump pulse energy 35 mJ and delay time 39 ns, the spontaneous line shifts to red and broadens. The red-shift of 24 nm (73 mev) displayed in Fig. 2 deduces a shock pressure of ~ 15 kbar, which agrees with our independent measurement using an x-cut quartz transducer. The observed line broadening was found to be proportional to the shock pressure (see Fig. 3) and attributed to the shock-wave-induced exciton collisions.[5] The larger red-shift of 36 nm and intensity decrease of the stimulated emission line under shock were well explained by the shock-wave-induced band gap shrinkage through the gain reduction mechanism based on exciton-exciton scattering process.[4]

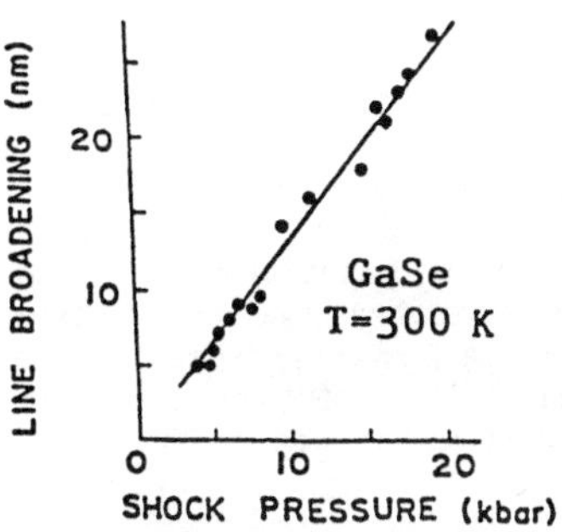

Fig. 3 Shock-induced line broadening

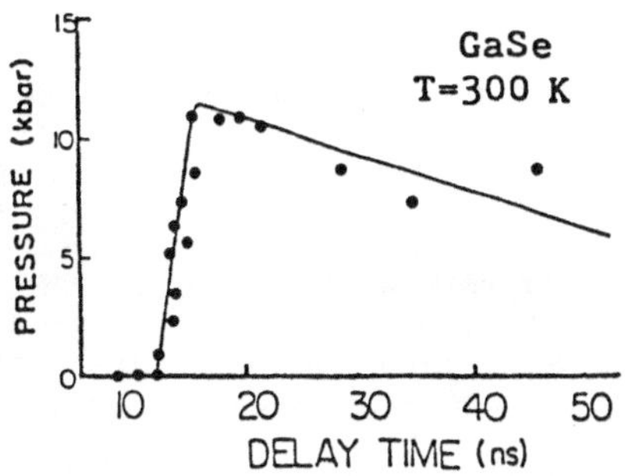

Fig. 4 Temporal profile of shock pressure

The time evolution of the shock pressure could be sketched out by measuring the red-shifts of the spontaneous emission line from GaSe at different delay times between the probe and pump beams. The results for a GaSe sample of 20 μm in thickness are shown in Fig. 4. The apparent pressure rise time displayed in Fig. 4 is 3 nsec which should be the upper limit of the real one because of the limited time resolution and unevenness of the Al-foil and sample thickness. The measured shock duration in GaSe is about 80 nsec.[3]

3.2 CdSe

Figure 5 shows shocked and unshocked PL from CdSe due to Γ_{7c} electron-Γ_{9v} hole recombination for a probe intensity of 1.2×10^7 W/cm² at T = 80 K. The shocked spectrum with a pump pulse energy 25 mJ and delay 35 ns shows a blue shift which deduces a shock pressure of 11 kbar. The carrier temperature T_e was obtained by fitting the high-energy tail of the data to

$$I(E) = C\,(E-\tilde{E}_g)^{1/2} \exp[-(E-\tilde{E}_g)/k_B T_e] \,.$$

The fit results are shown by the solid lines in Fig. 5. Significant carrier heating (126°K → 162°K) by the shock compression via the lattice was observed for the first time.[6]

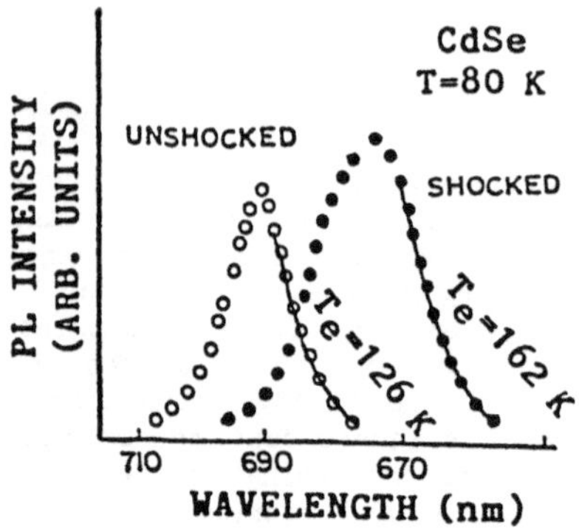

Fig. 5 Unshocked and shocked PL spectra from CdSe

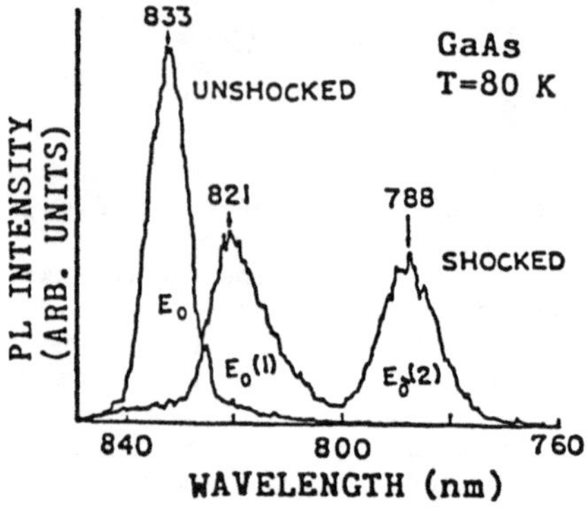

Fig. 6 Unshocked and shocked PL spectra from GaAs

3.3 GaAs

Figure 6 shows shocked and unshocked PL from GaAs due to Γ_{6c} electron-Γ_{8v} hole recombination for a probe intensity of 1.6×10^7 W/cm² at T = 80 K. The shocked spectrum with a pump pulse energy 25 mJ and delay 46 ns shows a blue shift due to the hydrostatic component of the pressure. It also splits into two components corresponding to the light hole transition $E_0(1)$ and heavy hole transition $E_0(2)$ due to the reduced symmetry by the uniaxial nature of the shock pressure.[7] Since the energy gap E_c-E_{v2} (heavy hole) expands linearly with the increasing pressure, the blue shift of the heavy hole line was chosen to calibrate the shock pressure which was ~ 10 kbar in GaAs.

4. REFERENCES

1. K. P. Leung, S. S. Yao, and R. R. Alfano: in Shock Waves in Condensed Matter 1983, ed. by J. R. Asay, R. A. Graham, and G. K. Straub (Elsevier, Amsterdam 1984) p. 343
2. K. P. Leung, A. G. Doukas, P. H. Jones, D. Paradimitriou and R. R. Alfano: Phys. Rev. B 31, 8329 (1985)
3. K. P. Leung, S. S. Yao, A. G. Doukas, and R. R. Alfano: Phys. Rev. B 31, 942 (1985)
4. X. Z. Lu, R. Rao, B. Willman, S. Lee, A. G. Doukas and R. R. Alfano: Phys. Rev. B 35, 7515 (1987)
5. X. Z. Lu, S. Lee, R. Garuthara and R. R. Alfano: Appl. Phys. Lett. 51, 1789 (1987)
6. R. Garuthara, X. Z. Lu, S. Lee, and R. R. Alfano: Shock Wave Effect on the Electron-Hole Plasma in CdSe, submitted to J. Appl. Phys.
7. X. Z. Lu, R. Garuthara, S. Lee and R. R. Alfano: Appl. Phys. Lett. 52, 93 (1988)
8. R. Rao, X. Z. Lu, A. G. Doukas, S. Lee, B. Willman, P. Harris and R. R. Alfano: in Proceedings of the International Conference on Lasers'85, ed. by C. P. Wang (STS Press, McLEAN, VA 1986), p. 171
9. X. Z. Lu, R. Rao, B. Willman, S. Lee, A. G. Doukas, and R. R. Alfano: Phys. Rev. B 36, 1140 (1987)

This work is supported by the City College of New York and Air Force Office of Scientific Research.

Hole Burning in the Free Exciton Line in GaSe

F. Minami, A. Hasegawa, K. Azuma, and K. Inoue

Research Institute of Applied Electricity, Hokkaido University, Sapporo 060, Japan

Exciton localization in semiconductors containing disorder has recently attracted much interest. Excitons localized by alloy fluctuations have been studied in various mixed compounds. Exciton confinement effects by stacking disorder are also predicted in layered materials such as GaSe, and the fine structure of exciton ground state, observed both in absorption and reflectivity, is interpreted in terms of anisotropic localization by the stacking disorder [1]. However, there has been no direct experimental evidence for the exciton localization by layer stacking disorder. In the present study, we performed "hole burning" experiments within the exciton line and confirmed the existence of the localized exciton states in layered compound GaSe.

The experiments were performed on platelets (∿20 μm) of ε-GaSe, cleaved from good quality single crystals perpendicular to the c axis. The excitation source was a synchronously mode-locked dye (rhodamine 6G) laser, whose pulse and spectral widths were ∿4 ps and ∿1 meV. The output laser energy was about 10^{-6}∿10^{-5} Jcm^{-2}/pulse. We made two different types of hole burning experiments: pulse propagation delay and pump-probe experiments [2]. Stress was laid on the propagation delay measurements. The propagation delay of the pulse through the sample was measured by cross-correlation technique. Delays in the linear regime were removed by monitoring the position of the "coherent spike". For pump-probe experiment, we used a one-laser system in which the pump and the probe pulses have the same frequency and only the absorption change at the peak of the hole is monitored. All measurements were made at 120 K.

Figure 1 shows the measured absorption spectrum $\alpha(E)$ of the n=1 free exciton in GaSe. The optical polarization was perpendicular to crystal c axis. The exciton absorption is centered at 2.086 eV and has a linewidth of ∿10 meV. Since polariton effects are not important for the present optical polarization, the exciton line is considered to be broadened inhomogeneously as a result of stacking disorder. The propagation delay of the pulse at the exciton resonance is also plotted in Fig. 1 as a function of photon energy. At energies higher than the peak of the exciton line, the propagation delay is too small to be observed. The delay increases quickly around E=2.083 eV and decreases gradually at lower energies. The maximum delay occurs ∿3 meV below the exciton absorption peak. When we decrease the pulse energy, the position of the maximum delay shifts to lower energy and the delay increases at low-energy side. Further, larger delays are observed for light pulses with longer durations. This behavior can be explained by hole burning by the pulse in the inhomogeneously broadened exciton line. Since the change in absorption coefficient $\Delta\alpha<10^3 cm^{-1}$ and the homogeneous width $\Gamma_h>1$ meV in GaSe [3], the observed delay cannot be interpreted in terms of the model proposed by Hegarty [4], where the delay is proportional to the ratio $\Gamma_h/\Delta\alpha$. We here consider the

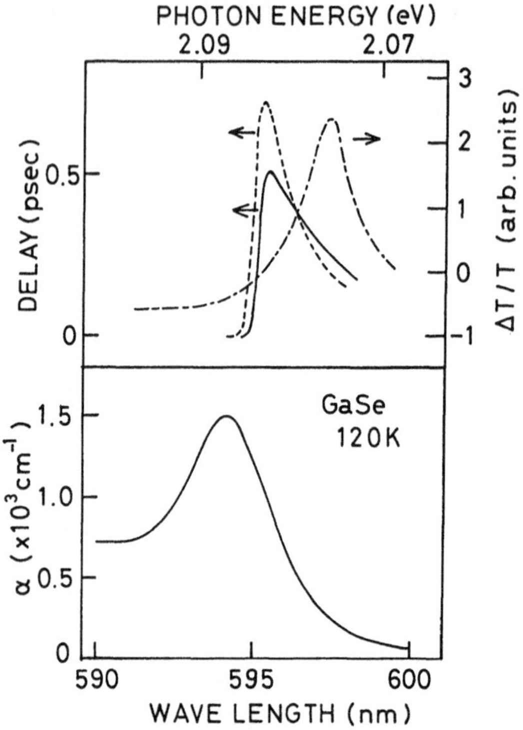

Fig. 1 Absorption coefficient α (lower trace) of the n=1 free exciton line at 120 K. Upper traces represent the pulse propagation delays for two different laser energies: 10^{-5} (dashed line) and 5×10^{-6} (solid line) Jcm^{-2}/pulse. Dot-dash curve represents the change in transmission $\Delta T/T$ for 5 ps delay.

propagation delay caused by pulse distortion effects due to incoherent bleaching [5]. The pulse propagation in this picture is envisaged as follows. The early part of the pulse burns a hole and is strongly attenuated, while the remainder of the pulse passes without loss through the sample. This leads to the propagation delay of the pulse.

To confirm the above picture, we compared the experimental data with predictions of the theory of saturable absorber transmission [5]. In the dynamical transmission model described in Ref. 5, the temporal variation of transmittance T is written as

$$d(\ln T)/dt=2I(t)(1-T)/E_S-(\ln T-\ln T_0)/\tau, \quad (1)$$

where I(t) is the input pulse function, E_S is the saturation energy for the absorber, T_0 is the low-power transmittance, and τ is the recovery time of the hole. If we assume that the hole recovery process is negligible on the time scale of the incident pulse, Eq. (1) can be solved analytically and T is obtained as

$$T^{-1}=1+(1-T_0)T_0^{-1}\exp[-\sqrt{\pi}\ I_p t_p(1+\mathrm{erf}(t/t_p))/E_S]. \quad (2)$$

Here we assume that the input pulse shape is gaussian:

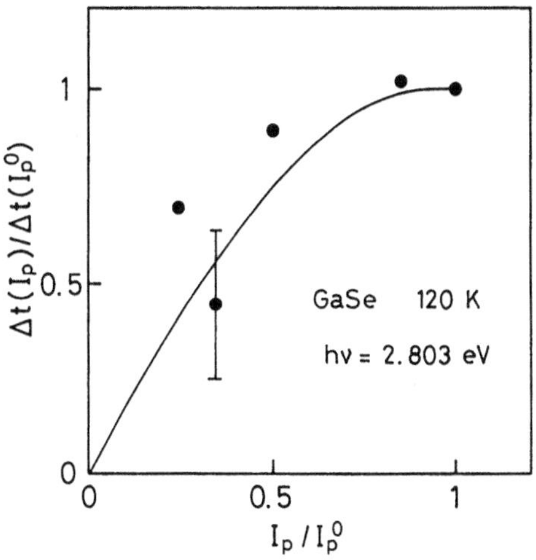

Fig. 2 Experimental (circles) and theoretical (solid line) laser power dependences of the propagation delay at the energy position of the maximum delay. The maximum laser energy $I_p{}^0 t_p$ is 10^{-5} Jcm^{-2}/pulse.

$$I(t)=I_p \exp(-t^2/t_p{}^2), \tag{3}$$

where I_p is the peak incident intensity and t_p is the pulse duration. Since the transmitted intensity is $T(t)I(t)$, the delay of the peak transmittance with respect to the input pulse, Δt, is given, for $t<t_p$, as

$$\Delta t=\beta t_p/(1+T_0(1-T_0)^{-1}\exp(\sqrt{\pi}\beta)), \tag{4}$$

where $\beta=I_p t_p/E_s$. It is found from Eq. (4) that the maximum delay occurs around $\beta \simeq 1$ and the delay decreases for larger and smaller values of β. Since $E_s \propto \alpha$, the observed energy dependence of the delay is explained well by Eq. (4). The delays are slightly overestimated in the model. We consider that this overestimation arises from neglect of the "spectral diffusion" effect [6,7], i.e., the second term in the right-hand side of Eq. (1). The intensity dependence is another check of the validity of the model. A comparison of the experimental and theoretical I_p dependences is made in Fig. 2. The agreement between theory and experiment is quite satisfactory in view of the experimental accuracy. It is thus concluded that the exciton confinement occurs in the low energy tail of the inhomogeneously broadened line, and the propagation delay of the light pulse is explained by pulse-reshaping effects due to bleaching of the exciton resonance.

The result of one laser pump-probe experiment is presented in Fig. 1. The change in transmission ΔT for 5 ps delay is plotted as a function of photon energy. It can be seen that excitation on low-energy side of the line burns a hole. This can be also used to support the ideas of localization of excitons by stacking disorder. The high energy extremity of the region where hole burning occurs (i.e., $\Delta T>0$), corresponds well to the peak of propagation delay data. This is other evidence for the fact that the propagation delay of the pulse is produced by hole burning. The difference between the peak position of both data is interpreted as follows. In propagation delay experiment, large delay is produced, when saturation occurs at the trailing part of the pulse. On the other hand, in pump-probe experiment, the saturation at the leading part of the pump pulse leads to large change in $\Delta T/T$. This is the reason that the peak of pump-probe data is situated at lower energy. The contribution of spectral diffusion would also be important in accounting for this difference.

In conclusion, we have presented the results of two different hole burning experiments in the n=1 free exciton line in GaSe. Both results support arguments for partial exciton localization by stacking disorder.

References

[1] J.J. Forney, K. Maschke, and E. Mooser, J. Phys. C**10**, 1887 (1977).
[2] J. Hegarty and M.D. Sturge, J. Opt. Soc. Am. B**2**, 1143 (1985).
[3] S. Asaka and F. Minami (unpublished).
[4] J. Hegarty, Phys. Rev. B**25**, 4324 (1982).
[5] A.C. Selden, Brit. J. Appl. Phys. **18**, 743 (1967); J. Phys. D**3**, 1935 (1970).
[6] X.C. Zhang, M. Gal, and A.V. Nurmikko, Phys. Rev. B**30**, 6214 (1984).
[7] F. Minami and K. Era, Phys. Rev. B**35**, 2509 (1987).

High Density Femtosecond Excitation of Hot Carrier Distributions in InP and InGaAs

W. Kütt, K. Seibert, and H. Kurz

Institute of Semiconductor Electronics,
Technical University Aachen, Sommerfeldstraße,
D-5100 Aachen, Fed. Rep. of Germany

Time resolved transmission and reflectivity measurements with fs-pulses have been performed to investigate the dominant electronic relaxation channels of highly excited InP and InGaAs on a subpicosecond time scale. While in transmission experiments only the scattering out from optically coupled states can be studied/1,2/, reflectivity transients are also sensitive to the temporary occupation of the satellite valleys /3/.

In this short contribution we would like to point out some specific differences between InP and InGaAs during and after excitation with amplified 50fs-pulses at 2 eV. The change of bulk reflectivity and of transmission through thin films are probed at the same wavelength with pulses as short as 50 fs. The probe beam is strongly attenuated and focussed to a third of the pump spot diameter to ensure sufficient spatial resolution. Combining boxcar and lock-in techniques optical changes could be detected with a sensitivity of 10^{-4}. Thin InP and InGaAs films grown by MOCVD are glued on a sapphire substrate. The substrate is removed by appropriate etching. For comparison InP bulk samples and thick InGaAs-epilayers are used in reflectivity measurements.

In Fig. 1 the transmission changes induced during the first picosecond after the excitation are shown for both samples and orthogonal polarization of pump and probe beams. At an excitation fluence of 80 $\mu J/cm^2$ generating $8x10^{18}$ carriers $/cm^3$ InP

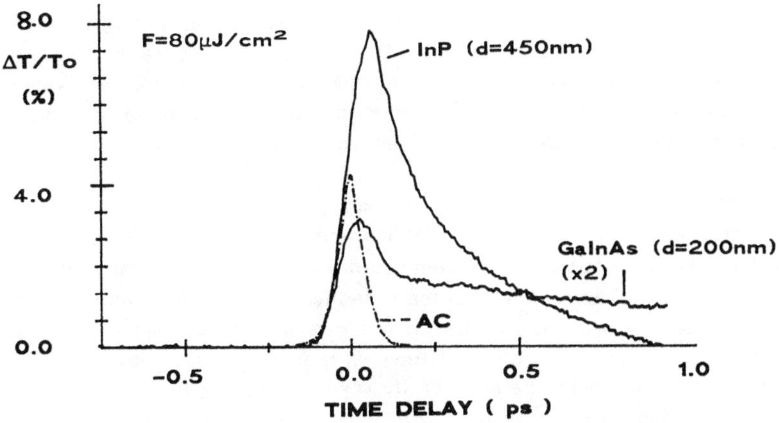

Fig.1: Transient transmission of thin InP- and GaInAs-films at $F=80\mu J/cm^2$

Springer Series in Chemical Physics, Vol. 48: **Ultrafast Phenomena VI**
Editors: T. Yajima · K. Yoshihara · C.B. Harris · S. Shionoya

shows much more instantaneous bleaching due to absorption saturation than InGaAs. The maxima of transmission are split in time, indicating different relaxation rates. After an ultrafast initial drop which follows closely the trailing edge of the excitation pulse the transmission of InGaAs merges into a monoexponential decay with a time constant of 1ps, characteristic for the cooling of hot thermalized carrier distributions.

This initial fast drop cannot be observed in InP. The major portion of the transmission change can be described by a monoexponential decay with a time constant of τ=0.34ps. These results lead to the conclusion that in InGaAs the absorption saturation is primarily limited by intervalley scattering of electrons with scattering times comparable to the laser pulseduration (50 fs). In InP, however, this ultrafast relaxation channel is blocked leaving the depopulation of optically coupled states to intercarrier collisions and intraband phonon interactions.

This suggestion is supported by transmission data of InP which we have performed with different probe energies up to 2.5 eV at an excitation energy of 2 eV. These data show a rapid onset of absorption saturation. The broad spreading of the distribution already during the pulse excitation demonstrates that ultrafast intercarrier scattering has to be taken into account in studies of various relaxation channels. The striking differences between InP and InGaAs are related to the probability of ultrafast intervalley scattering. Due to the high electron excess energy in InGaAs (1.12 eV) the transfer to the L- and X-valley dominates all other scattering events. The excess energy in InP (0.55 eV) is not sufficient to open the channel into the L-valley located 0.61 eV above the Γ-conduction band minimum.

Conclusive information on occupation of satellite valleys via intervalley scattering in InGaAs is provided by time resolved reflectivity measurements. As shown previously negative reflectivity changes indicate occupation of higher conduction band states /3/. As long as electrons relax within the central valley, only positive changes of the reflectivity are expected. As soon as a significant portion enter the satellite valleys the reflectivity is reduced due to the dominance of negative plasma contributions.

In Fig. 2 the relative changes of reflectivity are plotted for the same absorbed fluence of 45 J/cm^3 . While the reflectivity of InP rises continuously during the excitation pulse and approach a maximum after 3 picoseconds, the reflectivity of InGaAs drops during the pulse, passes through a minimum, recovers to the unpertubated value within a picosecond and continues to rise to a positive maximum after 5-10 ps, depending on the density of carriers created.

The negative reflectivity drop in InGaAs proves conclusively the ultrafast accumulation of carriers in the satellite valleys. The reflectivity minimum corresponds to a moment at which the internal thermalization is completed and the carrier density at the surface reaches a maximum. The recovery of the reflectivity and the following turn to positive values is consistent with the cooling of hot carrier distributions refilling the Γ-valley. The cooling of thermalized carrier distribution is found to slow down significantly in both compounds when the carrier density is raised from a 10^{18} cm^{-3} level to the maximum carrier densities around 10^{20} cm^{-3} . The decrease at long time delays is governed by diffusion rather than by Auger recombination processes.

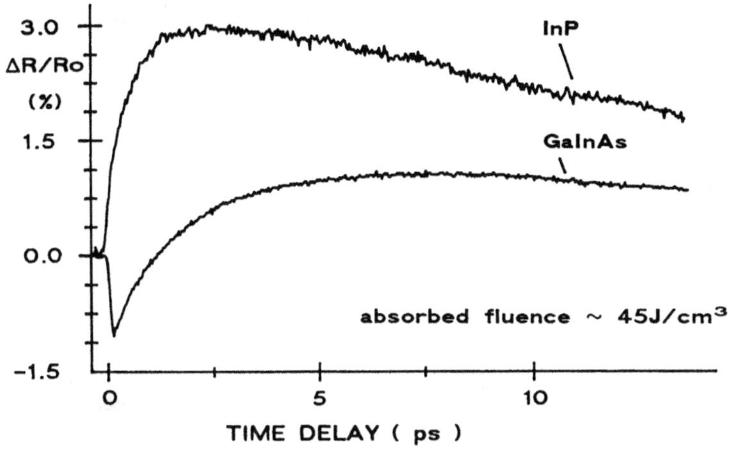

Fig.2: Reflectivity changes for InP and InGaAs for $F \times \alpha$ = 45J/cm^3

In summary we have presented several specific differences in the relaxation behavior of InP and InGaAs. Intervalley scattering is shown to be the dominant initial relaxation process in InGaAs while in InP the intercarrier scattering competes with the polar optical phonon emission. Reflectivity measurements provide the final evidence for intervalley scattering in InGaAs.

This work was entirely supported by the "Deutsche Forschungsgemeinschaft". We would like to thank G. Bosch and U. Lemmer for model calculations of carrier relaxation processes.

References:

1. A.J. Taylor, D.J. Erskine and C.L. Tang: JOSA B2, 663 (1985)
2. W.Z. Lin, R.W. Schoenlein, J.G. Fujimoto and E. Ippen: IEEE QE 24, 267 (1988)
3. H. Kurz, W. Kütt, K. Seibert and M. Strahnen: Solid State Electronics 31, 447 (1988)

Single-Shot Reflectivity Study of the Picosecond Melting of Silicon Using a Streak Camera

Juen-Kai Wang, P. Saeta, M. Buijs, M. Malvezzi[1]*, and E. Mazur*

Division of Applied Sciences and Department of Physics, Harvard University, Cambridge, MA 02138, USA

1. Introduction

Numerous investigations on the phase transition of silicon during picosecond laser annealing have been performed in recent years. It has been well established that the silicon surface melts during a picosecond laser pulse.[2] Because liquid silicon is a metal, the reflectivity increases on melting. This has indeed been observed using optical pump-and-probe techniques.[2-4] Standard picosecond pump-and-probe measurements, however, have some serious inherent drawbacks. First, they cannot resolve reflectivity changes that occur on a time scale of a few picoseconds, because they integrate over the duration of the probe pulse (typically 20 ps or more). Second, they determine the time profile of the reflectivity for every pump fluence in a step-wise manner by varying the delay between the pump and the probe pulse. This introduces a large amount of scatter in the data points, due to shot-to-shot variations in the pump fluence, and requires a large amount of data to be taken for every time profile of the reflectivity. Also, they provide no spatial information on the melting process.

To obtain spatial resolution, a better time resolution, and to measure the time profile of the reflectivity on a single-shot basis, we use a streak camera with a time resolution of 1.8 ps for the detection of the probe pulse.

2. Experimental setup

The frequency-doubled output (30 ps, 532 nm) of a mode-locked Nd:YAG laser is split into a probe and a pump pulse. The duration of the probe pulse is stretched by splitting it in four, delaying the four resulting pulses with respect to each other, and recombining them spatially to form a longer probe pulse of 120 ps duration. The probe pulse then images an area around the 100-μm wide melting area onto the entrance slit of a streak camera (Hamamatsu Photonics C1587).

The entrance slit of the camera is split into two parts. The larger part is used to image the probe pulse, the smaller part to image a fraction of the pump pulse. The latter acts as the timing reference.

To enhance the sensitivity of our measurements, the probe pulse is *p*-polarized and the probe angle of incidence is chosen to be 65°, close to Brewster's angle. At this angle the reflectivity of solid silicon is small (about 10%), leading to an increase in reflectivity on melting of a factor of 8.

Springer Series in Chemical Physics, Vol. 48: **Ultrafast Phenomena VI**
Editors: T. Yajima · K. Yoshihara · C.B. Harris · S. Shionoya

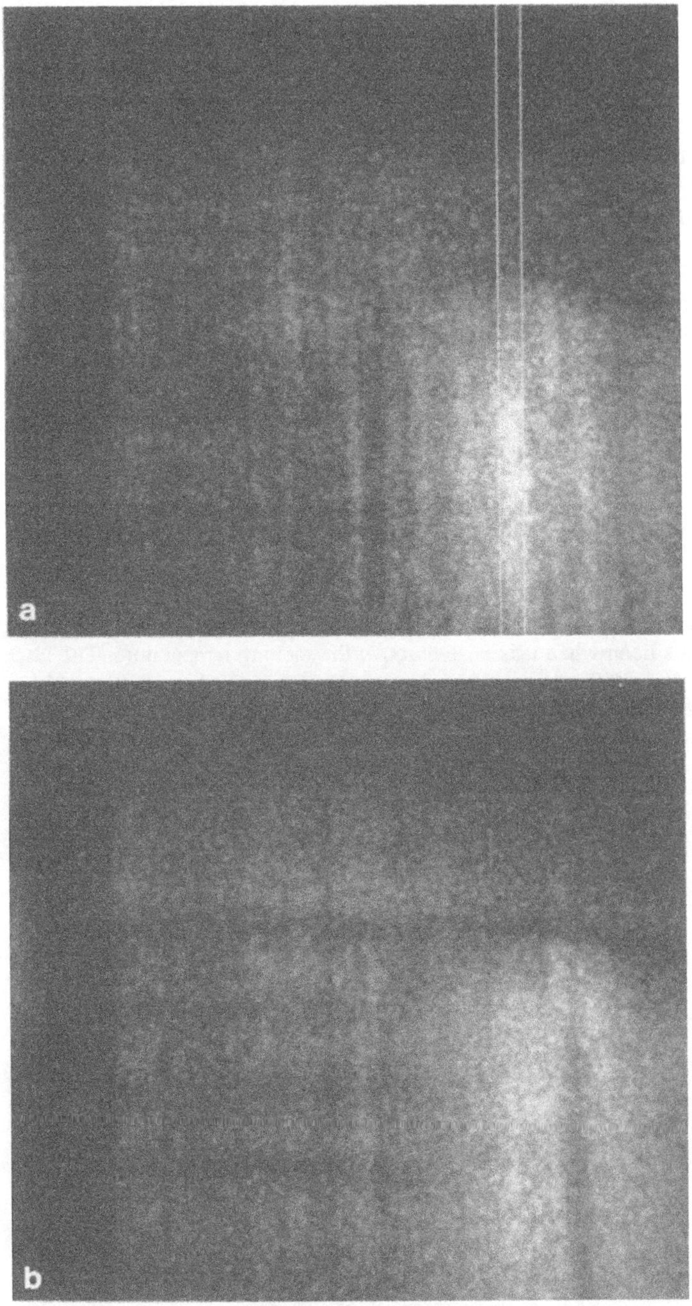

Fig. 1. Streak camera images of laser melting of silicon. Laser fluence is 470 mJ/cm^2. The white lines in (a) indicate the region used for analysis of the reflectivity.

3. Results and Discussion

The streak images of two measurements are shown in Figs. 1a and 1b. The time axis is displayed vertically, with time increasing from top to bottom. The full height of the image corresponds to 190 ps. The horizontal axis reflects the spatial profile. The pulse shown on the left side, with a width of about 30 ps, is the pump pulse. The stretched probe pulse covers most of the image. The bright part is where the silicon surface is melting. Spatially it reflects the Gaussian intensity profile of the pump pulse. The time profile allows one to study the melting dynamics.

Both measurements were performed at the same laser fluence. In the case shown in Fig. 1b, some surface irregularity leads to the formation of a plasma on the surface in the region where the absorbed energy is highest. The plasma scatters the incoming probe light, leading to the dark area in the center of the melting region. This clearly emphasizes the need for spatial resolution. An integrating detector would give rise to erroneous conclusions about the reflectivity.[5]

Fig. 2 shows the reflectivity profile at the center of the melting region, between the two white lines, in Fig. 1a. The reflectivity reaches the value for liquid silicon within the 30-ps duration of the pump pulse. The high time resolution of the streak camera enables us to confirm that the reflectivity follows the trend predicted by numerical simulations of heating above the melting temperature in silicon.[3] According to a Drude model one expects a decrease in reflectivity of molten silicon when it is heated above the melting temperature. The laser fluence in Figs. 1a and 1b is 470 mJ/cm^2 which is more than twice as large as the melting threshold for silicon (200 mJ/cm^2). A numerical solution of the one-dimensional heat equation shows that at this fluence the temperature of the liquid silicon exceeds the melting temperature by more than 1000 K.

Measurements with longer and more uniform probe pulses are currently in progress.

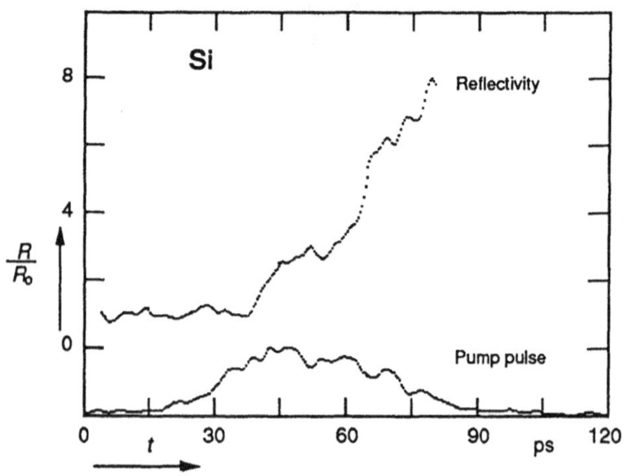

Fig. 2. Reflectivity of silicon during melting with a picosecond laser pulse. R_0 is the reflectivity of solid silicon. The lower trace shows the intensity of the pump pulse.

Acknowledgments

MB acknowledges a Fellowship from the Netherlands Organization for Scientific Research (NWO). PS acknowledges a Ph.D.-Fellowship from AT&T Bell Laboratories. This work was supported by Hamamatsu Photonics K.K., and by the Joint Services Electronics Program under contract with Harvard University.[6]

References

1. Permanent address: Istituto di Fisica Applicata, Pavia, Italy
2. J.M. Liu, H. Kurz and N. Bloembergen, *Appl. Phys. Lett.*, 41, 643 (1982)
3. P.M. Fauchet and K.D. Li, *Mat. Res. Soc. Symp. Proc.*, 100, 477 (1988)
4. D. von der Linde and N. Fabricius, *Appl. Phys. Lett.*, 41, 991 (1982)
5. I.A. Boyd, S.C. Moss, T.F. Bogess and A.L. Smirl, *Appl. Phys. Lett.*, 46, 366 (1985)
6. Contract number: N00014-84-K-0465

Subpicosecond Transient Grating Experiments in Amorphous Semiconductors

G. Noll, E. Göbel, and U. Siegner

Fachbereich Physik, Philipps-Universität Marburg, Renthof 5, D-3550 Marburg, Fed. Rep. of Germany

The study of the initial carrier relaxation in disordered and amorphous semiconductors has gained increasing interest within the last years /1/. In particular, the relaxation of carriers from extended into localized electronic states across the so-called "mobility edge" has attracted considerable attention. So far these experiments have been performed by excite and probe techniques which examine the photoinduced changes of the absorption. We report the first subpicosecond transient grating experiments, which according to our interpretation, provide a very convenient means to study extended state relaxation and localization in amorphous materials. Furthermore, we have performed nondegenerate transient grating experiments with a compressed Nd:YAG pulse (λ = 1.06 μm) for probing the photoinduced grating in order to eliminate the contribution of interband absorption.

The degenerate transient grating experiments are performed with a hybridly mode locked and cavity dumped linear dye laser system synchronously pumped by a frequency doubled (KTP) mode locked Nd:YAG laser. Rh6G was used as the gain medium and DODCI as the saturable absorber. Two Gires–Tournois interferometer (GTI) plates are employed for adjustable group velocity dispersion. A cavity dumper is used as the output coupler to reduce the repetition rate and enhance the peak power of the output pulses. The autocorrelation trace of the output pulses at 620 nm at a repetition rate of 1.5 MHz amounted typically to about 360 fs (FWHM). Pulse energy typically is about 5 nJ under these conditions. Pulses with a wavelength of 585 nm with similar characteristics could be produced as well by changing the GTI assembly conditions. The nondegenerate transient grating experiments are performed with visible pulses from a standard synchronously pumped and cavity dumped dye laser and synchronized 1.06 μm Nd:YAG pulses compressed in a 1 km single mode fiber and two 1700 /mm gratings used in a double pass configuration. Compressed pulses with a FWHM of the autocorrelation trace of 1.9 ps at a repetition rate of 76 MHz are obtained. The time resolution of the nondegenerate transient grating experiments is determined by a crosscorrelation of the 1.06 μm pulses with the visible dye laser pulses to 7 ps, which proves the good synchronization of the cavity dumped dye laser and the compressed Nd:YAG pulse train.

Hydrogenated amorphous silicon (a–Si:H) films with 0.25 μm to 1 μm thickness are deposited by glow discharge on quartz glass substrates. The transient grating experiments are performed at room temperature. Two parallel polarized pump beams are focussed on the sample without time delay to generate the light induced transient grating. The first order diffracted light intensity of either an orthogonally polarized weaker part of the dye laser output (degenerate transient grating) or the compressed IR pulses (nondegenerate) was measured as a function of delay time with respect to the pump pulses.

The time behavior of the transient grating signal in a degenerate experiment performed with subpicosecond time resolution is shown in Fig. 1. The laser photon energy is larger than the optical gap E_{04}. Two time regimes can be clearly identified: The initial fast component disappears within a few picoseconds whereas a slow component decays within several nanoseconds for a–Si:H films with low defect densities.

Springer Series in Chemical Physics, Vol. 48: Ultrafast Phenomena VI
Editors: T. Yajima · K. Yoshihara · C.B. Harris · S. Shionoya

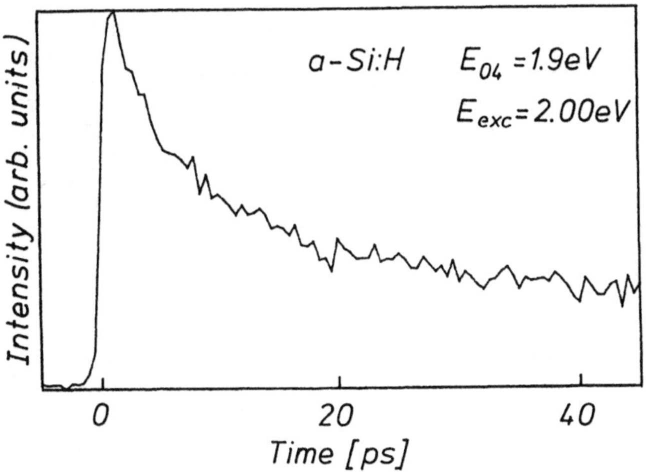

Fig. 1 Degenerate transient grating signal versus probe pulse delay time at 300K $(E_{04}) = E(\alpha = 10^{4}\ cm^{-1})$

The diffraction of the probe beam in the transient grating experiments can be caused by an absorption and/or refractive index grating induced by the photoexcited population grating. The initial fast component has to be attributed to a change of absorption and/or refractive index during relaxation of the photoexcited carriers, because diffusion and recombination can be neglected on this time scale in a–Si:H. We conclude that the diffraction is dominated by a refraction index grating, since the photoinduced absorption at the same photon energy ($E_{probe} > E_{04}$) does not exhibit a fast component /2,3/. In fact, a fast component in the photoinduced absorption has only been observed for lower energies ($E_{probe} < E_{04}$) due to bleaching of the interband transitions by relaxation of the photoexcited carriers into tail states /1,3/. In order to eliminate the possible effect of bleaching of the interband transitions in the transient grating signal we have performed the nondegenerate experiments, because interband absorption at 1.06 μm (1.16 eV) is very weak in a–Si:H. The experimental result of a nondegenerate transient grating experiment is shown in Fig. 2.

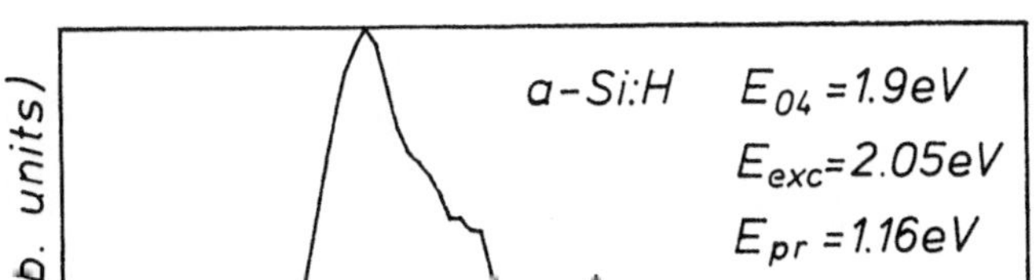

Fig. 2 Nondegenerate transient grating signal versus IR probe pulse delay at 300K

We observe again a fast component similar to the degenerate experiment. Thus this fast component cannot be attributed to bleaching of interband transitions, which confirms our interpretation that the fast component in the transient grating signal is due to refractive index changes caused by the inital relaxation of photoexcited carriers across the mobility edge. The constant background of the signal in the nondegenerate experiment is due to the high repetition rate of the IR probe pulse compared to the repetition rate in the degenerate case. The efficiency of the photoinduced diffraction grating obviously does not decay within 13 ns – apart from the fast changes – indicating that relaxation within the localized states does not change the diffraction efficiency opposite to relaxation between extended and localized states. A time constant for this initial relaxation of about 4 ps can be deduced from the data of Fig. 1 for a–Si:H.

In conclusion, we have demonstrated that subpicosecond transient grating experiments are a powerful tool to investigate the very fast carrier relaxation processes in amorphous and disordered semiconductors. By comparing the results of degenerate and nondegenerate experiments we have been able to show that the relaxation of carriers from extended into localized states causes a change of the refractive index. The time constant for this relaxation process is about 4 ps for a–Si:H.

Acknowledgement:

We would like to thank H. Mell for preparation of the samples. This work is supported by the Deutsche Forschungsgemeinschaft. One of the authors (G.N.) gratefully acknowledges the support by the "Stifterverband für die Deutsche Wissenschaft" to attend the meeting.

References:

1. Z. Vardeny, J. Tauc, in: Disordered Semiconductors (ed. M.A. Kastner et al.) Plenum Press, N.Y., London (1987) p. 339

2. G. Noll, E. Göbel, J. Non–Cryst. Solids 97&98 (1987) p. 141

3. P.M. Fauchet, D. Hulin, A. Migus, A. Antonetti, K. Kolodzey, S. Wagner, Phys.Rev.Lett. 57 (1986) p. 2438

Femtosecond Dynamics of Optical Nonlinearities in Wide-Gap II–VI Semiconductors

J. Puls [1], *W. Rudolf* [2], *F. Henneberger* [1], *D. Lap* [2], *V. Petrov* [2], *U. Stamm* [2], *and B. Wilhelmi* [2]

[1]Humboldt-Universität, Sektion Physik, Bereich Halbleiteroptik, Invalidenstr. 110, DDR-1040 Berlin, G.D.R.
[2]Sektion Physik der Friedrich-Schiller-Universität, Max-Wien-Platz 1, DDR-6900 Jena, G.D.R.

Previous studies [1] have shown that II-VI semiconducting compounds exhibit very strong optical nonlinearities in the vicinity of the fundamental absorption edge at room temperature. They are related to specific many-body effects (sreening, gap-shrinkage, bandfilling) in the carrier system excited and make these materials good candidates for various applications in the visible spectral range. In this paper we report on measurements of the femtosecond dynamics of these nonlinearities.

Using the amplified output of a CPM laser [2] at 618 nm with a pulse duration of 110 fs, we have studied three different situations in an excite-and- probe geometry:

(i) above-gap excitation of pure CdSe,

(ii) below-gap, exciton resonant excitation of CdS_xSe_{1-x} and, for comparison,

(iii) excitation of a commercial filter made from CdS_xSe_{1-x} microcrystallites embedded in glass.

In accord with earlier experiments on the ns- and ps-time scale [1] we have observed in all three situations a pronounced increase of the probe transmission, however, with a different time behaviour. For the filter (Fig. 1a) the transmission rise follows the pump and no recovery is seen within 1 ps. The excitation is here only slightly above the absorption-edge and the nonlinearity is caused by the filling and associated blocking of the low-energy states in the micro-crystallites. Thus, the recovery is controlled by the carrier recombination of some ten ps [3].

For exciton-resonant excitation (Fig. 1b) of bulk CdSSe we find a distinct coherence peak, followed by a plateau and subsequent recovery with a time constant of about 250 fs. To separate coherent interaction and that originating from real carriers we have studied the first-order diffraction from the transient grating produced by pump and probe together. A slight asymmetry of the efficiency (Fig.2) with respect to probe ahead or behind pump is found, which we attribute to the exciton dephasing with a characteristic time of about 50 fs. Substracting the coherence part in Fig. 1b we get a switch-on time for the nonlinearity via carriers somewhat below 100 fs. Physically, this is the time necessary for the ionization of excitons by LO-phonon absorption resulting in free carriers screening

Springer Series in Chemical Physics, Vol. 48: **Ultrafast Phenomena VI**
Editors: T. Yajima · K. Yoshihara · C.B. Harris · S. Shionoya

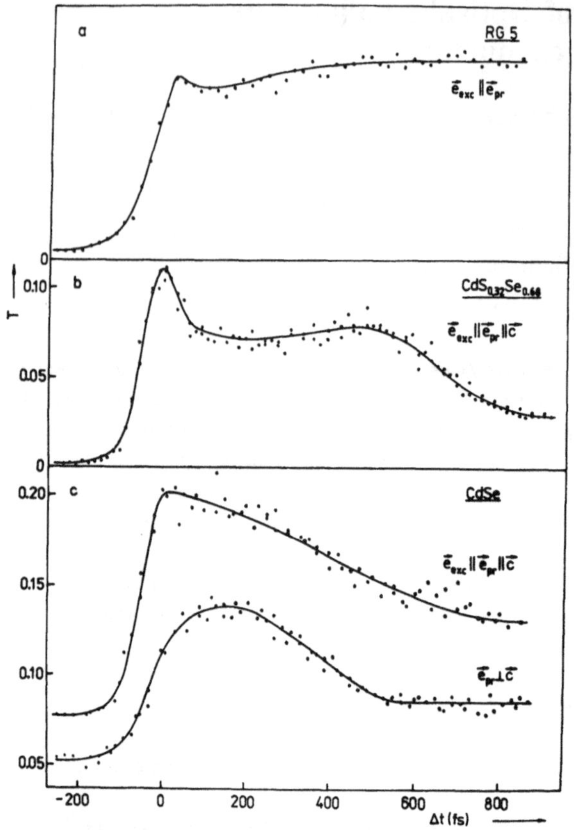

Fig. 1
Probe beam transmission versus pump-probe delay (pump: 1 mJ /cm^2 , probe: 0.01 mJ /cm^2)
a) Above- edge excitation of a commercial filter RG 5 (thickness d=1mm).
b) B-exciton resonant excitation of CdS_xSe_{1-x} (d=3 μm). A value of x = 0.32 is used to match the 618 nm laser output.
c) 250 meV above-gap excitation of CdSe (d=0.3 μm). All data are taken at room temperature

out the exciton resonance. Simultaneously the gap shrinks, so that the actual band edge is below the excitation frequency at larger delays and a complex carrier dynamics sets in leading to a nearly constant probe transmission. Currently, we attribute the very fast recovery to the decay of the non-thermal carriers at the excitation frequency down to the band bottoms (cf.Fig. 1c). Note, that at room temperature there is no essential difference between the linear levels of the below-gap exciton and above-gap continuum absorption.

For 250 meV above-gap excitation of CdSe (Fig. 1c) no clear coherence peak is seen , since we produce directly carriers with very rapid dephasing: the first-order efficiency follows instantaneously the pump pulse in this case (see Fig.2). The pump polarization was chosen parallel to the crystal axis c, so that primarily only B-holes were excited. We observe essentially the state-filling dynamics of the carriers studied at low temperature in [3] with a switching contrast much smaller than for exciton- resonant excitation in Fig. 1b. However, we find different results for probe polarization parallel and perpendicular to c. In the first case (upper curve in Fig. 1c) the absorption bleaching due to B-holes is probed and, accordingly, no delay of the respective transmission increase is seen. But, using probe polarization perpendicular c, we monitor the bleaching due to A-holes, which, in fact, shows up delayed to that of B-holes. Thus, we have directly observed the B- to A-hole conversion in CdSe which takes about 80 fs.

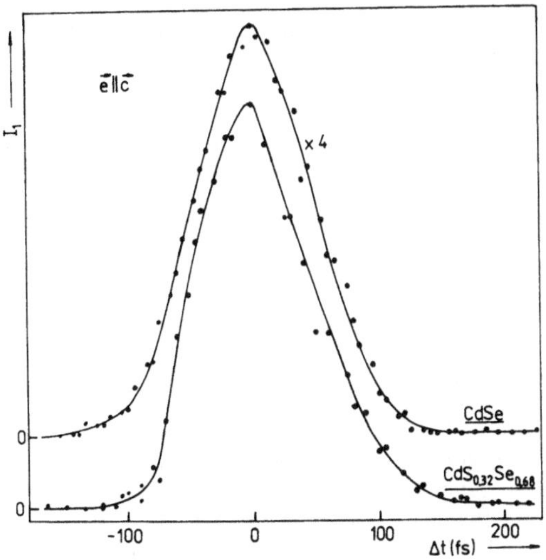

Fig.2
First-order diffraction signal versus pump-probe delay for b) and c) in Fig.1

In conclusion, we have observed a variety of basic processes of the light-matter interaction in the direct-gap II-VI's. The data presented in our paper directly demonstrate that these semiconductors can be used for optical switching and logic operations in the subpicosecond range, in particular, for the exciton- resonant nonlinearity of Fig. 1b. In this situation, high contrast, fs-recovery, and room temperature distinguish the wide-gap II-VI compounds from other materials.

References

1. F. Henneberger, J. Puls, H. Rossmann, Ch. Spiegelberg, M. Kretzschmar, and I. Haddad, Proc. 6th General Conf. Condensed Matter Division European Phys. Soc., Physica Scripta T13 (1986) 195.
   J. Puls, F. Henneberger, F. Fink, and I. Jüpner, Proc. Internat. Symp. Ultrafast Phenomena Spectroscopy, Reinhardsbrunn (GDR),1985.
2. W. Dietel, E. Döpel, G. Marowski, V. Petrov, C. Rempel, W. Rudolph, F.P. Schäfer, and B. Wilhelmi, Appl. Phys. B45 (1988).
3. N. Peyghambarian and S.W. Koch, J. Phys. Appl. 22 (1987).

Nonlinear Process-Induced Higher-Order Components of a Picosecond Transient Grating in CdS

H. Saito and A. Watanabe

Department of Applied Physics, Okayama University of Science, Ridai-cho 1-1, Okayama 700, Japan

Transient-induced grating experiments have been widely used to study the dynamics of diffusion, relaxation and recombination of photoexcited carriers in solids./1/ We demonstrate that the simultaneous observation of the decay of different diffraction orders is required to provide full information on the dynamics of excited carrier system./2/ Experiments were performed by monitoring simultaneously the temporal behavior of the first- and second-order diffraction intensity of transparent probe pulses due to a free-carrier index grating produced in CdS by interference of two picosecond light pulses. As the excitation light source, the third-harmonic pulses (355nm) of an active-passive mode-locked Nd:YAG laser (pulse width of 25ps) were used. As the probe pulses the second-harmonics (532nm) of the Nd:YAG laser, which is transparent for CdS, were used. Measurements were made for three different spacings at the temperatures from 10K up to 140K.

Experimental results for the diffraction intensity versus time delay measured at 140K are shown in Fig.1. The solid and open circles correspond to the first-order [$I_{dif}^{(1)}$] and second-order [$I_{dif}^{(2)}$] signals, respectively. It is noted that both signals decay exponentially, and that the decay constant of $I_{dif}^{(2)}$ is just half of that of $I_{dif}^{(1)}$. The ratio of $I_{dif}^{(2)}$ to $I_{dif}^{(1)}$ (squares in Fig.1), therefore, decays in the same manner as $I_{dif}^{(1)}$ which is represented by solid line. In Fig.2 are summarized experimental results measured at 10K. The

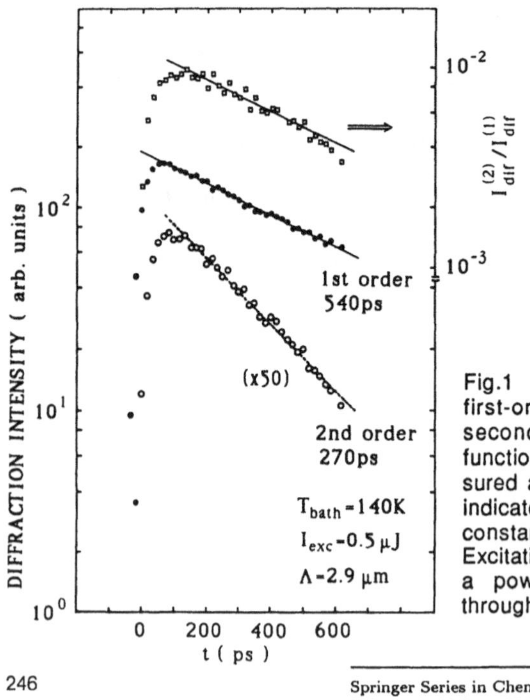

Fig.1 Observed diffraction intensity in the first-order, $I_{dif}^{(1)}$ (filled circles) and the second-order, $I_{dif}^{(2)}$ (open circles) as a function of time delay for Λ=2.9 μm measured at T_{bath}=140K. Solid and dotted lines indicate the exponential decay with the time constant of 540 and 270 ps, respectively. Excitation energy, I_{exc} of 1 μJ corresponds to a power density of about 10 MW/cm^2 throughout this work.

Springer Series in Chemical Physics, Vol.48: **Ultrafast Phenomena VI**
Editors: T. Yajima • K. Yoshihara • C.B. Harris • S. Shionoya

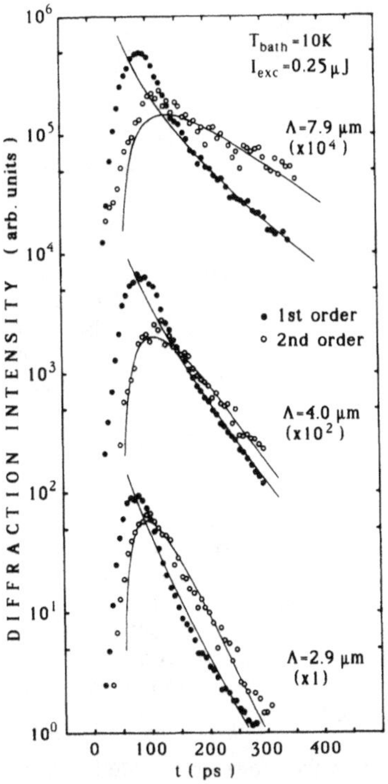

Fig.2 Observed diffraction intensity in the $I_{dif}^{(1)}$(filled circles) and $I_{dif}^{(2)}$ (open circles) at T_{bath}=10K. Solid curves are the calculated diffraction intensities.

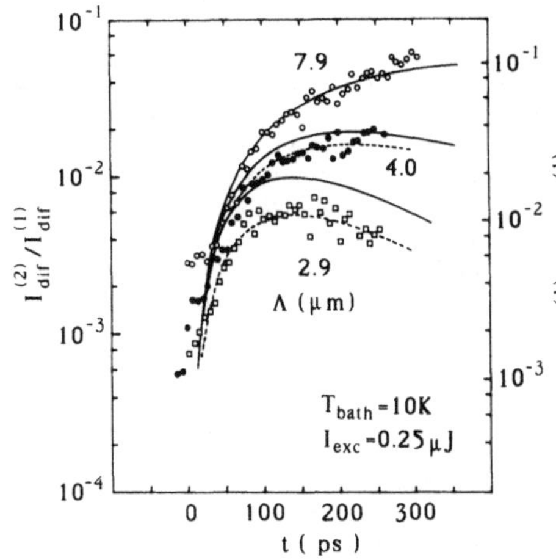

Fig.3 Observed ratio, $I_{dif}^{(2)}/I_{dif}^{(1)}$, as a function of time delay measured at T_{bath}=10K for three grating spacings; Λ=7.9 μm (open circles), Λ=4.0 μm (filled circles) and Λ=2.9 μm (squares). Solid and dashed curves indicate the calculated ratio, $I_{dif}^{(1)}(\Lambda/2)/I_{dif}^{(1)}(\Lambda)$, for $\beta=1.2\times10^{10}$ s^{-1} and D=8 cm^2/s.

intensity scale for $I_{dif}^{(2)}$ (open circles) is shifted with respect to $I_{dif}^{(1)}$ (solid circles), so that the data for $I_{dif}^{(1)}$ and those for $I_{dif}^{(2)}$ nearly coincide in the later time regime. In contrast to the results at high temperature (Fig.1), both $I_{dif}^{(1)}$ and $I_{dif}^{(2)}$ decay nonexponentially and shows roughly the same decay time in the later time regime. These decay characteristics can be clearly seen by the plots shown in Fig.3, where the experimental results for the ratio $I_{dif}^{(2)}/I_{dif}^{(1)}$ for the same data as shown in Fig.2 are depicted. It is found that the obtained ratio only for Λ=2.9 μm decays, but another two (Λ=4.0 and 7.9 μm) increase with time delay.

The diffraction of the probe beam is due to the spatially periodic change of the refractive index of the sample induced by free carriers or excitons, which is proportional to the carrier or the exciton density. The intensity of the n-th diffraction order due to a sinusoidal grating with the spacing of Λ in the Raman-Nath regime, $I_{dif}^{(n)}(\Lambda)$, becomes in proportion to 2n power of the difference of the free carrier density at the peak of the grating and its minimum./3/ Thus the decay time of $I_{dif}^{(2)}(\Lambda)$ should become just half of that of $I_{dif}^{(1)}(\Lambda)$. The results shown in Fig.1 are in accordance with this prediction. The results shown in Figs. 2 and 3, on the contrary, cannot be simply described by the second-order diffraction due to the a sinusoidal grating.

Nonlinear processes, such as bimolecular recombination of electrons and holes and free carrier-to-exciton Mott transition, result in a deformation of the shape of the grating, which leads to the occurrence of higher-order Fourier components, e.g., the doubled period in the diffraction grating. Calculation of diffraction intensities in the presence of a nonlinear process is performed by solving the nonlinear continuity equation which describes the decay of the grating due to the nonlinear relaxation process and the diffusion of carriers. An approximate solution assumed for the equation is expanded into a Fourier series. The j-th Fourier component corresponds to the grating with the period of Λ/j, and yields the diffraction intensity in the n-th order, i.e., $I_{dif}^{(n)}(\Lambda/j)$.

Solid curves in Fig.2 represent $I_{dif}^{(1)}(\Lambda)$ and $I_{dif}^{(1)}(\Lambda/2)$, the first-order intensity for Λ (j=1) and $\Lambda/2$ (j=2) components, calculated for β (bimolecular relaxation rate) $=1.2\times10^{10}$ s^{-1} and D (ambipolar diffusion coefficient)= 8 cm^2/s. [In this calculation the contribution of $I_{dif}^{(2)}(\Lambda)$ to the second-order diffraction intensity is neglected.] Solid and dashed curves in Fig.3 indicate the calculated ratio, $I_{dif}^{(2)}(\Lambda)/I_{dif}^{(1)}(\Lambda)$ for the same β and D values as in Fig.2. The dashed curves for Λ=4.0 and 2.9 µm are also the calculated results but are shifted so as to fit well to the respective data points. It is noted that agreement between the experimental and calculated results is good as far as the time-dependence is concerned. Similar results are obtained for the data measured under different excitation intensities and also those measured at 40K.

The obtained rate of $\beta=(1\sim1.5)\times10^{10}$ s^{-1} is much smaller than the bimolecular recombination rate for the degenerate plasma of about $10^{11}s^{-1}$./4/ At carrier temperatures approximately below 60K for CdS, exciton molecules are thermally stable. For nondegenerate plasma the recombination of free carriers is of minor importance and very fast recombination of the exciton molecule determines the decay of the free carrier and the exciton system. Taking these facts into consideration, we can conclude that the β values obtained above are determined mainly by the free-carrier to exciton Mott transition.

At temperatures above 60K, where the exciton molecule is thermally dissociated, thermal equilibrium is established between the exciton and the free carrier. Then the exciton density becomes in proportion to the square of the free carrier density. This nonlinear relation also brings about the exciton grating with doubled period in the free carrier grating. With further increase of the temperature up to about 140K, the exciton density becomes negligibly small because of thermal dissociation, and no higher-order Fourier components result in. Thus the observed results in Fig.1 are in perfect agreement with the theoretical predictions for an ideal sinusoidal grating.

These results demonstrate that the spatial resolution in transient-induced grating experiments can be extended to a microscopic scale by inclusion of higher diffraction orders, thus opening new areas for the study of carrier dynamics in solids in picosecond and subpicosecond time scales.

The authors acknowledge E.O. Göbel for helpful discussion. The experimental work was performed at the Max-Planck-Institut für Festkörperforschung, Stuttgart, F.R. Germany.

References

/1/ A.L. Smirl, in Semiconductors Probed by Ultrafast Laser Spectroscopy, edited by R.R. Alfano (Academic, Orlando, Fla., 1984), Vol.1, Sec.10, p329.
/2/ H. Saito and E.O. Göbel, Opti. Lett. **11**, 354 (1986); H. Saito and A. Watanabe, Phys. Rev. B **37** (to be published).
/3/ W.R. Klein and B.D. Cook, IEEE Trans. Sonics Ultrason. **SU-14**, 123 (1967).
/4/ H. Saito and E.O. Göbel, Phys. Rev. B **31**, 2360 (1985).

Dynamics of Exciton-Polariton Luminescence with High Repetition Tunable UV Picosecond Pulses

T. Kuga[1], *M. Kuwata*[2], *H. Akiyama*[2], *T. Hirano*[1], *and M. Matsuoka*[1]

[1]The Institute for Solid State Physics, University of Tokyo, Roppongi Minato-ku, Tokyo 106, Japan

[2]Department of Physics, Faculty of Science, University of Tokyo, Hongo, Bunkyo-ku, Tokyo 113, Japan

Recent advances in cw mode-locked picosecond lasers enable us to measure a time-resolved exciton luminescence under a weak excitation condition, which is desirable for the study of the dynamics of excitons and exciton-polaritons[1]. However, such measurements have been difficult in the UV region because of the limitation of light sources.

The lowest exciton state of CuCl, named Z_3 exciton, whose resonance energy lies in the UV region at 3.202 eV, has a simple electronic structure, and its polariton parameters are well established[2]. Thus, this exciton state is quite suitable for the basic study of the dynamics of exciton-polaritons. For this purpose, we have developed a high repetition rate tunable UV picosecond light source by means of sum frequency generation. By virtue of the high repetition rate (82MHz), one can measure the photo-luminescence under a weak excitation condition with high signal to noise ratio. In this report, we present our recent results on the time resolved exciton luminescence of CuCl at liquid helium temperature[3] and liquid nitrogen temperature.

Figure 1 shows the schematic diagram of the generation of UV pulses. The output of a cw mode-locked YAG laser is frequency doubled to pump a tunable dye (Rhodamine 6G) laser. The residual YAG laser pulse and the output pulse of

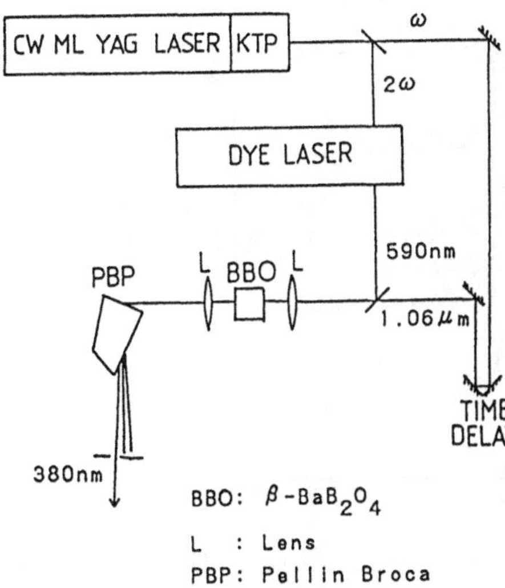

Fig.1. Block diagram of the UV pulse generation system.

Springer Series in Chemical Physics, Vol. 48: **Ultrafast Phenomena VI**
Editors: T. Yajima · K. Yoshihara · C.B. Harris · S. Shionoya

the dye laser are mixed in a β-BaB_2O_4 crystal (type I phase matching). The wavelength of the up-converted UV pulse is tunable from 375 to 395 nm by tuning the dye laser and the orientation of the crystal. The pulse duration and the spectral bandwidth of the UV pulse are 8ps and 3Å which are determined by the dye laser. The pulse duration is actually obtained by the autocorrelation measurement using biexciton two-photon excitation in CuCl. The average output power is about 5mW when the input power of the YAG laser and that of the dye laser are 2.5W and 250mW.

We measured the temporal behavior of the free exciton emission from high purity single crystals of CuCl at 4.2 K. The emission was analyzed with a 50 cm monochromator and a synchroscan streak camera. In the backward emission from a thick sample, it is found that the decay of the exciton luminescence has a very strong excitation frequency dependence. When the excitation frequency is in the transparent region of the upper branch polariton of Z_3 exciton, the free exciton emission shows a slow decay with a decay time of about 300 ps. When the excitation frequency is high and at resonance with the Z_{12} exciton where the absorption coefficient for the excitation beam is very large, the free exciton emission decays rapidly with the decay time shorter than the time resolution of our measurement system, about 80 ps. To clarify the origin of this difference, we measured the frequency-resolved temporal behavior of the forward and backward emission from a thin sample (2.6μm) under the excitation at the Z_{12} exciton resonance. Figure 2 shows the results[3]. Each curve does not show a type of a simple exponential decay but a pulse type. The delay times of these pulses

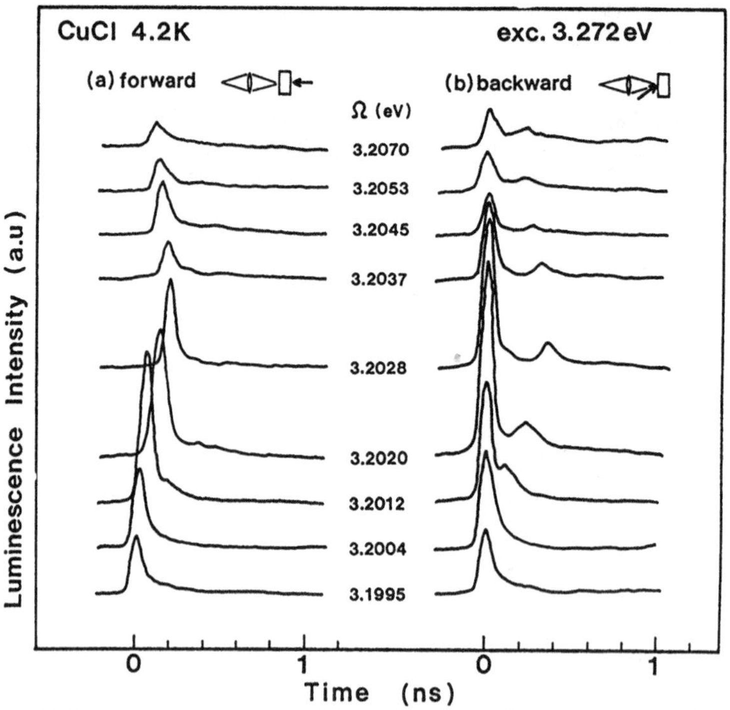

Fig.2. Temporal responses of the Z_3 exciton luminescence under the excitation of the Z_{12} exciton resonance at 4.2K for various photon energies of luminescence

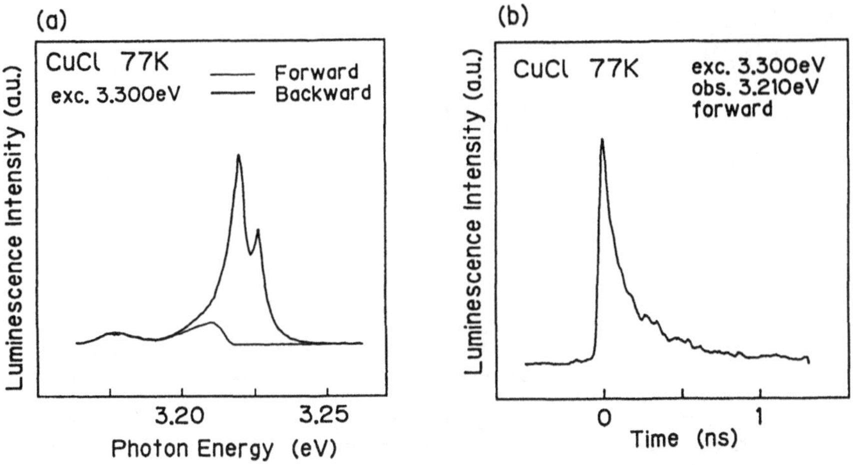

Fig.3. Time integrated luminescence spectra (a) and the temporal responses (b) of the Z_3 excitons at 77K in the forward and the backward directions

correspond to the propagation with the polariton group velocity[4]. These results show the excitation is strongly inhomogeneous along the beam direction, and that the Z_{12} excitons decay to the Z_3 exciton-band within a very short time. We also did the similar experiments at liquid nitrogen temperature. The forward and backward luminescence spectra show different line shapes as shown in Fig.3(a). Figure 3(b) shows a typical example of the temporal profile of the forward emission. It has rather long decay tail of about 200ps. This implies that at liquid nitrogen temperature the excitons are frequently scattered by phonons, and thus the exciton polariton concept fails.

In conclusion, under the weak excitation with UV picosecond pulses, the polariton propagation effects are found to play an essential role in the temporal profile of the exciton luminescence at 4.2K. It is also found that at 77K the exciton phonon interaction becomes important and the polariton picture loses its significance. The experiments between 4.2K and 77K are in progress.

References

1. U.Heim and P.Wiesner, Phys. Rev. Lett. **30**, 1205 (1973); P.Wiesner and U.Heim, Phys. Rev. **11**, 3071 (1975).
2. M.Ueta, H.Kanzaki, K.Kobayashi, Y.Toyozawa and E.Hanamura: *Excitonic Processes in Solids*, (Springer, Berlin, Heidelberg, New York, Tokyo 1986).
3. M.Kuwata, T.Kuga, H.Akiyama,T.Hirano and M.Matsuoka: to be published.
4. Y.Masumoto, Y.Unuma, Y.Tanaka and S.Shionoya, J.Phys.Soc.Jpn. **47** (1979) 1844.

Picosecond Dynamics of Exciton Polaritons in CuCl Single Crystals

T. Itoh, Jin Fashan, Y. Iwabuchi, and T. Ikehara*

Department of Physics, Faculty of Science, Tohoku University, Sendai 980, Japan

Evolution and decay kinetics of exciton polaritons photo-generated in semiconductors at low temperatures are governed by the following various processes: (1)initial formation of transient population of polaritons in real- and k-spaces inside a crystal, (2)propagation of polaritons with their group velocity, (3)intraband scattering by phonons or impurities resulting in the momentum relaxation along with the polariton dispersion curve, (4)trapping at impurities and defects inside the crystal and/or at the surface as non-radiative decay, and (5)escape from the surface as radiative decay.[1-3] Especially (2) and (5) are important in the polariton concept and together with (1) they influence the temporal behavior of exciton polariton resonant luminescence(Ex). On the other hand, its 2LO-phonon replica(Ex-2LO) is not affected by them since Ex-2LO is located at the transparent energy region and directly represents the energy distribution of the corresponding polaritons wherever they exist in the crystal.[2,4,5] In this paper picosecond temporal behaviors of Ex and Ex-2LO are studied for Z_3 exciton polaritons in pure single crystals of CuCl at 2K. The results show that when the excitation energy is located at a strongly absorptive region the polariton propagation has an important influence upon the temporal shape of Ex luminescence spectra in an early stage[6], but becomes less important in a late stage after the intraband inelastic scattering by phonons takes place. From the analysis the intraband relaxation rate is derived in the energy region of the Z_3 exciton.

Picosecond UV light source with average power of 5mW, width of 5ps and tunability from 350 to 390nm was obtained with use of a $LiIO_3$ crystal for doubling the frequencey of Pyridine 2 dye laser synchronously pumped by the second harmonic of a CW mode-locked YAG laser. Time-resolved spectra were analyzed with a double monochromator of subtractive dispersion followed by a synchroscan streak camera. Overall spectral and temporal resolutions of the system could be simultaneously set at 0.5meV and 15ps, respectively. CuCl single crystals of platelet shape grown from vapor phase were carefully selected in order to reduce possible impurity effects.

Figures 1(a) and (b) are the time-resolved luminescence spectra of Ex-2LO and Ex, respectively, at different delay times after the excitation at 3.273eV among the $Z_{1,2}$ exciton resonance for the 12μm sample. The uppermost spectra are the time-integrated ones. In (a) one can notice that the initial polariton distribution in energy is very broad and nearly proportional to the polariton density of states. With the lapse of time the intraband relaxation occurs and without much decrease of the total number of polaritons the piling up of the population is established around the energy of "bottle neck"[1] which is located slightly higher than the transverse exciton energy E_T. The decay time of the total number of the polaritons, that is, their lifetime, is about 4ns. On the other hand, in spectra (b) the resonant Ex luminescence

* On leave of absence from Yanbian University, China

Springer Series in Chemical Physics, Vol. 48: **Ultrafast Phenomena VI**
Editors: T. Yajima · K. Yoshihara · C.B. Harris · S. Shionoya

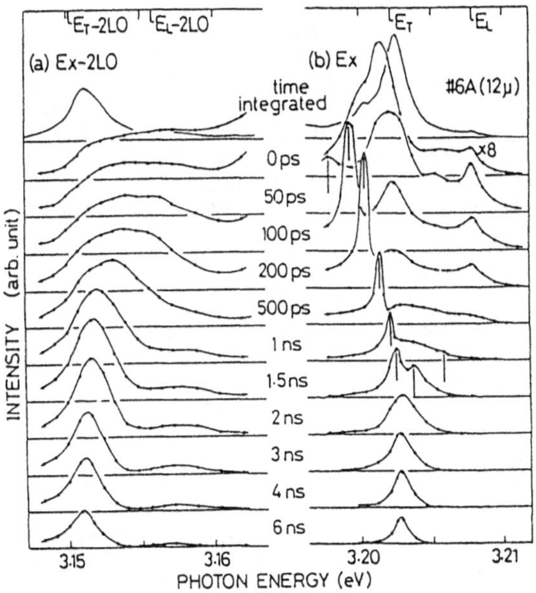

Fig.1 Time-resolved luminescence spectra of (a)Ex-2LO and (b)Ex in CuCl at different delay times together with the time-integrated ones under excitation among $Z_{1,2}$ exciton band at 2K. Note that the intensity of Ex at 0ps delay is reduced by a factor of 8

shows completely different behavior from those of (a) in an early stage. Just after the excitation an intense and broad luminescence band centered at 3.201eV is observed. The line shape seems to be mainly governed by the combination of the initial polariton distribution and the probability of polariton transmission through the surface. This broad band is transient and immediately decreases by more than one order in magnitude. Since Ex-2LO has no sharp decrease as seen in (a), the corresponding polaritons do not decay abruptly just after the excitation. The penetration depth of the excitation light at 3.273eV is less than 0.1μm and, therefore, the initial real-space distribution of the polaritons is generated very near to the front surface of excitation through interband scattering by multi phonons. The generated polaritons first propagate in all directions and half of them immediately impinge upon the front surface. However, due to their large refractive indexes roughly a thousandth of them with wavevectors almost normal to the surface can escape from the crystal as the initial Ex luminescence and the rest of them are reflected back. Then, the sharp drop is caused by the disappearance of the polaritons from the front surface on account of their inward propagation away from the surface. This interpretation is reasonable since the sharp drop of Ex completely disappears under excitation at the transparent energy region between the Z_3 and $Z_{1,2}$ exciton bands where the initial distribution of polaritons in real space is almost uniform.

With lapse of time the Ex luminescence becomes narrow, especially on the higher energy side, associated with the intraband relaxation. At the same time a sharp peak denoted by a vertical bar appears and shifts to the higher energy side. Another sharp peak becomes observable around 1ns on the higher energy side and both of them merge into one peak a little before 2ns. The delay time at which this secondary peak is observed is twice as much as the propagation time of the corresponding polaritons with their group velocity across the 12μm sample. Therefore, this structure is caused by the escape of the polaritons having wavevectors almost normal to the surface and once reflected back from the rear surface without any large deflection of their wavevectors during the propagation inside the crystal. After 3ns only one peak remains around 3.2025eV which coincides with the peak energy of the time-integrated spectrum. This fact means that the main peak in the time-integrated spectrum is caused by the polaritons at the bottle neck region. If the lifetime of the polaritons were much shorter than 3ns, the Ex luminescence would represent only the temporal behavior of polaritons with inhomogeneous distribution inside the crystal. It is also verified that the escape of the polaritons from the surfaces has minor influence on the total number of the polaritons since the corresponding sharp drop of intensity does not exist in Ex-2LO even when the secondary peak due to the polaritons once reflected back is observed in Ex.

The intensity ratio of the secondary peak to the initial luminescence at the same energy gives valuable information about the intraband scattering rate of polaritons. When one takes the probability of surface transmission, P, of a polariton having wave vector normal to the surface and its transmissivity, T, within one-way propagation across the sample, the intensity ratio, R, is equal to $(2-P)\cdot(1-P)\cdot T^2$. With use of theoretical value for P, T is obtained as $\sqrt{R/(2-P)\cdot(1-P)}$ and is plotted in Fig.2 by a closed circle for different polariton energies. The temporal behavior of the Ex luminescence has been also observed from the rear surface of the sample and the intensity ratio R' of initial luminescence in the forward spectrum to that in the backward one at the same energy is obtained. Here, $R'=(2-P)\cdot T$ and T is plotted by an open circle. The value T is nearly constant below E_T and the onset of decrease occurs at the bottle neck region. The transmissivity, T, is connected with the intraband scattering rate of polaritons, $1/\tau$, with which significant deflection of the polariton wavevector occurs, and T is given by $\exp(-d/Vg\cdot\tau)$. Here, Vg is the group velocity of polaritons and d the sample thickness. With use of theoretical value for Vg, $1/\tau$ is plotted by a triangle. Below E_T the rate is not shown because the decay is considered to be mainly governed by some extrinsic effect. Theoretical values for the acoustic-phonon scattering process[1] are calculated for Z_3 exciton polaritons in CuCl and drawn in Fig.2 by a solid curve, which shows a similar tendency to the experimental one and gives a right order of magnitude. The scattering time τ is about 3ns at the bottle neck. This value is reasonable since after this time the anisotropic propagation and the inhomogeneous distribution of the polaritons inside the crystal become almost obscured in the time-resolved Ex spectra. Therefore, the scattering rate experimentally obtained is considered to be mainly ascribed to the intraband scattering by acoustic-phonons.

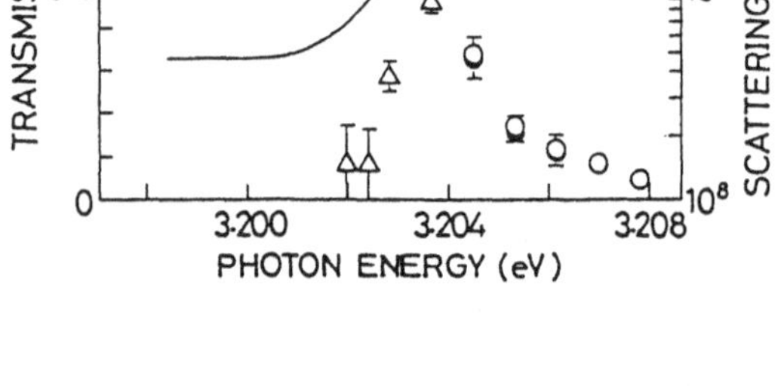

Fig.2 Polariton transmissivity, T, and its intraband scattering rate, $1/\tau$, versus energy, denoted by a circle (backward:closed and forward:open) and a triangle, respectively. Solid curve represents theoretical acoustic-phonon scattering rate for Z_3 excitons at 2K

References

1. Y. Toyozawa: Prog. Theor. Phys. 20, 53 (1958); ibid. Suppl. 12, 111 (1959)
2. H. Sumi: J. Phys. Soc. Jpn. 41, 526 (1976)
3. Y. Masumoto and S. Shionoya: J. Phys. Soc. Jpn. 51, 181 (1982)
4. Y. Masumoto and S. Shionoya: Phys. Rev. B30, 1076 (1984)
5. J. Aaviksoo, A. Freiberg, T. Reinot and S. Savikhin: J. Luminescence 35 267 (1986)
6. T. Kuga, M. Kuwata, H. Akiyama, T. Hirano and M. Matsuoka: in this conference, WD3

Dynamics of Free and Momentarily Localized Excitons in HgI_2 and PbI_2

J. Takeda[1], T. Goto[1], and M. Matsuoka[2]

[1]Department of Physics, Faculty of Science, Tohoku University, Sendai 980, Japan

[2]Institute for Solid State Physics, University of Tokyo, Roppongi, Minato-ku, Tokyo 106, Japan

The lower energy tail of the lowest exciton absorption band in various semiconductors and insulators is known to obey Urbach rule [1]. The exciton associated with this rule is assigned to be a momentarily localized exciton at a distorted lattice site by Toyozawa [2], and is called an Urbach exciton. Recently, we measured the excitation energy dependence of the luminescence quantum yield at 100 K in HgI_2 [3] and PbI_2 [4], and it has been concluded that each of the free and Urbach excitons is thermalized but they are not in the same thermal equilibrium. In this paper, we report the time response of these exciton luminescences in the picosecond region in order to observe the exciton dynamics more directly.

Platelet-like crystals and evaporated films of HgI_2 and PbI_2 were used in this work. HgI_2 and PbI_2 single crystals were excited by the second and third harmonics, respectively, of the 1.06 μm light from a mode-locked YAG laser. The luminescence from the crystal surface was passed through a monochromator and analyzed by a streak camera and an optical multi-channel analyzer. The time and spectrum resolution of this system were about 70 ps and 5 cm^{-1}, respectively.

Figure 1 shows the time response of the free and Urbach exciton luminescences of HgI_2 at 100 K together with the shape of the exciting laser pulse in a logarithmic scale. Notation F means the free exciton and U the Urbach exciton. The detected photon energy of the luminescence is indicated for each spectrum in a unit of kcm^{-1}. Broken lines show the calculated curves where the one-dimensional exciton diffusion, the suface recombination and the re-absorption of the luminescences are taken into account, as treated by Kurita et al. [5]. In this treatment, the exciton lifetime τ and the surface recombination velocity S are adjustable parameters for the free exciton luminescence, while τ, S and the diffusion coefficient of the exciton D are the ones for the Urbach exciton luminescence.

The temperature dependence of the free and Urbach exciton lifetimes denoted by circles and squares, respectively, is shown in Fig. 2 in a logarithmic scale. Above 100 K, the lifetime of the free exciton is shorter than that of the Urbach exciton. We also find that the lifetimes of the free and Urbach excitons are nearly independent of the detected photon energy in the respective energy region. If photo-generated excitons are immediately thermalized, the exciton lifetime should be constant in any energy region.

Springer Series in Chemical Physics, Vol.48: **Ultrafast Phenomena VI**
Editors: T. Yajima · K. Yoshihara · C.B. Harris · S. Shionoya

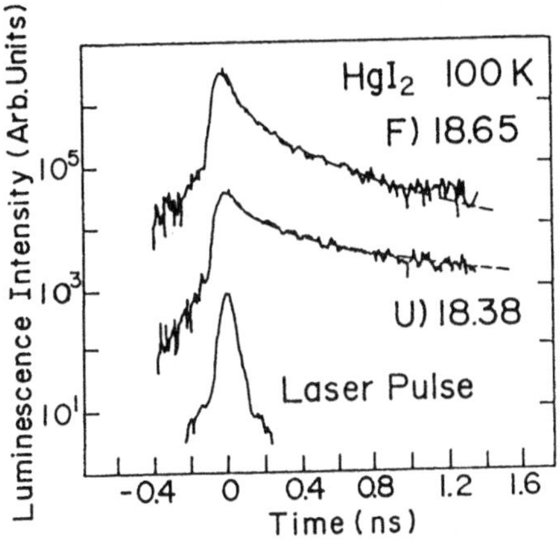

Fig. 1. Time response of the free and Urbach excitons of HgI_2 at 100 K

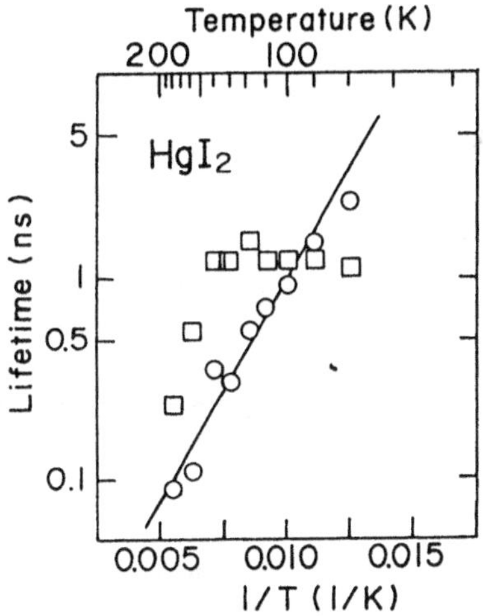

Fig. 2. Temperature dependence of the lifetimes of the free and Urbach excitons

Therefore, it is concluded that each of the free and Urbach excitons is thermalized, but they are not in the same thermal equilibrium. These conclusions agree well with the previous ones obtained from ref. [3].

The lifetime of the free exciton increases exponentially with increasing the inverse of temperature as shown by a straight line in Fig. 2. From the slope of this line the thermal activation energy is estimated to be 38±4 meV, which is almost equal to the exciton binding energy of HgI_2 [6]. This fact means that the decay of the free exciton is dominated by the process of the dissociation into an electron-hole pair. On the other hand, the lifetime of the Urbach exciton is almost constant between 77 and 140 K, and abruptly decreases above 140 K. This fact indicates that the decay process of the Urbach exciton is not dominated by the radiative one, which is independent of temperature. The Urbach exciton is considered to be decomposed to a free electron-hole pair or converted to a free exciton.

Figure 3 shows the diffusion coefficients of the free and Urbach excitons as a function of temperature. Open circles show the diffusion coefficients of the free exciton and closed circles those of the Urbach exciton. The diffusion coefficient of the Urbach exciton is obtained by the above mentioned method, while that of the free exciton is estimated from the line-width of the exciton absorption band in HgI_2 thin film. Solid lines represent the calculated curves for the diffusion coefficients of the free and Urbach excitons, D_f and D_u as a function of temperature:

$$D_f \propto \{ \exp(\theta / kT) - 1 \} , \qquad [1]$$

$$D_u \propto T^{1/2} , \qquad [2]$$

where θ is the LO-phonon energy. The former equation is obtained from the

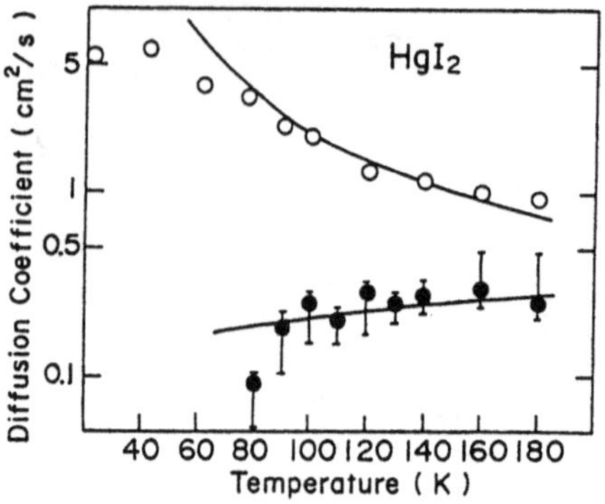

Fig. 3. Temperature dependence of the diffusion coefficients of the free and Urbach excitons

theory of the exciton scattering by the LO-phonons and the latter from that by acoustical phonons under the deformation potential. The solid curves can fit well to experimental points above 90 K. This fact suggests that the relaxation of free excitons is caused by the LO-phonon scattering and that of Urbach excitons by the acoustical phonons. Moreover, the value of D_u is smaller than that of D_f, which is reasonable because the Urbach exciton accompanied with the lattice distortion should be less mobile than the free exciton.

In PbI_2, the lifetime of the free exciton is also shorter than that of the Urbach exciton at 100 K. Thus, the same conclusion is obtained, that the free and Urbach excitons are not thermalized with each other.

References

1. F. Urbach: Phy. Rev. 92, 1324 (1953)
2. Y. Toyozawa: ISSP Tech. Rep. University of Tokyo, A119 (1964)
3. J. Takeda, T. Goto: Solid State Commun. 64, 1469 (1987)
4. J. Takeda, T. Goto: Solid State Commun. 59, 209 (1986)
5. A. Kurita, Y. Fujikawa, T. Kushida: J. Lumi. 38, 70 (1987)
6. T. Goto, A. Kasuya: J. Phy. Soc. Jpn. 50, 520 (1981)

Formation and Relaxation of Excitonic Magnetic Polarons in $Cd_{1-x}Mn_xTe$ and $Cd_{1-x}Mn_xSe$

Y. Oka, I. Souma, and Y. Kashiwagi

Research Institute for Scientific Measurements, Tohoku University, Katahira, Sendai 980, Japan

The dynamics of magnetic polarons is studied in $Cd_{1-x}Mn_xTe$ and $Cd_{1-x}Mn_xSe$ for x = 0.01 - 0.6 by picosecond time-resolved spectroscopy of the exciton luminescence. The luminescence shows the formation and relaxation kinetics of the magnetic polarons in localized exciton states, in the triplet exciton states as well as in a spin glass phase.

1. INTRODUCTION

Semimagnetic semiconductors show a number of magnetooptical properties which are induced by the exchange interaction of the band electrons with magnetic-ion spins.[1] Magnetic polaron, a magnetically bound state of the exciton by the exchange coupling with Mn ions, is such evidence.[2] In this paper we study the dynamics of excitonic magnetic polarons in $Cd_{1-x}Mn_xTe$ and $Cd_{1-x}Mn_xSe$ by time resolved spectroscopy of the luminescence. Under picosecond-pulse excitation the excitonic magnetic polarons show varieties of the fast energy relaxations which are related to the realignment of magnetic-ion spins.

Experiment was performed on single crystals of $Cd_{1-x}Mn_xTe$ and $Cd_{1-x}Mn_xSe$ (x = 0 - 0.6) grown by Bridgeman method. Luminescence measurement was made on a cleaved surface of a sample which was placed in a superconducting magnet. The optical excitation was made by mode locked pulses of an argon ion laser and also by second harmonics of a mode-locked YAG laser.[3,4] Time resolution of the system is 20 ps.

2. EXPERIMENTAL RESULTS AND DISCUSSION

2.1 Magnetic Polaron Formation in Localized Exciton Region

Figure 1 shows the time characteristics of the exciton luminescence in $Cd_{0.8}Mn_{0.2}Te$ at magnetic fields of 0 and 7 T. The arrows at the top of the spectra indicate the uppermost exciton energies which are determined by the reflection spectra of the exciton. As time increases, the energy of the luminescence peak in the time resolved spectrum decreases. Since microscopic variation of the Mn composition is significant in crystals of x > 0.1, the exciton energies are locally fluctuated in the lattice and localized states of the exciton are created in these crystals. The energy of the exciton is then decreased either by the formation of magnetic polarons or by the localization of excitons. Therefore it is significant to discriminate the magnetic-polaron formation from the localization effect of the exciton.[2]

The lower spectrum in Fig. 1 is the case at 7 T where the large Zeeman shift of 47 meV in the exciton state arises due to the giant g-value of the exciton which is also induced by the exchange effect. At 7 T the Mn spins are mostly polarized by the magnetic field as is known from the saturation of the Zeeman splitting. The lowest state of the magnetic polarons are

Springer Series in Chemical Physics, Vol. 48: **Ultrafast Phenomena VI**
Editors: T. Yajima · K. Yoshihara · C.B. Harris · S. Shionoya

realized immediately when the exciton is created, since the surrounding Mn spins are initially polarized by the external field. The remaining shift of the peak energy to the lower side is then caused by the localization of the exciton in the fluctuating band edge. To derive the magnetic polaron formation process from the transient exciton luminescence, we measure, at time t and at magnetic field H, the energy difference between the uppermost exciton energy (indicated by the allows) and the luminescence-peak energy, which we denote as $E_b(t,H)$. To subtract the localization effect we calculate $E_b(t,7T) - E_b(t,H)$ which corresponds to the energy for the magnetic polaron formation.

Figure 2 shows the magnetic polaron formation at 0 T, $E_b(t,7T) - E_b(t,0T)$, where the energy of the exciton decreases about 5 meV within 500 ps after the excitation. The exciton energy further decreases gradually with a time constant of 1 ns. This is due to subsequent magnetic formation during the localization process. Observed time dependence of the exciton luminescence displays the polarization of the Mn-ion spins within 500 ps to form the magnetic polaron and also shows the following localization.

Relaxation process of the excitonic magnetic polarons can be observed by the decay time and the rise time of the luminescence at each photon energy. Figure 3 shows the decay and rise time, τ_d and τ_r, derived from the transient luminescence by using the deconvolution calculation. At the uppermost energy of the exciton state (at 1.876 eV) τ_d is only 50 ps, while it increases to 1 ns at the lower edge of the luminescence (at 1.84 eV). The large variation of the decay time as a function of the exciton energy shows the localization of the exciton in the mixed crystals. The rise time of the luminescence shows the creation rate of the magnetic polarons at each exciton energy. In case that the excitonic magnetic polarons are well localized at the lower energy region of the luminescence, the rise time must coincide with the decay time owing to successive hopping to reach the lowest sites. This condition, $\tau_r = \tau_d$, is actually realized at the lower edge of the exciton luminescence at 7 T. However in Fig. 3 around

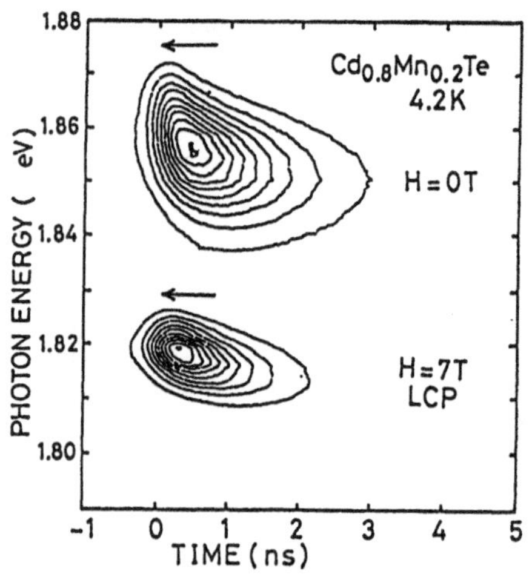

Fig. 1 Transient characteristics of the exciton luminescence in $Cd_{0.8}Mn_{0.2}Te$. The magnetic polaron formation takes place at H = 0 T during the luminescence decay while only the localization effect arises at H = 7 T.

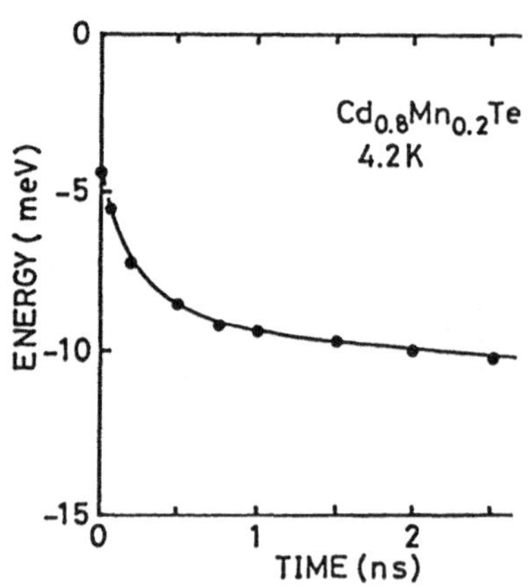

Fig. 2 Plot of $E_b(t,7T) - E_b(t,0T)$ vs t, which shows the formation energy of the magnetic polarons.

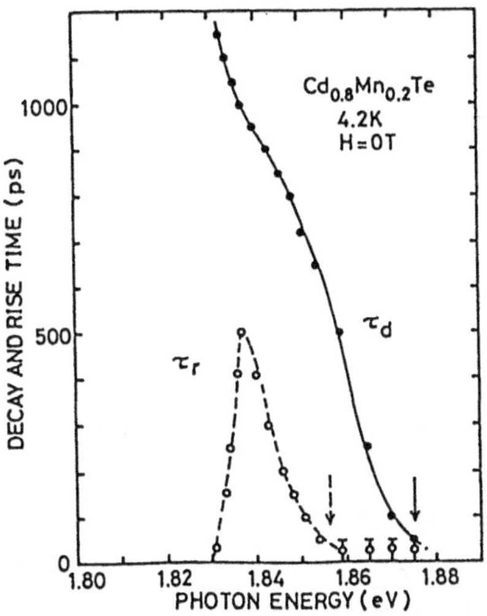

Fig. 3 The decay time and the rise time of the exciton luminescence. The solid arrow shows the uppermost exciton energy while the dashed arrow shows the peak energy of the time-integrated luminescence.

1.84 eV τ_r is only 500 ps while τ_d is 1 ns. The large difference between the two characteristic times indicates the faster creation of the magnetic polarons in the localized exciton regime, indicating the additional formation mechanism such as the tunneling-induced formation.

2.2 Triplet Magnetic Polarons and Magnetic Polarons in Spin Glass

The excitonic magnetic polarons have the singlet and triplet states with respect to the internal spin configuration of the electron and the hole. The triplet component can arise in the luminescence spectra if the dipole-allowed singlet excitons are discriminated using the polarization selection rule. Experimental results obtained by such conditions show the triplet magnetic polaron in $Cd_{1-x}Mn_xTe$ of x = 0.2-0.3, where the binding energy is about 10 meV larger than that of the singlet-magnetic polaron. The observed lifetime is 4 ns which is 2 times longer than the singlet magnetic polarons.[4]

For x > 0.2 the Mn-ion spins in semimagnetic semiconductors show the spin glass transition at temperatures below 20 K. Magnetic polaron states must be significantly affected by the spin glass phase of the Mn ions. Magnetic polarons are not formed in the spin glass phase if the spin-glass interaction among the Mn spins is stronger than the exchange interaction between the excitons and the Mn ions, while in the opposite case the excitons can destroy the spin glass and polarize the Mn spins to form the magnetic polarons. The temperature dependence of the exciton luminescence at zero magnetic field shows the restriction of the magnetic polaron formation below the spin-glass transition temperature. Time resolved spectra of the excitonic luminescence in the spin glass regime show the broadly-spread spin configuration of Mn ions. External high magnetic field is found to destroy the spin glass and assists the excitons to reach the ground states of the magnetic polarons.

[1] J.K. Furdyna, J. Appl. Phys. 53, 7637 (1982).

[2] Y. Oka, K. Nakamura, I. Souma and H. Fujisaki, Proc. Intern. Conf. on Physics of Semiconductors, Stockholm (World Scientific, Singapore 1986) p.1771.

[3] J.J. Zayhowski, R.N. Kershaw, D. Ridgley, K. Dwight, A. Wold, R.R. Galazka and W. Giriat, Phys. Rev. B35, 6950 (1987).

[4] Y. Oka, K. Nakamura, I.. Souma, M. Kido and H. Fujisaki, J. Lumines. 38, 263 (1987).

Space- and Time-Resolved Spectroscopy of the Ultrafast Exciton Motion at a Stacking Fault Interface in Layered BiI_3 Crystals

T. Karasawa, T. Kawai, I. Akai, and Y. Kaifu

Department of Physics, Osaka City University, Sumiyoshi-ku, Osaka 558, Japan

The exciton translational motion is a basic problem of energy transport in solids. However, its direct observation is rather difficult because intrinsic excitons usually have large absorption coefficients and short lifetimes due to various scattering processes. In this report, we present the observation of exciton motion in the picosecond time range on a specific two-dimensional (2D) lattice composed by a stacking fault in a layered crystal BiI_3.

Stacking faults distort the unit cell symmetry resulting in new exciton states at a 2D lattice. In BiI_3, characteristic sharp absorption lines, named R, S and T from the high energy side, appear near the indirect absorption edge and have been attributed to the exciton transition near the stacking fault (called stacking fault excitons, SFE)[1]. Their internal structure is well described by the cationic exciton model modified by deformed unit cells[2]. The SFE shows intense and sharp resonant luminescences without Stokes shifts[3]. The luminescences are observed to extend along the crystal plane beyond the exciting laser spot, which makes it possible to examine the exciton motion by space- and time-resolved spectroscopy.

Figure 1 shows the experimental configuration schematically. A laser beam from a mode-locked Ar^+ laser exciting the band to band transition region was focused to a spot of less than 40μm in diameter. on the sample surface. The

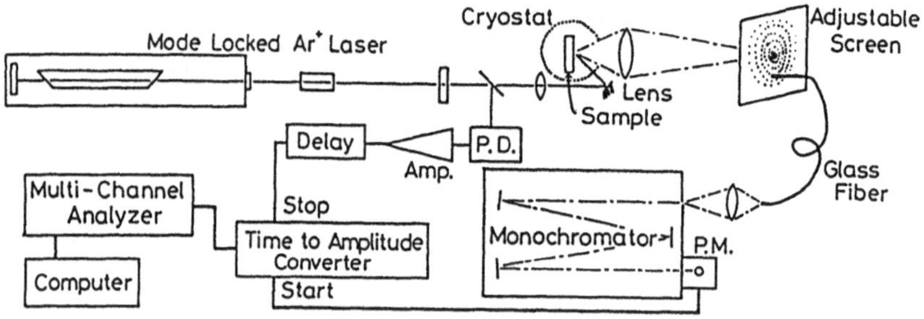

Fig.1 Experimental Configuration

Springer Series in Chemical Physics, Vol. 48: **Ultrafast Phenomena VI**
Editors: T. Yajima · K. Yoshihara · C.B. Harris · S. Shionoya

luminescence pattern was enlarged with a lens and resolved by a gradient index-type glass fiber set on an adjustable screen leading to a monochromator. The temporal behavior of luminescences was detected by a time-correlated single photon counting method. This system attained a spatial resolution of less than 30μm and a time resolution of less than 25ps by deconvolution analysis. Space-resolved transmission spectra were also obtained by putting a white light source behind the sample.

The space-resolved luminescence spectra are shown in Fig.2 for various distances r at 4.2K. Uppermost in Fig.2 is a space-resolved transmission spectrum at r=0, and the same ones are obtained at any point within more than r=1mm on the sample plane showing the macroscopically uniform occurrence of the stacking fault. The luminescence spectra consist of the indirect exciton recombination band L_C accompanied by the bulk phonon C, the resonance lines R, S and T, and an additional peak T'. The R, S and T lines can be detected with exact zero Stokes shifts even at r=400μm. While the band L_C fades out at larger r, the peak T' appears, depending on the sample, and increases intensity relative to that of T with increasing r. The T' is considered due to trapped excitons at some defects along the 2D lattice[4]. The stationary distributions of luminescence intensity along the lateral direction on the sample plane can be well described by the 2D diffusion equation including relaxation. For the T at 4.2K, an extraordinarily large diffusion constant of more than $10^5 cm^2/sec$ is estimated using the exciton lifetime of ≃1nsec[5]. The lateral distribution is seriously reduced below 4.2K, and the diffusion constant changes to $10^3 cm^2/sec$ at 2K.

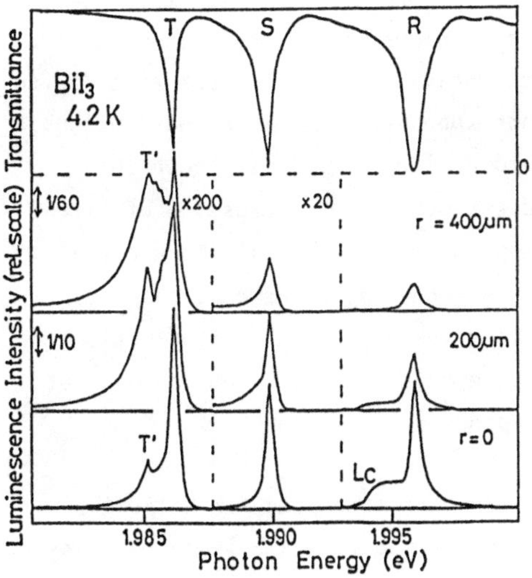

Fig.2 Space-Resolved Spectra

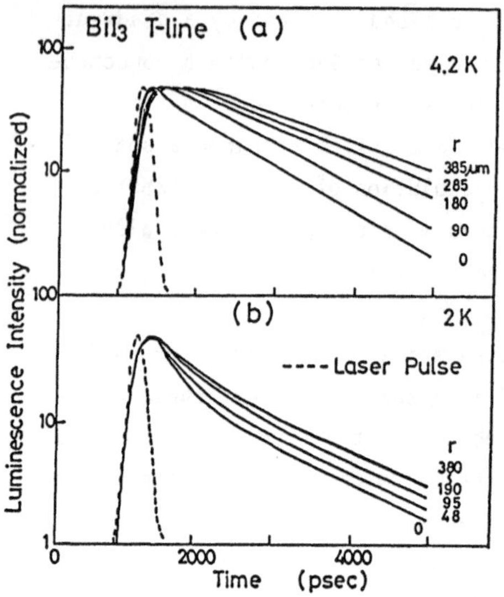

Fig.3 Temporal Behavior of Luminescence T

The temporal behavior of luminescence T at 4.2K is shown in Fig.3(a) at various r's with normalized peak intensity. At r=0, the luminescence T has a rising component characterized by the decay time constant of the upper lying state S, reflecting a cascade relaxation from S to T[5]. After that, it shows nonexponential rapid decay at an early stage. At r≠0, peak delay and a growing component after the peak can be seen with increasing r. This behavior can be explained qualitatively by diffusion. However, the mean velocity estimated from the peak delay is rather faster than that obtained from decay profiles calculated by the diffusion equation using the parameters of stationary distribution. Then, the details of time behavior cannot be explained quantitatively only by the diffusion process. The results suggest coexistence of rather faster transfer mechanisms of exciton than the diffusion in our system. At 2K, the temporal behavior shows the same rise due to the cascade process, the early stage nonexponential decay becomes obvious, and the peak delay is no longer observed at larger r. This change corresponds to the reduction of lateral distribution by suppression of diffusion at 2K. At lower temperatures, trapping centers inhibit the exciton diffusion, and the translational motion is governed by a faster transport mechanism such as resonance transfer. It gives ultrafast exciton motion along the stacking fault interface. The zero Stokes shift luminescence at large r suggests the existence of such process.

The authors are grateful to Prof. T.Iida and T.Komatsu for helpful discussions. This work is partially supported by the Yamada Science foundation.

References

1. Y.Kaifu, T.Komatsu: J.Phys.Soc.Jpn 40 1377 (1976)
2. T.Komatsu, Y.Kaifu, S.Takeyama, N.Miura: Phys.Rev.Lett. 58 2259 (1987)
3. K.Watanabe, T.Karasawa, T.Komatsu, Y.Kaifu: J.Phys.Soc.Jpn 55 897 (1986)
4. T.Karasawa: J.Lumin. 38 51 (1987)
5. I.Akai, T.Karasawa, Y.Kaifu, A.Nakamura, M.Shimura, M.Hirai: J.Lumin. to be published

Ultrafast Optical Dephasing of Two-Dimensional Excitons in BiI_3

A. Nakamura[1], Y. Ishida[2], T. Yajima[2], T. Karasawa[3], I. Akai[3], and Y. Kaifu[3]

[1]Department of Applied Physics, Nagoya University, Chikusa-ku, Nagoya 464-01, Japan

[2]Institute for Solid State Physics, University of Tokyo, Minato-ku, Tokyo 106, Japan

[3]Department of Physics, Osaka City University, Sumiyoshi-ku, Osaka 558, Japan

Coherent transient phenomena of localized and delocalized excitonic states in semiconductors have recently attracted much attention[1]. In a BiI_3 layered semiconductor, there exist two-dimensional excitons trapped by stacking faults and they give sharp absorption lines depending on temperatures and samples[2,3]. It is interesting to investigate the homogeneous line width and quantum beats of these stacking-fault excitons by transient spectroscopy in the ultrashort time region. Recent development of transient spectroscopy has allowed us to observe optical dephasing with ultrahigh resolution which is determined by the correlation time τ_c instead of time duration of light pulses[4]. Furthermore, it is suggested that one can derive some information on optical dephasing from the peak shift of the correlation traces of transient degenerate four-wave mixing (DFWM), even if the decay behavior is not observed. This paper reports optical dephasing of stacking-fault excitons in BiI_3 at low temperatures using DFWM in k-space with two incident beams.

Figure 1 shows typical absorption spectrum of BiI_3 at 4.2 K. Q, R, S, T lines are due to two-dimensional exitons trapped by stacking faults. These excitons originate from the distortion of bulk excitonic states ($\psi_{6,10}$, ψ_3, ψ_4) by stacking faults. The cascade relaxation among these states is observed by the luminescence experiments. The lifetimes for R, S and T excitons are about 200, 400 and 800 ps, respectively.

The DFWM experiments were made using a synchronously-pumped hybridly mode-locked cavity dumped R6G laser which generated 3.7 nJ sub-picosecond

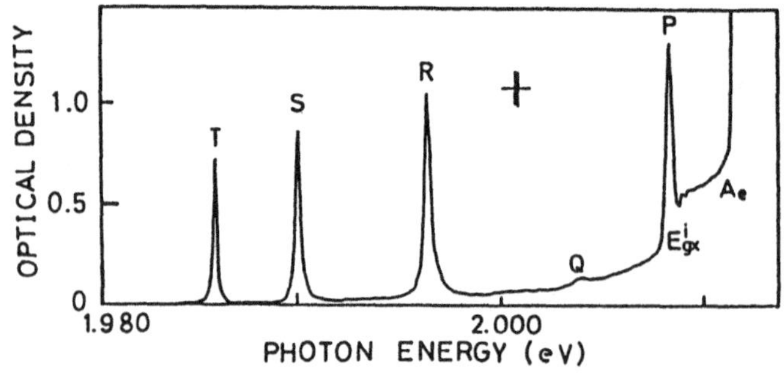

Fig. 1 Typical absorption spectrum of BiI_3 near the indirect exciton edge at 4.2 K.

Springer Series in Chemical Physics, Vol. 48: **Ultrafast Phenomena VI**
Editors: T. Yajima · K. Yoshihara · C.B. Harris · S. Shionoya

pulses. The spectral width and the pulse duration are 13 A and 0.9 ps ($sech^2$ shape), respectively. Since these pulses are not transform-limited, the time resolution of the DFWM experiments is governed by the correlation time τ_c (inverse of the spectral width ~0.38 ps in this experiment) instead of the pulse duration. When two light beams with wavevector $\vec{k}_1$, $\vec{k}_2$ are incident on the BiI_3 platelet crystal, output beams are produced in the new directions $\vec{k}_3 = 2\vec{k}_1 - \vec{k}_2$, $\vec{k}_4 = 2\vec{k}_2 - \vec{k}_1$.

Figure 2 shows correlation traces for the $\vec{k}_3$ and $\vec{k}_4$ directions, which were measured for the R-line excitation at 10 K. Both the decay behavior and the peak shift are observed. The decay time τ and the peak shift are 0.80 and 0.16 ps, respectively. The dephasing time T_2 is estimated for the limiting cases of the spectral broadening: T_2 is 4τ for the extremely inhomogeneous broadening and T_2 is 2τ for the homogeneous broadening. Assuming the extremely inhomogeneous broadening of the absorption line (~16 cm^{-1}), T_2 is estimated to be ~3.2 ps. Since T_1 of R-excitons of 200 ps, T_2 is dominantly determined by the pure dephasing process. Since it is not obvious that the above assumption is fulfilled for the R line of this sample, we should note that there exists an uncertainty in the estimated value of T_2 by a factor of 2 at maximum.

The peak shift is related to all relaxation processes of the system: longitudinal (T_1), transverse (T_2) and cross (T_3) relaxation processes. Moreover, the magnitude of the shift depends also on the incident pulse characteristics, for example, a transform-limited pulse and a phase modulated pulse. The model calculations[5] in the limiting cases where a

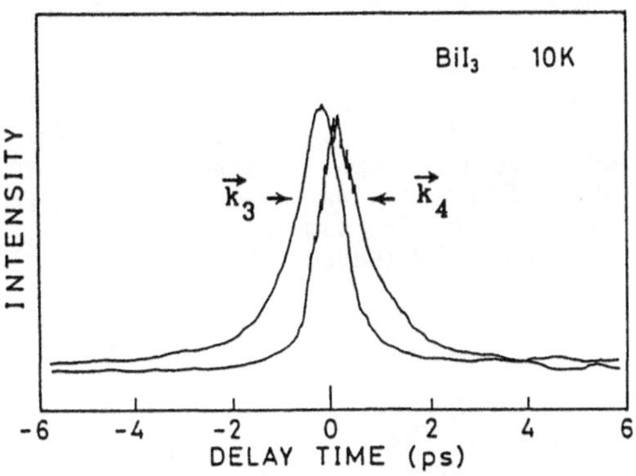

Fig. 2 Correlation traces for two output beams $\vec{k}_3$ and $\vec{k}_4$ of transient DFWM in BiI_3. The excitation wavelength is resonant with R-absorption line.

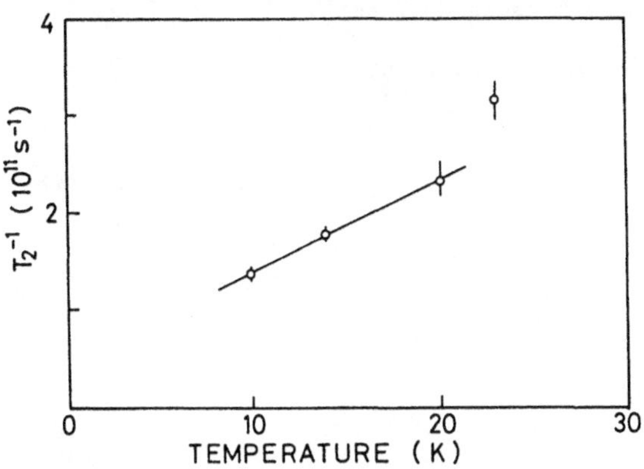

Fig. 3 Temperature dependence of inverse of T_2 for R-absorption line.

2-level system without cross relaxation processes and transform-limited pulses are assumed, provide the T_2 value comparable to the directly determined T_2 within the factor of 5.

In what follows, to investigate the mechanism of the optical dephasing of stacking-fault excitons, we have measured the temperature dependence of phase relaxation times between 10 and 23 K. We used the sample which exhibited a sharper absorption lineshape of the R line (spectral width $\sim 5\ cm^{-1}$ at 10 K). The decay time of the correlation trace is ~ 3.7 ps at 10 K. Since the decay time is comparable to the inverse of the spectral width, T_2 is rather approximate to $2\tau \sim 7.4$ ps. Figure 3 shows the inverse of T_2 as a function of temperatures. T_2^{-1} increases linearly with temperatures between 10 and 20 K. Watanabe et al.[3] have reported that the line width of the R line depends linearly on the temperatures between 6 and 25 K. Although the inhomogeneous broadening remains below 6 K in the high quality sample, such a temperature dependence is interpreted in terms of the acoustic phonon scattering with two-dimensional excitons trapped by stacking faults. Our results of T_2 is consistent with the line width variation and consequently we also conclude that optical dephasing is mainly induced by the phonon scattering.

The values of T_2, however, depend on the samples studied and T_2 is shorter in the sample which shows the broader absorption spectrum. Considering that stacking faults extend over ~ 100 µm in xy-plane, excitons trapped at these stacking faults can diffuse in xy-plane[6]. If there are small imperfections of stacking-fault plane, exciton may be scattered by them. Therefore, one of the possible interpretations of the sample dependence of the optical dephasing is the exciton scattering by imperfections.

In conclusion we have demonstrated both the decay behavior and the peak shift in the correlation trace of the DFWM in the sub-picosecond time region. The phase relaxation times of excitons trapped by stacking faults are determined for the BiI_3 crystal studied here at low temperatures. The values of T_2 at 10 K is $3.2 \sim 7.4$ ps depending on the samples. The optical dephasing of stacking-fault excitons is mainly induced by the phonon scattering at the temperatures 10 to 20 K.

References

1. L. W. Molenkamp and D. A. Wiersma: Phys. Rev. B32, 8180 (1985)
2. Y. Kaifu and T. Komatsu: J. Phys. Soc. Jpn. 40, 1377 (1976)
3. K. Watanabe, T. Karasawa, T. Komatsu and Y. Kaifu: J. Phys. Soc. Jpn. 55, 897 (1986)
4. T. Yajima, N. Morita and Y. Ishida: J. Opt. Soc. Am. B1, 526 (1984)
5. T. Yajima and N. Morita: in Method of Laser Spectroscopy, ed. by Yehiam Prior, A. Ben-Reuven and M. Rosenbluh (Plenum Publishing Corporation 1986) p.75
6. T. Karasawa, T. Kawai, I. Akai and Y. Kaifu: J. Lumin. 40/41, 431 (1988)

Ultrafast Relaxation of Localized Excitations in Solids

Y. Kayanuma

Department of Physics, Faculty of Science, Tohoku University,
Sendai 980, Japan

1. Introduction

In phenomenological theories of coherent or incoherent optical processes, the relaxation in the excited states is often treated by introducing two parameters, T_1 and T_2, which represent the population decay time and the phase relaxation time, respectively. The (homogeneous) phase relaxation is caused by the stochastic fluctuations in the electronic energy levels and is, in many cases, originated from the interaction with elementary excitations such as phonons in condensed media.

It should be remembered that the fluctuation-dissipation theorem tells us that wherever fluctuation is, dissipation is. In fact, the amplitude of the energy fluctuation D is generally related to the amount of the energy relaxation ΔE through the Einstein relation,

$$D^2 = 2kT^*\Delta E \tag{1}$$

where $T^* \equiv (\hbar\bar{\omega}/2k)\coth(\hbar\bar{\omega}/2kT)$ is the effective temperature and $\hbar\bar{\omega}$ is the representative energy of the quanta of the elementary excitations. Therefore, the phenomenological models or the stochastic models, in which the energy relaxation is neglected, are justified only in the weak coupling ($\Delta E \to 0$) and at high temperature ($kT^* \to \infty$) limit.

The purpose of the present work is to survey theoretically some aspects of the dynamical relaxation of localized centers in condensed media in the opposite , namely, the strong coupling limit through the optical responses.

2. Model

Let us consider a simple two-level system ($|1\rangle$ and $|2\rangle$) interacting with the bulk phonon modes of the crystal. The Hamiltonian is given by ($\hbar = 1$)

$$H = H_1|1\rangle\langle 1| + H_2|2\rangle\langle 2| , \tag{2}$$

where

$$H_1 = \sum_k \omega_k b_k^\dagger b_k , \tag{3}$$

$$H_2 = H_1 + \varepsilon + V , \tag{4}$$

$$V = \sum_k \alpha_k\omega_k(b_k+b_k^\dagger) . \tag{5}$$

Here, b_k ($b_k^\dagger$) is the annihilation (creation) operator for k-th phonon mode with energy ω_k, α_k is the coupling constant and ε is the electronic exci-

Springer Series in Chemical Physics, Vol. 48: Ultrafast Phenomena VI
Editors: T. Yajima · K. Yoshihara · C.B. Harris · S. Shionoya

tation energy. It is assumed that $|1\rangle$ and $|2\rangle$ are optically connected with each other by a constant dipole moment.

As for the excitation spectrum of phonons, we assume the following spectral density at low temperatures:

$$J(\omega) \equiv \frac{1}{2\pi}\int_{-\infty}^{\infty} \langle V(t)V(0)\rangle\, e^{i\omega t}\, dt$$
$$= \begin{cases} (2g\omega^2/\pi\bar{\omega}^2)(2\bar{\omega}\omega-\omega^2)^{1/2} & 0 \le \omega \le 2\bar{\omega}\,, \\ 0\,, & \text{otherwise.} \end{cases} \qquad (6)$$

Here, $\langle\cdots\rangle$ is the average over phonon coordinates and $g(=\sum_k \alpha_k^2)$ is the coupling constant by which ΔE(= half of the Stokes shift) and $D(\simeq$ width of the absorption spectrum) are given by $\Delta E = g\bar{\omega}$ and $D^2 = g\bar{\omega}^2$, respectively. The strong coupling condition $g \gg 1$ leads to the inequalities,

$$\Delta E \gg D \gg \bar{\omega}\,. \qquad (7)$$

For typical values, $\bar{\omega} \simeq 10$meV, $g \simeq 50$, we have three characteristic relaxation times; the phase relaxation time $\tau_{ph} \simeq D^{-1} \simeq 10^{-14}$sec, the energy relaxation time $\tau_{LR} \simeq \bar{\omega}^{-1} \simeq 10^{-13}$sec and the population decay time $T_1 \simeq 10^{-7}$sec.

3. Spontaneous and Coherent Optical Responses

By virtue of the assumed linear coupling, we can write down analytical expressions of any order of optical responses. The spontaneous Raman spectrum in this model was first investigated by HIZHNYAKOV et al.[1]. In Fig.1, the calculated spontaneous Raman spectrum is shown for $g = 40$ and for the incident frequency $\omega_2 = \varepsilon$ vs the normalized Raman shift. The characteristic structures — the Rayleigh scattering, the lower order Raman scattering,

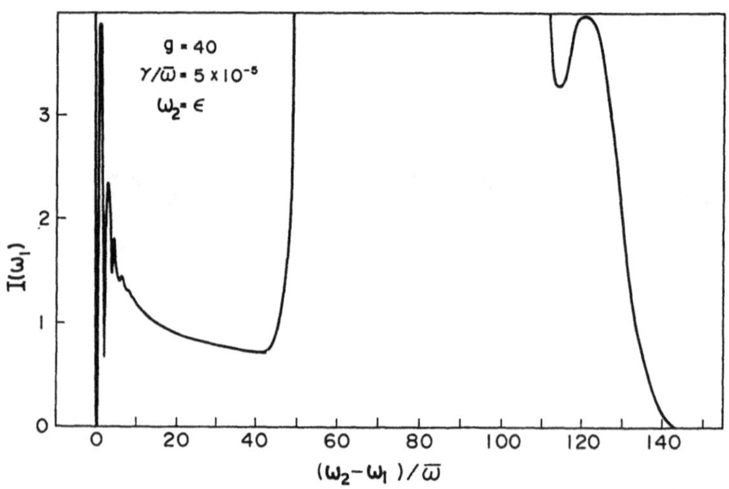

Fig.1 Spontaneous resonant Raman spectrum in strong coupling system for $g = 40$, $\omega_2 = \varepsilon$ and $\gamma \equiv 1/2T_1 = 5\times10^{-5}\bar{\omega}$

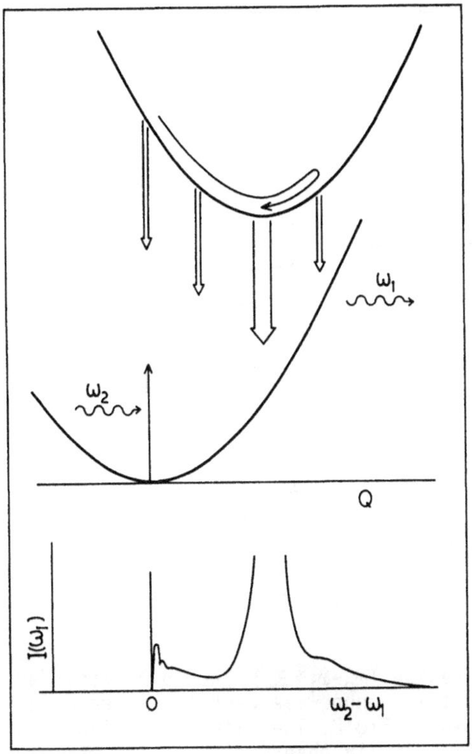

Fig.2 Schematic plot of the relaxation of strong coupling system(upper) and the spontaneous emission spectrum(lower)

the hot luminescence and the ordinary luminescence— are well resolved in the frequency domain. The analysis of the time resolved emission spectrum[2] clearly reveals how these structures are formed successively in accordance with the above mentioned three time scales, τ_{ph}, τ_{LR}, and T_1 as schematically shown in Fig.2.

The integrated amplitude of the n-th order Raman scattering $I_n^{(s)}$ is approximately given by[3]

$$I_n^{(s)} \propto |D^n \partial^n \Phi(\omega_2)/\partial\omega_2^{\,n}|^2/n! \tag{8}$$

where $\Phi(\omega_2)$ is the complex polarizability for the incident photon ω_2, the imaginary part of which is nothing but the broad Gaussian absorption spectrum. Thus, we see a duality peculiar to the strong coupling system: The quantum mechanical phonon structure is modulated by the semiclassical polarizability. The calculated line shape reproduces the essential features of that observed in the F-centers in alkali halides[4,5], a typical example of the strong coupling system.

The signals for coherent four wave mixing are also calculated exactly. In Fig.3, an example of the stimulated and the inverse Raman signal is presented for the pumping frequency $\omega_2 = \varepsilon - 10\bar{\omega}$ as a function of the probe frequency ω_1. (The gain is plotted upward of the dashed line.) The dominant contribution to the Raman gain signal is composed of two parts, the absorption saturation and the stimulated emission, which have broad Gaussian spectra widely separated (~ $2\Delta E$ ~ 1eV) from each other. In Fig.3, only the fine structures superimposed on the absorption saturation are plotted.

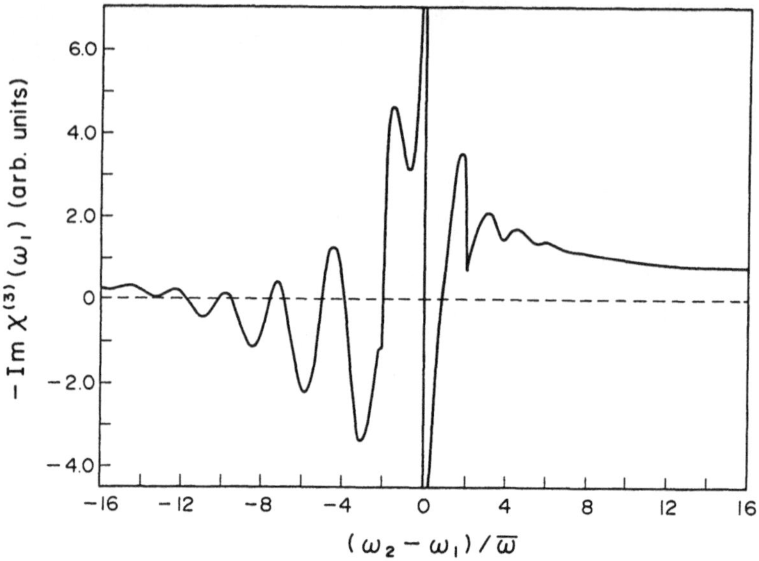

Fig.3 Stimulated resonant Raman signal $-\mathrm{Im}\,\chi^{(3)}(\omega_1)$ for g = 40 and $\omega_2 = \varepsilon - 10\overline{\omega}$

The smooth tail part in the Stokes region ($\omega_2-\omega_1 > 8\overline{\omega}$) should be regarded as the transient induced emission from the phonon wave packet sliding down the adiabatic potential surface as shown in Fig.2. The ultrafast relaxation dynamics is further clarified by the time domain observation of the stimulated Raman scattering using ultrashort light pulses for ω_1 and ω_2[6].

A peculiar feature in Fig.3 is that the inverse Raman signal ($\omega_1 > \omega_2$) changes its sign alternately according to the phonon numbers. This is explained by the duality, namely by the fact that the amplitude of n-th order inverse Raman signal $I_n^{(a)}$ is given by

$$I_n^{(a)} \propto -\{D^n \partial^n \Phi(\omega_2)/\partial \omega_2^{\,n}\}^2/n!. \tag{9}$$

(Note the difference from (8)!) Equations (8) and (9) are consistent with the semiclassical picture of the Raman scattering since $\partial^n\Phi(\omega)/\partial\omega^n \propto \partial^n\Phi(Q)/\partial Q^n$ where Q is the interaction mode coordinate (see Fig.2). This is a consequence of the ultrafast homogeneous phase relaxation in this system. The line shape of the inverse Raman signal depends sensitively on the pumping frequency ω_2 but its dependence is quite different from that of the weak coupling system[7].

In Fig.4, an example of the parametric mixing type signal is shown for the signal frequency $\omega_3 = 2\omega_2 - \omega_1$ against $\omega_2 - \omega_1$. The difference in the line shape between the region of CSRS ($\omega_1 > \omega_2$) and CARS ($\omega_1 < \omega_2$) can also be understood on the basis of duality and is related to the existence of the energy relaxation in this system.

4. Conclusion

In this article, we have seen how the ultrafast relaxation both in the phase and in the energy is reflected in the spontaneous and coherent optical processes. A remarkable feature is that, once excited by a resonant photon,

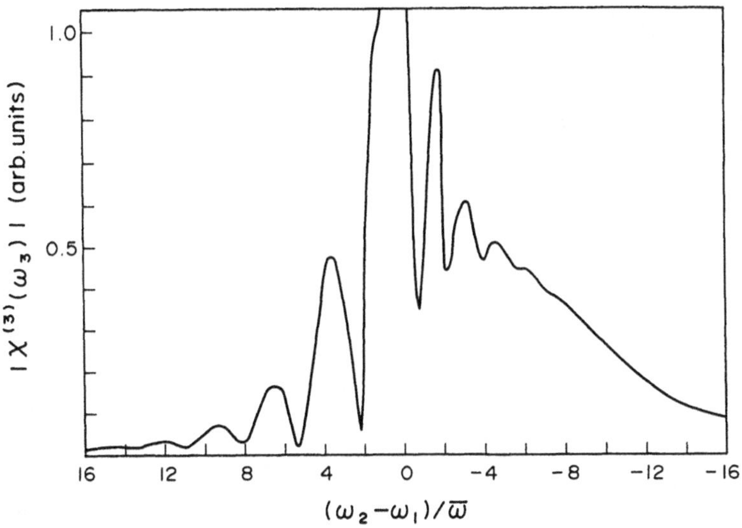

Fig.4 Signal of the third order optical response $|\chi^{(3)}(\omega_3)|$ for $\omega_3=2\omega_2-\omega_1$, g = 40 and $\omega_2 = \varepsilon$

the system undergoes a violent but coherent motion in the interaction mode coordinate space. This is most clearly exhibited by the hot luminescence peak appearing in the low energy tail of the ordinary luminescence band in Fig.1, which is the luminescence from the classical turning point of the oscillation of the phonon wave packet. This coherent motion sometimes induces a hot nonradiative transition across the quasicrossing region of the adiabatic potential surfaces instead of closing a normal optical cycle[8]. Furthermore, it is noticed that, in some strong coupling centers, the giant excess energy ΔE released to the lattice may even give rise to the defect reactions before dissipating away [9].

So far, the nonlinear optical techniques have mainly been developed in and applied to the research field of weak coupling systems with great successes. It will be a challenging target to make clear what is going on during the ultrafast relaxation processes of strong coupling systems by using sophisticated optical techniques.

References

1. V.Hizhnyakov and I.Tehver: Phys. Stat. Sol. 21, 755((1967)
2. Y.Kayanuma: J. Phys. Soc. Jpn. 57, 292 (1988)
3. Y.Toyozawa, A.Kotani and A.Sumi: J. Phys. Soc. Jpn. 42, 1495 (1977)
4. Y.Mori, R.Hattori and H.Ohkura: J. Phys. Soc. Jpn. 51, 2713 (1982)
5. Y.Kayanuma, Y.Mori and H.Ohkura: J. Lumin. 38, 139 (1987)
6. Y.Kayanuma: To appear
7. J.P.Haushalter and M.D.Morris: Anal. Chem. 53, 21 (1981)
8. Y.Kayanuma: J. Phys. Soc. Jpn. 51, 3526 (1982)
9. Y.Suzuki, H.Abe and M.Hirai: These Proceedings

Double Laser Excitation Spectroscopy on Picosecond Photochemical Reactions in Alkali Halide Crystals

Y. Suzuki, H. Abe, and M. Hirai

Department of Applied Physics, Faculty of Engineering, Tohoku University, Aramaki Aoba, Sendai 980, Japan

Photo-induced defect formation process in alkali halide crystals is one of the most extensively investigated photochemical reactions in condensed matters. It is established that the Frenkel pair consisting of an F center (a halogen(X) vacancy occupied by an electron) and an H center (an interstitial halogen atom with a molecular form of X_4^{3-}) is primarily generated by photo-irradiation through a relaxation process of a self-trapped exciton (STE). The STE is a halogen molecular ion (X_2^-) binding an extra electron:X_2^{--*}. However, the details of the branching mechanism from the STE to the pair of the F center and the H center (F-H pair, hereafter) have not been understood satisfactorily. Although a number of picosecond spectroscopic studies[1-3] were made on this process to obtain information on the branching mechanism, consecutive reaction steps included in this process, as shown in Fig.1, made the interpretation of the experimental results somewhat obscure. One group[1] attributed the observed growth time of the F center to an electron-trapping rate (k_2) of the self-trapped hole (X_2^-), another[2] to an internal relaxation rate (k_3) of the self-trapped exciton at higher excited states (STE_H), the other[3] to a branching rate (k_4) from STE_H to F-H pair($F+H_{nn}$). Recently, we have developed a spectroscopic system for picosecond laser photolysis to observe a chemical reaction from photo-excited "intermediates". This system allowed us to observe the branching step (k_4) of the F-H pair from the STE_H selectively without any interference from the other preceding steps. Experimental results of NaCl and RbCl crystals are presented in this paper.

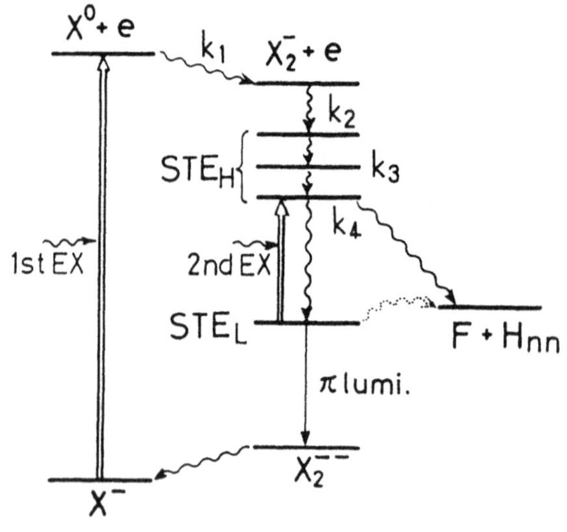

Fig.1

Schematic energy diagram for electron-hole recombination process to yield F-H pairs ($F+H_{nn}$) in alkali halide crystals. X^-:halogen ion, STE's:self-trapped excitons

Springer Series in Chemical Physics, Vol. 48: **Ultrafast Phenomena VI**
Editors: T. Yajima · K. Yoshihara · C.B. Harris · S. Shionoya

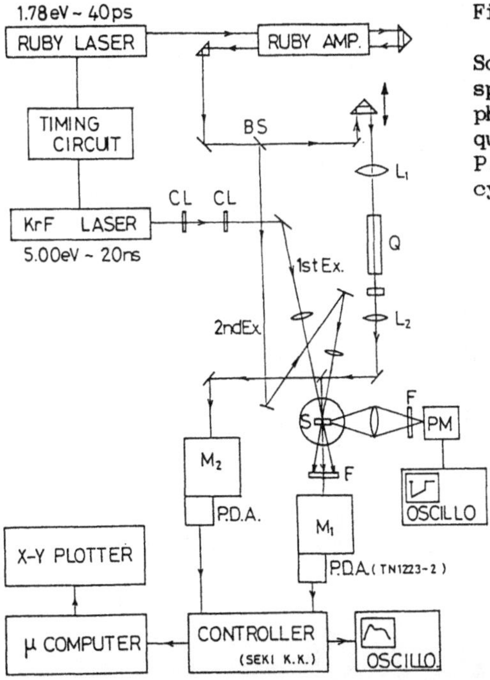

Fig.2

Schematic diagram of picosecond spectrometer for double laser photolysis. S:sample, Q:fused quartz, M_1,M_2:polychromator, P.D.A.:photodiode array, CL: cylindrical lens.

The picosecond spectroscopic system of the double laser photolysis is shown in Fig.2 and will be described in detail elsewhere[4]. Used were two pulsed laser systems connected to each other with a timing circuit. A nanosecond UV pulse from a KrF laser (5.00eV,20ns) was employed as the first exciting source to generate the self-trapped exciton at the lowest triplet state (STE_L:the intermediate). A picosecond red pulse from a mode-locked ruby laser (1.78eV,40ps) was used as the second excitation source to pump the STE_L to the STE_H before the STE_L decays radiatively (π lumi. in Fig.1). Since the decay times of the STE_L in NaCl and RbCl crystals are 0.34 and 3.0ms, respectively, the delay time of the ruby laser pulse to the KrF pulse was fixed to 0.10ms. A conventional pump & probe technique with picosecond continuum was utilized to obtain time-resolved absorption spectra in picosecond range under the second excitation. Ultrapure-graded NaCl and RbCl single crystals were grown in Utah University. A 11x16x3.5mm^3 block was kept at 15K in a cryostat.

Shown in the left frame of Fig.3 is a temporal change of absorption spectra of NaCl crystal at 14K before and after the second excitation. The time zero corresponds to the time when the peak of the ruby pulse reaches the crystal. The STE absorption obviously observed around 2.0eV in the spectrum at -58ps evidences the generation of the STE_L by the first excitation. The STE absorption decreases instantaneously with the second excitation as shown by curves from -25ps to 208ps, while the F absorption increases as compensation for the STE absorption. As shown in the right frame of Fig.3, RbCl crystal exhibited a similar absorption change. The STE absorption of RbCl peaking at 1.9eV was immediately substituted by the F absorption at 2.0eV under the second excitation. From the convolution analysis of the temporal changes of the STE and the F absorption, it was confirmed that the photo-conversion from STE_L to F center in both crystals occurs within the response limit (20ps) of the present detecting system.

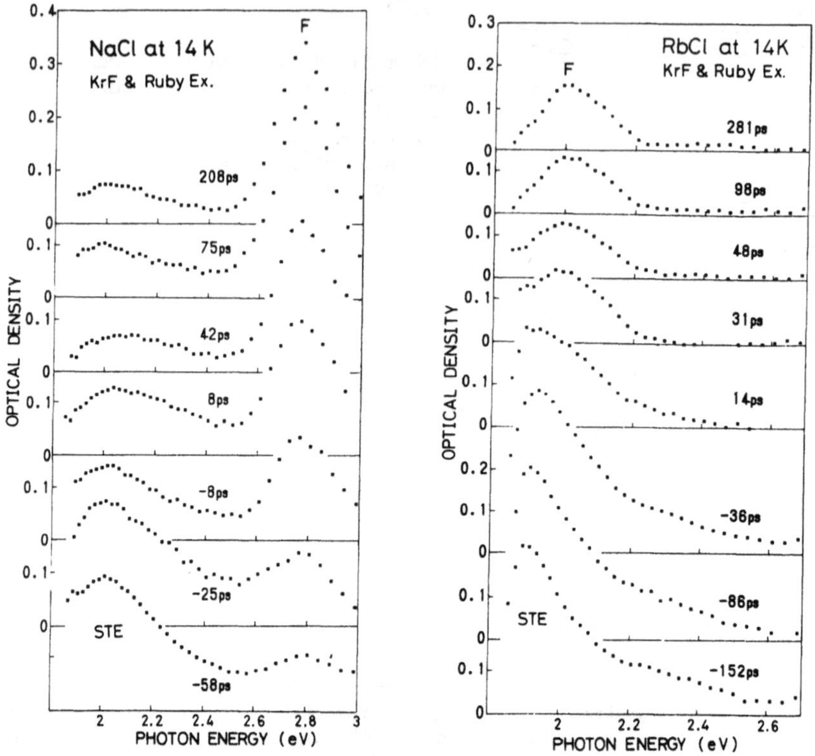

Fig.3 Time-resolved absorption spectra of NaCl(left frame) and RbCl (right) crystals at 14K under the second excitation with a mode-locked ruby laser 0.10ms after the first excitation with KrF laser.

From the pumping power dependence of the secondarily induced F absorption, the net branching efficiencies from STE_H to F center were estimated to be about 0.08 and 0.05 for NaCl and RbCl, respectively. These small values of the branching efficiency imply that the considerable conversion from STE_L to F center as seen in Fig.3 is caused by repetitional excitation of the STE_L restored from the STE_H during the pulse duration of the ruby laser. The branching times were also estimated to be 0.4 and 0.7ps for NaCl and RbCl, respectively. Since the averaged frequency of the vibration coupling to the STE is expected to be about $2\times10^{12}s^{-1}$, the extremely short branching times suggest that the conversion from STE to F center is completed within a few periods of lattice vibration. These results may be explained by the dynamical mechanism of the non-radiative transition from STE_H to F center and/or STE_L, as proposed by SONG and LEUNG[5], as far as alkali chloride crystals are concerned.

References

1. R.T.Williams, J.N.Bradford and W.L.Faust: Phys. Rev. B18, 7083 (1978)
2. J.M.Ortega: Phys. Rev. B19, 3222 (1979)
3. M.Hirai, Y.Suzuki, H.Hattori, T.Ehara and E.Kitamura: J. Phys. Soc. Jpn. 56, 2948 (1987)
4. Y.Suzuki, H.Abe, K.Tasaki and M.Hirai:to be submitted to J.Phys.Soc.Jpn.
5. K.S.Song and C.H.Leung: Solid State Commun. 32, 565 (1979)

Femtosecond to Microsecond Dynamics of Photoexcitations in a Polydiacetylene Film

T. Kobayashi, M. Yoshizawa, K. Ichimura, and M. Taiji

Department of Physics, Faculty of Science, University of Tokyo, Hongo 7-3-1, Bunkyo-ku, Tokyo 113, Japan

Among numerous polymers, polydiacetylenes are of special interest because of their large optical nonlinearities, ultrashort T_2 [1] and T_1 [2-4], and phase transitions with dramatic color changes [5]. We have performed a detailed study of time-resolved absorption spectrum of a polydiacetylene (poly-3BCMU) with the side groups of R=3BCMU (3-butoxycarbonylmethyl-urethane). In the present study time evolution from femtosecond to submillisecond was extensively investigated, including temperature (10-290 K), probe light polarization, and pulse energy dependences.

The experiment was performed using the amplified pulses of the CPM-ring dye laser (2.0 eV, 100 fs), the second harmonic pulses (2.33 eV) of a mode-locked Nd:YAG laser (30 ps) and a Q-switched Nd:YAG laser (5 ns), respectively. Cast film samples with thickness of the order of 100 nm were used. The temperature dependence was studied using a continuous-gas-flow cryostat.

Figures 1 and 2 show transient photoinduced absorption spectra when the polarization of pump pulse and that of probe light are parallel to each other. A photoinduced bleaching peak appears at the absorption peak of singlet excitons (1.93 eV at 10 K and 2.00 eV at 290 K) and broad-band absorption is observed below 1.9 eV both at 10 and at 290 K. This absorption is most probably caused by self-trapped singlet excitons. At the delay time between -0.1 ps and 0.1 ps, i.e. the probe pulse arrives at the sample simultaneously with the pump pulse, three minima are observed at 1.97, 1.78, and 1.70 eV (Fig. 2). The bleaching peak at 1.97 eV is the

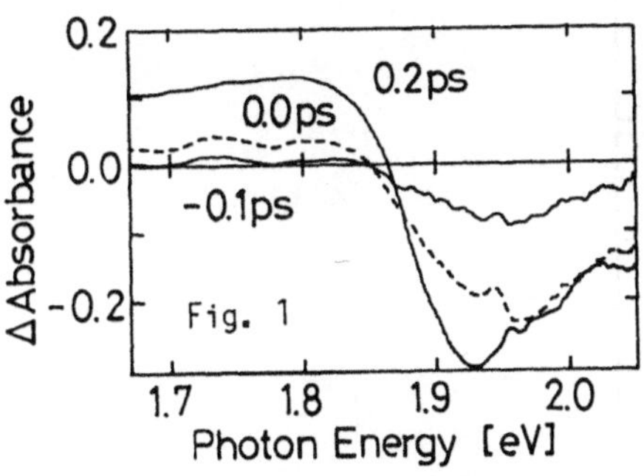

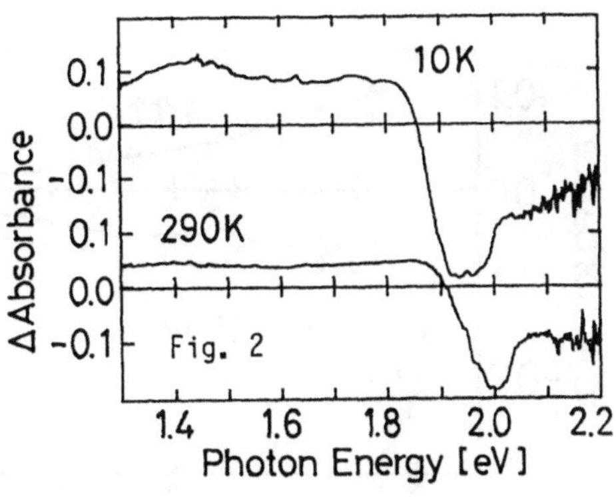

Fig. 1. Photoinduced absorption spectra at the delay time of 0.1 ps.

Fig. 2. Photoinduced absorption spectra at 10 K.

Springer Series in Chemical Physics, Vol. 48: **Ultrafast Phenomena VI**
Editors: T. Yajima · K. Yoshihara · C.B. Harris · S. Shionoya

hole burned by the 1.97-eV pump pulse and the other two minima can be explained in terms of Raman gain. The shifts are consistent with the reported phonon frequency [6].

Measurements carried out with crossed polarization direction of pump and probe field reproduce in effect the spectra depicted here except that the three negative peaks are vanished and the absorbance change is about 30% smaller. The corresponding anisotropy of the absorption is observed in the ground state. The excited states anisotropy remains for more than several picoseconds.

Figure 3 shows the temporal evolution of the absorbance change at the three different energies. The bleaching at 1.97 eV has a maximum at 0.0 ps, because the hole is observed at this photon energy. Both formation time constants for the absorption at 1.77 eV and the bleaching at 1.92 eV are less than the resolution time of the present study, 100 fs. Assuming the absorbance changes decay exponentially, the decay time constant is estimated to be 2.0±0.2 ps at 10 K and 1.5±0.2 ps at 290 K.

An exciton length in one-dimensional chains can be estimated using the phase space filling model [7]. The relative change in the absorption caused by excitons is given by

$$\Delta A(\lambda_{ex})/A(\lambda_{ex}) = -\rho/\rho_s, \tag{1}$$

where ρ is the photogenerated exciton density. The saturated exciton density ρ_s is obtained as 1.0×10^{20} cm^{-3} at 290 K. Using the value for the cross-sectional area for a polymer chain of 100 Å^2 [8], an exciton length is estimated to be 66 Å.

The intensity of the absorbance change was found to be saturated when the pump intensity is increased. The saturation photon densities are obtained to be $(1.2\pm0.3)\times10^{15}$ and $(1.6\pm0.4)\times10^{15}$ photons/cm^2 at 290 and 10 K, respectively. However, the shape as well as the temporal behavior of the photoinduced absorption are independent of the pump intensity. The saturation can be explained by the bleaching (saturation absorption) at the pump wavelength. The absorption of triplet excitons, which was observed at 1.41 eV in polydiacetylene-PTS by two photon excitation [4], is not observed in poly-3BCMU films at the peak power up to 10 GW/cm^2.

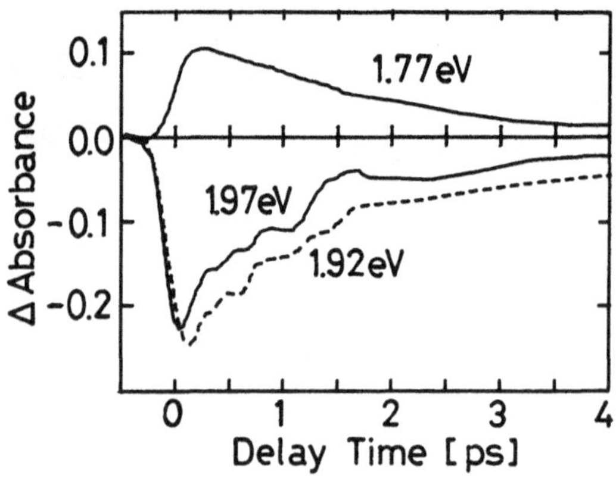

Fig. 3. The time dependences of the photoinduced absorbance changes.

Bleaching is still observed at 10 μs after 5-ns pulse excitation in 1.7-2.3 eV. Hence there is another intermediate with much longer lifetime than 1.5 ps. No difference in the spectral shape of the absorbance change between the picosecond and the nanosecond experiments can be observed. The decay kinetics in the nanosecond region can be represented by $\mathrm{erf}[(t/\tau)^{-d}]$ with τ = 80 ns and d=0.65±0.04. This absorbance change is discussed in terms of band depletion due to another long-lived intermediate, since the spectral change can be closely reproduced by the energy derivative of the absorbance spectrum. This type of decay function ($\tau \sim 1$ μs) was also observed for another polydiacetylene, PTS [9]. Triplet exciton absorption was observed around 1.4 eV with 18.2±0.5 μs lifetime at excitation energy well above the band gap.

In conclusion, the time-resolved photoinduced absorption spectrum of poly-3BCMU has been studied from femtoseconds to microseconds. The hole burning and the absorption of self-trapped excitons have been observed. An exciton length has been estimated as 66 Å.

REFERENCES

1. T. Hattori and T. Kobayashi: Chem. Phys. Lett. 133, 230 (1987)
2. S. Koshihara, T. Kobayashi, H. Uchiki, T. Kotaka, and H. Ohnuma: Chem. Phys. Lett. 114, 446 (1985)
3. G. M. Carter, J. V. Hryniewicz, M. K. Thakur, Y. J. Chen, and S. E. Meyler: Appl. Phys. Lett. 49, 998 (1986)
4. B. I. Greene, J. Orenstein, R. R. Millard, and L. R. Williams: Chem. Phys. Lett. 139, 381 (1987)
5. G. N. Patel, R. R. Chance, and J. D. Witt: J. Chem. Phys. 70, 4387 (1979)
6. Y. Tokura, T. Koda, A. Itsubo, M. Miyabayashi, K. Okuhara, and A. Ueda: J. Chem. Phys. 85, 99 (1986)
7. S. Schmitt-Rink, D. S. Chemla, and D. A. B. Miller: Phys. Rev. B32, 6601 (1985)
8. G. Lieser, B. Tieke, and G. Wegner: Thin Solid Films 68, 77 (1980)
9. T. Kobayashi, J. Iwai, and M. Yoshizawa: Chem. Phys. Lett. 112, 360 (1984)

Ultrashort Surface-Plasmon and Phonon Dynamics

M. van Exter and A. Lagendijk

Natuurkundig Laboratorium van de Universiteit van Amsterdam, Valckenierstraat 65, NL-1018 XE Amsterdam, The Netherlands

Kretschmann [1] was the first to demonstrate that the reflectivity of a thin metal layer (e.g. silver) evaporated on a transparent and optically dense substrate (e.g. glass) as measured by a light source coming from the glass side depends strongly on the angle of incidence. In the neighborhood of the so-called "plasmon angle", the reflectivity of the glass-metal-air system can display a sharp minimum. This decrease in reflectivity is called Attenuated Total Reflection (ATR) and is a result of surface plasmon (SP) generation. The electro-magnetic field penetrates into the metal and excites "leaky" SP at the metal-air interface.

With the use of two mode-locked dye lasers, producing "yellow" and "red" pulses of 5 ps FWHM we have observed the generation and decay of surface plasmons. [2] We direct both beams through a prism of BK7 onto a 45 nm film of silver evaporated on the prism. The two beams both reach the silver film at approximately the plasmon angle (Θ=43^0 in our case). The yellow pump beam creates SP propagating along the silver film at almost the speed of light. The energy dumped by the decaying "yellow" SP temporarily changes the dielectric constant of silver. After a variable delay the red probe pulse measures the, slightly changed, reflectivity of the glass-silver-air system in the same ATR configuration.

Under typical experimental conditions we focus two laser beams of 15 mW average power each on the film and the rise in temperature, corresponding to the energy dumped by one laser pulse, is about 1 K. The resulting gain/loss in the intensity of the reflected probe beam is typically 10^{-4}-10^{-5}. A special double-modulation technique [2] at 8.8 Mhz allows us to reach almost the quantum-noise limit and detect changes in the reflectivity of 10^{-7} (RC-time=0.4 s).

Conservation of momentum along the film (phase matching) restricts the propagation of the SP to one direction. One thus expects the heat distribution along the film, resulting from the decay of SP, to have an exponential tail. Experimentally we observe the spatial distribution of heat by moving the red probe spot with respect to the yellow pump spot over the film meanwhile monitoring the

Springer Series in Chemical Physics, Vol. 48: **Ultrafast Phenomena VI**
Editors: T. Yajima · K. Yoshihara · C.B. Harris · S. Shionoya

signal strength. A perfect exponential "heat" tail (more than two decades) was observed and a mean-free path of the optical SP of 12.8(7) μm could be determined. From the theoretical dispersion relation of the SP we know their group velocity to be 0.89c so the measured mean-free path reflects a momentum life time of 48(3) fs.

The previous spatially-resolved data were obtained for a probe pulse arriving about hundred picoseconds after the pump pulse. It is also very interesting to observe the change in reflectivity as a function of <u>time</u> for fixed angle and exact overlap between the laser foci. As we have a time resolution of 5 ps there is no possibility to observe the decay of the SP directly in the time domain, but we can look at their decay products.

In Fig.1 we have pictured a typical time-resolved measurement. The long-time decay (ns) of the reflectivity results from the diffusion of heat. Although the heat diffusion constant of silver is almost a thousand times larger than that of BK7 glass, the larger surface and larger thermal gradient perpendicular to the film make the heat diffusion into the glass dominant. This can also be concluded from the observed quasi-one-dimensional behavior of the heat diffusion; the ns decay is almost insensitive to the size of the focused laser spots.

In Fig.2 we present a measurement, concentrating on the short-time decay, that clearly shows an intriguing oscillation in the reflectivity. We will explain these small and fast (24 ps) oscillations in terms of acoustic phonons bouncing up and down the film. Similar oscillations have been recently observed by Thomsen et al. [3] in the absorption of a-As_2Te_3.

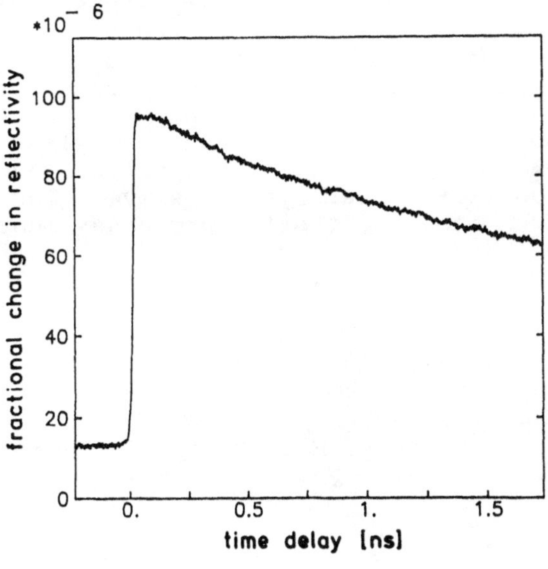

Fig.1 Reflectivity change of probe induced by pump pulse

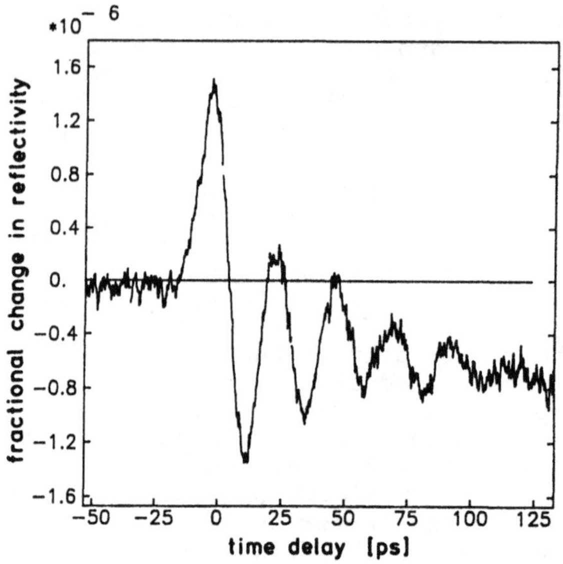

Fig.2 Reflectivity change of probe induced by pump pulse

The electric field associated with the surface plasmons has a penetration depth in the silver of only 12 nm, so most of the energy is released in a thin upper layer at the silver-air interface. The deposited energy first of all gives rise to an increase in the temperature but also creates an expansion of the lattice, part of it being stationary and part of it in the form of a "strain wave" travelling at the speed of sound perpendicular to the surface. From a microscopic point of view acoustic phonons are generated by multi-phonon decay. The experimental run of Fig.2 was fitted by an exponentially decaying sinusoidal curve. The oscillation period could be determined with 1 % accuracy to be 23.8(3) ps. Using the 45 nm thickness of the film we calculate a longitudinal speed of sound of v=3.78(5) km/s. The decay of the acoustic signal is almost completely due to transmission of phonons into the glass. The measured decay (50$\pm$5 %) agrees with the calculated acoustic reflection coefficient at the silver-glass boundary and losses due to other scattering processes must therefore be weak.

1. E. Kretschmann, Z. Physik 241, 313 (1971).
2. M. van Exter and A. Lagendijk, Phys. Rev. Lett. 60, 49 (1988).
3. C. Thomsen, J. Strait, Z. Vardeny, H.J. Maris, J. Tauc and J.J. Hauser, Phys. Rev. Lett. 55, 989 (1984).

Femtosecond Image-Potential Dynamics in Metals

R.W. Schoenlein[1], *J.G. Fujimoto*[1], *G.L. Eesley*[2], *and W. Capehart*[2]

[1]Department of Electrical Engineering and Computer Science, Research Laboratory of Electronics, Massachusetts Institute of Technology, Cambridge, MA 02139, USA

[2]Physics Department, General Motors Research Laboratories, Warren, MI 48090, USA

We report the first time-resolved studies of image-potential states in metals. Femtosecond ultraviolet pump and visible probe techniques are combined with two-photon photoemission energy spectroscopy to perform transient photoemission measurements with femtosecond resolution. The $n = 1$ image-potential state in Ag(100) is observed to decay within a few tens of femtoseconds.

The existence of image-potential states has been theoretically predicted[1] and experimentally investigated using inverse photoemission[2] as well as two-photon photoemission techniques[3,4]. They represent a unique class of surface states created by the Coulombic attraction between an electron existing outside a material surface and the electronic image charge in the solid. Electrons in these states are essentially trapped by the image-potential on one side and a potential barrier at the crystal surface created by a gap of available electronic states. This results in a two-dimensional electronic confinement in which the wavefunctions exist primarily outside the solid. The energy levels form a Rydberg-like series approaching the vacuum potential.

Energy levels of image-potential states may be predicted using a simple hydrogenic model which yields a binding energy $E_B \sim 1/n^2$ with a maximum binding energy of $\frac{1}{16}R_y = 0.85$ eV[2]. Echenique et al.[1,5] have developed a more detailed theoretical treatment based on wavefunction phase analysis. They predict binding energies and estimate the lifetime broadening of the states by considering the wavefunction overlap with the crystal. Two-photon photoemission measurements in Ag (100) have yielded binding energies of 0.53 eV and 0.16 eV $\pm$ 0.02 eV for the $n = 1$ and $n = 2$ states, respectively[4].

Our studies are performed on the (100) surface of Ag using a two-photon pump and probe photoemission technique employing both visible and ultraviolet femtosecond pulses. The laser system consists of a CPM ring dye laser with an 8 kHz copper vapor laser amplifier[6] which produces 55 fs pulses at 620 nm (2 eV). Ultraviolet pulses of $\sim$ 60 fs duration (at 4 eV) are generated in the pump arm by frequency doubling in a 100 μm KDP crystal. A prism pair is employed to compensate dispersion in the UV due to optical components.

The UV pulses populate the $n = 1$ image-potential state via nonresonant transitions from occupied bulk states below the Fermi energy. The transient dynamics of the image-potential state are probed by a delayed visible (2 eV) pulse which photoemits electrons to the vacuum. Both visible and UV beams were P-polarized. No photoemission was observed with S-polarized light, commensurate with selection rules. The two wavelength technique breaks the symmetry of standard two-photon photoemission, thereby separating effects of the pump from those of the probe. One-photon photoemission from image-potential states with visible photons is more effective than with UV photons since most of the UV energy is absorbed by d-band transitions. Thus, the effects of nonequilibrium heating are limited by working at the lowest possible UV flu-

Springer Series in Chemical Physics, Vol. 48: **Ultrafast Phenomena VI**
Editors: T. Yajima · K. Yoshihara · C.B. Harris · S. Shionoya

ences. This approach permits nearly background-free measurements of electron energy spectra on a femtosecond time scale.

The single crystal Ag sample was cut in a (100) orientation with an accuracy of $\pm 2°$, polished and sputter ion cleaned in a 10^{-10} torr vacuum chamber. Surface quality was verified using standard LEED and Auger spectroscopy. Photoemitted electrons are detected by a cylindrical mirror analyzer (CMA) with an energy resolution of better than 200 meV and a collection angle of $\sim 15°$.

Figure 1 shows a series of two-photon (4 eV and 2 eV) photoemission spectra taken at various time delays with UV and visible fluences of $\sim 5 \times 10^{-2}$ mJ/cm^2 and ~ 0.7 mJ/cm^2, respectively. Positive delay corresponds to visible pulses arriving after the UV pulses, and the zero energy was determined by the onset of the photoemission signal. The image-potential peak appears at ~ 1.5 eV which corresponds to the $n = 1$ image state lying ~ 0.5 eV below the vacuum potential and is consistent with previously reported results[4]. Photoemission from the $n = 2$ state is not observed since the 4 eV pump energy is not sufficient to populate this state. The 185 meV energy width is limited by the resolution of the energy analyzer.

The time decay of the observed signal is quite rapid, indicating an excited state lifetime which is comparable to the pulse duration. Population of the $n = 1$ state via nonresonant transitions results in a broad electronic distribution which extends over a range of energies and momentum states in the $k_{\parallel}$ direction. Because of the acceptance angle of the CMA, our measurements are sensitive to electrons with finite transverse momentum and this may affect the observed lifetime. There is no evidence of electron dynamics within the $n = 1$ band since the energy profile and width does not change substantially over the lifetime of the state.

The background (> 200 fs delay) photoemission spectra ($\times 100$) is shown for reference. We observe a peak at 1.5 eV resulting from three-(2 eV)photon photoemission via the $n = 1$ image-potential state. The peak at 3.5 eV corresponds to two-(4eV)photon photoemission from the $n = 1$ state.

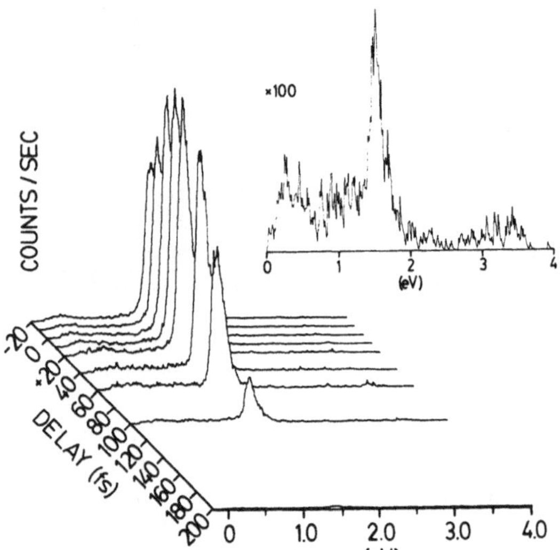

Fig. 1. Photoelectron energy spectra at various time delays showing the dynamics of the image-potential state.

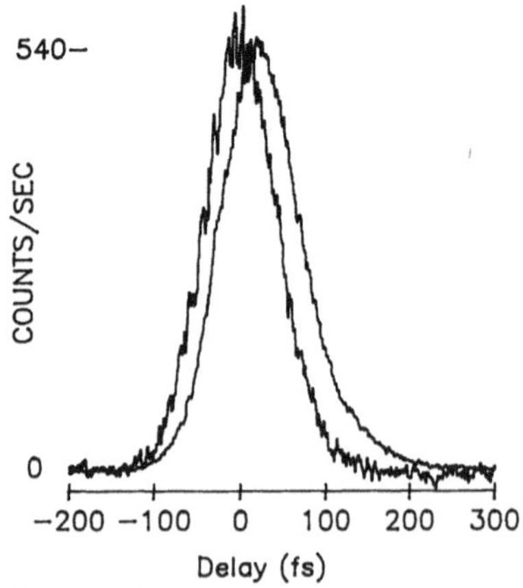

Fig. 2. Cross-correlation trace and femtosecond transient decay of the image-potential state.

Figure 2 shows a time-resolved photoemission measurement taken with the energy analyzer tuned to the peak of the image-potential signal at 1.5 eV. Also shown is a UV/visible cross-correlation measurement taken under identical experimental conditions by quenching the image-potential lifetime by adsorbing oxygen onto the sample surface. This is important in order to accurately determine the pulse duration at the sample as well as the zero delay. The finite lifetime of the image-potential state is evidenced by the asymmetry of the pump-probe trace and the shift toward positive delay. This clearly indicates that the process is not instantaneous. Theoretical fits performed by convolving the cross-correlation with single sided exponentials indicate a decay time in the range of 10 to 30 fs. However, the dynamics of these states may be sensitive to surface impurities.

In summary, transient dynamics of the $n = 1$ image-potential state in Ag (100) have been investigated using femtosecond ultraviolet pump and visible probe techniques combined with two-photon photoemission energy spectroscopy. This novel approach separates effects of the pump from those of the probe and permits nearly background-free measurements of photoelectron spectra with femtosecond resolution. Our results show the population of the $n = 1$ image-potential state and the subsequent relaxation of the electron gas on a time scale comparable to the pulse duration. High resolution pump-probe measurements indicate an excited state lifetime of a few tens of femtoseconds which is within the range predicted by theory[1,5]. The observed lifetime is relatively long for such high-energy electrons ($\sim$ 4 eV above the Fermi energy) indicating that the image-potential states are considerably decoupled from the solid. Additional studies incorporating a variable wavelength femtosecond UV pump should provide information about the dynamics of higher order states and their coupling to the crystal lattice. This technique should have applications for performing time-resolved surface studies in a variety of material systems.

We wish to thank E. P. Ippen for helpful discussions. This work was supported in part at MIT by the Joint Services Electronics Program Contract DAAL03-86-K-0002, the Air Force Office of Scientific Research, and a grant from the AT&T Foundation.

References

1. P. M. Echenique and J. B. Pendry, *Solid State Phys.* **11**, 2065 (1978).

2. D. Straub and F. J. Himpsel, *Phys. Rev. B* **33**, 2256 (1986).

3. K. Giesen, F. Hage, F. J. Himpsel, H. J. Riess, and W. Steinmann, *Phys. Rev. Lett.* **55**, 300 (1985).

4. K. Giesen, F. Hage, F. J. Himpsel, H. J. Riess, and W. Steinmann, *Phys. Rev. B* **35**, 971 (1987).

5. P. M. Echenique, F. Flores, and F. Sols, *Phys. Rev. Lett.* **55**, 2348 (1985).

6. W. Knox, M. Downer, R. Fork, and C. V. Shank, *Opt. Lett.* **9**, 552 (1984).

Part VI

Dynamical Processes in Semiconductor Quantum Wells and Microstructures

Femtosecond Luminescence Spectroscopy: Investigation of Semiconductors and Semiconductor Microstructures

*J. Shah, T.C. Damen, and B. Deveaud**

AT&T Bell Laboratories, Crawfords Corner Road,
Holmdel, NJ 07733, USA
*Permanent address: LAB/ICM, CNET, Lannion, France

We have developed a luminescence spectroscopy system with 65 fs time resolution and high sensitvity. Using this technique, we have investigated a number of phenomena of fundamental and device importance in semiconductors and their microstructures.

1. INTRODUCTION

Femtosecond spectroscopy is a very powerful technique for investigating properties of matter. Excite-and-probe techniques with time resolution as short as 6 fs [1] have been used for such studies. Luminescence spectroscopy provides complementary information but has been limited to a time resolution of about 400 fs [2]. We have recently demonstrated [3] that it is possible to measure luminescence spectra with 65 fs time resolution and with good sensitivity. We are therefore able to measure luminescence spectra with time resolution better than most excite-and-probe studies.

We describe important features of this 65 fs luminescence spectroscopy system and discuss how we have used ultrafast luminescence spectroscopy to investigate (i) transport of carriers in semiconductor superlattices, (ii) intervalley scattering in GaAs and (iii) exciton transfer between monolayer-flat regions in a single quantum well.

2. LUMINESCENCE SPECTROSCOPY WITH 65 fs TIME RESOLUTION

The technique we used is based on the principle of sum frequency generation [4]. The time resolution is determined by the laser pulsewidth and the group velocity dispersion in the nonlinear crystal used for upconversion [5]. We have discussed earlier [3] that for a given excitation density, the upconverted luminescence signal is proportional to the average power incident on the nonlinear crystal under the conditions of our experiments. It is therefore essential to have a laser with narrow pulsewidth and a reasonably large (> 100 mW) average power.

The starting point for the present experiments was a Rh 6G dye laser synchronously pumped with the second harmonic of a compressed, mode-locked YAG laser [6]. The dye laser output was typically 300 mW average power at 80 MHz repetition rate, with 300 fs pulsewidth. We compressed these pulses [3] using a fiber-prism compressor [7,8]. The fiber length was 12.7 cm and the separation between the 60 degrees fused quartz prisms was approximately 180 cm. Using 200 mW input power, we obtained 60 fs compressed pulses with an average power of 120 mW. Increasing the input power to 300 mW increased the pulsewidth to 65 fs and the average power to 150 mW. It may be possible to further improve these results by optimizing the fiber length and other parameters of the compressor. These results were obtained with the dye laser operating at 5900 Å . The tunability of the compressed pulses was not investigated.

Springer Series in Chemical Physics, Vol. 48: **Ultrafast Phenomena VI**
Editors: T. Yajima · K. Yoshihara · C.B. Harris · S. Shionoya

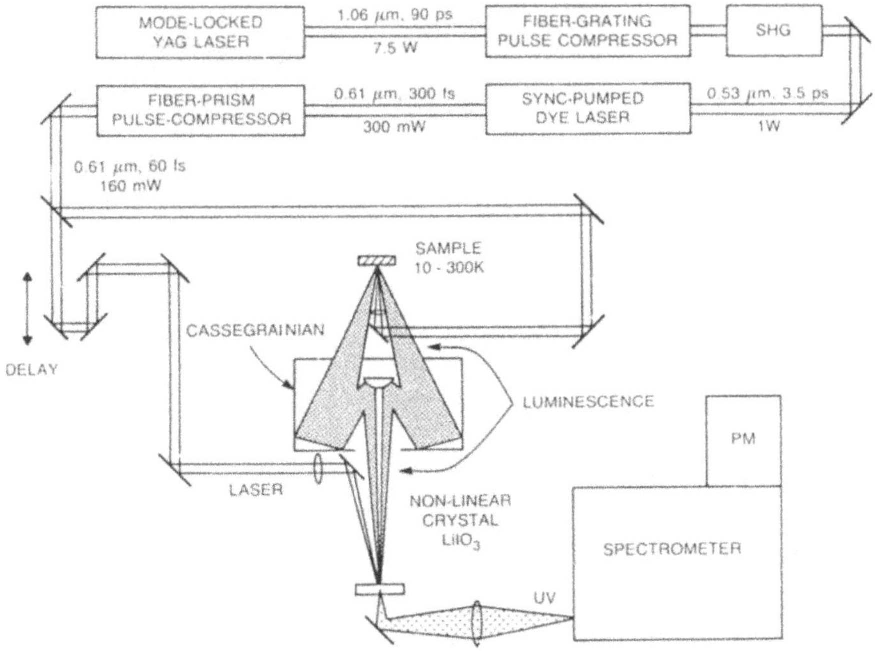

Fig. 1. Schematic of luminescence spectroscopy system for 65 fs time resolution.

Luminescence spectroscopy was performed by splitting this laser into two beams (Fig. 1). One beam excited the luminescence which was collected by dispersion free optics and focussed on a 0.3 mm thick nonlinear crystal of $LiIO_3$. The second compressed laser beam was passed through a variable delay stage and was also focussed at the same point in the nonlinear crystal. Sum frequency photons were generated only during the laser pulse, thus providing the femtosecond time resoultion. The phasematching restrictions and a spectrometer provided the necessary spectral resolution.

The time resolution of the system was determined by measuring the cross-correlation between the scattered light from the surface of the sample with the delayed laser beam at the nonlinear crystal. A typical cross-correlation trace is shown in Fig. 2. This demonstrates a time resolution of 65 fs. The time evolution of luminescence intensity at 1.78 eV from bulk GaAs at 300 K is also shown in Fig. 2. Compared to the earlier result [2], we can resolve a short delay in the onset of luminescence. Similar results for GaAs at other energies and for AlGaAs were reported in [3].

3. BLOCH TRANSPORT IN SUPERLATTICES

The formation of minibands in thin superlattices leads to a number of interesting transport phenomena, including the possibility of Bloch oscillations. While transport of carriers over large distances in the direction perpendicular to the superlattice planes (perpendicular transport) has been established, the nature of such transport is not well established. Using time resolved luminescence spectroscopy in stepwise graded gap superlattices (see inset of Fig. 3), we have achieved a better understanding of the nature of this transport.

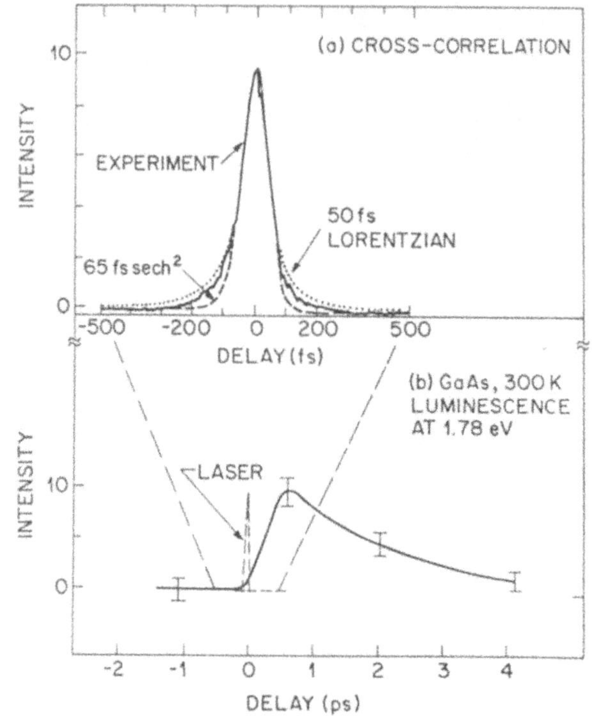

Fig. 2. (a) Cross-correlation showing system time resolution; (b) Luminescence from bulk GaAs showing the delay in onset [3].

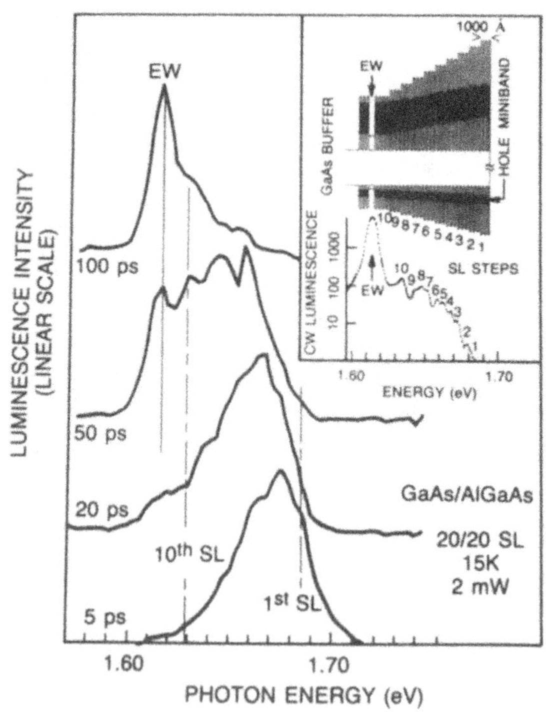

Fig. 3. Time resolved luminescence spectra from a 20/20 graded gap GaAs/AlGaAs superlattice; inset shows the sample structure and the cw luminescence spectrum [9].

The cw luminescence spectrum of a stepwise graded gap superlattice sample (inset of Fig. 3) shows a series of peaks, each corresponding to a step in the sample. Therefore, the peak of the time resolved luminescence spectrum at a given delay after the excitation pulse directly determined the position of photoexcited carriers in superlattices [9]. Fig. 3 shows the time resolved spectra for a 20/20 superlattice (i.e. superlattice with 20 Å GaAs well and 20 Å AlGaAs (30 % Al) barrier). These spectra clearly establish the motion of the carriers in the perpendicular direction.

By measuring such luminescence spectra for superlattices of different periodicities, we established that transport in thin superlattices proceeds via Bloch states (Bloch transport) and determined the mobility for such transport. We determined the variation of the mobility for perpendicular transport with superlattice period and showed that localization effects become important for barrier thicknesses > 30 Å . Our technique thus provides important information about transport in semiconductor superlattices.

4. INTERVALLEY SCATTERING IN GaAs

We have observed [10] that the luminescence intensity in GaAs photoexcited with 300 fs pulses, rises extremely slowly, taking nearly 10 ps to reach its peak value. This long risetime is in contrast to a much faster risetime observed in InP (Fig. 4). We have shown that this difference between two very similar semiconductors arises from the fact that, for our excitation photon energy (2.04 eV), photoexcited electrons are energetically allowed to make a transition to the L valley in GaAs but not in InP. Therefore, the slow rise in GaAs is primarily caused by a slow return of the carriers from the L valley to the Γ valley. By comparing these data with Monte Carlo simulations, we have determined that the average Γ to L transfer time is 100 fs and the average L to Γ return time is 2.5 ps. These are important parameters for understanding high field transport in GaAs.

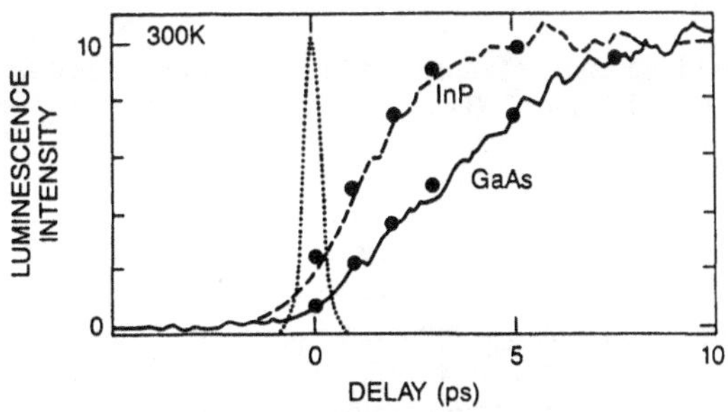

Fig. 4. Time evolution of luminescence from GaAs and InP [10].

5. EXCITON TRANSPORT IN A SINGLE QUANTUM WELL

Interruption of growth at hetero-interfaces leads to formation of large, monolayer-flat islands in quantum wells. We have investigated the transfer of excitons between these islands by time resolved luminescence spectroscopy.

Fig. 5 shows the time resolved spectra from a single quantum well whose nominal width is 25 Å. Two distinct luminescence features are observed, arising from islands with average thickness of 10 monolayers (higher energy feature) and 11 monolayers (lower energy feature). Initially the higher energy luminescence dominates, indicating that the islands with 10 monolayer thickness dominate within the excitation spot (≈ 30 μm diameter). With increasing time delay, the excitons transfer to islands with larger

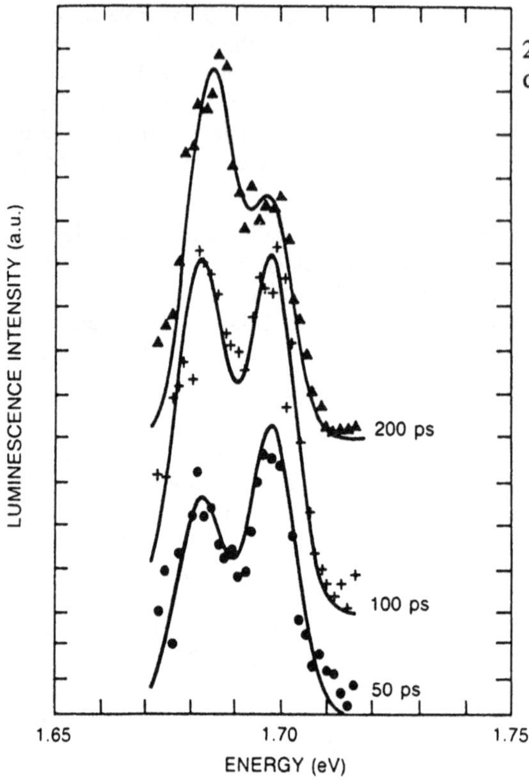

Fig. 5. Luminescence spectra of a 25 Å wide GaAs quantum well at three different delays [11].

thickness as shown by the decrease in the high energy feature and an increase in the low energy feature. We find that the transfer time is about 200 ps.

6. SUMMARY

We have built a system that allows measurement of luminescence spectra with 65 fs time resolution and with high sensitivity. We have discussed important features of this system and reviewed our recent work on semiconductors and semiconductor microstructures using the technique of ultrafast luminescence spectroscopy. This work sheds new light on many phenomena which are of fundamental interest in semiconductor physics and which form the basis for understanding many interesting and useful devices.

Acknowledgement: It is a pleasure to acknowledge collaborations with A. C. Gossard, B. Lambert, P. Lugli, A. Regreny, W. T. Tsang and C. W. Tu, and to thank A. E. DiGiovanni for technical assistance.

[1] R. L. Fork, C. H. Brito Cruz, P. C. Becker and C. V. Shank, Opt. Lett. 12, 483 (1986)

[2] Jagdeep Shah, T. C. Damen, B. Deveaud and D. Block, Appl. Phys. Lett. 50, 1307 (1987)

[3] T. C. Damen and Jagdeep Shah, Appl. Phys. Lett. 52, 1291 (1988)

[4] H. Mahr and M. D. Hirch, Opt. Commun. 13, 96 (1975)

[5] Jagdeep Shah, IEEE J. Quantum Electronics QE-24, 276 (1988)

[6] We thank Spectra Physics for the loan of the pulse compressor

[7] R. L. Fork, O. E. Martinez and J. P. Gordon, Opt. Lett. 9, 150 (1984)

[8] J. D. Kafka and T. Baer, Opt. Lett. 12, 401 (1987)

[9] B. Deveaud, Jagdeep Shah, T. C. Damen, B. Lambert and A. Regreny, Phys. Rev. Lett. 58, 2582 (1987)

[10] Jagdeep Shah, B. Deveaud, T. C. Damen, W. T. Tsang and P. Lugli, Phys. Rev. Lett. 59, 2222 (1987)

[11] B. Deveaud, T. C. Damen, Jagdeep Shah, and C. W. Tu, Appl. Phys. Lett. 51, 828 (1987)

Femtosecond Carrier–Carrier Scattering Dynamics in p-Type and n-Type Modulation-Doped Quantum Wells

W.H. Knox[1], *D.S. Chemla*[1], *G. Livescu*[1], *J.E. Henry*[1], *J.E. Cunningham*[1], *and S.M. Goodnick*[2]

[1]AT&T Bell Laboratories, Crawfords Corner Road, Holmdel, NJ 07733, USA
[2]Oregon State University

Optical studies of carrier thermalization dynamics in semiconductors [1-3] suffer from two basic ambiguities which have received very little attention in experimental and theoretical studies in the past. For each photon which is absorbed, one electron and one hole are created. Since the electrons and holes in general have different properties we should expect that their subsequent relaxation dynamics will be different. In studies of optical saturation, the experimental observable is the absorption, which is related indirectly to the <u>sum</u> of the electron and hole occupation functions. Thus, optical absorption saturation experiments cannot provide information on the separate contributions of the electrons and holes. Therefore, the separate contributions of electrons and holes to the thermalization rate and the separate contributions to the optical absorption saturation signals are of interest to study.

We discuss a new set of experiments in which the goal is to determine which scattering process (e-e, e-h or h-h) provides the dominant contribution to the thermalization rate of a non-thermal carrier distribution. Monte-Carlo simulations indicate that the electron population dynamics dominate the saturation signals. We use GaAs multiple quantum wells (MQW) which are doped at about 3×10^{11} cm^{-2} with either excess electrons or excess holes, thus breaking the symmetry imposed by charge neutrality in optical excitation of undoped semiconductors. We excite at less than one phonon excess energy so that carrier-carrier scattering is the only relaxation mechanism in our experiments.

We use an infrared femtosecond spectroscopic system which has been recently developed which generates intense 100 fs continuum pulses centered at 805 nm [4]. This provides excitation pulses of 100 fs duration which are continuously tunable around the GaAs bandedge with about 10-20 nJ pulse energies, obtained by a series of 10 nm bandpass filters. In a previous experiment [3], undoped 9.6 nm MQW were excited at a carrier density of about 2×10^{10} cm^{-2}, a limit imposed by the weak infrared continuum. A thermalization time of about 100 fs was obtained. Figure 1a shows the same sample excited at about $3\text{-}5\times10^{11}$ cm^{-2} carrier density. At this high density, the nonthermal distribution is twice as broad as the pump pulse bandwidth and we obtain about 30 fs thermalization time, by comparison with our relaxation time approximation model. At low densities, elastic scattering dominates, exchanging small amounts of energy per collision, At high densities inelastic scattering becomes important, however, screening reduces the scattering rate, so the net dependence on carrier density is weak. When we repeat the experiment in n-type QW (Figure 1b), we do not resolve a non-thermal distribution. This indicates that the thermalization is faster than our system can resolve (less than 10 fs by comparison with our relaxation time approximation model). Thus, the electron-electron scattering contribution to femtosecond carrier thermalization is apparently very important. We then repeat this in p-type QW doped at 2×10^{11} cm^{-2} (Figure 1c) and obtain a completely different result. In this case a non-

Springer Series in Chemical Physics, Vol. 48: **Ultrafast Phenomena VI**
Editors: T. Yajima · K. Yoshihara · C.B. Harris · S. Shionoya

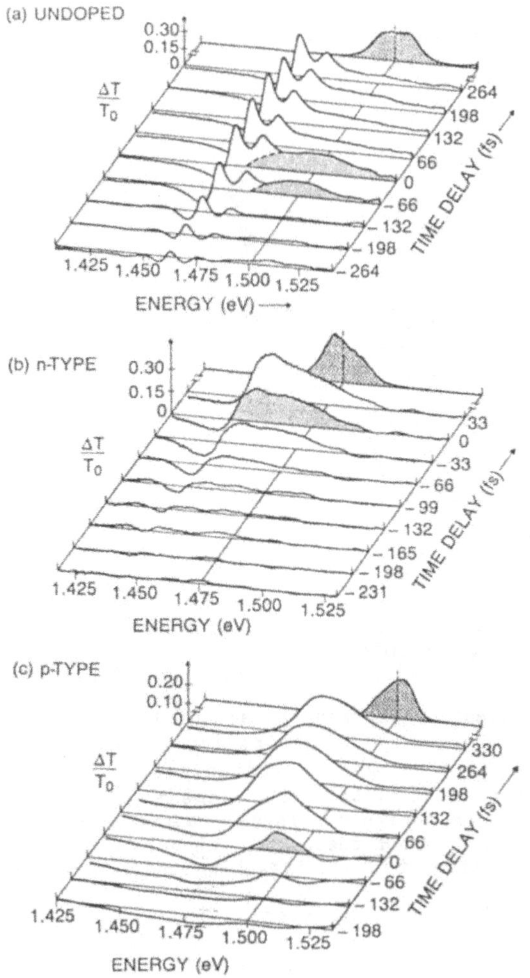

Figure 1. Differential transmission spectra for undoped and modulation-doped samples. Excitation density is about 2×10^{11} per cm^2. For the undoped case, broad holeburning is observed with about a 30 fs thermalization time. For the n-type case, no holeburning is obtained, indicating less than 10 fs thermalization time. P-type samples show strong holeburning with a 60 fs thermalization time and also a large dynamical edge broadening.

thermal distribution is clearly observed around t=0, but a large negative dip is observed around t=0. This is due to a large instantaneous renormalization of the Fermi sea. This effect appears to be much larger for the p-type case than the n-type case. The thermalization time is about 60 fs for this case, by comparison with our model. Optical excitation densities are about $2\text{-}3 \times 10^{11}$ cm^{-2} for MD samples.

We obtain thermalization times from this data, assuming that there is no loss of cariers to intervalley scattering, radiative or nonradiative decay and no phonon emission, using a relaxation time approximation [5] with a local equilibrium distribution which conserves carriers, and we set the initial distribution energy equal to the final Boltzmann distribution energy so that the relaxation conserves energy. We have compared the results of our relaxation-time approximation model with Monte-Carlo simulations which have been recently carried out [6]. We show in Figure 2 the electron occupation function for the conditions of the low density experiments (undoped,

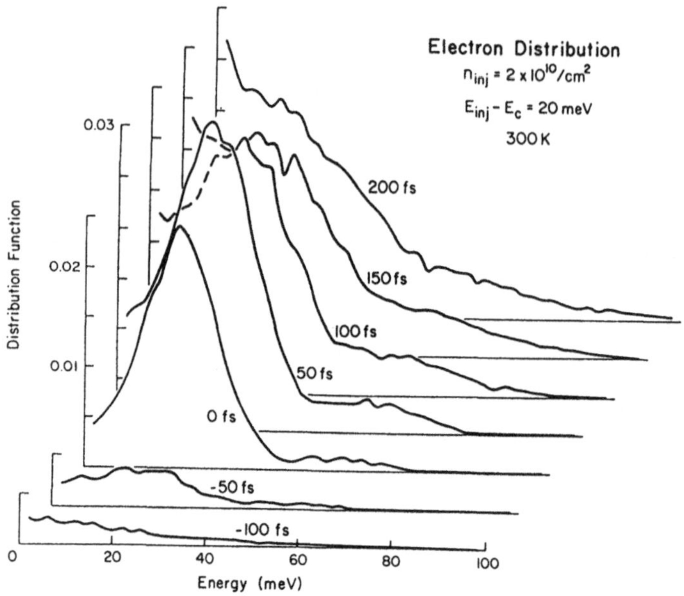

Figure 2. Monte-Carlo simulation of thermalization of electron distribution for undoped sample at $2x10^{10}$ cm^{-2}.

excitation at $2x10^{10}$ cm^{-2}) The connection between the sum of the electron and hole occupation functions and the absorption bleaching is complicated in regions of absorption in which excitonic or continuum resonances [7] modulate the absorption above the density of states significantly. In the present case, we excite non-thermal distributions in the n=1 continuum where the absorption is not resonantly enhanced so that we may make a reasonably direct correlation between bleaching and occupation.

In summary, we find that carrier thermalization in n-type doped samples is much faster than in p-type doped samples, suggesting that e-e collisions make the dominant contribution to femtosecond carrier thermalization in near bandedge excitation conditions. Thermalization times of less than 10 fs are obtained for n-type materials. Our relaxation time approximation model provides a good qualitative description of our results, and Monte Carlo simulations indicate that the electron population dominates the transient optical saturation signals. Further studies will be required to determine dependence of these relaxation rates on excess energy, doping and excitation density, well size, and other parameters.

[1] C.L. Tang and D.J. Erskine, Phys. Rev. Lett. 51, 840 (1983).

[2] J.L. Oudar, D. Hulin, A. Migus, A. Antonetti and F. Alexandre, Phys. Rev. Lett. 55, 2074 (1985).

[3] W.H. Knox, C. Hirlimann, D.A.B. Miller, J. Shah, D.S. Chemla and C.V. Shank, Phys. Rev. Lett. 56, 1191 (1986).

[4] W.H. Knox JOSA B4, 1771 (1987).

[5] N.W. Ashcroft and N.D. Mermin, "Solid State Physics", Holt Rinehart and Winston, N.Y. pp. 317-327.

[6] S.M. Goodnick and P. Lugli, Phys. Rev. B37, 2578 (1988).

[7] G. Livescu, D.A.B. Miller, D.S. Chemla, M. Ramaswamy, T.Y. Chang, N. Sauer, A.C. Gossard and J. English, IEEE JQE Special Issue on Quantum Wells and Superlattices, Aug. 1988.

Cooling of Hot Carriers in Three- and Two-Dimensional $Ga_{0.47}In_{0.53}As$

H. Lobentanzer, W. Stolz, and K. Ploog

Max-Planck-Institut für Festkörperforschung, Heisenbergstr. 1, D-7000 Stuttgart 80, Fed. Rep. of Germany

Cooling of hot carriers in undoped $Ga_{0.47}In_{0.53}As$ has been investigated systematically so far in bulk material with a free surface /1/ and in a thin $Ga_{0.47}In_{0.53}As/Al_{0.48}In_{0.52}As$ multiple-quantum-well (MQW) structure /2/. We report here a comparative study of the cooling behavior of hot carriers in a $Al_{0.48}In_{0.52}As/Ga_{0.47}In_{0.53}As/InP$ heterostructure (thickness of the $Ga_{0.47}In_{0.53}As$-layer: 0.75μm) and in $Ga_{0.47}In_{0.53}As/Al_{0.48}In_{0.52}As$ MQW's of a well thickness of 3.4, 8.0 and 13.8nm.

Our samples are directly excited with picosecond pulses of a synchronously pumped dye laser (repetition rate 80MHz, pulse duration 4ps). The laser pulses with a photon energy of 1.48eV are only absorbed in the $Ga_{0.47}In_{0.53}As$ layers. All experiments are performed at a temperature of 5K. The photoluminescence in the infrared spectral range between 1.0 and 1.55μm is spectrally dispersed by a monochromator and temporally resolved by a 2D Hamamatsu synchroscan streak camera, using a S1 photocathode. The time resolution amounts to 15ps and is limited by temporal broadening of the spectrometer and the trigger jitter of the streak camera system.

Transient photoluminescence spectra of the bulk sample are depicted in Fig. 1. Assuming fast thermalisation between electrons and holes, we can perform a linefit of the luminescence, using a band-band recombination model.

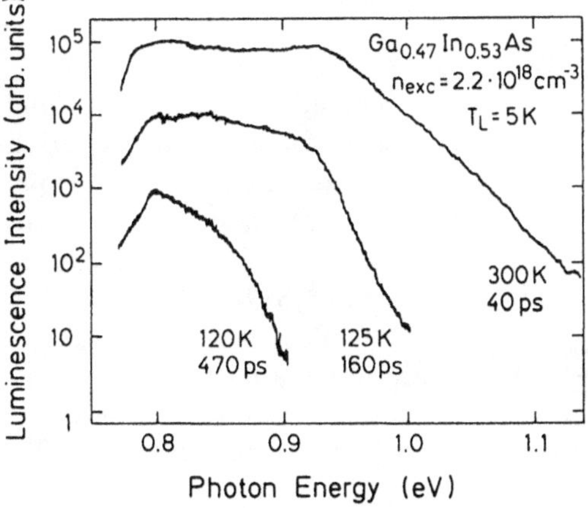

Fig.1 Transient photoluminescence spectra for bulk $Ga_{0.47}In_{0.53}As$ at three different delay times (40, 160, and 470ps) depicted for an excitation density of $2.2 * 10^{18} cm^{-3}$. The spectra have been normalized and multiplied by a factor of 10^5, 10^4, and 10^3 for delay times of 40, 160, and 470ps, respectively

Springer Series in Chemical Physics, Vol. 48: **Ultrafast Phenomena VI**
Editors: T. Yajima · K. Yoshihara · C.B. Harris · S. Shionoya

The nonequilibrium carrier temperature and the excitation density as a function of time delay after the excitation pulse are obtained in that way. These experimental results are compared with cooling curves calculated by a theoretical model /2/ taking into account energy loss by polar optical scattering , acoustic deformation potential scattering and carrier heating by Auger processes. We introduce a factor F, which describes the reduction of polar optical energy loss in comparison to the theory at high excitation densities because of LO phonon overpopulation and/or screening of the Coulomb interaction between carriers and lattice atoms. The other a priori unknown parameters which enter the model are the acoustic deformation potential constant E_{AC} of holes and Auger coefficient C_A.

Figure 2 shows the experimentally obtained cooling curves (symbols) and theoretically calculated curves (solid lines). The most important results from this fitting procedure concerns the Fröhlich interaction in $Ga_{0.47}In_{0.53}As$. We observe a polar optical energy-loss-rate (ELR) in the heterostructure for low excitation density ($5*10^{16}cm^{-3}$), which is close to theoretical expectations (F=3). Therefore the strength of the electron (hole)-londitudinal optical (LO) phonon interaction is well described by the theoretically predicted Fröhlich interaction. For higher excitation densities the polar optical ELR is strongly reduced in comparison to theory, for example F=100 for an excitation density of $2.2*10^{18}cm^{-3}$. This reduction is mainly caused by a nonthermal population of LO (and transverse optical (TO)) phonons.

A value of $E_{AC} = 3eV$ is obtained for the acoustic deformation potential of holes and a value of $7*10^{-29}cm^6sec^{-1}$ for the Auger coefficient.

Next we investigate the problem of carrier cooling in multiple quantum well (MQW) structures of well thickness 3.4, 8.0, and 13.8nm, thus studying the transition from bulk to quasi 2D systems and the well thickness dependence of the different energy loss rates. The above explained experiment and fitting procedure is performed on these MQW's. Again we concentrate first on polar optical energy loss of these structures. As depicted in Fig.3, the polar optical energy loss in all MQW's is close to theoretical expectations, indicating a strength of the Fröhlich

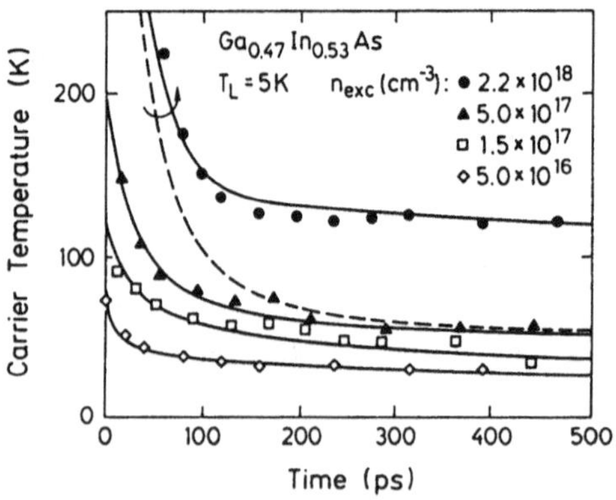

Fig.2 Cooling data for bulk $Ga_{0.47}In_{0.53}As$ for different excitation densities. Theoretical fits are depicted (solid curves) using E_{AC}=3eV and F=100,70,40, and 3 for excitation densities of $2.2*10^{18}$, $5*10^{17}$, $1.5*10^{17}$, and $5*10^{16}cm^{-3}$. The dashed curve shows a theoretical curve for the highest excitation density with the same parameters, but neglecting Auger recombination.

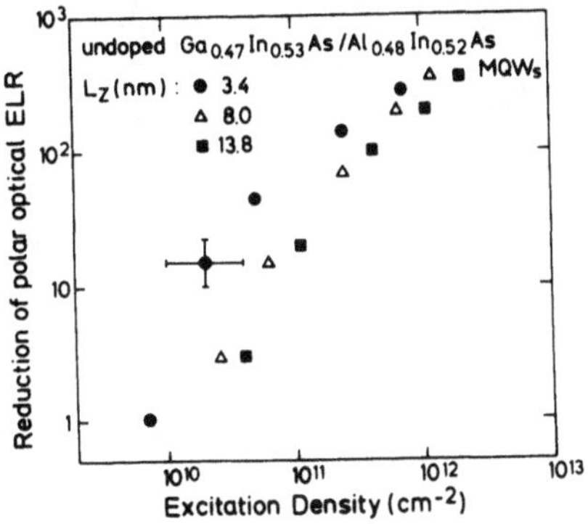

Fig.3 Reduction of polar optical energy loss as a function of (two-dimensional) excitation density for the 3.4nm thick MQW (circles), the 8.0nm thick MQW (triangles) and the 13.8nm thick MQW (squares)

interaction comparable to the 3D case. However, a small increase of the polar optical ELR in the low excitation limit is found with decreasing well thickness. This theoretically predicted effect/3/ has also been observed in a cw experiment/4/. For higher excitation density a reduction of the polar optical ELR is observed in the MQW's comparable to the bulk case. All MQW's show a similar reduction of polar optical ELR as a function of the 2D excitation density, which indicates the two-dimensional character of the carrier systems in these structures.

Further results concern the dependence of the ELR by acoustic deformation potential scattering of holes and the dependence of the Auger processes on quantum well thickness. We observe an increase of the ELR by acoustic deformation potential scattering of heavy holes with decreasing well thickness. The acoustic deformation potential scattering depends on the heavy hole effective mass m_h^* and the acoustic deformation potential E_{AC} via $(m_h^*)^{5/2} E_{AC}^2$. Therefore the observed increase of the acoustic deformation potential scattering can be attributed to an increase of the effective heavy hole mass as well as to an increase of the acoustic deformation potential. We also observe a decrease of the Auger coefficient with decreasing well thickness and therefore increasing energy of the first electron-heavy hole transition. A similar dependence of the Auger coefficient on the energy gap has been also reported for InGaAsP /5/.

Travel support by the Stifterverband für die deutsche Wissenschaft is gratefully acknowledged.

References

(1) K. Kash and J. Shah, Appl. Phys. Lett. 45, 401 (1984)
(2) H. Lobentanzer, H.-J. Polland, W.W. Rühle, W. Stolz, and K. Ploog, Phys. Rev. B36, 1136 (1987)
(3)F.A. Riddoch and B.K. Ridley, J. Phys. C, Solid State Phys. 16, 6971 (1983)
(4) H. Lobentanzer, W.W. Rühle, W. Stolz, and K. Ploog, Solid State Commun. 62, 53 (1987)
(5) E. Wintner and E.P. Ippen, Appl. Phys. Lett. 44, 999 (1984)

Tunneling Processes in AlAs/GaAs Double Quantum Wells

T. Tada[1], *A. Yamaguchi*[1], *T. Ninomiya*[1], *H. Uchiki*[1], *T. Kobayashi*[1], *and T. Yao*[2]

[1]Department of Physics, University of Tokyo, Bunkyo-ku, Tokyo 113, Japan
[2]Electrotechnical Laboratory, Tsukuba-shi, Ibaraki 305, Japan

1. INTRODUCTION

Quantum transport of photogenerated carriers through a barrier layer with an indirect band gap was studied. With the use of AlAs/GaAs double quantum well (DQW) structures in which two wells have different widths, the tunneling of carriers from one well to the other can be observed directly by measuring the photoluminescence (PL) from the two wells with different recombination energies. For the barrier layers, we choose AlAs to avoid possible uncertainty of barrier-height energy coming from compositional fluctuation when we use a barrier made of $Al_xGa_{1-x}As$. Two wells of different widths (60 and 80Å) are separated by the AlAs barrier of the thickness of L_B. (Fig. 1) Several samples with different L_B(15~120Å) were used for the study of the dependence of tunneling rate on L_B. The samples were grown on undoped GaAs substrates by molecular beam epitaxy.

2. EXPERIMENTAL

The 514.5-nm (2.41eV) line from Ar^+ laser (Spectra Physics Model 171-09) was used for excitation in the steady-state PL measurements. For the time-resolved PL measurements, a dye laser synchronously pumped by the second harmonics of a cw mode-locked Nd^+:YAG laser (Spectra Physics, Model 3460) was used. The pulse width and wavelength of the dye laser were about 5ps and 575nm

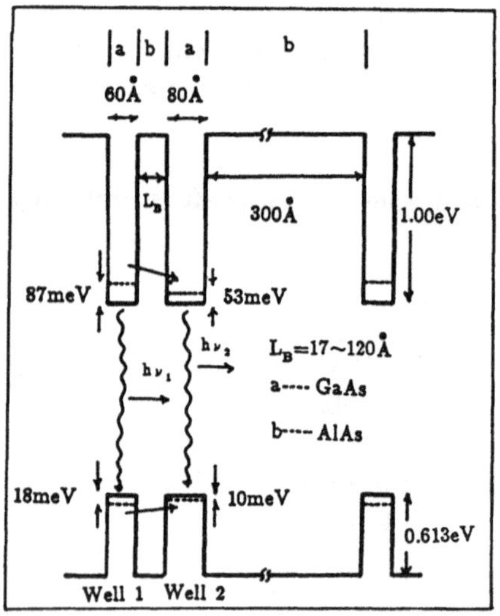

Fig.1 The schematic band diagram of the DQW structure.

Springer Series in Chemical Physics, Vol. 48: **Ultrafast Phenomena VI**
Editors: T. Yajima · K. Yoshihara · C.B. Harris · S. Shionoya

(2.16eV), respectively. The time dependence of the PL intensity was measured by a time-correlated single-photon counting method.

3. RESULTS AND DISCUSSION

The steady-state PL spectrum at 4.2K for a GaAs/AlAs DQW sample with the barrier width of 28Å is shown in Fig. 2. It has two peaks 1 and 2 with the peak energies of 1.63eV and 1.59eV, corresponding to Well 1 (L_z=60Å) and Well 2(L_z=80Å), respectively. The peak energies are consistent with a simple envelope-function approximation.

Figure 3 shows the ratio of the spectrally integrated intensity of PL at 4.2K between Well 1 (I_1) and Well 2 (I_2) as a function of the barrier width. Tunneling process plays an important role for the barriers narrower than 40A. We analyze this tunneling process by the following rate equations [1] :

$$dn_1/dt=G_1-n_1/\tau_1-\nu n_1 \tag{1}$$

$$dn_2/dt=G_2-n_2/\tau_2+\nu n_1 \tag{2}$$

where $n_{1(2)}$ is the population of the photogenerated carriers, $G_{1(2)}$ is the generation rate, and $\tau_{1(2)}$ is the life time of the photogenerated carriers in Well 1(2), respectively. Tunneling rate ν from Well 1 to Well 2 is given by [1]

$$\nu=\nu_0\cdot 4kE(V-E)/[4kE(V-E)+\{(k-1)E+V\}^2\sinh^2(L_B/\lambda)] \tag{3}$$

where λ is the 1/e penetration depth of the wave function into the barrier, and ν_0 corresponds to the classical period of the electron motion in the well, and k is the ratio of the electron mass in the barrier to that in the well, and V is the barrier height energy, and E is the energy of the electron. V is calculated from the difference in the energy of Γ-valleys between AlAs and GaAs by using Kroemer's 62:38 rule. [2]

Thus by using the steady state solutions of the rate equations, the ratio I_1 to I_2 is calculated as

$$I_1/I_2=(\tau_{r2}/\tau_{r1})/[(1+G_2/G_1)\tau_2\nu+G_2\tau_2/G_1\tau_1], \tag{4}$$

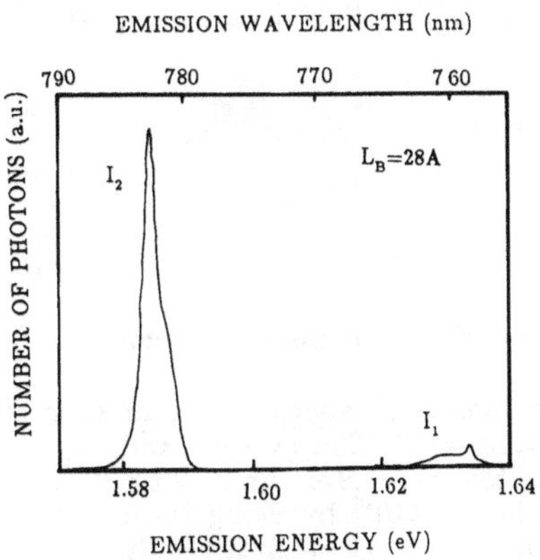

Fig.2 The steady state PL spectrum for L_B=28Å

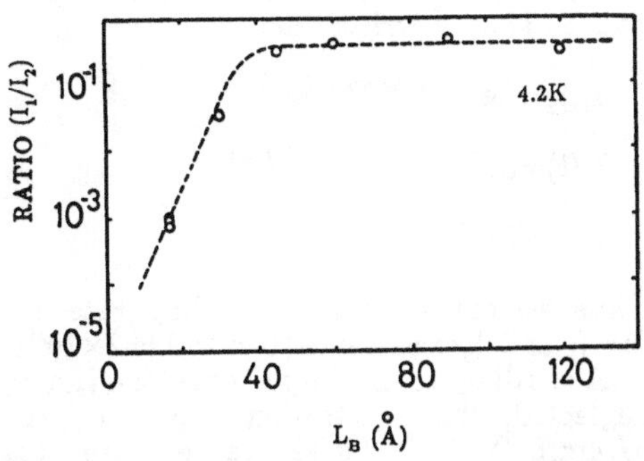

Fig.3 The ratio of I_1 to I_2 is plotted as a function of L_B. The dashed line represents the fitted curve.

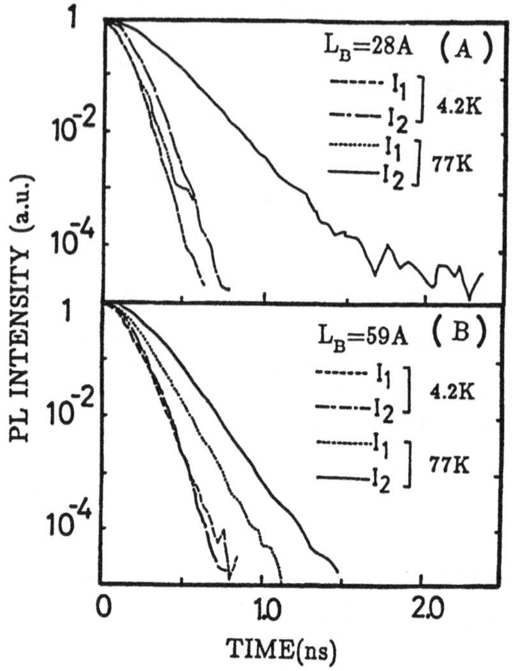

Fig.4 The temporal profile of the PL intensity.

where $\tau_{r1(2)}$ is the radiative recombination time for Well 1(2). τ_{r1}/τ_{r2} is estimated as 1.1 from a simple quantum mechanical calculation. By fitting the experimental results to eqs (3) and (4), we estimated λ and ν_0 as 6A and $4\times10^{14}s^{-1}$, respectively. The dashed line in fig.3 is the fitted curve.

Figure 4 shows the decays of PL from the sample with L_B=28A (Sample A) and 59A (Sample B) at 4.2K and 77K. We estimate the decay time by the convolution of an exponential decay function and instrument function. The estimated decay times of I_1 and I_2 for L_B=28A are 30ps and 50ps at 4.2K, respectively, and 40ps and 160ps at 77K, respectively. As for L_B=59A, those of I_1 and I_2 are 50ps and 50ps at 4.2K, respectively, and 100ps and 130ps at 77K, respectively. The decay times of I_1 are shorter than that of I_2 in Sample A, but the difference is small in Sample B. This is because an electron can tunnel from Well 1 to Well 2 in Sample A (L_B=28A) but can not in Sample B (L_B=59A). The decay of I_1 and I_2 can be simply written as:

$$I_1(t)=Ae^{-(1/\tau_1+\nu)t} \tag{5}$$

$$I_2(t)=Be^{-t/\tau_2}-C\cdot e^{-(1/\tau_1+\nu)t}$$
$$\sim Be^{-t/\tau_2}\,. \tag{6}$$

Thus we can estimate tunneling rate ν from the difference between decay times for I_1 and I_2 and get $\nu(L_B=28A)=2\times10^{10}s^{-1}$.

According to a simple envelope-function approximation where $\Gamma-X$ mixing is neglected, the penetration depth λ can be written as $\hbar/\sqrt{2m_e(V-E)}$ and ν_0 as $h/(4m_eL_z^2)$ [3] . These values are calculated to be 5.8A and $7.6\times10^{13}s^{-1}$, respectively. The tunneling rate are calculated to be 0.9×10^{10} by using these values and eq.(3). These values are in good agreement with the experimental results.

Recently Ando suggested that the mixing of the Γ valley with X-valley in $GaAs/Al_xGa_{1-x}As$ system has little effect on tunneling property even when the X-valley is lower than the Γ-valley. [4] Therefore it is valid to analyze our experimental results with the simple envelope-function approximation because Γ−X mixing has little effects on the DQW system in our study although the energy of X-valley of AlAs is lower than that of the Γ-valley.

4. SUMMARY

We observed the tunneling process in AlAs/GaAs DQW structures directly with time-resolved and steady-state PL measurements, and obtained the tunneling rate of $2\times10^{10}s^{-1}$ for L_B=28A and penetration depth of the wave function into the AlAs barrier is 6A. This is consistent with the simple envelope function approximation.

5. REFERENCES

1. Tetsuya Tada, Atsushi Yamaguchi, Toshiyuki Ninomiya, HisaoUchiki, Takayoshi Kobayashi, and Takafumi Yao: J. Appl. Phys **63** (1988) 5491.
2. H.Kroemer: Surf. Sci. **174** (1986) 299.
3. Y.Masumoto, S.Tarucha, and H.Okamoto: Phys. Rev. **B 33**, (1986), 5961.
4. T.Ando: Third Brazilian School of Semiconductor Physics, edited by C.E.T. Goncalves da Silva (World Scientific, Singapore,1987) p.23.

Picosecond Laser Study of Electron Dynamics in Resonant Tunneling Structures

M. Tsuchiya, T. Matsusue, and H. Sakaki

Institute of Industrial Science, University of Tokyo, Roppongi 7-22-1, Minato-ku, Tokyo 106, Japan

Resonant tunneling (RT) phenomenon in double-barrier (DB) heterostructure [1,2] has a conceptual similarity to a transmission of optical waves in Fabry-Perot (FP) resonator and involves time delay. Its dynamics should be investigated since they limit the ultimate speed of RT devices. Such a study will also clarify both the validity and the limitation of FP analogy of RT. In this paper, we report our study on the electron dynamics in AlAs-GaAs-AlAs DB structures by measuring the time resolved photoluminescence (PL). Picosecond pulses of a mode-locked dye laser were used to generate electron hole pairs in quantum wells (QW) in the structures, and the subsequent PL from particular QWs was monitored by a streak camera to determine the time variation of electron density in the QWs, which is lost by recombination (radiative[3] and nonradiative) and/or tunneling. Since the heavy hole mass is quite heavy, hole tunneling can be neglected at least in the initial phase of tunneling.

1. Escape of Electrons from Double-Barrier Resonant Tunneling Structures[4]

RT in DB structures is a basic and key process which governs the ultimate speed of RT diodes. Electron build up into QW and escape from QW causes intrinsic time delay due to multiple reflection. In ideal case, build up time and escape time are $\sim\hbar/\Delta E$, where ΔE is the width of transmission peak. To investigate this process, we prepared a series of AlAs/6.2nm-GaAs/AlAs DB structures with the barrier width L_B = 2.8~6.2nm. The tunneling escape rate $1/\tau_T$ of optically generated electrons in GaAs QW through thin AlAs barriers was determined by measuring the decay time of QW-PL. The laser is set to generate electron heavy-hole pairs only in the ground subband. The measured decay time was ~50ps when L_B = 2.8nm and increased systematically as L_B was increased, approaching asymptotically that of multi QW, which is equal to the recombination-limited lifetime τ_R. τ_T was estimated to be $1/\tau - 1/\tau_R$, and is plotted in Fig.1 as a function of L_B. Note that τ_T increases as $\exp(L_B/\text{const})$. The data for $L_B<4$nm fit well with the theoretically predicted tunneling lifetime (broken line and chained line) $\tau_T=2\hbar/\Delta E$, indicating that the tunneling escape process is well described by the simple FP-like model. In $L_B>5$nm, however, a discrepancy appears. This seems due to an un-canceled portion of non-radiative recombination process, which is known to be less in multi QW than in single QW structures.

2. Electron Dynamics in Coupled Resonant Tunneling Structures

Next, we investigated the effect of resonant coupling on tunneling process in coupled QWs of Fig.2, where L_B=3.1nm, L_{W1}=7.1nm, and L_{W2}=5.1nm. We measured the variation of tunneling rates at different bias voltage Va applied to the semi-transparent Schottky contact formed at the surface. The photon energy of laser was chosen to be 1.771eV to generate carriers in both wells. Under the flat band condition (Va~+0.6V), the electrons in QW2 have the highest energy, and therefore, transfer to QW1 or escape rightwards to outside. When the ground levels of QW1 and QW2 are in resonance(Va~-0.6V),

Springer Series in Chemical Physics, Vol. 48: **Ultrafast Phenomena VI**
Editors: T. Yajima · K. Yoshihara · C.B. Harris · S. Shionoya

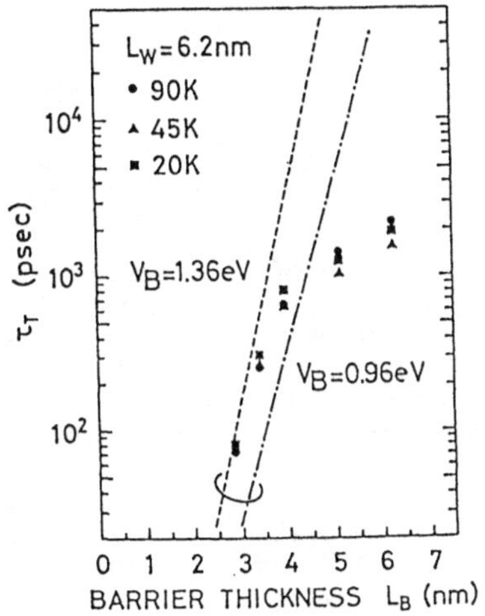

Fig.1 The measured and calculated tunneling escape time in AlAs/GaAs/AlAs DB structure vs. barrier thickness, where V_B is barrier height

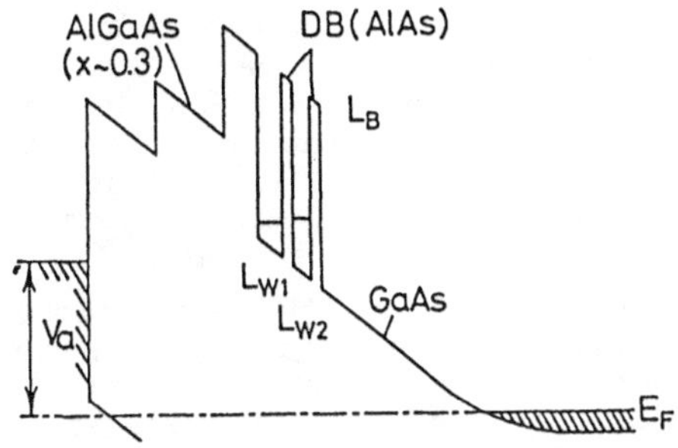

Fig.2 Band diagram of coupled RT structure

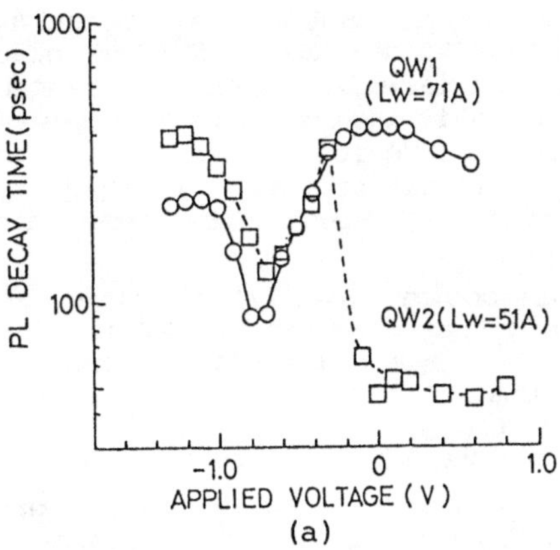

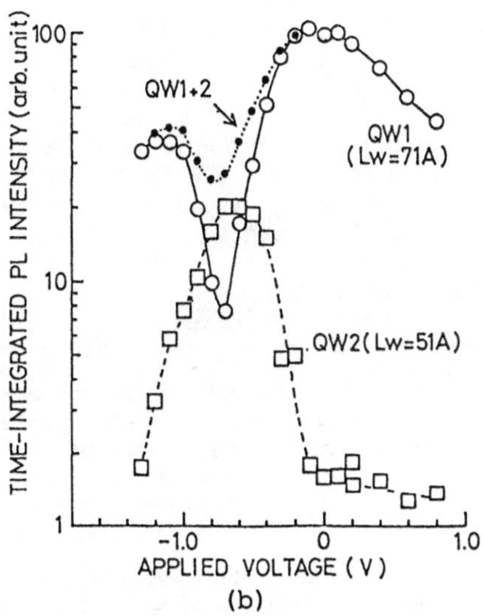

Fig.3 PL decay time(a) and time-integrated PL intensity(b) as a function of applied voltage

interwell tunneling is enhanced. When Va is made further negative, resonance breaks down, and tunneling from QW1 to QW2 comes to be allowed. Measured PL decay time and time-integrated PL intensity I at 15K are shown in Fig.3 as a function of Va. Their variations with Va indicate large change in resonant tunneling dynamics as will be described below.

Before resonance(Va>-0.1V), the ground level E1(QW1) of QW1 is far less than that E1(QW2) of QW2. The PL decay time of QW1 (τ_1) in this case is ~400ps, which is nearly equal to radiative recombination lifetime τ_R. In contrast, the decay time of QW2 (τ_2) is ~50ps. This decay rate $1/\tau_2$ is equal to the sum of two escape rates $1/\tau_{23}$, $1/\tau_{21}$ which correspond to the tunnel escape from QW2 to outside and to QW1, respectively. The former

(τ_{23}) is estimated to be ~100ps by the measured data of Fig.1 with the use of relation $\tau_T \propto L_W^4$. Hence the latter τ_{12} is found to be also ~100ps. Note that this tunneling process was clearly observed also in excitation spectrum of QW1, where the structure associated with QW2 exciton appeared. With these time constants, PL efficiency η_1 of QW1 and that η_2 of QW2 are expected to be ~1 and ~0.1, respectively, since $\eta_2=\tau_2/\tau_R$. The expected values agree well with the data.

When E1(QW1) and E1(QW2) are close to but slightly off the resonance(-0.4V<Va<-0.1V), QW1 and QW2 get slightly coupled, which allows a slow electron transfer (injection) from QW1 to QW2. This leads to the increase in τ_2. When the two levels are at resonance(-0.7V<Va<-0.4V), interwell tunneling gets frequent and large variations in decay time and PL intensity appear. The fact that $\tau_1 \sim \tau_2$ and $I_1 \sim I_2$ in this region indicates that the interwell tunneling time τ_{12} is much faster than τ_{23}(~100ps). This is confirmed by the measured delay time of the PL peak, which shows that τ_{12} is less than the time resolution(~30ps) of the detection system. Indeed, when barrier height is 1.13eV, τ_{23} is calculated to be 100ps for L_W=7.1nm and L_B=3.1nm, while τ_{12} is predicted to be 0.62ps from the calculated energy splitting $\Delta E(\sim \hbar/\tau_{12})$ for coupled QW with $L_{W1}=L_{W2}$=6.2nm and L_B=3.1nm. Hence, quasi-equilibrium exists between N1 and N2, the electron density of each well. The total density N(=N1+N2) is then written as $dN/dt=-N2/\tau_{23}=-N/(1+R)\tau_{23}$, where R is N1/N2. This indicates that the lifetimes τ_1 and τ_2 get equal and are $(1+R)\tau_{23}$, while the PL efficiencies are $R\tau_{23}/2\tau_R$ for QW1 and $\tau_{23}/2\tau_R$ for QW2. Therefore, one expects that τ_1 equals τ_2, and η_1/η_2(~R) decreases as Va decreases, which agrees with the data. Since τ_{23} is expected to change with Va less sensitively than R does, τ_1, τ_2 and η_1 should decrease, and η_2 is nearly constant as Va decreases. Note here also that τ_R and τ_{NR} (nonradiative recombination lifetime) have not changed appreciably with Va in this region for the following reasons; the measured sum of PL intensity from QW1 and QW2 in this region varies with Va in proportion to $\tau_1(\sim\tau_2)$, indicating that τ_R is nearly constant. In addition, the remarkable increase in PL intensity of QW2 at resonance indicates that the non-radiative path remained small here.

When Va gets more negative beyond the resonance (Va<-0.7V), the tunneling rate from QW2 to QW1 gets negligible and that from QW1 to QW2 is also somewhat reduced because of the breakdown of resonance. This mechanism explains the increase in τ_1 and τ_2, and change in PL intensity. The data indicate τ_{12} increases to ~540ps since $\tau_1(=(1/\tau_R+1/\tau_{12})^{-1})$ goes up to ~230ps. The value is reasonable considering the dependence of τ_T on L_W.

In summary, the escape rate of electrons from DB-RT structures was directly measured and well explained by FP-like model. Electron dynamics in coupled RT structures were also investigated. Under the resonance condition, large enhancement of interwell tunneling was observed and the decay process with two QWs under quasi-equilibrium was detected. Tunneling into the other well was observed under the off-resonant condition, and well explained by FP-like model.

We thank Dr. H. Onose for his help and the Grant-in-Aid from the Ministry of Education, Science and Culture.

References

1. L.L. Chang, L.Esaki, and R.Tsu, Appl.Phys.Lett. 24, 593 (1974)
2. T.C.L.G.Sollner, W.D.Goodhue, P.E.Tannenwald, C.D.Parker, and D.D.Peck, Appl.Phys.Lett. 43, 588 (1983)
3. T.Matsusue and H.Sakaki, Appl.Phys.Lett. 50, 1429 (1987)
4. M.Tsuchiya, T.Matsusue, and H.Sakaki, Phys.Rev.Lett. 59, 1934 (1987)

Phase Relaxation of Two-Dimensional Excitons in a GaAs Single Quantum Well

A. Honold[1], *L. Schultheis*[2], *J. Kuhl*[1], *and C.W. Tu*[3]

[1]Max-Planck-Institut für Festkörperforschung, Heisenbergstr. 1, D-7000 Stuttgart 80, Fed. Rep. of Germany
[2]ASEA Brown Boveri Corporate Research, CH-5405 Baden, Switzerland
[3]AT&T Bell Laboratories, Murray Hill, NJ 07974, USA

The phase relaxation of excitons in a solid, characterized by the phase coherence time T_2, which can be determined by time-resolved degenerate four-wave mixing[1], or the corresponding homogeneous linewidth Γ_h, is a very sensitive sensor for the interactions of the excitons with their environment. Measurements of the phase coherence time T_2 of three-dimensional (3D) excitons in GaAs subjected to collisions with incoherent excitons or free carriers[2], and acoustic phonons[3] at variable densities have been successfully applied to study the respective scattering mechanisms and to determine the dephasing efficiency of the corresponding collision processes. The experimentally found dependence of T_2 on the density of the scattering particles revealed strong exciton-exciton interaction and even ten times more effective exciton-free carrier scattering. In contrast, much less is known on the phase relaxation of two-dimensional (2D) excitons in GaAs, and in particular on the influence of the reduced dimensionality on the interactions of these excitons with other quasi-particles[4].

We report two-pulse selfdiffraction experiments on free 2D excitons in a 12nm GaAs single quantum well. The phase coherence of excitons, generated by a first pulse, is probed by a second, delayed pulse via coherent interaction and subsequent selfdiffraction. The variation of the diffracted intensity versus the delay of the two pulses reveals the phase relaxation. The phase coherence time T_2 is determined by solving numerically the optical Bloch equations of the two level system in the small signal regime and fitting the theoretical to the experimental diffraction curves. We fit the data assuming homogeneous broadening of the exciton transition, therefore, the determined phase coherence times represent the lower limit of the possible values with a maximum correction for strong inhomogenous broadening by a factor of 2 longer times.

We used in the experiments pulses from two synchronously pumped dye lasers with 2.6ps duration, 0.9meV spectral width, and less than 2ps jitter between the pulse trains of the two lasers. The pulses of one laser are tuned into the center of the exciton resonance to determine the phase relaxation of the excitons. The pulses of the second dye laser can be tuned independently to create either resonant excitons or free carriers within the sample.

The phase coherence time at low excitation densities ($N_x < 10^9 cm^{-2}$) without pumping excitons or free carriers with the second dye laser and at a temperature of 2K is $T_2 = 6 \pm 0.5$ps corresponding to a homogeneous linewidth of $\Gamma_h = 0.22 \pm 0.02$meV. These values represent the low density limits for T_2 and Γ_h of this sample.

Springer Series in Chemical Physics, Vol. 48: **Ultrafast Phenomena VI**
Editors: T. Yajima · K. Yoshihara · C.B. Harris · S. Shionoya

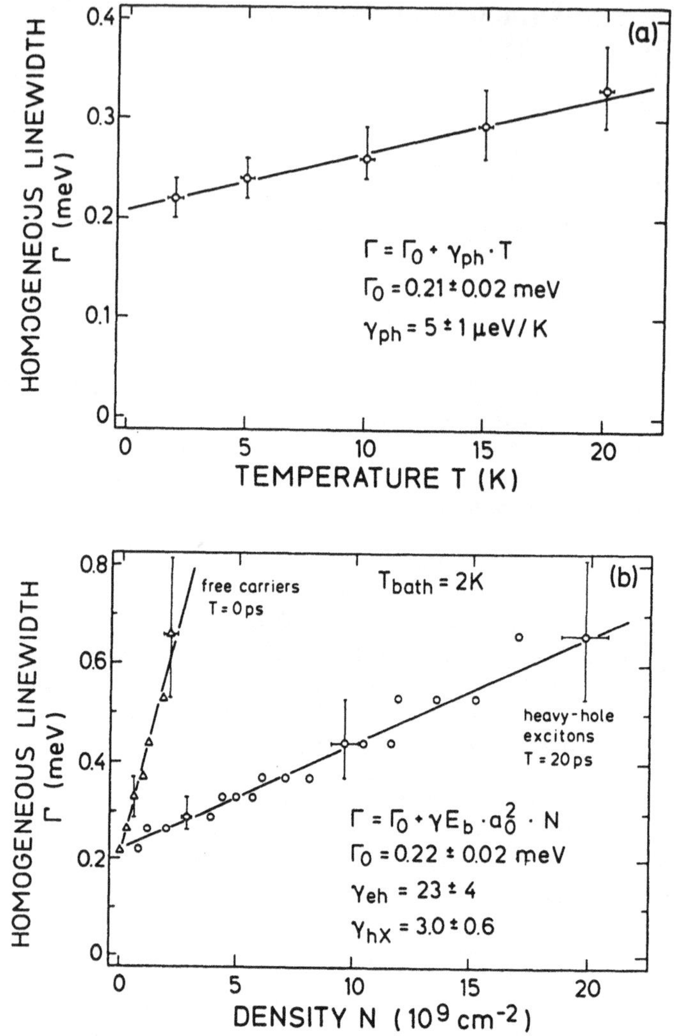

Figure 1. Dependence of the homogeneous linewidth Γ_h of 2D excitons in a 12nm GaAs single quantum well on the temperature (a) and the excitation density (b). The solid lines are fits assuming a linear dependence.

The exciton-acoustic phonon interaction is studied by measuring the temperature dependence of the homogeneous linewidth. Absorption of one acoustic phonon in the center of the Brillouin zone by an exciton leads to a linear dependence of the homogeneous linewidth on the temperature. Figure 1(a) depicts the observed temperature dependence of the homogeneous linewidth, which can be fitted very well by the expected linear dependence for one-phonon absorption. The slope of $\gamma_{ph} = 5 \pm 1\mu eV/K$, which gives a measure for the exciton-acoustic phonon interaction strength, agrees fairly well with the value of $\gamma_{ph} = 5\mu eV/K$ determined in previous transmission experiments on excitons in a 13.5nm GaAs single quantum well[4]. The comparison with the value of $\gamma_{ph} = 17\mu eV/K$ determined for 3D excitons[3] demonstrates a tremendous weakening of the exciton-acoustic phonon interaction in 2D compared with 3D systems.

The interaction with incoherent excitons and free carriers is studied by measuring the phase coherence time T_2 at a temperature of 2K in the presence of excitons or free carriers injected independently by the second dye laser. The excitons are injected 20ps before the first pulse of the selfdiffraction experiment arrives to ensure that the injected excitons have lost their phase coherence completely and act only as a background of incoherent scatterers. The free carriers are created in temporal overlap with the first pulse of the selfdiffraction experiment to prevent the condensation of the free carriers to excitons.

The determined homogeneous linewidth Γ_h as a function of the excitation density is shown in Figure 1(b). We find a linear broadening with increasing excitation density, starting at surprisingly low densities. The broadening is described by the function

$$\Delta\Gamma(N) = \Gamma(N) - \Gamma(0) = \gamma a_0^2 E_b N \quad ,$$

with a_0 the exciton Bohr radius, E_b the exciton binding energy, N the excitation density, and γ a dimensionless broadening parameter. The experimental values are $\gamma_{hX} = 3 \pm 0.6$ and $\gamma_{eh} = 23 \pm 4$ for exciton-exciton and exciton-free carrier interaction, respectively. The exciton-free carrier collisions are more effective than exciton-exciton collisions by a factor of 8. The exciton-free carrier interaction is due to the long-ranged Coulomb interaction while the exciton-exciton interaction is composed of an attractive part due to screening of the Coulomb interaction and a repulsive part due to the Pauli exclusion principle for identical fermions, i.e. the electrons and holes which form the excitons. The long-ranged Coulomb interaction is much more efficient in scattering at larger particle distances than the exciton-exciton interaction and explains the higher broadening coefficient for exciton-free carrier collisions.

The comparison with the broadening of 3D exciton transitions[2] yields at comparable particle distances a much stronger interaction of the 2D excitons with both free carriers and incoherent excitons. This result is in agreement with the theoretical prediction[5] that screening effects are less important in 2D and therefore, the interaction of 2D excitons with incoherent excitons and free carriers is enhanced compared with 3D excitons.

The residual homogeneous linewidth at low temperature and low excitation density of $\Gamma_h = 0.22 \pm 0.02$meV is attributed to interactions with impurities, the imperfect interfaces of the quantum well, and additional scattering inside the barrier material due to the imperfect confinement of the exciton wave function in the quantum well with finite barrier height.

In conclusion, we studied the phase relaxation of free 2D excitons and their interaction with acoustic phonons, free carriers, and incoherent excitons by means of two-pulse selfdiffraction. We find weaker interaction of the 2D excitons with acoustic phonons but stronger interaction with free carriers and incoherent excitons compared to 3D excitons. As a consequence, the phase relaxation of 2D excitons is extremely sensitive to the excitation density.

REFERENCES

1. L. Schultheis, J. Kuhl, A. Honold, and C.W. Tu: Phys. Rev. Lett. 57, 1797 (1986)
2. L. Schultheis, J. Kuhl, A. Honold, and C.W. Tu: Phys. Rev. Lett. 57, 1635 (1986)
3. L. Schultheis, J. Kuhl, A. Honold, and C.W. Tu: in Proceedings of the 18th International Conference on the Physics of Semiconductors, ed. by O. Engström (World Scientific, Singapore 1987) p. 1397
4. L. Schultheis, A. Honold, J. Kuhl, K. Köhler, and C.W. Tu: Phys. Rev. B 34, 9027 (1986)
5. S. Schmitt-Rink, D.S. Chemla, and D.A.B. Miller: Phys. Rev. B 32, 6601 (1985)

Intra-Well and Cross-Well Transport Measurements in Multiple Quantum Wells Using Transient Gratings

R.J. Manning [1], *A. Miller* [1], *D.W. Crust* [2], *D. Herbert* [1], *and K. Woodbridge* [3]

[1]Royal Signals and Radar Establishment, Great Malvern, Worcs., WR14 3PS, UK
[2]University of St. Andrews, St. Andrews, Fife, KY16 9SS, UK
[3]Philips Research Labs., Redhill, Surrey, RH1 5HA, UK

In this paper we describe experiments, using sub-picosecond pulses, which exploit the large excitonic refractive non-linearity [1] in quantum wells in two degenerate four wave mixing configurations. These enable us to study photo-carrier transport along and across wells in a multiple quantum well sample. The two arrangements, forward travelling and counter propagating (Fig 1), produce transient phase gratings parallel to and across the plane of the wells respectively. By measuring the temporal decay of these gratings, the intra-well and cross-well contributions to carrier diffusion may be determined. The diffusion is highly anisotropic, and the decay rates are dependent on the angle between the grating and the wells.

The sample (KLB 269) consisted of 120 periods of 6.5 nm thick GaAs quantum wells, separated by 21.2 nm $Al_xGa_{1-x}As$ ($x \sim 0.4$) barriers, and was grown by molecular beam epitaxy. The substrate was removed by selective etching, and the sample had anti-reflection coatings deposited front and back before being mounted on sapphire. We used a tunable, synchronously

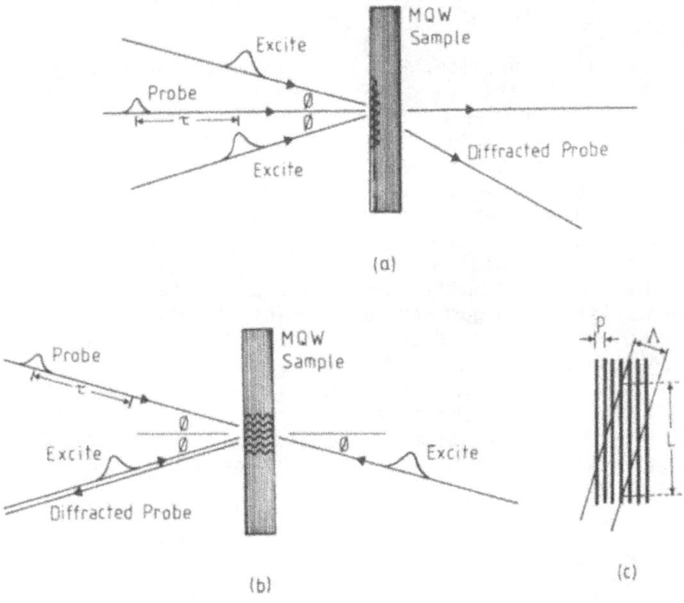

Figure 1. a) Forward travelling configuration, b) Counter propagating, c) relative position of wells to grating for a rotation of the sample in (b)

Springer Series in Chemical Physics, Vol. 48: **Ultrafast Phenomena VI**
Editors: T. Yajima · K. Yoshihara · C.B. Harris · S. Shionoya

mode-locked Styryl 9 dye laser with intracavity saturable absorber, giving cavity dumped pulses ~ 600 fs long, separated by ~ 65 ns. The output was split into two excite and one probe beam, the latter being delayed with respect to the two excite pulses via a stepper motor driven delay stage. The first order diffracted light from the probe beam was measured using phase sensitive detection, giving efficiencies and lifetimes.

The forward travelling geometry (Fig 1a) was used to determine the size of the nonlinearity and the ambipolar diffusion coefficient for carriers moving in the plane of the wells. This geometry employs two excite pulse, incident at ϕ either side of the normal to the sample surface, which are temporally and spatially coincident on the sample (spot size ω_o = 25 μm). They interfere to produce, within a gaussian envelope, a sinusoidal intensity pattern of period Λ, along the wells, determined by ϕ. The experiments were performed at the peak diffraction efficiency, which was to the long wavelength side of the 830 nm heavy hole exciton feature, consistent with a refractive grating. The refractive index change per electron-hole pair per unit volume was determined to be $n_{eh} \sim 0.4 \times 10^{-19}$ cm^3 [2].

For small diffraction efficiencies, the decay rate of the diffracted signal, r, is twice the relaxation of the carrier density modulation, ie

$$r = \frac{8\pi^2 D_m}{\Lambda^2} + \frac{2}{\tau_R} \tag{1}$$

where D_m is the ambipolar diffusion coefficient and τ_R is the carrier recombination time, and we have ignored the diffusion rate out of the excited area. By measuring the grating decay rates at different angles, we deduced an in-well ambipolar diffusion coefficient, D_m, of 16.2 cm^2/s, similar to that measured for bulk GaAs [2].

The counterpropagating configuration, Fig 1(b), accessed the short period grating perpendicular to the wells. This grating was generated by interference at the two excite pulses entering the sample from opposite directions at equal angles to the wells. The time delayed probe was counterpropagating to one of the excite beams. The grating period, $\Lambda \sim 120$ nm, covered approximately 4 wells. The front excite beam was polarisation modulated for phase sensitive detection of the diffracted probe. The diffracted probe signal as a function of time for several angles of rotation about a horizontal axis is shown in Fig 2. θ is the angle of rotation of the sample from the normal (where the grating is perpendicular to the wells). From geometrical considerations, Fig 1(c), we would expect intra-well diffusion to contribute to the observed decay rate when the grating is angled with respect to the wells. For a horizontal axis of rotation, the separation of carrier density maxima along the wells is given by $L = n\Lambda/\mathrm{Sin}\,\theta$ (where n is the average refractive index), up to the point where L becomes comparable with the spot size ($\theta < 1°$). We assume that the diffusion of the carriers parallel and perpendicular to the wells to be independent, and modify (1) to

$$r = \frac{8\pi^2 D_m \mathrm{Sin}^2\theta}{n^2 \Lambda^2} + \frac{2}{\tau_\perp} + \frac{2}{\tau_R} \tag{2}$$

where $\tau_\perp$ is the grating decay time due to cross-well diffusion. The measured decay rates are plotted in Fig 3 as a function of $\mathrm{Sin}^2\theta/n^2$. At or

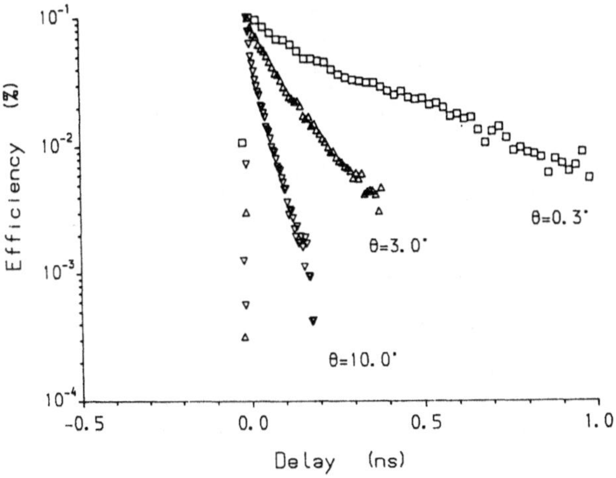

Figure 2. Diffracted signal decays for configuration 1(b), for different rotations, θ, of the sample from the normal.

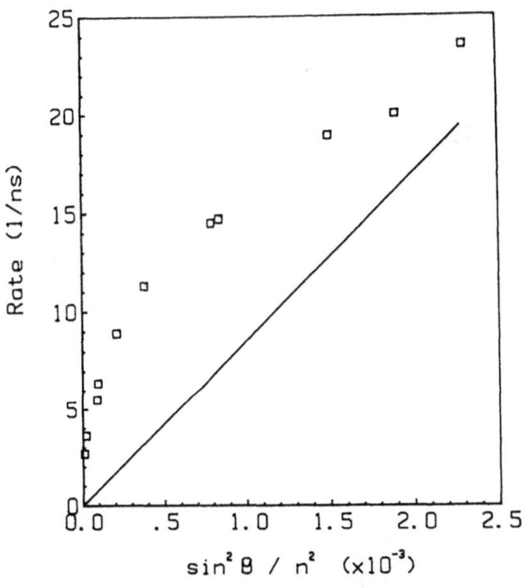

Figure 3. Signal decay rates as a function of angle. Purely intra-well diffusion is given by the solid line.

near zero angle ($\theta < 1°$), there is little or no carrier density modulation along the well, and this limiting decay rate gives the cross-well diffusion time ($\sim$ 1 ns from Fig 3). Diffusion out of the spot area and carrier recombination, both take place on timescales > 50 ns, and cannot account for this limiting value.

For barriers of 21.2 nm thickness, tunnelling should be negligible, and this timescale arises from the time to eject sufficient carriers out of the well via thermionic emission to balance the population gradients. We plot (solid line, Fig 3) the anticipated relaxation rate due to intrawell diffusion using the measured diffusion coefficient. There is an enhancement above the expected rate at small angles, possibly due to a differential emission rate over the barriers for electrons and holes (the holes have a lower barrier to overcome). This leads to some degree of charge separation which is then rapidly neturalised by electron movement along individual wells, causing an enhancement of the along-well diffusivity. At larger angles, the "wash-out" along the wells occurs significantly rapidly that thermionic emission is too slow to contribute to any enhancement, and the points approach the line given by (2).

In conclusion, we have shown that the highly anisotropic nature of carrier diffusion affects the time constants measured using a phase conjugate geometry, and that the grating decay rate may be controlled by sample rotation. The timescale at very small angles ~ 1 ns, indicates the time taken for carriers to be ejected out of the wells, assuming the movement across the AlGaAs and subsequent trapping is extremely rapid. This gives the time needed to build up a photocurrent in photoconductive devices, and as such gives the intrinsic limit to their speed of response.

References

1. D. S. Chemla, D. A. B Miller, P. W. Smith, A. C. Gossard and W.Wiegmann: IEEE J. Quantum Electron. QE-20, 265 (1984).

2. R. J. Manning, D. W. Crust, D. W. Graig, A. Miller and K. Woodbridge: J. Mod. Opt. 35, 541 (1988).

Pulse Propagation in GaAs Quantum Wells

Y. Masumoto and M. Adachi

Institute of Physics, University of Tsukuba, Tsukuba, Ibaraki 305, Japan

There are many interesting phenomena in the physics of pulse propagation in the condensed matter. Especially, associated with the excitons in semiconductors, pulse distortion occurs as a result of the exciton polariton effect or the hole burning effect. The latter is considered to be important in excitons in semiconductor quantum wells[1]. Nowadays, the form of laser pulses is controllable by means of the pulse handling technique, such as the pulse compression. Therefore the technique has a possibility to explore new aspects in the physics of pulse propagation associated with the excitons in semiconductors. In this work, we have studied the pulse propagation in GaAs quantum wells by comparing the experimental results with the computational simulation.

The excitation laser pulses are generated by a dye laser synchronously pumped by a mode-locked Kr ion laser. Temporal and spectral widths of laser pulses are 9.0 ps and 2.9 meV, respectively. The laser beam was split in two parts. One beam was focused on the 200 periods of GaAs (43 Å)-AlAs (62 Å) multi-quantum-wells in a cryostat. The transmitted beam and another beam were mixed in an RDP crystal to generate SHG pulses. Figure 1 shows two SHG correlation traces, bold and fine lines, corresponding to the cases that the sample is present or absent, respectively. Here, the excitation photon energy hits the low energy tail part of the lowest exciton (1s heavy exciton) in GaAs quantum wells. The fine correlation trace is composed of a coherent spike (correlation width = 520 fs) and a symmetrically broad correlation base (correlation width = 19 ps). On the other hand, the bold correlation trace is composed of asymmetric correlation base around the coherent spike. The leading edge of the pulses was strongly absorbed, while the trailing edge was absorbed little. The pulse deformation disappeared when the excitation density is reduced to 10^{-8} $J/cm^2 \cdot$pulse at 2.1 K. The pulse deformation is reduced when the excitation photon energy increases under the common excitation density. It completely diappeared when the excitation laser hits the transparent energy region below exciton resonance. Above mentioned results are clearly understood in terms of hole-burning of exciton states in GaAs quantum wells, as described below.

Firstly, we confirmed that the shift of the coherent spike does not depend on both the excitation photon energy and excitation density and is 30 fs. The fact indicates that there is not any detectable anomalous dispersion effect due to excitons. The shift of the coherent spike probably comes from the linear refractive index of the sample 1.05 μm thick. Therefore, we could accurately decide the peak delay of the base correlation trace by regarding the coherent spike as the zero time standard. The peak delay varies with the change of the excitation density and the excitation photon energy. In Fig. 2, the peak delay is plotted as a function of the excitation density. With the increase of the excitation density, the peak delay increases once, afterward it decreases. With the increase of the excitation photon energy, the maximum peak delay shifts to the higher

Springer Series in Chemical Physics, Vol. 48: **Ultrafast Phenomena VI**
Editors: T. Yajima · K. Yoshihara · C.B. Harris · S. Shionoya

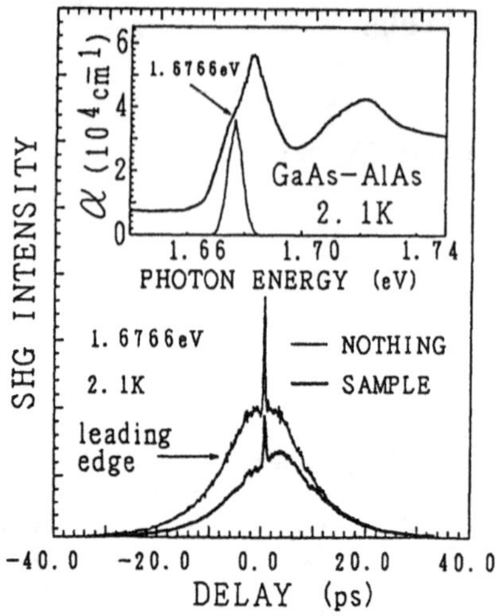

Fig.1 SHG correlation traces. Bold and fine lines correspond to cases that the sample is present or absent, respectively. In the inset, absorption spectrum of the sample and excitation laser spectrum are shown.

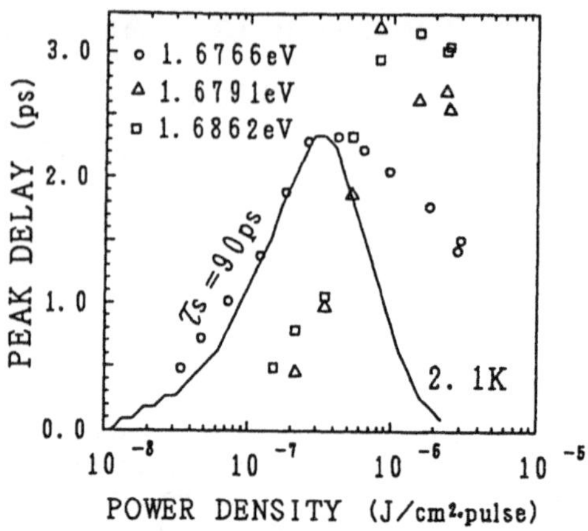

Fig.2 Peak delay of the base correlation traces relative to the coherent spike as a function of the excitation density. Experimental results are shown by symbols together with the calculated result (line) for 1.6766 eV .

excitation density. Selden has presented an equation which describes the delay of pulses propagating through saturable absorbers[2]. On the basis of the equation, we simulated the temporal change of transmittance T(t) with a parameter, relaxation time τ_s defined as the decay time for the excited state of the absorber. Then, the correlation traces, $\int_{-\infty}^{\infty}f(t+\tau)T(t)f(t)dt$, are calculated with the Lorentzian-type pulse, f(t). Thus, the peak shift of the correlation traces is calculated. The calculated peak shift for 1.6766 eV is described as a function of the excitation density in Fig.2, together with the experimental peak shift of the correlation trace (base) relative to the coherent spike. The calculation qualitatively describes the experimental feature. In particular, the calculation describes that the peak delay attains its maximum at the certain excitation density and is reduced at both sides. Thus the relaxation time τ_s = 90 ps is now obtained. With the increase of the excitation photon energy, the relaxation time τ_s was found to decrease. It is known that the relaxation time of the excitons in GaAs quantum wells decreases with the increase of the exciton energy[3]. Therefore the result is reasonable. Hegarty explained his experimental data on the basis of his approximation. Hegarty's approximation can not explain that the peak delay attains its maximum at the certain excitation density. In his approximation, the higher the excitation density is, the larger the peak shift is. Therefore Hegarty's approximation is not good. The agreement between the experimental delay and the calculated one based on Selden's model is not good enough. To improve the agreement, the extension of Selden's model to the inhomogeneously broadened excited states may be necessary.

The coherent spike shows small change with the change of the excitation photon energy. In the transparent energy region, the peak amplitude ratio of the coherent spike to the base correlation trace does not depend on

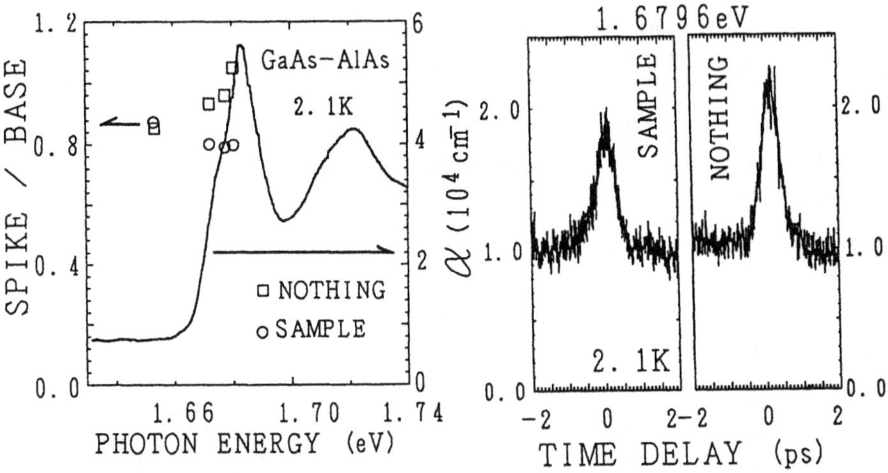

Fig.3 The peak amplitude ratio of the coherent spike to the base correlation trace as a function of the excitation photon energy. In the right part, the typical coherent spikes corresponding to cases that the sample is present or absent are shown.

whether the sample is absent or present. It is almost unity, which is a characteristic feature of the Gaussian random field. In Fig.3, the ratio is plotted as a function of photon energy. It is reduced as the incident photon energy hits the exciton energy. It was found to be independent of the excitation intensity. The phenomena are explained as the coherent propagation effect of the incoherent pulses observed in Na vapor[4]. The effect was observed in a semiconductor for the first time. It is considered that energy dependence of the linear absorption or refractive index works to change the ratio. To identify which essentially works to change the ratio, energy dependence of absorption or refractive index, some calculations are necessary.

In summary, we observed the hole burning effect due to excitons in the pulse propagation in GaAs quantum wells. The observed pulse distortion is qualitatively explained by Selden's model. However, it is suggested that the extension of Selden's model is necessary. In addition, coherent propagation effect of the incoherent pulses was observed in GaAs quantum wells.

References

1. J. Hegarty: Phys. Rev. B 25, 4324 (1982)
2. A. C. Selden: J. Phys. D 3, 1935 (1970)
3. J. Hegarty and M. D. Sturge: J. Opt. Soc. Am. B 2, 1143 (1985)
4. N. Morita, K. Torizuka and T. Yajima: J. Opt. Soc. Am. B 3, 548 (1986)

Excitonic-Polariton Propagation in a GaAs/AlGaAs Quantum Well

K. Ogawa, T. Katsuyama, and H. Nakamura

Central Research Laboratory, Hitachi Ltd., Kokubunji, Tokyo 185, Japan

1. Introduction

Electromagnetic waves in semiconductor crystals are coupled to excitons and are transformed into excitonic polaritons in the spectral region around excitonic resonance [1, 2]. The optical properties of excitonic polaritons characteristically depend on the wavevector. Time-of-flight measurements with short light pulses have revealed a remarkable decrease in light group velocity in bulk CuCl and GaAs crystals at low temperatures, which is induced by excitonic polaritons [3-5].

Two-dimensional light propagation caused by surface excitonic polaritons in ZnO crystal has been observed [6, 7]. In semiconductor quantum wells, however, no experimental evidence for excitonic polaritons has been obtained. As far as we know, this is because all low-temperature optical measurements of quantum wells have been performed with incident light propagating perpendicular to the quantum-well layers. Quantum-well excitonic polaritons (QWEPs) will form in light propagation parallel to the layers [8], since translational motion of the quantum-well excitons is allowed along the layers.

2. Experimental Procedure

We used a picosecond time-of-flight method to measure the propagation delay time of the light transmitted along the quantum-well layer. For optical measurements in the direction parallel to the quantum well, a waveguide-type sample containing a GaAs single quantum well was grown by molecular-beam epitaxy on a semi-insulating GaAs substrate. The sample formed a so-called leaky waveguide, in which the refractive index of the core was smaller than that of the cladding [9]. A 50-A GaAs quantum well was sandwiched between 1.8-μm $Al_{0.3}Ga_{0.7}As$ barrier layers. These layers constituted the core of the waveguide sample. The cladding was made of GaAs. The sample was mounted on Si substrate and set in an optical cryostat (Air Products LT-3-110). The delay time was measured as a function of the incident laser.

3. Experimental Data and Discussion

To gain information on excitonic-resonance energy of the quantum-well waveguide at low temperatures, the optical-

Springer Series in Chemical Physics, Vol. 48: **Ultrafast Phenomena VI**
Editors: T. Yajima · K. Yoshihara · C.B. Harris · S. Shionoya

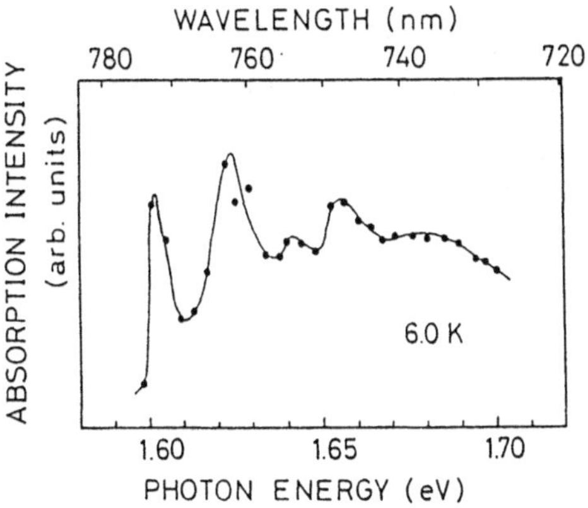

Fig. 1 Optical-absorption spectrum of the quantum-well waveguide at 6.0 K. The y-axis is absorption intensity; the x-axis corresponds to incident photon energy or wavelength. Dots denote experimental data. The solid line is drawn as a guide.

absorption spectrum was taken. This spectrum is shown in Fig. 1. Several lines are observed in the spectrum. The three lines at the lower photon energies, 1.602 eV, 1.625 eV and 1.642 eV are considered to be associated with two-dimensional excitons in the quantum well. This is because these lines are also observed in a GaAs quantum-well waveguide containing no superlattice layers. The rest of the absorption lines at the photon energies higher than 1.65 eV are probably caused by excitons in the superlattice layers.

Identification of the three absorption lines has been performed by polarization-sensitive measurements [9] using transverse-electric (TE) and transverse-magnetic (TM) light polarization. The absorption lines at 1.602 eV and 1.642 eV are identified as heavy-hole (hh) and light-hole (lh) exciton lines, respectively. The broad line lying at 1.625 eV between each line is probably composed of hh- and lh-exciton absorption lines. It is suggested that both hh and lh excitons have doublet-line structures in the optical-absorption spectrum and that the higher-energy part of the hh-exciton doublet unexpectedly overlaps with the lower-energy part of the lh-exciton doublet.

Two tentative explanations for the occurrence of the doublet structures are proposed. One is that the doublets are caused by fluctuation of the width of the quantum well. In this interpretation, the quantum-well width will fluctuate with binary values. The other tentative explanation is that the doublets are produced by the longitudinal-transverse (LT) splitting between the lower- and upper-polariton branches [2]. Since the energy spacing between each component of the doublets is nearly 20 meV, we may expect a giant LT splitting of QWEPs. Theoretical work has predicted that the LT splitting is remarkably large for QWEPs on account of the enhancement of the oscillator strength of quantum-well excitons [8].

In Fig. 2, the propagation delay of the light pulse obtained by the time-of-flight measurements is plotted against the incident photon energy. The delay time increases drastically at almost the same photon energies as those of the excitonic-absorption lines in Fig. 1. This shows that the

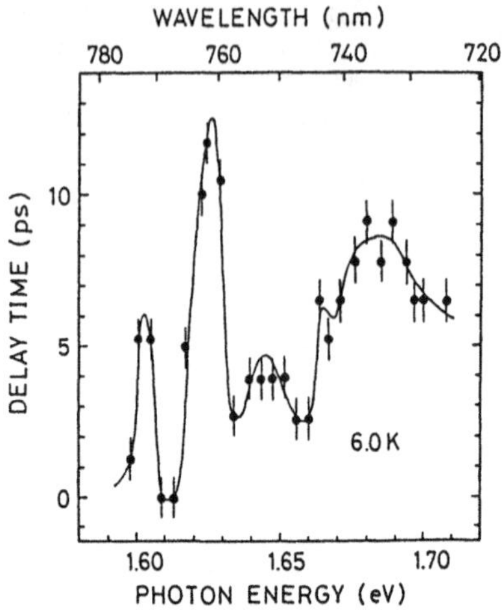

Fig. 2 Propagation delay time of the probe pulse transmitted through the waveguide as a function of photon energy or wavelength. Dots are experimental data taken at 6.0 K. The solid line is a guide.

group velocity of the light pulse decreases at around the photon energy resonant to the exciton lines. In Fig. 2, the delay time is about 12 ps at the photon energy of the hh-exciton line, 1.625 eV. The filling factor of the quantum well in the waveguide is only 0.1%. Simply multiplying the actual length of the waveguide, 650 μm, by the filling factor, we obtain the effective length of the waveguide, which contributes to the decrease in group velocity, as 0.9 μm. Thus the group velocity is given as 7×10^4 m/s at 6.0 K. This value is less than 1/1000 of the light velocity. Almost the same value was reported for bulk GaAs crystal at 1.3 K [4]. Such a considerable decrease in group velocity, as well as the coincidence between the photon energies of the absorption and the delay-time lines, indicates that the QWEPs exist in the quantum well.

In conclusion, we have presented the first experimental evidence for quantum-well excitonic polaritons.

References

1. J. J. Hopfield: Phys. Rev. 112, 1555 (1958)
2. J. J. Hopfield, D. G. Thomas: Phys. Rev. 132, 563 (1963)
3. Y. Segawa, Y. Aoyagi, S. Namba: Solid State Commun. 32, 229 (1979)
4. R. G. Ulbrich, G. W. Fehrenbach: Phys. Rev. Lett. 43, 963 (1979)
5. Y. Masumoto, Y. Unuma, Y. Tanaka, S. Shionoya: J. Phys. Soc. Jpn. 47, 1844 (1979)
6. J. Lagois, B. Fisher: Phys. Rev. Lett. 36, 680 (1976)
7. F. DeMartini, M. Colocci, S. E. Kohn, Y. R. Shen: Phys. Rev. Lett. 38, 1223 (1977)
8. M. Nakayama: Solid State Commun. 55, 1053 (1985)
9. J. S. Weiner, D. S. Chemla, D. A. B. Miller, H. A. Haus, A. C. Gossard, W. Wiegmann, C. A. Burrus: Appl. Phys. Lett. 47, 664 (1985)

Investigation of Intersubband Relaxation in GaAs/$Al_xGa_{1-x}As$ Quantum Well Structures by an Infrared Bleaching Technique

A. Seilmeier[1], *M. Wörner*[1], *H.-J. Hübner*[1], *G. Weimann*[2], *and W. Schlapp*[3]

[1]Physik Department, Technische Universität München, Arcisstraße 21, D-8000 München 2, Fed. Rep. of Germany
[2]Walter Schottky Institut, Technische Universität München, Am Coulombwall, D-8046 Garching, Fed. Rep. of Germany
[3]Forschungsinstitut der Deutschen Bundespost, D-6100 Darmstadt, Fed. Rep. of Germany

Carrier relaxation is more complex in 2D-heterostructures than in 3D-semiconductor material. The carrier confinement generates two dimensional subbands and two different relaxation mechanisms have to be considered in the investigation of carrier cooling: intrasubband and intersubband transitions. There exists a substantial amount of data on carrier cooling in bulk semiconductors and on intrasubband relaxation in GaAs/$Al_xGa_{1-x}As$ quantum well structures. Time constants in the order of several 10^{-12} s to 10^{-13} s have been measured via transient absorption changes or time resolved photoluminescence. In contrast, our knowledge of intersubband relaxation is very poor. This is due to the fact that the above mentioned techniques do not give direct information on intersubband processes.

In this paper, intersubband transitions at 300K are studied by an ultrafast infrared bleaching technique /1/ which we used for the first time. An intense tunable picosecond pulse bleaches the intersubband absorption in the infrared. A second light pulse of the same frequency measures the recovery of the absorption as a function of time. The infrared light pulses are generated with a mode-locked Nd:glass laser system. Difference frequency generation between the fundamental of the laser and the emission of a traveling-wave infrared dye laser supplies picosecond pulses tunable between 5 µm and 12 µm. The pulse duration amounts to 2 ps.

The quantum well samples investigated here are grown by molecular beam epitaxy on (100) semi-insulating GaAs substrates of 350 µm thickness. They consist of fifty thin undoped GaAs layers with thicknesses between 47Å and 73Å, which are embedded in 400Å thick $Al_xGa_{1-x}As$ layers with an Al concentration between x = 0.3 and x = 0.35. The central 100 Å of the barriers are doped with Si, which leads to a two-dimensional carrier concentration of about 4 to 5×10^{11} cm^{-3} per GaAs quantum well. The total multi-quantum well structures are cladded in between of 0.2 µm thick $Al_xGa_{1-x}As$ layers to avoid surface depletion and substrate effects.

It is well-known that intersubband transitions in the infrared exhibit strong absorption cross-sections for light polarized normal to the layers. Therefore, we have to prepare our samples in a special prism geometry (see inset in Fig.1) in which a large component of the electric field vector is normal to the layers. In these samples the expected strong infrared absorption is clearly observed (Fig.1).

Springer Series in Chemical Physics, Vol. 48: **Ultrafast Phenomena VI**
Editors: T. Yajima · K. Yoshihara · C.B. Harris · S. Shionoya

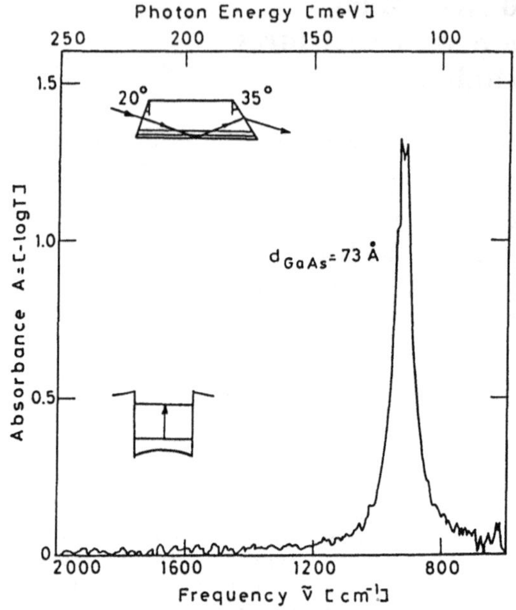

Fig. 1 Infrared absorption of a multi-quantum well sample. The strong absorption band is due to the intersubband transition. Inset: sample geometry.

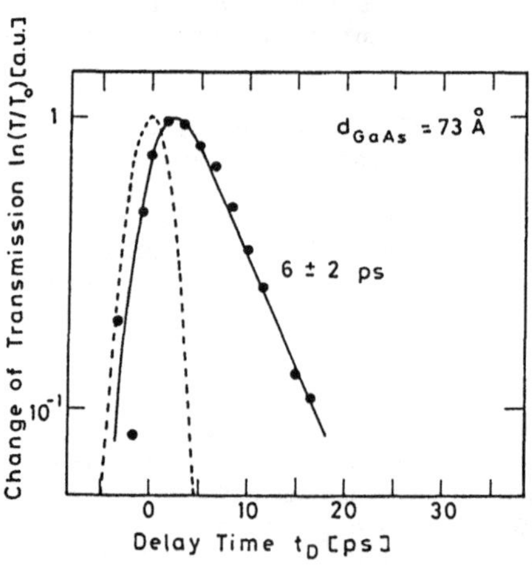

Fig. 2 Bleaching of the probe pulse as a function of the delay time. T_0 is the transmission without excitation.

Fig.2 shows time-resolved data measured on a 73Å sample. The frequency of our infrared pulses is tuned to the center of the absorption line at $\tilde{\nu}$=920cm^{-1} The pump pulse transfers a considerable amount of electrons to the upper subband. The absorption saturates and the transmission rises. The transmission signal in Fig.2 is a direct measure of the number of excited electrons. From the experimental data we deduce a time constant of 6 ps for the absorption recovery. The broken curve represents the autocorrelation of the infrared pulses. Experimental results for other well thicknesses are summarized in Table 1. In thinner quantum wells longer time constants are observed.

Table 1 Sample parameters, subband splitting, relaxation time

well thickness d [Å]	Al concentration x	subband separation ΔE [cm^{-1}]	absorption recovery τ [ps]
47±3	0.35	1300	14±2
52±2	0.35	1225	11±2
59±2	0.33	1140	8±2
59±1	0.29	1015	8±2
73±1	0.31	920	6±2

The mechanism of the intersubband relaxation is of special interest. In our experiment the excited carriers first have to be transferred with finite $k_{\parallel}$ to the ground subband. Subsequently, the carriers are cooled via intrasubband relaxation. At room temperature polar LO-phonon scattering is believed to be the most efficient relaxation mechanism in both steps.

There is strong indication that the observed time constants represent the intersubband relaxation time: (i) The theoretical treatment of polar LO-phonon interaction shows that intrasubband scattering at 300K is faster than intersubband scattering. (ii) Hot carriers in the lower subband exhibit approximately the same intersubband absorption cross-sections as carriers at 300K. Excited carriers contribute again to the absorption immediately after their transfer from the upper state to the ground subband. Intrasubband relaxation should not be observable by the bleaching technique. (iii) The experimental data in Table 1 are in favorable agreement with a relation $\tau \propto d_{GaAs}^{-2}$. An increase of the intersubband relaxation time τ with decreasing d_{GaAs} is also expected from theoretical considerations, directly reflecting the characteristic weakening of the polar interaction with increasing wave vector change /2/.

The observed intersubband time constants are surprisingly long. Theoretical calculations for infinitely deep wells give values of the order of approximately 2 ps. However, longer time constants may be expected from more detailed theories, which take into account finite barrier heights and the phonon modes of quasi-2D systems /3/. In addition, the time constants may be reduced by the fact that in our samples with d ~ 50Å the upper subband is close to the band minimum in the barrier generated by the doped layer. Interactions with barrier states result in a penetration of the wave function into the barrier. The overlap integral between the wide-spread wave function of the upper subband with the strongly confined wave function of the lower subband is reduced and longer time constants are expected.

References

1 A. Seilmeier, H.-J. Hübner, G. Abstreiter, G. Weimann, W. Schlapp: Phys. Rev. Lett. 59, 1345 (1987); A. Seilmeier, H.-J. Hübner, M. Wörner, G. Abstreiter, G. Weimann, W. Schlapp: Solid State Electron. 31, 767 (1988)

2 B.K. Ridley: In Optical Properties of Narrow-Gap Low-Dimensional Structures, Eds. C.M. Sotomayor et al. (Plenum, New York 1987) p.177

3 M. Babiker, B.K. Ridley: Superlattices and Microstructures 2, 287 (1986)

Time-Resolved Photoluminescence Spectroscopy of GaAs Quantum Wells with 1 W Picosecond Light Pulses Generated from a Visible Diode Laser

H. Yokoyama[1], *M. Fujii*[2], *M. Sugimoto*[1], *H. Iwata*[1], *K. Onabe**, *and T. Suzuki*[1]

[1]Opto-Electronics Research Labs., NEC Corporation,
4-1-1, Miyazaki, Miyamae-ku, Kawasaki 213, Japan
[2]Microelectronics Reseach Labs., NEC Corporation,
4-1-1, Miyazaki, Miyamae-ku, Kawasaki 213, Japan
*Permanent address: Department of Applied Physics,
University of Tokyo, 7-3-1 Hongo, Bunkyo-ku, Tokyo 113, Japan

Time-resolved photoluminescence (PL) measurements bring us important information about the dynamics of elementary excitations in many materials. From the view point of electronic and opto-electronic semiconductor device research, measurement in the region of subnanoseconds to several tens of nanoseconds is especially important for learning the nonradiative carrier recombination lifetime, which is a good indicator of the quality of semiconductors[1-3]. In such measurements, mode-locked gas, solid state, and dye lasers are generally used to generate excitation picosecond light pulses. To date, these lasers have been indispensable for generating visible picosecond light pulses which can efficiently excite the carriers in GaAs, or AlGaAs, although the operation of these lasers is not so easy.

We have recently demonstrated that the time-resolved PL of GaAs/AlGaAs quantum wells (QWs) are successfully measured by using 660nm picosecond light pulses (100mW peak power) generated from an AlGaInP visible diode laser[4]. In this paper, we report the results of time-resolved PL study on the nonradiative carrier recombination velocity in QWs and double-heterostructures (DHs) with 1W peak picosecond light pulses from a visible diode laser.

The AlGaInP diode laser used is a 70µm wide stripe double-heterostructure gain-guiding one grown by two step metalorganic vapor phase epitaxy[5]. As is known, short pulse current driven diode lasers can produce light pulses of ten to hundred picosecond time duration without any modelocking technique[6]. In the present experiment, the diode laser was driven with one to several nanosecond pulse current at 100kHz repetition rate. The samples measured were undoped single quantum wells (SQWs), multiple quantum wells (MQWs), and DHs grown by molecular beam epitaxy. An ultrafast streak camera (Hamamatsu Photonics, OOS-3085) and an ultrafast photon-counting system (Horiba, NAES 550) were used for time-resolved measurements of laser pulses and PLs.

Figure 1 shows the streak camera traces of red light pulses from the AlGaInP diode laser. Compensating the trigger jitter (50ps) and the streak camera resolution (10ps), actual FWHM of light pulses were respectively obtained to be 85ps and 93ps for Fig. 1(a) and (b). The light pulse energy was respectively measured as 31pJ and 150pJ for Fig. 1(a) and (b), and the peak light pulse power was estimated to be 360mW and 1.1W for each case. This rather high peak power was achieved because of the wide stripe diode laser structure. Lasing spectrum under a pulsed excitation had about 3nm half width centered at 657nm indicating multimode operation and spectral chirping. In the PL measurement, the laser beam illuminates the sample surface of $0.9\times0.3\text{mm}^2$ area.

Springer Series in Chemical Physics, Vol. 48: Ultrafast Phenomena VI
Editors: T. Yajima · K. Yoshihara · C.B. Harris · S. Shionoya

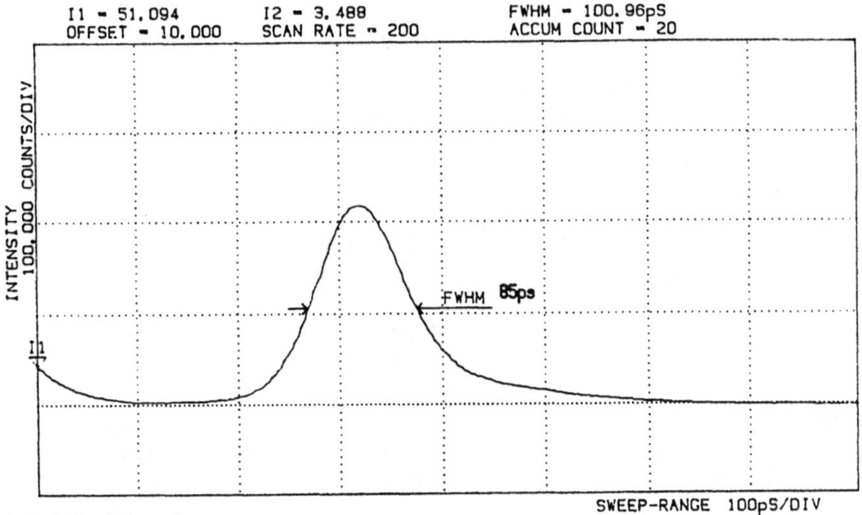

(a) 360mW peak

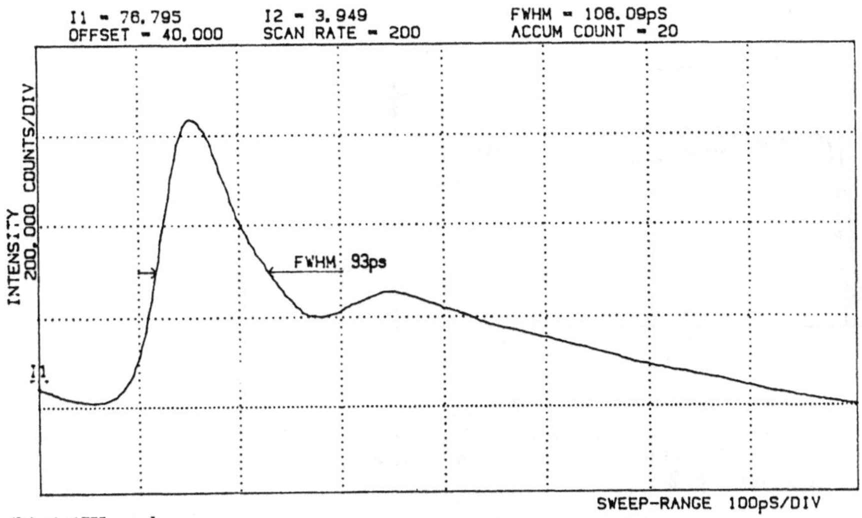

(b) 1.1W peak

Fig. 1 Streak camera traces of light pulses generated from an AlGaInP visible diode laser driven by short pulse current. Electric pulser output; (a): 1.8ns duration, 25V, (b): 2.4ns duration, 25V.

Figures 2 and 3 show the PL decay traces for QWs and DHs measured with the ultrafast photon counting system. In this measurement, central 10nm width spectral area were extracted from whole spectral envelopes. These figures prove that high S/N timeresolved PL measurement is performed with visible diode laser picosecond pulses. Although a few picojoule light pulse energy was enough for PL decay detection[4], the light pulse energy over 100pJ markedly shortened the measurement time to practical use (a few minutes per sample). Therefore, a simple, reliable, and practical PL decay measurement of GaAs semiconductors including QWs has been realized by utilizing a wide stripe AlGaInP diode laser, in place of modelocked lasers.

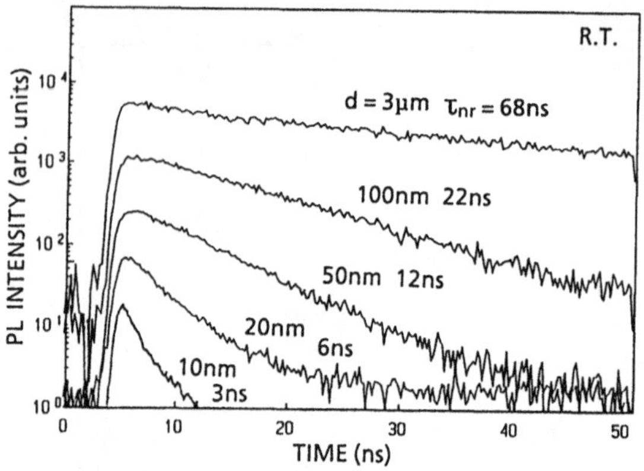

Fig. 2 PL decay traces for GaAs/$Al_{0.4}Ga_{0.6}As$ DHs with different GaAs thickness. AlGaAs thickness is 300nm. Laser pulse energy density: ~40nJ/cm^2, sampling frequency: ~8kHz, sampling time: 240s.

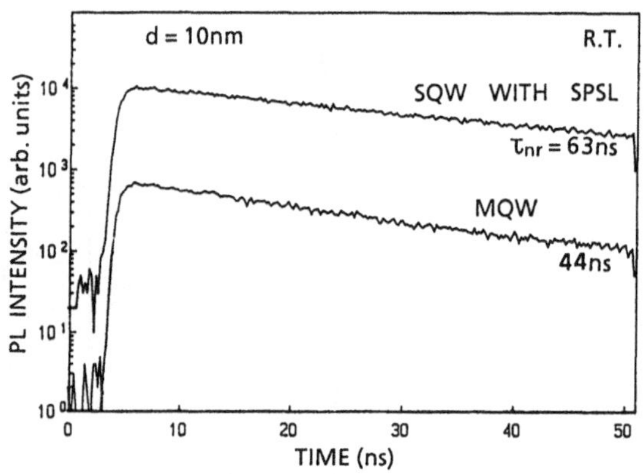

Fig. 3 PL decay traces for an SQW and an MQW (60 wells). The SQW has GaAs/AlAs SL barriers in place of AlGaAs alloy barriers. The well width is about 10nm for both samples. Measurement conditions are the same as those for Fig. 2.

Figure 2 shows the PL decay traces for a series of GaAs/$Al_{0.4}Ga_{0.6}As$ DHs with different GaAs thicknesses. In our measurement condition, the carrier recombination is dominated by the monomolecular nonradiative process in the exponential PL decay region, then we define overall nonradiative carrier recombination lifetime τ_{nr} by $\tau_{nr}=2\tau$, where τ represents the PL decay time constant. It is apparent that the decrease in GaAs thickness resulted in the τ_{nr} reduction. This result indicates that the τ_{nr} of an GaAs/AlGaAs SQW is dominated by the well/barrier interface recombination. Using a simple expression for the relation between τ_{nr} and interface recombination[1)], the interface recombination velocity and the intrinsic nonradiative lifetime of GaAs layers are respectively estimated to be ~300 cms^{-1}, and ~70ns.

On the other hand, the PL decay traces, shown in Fig. 3, for an MQW (10nm GaAs/10nm $Al_{0.3}Ga_{0.7}As \times 60$periods) and an SQW with short-period superlattice (SPSL) barrier layers (10nm GaAs, 1nm AlAs/2nmGaAs×33periods for each barrier) have decay times which are much longer than that for the SQW with AlGaAs alloy barriers. This clearly shows that the interface recombination velocity of QWs, from which we observed the PL, were markedly reduced. This result may be because of residual impurity trapping by many heterointerfaces involved in SPSLs or an MQW structure.

References

1) P. Dawson and K. Woodbridge, Appl. Phys. Lett. 45, 1227 (1984).
2) J. E. Fouquet and R. D. Burnham, IEEE J. Quantum Electron. QE-22, 1799 (1986).
3) H. Yokoyama, H. Iwata, M. Sugimoto, K. Onabe, and R. Lang J. Appl. Phys. 63, 4755 (1988).
4) H. Yokoyama, H. Iwata, K. Onabe, and T. Suzuki, Rev. Sci. Instrum. 59, 663 (1988).
5) K. Kobayashi, S. Kawata, A. Gomyo, I. Hino, and T. Suzuki, Electron Let. 21, 931 (1985).
6) T. Kobayashi, A. Yoshikawa, A. Morimoto, Y. Aoki, and T. Sueta, Proceedings of XIth Int. Quantum Electron. Conf., Boston, U.S.A., W1, 1980.
7) H. Yokoyama, H. Iwata, M. Sugimoto, K. Onabe, and R. Lang, Proceedings of XVIth Int. Quantum Electron. Conf., Tokyo, Japan, TUG2, 1988.

Light-Induced Selection Rules in Semiconductors Using Ultrashort Pulses

M. Joffre[1,2], *D. Hulin*[1,2], *A. Migus*[1], *A. Antonetti*[1], *and M. Combescot*[2]

[1]Laboratoire d'Optique Appliquée, Ecole Polytechnique, ENSTA, F-91120 Palaiseau, France

[2]Groupe de Physique des Solides de l'Ecole Normale Supérieure, Université Paris VII, 2 place Jussieu, F-75005 Paris, France

It has been recently shown[1,2,3] that the interaction of semiconductors with light is not only restricted to the creation of real excitations (electrons, holes, excitons...) but that photons with energies smaller than any absorption edge (therefore not absorbed in the semiconductor) perturb the medium's optical properties during a time limited by the light pulse duration. This so-called optical Stark effect shows up as a modification of the optical absorption spectrum, mainly a shift towards higher energy of the different absorption lines or bands. The magnitude of this shift is proportional to the light intensity and, for a given transition, inversely proportional to the detuning between the photon energy and the absorption line. Such an effect had already been reported in atomic vapors but in order to be noticeable in semiconductors, it requires strong light intensities, which can only be provided through the use of ultrashort and intense pulses.

Beyond its fundamental interest[4,5] and the potential applications[6,7,8], the optical Stark effect presents a very interesting property. Since the coupling between the light and an energy level is proportional to the square of the optical matrix element, the amplitude of the shift will differ for the various absorption lines involved. This situation turns out to be particularly fruitful in the case of degenerate transitions since it allows the separation of the different contributions. To split a degeneracy, one usually uses magnetic or electric fields, uniaxial stress, here this is the first case in semiconductors where light is used to provide such an effect.

M. and R. Combescot[5] have recently shown that the exciton shifts can be calculated without the Coulomb interaction when the detuning is large compared with the exciton binding energy. In these conditions the detuning is almost the same for all transitions (towards bound and unbound electron-hole states) and one expects rigid shifts of the valence and conduction bands. This remark is quite useful as it avoids many-body treatments and allows one to consider simply transitions from one free valence state to one free conduction state. The situation is then similar to the two-level system of atomic physics, except that we now have a multi-level atom. The conduction band being spin degenerate and the valence band resulting from a p-like spin orbit coupling in III-V compounds, one expects to find[9] at large detuning the shifts of a (2 + 2x3) i.e. 8-level dressed atom.

In the case of GaAs MQWS both the heavy and light hole excitons are fourfold degenerate. The laser beam splits each exciton into four different levels. However, when the beams propagate along the growth axis, only two excitons instead of four can be created by probe photons, since only σ transitions are allowed. These two excitons are shifted by

Springer Series in Chemical Physics, Vol. 48: **Ultrafast Phenomena VI**
Editors: T. Yajima · K. Yoshihara · C.B. Harris · S. Shionoya

$$\Delta E_{H\pm} = 2\delta_{H\pm} + \delta_{L\pm} \tag{1}$$

for the heavy-hole transitions and

$$\Delta E_{L\pm} = 2\delta_{L\pm} + \delta_{H\pm} \tag{2}$$

for the light-hole transitions. In these equations we have used

$$\delta_{H\pm} = \frac{\lambda_{\mp}^2}{\Omega_H} \tag{3}$$

$$\delta_{L\pm} = \frac{\lambda_{\pm}^2}{3\Omega_L}. \tag{4}$$

Ω_H and Ω_L are respectively the detuning of the pump beam to the heavy and light hole resonances and $\lambda_{\pm}$ is related to $\sigma\pm$ field amplitudes. These equations take into account the fact that both heavy- and light-hole transitions use the same final state, namely the conduction band. In the particular case of linearly polarized light, the two shifted exciton transition energies reduce to only one as $\lambda_{+}^2 = \lambda_{-}^2$.

The experiments have been performed in GaAs-GaAlAs MQWS. Short tunable pulses[10] have been used to get sufficiently high intensity. The pump beam is an amplified part of the wavelength continuum generated in water by 60 fs pulses, leading to intensities as high as $10 GW/cm^2$. Pump and probe are both propagating along the growth axis. The pump is polarized either linearly or circularly while the probe polarization is analyzed after the sample. The intensity of the pump is kept as low as possible to stay in the small signal(perturbative) regime.

Figure 1 exhibits the experimental results obtained in MQWS for circular polarizations. Because of the small value of the shift at relatively low pump intensity, it is more instructive to show the differential transmittance spectra. In this very sensitive method the magnitude of the signal is proportional to the line shift and a form factor deduced from the unperturbed lineshape. In Figure 2 we present the

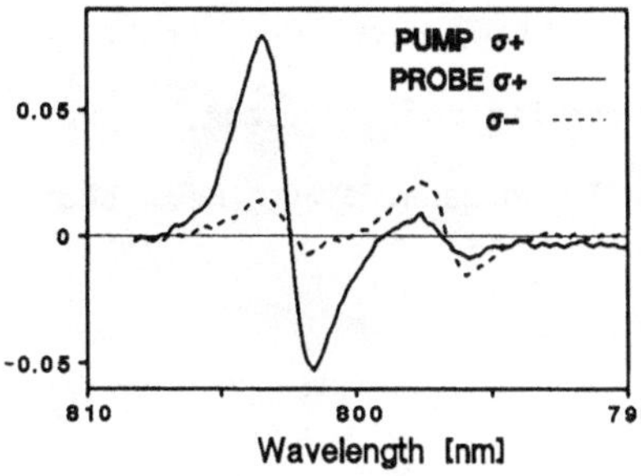

Fig. 1: Differential absorption spectra recorded at the maximum of the $\sigma+$ pump pulse using a $\sigma+$ (solid line) or a $\sigma-$ (dashed line) probe pulse

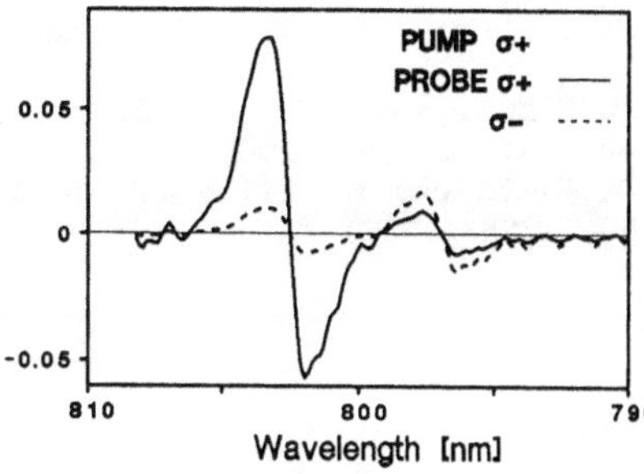

Fig. 2: Theoretical spectra for a $\sigma+$ (solid line) or $\sigma+$ (dashed line) probe polarization. These "theoretical" curves are noisy since they are deduced from the derivative of the experimental absorption spectrum.

corresponding theoretical curves. These curves are obtained by multiplying the derivative of the experimental curves by the calculated shifts. As can be seen the agreement is good. The only difference comes from the associated reduction of absorption during the optical Stark shift in the experimental curve and also from the band-mixing which was not accounted for in the theory.

As we stated above, the two exciton shifts are identical in the case of linear pump polarization. Consequently there can be only one measured shift, whatever the probe polarization is. This fact has been previously observed[1]. The question of the non-dependence in MQWS of the relative polarizations of the probe and pump beams (linear in plane polarizations) has been raised recently by Fröhlich et al [11]. They argue that one should observe an effect only if the probed component of the exciton is the same as the one which has been perturbed by the pump (parallel polarizations). They suggest that the existence of a very fast relaxation process between the orthogonal exciton components can explain this non-dependence of the relative polarizations. We have shown that within the framework of our theory the experimental result can be explained without any relaxation between the virtually excited states.

Part of this work is supported by Direction des Recherches et Etudes Techniques (grant 86/089). Groupe de Physique des Solides is Laboratoire Associé CNRS LA17.

REFERENCES

1. A. Mysyrowicz, D. Hulin, A. Antonetti, A. Migus, W.T. Masselink and H.M. Morkoç, Phys. Rev. Lett. 56, 2748 (1986)
2. A. von Lehmen, D.S. Chemla, J.E. Zucker and J.P. Heritage, Opt. Lett. 11, 609 (1986)
3. K. Tai, J. Hegarty and W.T. Tsang, Appl. Phys. Lett. 51, 152 (1987)
4. S. Schmitt-Rink and D.S. Chemla, Phys. Rev. Lett. 57, 2752 (1986)
5. M. Combescot and R. Combescot, Phys. Rev. Lett. 61, 117 (1988)
6. M. Combescot, Solid. State Comm. (in press)
7. D. Hulin, A. Mysyrowicz, A. Antonetti, A. Migus, W.T. Masselink, H.M. Morkoç, H.M. Gibbs and N. Peyghambarian, Appl. Phys. Lett. 49, 749 (1986)
8. M. Yamanishi, Phys. Rev. Lett. 59, 1014 (1987)
9. D.S. Chemla, D.A.B. Miller and S. Schmitt-Rink, Phys. Rev. Lett. 59, 1018 (1987)
10. A. Migus, A. Antonetti, J. Etchepare, D. Hulin and A. Orszag, J. Opt. Soc. Am. 2, 584 (1985)
11. D. Fröhlich, R. Wille, W. Schlapp and G. Weimann, Phys. Rev. Lett. 59, 1748 (1987)

Simultaneous Virtual and Two-Photon Femtosecond Excitations in GaAs MQWS

W.H. Knox [1], *J.B. Stark* [2], *D.S. Chemla* [1], *D.A.B. Miller* [1], *and S. Schmitt-Rink* [3]

[1]AT&T Bell Laboratories, Crawfords Corner Road, Holmdel, NJ 07733, USA

[2]Massachusetts Institute of Technology, Cambridge, MA 02139, USA

[3]AT&T Bell Laboratories, Murray Hill, NJ 07974, USA

When intense excitation is applied to a semiconductor at energies below the band gap, AC Stark shifts of the band edge states are obtained. This effect, which is well-known for the case of atomic systems, has been recently observed in semiconductors [1,2]. At large detunings which are required for ultrashort (less than 100 fs) excitation, two-photon carrier generation becomes more important relative to the AC Stark shift, because the AC Stark shift decreases inversely in proportion to the detuning. We study the effects of two-photon carrier generation under the conditions of intense virtual excitation and we find that two-photon absorption is strong and unavoidable and is a natural consequence of excitation below the band edge with ultrashort optical pulses. We also find that at very low pump and probe intensities the exciton shifts with no loss of exciton oscillator strength, a result which is different than the high-intensity regime.

We use a newly developed laser system which generates 100 fs optical pulses of microjoule energies at 8 kHz repetition rate with center at 805 nm [3]. The quantum wells are 7.5 nm thick, yielding an exciton absorption line at 780 nm. Since the continuum center is 805 nm, we now have available extremely intense (up to 1000 GW/cm^2) continuously tunable excitation pulses below the band gap of GaAs quantum wells. Kilohertz repetition rates allow us to use differential detection techniques for small signal analysis.

In Figure 1 we show the absorption spectrum of the sample at 35K measured with 100 fs continuum pulses of intensity 10 kW/cm^2. The derivative of the absorption spectrum with respect to frequency is compared with the measured differential spectrum at zero time delay. In the limit of a pure shift, the measured differential spectrum should have the same lineshape as the derivative of the linear absorption. We find that this is indeed the case for the small signal regime where the transmission change is only a few percent, obtained at about 30 MW/cm^2. Recent numerical calculations [5,6] based on the theory of Ref. [4] show that the low-density regime in semiconductors is qualitatively different from the atomic system: there is essentially no loss of oscillator strength with shift. Theoretically, this can only be explained if the shift of the band edge states is larger than the exciton shift and compensates the exciton oscillator strength loss due to phase space filling [6]. Experiments with weak pump but at higher probe intensities indicate a departure from this pure shift regime, presumably because of an interaction between the real excitons created by the probe beam (and the virtual excitons created by the pump beam).

At higher pump intensities we study two-photon carrier generation by monitoring the photocurrent in a PIN structure (Figure 2). A quadratic dependence is obtained. We estimate, using the known two-photon absorption coefficient of GaAs [7], that at 10 GW/cm^2 we expect a photocurrent of 6.7 microamps, about 10 times higher than we measure. The photoconduction quantum efficiency in perpendicular transport at low temperature in quantum wells is difficult to estimate, however, our estimates of the expected two-photon photocurrent are in reasonable agreement with our measured values.

Springer Series in Chemical Physics, Vol. 48: **Ultrafast Phenomena VI**
Editors: T. Yajima · K. Yoshihara · C.B. Harris · S. Shionoya

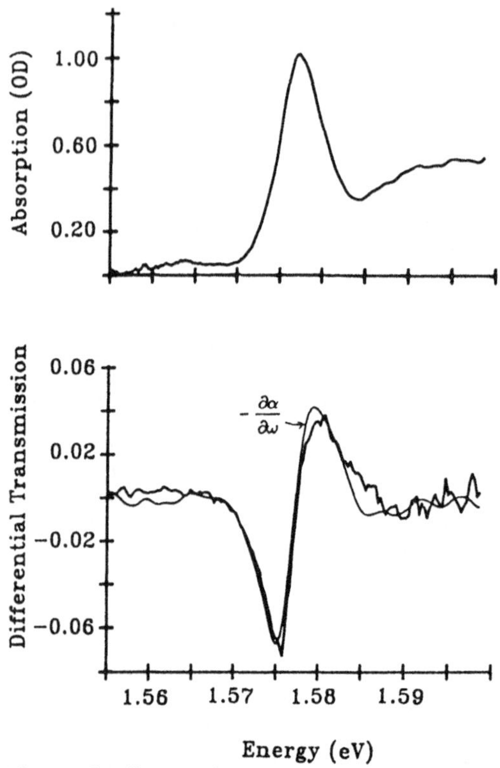

Figure 1. Absorption spectrum and derivative of absorption compared to measured differential transmission at delay t=0.

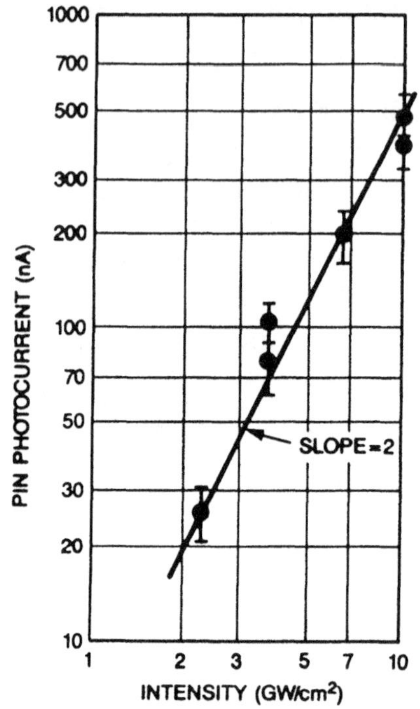

Figure 2. Photocurrent obtained in DC-biased PIN structure at 10 volts DC bias. A quadratic behavior indicates two-photon carrier generation.

When carriers are excited by two-photon absorption in GaAs in our case they are excited at 3.0 eV, or 1.5 eV above the bandgap. These carriers instantaneously create a weak perturbation of the excitonic absorption through a weak screening and three-photon processes. This effect prevents the full recovery of the excitonic absorption and becomes stronger in time as the carriers cascade down to the bandedge by phonon emission and intervalley scattering. Of course, after the carriers recombine in a few ns the excitonic absorption fully recovers. Figure 3 shows the full time course of the differential absorption with 100 fs excitation pulses at 80 meV below the exciton energy at a pump intensity of a few GW/cm^2. This plot is obtained by overlaying a grid on top of our measured differential spectra. We observe a rapid initial transient at both the heavy-hole and light-hole excitons with an instrument-limited response. This transient in the high-density regime contains excitonic shift and bleaching contributions. After the rapid transient the signals do not recover to zero in the high-density regime. The heavy-hole exciton differential signal at +264 fs indicates line broadening with precise area conservation. The relative transmission change at the heavy-hole exciton at t=0 is about 50% for this case.

We note that the oscillations in the spectra at negative time delays are clearly observed in this data. These are due to the interaction of the pump pulse with the polarization which is induced by the probe field [8].

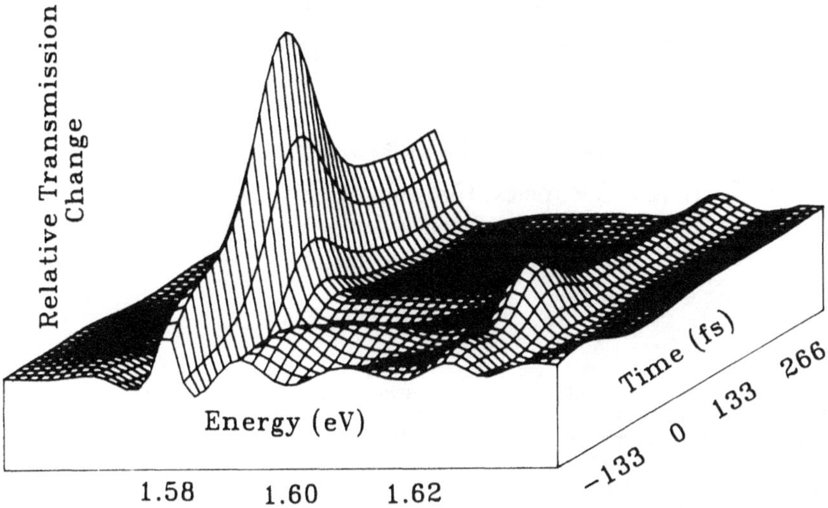

Figure 3. Measured differential transmission spectra for high pump intensity conditions at large detunings. The long time transient due to two-photon carrier generation is clearly observed.

In conclusion, we show that two-photon carrier generation is important when making intense virtual excitations in the large detuning limit. This causes power dissipation and long (ns) lifetime transients and is undesirable in general. Use of a reverse-biased PIN structure provides an independent measure of the two-photon generated photocarrier density. In the small detuning limit, large shifts of the excitons can be obtained at lower pump powers, so that two-photon effects are weaker, however, very short optical pulses (less than 100 fs) because of their large bandwidth inherently violate the small detuning limit. Virtual excitation and two-photon carrier excitation effects are connected in a fundamental way and a full theoretical analysis of their inter-relation is required.

[1] A. Mysyrowicz, D. Hulin, A. Antonetti, A. Migus, W.T. Masselink and H. Morkoc, Phys. Rev. Lett. 56, 2748 (1986).
[2] A. von Lehmen, D.S. Chemla, J.E. Zucker and J.P. Heritage, Opt. Lett. 11, 609 (1986).
[3] W.H. Knox, JOSA B4 1771 (1987).
[4] S.Schmitt-Rink and D.S. Chemla, Phys. Rev. Lett. 57, 2752 (1986). S. Schmitt-Rink, D.S. Chemla and H. Haug, Phys. Rev. B37, 941 (1988).
[5] W. Schaefer, unpublished.
[6] C. Ell, J.F. Muller, K. El Sayed and H. Haug, unpublished.
[7] B.S. Wherrett, JOSA B1, 67 (1984).
[8] B. Fluegel, N. Peyghambarian, G. Olbright, M. Lindberg, S.W. Koch, M. Joffre, D. Hulin, A. Migus and A. Antonetti, Phys. Rev. Lett. 59, 2588 (1987); C.H. Brito-Cruz, J.P. Gordon, P.C. Becker, R.L. Fork and C.V. Shank, IEEE JQE 24, 261 (1988)

Ultrafast Control of Quantum Interference Currents by Virtual Charge Polarizations in Biased Quantum Well Structures

M. Yamanishi[1], M. Kurosaki[1], Y. Osaka[1], and S. Datta[2]

[1]Department of Physical Electronics, Hiroshima University, Saijyocho, Higashi-Hiroshima 724, Japan
[2]School of Electrical Engineering, Purdue University, W. Lafayette, IN47907, USA

1. Introduction

Recently, ultrafast optical phenomena based on virtual pair excitations in semiconductor quantum well (QW) structures have been attracting a great deal of attention. For instance, ultrafast optical nonlinearity, so called optical Stark effect, in GaAlAs QW structures has been experimentally demonstrated [1][2]. A new modulation scheme of optical properties by virtual charge polarization in DC biased QW structures has been theoretically proposed [3],[4]. In both cases, virtual pair excitation by an off-resonant pump light is essential to realize the ultrafast switches of the optical properties of the structures. In addition, in the latter case, one may expect a generation of ultrafast voltage pulse resulting from the virtual charge polarization caused by an asymmetry of the potential profiles in the biased QWs. In this paper, the dynamics of the virtual population for a short pump light will be clarified within a frame work of density matrix formalism. A possibility of ultrafast control of quantum interference currents through the voltage modulations by the virtual charge polarizations will be discussed, based on the result of the transient response of the virtual excitations.

2. Dynamics of virtual excitations

There exist many quantum states involving excitons and subband states in actual QW structures. In order to avoid inessential complications, we postulated a simple two level system interacting with a classical radiation field, $E_p(t)\cos(\omega_p t)$. The assumption would be justified if the detuning energy of the pump light with respect to the lowest excitonic gap, $\hbar\Delta_{1e\text{-}1hh}$ is smaller than or comparable to the binding energy of the exciton. Even if the detuning energy is larger than the binding energy, the assumption may be still meaningful because the virtual excitation regarding the lowest exciton associated with the smallest detuning energy may be a bottle neck of the response of the overall excitations.

For the two level system, density matrix formalism may result in the following normalized relations, under the rotating wave approximation, ignoring the T_1-relaxation,

$$\left.\begin{aligned} &\frac{d(\sigma_{21})}{d(|\Delta|\cdot t)} = (\pm i - 1/(|\Delta|\cdot T_2))\sigma_{21} + i\,\frac{\Omega(|\Delta|\cdot t)}{|\Delta|}(1-2\rho_{22}) \\ &\frac{d\rho_{22}}{d(|\Delta|\cdot t)} = 2\,\frac{\Omega(|\Delta|\cdot t)}{|\Delta|}\,\mathrm{Im}(\sigma_{21}) \\ &\rho_{21} = \sigma_{21}\,e^{-i\omega_p t} \end{aligned}\right\} \quad (1)$$

where $\Omega(|\Delta|\cdot t) = \mu E_p(t)/2\hbar$, $\Delta = \omega_p - (\omega_2 - \omega_1)$ and T_2 are the Rabi frequency, the detuning frequency and the dephasing time, respectively. Figure 1 shows the transient responses of the upper level population ρ_{22} for pump pulses described by Gaussian

Springer Series in Chemical Physics, Vol. 48: **Ultrafast Phenomena VI**
Editors: T. Yajima · K. Yoshihara · C.B. Harris · S. Shionoya

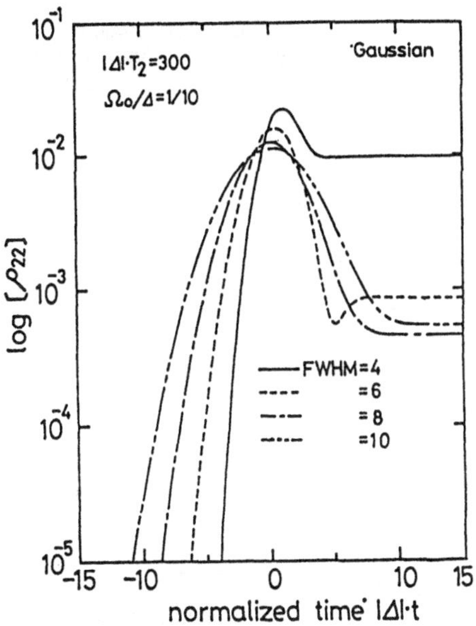

Fig.1 Estimated responses of carrier population ρ_{22} for Gaussian pump pulses.

functions, $\Omega(|\Delta|\cdot t) = \Omega_0 \exp(-|\Delta|^2 t^2/\delta^2)$ with full width at half maximum (FWHM), 4~10 and a normalized peak value $\Omega_0/|\Delta|$ of 1/10, for a fixed normarized dephasing time, $|\Delta|\cdot T_2 = 300$.

As a result of the numerical estimations, a criterion for quasi pure virtual excitations is given as follows, $6/|\Delta| \le \text{FWHM} \le T_2/10$. For detuning energies, $\hbar\,\Delta$ =5meV and 30meV, the unity normalized time, $|\Delta|\cdot t = 1$ gives actual times 125f·s and 21f·s, respectively. Therefore, the virtual population can almost follow up the pump pulse with a FWHM of 750f·s (126f·s) and a detuning energy of 5meV (30meV) if the dephasing times are ten times longer than the FWHMs. With experimental data on the dephasing time in GaAs QWs at very low temperature ~10K [5] and absorption tail at 80K [6], it can be confirmed that such a large value of the dephasing time is realizable for a detuning energy of 5meV (30meV) at 10K (80K), Particularly, it should be noted that the dephasing time tends to increase even more with the increasing detuning energy because of the non-Markovian nature of the relaxation processes.

3. Control of quantum interference currents

In addition to proposed optical nonlinearity, an extremely short voltage pulse would be generated by the virtual charge polarization in the biased QW structure [3],[4]. In the device structure shown in Fig.2 the quantum interference current [7] flowing in the channels A and B could be controlled through the modulation of voltage drop V_{AB} between the channels due to virtual charge polarization in the multiple quantum well (MQW) structure inside the channels. The thickness L_{ZQ} of the channels should be smaller than that L_{ZC} of the MQW to realize selective virtual populations in the MQW. The screening field E_s may induce a voltage pulse between the channels A and B in Fig.2, given by

$$\Delta V_{AB} = (N \cdot E_s \cdot L_{ZC})$$

where N is the period of the multiple quantum wells. In general, the modulation depth ΔV_{AB} of gate voltage required for 100% modulation of quantum interference currents is fairly small [7], for instance,

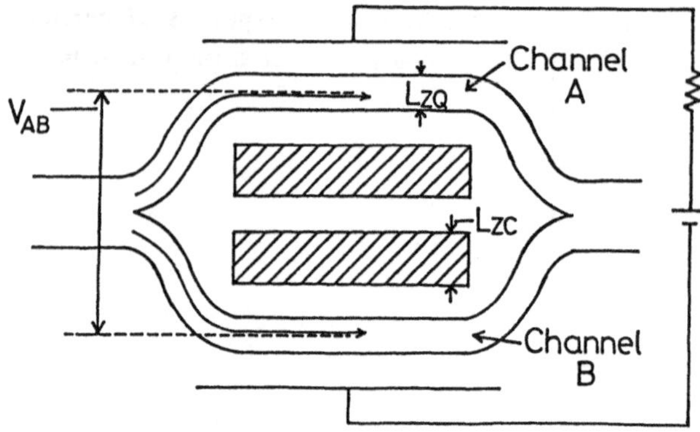

Fig.2 Cross sectional view of the proposed device for the ultrafast control of quantum interference currents.

$$\Delta V = \frac{h}{2e}\left(\frac{V}{L}\right) = \frac{h}{2e}\left(\frac{1}{\tau_t}\right) \sim 2mV$$

for a channel length, L=2000Å and an average electron velocity (Fermi-velocity), v=1.86x10^{7}cm/sec, i.e., a transit time, τ_t ~1.1psec. Such a small modulation of the voltage would be realized by a small pump power density, ~3MW/cm^2 and a detuning energy of 5meV in the device with 10 period graded gap quantum wells without external bias voltages. The internal voltage modulation without bias voltages [3] is extremely important because the modulation depth of the quantum interference currents becomes shallow for a large bias gate voltages [7]. Also, it is extremely important that the switch of the internal voltage due to virtual charge polarizations is expected to be free from C.R time constant, resulting in a fairly short switching time, ~1psec and, consequently, a small power delay product of the device, 30 femtojoule. Many devices stacked on a single chip can be simultaneously driven by a single pump pulse because the virtual process is at least in principle loss free. This is an important advantage of this kind of modulation scheme based on virtual excitations over those based on real ones.

This work was partially supported by a Scientific Research Grant-In-Aid for Specially Promoted Research (Project No. 61065006) from the Ministry of Education, Science and Culture of Japan.

References

1. A.Mysyrowicz, D.Hulin, A.Antonetti, A.Migus, T.Masselink, H.Horkoc, Phys. Rev. Lett. **56**, 2748 (1986)
2. A.Von Lehmen, D.S.Chemla, J.E.Zucker, J.P.Heritage, Optics Lett. **11**, 609 (1986)
3. M.Yamanishi, Phys. Rev. Lett. **59**, 1014 (1987)
4. D.S.Chemla, D.A.B.Miller, S.Schmitt-Rink, Phys. Rev. Lett. **59**, 1018 (1987)
5. L.Schultheis, A.Honold, J.Kuhl, K.Kohler and C.W.Tu, Phys.Rev. **B-34**, 9027 (1986)
6. A.Von Lehmen, J.E.Zucker, J.P.Heritage and D.S.Chemla, Phys.Rev. **B-35**, 6479 (1987)
7. S.Datta, M.R.Melloch, S.Bandyopadhyay and M.S.Lundstrom, Appl.Phys.Lett. **48**, 487 (1986)

Ultrafast Inter-Subband Relaxation of Photoexcited Carriers in Semiconductor Quantum Dots

T. Takagahara

NTT Electrical Communications Laboratories,
Musashino-shi, Tokyo 180, Japan

1. Introduction

Ultrafast inter-subband relaxation processes of photoexcited carriers in semiconductor quantum dots are now attracting much attention from the viewpoint of fundamental physics and also from the interest of applied physics. Recently, ultrafast absorption recovery was observed in semiconductor-doped glasses employing femtosecond pump-probe techniques[1,2]. Along with this, the degenerate four-wave mixing in semiconductor-doped glasses was measured and the ultrafast dephasing time was estimated[3]. These measurements yielded a relaxation time of about 200 femtoseconds. The most important mechanism of this ultrafast relaxation is considered as phonon-mediated inter-subband transition. So in this paper, we will formulate the electron-phonon interactions in semiconductor quantum dots and estimate the inter-subband relaxation rate. First of all, it is to be noted that in semiconductor microcrystallites, the translational symmetry is totally lost due to three-dimensional confinement and thus the concept of wavevector is not applicable for both electrons and phonons. Instead the electronic states become discrete due to the quantum confinement effect and form a series of subband states. Similarly the phonon spectra become discrete due to the boundary condition at the microcrystallite surface.

2. Electron-Phonon Interactions in Semiconductor Microcrystallites

In this section the electron-phonon interactions in semiconductor microcrystallites will be derived. First we will discuss the optical phonon mode. The optical phonon modes in a finite crystal can be classified into longitudinal , transverse and surface modes[4]. As for the longitudinal optical mode in a spherical semiconductor microcrystallite, the electron-phonon coupling occurs through the longitudinal depolarization field and the electron-LO phonon interaction Hamiltonian is given as

$$H_{LO} = \sqrt{\frac{4\pi e^2 \hbar\omega_{LO}}{R}\left(\frac{1}{\epsilon_\infty} - \frac{1}{\epsilon_0}\right)} \sum_{l,m,j} N_{lmj} Y_{lm}(\theta_e, \varphi_e) j_l(k_{lj} r_e / R), \qquad (1)$$

where R is the radius of a quantum dot,$\epsilon_0(\epsilon_\infty)$ the static(high frequency) dielectric constant, j_l and Y_{lm} are a spherical Bessel function and a spherical harmonics, respectively, k_{lj}'s are zeros of the j_l function and N_{lmj} is the normalization constant. The coupling strength is inversely proportional to the square root of the particle radius. This implies that the confinement of electrons and phonons in a small crystallite enhances their interaction.

The electron-phonon interaction associated with the surface optical phonon mode will be discussed. The amplitude of surface phonon mode is in general localized in the vicinity of the crystal surface and on the contrary the electronic wavefunction has a small amplitude at the crystal surface. Thus the electron-surface phonon coupling is considered to be small. However, the lowest-order surface optical phonon mode which is usually called Fröhlich mode[5] has a uniform lattice polarization in a spherical crystallite and the coupling with electrons is expected to be large. In this case also, the electron-phonon coupling occurs through the longitudinal depolarization field and the electron-surface phonon interaction Hamiltonian is given as

Springer Series in Chemical Physics, Vol. 48: **Ultrafast Phenomena VI**
Editors: T. Yajima • K. Yoshihara • C.B. Harris • S. Shionoya

$$H_{SF} = \sqrt{\frac{2\pi\hbar e^2\omega_{LO}^2}{9R^3\omega_F}\left(\frac{1}{\epsilon_\infty} - \frac{1}{\epsilon_0}\right)} \cdot r_e Y_{1m}(\theta_e, \varphi_e), \tag{2}$$

where m is ±1 or 0. The coupling strength is inversely proportional to the square root of the quantum dot radius R.

As for the acoustic phonon modes in a semiconductor microcrystallite, the mode frequencies and mode eigenfunctions are determined by solving the elastic wave equation with an appropriate boundary condition[6]. As one of these acoustic modes, we will consider the radial vibration modes. These modes are longitudinal acoustic modes and couple with electrons through the deformation potential. The explicit form of the electron-LA phonon interaction Hamiltonian is given as

$$H_{DF} = E_d\sqrt{\frac{\hbar}{4\pi\rho\omega_{LA}R^3}}\frac{hj_0(hr_e)}{\sqrt{j_1^2(hR) - j_0(hR)j_2(hR)}}, \tag{3}$$

where E_d is the deformation potential and h is a constant determined by the boundary condition and is proportional to R^{-1}. Then the coupling strength is inversely proportional to $R^{5/2}$. As typical transverse acoustic modes, we can take up rotatory vibration modes. These acoustic modes have no dilatation and thus do not have deformation potential coupling. However, the piezoelectric coupling with electrons through the strain field is possible. The Hamiltonian of piezoelectric coupling between electrons and TA phonons is given as

$$\begin{aligned} H_{PZ} = &\frac{2\pi e}{\epsilon_0 R^{3/2}}\sqrt{\frac{3\hbar}{8\pi\rho\omega_{TA}}}\frac{1}{\sqrt{j_1^2(\kappa R) - j_0(\kappa R)j_2(\kappa R)}} \\ &\left[\frac{2}{5}\sqrt{\frac{2\pi}{3}}(3e_{15} + e_{33} - e_{31})\Big(j_1(\kappa r_e) - \frac{r_e}{R}j_1(\kappa R)\Big)(Y_{11}(\theta_e,\varphi_e) + Y_{1-1}(\theta_e,\varphi_e))\right. \\ &\left. + \frac{8}{15}\sqrt{\frac{3\pi}{7}}(2e_{15} - e_{33} + e_{31})j_3(\kappa r_e)(Y_{31}(\theta_e,\varphi_e) + Y_{3-1}(\theta_e,\varphi_e))\right], \end{aligned} \tag{4}$$

where e_{15}, e_{33} and e_{31} are the piezoelectric constants and κ is a constant determined by the boundary condition. The coupling strength is inversely proportional to $R^{3/2}$.

We have formulated three types of the electron-phonon coupling in semiconductor microcrystallites. The dependences of the coupling strength on the quantum dot radius are different for the three coupling schemes. The Fröhlich coupling with LO phonons or Fröhlich surface modes has a matrix element of the order of a few meV for a quantum dot radius of $100\mathring{A}$. On the other hand, the matrix elements of the deformation potential coupling and the piezoelectric coupling are two orders of magnitude smaller than that of the Fröhlich coupling for the particle radius of $100\mathring{A}$. However, the dependence on the particle size is stronger for the acoustic phonon coupling. Thus for smaller size particles, the matrix elements of the acoustic phonon coupling can be comparable to or larger than those of the optical phonon coupling.

3. Inter-Subband Relaxation Rate

On the basis of these electron-phonon interaction Hamiltonians, we can estimate the inter-subband relaxation rate. The inter-subband relaxation can occur between the energy levels with the same angular momenta and also between the energy levels with different angular momenta. The inter-subband transition matrix elements are not very strongly dependent on the angular momenta of energy levels and also on the subband indices. For a one-phonon emission process in which the inter-subband energy happens to be equal to the LO phonon energy, the relaxation time is estimated as follows. For CdS, using the parameters $\omega_{LO} = 305cm^{-1}$, $\omega_{TO} = 240cm^{-1}$ and $\epsilon_\infty = 5.3$, we obtain

$1/\Gamma_{ph} = 66.2 \cdot \sqrt{R}(fs)$, where R is measured in unit of $100\mathring{A}$. For CdSe, with the parameters $\omega_{LO} = 212cm^{-1}$, $\omega_{TO} = 172cm^{-1}$ and $\epsilon_\infty = 6.1$, a little longer relaxation time is obtained as $1/\Gamma_{ph} = 88.9 \cdot \sqrt{R}(fs)$. However, these values are a few times as short as the observed relaxation time. The exact matching between the inter-subband energy and integer times the LO phonon energy is rather fortuitous. In general there is a mismatch between the two and this energy mismatch is compensated for by acoustic phonons. Participation of acoustic phonons reduces the inter-subband relaxation rate drastically as can be seen from the previous argument on the magnitude of matrix elements. In fact, for a two-phonon process involving one LO phonon and one LA or TA phonon, we found that the relaxation rate is inversely proportional to R^6. It is estimated that for $R = 50\mathring{A}$, $1/\Gamma_{ph} \cong 2.3ps$ and for $R = 33\mathring{A}$, $1/\Gamma_{ph} \cong 200fs$. This result may suggest that the radius of microcrystallites contained in semiconductor-doped glasses is about $30\mathring{A}$.

4. Discussion

On the basis of these calculations, we can describe the relaxation processes of photoexcited carriers in a semiconductor microcrystallite. When the excess energy of carriers is more than a few times the LO phonon energy, the relaxation occurs mainly through LO phonon emission. As the carrier energy decreases, the acoustic phonons begin to participate in the relaxation process. Furthermore, when the excess energy is less than one LO phonon energy, the carrier relaxation occurs through emission of acoustic phonons alone and the relaxation rate slows down drastically. In this region the radiative recombination and the trapping into shallow traps come into play as another relaxation channels in competition with the acoustic phonon emission. The radiative recombination lifetime is calculated from the oscillator strength of the excitonic transition[7]. The radiative lifetime decreases with increase in the particle size. When the particle radius is assumed to be about $30\mathring{A}$ which yielded a reasonable value of about 200 femtoseconds for the inter-subband relaxation time, the radiative lifetime is estimated to be about 100 picoseconds. This value is also in reasonable agreement with the observed decay time.

In summary, we have developed a theory of electron-phonon coupling in a semiconductor microcrystallite and clarified the dependence of the coupling strength on the particle radius for the first time. On the basis of this theory, we have estimated the inter-subband relaxation time and obtained a reasonable agreement with the experimental values.

References

1. M. C. Nuss, W. Zinth and W. Kaiser, Appl. Phys. Lett. **49**, 1717(1986).
2. N. Peyghambarian, G.R. Olbright and B.D. Fluegel, Proceedings of XIVth IQEC, San Francisco, 1986, PD20.
3. T. Tokizaki, Y. Ishida and T. Yajima, this volume.
4. R. Engleman and R. Ruppin, J.Phys. C **1**, 614(1968).
5. H. Fröhlich, Theory of Dielectrics(Clarendon, Oxford, 1949).
6. A.E.H. Love, A Treatise on the Mathematical Theory of Elasticity, Fourth Edition(Cambridge University Press, 1959).
7. T. Takagahara, Phys. Rev. B **36**, 9293(1987).

Ultrafast Optical Nonlinearity in Semiconductor-Doped Glasses Controlled Through the Trapping State

M. Tomita[1], *T. Matsumoto*[2], *and M. Matsuoka*[2]

[1]Department of Physics, Faculty of Science, Shizuoka University, 836 Ohya, Shizuoka 422, Japan

[2]Institute for Solid State Physics, University of Tokyo, Roppongi, Minato-ku, Tokyo 106, Japan

Recently, semiconductor-doped glasses have attracted increasing attention to their quantum size effect and large nonlinear optical susceptibilities. Ultrafast relaxation times as well as other various properties of these glasses have been measured [1]. The results reported by many authors are, however, often diversified. In order to understand and resolve these differences in part, we discuss here a nonlinear dynamical aspect of these material in relation to nonlinear photoluminescence at 77 K.

Photoluminescence spectra of a Toshiba R-63 color glass filter show a low energy broad band peak at 650 nm for a weak excitation intensity, and a rather sharp peak at 605 nm for increased intensities. The low energy peak has a lifetime of about 2 μsec, therefore it is attributed to the emission from a surface trapping state. The radiative lifetime of 605 nm emission depends strongly on the excitation power (see Fig.1). Under the weak excitation of 532 nm pulses, the lifetime is less than 20 psec. Then with increasing intensity the radiative decay becomes slower, and the decay curve approaches a doubly exponential type with decay times 440 psec and 1250 psec. In accordance with a sufficient decrease of the photo-darkening effect at this temperature, the nonlinear behavior is reversible as a function of the excitation intensity.

Considering a four-level system as shown in Fig.2, the above experimental results are explained well in terms of a change of carrier-recombination route due to population saturation of a trapping state lying at lower energy. With this model, the photo-excited carriers firstly relax into the low-lying trapping state with a very fast non-radiative decay rate γ_{Ca}, then relax radiatively from this trapping state with the 650 nm luminescence. When the trapping states of a finite number were filled up, the carriers now decay directly from the band edge state with the 605 nm luminescence.

The above interpretation was confirmed by an experiment with superposed two mode-locked pulse trains, one for a strong pre-excitation and the other for a weak probe, delayed with respect to the other (Fig.3). The weak pulses went through the mechanical chopper, so that only the photoluminescence excited by the weak pulses could be detected by a lock-in amplifier. It turns out that, even though the weak probe pulses cannot excite the 605 nm luminescence by themselves (Fig4.a), they can excite it, if the pre-excitation pulses enter beforehand within the lifetime of the trapping state (Fig4.b).

This experimental result suggests that it is possible to observe the ultrafast nonlinear response of the interband

Springer Series in Chemical Physics, Vol. 48: **Ultrafast Phenomena VI**
Editors: T. Yajima · K. Yoshihara · C.B. Harris · S. Shionoya

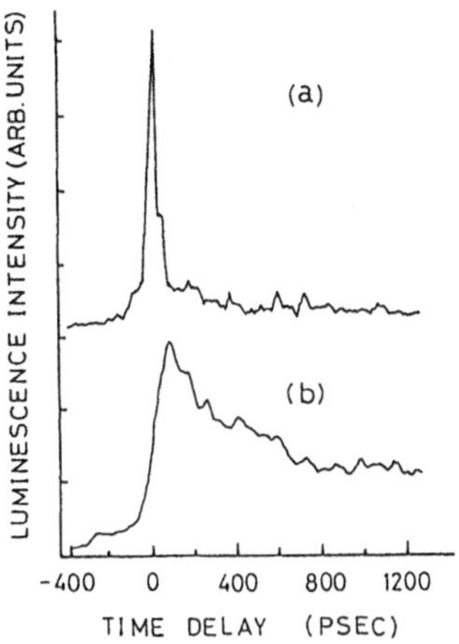

Fig.1 Time dependences of the 605 nm luminescence under the excitation intensity of (a) 5 mW/mm^2, (b) 500 mW/mm^2

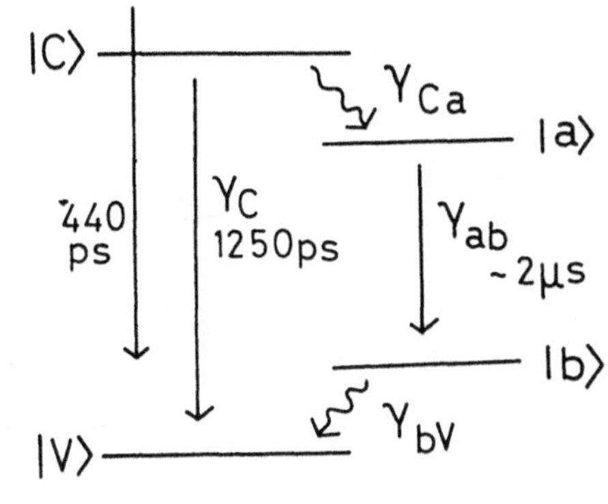

Fig.2 Four-level model of the semiconductor-doped glass

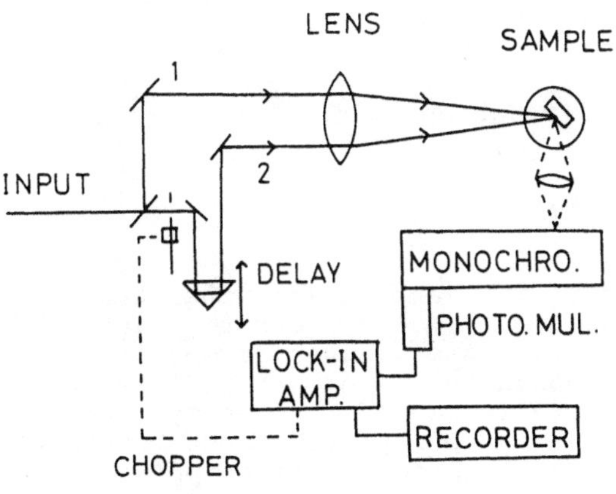

Fig.3 Schematic diagram of the experiment for induced luminescence

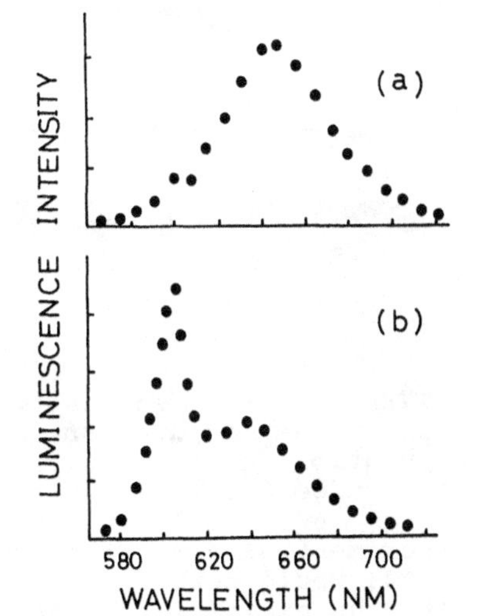

Fig.4 Photoluminescence spectra of the color filter without (a) and with (b) pre-excitation pulse

transition selectively, by saturating the trapping state within 2 μsec before the signal excitation pulses by the pre-excitation pulses. For example, the response of the transient grating was examined with the configuration as shown in Fig.5. The beams k_1, k_2, k_3 are from a cw mode-locked dye laser, and the pre-excitation beam is from the second harmonic light of the YAG laser which pumped the dye laser. Fig.6 shows the diffraction signals as a function of the delay time of the k_3 pulse. The background component of the diffraction signal corresponding to the slow nonlinear response is reduced by the pre-excitation pulses (open circle).

Our results consistently explain the experiments on the optical nonlinear response reported by ROUSSIGNOL et al.[2] and

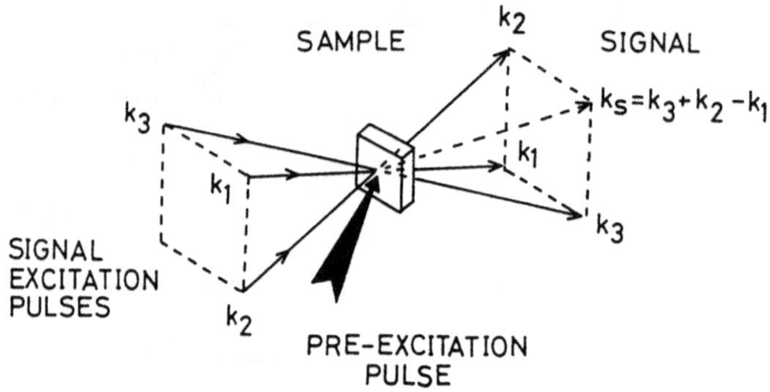

Fig.5 Schematic diagram of four-wave mixing experiments

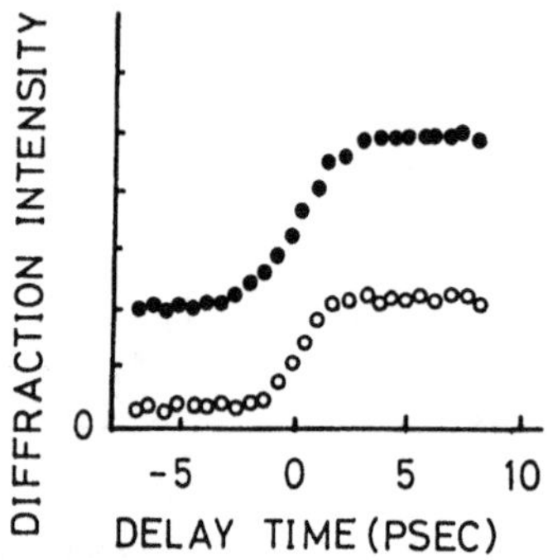

Fig.6 Four-wave mixing signals in Schott RG-645 filter with (open circles) and without (filled circles) pre-excitation pulse

COTTER[3]. For samples with a small amount of the trapping state, the population saturation easily occurs, and the population grating with fast response is then formed in the band edge state. This corresponds to the growth of the high energy 605 nm luminescence with increased intensity in our experiment. A 4.4 kHz narrow resonance reported by REMILLARD et al.[4] would correspond to the decay rate of the trapping state with the excitation intensity below the saturation point.

Some characteristics of the semiconductor-doped glasses were revealed. The new method was proposed which may be used to analyze the size effects in these glasses. This technique may also be used to control an optical switch or make it operable by a weak driving power.

References

1. R.K. Jain, R.C. Lind: J. Opt. Soc. Am. 73, 647 (1983); M.C. Nuss, W. Zinth, W. Kaiser: Appl. Phys. Lett. 49, 1717 (1986) ; G.R. Olbright, B.D. Fluegel, S.W. Koch, N. Peyghambarian: In Ultrafast Phenomena V, (Springer 1986) p.270
2. P. Roussignol, D. Ricard, K.C. Rustagi, C. Flytzanis: Opt. Commun., 55, 143 (1985)
3. D. Cotter: In Ultrafast Phenomena V, (Springer, 1986) p.274
4. J.T Remillard, D.G. Steel: Opt. Lett. 13, 30 (1988)

Part VII

Nonlinear Optics, Coherent Spectroscopy and Spectroscopic Methods

Femtosecond Photon Echoes

C.V. Shank, P.C. Becker, H.L. Fragnito, and R.L. Fork

AT&T Bell Laboratories, Crawfords Corner Road,
Holmdel, NJ 07733, USA

Use of the photon echo technique to study dephasing in semiconductors and large molecules in solution has been frustrated by the very rapid dephasing in such systems. Recently advances in pulse compression techniques have permitted the generation of optical pulses as short as 6 femtoseconds [1] making the observation of coherent transients in such systems possible.

Treacy [2] showed in his original work on pulse compression that a pair of gratings could be used to compensate a linear frequency chirp or quadratic phase distortion. He also pointed out that when compressing optical pulses with a large bandwidth, a grating pair impresses an unwanted cubic phase distortion on the optical pulse. This has provided a limit to pulse compression using this approach to just under 10 femtoseconds.

A pair of prisms can be used in much the same manner as a grating pair for pulse compression. A prism pair also induces a cubic phase distortion but amazingly it is of the opposite sign of that caused by a grating pair. Correction to third order in the phase can then be accomplished by using a grating pair followed by a prism pair to cancel out the third order distortions in the phase. Using this approach an optical pulse of 6 femtoseconds has been generated and used for the experiments described here.

Recent hole burning experiments in large molecules in solution [3] suggest that it may be possible to observe photon echoes in these systems. Hole burning experiments have shown that T_2 is on the order of 70 femtoseconds for a single vibronic transition. With a six femtosecond optical pulse the entire vibronic spectrum is excited so the echo will be the sum of all the echoes from the manifold of vibronic states.

Several attempts have been made to observe coherent transient processes in organic dyes including four wave mixing, three pulse echoes, and four wave mixing with incoherent pulses. In the experiments reported here we attempt to observe echoes using a two pulse sequence. The two pulses having wave vectors K_1 and K_2 make a small angle and generate an echo in the momentum matched direction $2K_2 - K_1$. The echo is then separated spatially from the exciting pulses. The energy of the generated echo is measured as a function of relative time delay between the exciting pulses.

The primary utility of a photon echo is to determine the dephasing time, T_2. For a purely inhomogeneously broadened two level transition the echo energy will decay exponentially with the relative time delay, τ, between the exciting pulses:

Springer Series in Chemical Physics, Vol. 48: **Ultrafast Phenomena VI**
Editors: T. Yajima · K. Yoshihara · C.B. Harris · S. Shionoya

$$E_{echo} = \exp(-4\tau/T_2) \quad . \tag{1}$$

For the case of a homogeneously broadened two level transition a polarization free decay will be observed to relax as $\exp(-2\tau/T_2)$. These relations show that an optical pulse much shorter that T_2 must be used to time resolve the echo decay.

The experiments were performed using compressed pulses corrected to third order in a manner described previously. The pulse repetition rate was 10 kHz and the pulse energy was 20 nJ. The energy of the pulse was less than that needed for a π pulse so the echoes observed in the experiments described here are in the small signal perturbation limit. The pulses were split into two parts using a beam splitter in a modified Michelson configuration to form the two excitation pulses. The two pulses were focused with a 5 cm focal length lens into a flowing stream containing the molecule under study. One pulse was delayed with respect to the other using a stepper motor controlled delay.

The signal detected in the echo direction was directed into a photomultiplier. The detection electronics consisted of a boxcar integrator followed by a phase lock detector. The signal was measured as a function of relative time delay between the pulses.

The measured echo energy for the oxazine dye Nile Blue is plotted in Figure 1. Note that a very rapid 4 femtosecond time decay is observed. The dynamic range of the experiment is limited by the lack of perfect phase compensation for this very short 6 femtosecond optical pulse. The action of imperfect phase compensation is to produce energy in the trailing edge of the pulse.

How do we reconcile this rapid polarization free decay with the much longer $T_2 \approx 70$fs reported in the hole burning experiments? To understand this data we need to consider the photon echo from a system of vibronic levels. Each level in the ground state is coupled to a manifold of vibronic levels having a Franck-Condon overlap with

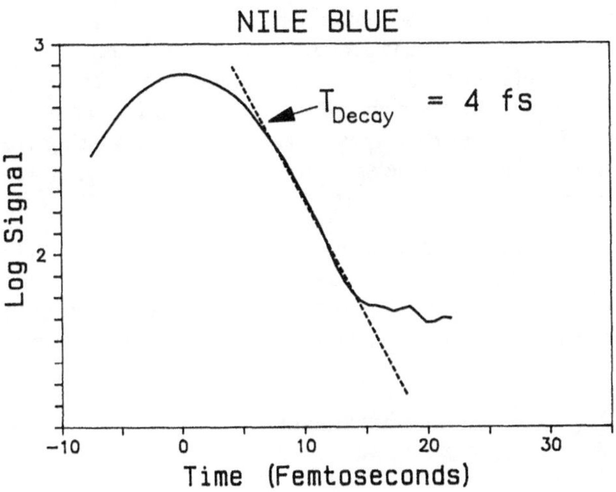

Figure 1. Plot of the log of the echo energy versus time delay.

the ground state. For purposes of our discussion let us assume delta function in time excitation pulses and strong inhomogeneous broadening. The echo energy as a function of time delay is given by

$$S(\tau) \approx \left[\sum_{ij} \mu_{oi}^2 \mu_{oj}^2 \cos(w_{ij}\tau)\right]^2 \exp(-4\tau/T_2)$$

where |i> and |j> are excited state vibronic levels and ω_{ij} is the frequency difference between these levels. In such a multilevel system the exponential term can be dominated by the term in the brackets which comes about from the dephasing due to the sum of the echoes coming from the manifold of vibronic levels. Obviously this consideration precludes the determination of T_2 from a manifold of levels using the photon echo technique.

A number of conclusions can be drawn from this discussion. First it is necessary to take a system point of view in calculating the echo response. Even though hole burning indicates that T_2 is ≈ 70 femtoseconds the echo from a manifold of vibronic levels is dephased much more rapidly. The measured dephasing rate of the manifold of vibronic levels is 4 femtoseconds and is at the limit determined by the spectral linewidth of the S_1 electronic transition.

A more interesting system to study is the polarization dephasing rate of band to band transitions in the direct band semiconductor GaAs. The dephasing rate provides a direct measure of the process of momentum relaxation. At high carrier densities the carrier momentum loses phase coherence primarily through the screened Coulomb interaction between carriers. Both elastic and inelastic carrier-carrier collisions contribute to the momentum dephasing. At low carrier densities electron-phonon interactions are the dominate dephasing process. The density dependence of the polariztion rate provides important information concerning the carrier-carrier interaction.

The experiment was performed in the manner just described for molecules. The sample was a thin .1 micron thick sample of GaAs grown by molecular beam epitaxy. Both faces of the sample were antireflection coated. The excitation pulse energy at the sample ranged from .1-.01 nJ per pulse which corresponds to carrier densities ranging from 10^{17} to 10^{18} cm^{-3}. The carrier density was estimated by measuring the number of photons absorbed in the material. The spot size of the focused beam was measured to be 30 microns in diameter.

In Figure 2 we have plotted the log of the echo energy versus the relative time delay between the pulses. At the highest density an exponential decay with a time constant of 3.5 femtoseconds is measured. This is very close to the system response limit. As the density is reduced the echo decay time lengthens and is very clearly resolved.

The echo decay constant, T_{echo} is plotted as a function of density in Figure 3. The points are experimental and the solid curve is a power law fit to the data and is given by the expression

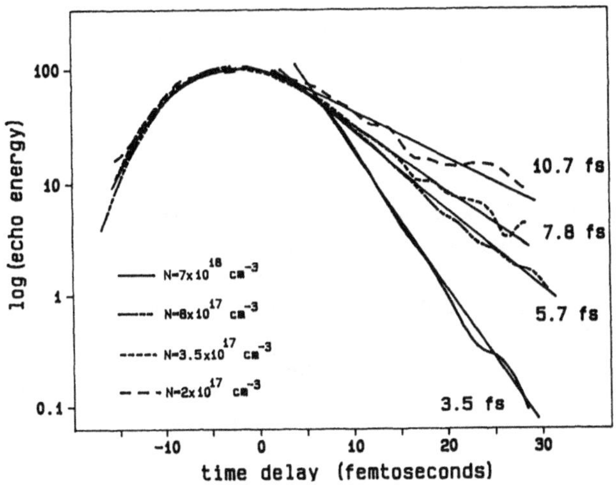

Figure 2. Plot of the log of the photon echo energy as a function of relative time delay between excitation pulses. The time constant of the exponential decay, T_{echo}, is indicated for each carrier density.

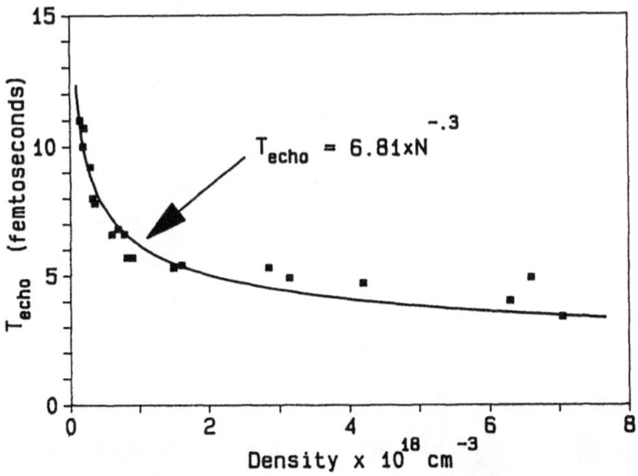

Figure 3. Echo decay time constant as a function of the carrier density for GaAs.

$$T_{echo} = 6.81 \times N^{-.3} .$$

According to expression (1) the polarization dephasing time is given by the relation $T_2 = 4T_{echo}$. The data reveal that the dephasing time T_2 ranges from 14 femtoseconds to 44 femtoseconds while the density has been changed from 7×10^{18} to 1.5×10^{17} cm^{-3}.

The dependence of the dephasing rate on density clearly suggests that the dephasing process is dominated by carrier-carrier interactions. The momentum relaxation rate T_m in the limit of large carrier densities is given by

$$T_m\ N^{-1} S(N)$$

where N is the carrier density and S(N) is a density dependent factor representing the effect of Coulomb screening. In the Thomas-Fermi approximation $S(N) \approx N^{2/3}$ which is close to the experimentally measured $S(N) \approx N^{0.7}$. However, the Thomas-Fermi approximation is not valid in this situation because the excited carrier distribution is nonthermal. Currently a good theory does not exist to describe the process of screening in this highly nonequilibrium population distribution. These experiments provide the first information on nonequilibrium screening processes on such a rapid time scale.

1. R.L. Fork, C.H. Brito Cruz, P.C. Becker, and C.V. Shank Opt. Lett. 12 , 483 (1987).
2. E.P. Treacy, IEEE J. Quantum Electron. QE-9 , 454 (1969).
3. C.H. Brito Cruz, R.L. Fork, W.H. Knox and C.V. Shank, Chem Phys. Lett. 132 , 331 (1986).

Fourier-Transform Spectroscopy in Dye-Doped Polymer Films Using Femtosecond Accumulated Photon Echo

S. Saikan

Department of Physics, Faculty of Science, Osaka University, Toyonaka, Osaka 560, Japan

Fourier-transform spectroscopy using femtosecond accumulated photon echo has been applied to samples of dye-doped polymer. The material dependence of linear electron-phonon interaction and the origin of the quantum beat have been investigated.

1. INTRODUCTION

The persistent hole-burning effect in dye-doped polymers has recently attracted considerable attention because of the promising application to the frequency-domain optical storage. To accomplish this technique at higher temperature, it needs to find a material which exhibits a weak electron-phonon coupling. The study of the electron-phonon interaction in dye-doped material has usually been performed by the persistent hole-burning spectroscopy. However, since the hole spectrum varies dependent on the burning fluence, the detailed investigation on the burning kinetics /1/, in particular, on the time evolution of the hole spectrum, is indispensable to attain the material parameters such as the homogeneous width of the zero-phonon line and Debye-Waller factor. As an alternative method to investigate the electron-phonon interaction, we have recently developed Fourier-transform spectroscopy using femtosecond accumulated photon echo /2,3/. In this technique, the accumulated photon echo is excited by an extremely broadband incoherent light, and then the observed echo signal is Fourier-cosine transformed using computer. Because of the low power density per unit frequency of the broadband dye laser, the obtained spectrum is considered to be rather close to the hole spectrum at relatively short burning time. In this paper, we review the recent experimental results on the Fourier-transform spectroscopy in dye-doped polymers /2-4/.

2. ACCUMULATED PHOTON ECHO

Based on the density matrix incorporating the fluctuation of the electronic transition frequency, we have recently clarified several characteristic features of the accumulated photon echo /2,4/. The phase relaxation process is generally related to the fluctuation of the electronic transition frequency, and the temporal behaviour of the echo can be described by the relaxation function /5/ which is derived from the statistical averaging of the frequency fluctuation. The relaxation function for the accumulated photon echo can be factorized as a product of the corresponding first-order relaxation function. This is the most striking difference between the transient- and accumulated photon echo. In addition, it can be proved that the Fourier-cosine transform of the heterodyne-detected accumulated photon echo provides the hole-burning spectrum. Therefore, using a broadband dye laser, the hole-burning spectrum can be obtained over a wide spectral range.

Springer Series in Chemical Physics, Vol. 48: **Ultrafast Phenomena VI**
Editors: T. Yajima · K. Yoshihara · C.B. Harris · S. Shionoya

Under the excitation of such a broadband laser, many spectral components including zero-phonon line, phonon sideband and vibronic lines are simultaneously excited, and the quantum beats appear in the echo signal. However, it should be noted that the quantum beat in the accumulated photon echo is not induced by the sublevel coherence which is responsible for the quantum beat in transint pump-probe experiment /6/, but is induced by the interference between two single-photon transitions. This is because the accumulated photon echo originates from the accumulation of the saturation hole in the ground state, and this signal overwhelms the transient two-photon signal which is associated with the sublevel-coherence.

3. EXPERIMENTAL

The experimental arrangement for the heterodyne-detected accumulated photon echo is depicted in Fig.1 /7,8/. The synchronously pumped rhodamine-6G dye laser was used as the excitation light. When the dye laser was operated without birefringent filters, the bandwidth results in about 300 cm^{-1} around the centre wavelength of 590 nm. This bandwidth has given the time resolution of 40 fsec in the present experiment of accumulated photon echo. An acousto-optical modulator (ISOMET 1205C) was employed to intensity-modulate the pump-beam at 5 MHz. The signal has been amplified by a tuned preamplifier and then fed into a high frequency lock-in amplifier (PAR 5202). The samples we have employed are rhodamine, oxazine, cyanine, triphenylmethane and porphine dyes which were doped in several polymers including polyvinyl alcohol (PVOH), polyacrylic acid, polymethyl-methacrylate (PMMA) and polystyrene. The samples were kept at about 2 K in an immersion type superfluid helium cryostat. For the measurement of temperature dependence, a flow cryostat was used.

4. RESULTS AND DISCUSSION

As the typical examples, the observed echo signals in rhodamine-640, DODCI and tetraphenyl-porphine (TPP) are shown in Fig.2(A). In the former two samples, the low frequency side of the absorption spectrum was excited.

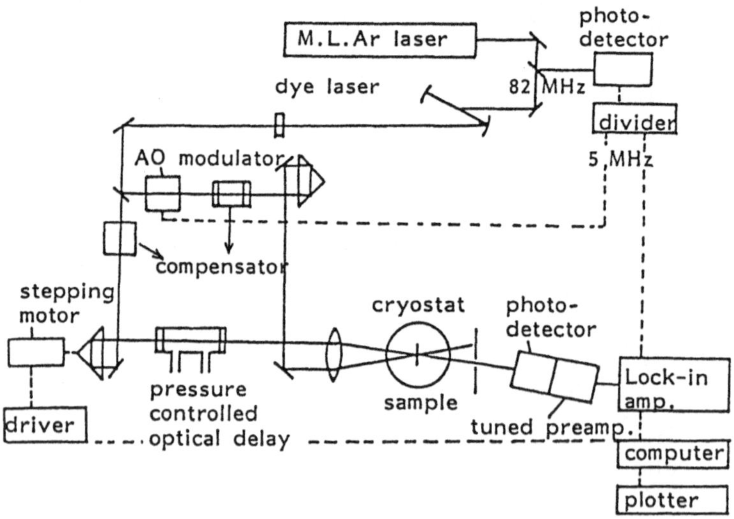

Fig.1 Experimental set-up for the high-frequency modulated heterodyne detection of accumulated photon echo.

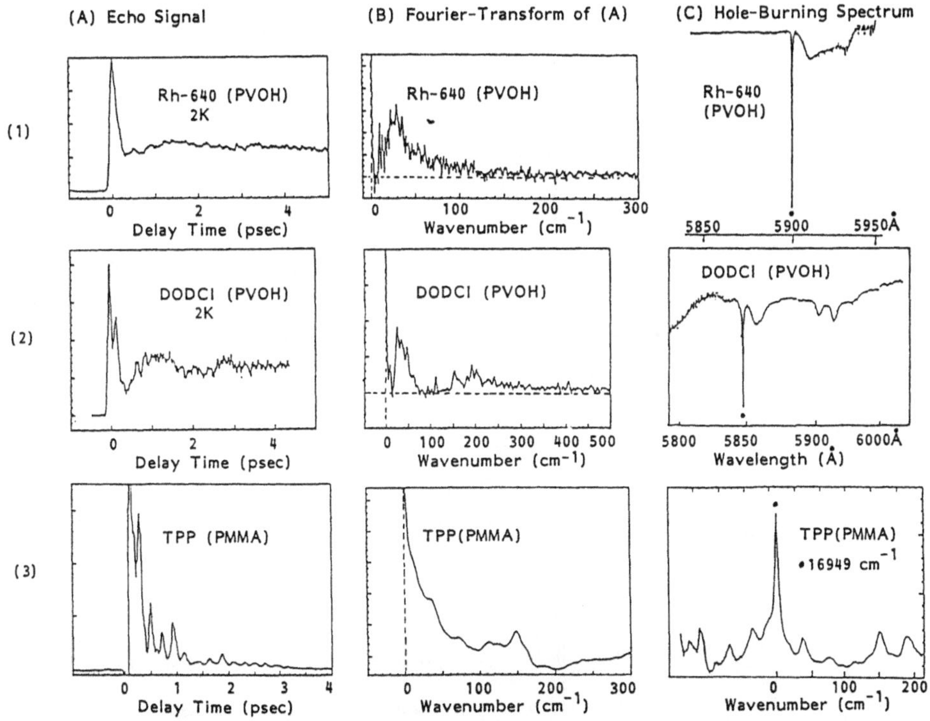

Fig.2 (A) Accumulated photon echoes in rhodamine-640, DODCI and TPP doped in polyvinyl alcohol (PVOH).
(B) Fourier-cosine transform of the echo signal in (A).
(C) Persistent hole-burning spectra. Closed circles indicate the burning laser wavelength.

On the other hand, in TPP the isolated vibronic band $Q_x(1,0)$ was excited. To assign the spectral origin of the quantum beats, the echo signals were Fourier-cosine transformed, and then compared with the persistent hole-burning spectrum that has been obtained by means of the spatial modulation method /9/. As seen in Figs.2B and 2C, the Fourier transformed spectrum well reproduces the hole-burning spectrum. The above comparison enabled us to assign the spectral lines associated with the quantum beats. The observed quantum beats in dye molecules of rhodamine-640, DODCI and TPP are caused, respectively, by the interference due to zero-phonon line vs. phonon sideband, zero-phonon line vs. vibronic line and vibronic line vs. vibronic line. Using this method, we have investigated the electron-phonon interaction in dye-doped polymers.

4.1 Electron-Phonon Interaction in Dye-Doped Polymers

The host polymer dependence of the temporal behaviour of echo was investigated for rhodamine-640 which was doped in four different host polymers. As shown in Fig.3(A), the echo signal consists of a sharp spike and a slowly decaying component. The relative weight of the two components was found to depend on the host polymer. We have previously assigned that the zero-phonon line and the phonon sideband are associated with the two components, and the dip observed after the spike reflects the quantum beat between them

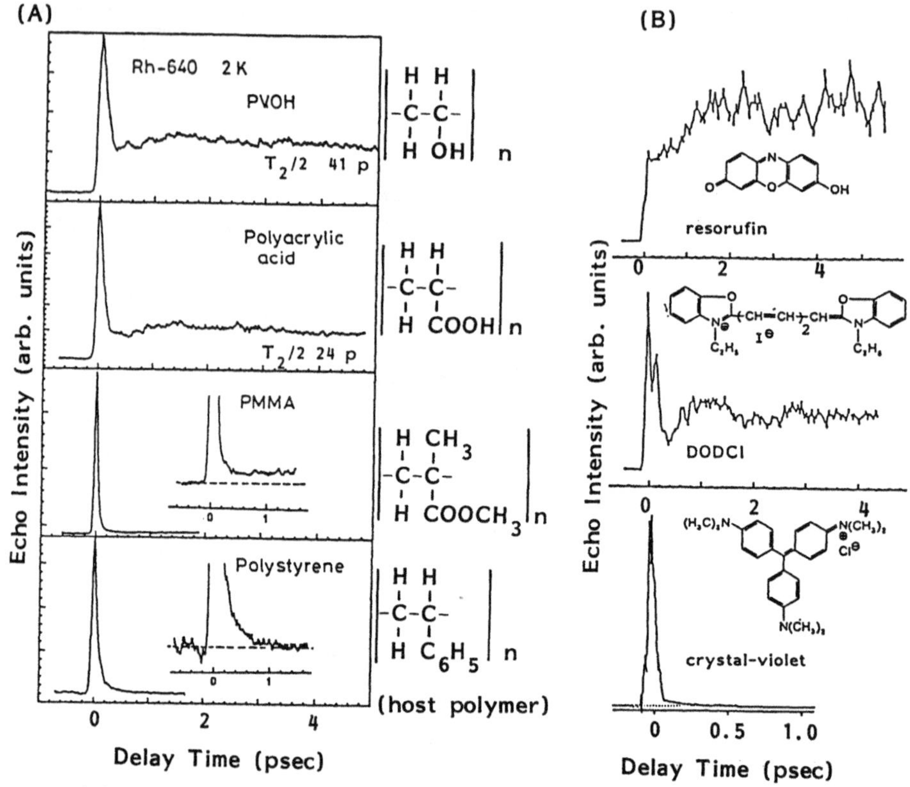

Fig.3 (A) Polymer dependence of accumulated photon echo in rhodamine-640.
(B) Accumulated photon echoes in resorufin, DODCI and crystal-violet doped in PVOH.

/9/. The Fourier-transformed spectra indicate that the electron-phonon interaction strength in the samples of rhodamine-640 decreases in the order of polystyrene, PMMA, polyacrylic acid and PVOH as the host polymer. Namely, there is a correlation between the electron-phonon coupling strength and the size of sidegroups on the polymer chain. This result has been interpreted in terms of the freedom of molecular motion in the polymer environment /9/. The relative intensity of the phonon sideband to the zero-phonon line was also investigated for three different dyes of resorufin, DODCI and crystal-violet. These dyes were doped in the same polymer of PVOH. As shown in Fig.3(B), since the echo signal in resorufin has no spike around the time origin, we can say that the electron-phonon interaction is very weak in this sample. Further, since this dye has a rigid skeleton in the molecular structure, we speculate that the electron-phonon interaction is weak in a non-flexible molecule. The appearance of the strong spike in DODCI is considered to be consistent with the above speculation because this dye has a flexible polymethine chain. Furthermore, in crystal-violet, which is known to exhibit a comformational change in the excited state due to the twisting of the phenyl ring, the sharp spike due to the phonon sideband predominates as shown in Fig.3(B). This result gives further support to our speculation. In a flexible dye without a rigid skeleton, bond rearrangement can occur after photo-excitation. Since the linear electron-phonon interaction is associated with the change in the equlibrium configuration between the electronic ground and excited state, it is reasonable that the electron-phonon coupling strength should be related to the molecular flexibility.

4.2 Vibronic Quantum Beat and Its Temperature Dependence

In the above experiments, we have found that the vibronic quantum beat becomes undetectable above the liquid nitrogen temperature. It makes a striking contrast with the previous observation of the quantum beat in transient pump-probe experiment performed at room temperature /10/. To elucidate the cause of the difference, we have derived the expressions for the signal strength of the transient- and accumulated-pump-probe experiments in three level system /4/. The three-level system has been assumed to consist of the upper two levels $|1\rangle$ and $|3\rangle$ with the decay rates γ_1 and γ_3, and the ground state $|2\rangle$ with the recovery rate γ_2. The expressions for the pump-probe signal can be described as

$$S(t_{21}) = (\mu_2^2 + \mu_3^2)^2 e^{-\gamma_2 t_{21}} + \mu_2^4 e^{-\gamma_1 t_{21}} + \mu_3^4 e^{-\gamma_3 t_{21}} + 2\mu_2^2\mu_3^2 e^{-\gamma_{31} t_{21}} \cos \omega_{31} t_{21} \qquad (1)$$

for the transient case, and

$$S(t_{21}) = e^{-2\gamma t_{21}} \left|\left\langle e^{i\int_0^{t_{21}} \delta\omega(t)dt} \right\rangle\right|^2 (\mu_2^4 + \mu_3^4 + 2\mu_2^2\mu_3^2 \cos \omega_{31} t_{21}) \qquad (2)$$

for the accumulated case, where the angular brackets denote the average over the modulation process of the transition frequency and t_{21} is the delay time between the pump and probe pulses. From eq.1, we noticed that the signal in the transient case gives no information about the phase relaxation except the sublevel coherence γ_{31}, and the quantum beat originates from this sublevel coherence. This agrees with the previous result obtained by MITSUNAGA and TANG /6/. On the contrary, in the accumulated case, the population relaxation rates in the excited state as well as the sublevel coherence do not contribute to the signal, and the quantum beat is induced by the interference between the two single-photon transitions. In this case, since the envelope of the quantum beat is determined by the relaxation function, the beat becomes obscure at higher temperature due to the line broadening of the vibronic state. Therefore, different temperature dependence of the vibronic quantum beat between the transient and accumulated pump-probe experiments is understood, if we can assume that the temperature dependence of the sublevel coherence is relatively small.

References

1. Y.Kanematsu, R.Shiraishi, S.Saikan and T.Kushida: Chem.Phys.Lett. 147 53 (1988)
2. S.Saikan, T.Nakabayashi, Y.Kanematsu and N.Tato: Phys.Rev.B in press
3. S.Saikan, T.Nakabayashi, Y.Kanematsu and A.Imaoka: J.Chem.Phys. in press
4. S.Saikan: Phys.Rev.A in press
5. R.Kubo: Fluctuation, Relaxation and Resonance in Magnetic Systems, edited by D.terHaar (Edinburgh,1962)
6. M.Mitsunaga and C.L.Tang: Phys.Rev. A35 1720 (1987)
7. S.Saikan, A.Fujiwara, T.Kushida and Y.Kato : Jpn.J.Appl.Phys. 26 L941 (1987)
8. S.Saikan, H.Miyamoto, Y.Tosaki and A.Fujiwara: Phys.Rev.B36 5074 (1987)
9. S.Saikan, Y.Kanematsu, R.Shiraishi, T.Nakabayashi and T.Kushida: J.Luminescence 38 15 (1987)
10. M.J.Rosker, F.W.Wise and C.L.Tang: Phys.Rev.Lett. 57 321 (1986)

Ultrafast Dynamics of Excitons and Polarons in Molecular Aggregates

S. de Boer and D.A. Wiersma

Department of Chemistry and Ultrafast Laser and Spectroscopy Laboratory, University of Groningen, Nyenborgh 169747AG, Groningen, The Netherlands

1. Introduction

Aggregates of dye molecules play, as light-absorbing species, an important role in nature and technology, and are therefore worth a detailed optical study. In this paper we report results of photon echo and dual-colour pump-probe experiments on two different types of aggregates. In the first type, the optically prepared state can be described as the optically allowed exciton, which carries a large oscillator strength resulting in a picosecond radiative lifetime! With increasing temperature the optical phase of this excitonic wave-packet is disturbed by collisions with phonons localized at the aggregate. In the second type of aggregate, initially, a super-emissive excitonic state is also formed, but this excitation is not stable and decays into a polaron state that becomes trapped. The driving force for this process is a combination of strong electron-phonon coupling and site inhomogeneity.

2. Theoretical Background

The Hamiltonian for the description of excitations in molecular aggregates has the well-known form [1]

$$\mathbf{H} = \mathbf{H}_{ex} + \mathbf{H}_{ph} + \mathbf{H}^{(1)}_{ex\text{-}ph} + \mathbf{H}^{(2)}_{ex\text{-}ph} \tag{1}$$

where

$$\mathbf{H}_{ex} = \sum_{n}[\varepsilon + D(0) + \Delta_n]a_n^+ a_n + \sum_{nm} J_{m-n}(0)a_n^+ a_m; \quad \mathbf{H}_{ph} = \sum_{q} \hbar\omega_q(b_q^+ b_q + 1/2) \tag{2,3}$$

$$\mathbf{H}^{(1)}_{ex\text{-}ph} = \sum_{qnm} F^q_{nm}\hbar\omega_q(b_q^+ + b_{-q})a_n^+ a_m; \quad \mathbf{H}^{(2)}_{ex\text{-}ph} = \sum_{qn} \chi^q_n \hbar\omega_q(b_q^+ + b_{-q})a_n^+ a_n. \tag{4,5}$$

In (2) through (5) the Bose operator $a^+(a)$ creates (annihilates) an electronic excitation of energy $[\varepsilon + D(0) + \Delta_n]$ at site n, where $D(0)$ represents the van der Waals shift that a monomer transition undergoes through interaction with the other (rigid) molecules and Δ_n represents the inhomogeneous site shift from the average energy ε. J_{m-n} stands for the intermolecular coupling term of site n to m. The Bose operator $b_q^+(b_q)$ creates (annihilates) a phonon of wavevector q and frequency ω_q. $\mathbf{H}^{(1)}_{ex\text{-}ph}$ is the electron-phonon coupling term that is associated with the lattice-vibrational dependence of J_{m-n}, while $\mathbf{H}^{(2)}_{ex\text{-}ph}$ expresses the dependence of the van der Waals shift on a local lattice distortion. In many cases one of these electron-phonon terms dominates. When the Hamiltonian (1) is considered in more detail one notes that the driving term for delocalization of the excitation is J, the intermolecular electronic coupling term. A counterforce against delocalization is provided by the local inhomogeneity Δ_n. It is clear however that in the case $J >> \Delta$, the optical excitation on the aggregate will be delocalized, leading to an optically allowed excitonic transition that carries all the oscillator

Springer Series in Chemical Physics, Vol. 48: **Ultrafast Phenomena VI**
Editors: T. Yajima · K. Yoshihara · C.B. Harris · S. Shionoya

strength of the coherently coupled monomers and which attains a very short radiative lifetime. We call these states superemissive. When the lattice vibrations are considered, the situation becomes more complex. In the case where J >> Δ and the coupling of the electronic excitation with the phonons is weak (case A) the inequality $\mathbf{H}^{(1)}_{ex-ph} \gg \mathbf{H}^{(2)}_{ex-ph}$ is obeyed. In this situation delocalized excitations (excitons) are created which undergo elastic and inelastic scattering through collisions with phonons. In the other limit where $\mathbf{H}^{(2)}_{ex-ph} \gg \mathbf{H}^{(1)}_{ex-ph}$ and J >> Δ (case B) we recover the classical Fröhlich-Holstein (FH) Hamiltonian [2] which has been shown to lead to excitations known as polarons and solitons. For this discussion we will focus only on the polaron-type solutions of the FH-Hamiltonian, which is achieved by a canonical transformation of the Hamiltonian (1). The most important result for our purposes pertains to the fact that the intermolecular coupling term J_{m-n} in the transformed Hamiltonian becomes

$$\tilde{J}_{m-n}(T) = J_{m-n}(0) \langle e^{\gamma_m^+} e^{\gamma_n} \rangle \tag{6}$$

where the term between brackets denotes a thermal average of the Franck-Condon factor, between a phonon-dressed and a phonon-relaxed state, which is usually small. One therefore concludes that 'phonon-dressing', besides lowering the optical excitation energy also results in an effective reduction of the intermolecular coupling term. However, if the electron-phonon coupling strength is such that the inequality $\Delta_m - \Delta_n \gg \tilde{J}_{m-n}$ holds, the polaron will be trapped at site m! This self-trapping will have a dramatic effect on the optical dynamics of the aggregate, the most important aspect being that the dynamic behaviour of the trapped polaron and its optical spectrum will mimic that of a perturbed monomer.

3. Results and Discussion

The aggregates we have under study derive from the monomers of the pseudoisocyanine (PIC) and thiapyrylium (TPY) dyes. Figure 1 shows the optical spectra of these species in monomeric and aggregated form embedded in the polymer polycarbonate [3]. In both cases a substantial red-shift and narrowing of the aggregates' spectrum is observed.

The emission spectrum of the PIC aggregate is also very simple: just one line in near coincidence with the origin absorption. A femtosecond stochastic accumulated photon echo study for excitation into the origin of

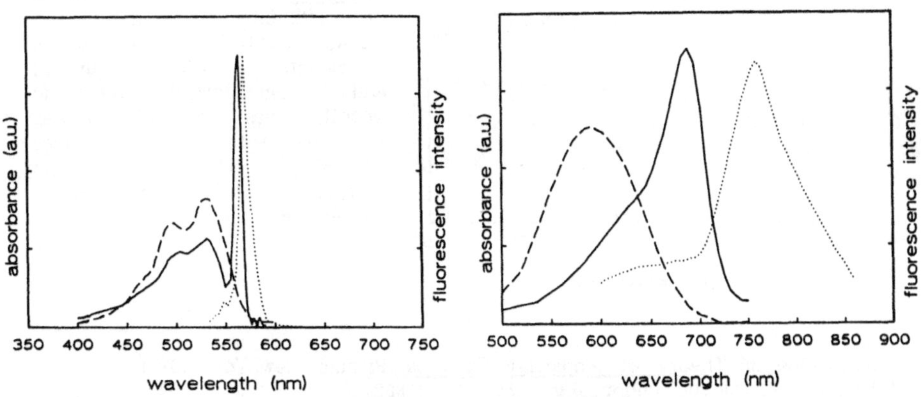

Figure 1. Low-temperature absorption and emission spectra of PIC-Br (left) and TPY-ClO_4 (right) in polycarbonate; monomer absorption (---), aggregate absorption (——) and aggregate emission (....).

a PIC–Br sample held at 1.5 K yields an echo lifetime of 8 ps. In addition a mild temperature dependence of the echo decay is found [4], indicative of weak electron–phonon coupling. The conclusion is that optical excitation in the origin region creates an exciton with a small wavevector, which is scattered by lattice phonons. The short lifetime is due to a combination of an ultrafast radiative [4] and, possibly, a radiationless (internal conversion) process. When the fluorescence quantum yield is known, the radiative lifetime of the aggregate can be calculated, from which, in turn, the number of coherently coupled monomers can be obtained.

The spectroscopy and dynamics of the TPY aggregate are very different. The emission spectrum is as broad as the absorption one, but red–shifted by about 1500 cm^{-1}. This observation indicates a case (B) of strong electron–phonon coupling. In complete agreement with this we found a photon echo signal only near zero probe delay time. This signal is assigned to the decay of the coherently excited phonon side–band in the TPY aggregate. Fluorescence lifetime measurements hint at the presence of two components in the emission: one fast (< 0.5 ns) and the other slow (ca. 2 ns). Pump–probe experiments at different wavelengths confirmed the presence of these two components; one decaying with a 45 ps the other with a 2 ns lifetime. Figure 2 shows two typical pump–probe signal traces. We propose that the fast decay in the upper trace is due to decay of the optically prepared excitonic **k**–state in the TPY aggregate and that the slow decay is due to a polaron formed by trapping of the exciton. It is clear that this exciton–polaron–transition model requires the 'slow' polaron emission to exhibit a rise time concurrent with the decay time of the 'fast' excitonic one. The lower trace in Fig.2 shows a pump–probe signal that fits this requirement at an excitation wavelength where the exciton's pump–probe signal cancels due to the combined effect of absorption loss by the ground state and a corresponding gain by the excited state.

In conclusion, we suggest that dynamical exciton–polaron transitions in aggregates will be regularly found in systems where optical excitation leads to a substantial change of the electron distribution in the excited state.

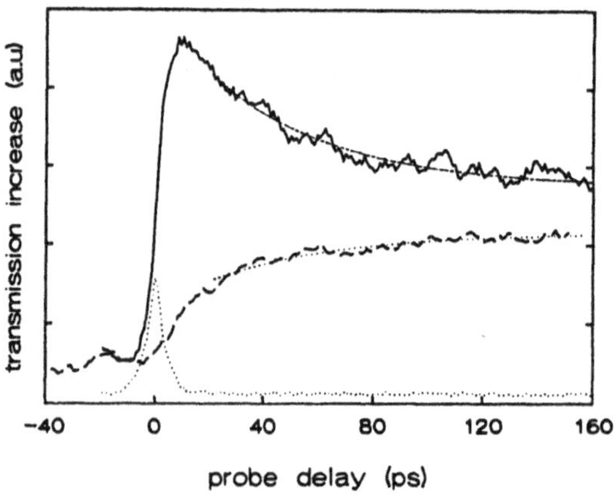

Figure 2. Pump–probe response of TPY–ClO_4 aggregates at 1.5 K; excitation and probe wavelength is 675 nm (upper) and 680 nm (lower) trace. The smooth lines in both traces are fits with response times of 45 ps. The dotted line represents the pulse auto–correlation trace.

References

1. A.S.Davydov in: Theory of Molecular Excitons (Plenum, New York 1971)
2. H.Fröhlich, Proc.R.Soc.London, Ser. A215, 291 (1952)
   T.Holstein, Ann.Phys. (N.Y.) 8, 325 (1959)
3. W.J.Dulmage, W.A.Light, S.J.Marino, C.D.Salzberg, D.L.Smith and W.J.Staudenmayer, J.Appl.Phys. 49, 5543 (1978)
4. S.de Boer, K.J.Vink and D.A.Wiersma, Chem.Phys.Lett. 137, 99 (1987)

Femtosecond Relaxation Studies of Semiconductors and Large Molecules

C.L. Tang, F.W. Wise, I.A. Walmsley, D. Edelstein, and E. Wachman

Cornell University, Ithaca, NY 14853, USA

Abstract - Results on the ultrafast relaxation dynamics of large molecules and nonequilibrium carriers in GaAs and related compounds and structures obtained using optical correlation spectroscopy and time-resolved hot-luminescence spectroscopy are reviewed. This is followed by discussions of recent experiments on wavelength-resolved quantum-beats in dye molecules and the generation of femtosecond laser pulses in the ultraviolet.

I. INTRODUCTION

There are two basic approaches to the study of ultrafast relaxation dynamics in condensed matter: polychromatic experiments and monochromatic experiments. In the monochromatic experiments, pump and probe pulses at the same wavelength are used. The delayed probe pulse measures the relaxation dynamics of the carriers out of the initially excited states. In polychromatic experiments, typically the sample is first excited by a short femtosecond laser pulse. The final states the system relaxes to are then studied with a wavelength-shifted probe continuum. These two basic approaches provide complementary information needed to completely characterize the relaxation dynamics of the system following femtosecond photoexcitation. The monochromatic experiments provide complete information on the relaxation dynamics of the initially excited states whereas the polychromatic experiments provide information on the final states that are accessible to optical probing. Not all the final states the system relaxes to initially are accessible to optical probing, however. In GaAs excited by 2 eV photons, for example, the dominant scattering mechanism from the initially excited states is deformation-potential scattering into the satellite valley states that are not directly connected by optical transition from the valence band states. The dynamics of this relaxation channel can only be studied optically by probing the initially excited states through monochromatic experiments.

II. RELAXATION DYNAMICS OF NONEQUILIBRIUM CARRIERS IN SEMICONDUCTORS

In the case of GaAs and related compounds, a fairly consistent and complete picture of the relaxation dynamics of carriers out of the initially excited states is now available based upon extensive optical transmission-correlation spectroscopy and pump-probe studies[1-5]. Much of these results have now been substantiated by other independent experiments and by extensive Monte-Carlo calculations[6].

In transmission spectroscopy the measured correlation trace $y(t)$ beyond the range where a coherent-artifact might be present is expressed as a sum of damped exponentials with amplitudes a_i and time constants T_i. The number of terms and the amplitude and time constant of each term are then determined using the linear-prediction least-squares fitting procedure[7] with no preprocessing of the data required. The results are summarized in

Table I - Summary of decay parameters for photoexcited carriers in semiconductors. Amplitudes are normalized so that $y(t=0)=1$.

$$y(t) = \sum a_i \exp(-t/T_i)$$

Material	Polarization	a	T (fs)
$Al_{0.34}Ga_{0.66}As$	parallel	0.85	40-45
		0.10	120-150
		0.05	1300-1700
GaAs	parallel	1.02	32-37
		0.09	140-260
		-0.11	1600-1900
MQW	parallel	1.67	43-48
		-0.67	1500-3000

Table I. The relaxation times depend weakly on carrier density. The range of values given in the table corresponds to carrier densities between 7×10^{17} cm^{-3} and 10^{19} cm^{-3}.

There are several notable features. In all the cases studied, there is a fast relaxation component on the order of 40 fs and a long component on the order of a picosecond or longer. This fast component is due mainly to inter-valley scattering to the satellite valleys and to a lesser extent to carrier-carrier scattering. The ratio is roughly 2:1 based on experiments involving temperature-tuning the band-gap of AlGaAs[3]. The precise physical origin of the long component is not yet clear.

In addition, in GaAs or $Al_xGa_{1-x}As$ bulk materials, there is an intermediate component on the order of 150 to 200 fs (see Table I). This is due to optical-phonon emission in the central valley. There is no such intermediate component in the quantum well samples that we have studied. A suggested explanation[8] for the absence of this component in quantum wells is that the corresponding valence band structure is very complicated. The net result is that the band of optically excited conduction band states is much wider than the optical phonon energy of 36 meV; in fact it can be as wide as 200 meV. Thus, optical phonon emission within the central valley does not remove the carrier from the optically-coupled conduction band states.

Finally, in bulk GaAs and GaAs/AlGaAs multiple quantum wells, there is a component with a negative amplitude corresponding to a "rising-wing" component at high carrier concentrations. This is due to the dynamic band-filling effect (or Burstein-Moss shift), which leads to a saturation of the optical transition from the split-off band[1]. This interpretation was confirmed by the fact that the rising wing component is missing in the data for $Al_{0.35}Ga_{0.65}As$ for 2 eV photons. In this case, direct optical transition from the split-off band is forbidden. The rate of rise of the wings is 1.65 to 2 ps corresponding to the return of the carriers from the L and X valleys. It peaks 3 to 5 picoseconds later, depending on the carrier density, corresponding to the eventual establishment of a quasi-equilibrium distribution near the bottom of the central valley. Similar results were obtained for quantum wells, but the carrier dependence is stronger.

To confirm some of these conclusions based upon the study of the initially excited states, we have also studied the time dependence of the hot luminescence emitted by the carriers in the central valley as they relax down to the bottom of the valley. A number of techniques are available for time-resolved detection of luminescence; among these, sum-frequency generation, or up-conversion, is best suited for the femtosecond time domain. This method was originally conceived for the study of picosecond dynamics[9], and was recently adapted to subpicosecond work by Shah et al.[10]. We have studied the time dependence of hot luminescence with an experimental arrangement which is conceptually similar to that described in Ref.10. In place of a synchronously-pumped dye laser and one or two stages of pulse compression, we use the CPM dye laser, however. In addition, we employ a 1 mm barium metaborate crystal for the up-conversion, rather than lithium iodate.

The dye laser routinely produces pulses of duration 35-40 fs at 635 nm, 10^8 Hz, and 15-20 mW in each arm. Figure 1 shows some data recorded. Data shown in Fig. 1(A) were obtained from an LPE grown GaAs layer for which carrier recombination is not quite complete in 11 nsec, the laser period. A background of thermal carriers is thus present at zero delay, which may be responsible for the unusually fast rise of luminescence observed at 1.65 eV.

III. WAVELENGTH-RESOLVED QUANTUM BEATS IN LARGE MOLECULES

Optical correlation spectroscopy has also been used to study the relaxation dynamics of photoexcited dye molecules. As previously reported[11-16], light pulse induced coherent molecular vibrations in a number of dye molecules have been seen for the first time directly in the time domain in the form of femtosecond quantum beats. Table II summarizes the corresponding amplitudes, frequencies, time constants, and phases for the dye molecules studied.

Coherent vibrations of molecules in the electronic ground state have also been seen in impulsive stimulated Raman scattering (ISRS)[12]. There are some major differences between the two effects[15,17-19], however. This is particularly striking in the case when the quantum-beat and ISRS experiments are wavelength-resolved.

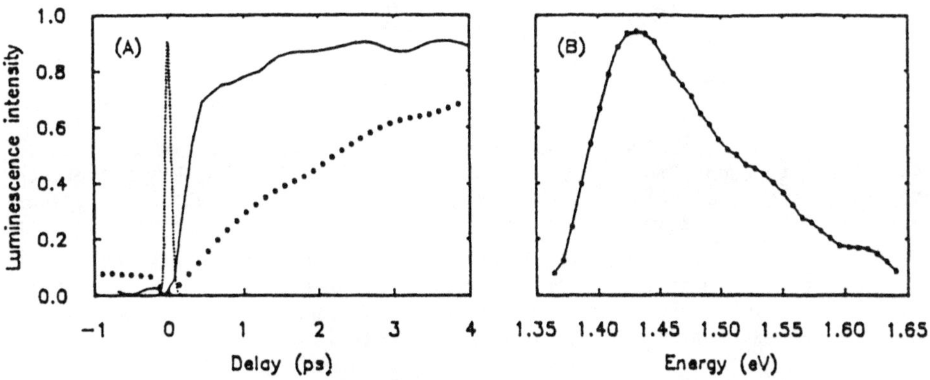

Fig. 1 - (A) Time-evolution of hot-luminescence from LPE GaAs at photon energies, 1.43 eV (symbols) and 1.56 eV (line). The curves are scaled so that the peak luminescence has a value 1 at each energy. The dotted line is the cross-correlation of the scattered laser light from the sample and the pump beam for the up-conversion process. (B) Luminescence from MOCVD grown GaAs sample at 500 fs delay. Initial carrier density is 7×10^{17} cm^{-3}.

Table II - Summary of decay parameters for photoexcited dye molecules. The amplitudes are normalized so that y(t=0)=1.

$$y(t) = \sum a_i \exp(-t/T_i) \cos(2\pi\nu_i t + \phi_i)$$

Molecule	a	T (fs)	ν(THz)	φ(deg)
Malachite Green	0.58	75		
	0.36	4800		
	0.06	205	6.60	11
Ethyl Violet	0.60	120		
	0.33	1700		
	0.07	380	6.45	7
Methyl Violet	0.55	39		
	0.23	2650		
	0.19	150		
	0.03	280	6.40	7
Victoria Blue	0.57	30		
	0.31	10000		
	0.11	105		
	0.01	190	6.14	13
Methylene Blue	0.98	27		
	0.01	1470		
	0.002	380	7.7	2
DODCI	0.47	185	0.87	5
	0.32	1050		
	0.09	370	6.35	3
	0.07	300	2.61	9
	0.03	180	8.17	93
	0.02	500	5.00	8
Nile Blue	0.72	43		
	0.20	400		
	0.04	180	4.86	140
	0.02	490	3.06	87
	0.006	560	5.00	12
	0.005	405	8.39	144
	0.005	4300	2.16	21
	0.0007	4700	6.96	73
	0.0004	3900	9.12	93
	0.0003	11500	10.6	27

In wavelength-resolved transmission spectroscopy, the correlation trace is taken through a monochromator instead of directly. The key difference observed is the following. In the ISRS, the molecular vibration is sine-like and the modulation of the Raman scattered light on the Stokes side is 180^0 out of phase from that on the anti-Stokes side. In the case of quantum beats, the modulation on the probe-pulse due to the molecular vibrations is always nearly cosine-like and in phase as shown in Figure 2. A theoretical explanation for this is given in Ref's 18 and 19.

IV. FEMTOSECOND UV PULSE GENERATION

Almost all current results on ultrafast processes are derived from femtosecond lasers in the visible or near infrared. There is a clear need to extend the wavelength range of high repetition rate femtosecond pulses into the ultraviolet. We report here an intra-cavity doubling experiment[20] using the new nonlinear optical crystal beta-barium metaborate (BBO) that

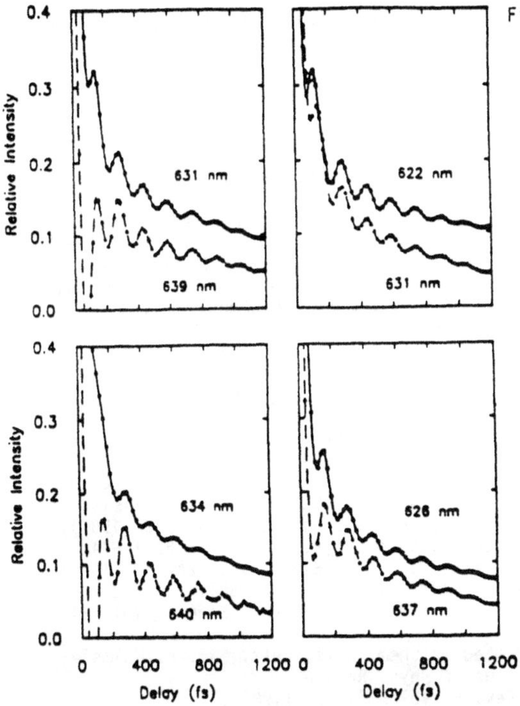

Fig. 2 - Comparisons of measured transmission correlation traces for ethyl violet resolved at various pairs of wavelengths separated by 215 cm^{-1}. All traces are cosine-like.

results in nearly complete recovery of the normal visible fundamental output power in the form of femtosecond uv pulses at the second harmonic.

The basis for our work is a Rh6G-DODCI dispersion-compensated CPM ring dye laser with the output coupler replaced by a high reflector. A third intracavity focus is obtained by using two r=10 cm dichroic mirrors in a "Z" configuration. A 55 u thin BBO crystal was used for intracavity second-harmonic generation and a 85 u BBO crystal was used for cross-correlation measurement between the generated uv pulse and the 630 nm fundamental pulse. Figure 3 shows the measured cross-correlation trace. 43 fs uv pulses at 10^8 Hz rate with 3 mW per laser arm were obtained using 49 fs $sech^2$ red pulses. This technique should be applicable to frequency-doubling of any existing femtosecond dye laser, thus expanding the available wavelength range of ultrashort pulses.

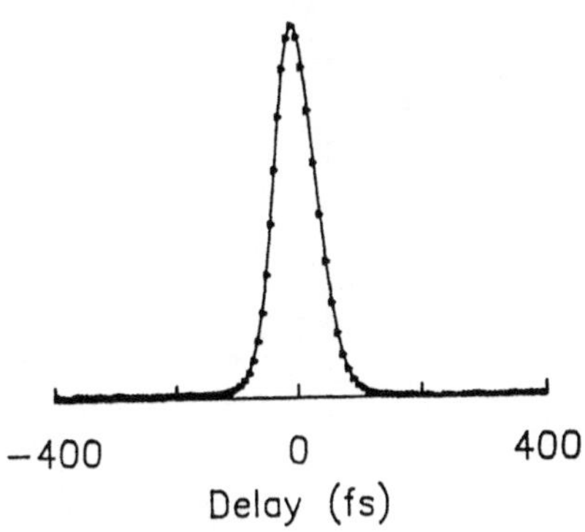

Fig. 3 - Cross-correlation trace of a 43 fs uv pulse with a 49 fs red pulse.

This work was supported by the National Science Foundation, the Joint Services Electronics Program, and the Naval Research Laboratory.

REFERENCES

1. A. J. Taylor, D. J. Erskine, and C. L. Tang, J. of Opt. Soc. of Am. B, Vol.2, 663(1985).
2. M. J. Rosker, F. W. Wise, and C. L. Tang, App. Phys. Lett., Vol. 49, 1726(1986).
3. F. W. Wise, I. A. Walmsley, and C. L. Tang, App. Phys. Lett., Vol. 51, 605(1987).
4. W. H. Knox, D. Chemla, and G. Livescu, Solid State Elect., Vol.31, 425(1988).
5. W. Lin, R. Schoenlein, J. Fujimoto, and E. Ippen, J. Quant. Elect. Vol. QE-24, 267(1988); [see also C. L. Tang, I. A. Walmsley, and F. W. Wise, Comment, App. Phys. Lett., Vol. 52, 850(March, 1988)].
6. D. Bailey, C. Stanton, M. Artaki, K. Hess, F. W. Wise, and C. L. Tang, Sol. State Elect., Vol. 31, 467(1988).
7. F. W. Wise, M. J. Rosker, G. L. Millhauser, and C. L. Tang, J. Quant. Elect., Vol. QE-23, 1116(1987).
8. C. J. Stanton, D. W. Bailey, K. Hess, Y. C. Chang, Phys. Rev. Vol. B37, 6575(1988).
9. H. Mahr and M. O. Hirsch, Opt. Comm., Vol. 13, 96(1975).
10. J. Shah, T. C. Damen, B. Deveaud, D. Bloch, App. Phys. Lett., Vol. 50, 1307(1987).
11. M. J. Rosker, F. W. Wise, and C. L. Tang, Phys. Rev. Lett., Vol. 57, 321(1986); F. W. Wise, M. J. Rosker, and C. L. Tang, J. Chem. Phys., Vol. 85, 2827(1987).
12. S. Ruhman, A. G. Joly, and K. A. Nelson, J. Chem. Phys., Vol.86, 6563(1987).
13. D. McMorrow, W. Lotshaw, G. Kenny-Wallace, J. Quant. Elect. Vol. 24, 443(1988).
15. A. Mokhtari and J. Chesnoy, Europhysics Lett., Vol. 5, 523(1988).
16. W. Zinth, R. Leonhardt, W. Holzpfel, and W. Kaiser, J. Quant. Elect., Vol. 24, 455(1988).
17. I. A. Walmsley and C. L. Tang, submitted to Chem. Phys. Lett.; I. A. Walmsley, M. Mitsunaga, and C. L. Tang, submitted to Phys. Rev. A.
18. M. Mitsunaga and C. L. Tang, Phys. Rev. A, Vol. 35, 1720(1987).
19. C. L. Tang, F. Wise, and I. A. Walmsley, Rev. of Modern Optics Special issue on Ultrafast Phenomena, A. Miller and Sibbett, editors, (scheduled for December, 1988)
20. D. C. Edelstein, E. S. Wachman, L. K. Cheng, W. R. Bosenberg, and C. L. Tang, App. Phys. Lett., Vol. 52, 2211-2213(June 27, 1988).

Coherent Time- and Frequency-Domain Spectroscopy with a Picosecond Distributed Feedback Dye Laser

G.M. Gale [1], *P. Schanne* [1], *and P. Ranson* [2]

[1]Laboratoire d'Optique Quantique du CNRS,
Ecole Polytechnique, F-91128 Palaiseau Cedex, France
[2]Département de Recherches Physiques, LA71,
Université Pierre et Marie Curie, 4 Place Jussieu,
F-75232 Paris Cedex 05, France

A broadly and continuously tunable picosecond distributed feedback dye laser system (DFDL), specifically designed for coherent time-domain spectroscopy, has been developed in our laboratory [1]. The dye laser source produces a single, Fourier-transform limited pulse with an average duration adjustable between 1.6 and 35 ps when pumped by a 20 ps second harmonic pulse produced by a mode-locked Nd/YAG laser. DFDL pulse width is determined by the length of the pumped region. The very low background and duration/frequency agility of this source make it ideally suited to the investigation of weak or drowned Raman features inaccessible to classical time-resolved CARS methods. Dynamic measurement ranges (> 10^{10}), unprecedented in picosecond coherent spectroscopy, are observed for strong Raman lines.

We have applied our new laser system to the study of the coherent behavior of the carbon - 13 isotopic satellites (I_- and I_+ at 1252 cm^{-1} and 1369 cm^{-1} respectively) of the two-vibron bound states (Ω_- and Ω_+ at 1277 cm^{-1} and 1384 cm^{-1} respectively) in crystalline carbon dioxide. The principal transitions of this Fermi doublet have been extensively investigated [2,3] but until now their isotopic companions have remained out of reach, due to their proximity to the parent states ($\approx$ 20 cm^{-1}) and their relatively feeble Raman cross-section (1% in the natural crystal). Figure 1 shows a time-delayed CARS excitation spectrum of the Ω_+ region of CO_2 solid at a temperature of 78K. Excitation frequencies (durations) were 532 nm (20 ps) and $\approx$ 571 nm tunable (35 ps), and probing was at 545 nm fixed (15 ps) delayed by 130 ps. Anti-Stokes frequency selection was also employed [1]. This spectrum shows that the isotopic satellite, I_+, of Ω_+, is clearly isolated from the main transition in our experiment and that its dephasing times (T_2) may hence be measured without fear of contamination by its intense neighbor, whereas in a CARS spectrum the isotopic line, with a strength of only 10^{-4} of its neighbor, would be drowned in the non-resonant background.

Exploiting the spectral isolation provided by the DFDL has enabled us to measure coherent relaxation times of both I_- and I_+ in the 5 - 130K temperature range. The results for I_+, compared to earlier [2] and new measurements for Ω_+ show that I_+ relaxes more than twice as slowly as Ω_+ over the whole temperature range investigated and that both states possess a large residual linewidth at low temperature.

Springer Series in Chemical Physics, Vol. 48: **Ultrafast Phenomena VI**
Editors: T. Yajima · K. Yoshihara · C.B. Harris · S. Shionoya

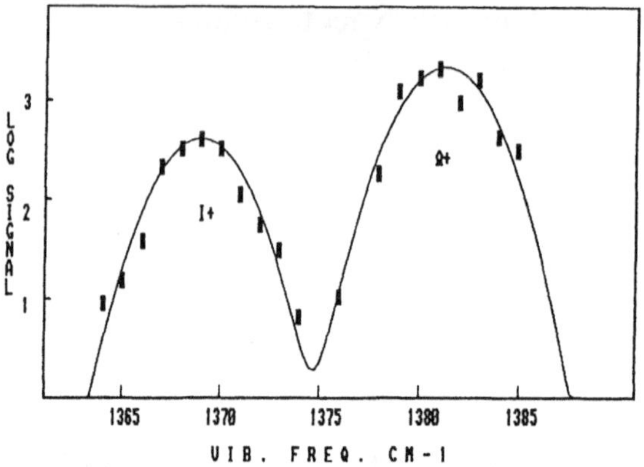

Fig.1 - Time-delayed (130ps) coherent excitation spectrum of I_+ region in CO_2 solid at 77K (with anti-Stokes selection)

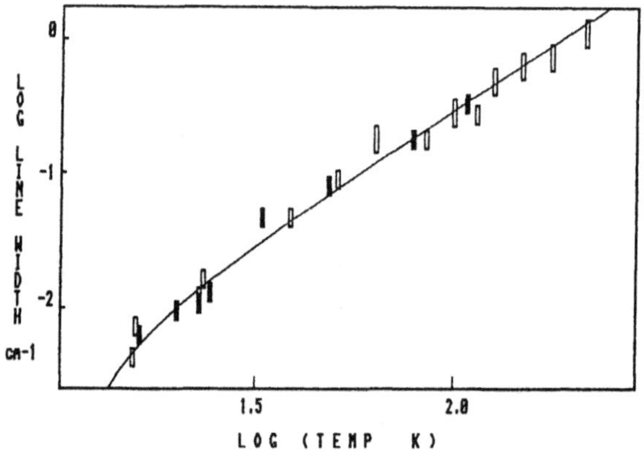

Fig.2 - Log (linewidth) versus log (temperature) for the Ω_- (open rectangles) and I_- (full rectangles)states.

In striking contrast (fig.2), the relaxation times for I_- are identical to those of the parent Ω_- transition, exhibiting a remarkable quadratic temperature dependence of relaxation rate over two orders of magnitude in decay time.

These results for the isotopic satellites provide new insight into the relaxation mechanisms of two-vibron bound states in crystals. The decay of the states above the free two-phonon quasi-continuum, which stretches from 1305 to 1360 cm^{-1}, can be simply described in terms of a population relaxation mechanism whereby the Ω_+ or I_+ state emits an acoustic phonon to access the free two-phonon band, leading to a temperature dependent linewidth $\Gamma = \sigma\ (1 + n)$, where σ is an overlap term and n is the Bose factor for high density of state phonons of energy $\approx 60\ cm^{-1}$. This expression describes well the observed temperature dependence, including the low-temperature intersection, and explains nicely the difference in absolute linewidth as being due to the overlap σ, which is very sensitive to the energy position of the state with respect to the two-phonon quasi-continuum.

A population-change decay channel cannot explain the observed quadratic temperature dependence for the states below the free two-phonon band and we are led to consider the pure dephasing mechanism of [2], which gives an n(1 + n) factor in the linewidth expression. Detailed calculation, in the framework of impurity relaxation theory using the Debye model for the density of states of acoustic phonons, gives the full line in fig.2, in excellent agreement with experimental results. This pure dephasing calculation, which is insensitive to the frequency of a given state, also explains, qualitatively at least, the identical relaxation times observed for Ω_- and I_-.

We conclude empirically that bound state and isotopic impurity relaxation in crystalline carbon dioxide can be treated on the same footing. This can be understood if one recalls that the Ω_- and Ω_+ levels are practically undispersed and hence the bound state can be represented to a good approximation as a sum over ω_1 and $2\omega_2$ molecular states. Otherwise expressed this means that the bound states in CO_2 are "impurity-like". This approach also allows one to hypothesise that the dramatic difference in relaxation mechanism between the "plus" and "minus" states may be due to their different composition.

We further conclude that our DFDL system allows the achievement of unprecedented sensitivity and selectivity in time and frequency-domain spectroscopy and anticipate numerous applications and developments of this technology.

1 - G.M Gale, P. Ranson and M.Denariez-Roberge
App. Phys. B44, 221 (1987)

2 - G.M. Gale, P. Guyot-Sionnest, W.Q. Zheng and C. Flytzanis
Phys. Rev. Lett. 54, 823 (1985)

3 - P. Ranson, R. Ouillon and S. Califano
J. Raman Spectr. 17, 155 (1986).

Raman Quantum Beats Obtained by Impulsive Stimulated Raman Scattering Close to an Electronic Resonance

J. Chesnoy and A. Mokhtari

Laboratoire d'Optique Quantique du CNRS,
Ecole Polytechnique, F-91128 Palaiseau Cedex, France

Coherent excitation of a Raman active vibration can be achieved with a single light pulse when its duration is shorter than the vibrational period of the mode. This so-called impulsive stimulated Raman scattering (ISRS) and the ensuing Raman quantum beats were first observed by the transient induced grating technique [1]. More recently, vibrational quantum beats were also observed in dye molecules by transient induced absorption measurements [2]. Here, we report additional features of these quantum beats in a similar dye molecule, malachite green in solution, using different pump-probe techniques.

The experiments were performed with a femtosecond dye laser described previously [3]. Following a single beam excitation, we have analysed the polarisation rotation and ellipticity change experienced by a delayed probe beam. In addition to these measurements of transient induced dichroism and birefringence, we have developed an experimental configuration to analyse the frequency modulation (spectral shift) of the probe beam as well as its amplitude modulation (time delay shift). See Fig.1.

In all these experimental configurations, we observe small amplitude oscillations superimposed on a large background originating from saturated absorption and dispersion decays (Fig.2). A new technique was developed to analyse these damped oscillations with improved signal to noise ratio [4].

We can interpret these observations in terms of ISRS on the 220 cm^{-1} breathing mode of the thriphenylmethane dye close to an electronic resonance. Our

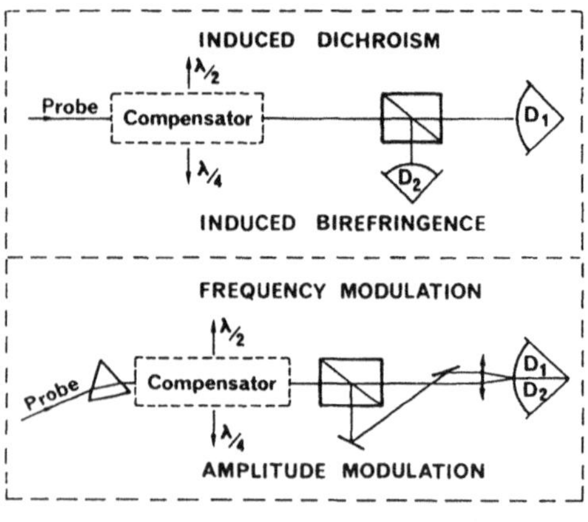

Fig.1
The probe pulse is focused on the excited spot after a variable delay with a polarisation at 45 degrees from the pump polarisation. The polarisation sensitive analysis of the probe beam is performed following the four schemes detailed beside. Analysis of the rotation and ellipticity of the polarisation gives access to the induced birefringence and dichroïsm (top). Polarisation analysis of the dispersed beam gives access to the differential frequency modulation and amplitude modulation (bottom)

Springer Series in Chemical Physics, Vol.48: **Ultrafast Phenomena VI**
Editors: T. Yajima · K. Yoshihara · C.B. Harris · S. Shionoya

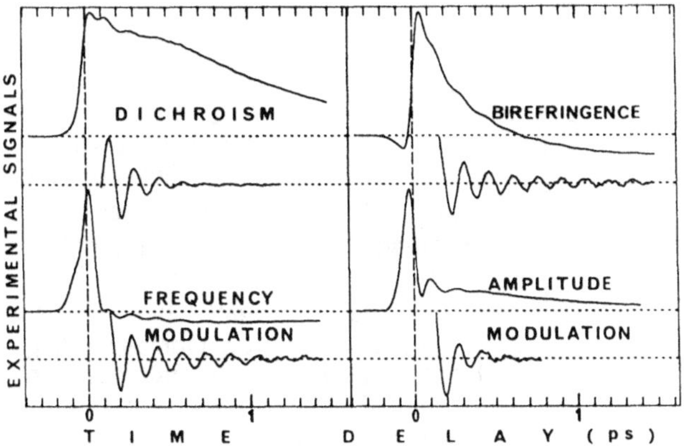

Fig.2
Time-resolved dichroism and birefringence (top) are observed on malachite green dissolved in water. Ultrafast beats are superposed on the relaxational decays, and are experimentally extracted for each observation. At the bottom, similar results are obtained by detecting frequency modulation as well as amplitude modulation of the probe beam. These four observations follow the detection schemes depicted in Fig.1

observations show the main trends of resonant ISRS such as the joint observation of an absorptive effect in addition to pure dispersive manifestations. Vibrations in the ground electronic state are involved to the same extent as vibrations in the excited state and both are experimentally separated as is seen qualitatively in Fig.2 in view of the different damping constants of the observed oscillations. Identical behaviours of the observed vibrational decays are observed for malachite green in different solvents.

The origin of our experimental signal is the modulation of molecular polarisability α and absorption cross-section σ by the vibrational oscillation with amplitude Q, in the ground (g) and excited (e) electronic states. The variation of the absorption coefficient $\Delta\Gamma$ and index of refraction Δn are given by

$$\Delta\Gamma(t) = \frac{\delta\sigma_g}{\delta q} Q_g(t) + \frac{\delta\sigma_e}{\delta q} Q_e(t)$$

$$\Delta n(t) \cong \frac{\delta\alpha_g}{\delta q} Q_g(t) + \frac{\delta\alpha}{\delta q} Q_e(t)$$

where q is the individual vibrational amplitude and Q the macroscopic vibrational coherence.

Our observations are interpreted in the framework of a time dependent density matrix expansion to third order in the fields that allows a description of the ISRS excitation and probing of a simple vibronic system [5,6] : far from resonance, ISRS can be described in a way analogous to the percussional excitation of a classical oscillator (Q_g oscillating in time as sin ωt). In contrast, when the resonant regime is approached, vibrational oscillations occur both in the ground and excited electronic

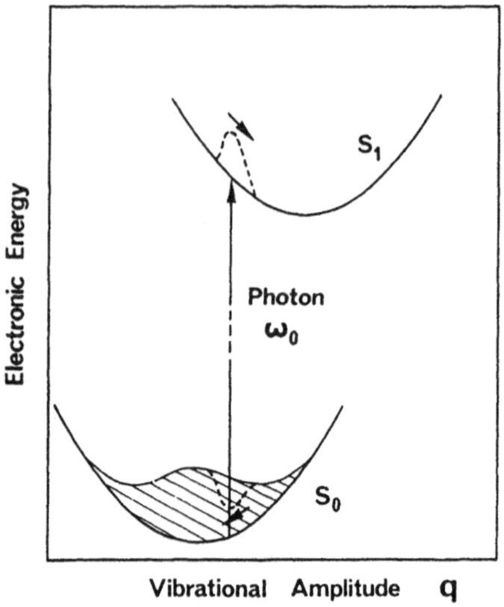

Fig.3
Classical picture of the impulsive stimulated Raman excitation of a vibronic system at resonance. Following the ultrafast optical excitation at energy $h\omega_0$, an off-equilibrium vibrational amplitude is coherently excited in S_1 and begins to oscillate in $\cos(\omega t)$ where ω is the vibrational frequency. Correspondingly a hole is burnt in the vibrational amplitude for S_0 and begins to oscillate in the same way. A treatment of the quantum system by density matrix expansion at third order in the fields permits the picture of ISRS to be refined

states. When the resonance is approached, Q_g (in the ground state) evolves from $\sin \omega t$ to $\cos \omega t$ at exact resonance. Q_e (in the S_1 electronic state) is excited only close to resonance and evolves with time following $\cos \omega t$. (Fig. 3)

References

1 - Y.X Yan, E.B. Gamble and J.A. Nelson, J. Chem. Phys. 83, 5391 (1985)
2 - M.J. Rosker, F.W. Wise and C.L. Tang, Phys. Rev. Lett. 57, 321 (1986)
3 - J. Chesnoy and L. Fini, Opt. Lett. 11, 635 (1986)
4 - J. Chesnoy and A. Mokhtari, Rev. Phys. App. 22, 1743 (1987)
5 - M. Mitsunaga and C.L. Tang, Phys. Rev. A35, 1720 (1987)
6 - A. Mokhtari and J. Chesnoy, Europhys. Lett. 5, 523 (1988)
J. Chesnoy and A. Mokhtari, accepted by Phys. Rev. A.

Picosecond Pump-Probe Interferometry of Nonlinear-Refractive Materials

D. Cotter, C.N. Ironside, B.J. Ainslie, and H.P. Girdlestone

British Telecom Research Laboratories, Martlesham Heath, Ipswich, Suffolk IP5 7RE, UK

Materials with significant refractive nonlinearity and ultrafast response are receiving intensive investigation currently. This is motivated by the interest in optical bistability [1] and nonlinear guided-wave optics, [2] both of which are aimed towards the development of all-optical switching devices. In some materials, such as semiconductors and organics, the nonlinearity is enhanced by utilising resonances, and in this case the nonlinear refraction is accompanied by an absorptive nonlinearity. [1]

Two types of measurements are commonly used to investigate the resonant optical nonlinearity of materials with ultrafast time resolution: Pump-probe transmission experiments directly measure the induced absorption change $\Delta\alpha$, whilst degenerate four-wave mixing yields a signal proportional to the squared modulus of the third-order susceptibility $\chi^{(3)}$ so that the absorptive and refractive nonlinearities are indistinguishable. [3] This ambiguity is significant in the case of resonant nonlinearities because the argument of the complex quantity $\chi^{(3)}$ is dependent on the detuning from resonance, [3,4] and moreover the relaxation dynamics for the real and imaginary parts of the nonlinear response need not be directly equivalent.

Here we describe a new technique which allows the induced refractive-index change to be measured unambiguously, with ultrafast time resolution and with absolute calibration, in both magnitude and sign. The technique can be used with bulk materials, and is also well suited to monomode waveguide geometries. We report measurements on semiconductor-doped fibres, which exhibit large optically-induced phase shifts with relaxation times as fast as ~10 ps. Comparison with pump-probe absorption measurements made under identical conditions reveal marked differences between the relaxation behaviour for the absorptive and refractive components of the optical nonlinearity. The phase of the material response is thus shown to be time-dependent, as well as frequency-dependent.

In the experimental arrangement (shown in Fig.1) the output from a mode-locked laser is divided into three beams. The probe and reference beams form a modified Mach-Zehnder interferometer, in the probe arm of which the sample is located. The optical path lengths of these two beams are equalised and fixed so that spatial interference fringes are produced at the final beam-splitter. The fringe positions are detected using bi-cell detectors together with appropriate sum- and difference-amplifiers and analogue-divider circuitry. The pump beam, which is at least ten times more intense than the probe, also passes through the sample. The refractive-index change so induced causes a fringe shift, which is measured as a function of pump-probe pulse time delay $\Delta\tau$. The pump beam is polarised

Springer Series in Chemical Physics, Vol. 48: **Ultrafast Phenomena VI**
Editors: T. Yajima · K. Yoshihara · C.B. Harris · S. Shionoya

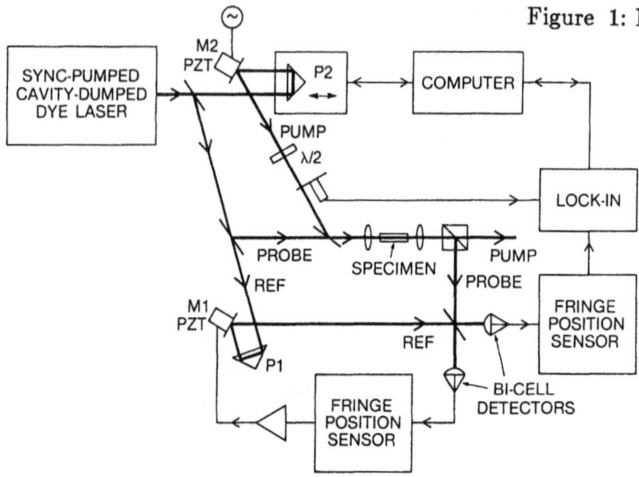

Figure 1: Experimental arrangement

orthogonally to the probe and reference beams and makes no direct contribution to the observed fringe pattern — any residual coherent artifacts at $\Delta\tau \simeq 0$ are nulled by rapid pump-beam phase modulation produced by vibrating mirror M2. The interferometer is stabilised actively to compensate for long-term drift and to suppress low-frequency fluctuations of the fringe pattern. The limit of resolution, using lock-in detection, is equivalent to an optically-induced fringe shift of $\sim\lambda/1000$. The absolute calibration of the induced refractive-index changes, in sign and magnitude, is made by direct comparison with the fringe shift produced by a mechanically-controlled path-length change in the interferometer. The technique has been tested by observing the induced refractive-index change in short (~25-50 mm) lengths of monomode silica fibre; the measured nonlinear refraction coefficient $n_2 = +2.5\times10^{-20}\ \mathrm{m^2W^{-1}}$ is in good agreement with previously-reported values.[5]

The lower part of Figure 2 shows the measured refractive-index change Δn in a 10 mm-long monomode fibre doped with CdS_xSe_{1-x} semiconductor microcrystals.[6,7] The measured values are shown as a function of time after excitation by a 4 ps pulse of energy 120 pJ at a wavelength (610 nm) in the absorption band-tail. The peak Δn is -1.5×10^{-5}, and the negative sign is consistent with the model of electronic band-filling. The initial relaxation time is typically 10–20 ps, and is followed by a slower decay of several hundred picoseconds.

By a minor rearrangement, the apparatus shown in Fig.1 can be used to measure the sample transmission change ΔT as a function of pump-probe time delay $\Delta\tau$. Thus the relaxation behaviour of the absorptive and refractive nonlinearities of a particular sample can be directly compared under identical conditions. The upper part of Fig.2 shows ΔT normalised to the low-intensity transmission T_0. A small increase in transmission at $\Delta\tau \simeq 0$ (perhaps due to band-filling) is dominated subsequently by a larger increase in absorption which exhibits a relatively slow recovery (pulse repetition period 0.26 μs). These observations suggest that, under the conditions of the measurements, the nonlinear refractive index (unlike the induced absorption) is dominated by the faster electronic contribution.

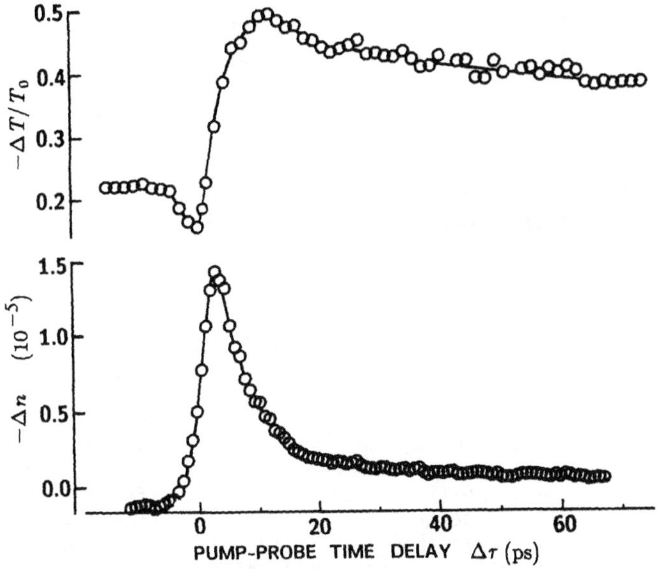

Figure 2: Measured transmission change ΔT and induced refractive index Δn in a 10 mm-long semiconductor-doped optical fibre as a function of time delay after excitation with a 4 ps pulse

Under different conditions we have observed unusual time-dependent effects which are believed to arise from the interference between the refractive nonlinearity of the semiconductor microcrystals and the much faster (and opposite polarity) Kerr nonlinearity of the host glass. In continuing work the pump-probe interferometric technique is being applied to other nonlinear materials in both bulk and waveguide form.

We thank R Cecil for valuable technical assistance with fibre fabrication. C N Ironside held a BT Short-Term Fellowship (permanent address: Department of Electronics & Electrical Engineering, University of Glasgow, Glasgow G12 8QQ, UK). Publication is by permission of the Director of Research & Technology, British Telecommunications plc.

[1] H M Gibbs: *Optical Bistability: Controlling Light with Light* (Academic, 1985)
[2] G I Stegeman and C T Seaton: J Appl Phys **58**, R57-R78 (1985)
[3] P Roussignol, D Ricard and C Flytzanis: Appl Phys **A 44**, 285 (1987)
[4] G R Olbright and N Peyghambarian: Appl Phys Lett **48**, 1184 (1986)
[5] R H Stolen and Chinlon Lin: Phys Rev **A 17**, 1448 (1978)
[6] B J Ainslie, H P Girdlestone and D Cotter: Electron Lett **23**, 405 (1987)
[7] D Cotter, B J Ainslie, H P Girdlestone and C N Ironside: Paper WG-3 in *Proc. XVI Int. Conf. on Quant. Electr.* (Tokyo, July 1988)

Measurement of Ultrashort Phase Relaxation Time of Semiconductor-Doped Glasses with Chirped Pulses

T. Tokizaki, Y. Ishida, and T. Yajima*

Institute for Solid State Physics, University of Tokyo,
Roppongi, Minato-ku, Tokyo 106, Japan
*Present address: Dept. of Applied Physics, Nagoya University,
Nagoya 464, Japan

It is known that semiconductor-doped glasses have large nonlinear susceptibility due to the quantum size effect of semiconductor micro-crystals [1]. Energy relaxation time in these materials has already been measured under various conditions of excess carrier energies, pump intensities and probe wavelengths, and the associated carrier dynamics are being clarified [2,3]. But the phase relaxation time T_2 has not been measured. Because of the ultrashort nature of T_2 in condensed matters, its study could effectively be made by a new approach of ultrafast spectroscopy with non-transform-limited broadband light source [4] which provide high time-resolution even with long pulse widths. Chirped pulses as well as incoherent light have been shown to provide a time resolution of the order of light correlation time (nearly the inverse bandwidth) in the T_2 measurement by degenerate four-wave mixing (DFWM) [5]. We report here the first measurement of T_2 in semiconductor (Cd_xSe_{1-x})-doped glasses by DFWM with broadband chirped pulses, and its dependence on the temperature and the excitation photon energy.

Tunable broadband chirped pulses were produced by self-phase modulation in a 6-cm polarization preserving optical fiber from high power transform-limited input pulses, that were generated by a tunable subpicosecond dye laser and amplifier system both pumped with a CW mode-locked YAG laser and YAG regenerative amplifier system [6]. The chirped pulses have typically a pulse width of 2 ps, but a correlation time of 30 fs corresponding to the bandwidth of 20 nm. The pulses divided into two equal-energy beams with wave vectors $\mathbf{k}_1$ and $\mathbf{k}_2$ were delayed mutually, and focused into the sample by a f=200 mm lens. Two output signals with wave vectors $\mathbf{k}_3 = 2\mathbf{k}_1-\mathbf{k}_2$ and $\mathbf{k}_4 = 2\mathbf{k}_2-\mathbf{k}_1$ were measured simultaneously as functions of the delay time. The result is called the correlation curve, whose decay and peak shift provide information on T_2 [4,5].

First, the correlation curves of DFWM were compared for two types of incident light, the chirped pulses of 2 ps width and the nearly transform-limited pulses of 30 fs width obtained by compressing the former. They have nearly the same spectrum with a central photon energy of 2.03 eV. A color filter glass (HOYA R-64) at a temperature of 8 K was used as the sample. The results shown in Fig. 1(a) and (b) show that the correlation curves for the two types of light sources exhibit the same features with the same peak shift between two curves for the $\mathbf{k}_3$ and $\mathbf{k}_4$ output beams and the same decay constant of 0.2 ps, corresponding to the T_2 value of 0.8 ps. It is therefore confirmed that chirped pulses provide essentially the same correlation curves with the same time resolution as those with transform-limited pulses having the same spectrum. Further, chirped pulses are advantageous compared with transform-limited pulses, because they are easier to generate, have much higher power giving better signal to noise ratio, and are not seriously affected by material dispersion.

Springer Series in Chemical Physics, Vol. 48: **Ultrafast Phenomena VI**
Editors: T. Yajima · K. Yoshihara · C.B. Harris · S. Shionoya

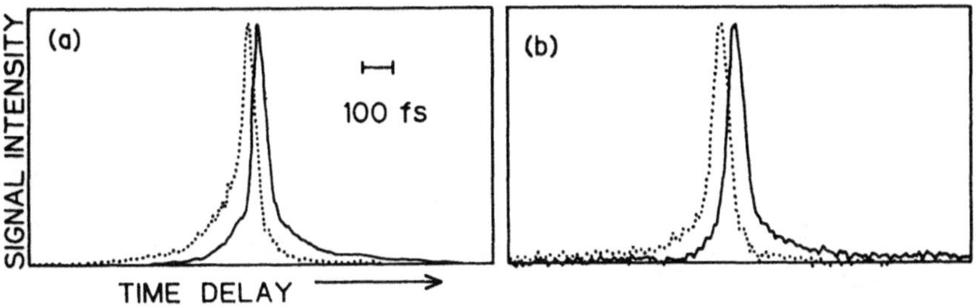

Fig. 1 Correlation curves of DFWM using (a) chirped pulses (2 ps) and (b) nearly transform-limited pulses (30 fs) with an excitation photon energy Ep of 2.03 eV. Solid line: $\mathbf{k}_4$ beam; Dotted line: $\mathbf{k}_3$ beam

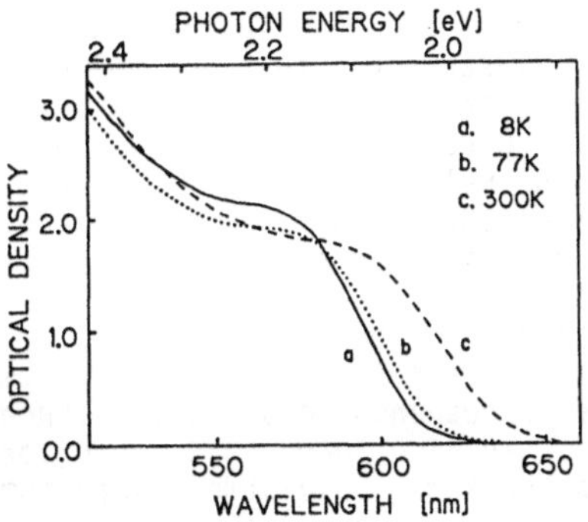

Fig. 2 Absorption spectra of a semiconductor-doped glass (HOYAR-64) at different temperatures.

Then, phase relaxation characters at various temperatures and excitation photon energies (Ep) were studied. Relative position between Ep and the absorption edge can be found from the absorption spectra shown in Fig. 2.

Fig. 3 shows the correlation curves for HOYA R-64 glass at different temperatures with a fixed excitation photon energy of Ep = 2.07 eV. The peak shift increases as the temperature decreases, being 13 fs at 300 K, 21 fs at 77 K and 34 fs at 30 K. Although the peak shift does not directly give the T_2 value due to appreciable contributions of other relaxation parameters, it still provides a rough measure of T_2. The result of Fig. 3, therefore, indicates the decrease of T_2 with temperature. This dependence is considered to be attributed to the increasing scattering of excited electrons with phonons, and also to the temperature dependence of the band gap particularly above 77 K.

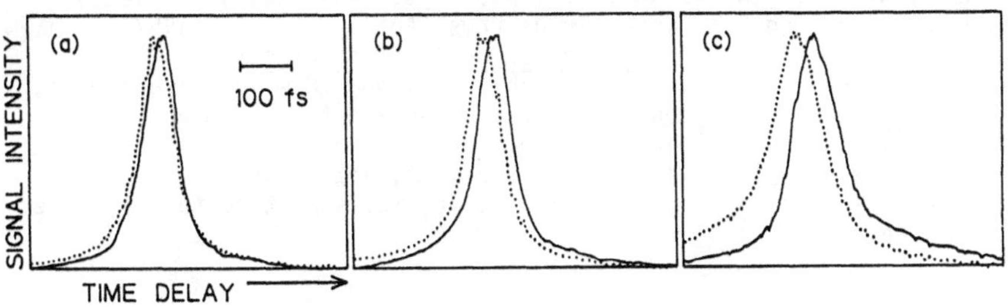

Fig. 3 Correlation curves of DFWM using chirped pulses with Ep = 2.00 eV, showing the temperature dependence. (a) 300 K, (b) 77 K, and (c) 8 K

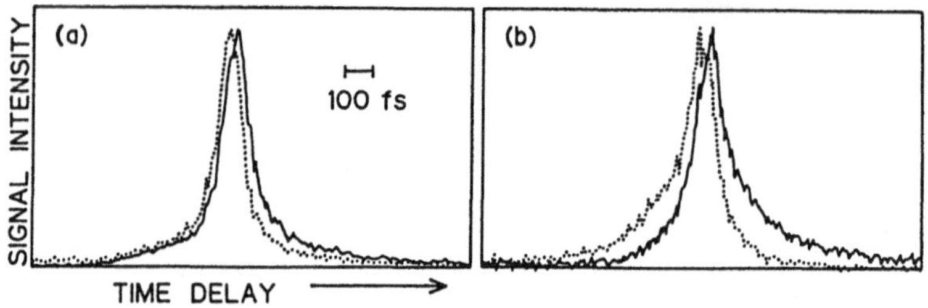

Fig. 4 Correlation curves of DFWM showing excitation photon energy (Ep) dependence. (a) Ep = 2.14 eV, (b) Ep = 2.00 eV

Experimental results with different excitation photon energy (Ep = 2.14 eV and 2.00 eV) at a fixed temperature of 8 K are shown in Fig. 4. Disappearance of clear exponential decay in Fig. 4(a) suggests that the value of T_2 for E = 2.14 eV is much shorter than that for Ep = 2.00 eV. In the latter, the value of T_2 is determined to be 1.0 ps from the decay. The result is interpreted as follows. The phase relaxation of high energy carriers becomes fast due to several kinds of perturbation, such as carrier-carrier and carrier-phonon interactions and reactions in surface states of micro-crystals. On the other hand, at low excitation-energies many electrons seem to be excited to some trap states and to be subject to few perturbations giving much longer T_2. This interpretation will also be supported by the fact that the energy 2.00 eV corresponds to a luminescence peak originated probably from the trap states. We also found that two color filters from different manufacturers having the same absorption spectrum showed much different correlation curves of DFWM accompanied by much different luminescence properties. This may also be attributed to different situations of the trap states.

In summary, a tunable broadband chirped-pulse source for femtosecond spectroscopy is developed, and is applied to the measurement of T_2 in semiconductor-doped glasses by DFWM. The chirped pulses are confirmed to give the same information in DFWM as that with transform-limited pulses and to be more advantageous. The temperature and excitation-photon-energy dependences of T_2 are clearly found, and their mechanisms are discussed.

References

1. R.K. Jain and R.C. Lind: J. Opt. Soc. Am. 73, 647 (1983)
2. M.C. Nuss, W. Zinth and W. Kaiser: Appl. Phys. Lett. 49, 1717 (1986)
3. N. Peyghambarian and S.W. Koch: Revue Phys. Appl. 22, 1711 (1987)
4. T. Yajima and N. Morita: In Methods of Laser Spectroscopy, ed. by Y. Perior, A. Ben-Reuven and M. Rosenbluh (Plenum, New York 1986) p.75; In Ultrafast Phenomena V, ed. by G.R. Fleming and A.E. Siegman (Springer, Berlin, Heidelberg 1986) p.536
5. Y. Ishida and T. Yajima: Revue Phys. Appl. 22, 1629 (1987)
6. T. Tokizaki, N. Morita, Y. Ishida and T. Yajima: submitted to Rev. Laser Eng. (in Japanese)

Ultrafast Fluorometry Using Temporally Incoherent Light

S. Asaka and K. Watanabe

Research Institute of Applied Electricity, Hokkaido University,
Sapporo 060, Japan

In recent years techniques of fast transient spectroscopy by using temporally incoherent light have been developed, and they have been successfully applied to the study of ultrafast dephasing [1] and population recovery [2] without ultrashort light pulses. As to the time resolved luminescence measurements, however, they are not directly applicable since the above techniques are based on the four-wave-mixing or saturation of absorption primarily irrelevant to emission processes. Thus the possibility of ultrafast fluorometry with an incoherent light is currently of great interest. Here we present a novel method of fluorometry with a high time resolution using temporally incoherent light source.

The experimental set-up is shown in Fig.1. This scheme is basically of the optical sampling method by frequency mixing of the sample fluorescence and the "gating" pulse. But here the exicitation and the gating pulses are not conventional single short pulses but long pulses with a fast intensity fluctuation. Then the time resolution is determined by the intensity correlation time of the fluctuation much shorter than the pulse width. Let the intensity autocorrelation of the incident laser pulses be expressed by

$$G(\tau)=G_o(\tau)+R(\tau), \tag{1}$$

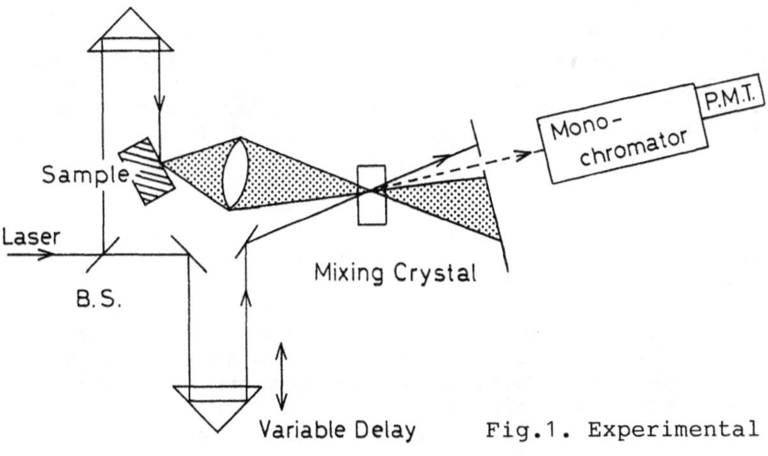

Fig.1. Experimental set-up

Springer Series in Chemical Physics, Vol. 48: **Ultrafast Phenomena VI**
Editors: T. Yajima · K. Yoshihara · C.B. Harris · S. Shionoya

where $G_o(\tau)$ is a slowly varying part with a width corresponding to the laser pulse width τ_p, and $R(\tau)$ is the so-called "noise burst" with a width of the fluctuation correlation time τ_c. The peak of $R(\tau)$ is at $\tau=0$. After some calculation, the energy E of the frequency-mixed pulse emitted from the nonlinear crystal is found to be written as

$$E \propto \exp(-\tau_{12}/T_f)\int_{-\infty}^{\tau_{12}} d\tau \exp(\tau/T_f)\{G_o(\tau)+R(\tau)\}, \qquad (2)$$

where T_f is the fluorescence decay time of the sample, and τ_{12} is the delay time of the gating pulse with respect to the excitation pulse. If we assume that $\tau_c \ll T_f \ll \tau_p$, (2) is rewritten as

$$E \propto T_f G_o(\tau_{12})+\tau_c R(0)\exp(-\tau_{12}/T_f). \qquad (\tau_{12}>0) \qquad (3)$$

In (3) the fluorescence decay curve directly appears on a broad background G_o. This means that if τ_c of the excitation light is short, we can obtain a high time resolution regardless of the excitation pulse width.

In the experiment we used a second harmonic light (532nm) of a Q-switched Nd:YAG laser as an excitation source. The repetition rate and pulse widths were 20 Hz and 7 nsec, respectively. The sample was DTC dye (3,3'-diethylthiacarbocyanine iodide) dissolved in water, whose fluorescence peaks at 580 nm. Frequency mixing was done in a beta barium borate (BBO) crystal. The signal light was distinguished from the intense second harmonics of the gating pulse at 266nm by a double monochromator. Then it was detected by a photomultiplier and fed into a gated integrator while τ_{12} was slowly scanned.

The autocorrelation $G(\tau)$ was first measured by replacing the sample with a plastic diffuser. The result is shown in Fig.2. From the figure the correlation time is found to be 50 psec, much shorter than the pulse width. The ratio of $G_o(0)$ to $R(0)$ should be 1:5 if the fundamental wave of the laser obeys the Gaussian random process. Figure 2 reproduces this ratio well.

A recorded decay curve for the DTC dye as a function of τ_{12} is shown in Fig.3. The decay curve above the nearly flat background is an exponential form with the decay time of 110 psec. This value is in good agreement with the one determined from a streak camera measurement (T_f=100psec) [3]. In Fig.3 the signal-to-background ratio is nearly equal to $\tau_c R(0):T_f G_o(0)$, which is expected according to (3).

In conclusion we have presented a new method of time resolved fluorescence measurement using temporally incoherent light pulses. Here the resolution time is determined by only the intensity correlation time of the excitation source, and not by the pulse width. This means that a resolution time much shorter than the excitation pulse width can be obtained also in a fluorescence decay measurement if we use a temporally incoherent light and frequency mixing technique. This feature was confirmed in an experiment in a dye solution as a sample mate-

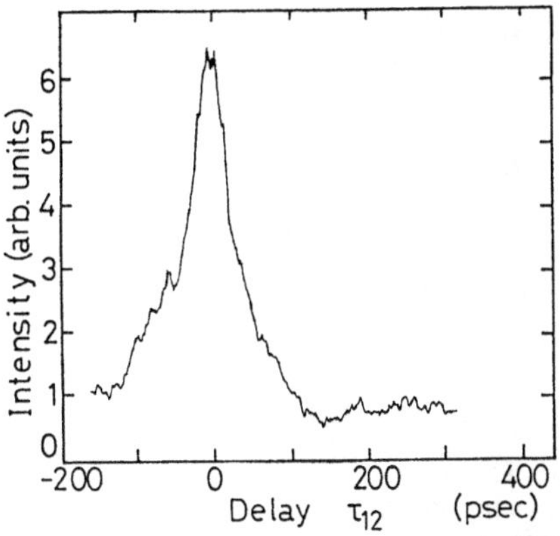

Fig.2. Autocorrelation trace of the excitation laser

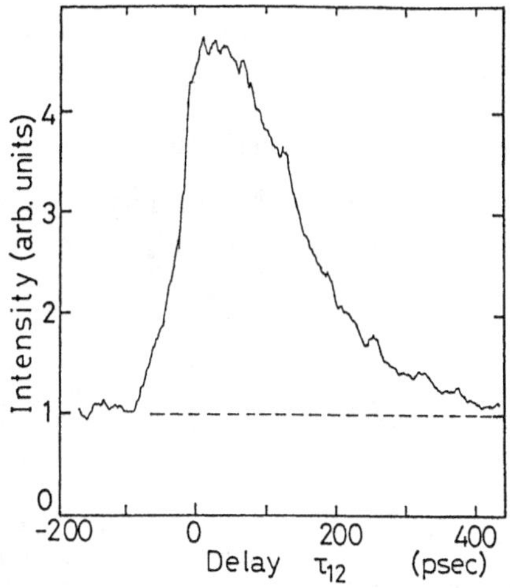

Fig.3. Measured fluorescence decay curve. The broken line is the level of $T_f G_o(\tau_{12})$

rial, where 50 psec resolution time was attained with 7 nsec laser pulses.

References

[1] S.Asaka, H.Nakatsuka, M.Fujiwara, and M.Matsuoka: Phys. Rev., A29, 2286(1984).
[2] M.Tomita and M.Matsuoka: J.Opt.Soc.Am., B3, 563(1986).
[3] H.Sato, M.Kawasaki, K.Kasutani, Y.Kusumoto, N.Nakashima, and K.Yoshihara: Chem.Lett., 1529(1980).

Femtosecond Kerr Dynamics and Three-Beam Degenerate Four-Wave Mixing with Incoherent Light

T. Hattori, A. Terasaki, Xusan Cheng, and T. Kobayashi

Department of Physics, Faculty of Science, University of Tokyo, Hongo, Bunkyo, Tokyo 113, Japan

Molecular dynamics in transparent liquids, such as orientational motion, intermolecular or intramolecular vibration, can be observed by time-resolved experiments such as a transient grating or a Kerr gating measurement with femtosecond optical pulses [1,2] as well as by frequency-domain Rayleigh scattering measurements. We have shown that incoherent light, or light with broad bandwidth, can be used for the measurements of Kerr dynamics of transparent liquids by phase conjugation generation by degenerate four-wave mixing (DFWM) as well as ultrashort pulses and that the time resolution in these measurements is determined by the correlation time of the excitation incoherent light [3]. These methods, i.e., the transient grating, the Kerr shutter, and the phase conjugation, can be treated in a single point of view and can be used for the measurement of Kerr dynamics since each of these techniques is one of the variety of transient DFWM [4].

As a general case for the observation of the Kerr response of liquids with incoherent light, we consider a three-beam scheme (folded boxcars configuration). The three beams are obtained from a single incoherent light source and have linear polarization with arbitrary directions. Two of them make a transient phase grating in the transparent liquids and the third probes it which arrives at the sample with a variable delay t_d. The intensity of the emitted light is detected after an analyzer with an arbitrary polarization direction. Measurements performed in the Kerr-shutter configuration, with one pump and one probe beam, can be regarded as a special case of the triple-beam configuration, where the two pump beams become the same one.

The third-order polarization which is responsible for the DFWM signal is expressed as

$$p^{(3)}(t) = \int_{-\infty}^{t} dt' r(t-t')[\ E(t-t_d)E^*(t')E(t') + gE(t)E^*(t')E(t'-t_d)\ ].\quad(1)$$

Here, r(t-t') is the temporal response function of the Kerr effect. The second term in the equation describes the coherent coupling of the pump and the probe beams, and the factor g is determined by the tensor elements of the response function and the polarization planes of the three excitation beams and the analyzer between the sample and the detector.

When the correlation time of the incoherent light, which determines the time resolution, is much shorter than the relaxation time of the Kerr response, the signal intensity, which is the square of the polarization amplitude, is expressed as a function of the delay time t_d as

Springer Series in Chemical Physics, Vol. 48: **Ultrafast Phenomena VI**
Editors: T. Yajima · K. Yoshihara · C.B. Harris · S. Shionoya

$$I(t_d) = 1 + g(g+2)\,|f(t_d)|^2 + (2t_c/T^{Kerr})\exp(-t_d/T^{Kerr}), \qquad (2)$$

with the third term vanishing for negative t_d. Here $f(t)$ is the autocorrelation function of the field amplitude normalized at $t_d=0$, t_c is the correlation time of the light, and T^{Kerr} is the relaxation time of the Kerr response when it is assumed as exponential. The first term in the r.h.s. gives a constant background, the second one is a coherence spike, which only reflect the coherence of the light, and the third term gives information about the dynamics of the liquid molecules.

Owing to the large spike which appears due to the coherent coupling, the fast behavior of the Kerr response may be obscured. We find, however, the coherence spike can be eliminated or even made negative by controlling the polarization of the excitation beams. Data obtained with carbon disulfide are shown in the figure. The spike intensity is largest ($g=1$ for Fig. a) when the polarization directions are all parallel. On the other hand, with Kerr shutter configuration the ratio of the spike intensity to the decaying part becomes much smaller than that with parallel polarization ($g=1/6$ for Fig. b), and negative spike appears with a certain combination of the polarization ($g=-2/3$ for Fig. c).

We also calculated the transient DFWM with three incoherent beams for the dephasing-time measurement, which has the advantage of flexible opti-

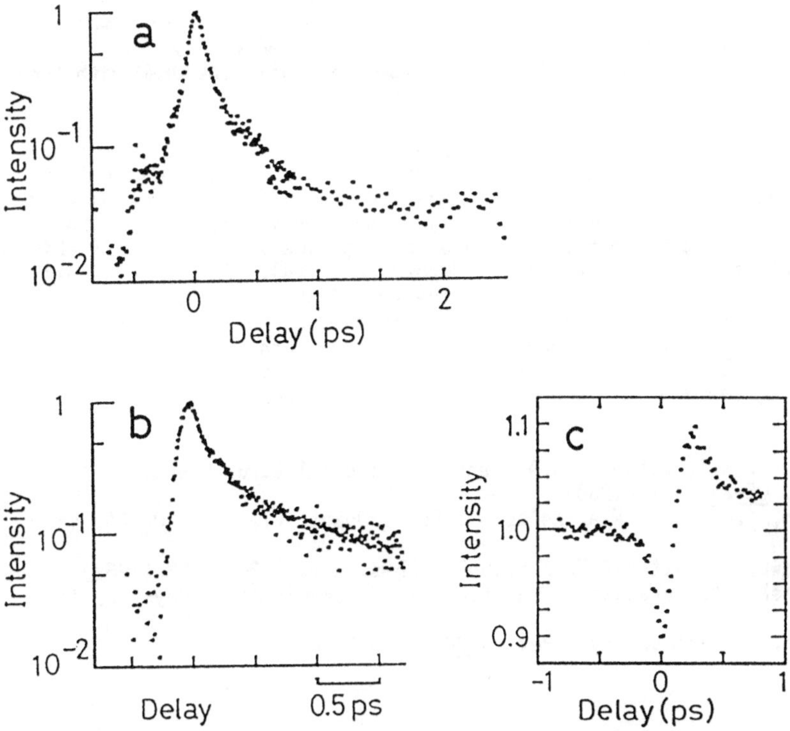

Delay time dependence of signal intensity with carbon disulfide. (a) Data obtained with all parallel polarization of three beams. (b) Data by Kerr-shutter configuration. (c) Data obtained with the polarization of two pump beams perpendicular to each other, and the polarization of the probe beam parallel to one of the pumps and that of the analyzer to the other cal arrangement for the detection of signal in a special direction of phase-matching with $\mathbf{k}_s=\mathbf{k}_1-\mathbf{k}_2+\mathbf{k}_3$. In calculation we assume the inhomogeneous function is Gaussian with a width δw, and the incoherent beams are in δ-function correlation [5]. In rotation wave coordinate, the three consequent pump fields are given as

$$E(t,\mathbf{r}) = E_1(t)\exp(i\mathbf{k}_1\cdot\mathbf{r})+E_2(t+\tau)\exp(i\mathbf{k}_2\cdot\mathbf{r})+E_3(t-T)\exp(i\mathbf{k}_3\cdot\mathbf{r}). \quad (3)$$

Here the fixed delay time T is comparatively larger than T_2. This means the nonlinear diffraction of E_3 through the DFWM process is produced after the transverse relaxation of the transient grating formed by the other two beams (E_1 and E_2) with a variable delay time τ is completed.

The main calculated results are shown as follows:
(1) The analytical solution shows that the total DFWM signal consists of five groups as

$$I(\tau,T) = I_1(\tau,T)+I_2(\tau,T)+I_3(\tau)+I_4(T)+I_5. \quad (4)$$

Here $I_4(T)$ (independent of τ) and I_5 (independent of τ and T) are background for T_2 detection, and $I_2(\tau,T)$ is a continuous function of τ while $I_1(\tau<0,T)$ and $I_1(\tau>0,T)$ have different forms. When T=0 we have $I_1(\tau<0,0)=0$, the total signal $I(\tau,T=0)$ is something like the DFWM with two incoherent beams as discussed in reference [5].

(2) Numerical result shows the peak position for the curve of the total intensity $I(\tau,T)$ expressed by (4) versus delay time τ may be shifted from the zero delay point. In a practical case the incoherent pump spectrum is not wide enough and thus the curve $I(\tau,T)$-δ shows only a coherent spike and no tail can be used to resolve the dephasing time T_2 in DFWM. At this time the peak shifting can be used to determine T_2 [4]. In finite inhomogeneous broadening case, the shifted value $\Delta\tau$ of the coherent spike is found to be related to $T_2\delta w$ and therefore can be used to estimate the dephasing time T_2 if the inhomogeneous broadening width δw is known. The peak shift will disappear in two extreme cases, the extremely inhomogeneous broadening ($\delta w \rightarrow \infty$) and the homogeneous broadening.

1. D. McMorrow, W.T. Lotshaw, G.A. Kenney-Wallace: IEEE J. Quant. Electron. QE-24, 443 (1988)
2. S. Ruhman, A.G. Joly, K.A. Nelson: IEEE J. Quant. Electron. QE-24, 460 (1988)
3. K. Kurokawa, T. Hattori, T. Kobayashi: Phys. Rev. A 36, 1298 (1987)
4. T. Kobayashi, A. Terasaki, T. Hattori, K. Kurokawa: Appl. Phys. B, in press
5. N. Morita, T. Yajima: Phys. Rev. A 30, 2525 (1984)

Fluorescence Lifetime Measurement by Optical Kerr Shutter Gated with Incoherent Light

H. Nakatsuka[1], *Y. Katashima*[1], *K. Inouye*[1], *and R. Yano*[2]

[1]Institute of Applied Physics, University of Tsukuba, Tsukuba Ibaraki 305, Japan
[2]Department of Physics, Faculty of Science, Kyoto University, Kyoto 606, Japan

Although the use of incoherent light in the measurement of dephasing time T_2 has become rather common in recent years [1], only a few papers have reported on the measurement of the population relaxation time T_1 by using incoherent light [2,3]. In this paper we report a new method of measuring the fluorescence lifetime by using an optical Kerr shutter which is gated with incoherent light. In this experiment the resolution time is determined by the intensity correlation time τ_c of the excitation laser pulse and is not limited by the pulse width.

A schematic diagram of the experiment is shown in Fig.1. The sample cell which contained an ethanol solution of cresyl violet was excited by second harmonic light of a Q-switched YAG laser pulse of 7 nsec duration (Spectra Physics, DCR-11). The fundamental light of the same pulse was used as the gate pulse to the Kerr cell which contained CS_2, and was plane polarized at 45° with respect to both the polarizer and the analyzer whose polarizations were orthogonal with each other. The intensity of the gate pulse at the Kerr cell of 1 cm pass length was 100 MW/cm^2, and it induced 0.4 rad of phase shift to the signal light passing through the cell. This corresponds to a signal transmission of about 4 % through the Kerr shutter.

As the first step to examine the resolution time of the system, we measured the intensity correlation $S(\tau) \propto \langle I_S(t+\tau) I_0(t)^2 \rangle$ between the second harmonic light $I_S(t)$ and the fundamental light $I_0(t)$ of the laser pulse by simply removing the sample cell and the red filter in Fig.1. The measured curve of $S(\tau)$ is shown in Fig.2 together with a theoretical curve where the fundamental light was assumed to be a Gaussian random noise with 0.5 cm^{-1} of spectral width. Although the calculated ratio between the peak height to the background height is 5 to 1, the measured ratio is about 3 to 1. We think this

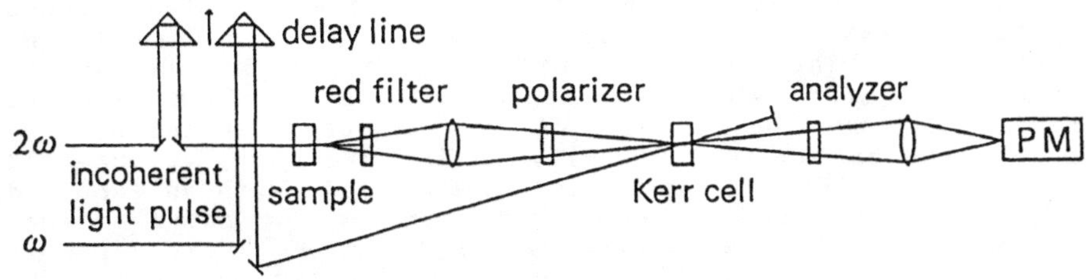

Fig.1 Schematic diagram of the experiment

Springer Series in Chemical Physics, Vol. 48: **Ultrafast Phenomena VI**
Editors: T. Yajima · K. Yoshihara · C.B. Harris · S. Shionoya

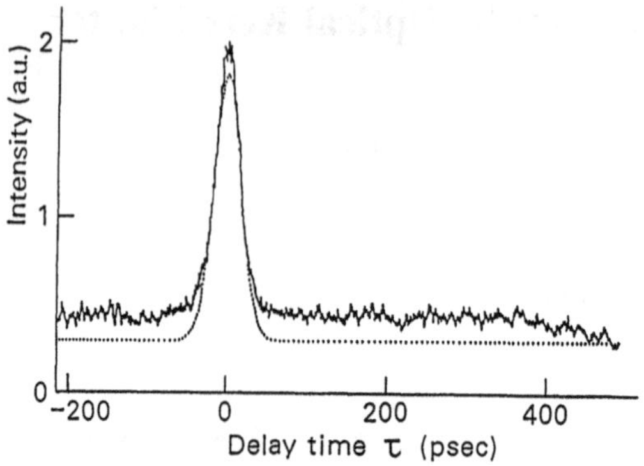

Fig.2 The intensity correlation $S(\tau) \propto \langle I_S(t+\tau) I_0(t)^2 \rangle$ between the fundamental light $I_0(t)$ and the second harmonic light $I_S(t)$ of the laser. The dotted curve is the calculated trace.

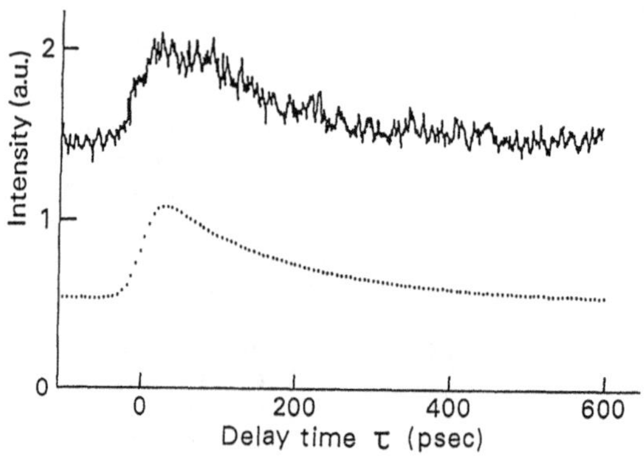

Fig.3 Fluorescence decay curve of cresyl violet in ethanol. The dotted curve is the calculated trace with 160 psec for T_1.

discrepancy occurred because the output of the YAG laser $I_0(t)$ might deviate from a Gaussian random statistics. Apart from this background ratio, Fig.2 clearly shows that we can obtain a resolution time of about 30 psec even by using a Q-switched laser pulse of 7 nsec duration.

A decay curve of the fluorescence from the dye cell is shown in Fig.3 together with a theoretical curve where fluorescence lifetime T_1 was assumed to be 160 psec. In this case lasing occurred from the dye cell upon excitation by the second harmonic light. The fluorescence lifetime was reduced considerably by the effect of the induced emission. By reducing the excitation intensity to the dye cell, we obtained a fluorescence lifetime of several nanoseconds with an enhanced background level in the decay curve. Since the ratio between the peak height to the background height in the decay curve is proportional to τ_c/T_1, it is desirable to use a laser whose intensity correlation time τ_c is not too small compared with the measured fluorescence lifetime T_1, so that a good signal to noise ratio can be obtained. In general the intensity correlation time of a laser can be easily controlled by changing the spectral width of the output.

We have demonstrated that incoherent light can be successfully used together with an optical Kerr shutter for the measurement of the fluorescence lifetime. The most important feature of this method is that we can very easily obtain a high time resolution by simply broadening the spectral width of the excitation laser pulse. In contrast to the previous methods of measuring the population decay time T_1 by using incoherent light, this method is free from the coherent artifact which may obscure the decay curve, and moreover it facilitates the simultaneous observation of the fluorescence decay behavior of the whole spectral region.

References

1. S. Asaka, H. Nakatsuka, M. Fujiwara, and M. Matsuoka: J. Phys. Soc. Jpn. 56, 2007 (1987).
2. M. Tomita and M. Matsuoka: J. Opt. Soc. Am. B 3, 560 (1986)
3. N. Morita, T. Tokizaki, and T. Yajima: ibid. 4, 1269 (1987)

New Method for the Measurement of Dephasing Time Using Incoherent Light with Reduced Noise and Its Application to CdS Fine Particles

K. Misawa[1], T. Hattori[1], Y. Ohashi[2], H. Itoh[2], and T. Kobayashi[1]

[1]Department of Physics, University of Tokyo, 7-3-1 Hongo, Bunkyo-ku, Tokyo 113, Japan

[2]Central Research Institute, Mitsui Toatsu Chemicals Inc., 1190 Kasama-cho, Sakae-ku, Yokohama 247, Japan

Recently a new spectroscopic technique using incoherent light has been developed, and it has been applied to measure the dephasing time T_2 in several materials by degenerate four-wave mixing (DFWM) [1-3]. The time resolution of the method is limited only by the correlation time (τ_c). This technique is a type of accumulated photon echo which can be heterodyne-detected in the same configuration as a pump-probe experiment [4,5]. This configuration has an advantage of easier alignment because the signal and the probe beams are collinear. Heterodyne-detected accumulated photon echoes usually require phase-sensitive detection using a mode-locked dye laser with a high repetition rate, and as a result noise is efficiently eliminated. However, methods using a nanosecond dye laser with a low repetition rate in the ordinary forward DFWM configuration tend to suffer from scattered noise.

Here we propose a simple method to study phase relaxation in the Kerr-shutter configuration (Fig. 1). A single incoherent light is divided into two beams. One of them (probe beam) passing through a polarizer is focused onto the sample and is directed to a detector through a crossed analyzer. The other (pump beam) is polarized 45° with respect to the probe beam and focused at the same point on the sample. Transmitted signal is detected as a function of the delay time between two beams. So far the Kerr-shutter configuration has been applied to investigate the molecular dynamics in several liquids [6]. For the measurement of T_2, elimination of the excitation pulses by polarization was suggested in forward three-pulse echo [7]. We calculated the signal intensity profile as a function of the delay time (τ) of the probe beam with respect to the pump beam and demonstrated that T_2 can be obtained from the profile.

Here, the following three assumptions are made for the incoherent light and the system to be studied:

(i) extremely large inhomogeneous broadening, and
(ii) $T_1 \gg \tau_c$, T_2, and
(iii) $\tau_c \ll T_2$.

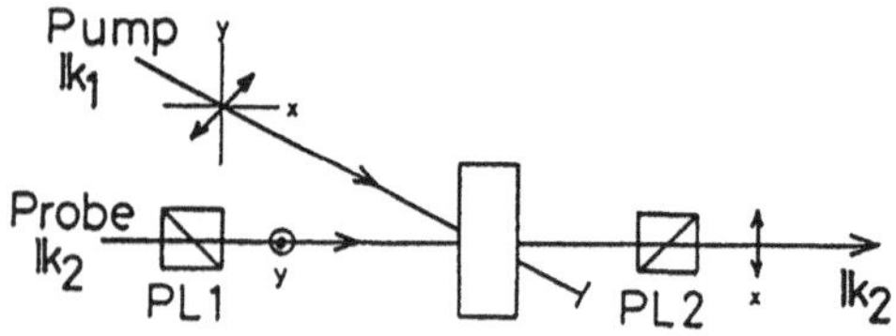

Fig. 1: DFWM in the Kerr-shutter configuration

Springer Series in Chemical Physics, Vol. 48: Ultrafast Phenomena VI
Editors: T. Yajima · K. Yoshihara · C.B. Harris · S. Shionoya

We implicitly assumed that the temporal pulse duration is long enough not to vary the population distribution of the material too much. Using ultrashort pulses, we cannot measure T_2 by this method.

On assumptions (i) and (ii), the $\mathbf{k}_2$ component of the third-order polarization is written with the correlation function G(u) as

$$P^{(3)}(t) \propto \int_{-\infty}^{t} dt' \exp[-2(t-t')/T_2] \Big(E(t'-\tau)G(t-t') + E(t')G(t-t'-\tau) \Big) . \quad (1)$$

The signal intensity is the statistical average of $|P^{(3)}|^2$:

$$\begin{aligned} I^{(3)}(t) \propto \int_{-\infty}^{t} dt' \int_{-\infty}^{t} ds' \exp[-2(2t-t'-s')/T_2] \\ \Big(\; & G(t'-s')G(s'-t)G(t-t') && \leftarrow \text{background} \\ & + G(t'-s'-\tau)G(s'-t+\tau)G(t-t') && \\ & && \leftarrow \text{cross terms} \\ & + G(t'-s'+\tau)G(t-t'-\tau)G(s'-t) && \\ & + G(t'-s')G(t-t'-\tau)G(s'-t+\tau) \; \Big) . && \leftarrow \text{decay} \end{aligned} \quad (2)$$

On assumption (iii), G(u) can be replaced with $\delta(u)$. After some calculations, we get the normalized signal intensity as follows:

$$\begin{aligned} I(\tau) &= 1 && (\tau < 0) \\ I(\tau) &= [1 + 2\exp(-2\tau/T_2)]^2 && (\tau > 0) . \end{aligned} \quad (3)$$

The applicability of this method was verified by measuring T_2 in cresyl violet doped in a PMMA film at 2K. Figure 2 shows the experimental setup. An incoherent light source was a broad-band dye laser around 620 nm pumped by a ns excimer laser. The dye laser had no tuning elements [3], and τ_c was about 170 fs. A Babinet compensator was used to suppress the background. By setting a focusing lens and a pinhole we can detect selectively the signal with high sensitivity. The data in Fig.3 have two decay components. The slower component corresponds to the zero-phonon line and can be clearly seen only when the excitation intensity is low enough as expected from the low saturation level of the zero-phonon line.

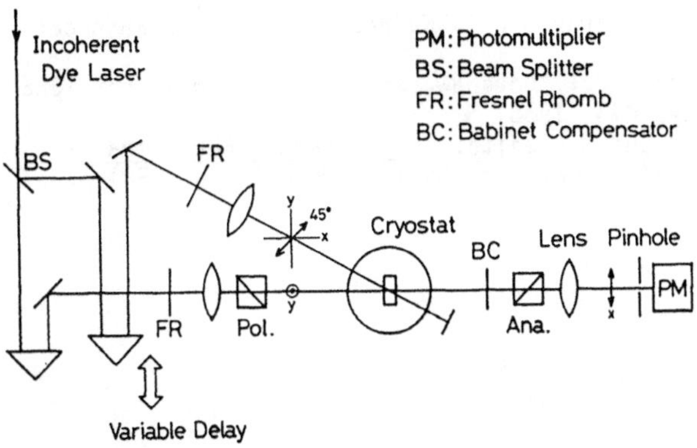

Fig. 2: Experimental setup for T_2 measurement under the Kerr-shutter configuration

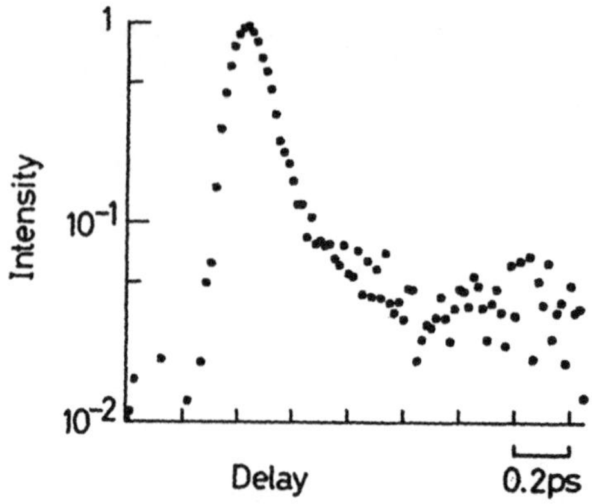

Fig. 3: Delay time dependence of the signal intensity obtained from cresyl violet doped in a polymer film at 2K

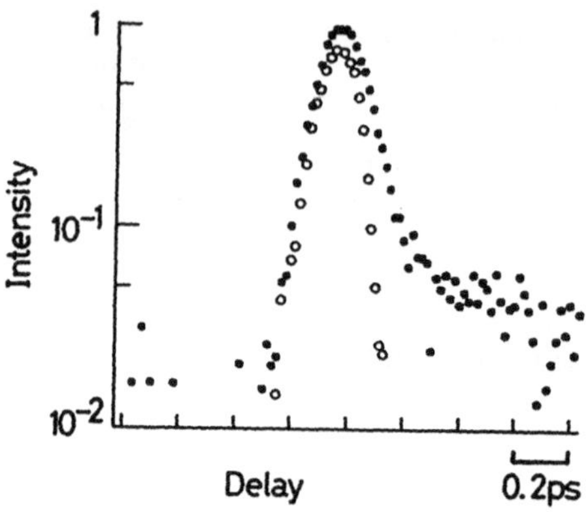

Fig. 4: Delay time dependence of signal intensity at 2K obtained from CdS microcrystallites: signal(●); autocorrelation(○)

We investigated dephasing in CdS microcrystallites with a mean diameter of 50A embedded in a polymer film. The excitation was resonant on the exciton absorption around 420 nm. We successfully obtained the signal with high S/N shown in Fig. 4, while failing to detect the signal by ordinary two-beam forward DFWM with similar intensity excitation. Clearly the signal trace has a decay tail which represents the dephasing process, but determination of T_2 is in progress. Making use of the present method, we are going to discuss the dependence of the dephasing in CdS microcrystallites on the excitation wavelength, temperature, and particle size.

1. N. Morita and T. Yajima: Phys. Rev. A 30, 2525 (1984)
2. S. Asaka, H. Nakatsuka, N. Fujiwara, and M. Matsuoka: Phys. Rev. A 29, 2286 (1984)
3. T. Hattori and T. Kobayashi: Chem. Phys. Lett. 133, 230 (1987)
4. W.H. Hesselink and D.A. Wiersma: Phys. Rev. Lett. 43, 1991 (1979)
5. S. Saikan, Y. Kanematsu, R. Shiraishi, T. Nakabayashi, and T. Kushida: J. Lumin. 38, 15 (1987)
6. See for example, D. McMorrow, W.T. Lotshaw, and A. Kenney-Wallace: IEEE J. Quantum Electron. QE-24, 443 (1988)
7. M. Matsuoka, H. Nakatsuka, and M. Fujita: In Picosecond Phenomena II, ed. by R. Hochstrasser et al. (Springer, Berlin, Heidelberg 1980) p.357

Estimation of Ultrafast Relaxation Parameters from Excitation Spectra for Second-Order Optical Processes

S. Kinoshita and T. Kushida

Department of Physics, Osaka University,
Toyonaka, Osaka 560, Japan

Although recent progress in ultrafast laser techniques has enabled us to obtain much information about very fast relaxation phenomena in various materials, there still exist problems in determining the relaxation parameters in the ultrashort time range from both experimental and theoretical points of view. For example, it is difficult to determine a very short lifetime of an excited state under weak excitation. For direct fluorescence decay measurements and also for indirect ground-state population recovery measurements, high-density excitation is needed, which often modifies relaxation characteristics. Further, it is necessary to examine the meaning of the phase relaxation time T_2, when it is very short, because the non-Markovian effect becomes important.

In the present paper, we show that the conventional resonance Raman and luminescence spectroscopy, which needs only weak stationary light for the excitation, provides detailed information about the relaxation parameters in the pico- and femtosecond time ranges. Namely, it is shown that a very short population relaxation time is obtainable from the intensity ratio of the Raman scattering to the total secondary emission (R/T) under resonant excitation. Further, the amplitude and the correlation time of the perturbation are shown to be determined from the analysis of the excitation profiles for the Raman and luminescence spectra. The values thus determined are compared with the relaxation parameters obtained by the ultrafast laser spectroscopy.

Let us consider dye molecules in solution, in which the solute molecules are perturbed by a number of surrounding solvent molecules. A stochastic approach, in which the perturbation is treated as a random energy modulation, is considered to be well applicable in this case. Then, under the assumption of a Gaussian-Markovian process for the modulation, the excitation profiles for the sharp Raman line and the broad fluorescence band can be expressed using four parameters; the lifetime of the excited state T_1, inhomogeneous broadening D_s, the amplitude D and the correlation time τ_c of the energy modulation [1].

From the calculation based on this model, it has been found that the intensity ratio of the first-order Raman line and the sum of the Raman line and the corresponding luminescence band under resonant excitation is given by $(R/T)_{res} = 2S/(\Delta E_h T_1)$, where ΔE_h is the half-width at half-maximum (HWHM) of the absorption spectrum without the inhomogeneous broadening and S is a numerical factor which varies monotonically from 1.0 in the fast-modulation limit ($D\tau_c \ll 1$) to 1.48 in the slow-modulation limit ($D\tau_c \gg 1$) [2]. In deriving this relation, the intensity of the Raman component has been assumed to be proportional to the square of the average polarizability. This assumption is valid so long as $D\tau_c$ is not much larger than unity, as is usually the case for the dye molecules in solution. The

Springer Series in Chemical Physics, Vol. 48: **Ultrafast Phenomena VI**
Editors: T. Yajima · K. Yoshihara · C.B. Harris · S. Shionoya

above relation can be interpreted in terms that under resonant excitation, the Raman process takes place within the reciprocal homogeneous absorption bandwidth irrespective of the rate and amplitude of the energy modulation concerned. It has also been found from our calculation that the ratio $(R/T)_{res}$ is rather insensitive to the inhomogeneous broadening D_s as well as $D\tau_c$. Therefore, it is considered to be possible to determine T_1 from the measured intensity ratio and the homogeneous width of the absorption band.

In order to confirm this, we have performed experiments in crystal violet and malachite green in water at room temperature and also in β-carotene in isopentane at 177 K. The Raman intensity has been obtained by integrating the first-order Raman peaks of the dye molecule and the corresponding fluorescence intensity by decomposing the observed fluorescence spectrum into the sum of Gaussian vibronic transitions. The homogeneous width of the absorption band has been estimated by assuming the same widths for the homogeneous and inhomogeneous broadenings. This assumption is based on the experimental results that the ratio of the homogeneous and inhomogeneous widths is almost unity in various dyes in solution [1,3]. The lifetimes estimated from these data of 1.2, 0.8 and 0.15 ps agree well with those of the values in the literature which were determined by ultrafast spectroscopic methods, i.e., 1.1, 0.7-1.2 and 0.1-0.265 ps, respectively [2]. This result shows that a very short population relaxation time can be estimated from the Raman-to-total emission intensity ratio under resonant weak-light excitation.

When the excitation frequency is moved away from resonance, the intensity ratio R/T increases remarkably for $D\tau_c > 1$, though it remains unchanged for $D\tau_c \ll 1$. The fact that R/T varies with the excitation energy is one of the typical examples of the non-Markovian behavior in the optical response. Our calculation has revealed that the inhomogeneous broadening affects the R/T ratio only weakly even for the off-resonant excitation. Therefore, it will be possible to obtain D and τ_c from the intensity ratio R/T under off-resonant excitation, provided T_1 is known.

Generally, the four parameters T_1, τ_c, D and D_s can be determined from the analysis of the excitation profiles for the Raman lines and the fluorescence band as well as of the absorption and fluorescence band-shapes. In fact, we have analyzed the Raman and fluorescence excitation profiles for β-carotene in isopentane, and determined the parameter values as T_1 = 170 fs, τ_c = 53 fs, D = 300 cm^{-1} and D_s = 360 cm^{-1} at 177 K [1]. These values have been found to reproduce the absorption and fluorescence band-shapes. Since $D\tau_c$ is 3, the energy modulation in this system is relatively slow. In such a case far from the fast-modulation limit, it is not the dephasing time T_2, but the amplitude D and the correlation time τ_c of the energy modulation, that characterize the linear and nonlinear spectroscopic properties. For example, the homogeneous width (HWHM) of the absorption band is expressed as $1/T_2 = D^2\tau_c + 1/2T_1$ in the fast-modulation limit, while it is $D\sqrt{2\ln 2}$ in the slow-modulation limit when $DT_1 \gg 1$. Namely, the pure dephasing rate which is given as $\Gamma = D^2\tau_c$ in the fast-modulation limit should be replaced, in the slow-modulation case, by the energy fluctuation amplitude D multiplied by some numerical factor that depends on the experimental method to measure it. In this case, various transient responses exhibit non-exponential behavior and in addition the lineshapes of the absorption and emission spectra become Gaussian even when the inhomogeneous broadening is absent.

In order to compare the parameters estimated from the excitation spectra for the second-order optical processes with those obtained by ultrafast nonlinear spectroscopy, we have calculated the time characteristics of photon echo and hole burning. A calculation of the $D\tau_c$-dependence of the time behaviors of the two-pulse photon echo by the use of Aihara's theory [4] has revealed that the echo intensity decays non-exponentially for $D\tau_c > 1$ as a function of the time separation t_s between the first and the second exciting pulses, though it exhibits a single exponential decay as $\exp(-4\Gamma t_s)$ for $D\tau_c \ll 1$. When T_1 is sufficiently long, the time response of the echo signal in the slow-modulation limit is expressed as $\exp[-(t_s/\tau_{pe})^3]$, where $\tau_{pe} = D^{-1}(3D\tau_c/4)^{1/3}$. For the case of $D\tau_c > 1$, the 1/e-decay time of the echo intensity has been found to be approximated by τ_{pe}, which is estimated to be 18 fs, if we use the parameter values obtained for malachite green in water at room temperature. This is much longer than $1/4\Delta E_h \sim 3$ fs and $1/4D^2\tau_c \sim 1$ fs. Since the absorption spectrum of organic dye molecules in solution usually has a shoulder due to the vibronic transitions, it is necessary to take into account the effect of the quantum beats due to the contribution of a pair of vibronic levels in the excited state, when the light pulses of the duration of 10 fs or less are employed for the experiment [5]. It has been found that the t_s-dependence of the echo signal calculated for the 1300-cm^{-1} separated vibronic levels shows an apparently fast decay with a 1/e-decay constant of $\sim$ 7 fs which is accompanied by a small peak separated by $\sim$ 20 fs from the main peak. On the other hand, the width (HWHM) of the spectral hole and the rapid recovery time of the burned hole observed in a time-resolved hole-burning experiment are expected, in the slow-modulation case, to be comparable to the spectral width of the pump pulse and τ_c, respectively. In fact, the values of $\sim$ 100 cm^{-1} and $\sim$ 50 fs reported recently for solutions of dye molecules such as cresyl violet [6] are comparable to the HWHM of the exciting light of $\sim$ 100 cm^{-1} and $\tau_c = 53$ fs estimated for β-carotene, respectively. These results demonstrate that the analysis of the excitation profiles of Raman scattering and luminescence offers detailed information on the ultrafast relaxation parameters.

References

1. J. Watanabe, S. Kinoshita and T. Kushida: J. Chem. Phys. 87, 4471 (1987); J. Watanabe: Dissertation in Osaka University (March, 1986); T. Kushida and S. Kinoshita: In Proc. Workshop on Advanced Raman Spectroscopy, Kanpur (1987), in the press.
2. S. Kinoshita and T. Kushida, Chem. Phys. Lett. in the press.
3. S. Kinoshita, N. Nishi and T. Kushida, Chem. Phys. Lett. 134, 605 (1987); S. Kinoshita, N. Nishi and T. Kushida, J. Lumin. 40/41, 581 (1987); S. Kinoshita and N. Nishi, J. Chem. Phys. to be published.
4. M. Aihara, Phys. Rev. B25, 53 (1982).
5. C. V. Shank, Papers presented at this conference and 1987 Intern. Conf. on Luminescence (Beijing).
6. C. H. Brito Cruz, R. L. Fork, W. H. Knox and C. V. Shank, Chem. Phys. Lett. 132, 341 (1986).

Rise–Fall Ambiguities and Their Removal from Frequency-Domain Ultrafast-Measurement Techniques

R. Trebino[1], *C.E. Barker*[2], *and A.G. Kostenbauder*[2]

[1]Combustion Research Facility, Sandia National Laboratories, Livermore, CA 94550, USA

[2]Stanford University, Stanford, CA 94305, USA

In the past decade, nonlinear-optical frequency-domain techniques have provided a useful alternative to ultrashort-pulse (time-domain) techniques for the measurement of a wide range of ultrafast events.[1-4] While, in principle, frequency-domain techniques measure the Fourier transform of the temporal response of a material, nonlinear-optical techniques often yield only its magnitude. As a result, the temporal response may not be uniquely determined. Obvious (and trivial) ambiguities include a constant phase factor, displacement of the time origin, and simultaneous inversion and complex conjugation.[5] Less obvious, and far less trivial, ambiguities also exist, even if the possible temporal responses are restricted.[5-7] The study of such ambiguities, known as the phase retrieval problem, has received much attention in other contexts.[7]

In this paper, we consider techniques that measure the magnitude of the frequency response and study a commonly used temporal response, the sum of two exponentials:[1,8]

$$h(t) = \theta(t)\,\{A\exp(-t/\tau_f) + B\exp(-t/\tau_s)\} \tag{1}$$

where τ_f and τ_s are fast and slow time scales, $A \equiv \alpha/\tau_f$, $B \equiv (1-\alpha)/\tau_s$, α is a real-valued (possibly negative) relative weight, and $\theta(t)$ is the unit step function. When α is between zero and one, h(t) represents a rapid fall followed by a slower fall, while other values of α correspond to a rapid rise and then a slower fall. The measured line shape is the squared magnitude of the sum of the Fourier transform (with respect to $\omega \equiv \omega_1-\omega_2$) of Eq. (1) and, potentially, some constant coherent background, $\beta+i\gamma$:

$$\left|\Gamma_{\alpha\beta\gamma}(\omega)\right|^2 = \left|\frac{1+i\omega\tau_\alpha}{(1+i\omega\tau_f)\,(1+i\omega\tau_s)} + \beta + i\gamma\right|^2 \tag{2}$$

where $\tau_\alpha \equiv \alpha\tau_s + (1-\alpha)\tau_f$. We point out that knowledge of only the magnitude of the frequency response of this decay is not sufficient to specify uniquely the parameter α. Specifically, there exists a nontrivial set of parameters, α', β', and γ', that satisfies for all ω the equation, $|\Gamma_{\alpha'\beta'\gamma'}(\omega)|^2 = |\Gamma_{\alpha\beta\gamma}(\omega)|^2$:

$$\alpha' = -\frac{2\tau_f}{\tau_s-\tau_f} - 2(\beta-i\gamma)\frac{\tau_s+\tau_f}{\tau_s-\tau_f} - \alpha \tag{3}$$

with $\beta'=\beta$ and $\gamma'=-\gamma$. Thus, for every line shape, there are two different possible values for the parameter, α. [Equation (3) can also be obtained by writing $\Gamma_{\alpha\beta\gamma}(\omega)$ as a single quotient and taking the complex conjugate of the numerator only. This produces a nontrivial change in h(t), but does not change $|\Gamma_{\alpha\beta\gamma}(\omega)|^2$.] We refer to this ambiguity as a rise-fall ambiguity because it often results in a temporal response involving a rapid fall and then a slower fall having an identical line shape to that of a very different temporal response, one involving a rapid rise and then a slower fall. This type of ambiguity could be particularly troublesome, especially when each of the above responses corresponds to a different physical phenomenon or when rises and falls both potentially occur in the same phenomenon.[9-12]

Springer Series in Chemical Physics, Vol. 48: **Ultrafast Phenomena VI**
Editors: T. Yajima · K. Yoshihara · C.B. Harris · S. Shionoya

The introduction of quadrature-phase coherent background, $i\gamma$, removes the ambiguity because it forces the value of α' off the real line. Since a complex relative weight violates the assumption of a real decay, it is no longer physically acceptable. Even if complex weights are allowed, quadrature phase may remove the ambiguity because the alternate solution for α is necessarily accompanied by ambiguity in γ: $\gamma' = -\gamma$. Thus, if only its sign is known, coherent, quadrature-phase background removes rise-fall ambiguities for the general case of complex decays. Note that any nonzero amount of quadrature-phase background is sufficient to remove the ambiguity, although in practice, experimental noise implies a minimum value.

This ambiguity can be seen in experiments on the optical-Kerr effect in carbon disulfide using an induced-moving-grating technique. This technique involves measuring the dispersion of the four-wave-mixing process, $\omega_0 = \omega_1 - \omega_2 + \omega_3$, and is described in detail in ref. 1. The optical-Kerr response in carbon disulfide has been measured[9-12] and fit with sum-of-two-exponentials decays. The parameters obtained from these measurements are: $\tau_f = 100 - 360$ fs, $\tau_s = 1.3 - 2.2$ ps, and $\alpha = 0.16 - 0.51$. Other time scales (e.g., the electronic component and the recently observed fsec rise[9,10]) are beyond the resolution of our experiment.

Experimental wavelengths are $\lambda_1 = 572$-8 nm, $\lambda_2 = 575$ nm, and $\lambda_3 = 570$ nm, intensities are ~1 GW/cm^2, and polarizations are x, y, and x, respectively. Figure 1 shows an experimental spectrum (in the absence of quadrature-phase background). Fitting these data, we obtain two solutions, which both comprise the solid line in Fig. 1. All of the parameters of these two fits are nearly identical, except the relative-weight parameters, which differ by a large amount. That these two solutions represent the predicted ambiguity is also demonstrated in Fig. 1, which displays a plot of $\chi^2(\alpha)$, the value of the normalized sum of the squares of the weighted residuals for a given value of α, with all other parameters fixed at their optimum values. Observe that there are two very steep minima, representing the two possible solutions, which satisfy Eq. (3) to within a few percent.

Removal of the above ambiguity is achieved simply by tuning the probe frequency to ~25 cm^{-1} ($\omega_3 - \omega_2 \sim 5 \times 10^{12}$ s^{-1}) from the fixed-frequency excitation beam. An $\omega_0 = \omega_3 - \omega_2 + \omega_1$ process, which adds coherently to the $\omega_1 - \omega_2 + \omega_3$ process, provides the background. The quadrature-phase component of this background is significant when the probe frequency, ω_3, is approximately equal to $\omega_2 \pm 1/\tau_f$. Figure 2 shows data obtained in the presence of this coherent background, in which a positive value of α (+0.26) achieves a good fit, and other values do not. The dashed curve, which does not fit the data well, results from the use of all of these parameters, except for α, for which we use the negative solution obtained previously (–0.91). Allowing all of the parameters to float does not improve the fit for any negative value for α. To verify that no other solution exists for α, we plot in Fig. 2 the function $\chi^2(\alpha)$ for these data, showing that the only minimum occurs at the correct positive-α location. Thus, we confirm that the introduction of quadrature-phase coherent background removes the ambiguity in α.

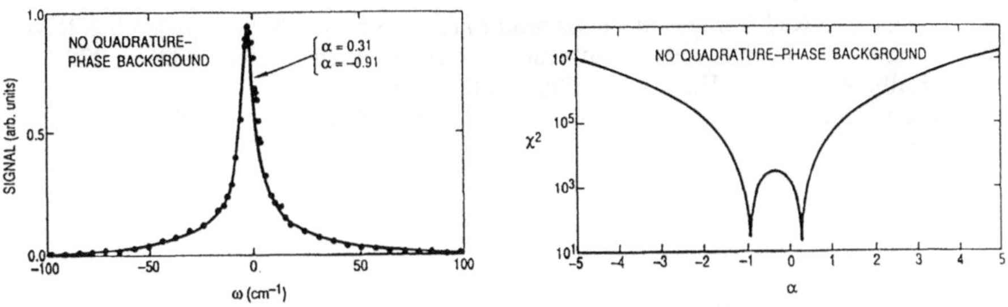

Fig. 1. Experimental results with negligible coherent quadrature-phase background (γ). Left: Diffraction efficiency vs. frequency difference, $\omega \equiv \omega_1 - \omega_2$. The solid line indicates the best fit, obtained for *two* different values of α: 0.31 and –0.91. [Fast and slow lifetimes obtained in these fits are 0.19 ps and 1.32 ps, respectively. Some in-phase background (β) exists in this scan: $\beta = 0.12$.] Right: Plot of χ^2 as a function of α. Observe the presence of two minima, indicating the ambiguity in the fit for α.

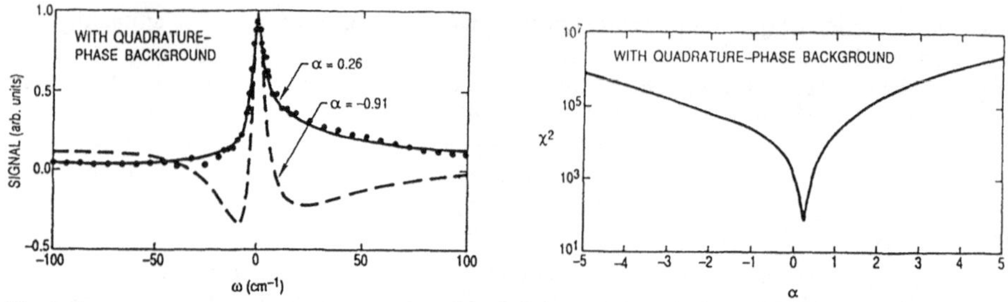

Fig. 2. Experimental results with substantial quadrature-phase coherent background. Left: Diffraction efficiency vs. frequency difference, $\omega \equiv \omega_1 - \omega_2$. The solid line indicates the best fit, obtained for a single value for α of +0.26. (Fast and slow lifetimes obtained in this fit are 0.19 ps and 1.42 ps, in agreement with Fig. 1 and short-pulse results.[9-12] Background parameters are $\beta = 1.40$ and $\gamma = -1.04$.) Using the value $\alpha = -0.91$ yields a visibly poor fit for all possible values of all other parameters. Right: Plot of χ^2 as a function of α. Observe the presence of only one minimum, indicating that the ambiguity in the fit for the parameter, α, has been removed.

We have limited discussion of these ambiguities to processes involving zero-frequency resonances, but our results are applicable to other coherent nonlinear methods, such as coherent anti-Stokes Raman scattering, where it is clear that ambiguities exist. It is not clear, however, whether such ambiguities will be as physically interesting as that discussed here.

Acknowledgements

This work was supported by the U.S. Department of Energy, Office of Basic Energy Sciences, Chemical Sciences Division, and the Air Force Office of Scientific Research.

References

1. R. Trebino, C.E. Barker, and A.E. Siegman, IEEE J. Quantum Electron. **QE-22**, 1413 (1986).
2. J.J. Song, J.H. Lee, and M.D. Levenson, Phys. Rev. A **17**, 1439 (1978).
3. T. Yajima, H. Souma, and Y. Ishida, Phys. Rev. A **17**, 324 (1978).
4. F. Keilmann, Appl. Phys. **14**, 29 (1977).
5. J.R. Fienup, J. Opt. Soc. Amer. **4**, 118 (1987).
6. E.J. Akutowics, Trans. Amer. Math. Soc. **83**, 179 (1956).
7. H. Stark, ed., *Image Recovery: Theory and Applications* (Academic Press, Inc., Orlando, 1987).
8. W. Yu, F. Pellegrino, M. Grant, R.R. Alfano, J. Chem. Phys. **64**, 2648 (1977).
9. C. Kalpouzos, W.T. Lotshaw, D. McMorrow, and G.A. Kenney-Wallace, J. Phys. Chem. **91**, 2028 (1987).
10. B.I. Greene and R.C. Farrow, in *Picosecond Phenomena III*, eds. K.B. Eisenthal, R.M. Hochstrasser, W. Kaiser, and A. Lauberau, (Springer-Verlag, Berlin, 1982), p. 209.
11. J.-M. Halbout and C.L. Tang, Appl. Phys. Lett. **40**, 765 (1982).
12. S. Ruhman, L.R. Williams, A.G. Joly, B. Kohler, and K.A. Nelson, J. Phys. Chem. **91**, 2237 (1987).

Time-Resolved Resonant Light Scattering of an Electron–Hole System in an Intense Laser Field

T. Iida [1] *and T. Higashimura* [2]

[1] Department of Physics, Osaka City University, Sumiyoshi-ku, Osaka 558, Japan
[2] Department of Natural Science, Osaka Women's University, Daisen-cho, Sakai, Osaka 590, Japan

The electron-hole(e-h) system in semiconductors highly excited by an intense laser-field(Ω_L-field) is described by the Bogolyubov quasiparticles[1,2]. In usual cases, the lifetime of the quasiparticle is the order of magnitude of collision lifetime, typically 10-10^3fs. In optical processes, a pair of the quasiparticles plays an important role. The pair state has the induced energy gap at which the density of states has divergent character. These features are expected to be clarified in the time-resolved resonant light scattering of probe light. We derive the general expression of the time-resolved spectrum of secondary emission in the same sprit of Aihara's theory[3], and apply the result to a simple two-band model of semiconductor quantum wells; we assume two-dimensional parabolic conduction ($E_c(k)$) and valence ($E_v(k)$) bands with extrema at k=0 and the band gap Eg.

We treat the Ω_L-field as a classical monochromatic field with the frequency Ω_L, since the coupling with the e-h system (g_L) is so strong that the perturbation theory cannot be used. Then, the Hamiltonian of the system depends on time explicitly. By the procedure given in [1], we eliminate the t-dependence by a canonical transformation and diagonalize the Hamiltonian by the Bogolyubov transformation. Neglecting the many-body effects, up to the constant energy shift, the Hamiltonian of e-h-Ω_L-coupled system is given as $H=\sum_k[e_\alpha(k)\alpha_k^+\alpha_k+e_\beta(k)\beta_k^+\beta_k]$ where $e_\alpha(k)+e_\beta(k)=\omega(k)$; $\omega(k)=2[g_L{}^2+x^2(k)]^{1/2}$ and $x(k)=[E_c(k)-E_v(k)-\Omega_L]/2$. The ground state of this system at T=0K is the vacuum state of the Bogolyubov quasiparticles, $|\tilde{0}\rangle$[2].

By representing the incident photon wavepacket and the resolution function of measurement system by F_1 and F_2 in energy domain, respectively, and denoting the mean incident(scattered) photon energy by $\Omega_{1(2)}$, the time-resolved spectrum, in the quasiparticle representation, is given as

Springer Series in Chemical Physics, Vol. 48: **Ultrafast Phenomena VI**
Editors: T. Yajima · K. Yoshihara · C.B. Harris · S. Shionoya

$$S(\Omega_1;\Omega_2;t)=\sum_\xi \left|\int dE\ e^{-iEt} \langle\xi|M(Q_2)G(z)M(Q_1)^+|\mathcal{O}\rangle\ F_2(\Omega_2-\Omega_L-E+E_\xi)\ \times \right.$$
$$\left. \times\ F_1(\Omega_1-\Omega_L-E)\right|^2 \tag{1}$$

with

$$M(Q)=\sum_k [u_k{}^2\beta_{-k}\alpha_{k+Q}-v_k{}^2\alpha_k^+\beta_{-(k+Q)}^+-u_kv_k\alpha_k^+\alpha_{k+Q}+u_kv_k\beta_{-k}\beta_{-(k+Q)}^+] \tag{2}$$

apart from a constant factor, where $|\xi\rangle$ is the excited state of H with energy E_ξ; $G(z)=1/[E+i\gamma-H]$, $\gamma>0$; t is the mean photon emission time being detected. In (2), the dependence on photon wave vector Q is ignored in u_k and v_k; $u_k{}^2=[(1/2)+x(k)/\omega(k)]$, $v_k{}^2=[(1/2)-x(k)/\omega(k)]$ and $u_kv_k=-g_L/\omega(k)$. Note that, in (1), the incident and scattered photon energies are shifted by Ω_L, since the system is driven by the Ω_L-field. The secondary emission (1) consists of three processes depicted in Fig.1; (a) is Rayleigh process with respect to the quasiparticle, (b) and (c) are the Stokes processes involving the scattering of one- and two-pairs of quasiparticles, respectively. Here, as an example, we take the (b)-process in the case of $\Omega_L>E_g$. The intermediate states are $\alpha_k^+\beta_{-k}^+|0\rangle$ with the energy $E_\xi=\omega(k)$ which has the induced energy gap $2g_L$. The density of states diverges at $2g_L$. Such a divergent character will be revealed more clearly in the quantum well due to its two dimensional nature. We can expect that the secondary emission has two peaks; one is given by $\Omega_2-\Omega_L+2g_L\cong\Omega_1-\Omega_L$ associated with the virtual process combining the peaks of F_1 and F_2, and the other is given by $\Omega_2-\Omega_L+2g_L\cong 2g_L$ associated with the process combining the pole of G(z) and the peak of F_2. We carry out the numerical computations for a typical case of GaAs/AlAs; the band gap is E_g= 1.5eV, the lifetime of pair state is assumed to be constant, $\gamma/E_g=0.004$. For the Ω_L-field, we put $\Omega_L/E_g=1.04$ and $g_L/E_g=0.04$. We assume a Lorentzian shape for $F_{1,2}$. The computed spectra are shown in Fig.2 where $\Omega_1/E_g=1.14$, the very narrow widths of $F_{1,2}$, $w_{1,2}/E_g=0.0007$, are adopted to see the spectral structure. Times are measured in units of $\hbar$/1meV=660fs.

More characteristic features are seen in the processes involving the LO-phonon(ω_0) scattering. In Fig.3 and 4, we show the spectra in energy and time domain for the 2LO-scattering in the (a)-process computed at T=0K, where we put $\Omega_1/E_g=1.13$, $w_1/E_g=0.008$, $w_2/E_g=0.016$ and $\omega_0/E_g=0.013$, and assume, for simplicity, the imaginary part of the self-energy to be constant, $\Gamma/E_g=0.004$. With increasing time, the peak changes from $\Omega_2\cong\Omega_1-2\omega_0$ to $\Omega_2\cong\Omega_L+2g_L-2\omega_0$ and to $\Omega_2\cong\Omega_L+2g_L$, reflecting the virtual and real processes. The divergent behavior of density of states at $2g_L$ plays an important role to give

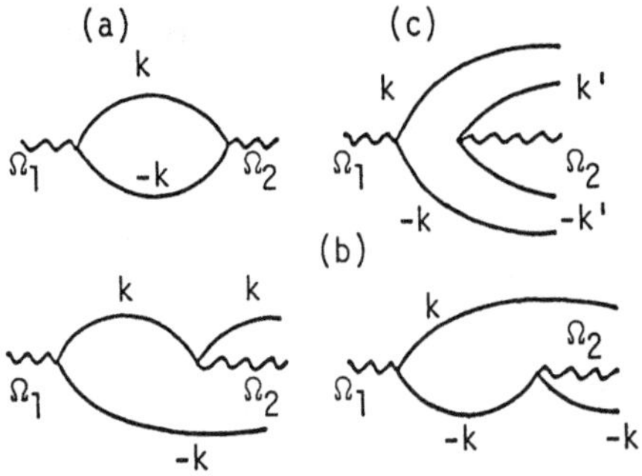

Fig.1 Secondary Emission Processes

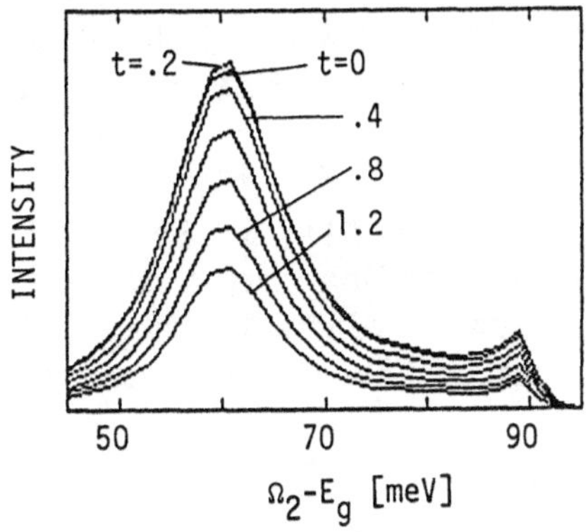

Fig.2 Time-Resolved Spectra

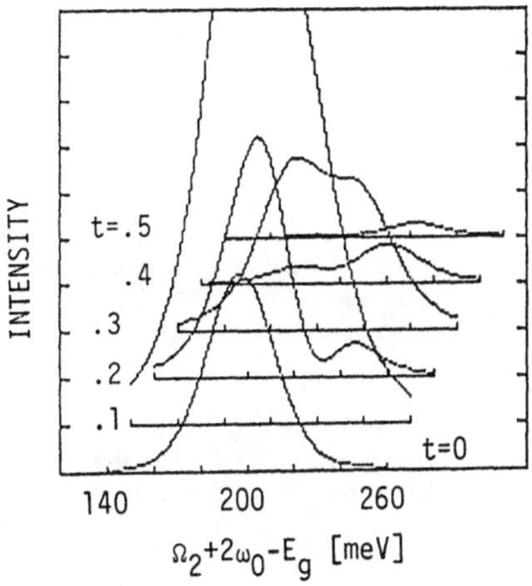

Fig.3 Time-Resolved Spectra

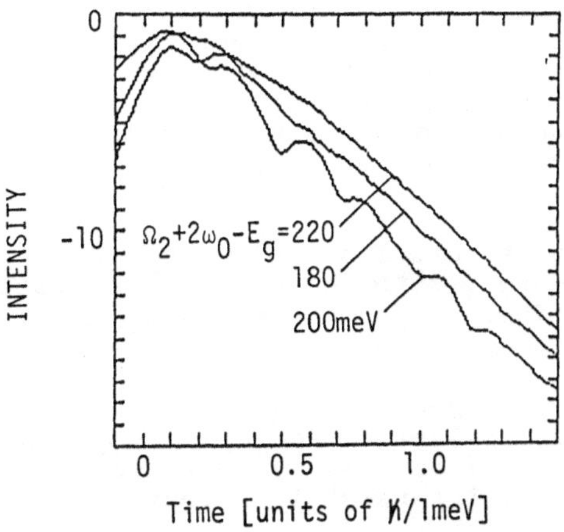

Fig.4 t-Dependence of Intensity

sufficient intensity at $\Omega_2 \cong \Omega_L+2g-2\omega_0$. Such phenomena cannot be expected in usual cases as three dimensional exciton-phonon systems. In Fig.3, the intensity decays with the time constant of system. The oscillatory behavior with period $2\pi/2\omega_0$ comes from the interference between the two processes through intermediate states with $E_\xi \cong E_g+2g_L$ and $E_\xi \cong E_g+2g_L+2\omega_0$.

References

1. Th. Harbich and G. Mahler: Phys. Stat. Sol. (b)104, 185 (1981)
2. Th. Harbich and G. Mahler: Phys. Stat. Sol. (b)117, 635 (1983)
3. M. Aihara and A. Kotani: Solid State Commun. 46, 751, (1983)

Microscopic Theory of Ultrafast Nonlinear Optical Phenomena in an Electron–Phonon System

M. Hama[1], *M. Aihara*[2], *and M. Yokota*[1]

[1]Department of Physics, Faculty of Engineering, Osaka City University, Osaka 558, Japan
[2]Department of Physics, Faculty of Liberal Arts, Yamaguchi University, Yamaguchi 753, Japan

With the development of laser pulse technique, transient nonlinear optical phenomena have become a subject of much investigation. YAJIMA and TAIRA [1] investigated the third-order transient optical parametric effect caused by two-pulse excitation in the Markovian limit. In ref. [1] it is shown that correlation trace (output light energy vs incident pulse separation) provides a general means of obtaining the transverse relaxation time T_2. The photon echo and the non-Markovian relaxation phenomena caused by two-short-pulse excitation in a localized-electron-phonon system were investigated by AIHARA from a microscopic viewpoint [2],[3]. In order to clarify the dynamics of lattice relaxation both in the spectral and time regions, we extend the results obtained in ref. 3 to the case of excitation pulses with finite temporal width. This enables us to study the dependence of signal intensity on the excitation photon energy.

Using the formalism of Ref. 3, apart from the unimportant multiplicative factor, we obtain the formula for off-diagonal element of density matrix corresponding to the third-order optical parametric effect;

$$
\begin{aligned}
\langle e|\rho^{(3)}(t)|g\rangle = & -i\int_{-\infty}^{t}dt_1\int_{-\infty}^{t_1}dt_2\int_{-\infty}^{t_2}dt_3 \\
& \times\{E_2(t_1-\tau_s)E_2(t_2-\tau_s)E_1^*(t_3)e^{-i(\Omega_i-\varepsilon)(t_1+t_2-t_3)-\gamma(t_1-t_2)} \\
& \times e^{-iL_{eg}(t-t_1)}(e^{-iL_{gg}(t_1-t_2)}+e^{-iL_{ee}(t_1-t_2)})e^{-iL_{ge}(t_2-t_3)} \\
& +E_2(t_1-\tau_s)E_1^*(t_2)E_2(t_3-\tau_s)e^{-i(\Omega_i-\varepsilon)(t_1-t_2+t_3)-\gamma(t_1-t_2)} \\
& \times e^{-iL_{eg}(t-t_1)}(e^{-iL_{gg}(t_1-t_2)}+e^{-iL_{ee}(t_1-t_2)})e^{-iL_{eg}(t_2-t_3)}\} \\
& \times \rho_R. \qquad (1)
\end{aligned}
$$

In the above equation, $E_j(t)(j=1,2)$ is temporal envelope function of j-th pulse which is peaked at t=0. τ_s is pulse separation. Ω_i is mean frequency of incident light, and ε is Franck-Condon energy. γ is lifetime of excited state. The hyper-operator L_{xy} (x,y=g,e) is defined by $L_{xy}A=H_xA-AH_y$, where $H_e(H_g)$ denotes the Hamiltonian associated with the ground (excited) state manifold. ρ_R is density matrix for the thermal equilibrium.

Springer Series in Chemical Physics, Vol. 48: **Ultrafast Phenomena VI**
Editors: T. Yajima • K. Yoshihara • C.B. Harris • S. Shionoya

The induced electric-dipole moment $\langle d(t)\rangle$ is expressed by

$$\langle d(t)\rangle = Tr[d\langle e|\rho^{(3)}(t)|g\rangle] + c.c., \tag{2}$$

where Tr denotes the trace operation over the reservoir variables. The electron-lattice interaction Hamiltonian, $V(=H_e-H_g-\varepsilon)$, is expressed, up to the quadratic interaction, by

$$V=\sum_k h_k e_k(b_k+b_k^\dagger)+\frac{1}{2}\sum_k\sum_q h_{kq}(e_k e_q)^{1/2}(b_k+b_k^\dagger)(b_q+b_q^\dagger). \tag{3}$$

Thermal average in Eq. (2) is evaluated by the cumulant expansion method, and we consider the second cumulant. We assume the density of state of phonon system to be the Gaussian profile with the width γ_p.

In the numerical evaluations, we assume that h_k and h_{kq} are constants ($S_L=h_k^2$ and $S_Q=h_{kq}^2$), and neglected the longitudinal relaxation time because it is usually much longer than transverse one in condensed matter. At first, we show in Fig. 1 the absorption spectrum of this two-level system. The origin of energy is set at that of the zero-phonon line. The lattice relaxation rate γ_p is $0.4\omega_p$, where ω_p is the mean phonon frequency. The coupling strength S_L of linear electron-phonon interaction is 5. The ratio of electron-phonon coupling strength of the linear interaction to the quadratic one (S_L/S_Q) is 100. The inverse of temperature $\beta=1/kT$ is $5\omega_p^{-1}$. The dephasing time T_2 is then evaluated as $812\omega_p^{-1}$. As is shown in Fig. 1, a sharp zero-phonon line and multiphonon structure appear in this system with intermediate coupling strength.

In Fig. 2, we show the logarithmic diplay of time-integrated intensity $I(\tau_s)$ for several values of excitation energy Ω_i, which is indicated by arrows in Fig. 1. Pulse envelope functions are assumed to have the same Gaussian profile $(E_1(t)=E_2(t)=(\delta^2/\pi)^{1/2}\exp[-\delta^2 t^2])$. The spectral width δ of

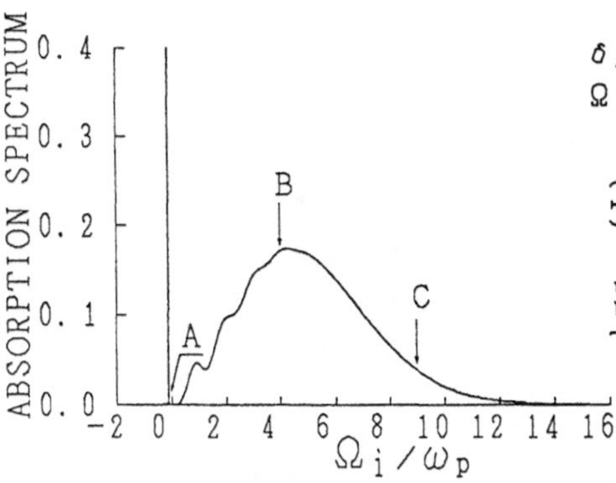

Fig. 1. Absorption spectrum of electron-phonon system with intermediate coupling strength. The arrows indicate the mean excitation energy where $I(\tau_s)$ in Fig. 2 is evaluated.

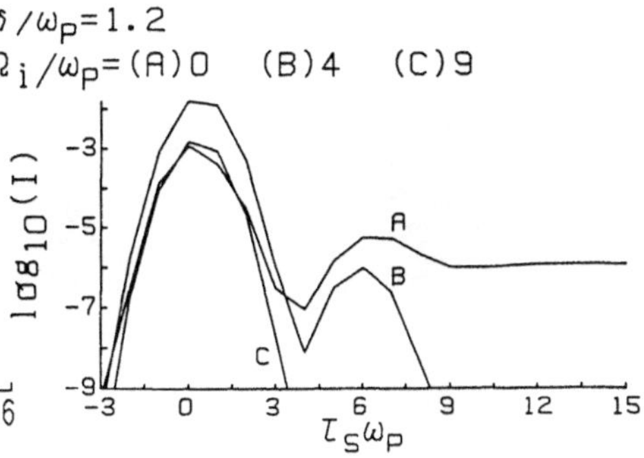

Fig. 2. The logarithmic display of time-integrated intensity as a function of pulse separation $\tau_s\omega_p$ for several values of excitation energy.

excitation pulses is set at $1.2\omega_p$. Other parameters are the same as in Fig. 1. From curve A (Ω_i=0) in Fig. 1, we find that $I(\tau_s)$ shows the damped oscillation with period $2\pi/\omega_p$ which reflects the dynamical nature of the system in the ultrashort time region, and that it is followed by the slow exponential dephasing with T_2 for $\tau_s >> \gamma_p^{-1}$. As is shown in curve B ($\Omega_i=4\omega_p$) and C ($\Omega_i=9\omega_p$), the initial rapid decay which is mainly determined by the excitation pulse profile becomes predominant as the excitation energy becomes high. This is explained by the fact that damping time of vibronic levels becomes shorter with increasing energy. That is, if $(n^{1/2}\gamma_p)^{-1}$ becomes shorter than pulse duration ($(n^{1/2}\gamma_p)^{-1}$ is the damping time of n-th vibronic level), dynamical nature of the system cannot be resolved. The behavior of curve C is related to the irreversibility of dipole moment discussed by YEH and EBERLY [4]. We should note that the dynamics of lattice relaxation is reflected in the τ_s-dependence of the time-integrated intensity. This is the distinctive advantage of the transient nonlinear optical method over the linear one, such as the time-resolved resonant light scattering.

References

1. T. Yajima and Y. Taira, J. Phys. Soc. Jpn. 47, 1620 (1979).
2. M. Aihara, Phys. Rev. B 21, 2051 (1980).
3. M. Aihara, Phys. Rev. B 25, 53 (1982).
4. J. J. Yeh, and J. H. Eberly, Phys. Rev. A 22, 1124 (1980).

Weak Localization of Femtosecond Laser Pulses by Random Media

R. Vreeker [1], *M.P. van Albada* [1], *R. Sprik* [1], *and A. Lagendijk* [1,2]

[1]Natuurkundig Laboratorium der Universiteit van Amsterdam, Valckenierstraat 65, NL-1018 XE Amsterdam, The Netherlands
[2]FOM Instituut voor Atoom- en Molecuulfysica, Kruislaan 407, NL-1098 SJ Amsterdam, The Netherlands

One of the interesting aspects of light propagation in disordered media is the possibility of photon localization in analogy with localization of electrons in random metallic systems [1]. A first indication of such behaviour is the recent observation of weak localization of light [2]. Weak localization manifests itself as a narrow peak in the intensity of light scattered from a disordered medium in backward direction. The effect arises from constructive interference between light waves scattered by the same particles, but in reversed sequence, and leaving the sample in the backscattering direction with the same phase. At an angle θ with respect to the backscattering direction the interference effect is reduced, since a given lightpath and its time-reversed counterpart now have a phase shift proportional to θd, where d is the distance between the first and last scattering center in the light path [3]. At angles θ larger than λ/d, where λ is the wavelength of the light, the interferences average out to zero. It follows that the coherent contribution of a light path to the backscattering peak is within an angular range related to the size of the path, i.e. the cone associated with the light path becomes narrower as its length, and hence d, becomes larger. We have performed pulsed light scattering experiments and were able to show the variation of the lineshape of the backscattering peak with the path length of the scattering events directly in the time domain.

The sample used in the time-resolved backscattering experiments was a concentrated suspension of TiO_2 particles (mean diameter 0.22 μm) in 2-methylpentane-2,4-diol (mean free pathlength 2 μm). Laser pulses with a duration of 100 fs were delivered by a colliding-pulse mode-locked ring laser operating at 620 nm. Time-resolved detection of the laser pulses scattered from the sample was performed using a light-gating technique, based on second-harmonic generation, with a temporal resolution of 30 fs. In Fig.1 experimentally obtained backscattering peaks at various delay times after

Springer Series in Chemical Physics, Vol. 48: **Ultrafast Phenomena VI**
Editors: T. Yajima · K. Yoshihara · C.B. Harris · S. Shionoya

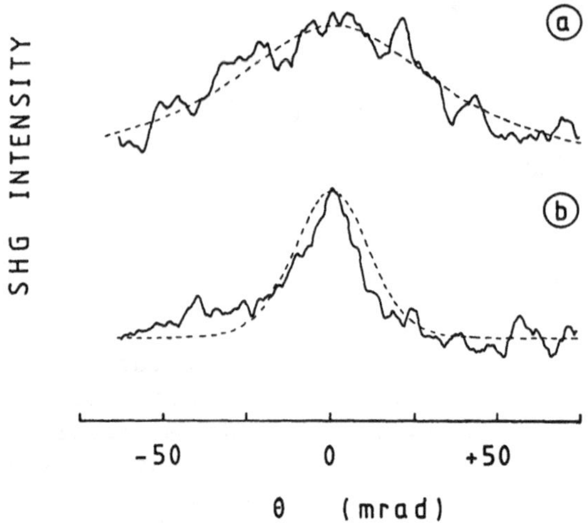

Fig. 1. Enhanced backscattering peaks measured for a 6.8 vol.% suspension of TiO_2 particles at different delay times after the laser pulse: (a) 30 fs; (b) 330 fs. Dashed curves represent results from calculations based on diffusion theory

the laser flash are shown. The backscattering peaks were detected for light polarized parallel to the polarization of the incoming laser pulse. We found that the width of the peak is reduced from approximately 80 mrad at 30 fs delay to approximately 25 mrad at a delay time of 330 fs. In Fig.2 time-resolved intensity profiles of light pulses scattered from the TiO_2 sample are shown. The profiles were detected at fixed angles with respect to the backscattering direction. The backscattered pulses have a short rise time, determined by the laser pulse duration and a slower long-time decay. The decay could be monitored for delay times up to 10 ps. Photons detected at these long delay times have traveled more than 2 mm in the sample and are elastically scattered over more than one thousand times.

As is seen from Fig.2, the scattered light intensity is largest at exact backscattering conditions, as could be expected. Under exact backscattering conditions all time-reversed light paths interfere constructively and the intensity equals almost twice the average intensity. At small angles with the backscattering direction, the phase difference between time reversed paths will increase with pathlength, and a coherent contribution to the scattered intensity occurs only during a short time after the laser flash. A quantitative analysis of the results was given on

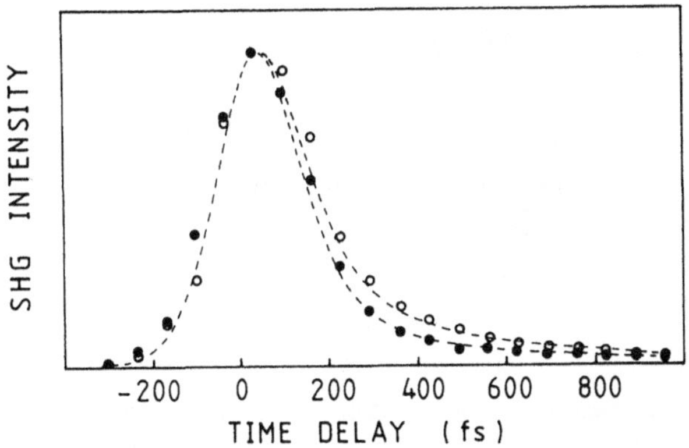

Fig. 2. Backscattered light pulses from a 6.8 vol.% suspension of TiO_2 particles measured for θ = 0 mrad (open circles) and θ = 27 mrad (full circles). Dashed curves represent results of calculations based on diffusion theory

basis of simple diffusion theory [4]. The results of this analysis are also given in Fig.1 and 2. Both the width of the coherent backscattering peak and the temporal intensity profiles are in good quantitative agreement with the theoretical predictions.

References

1. P.W. Anderson, Phil. Mag. B 52, 505 (1985).
2. M.P. van Albada and A. Lagendijk, Phys. Rev. Lett. 55, 2692 (1985).
3. A. Lagendijk, M.P. van Albada, and M.B. van der Mark, Physica 140 A, 183 (1986).
4. E. Akkermans, P.E. Wolf, and R. Maynard, Phys. Rev. Lett. 56, 1471 (1986).

Measurements of the Electronic Wave Function in the Time Domain

L.D. Noordam, A. ten Wolde, and H.B. van Linden van den Heuvell

FOM Instituut voor Atoom- en Molecuulfysica, Kruislaan 407, NL-1098 SJ Amsterdam, The Netherlands

For a one-electron atom, the characteristic time of the electronic wave function is the classical round trip time of the electron. The classical round-trip time of an electron scales with n^3, and is 0.15 fs for the ground state of a hydrogen atom. This time is about ten times shorter than an oscillation of visible light. For Rydberg states the characteristic time increases quickly: for the state with principal quantum number n = 40 the orbit time is already 10 ps. In this contribution a study of the dynamics of the electronic wave functions of Rydberg states with picosecond laser pulses is presented.

The effect of an intermediate resonance on multiphoton ionization is studied, in particular in the situation that the classical orbit time of the intermediate state is of the same order of magnitude as the time during which the atom is irradiated. Although the wave function of a Rydberg state is quite extended, only the part close to the core contributes to absorption of photons in the visible. In addition to this, the production of the Rydberg state also takes place in a small area, given by the dimensions of the ground state. If an atom is ionized in a two-step process, in which the involved intermediate states are Rydberg states, the ionization is resonantly enhanced only if the wave function can return to the core during the laser pulse. If the classical orbit time of the involved Rydberg states is longer than the duration of the pulse, the ionization probability will not be enhanced.

This enhancement can be measured experimentally in the following way. An atom is excited to a Rydberg state or a coherent superposition of states. In the present experiment this is done by three-photon excitation of xenon. The Rydberg atom is ionized by absorption of a subsequent photon. By scanning the photon energy the principal quantum number is varied, and therefore the classical orbit time of the intermediate state. Detection of electrons, resulting from ionization, as a function of the photon energy should show the transition from non-resonant direct four-photon ionization, to resonantly enhanced three-plus-one photon ionization.

The pulses of the dye laser (614 nm) are two times bandwidth limited. With a scanning autocorrelator the duration of the dye laser pulses is measured to be 3.5 ps. The duration of the UV pulses is estimated to be 3 ps, and the energy was 22 μJ. The light intensity in the focus is calculated to be 900 GW/cm^2. The excitation/ionization takes place in a metal box to minimize the electric fields. Outside this box the photoelectrons are decelerated to prevent background electrons with less than 1.33 eV kinetic energy from reaching a set of channel plates. The remaining electrons are accelerated in the direction of the channel plates. The signal of the channel plates is amplified, and integrated with a boxcar. With this experimental set-up, the ionization signal as a function of the photon energy can be measured.

The result of the experiment is presented in Figure 1. Firstly, a transition from an enhanced to a non-enhanced ionization signal can be seen. Secondly, this transition is not at the ionization threshold but at states with a finite principal quantum number. The sudden increase of

Springer Series in Chemical Physics, Vol. 48: **Ultrafast Phenomena VI**
Editors: T. Yajima · K. Yoshihara · C.B. Harris · S. Shionoya

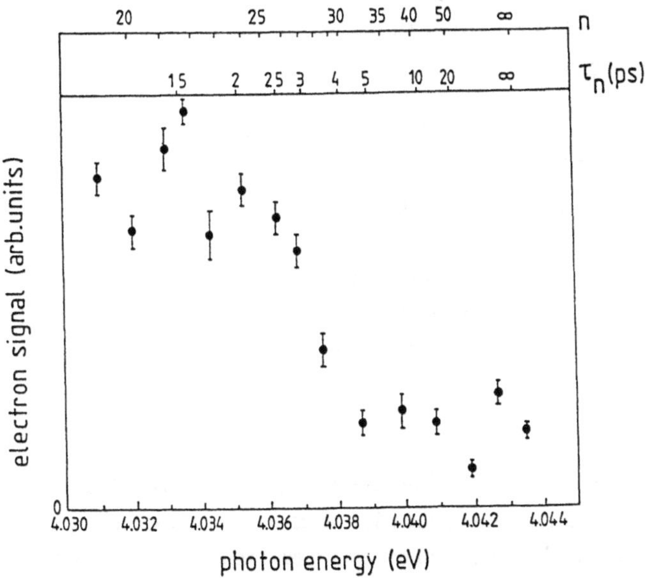

Fig. 1. The observed electron signal as a function of the energy of the ionizing light. In addition to the energy scale, a scale is given of the principal quantum number of the intermediate state, and the corresponding classical round-trip time of the n-state. The transition from resonant to non-resonant ionization can be seen at $\tau_n = 3$ ps. The pulse duration was of 3.0 ps and the calculated laser intensity 900 GW/cm^2.

ionization probability is observed as n=27 is excited as an intermediate resonance. The classical orbit time (τ_n) of this state is 3 ps. This time agrees with the pulse duration of the laser ($\tau_p = 3$ ps). For $\tau_n > \tau_p$ no wave function returns to the core during the laser pulse: only direct four-photon ionization can occur. The ionization probability is independent of n. For $\tau_p < \tau_p$ the wave function does return to the core within the laser pulse. Besides direct ionization, now resonantly enhanced three-plus-one photon ionization takes place. Due to the latter, the ionization probability increases enormously. In Figure 1 the increase can be seen for Rydberg states lower than n = 28.

In conclusion, for sufficiently high Rydberg states the wave function does not return to the core during the laser pulse. In this case several Rydberg states are coherently excited adding up to a wave packet [2], describing a radially localized moving electron. This wave packet can be detected by applying a second laser pulse.

References

[1] Liberman, J. Pinard and A. Tabel; *Phys.Rev.Lett.* **50**, 888 (1983).

[2] Parker and C.R. Stroud Jr.; *Phys.Rev.Lett.* **56**, 716 (1986); G. Alber, H. Ritsch and P. Zoller; *Phys.Rev.* **A34**, 1058 (1986).

Above-Threshold Ionization Observed in the Femtosecond Regime

H.B. van Linden van den Heuvell[1], H.G. Muller[1], P. Agostini[2], G. Petite[2], A. Antonetti[3], M. Franco[3], and A. Migus[3]

[1]FOM Instituut voor Atoom- en Molecuulfysica, Kruislaan 407, NL-1098 SJ Amsterdam, The Netherlands
[2]Service de Physique des Atomes et des Surfaces, CEN Saclay, F-91191 Gif-sur-Yvette, France
[3]Laboratoire d'Optique Appliquée, Ecole Polytechnique, ENSTA, F-91120 Palaiseau, France

Above-threshold ionization (ATI) is investigated in xenon using high-intensity light pulses at 620 nm with a pulse duration shorter than 100 fs, the shortest used to investigate ATI so far. The investigations extend existing measurements [1,2,3] into the femtosecond regime. The aim was to observe ATI, not pertured by ponderomotive effects, that, in the case of longer pulse durations, increase the kinetic energy of the electrons after the ionization.

The laser source was a colliding-pulse mode-locked ring laser working at 620 nm and four dye amplifiers pumped by a frequency doubled Q-switched Nd:YAG laser. This system generates 80-fs pulses in the mJ range at a 10-Hz repetition rate. The pressure of xenon gas was varied in such a way that no more than 20 ionization events per laser shot took place. Of these 50% were detected since the electron spectra were recorded with a magnetic-bottle spectrometer [4] with an acceptance angle of 2π sr.

A number of electron-energy spectra at pulse energies varying from 11 to 35 μJ are displayed in figure 1. The structure of ATI is clearly observable in all spectra [5], comparable with the situation where longer pulses are used: discrete peaks are spaced by the photon energy of 2.00 eV. Figure 1 shows that all peaks are shifted. The shifts are caused by the difference in AC-Stark shift of the neutral ground state and the ionic ground state, and the fact that the ionized electrons have a 'quiver energy' in the laser field of $\Delta E = e^2 I/4m\omega^2$ (m and e are the mass and the charge of the electron, ω and I are the frequency and intensity of the laser). Since the shift is dominated by this quiver energy, the relation between the light intensity and the shift of the peaks is linear. Such a linear dependence is in agreement with the observations in figure 1 and leads to very accurate measurements of light intensities. In experiments where longer pulses are used, the shift of the ionization limit is not detectable directly on the basis of the kinetic energy of the ionized electrons, since electrons are accelerated by the ponderemotive force. For short pulses the product of the ponderomotive force and the time that this force acts on the electron is too short

Springer Series in Chemical Physics, Vol. 48: **Ultrafast Phenomena VI**
Editors: T. Yajima · K. Yoshihara · C.B. Harris · S. Shionoya

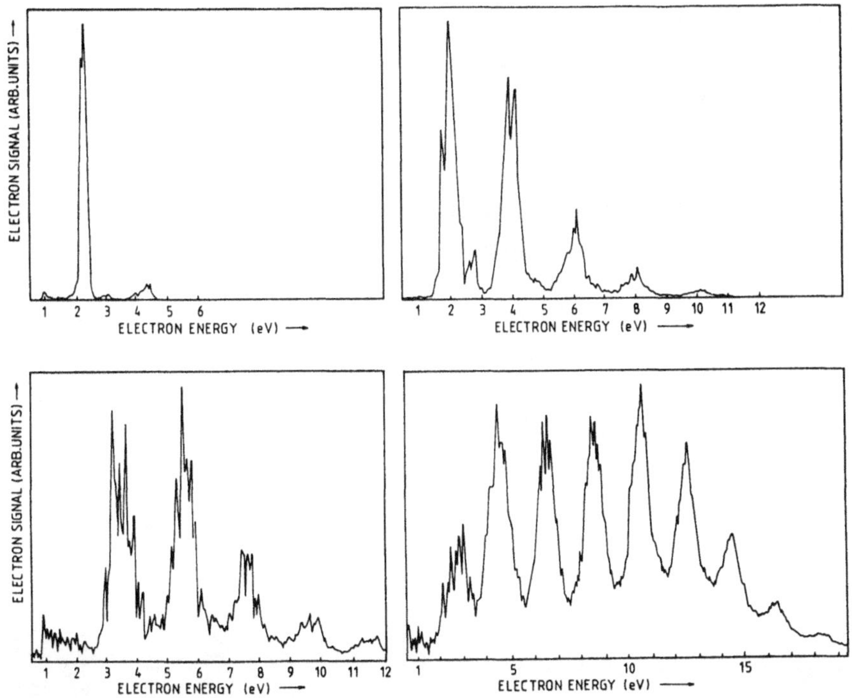

Figure 1 Electron-energy spectrum measured after multiphoton ionization of xenon with 620 nm light pulses of 100 fs duration. The energy of the pulses is 11 μJ for spectrum (a), 24 μJ for spectrum (b), 29 μJ for spectrum (c) and 35 μJ for spectrum (d). The assumed shifts of the spectra to lower energy are 1.7, 4.0, 4.5 and 5.6 eV, respectively.

to change the momentum of the electron, and the shift of the ionization limit is observed directly. An equivalent requirement is that the light is switched off, on a time scale short compared to the time that an electron moves a distance of the focus size. According to this interpretation, and neglecting the AC-Stark shifts of the ground state of the initial atom and the final ion, the spectrum given in figure 1d is measured at a light intensity of $1.7 \cdot 10^{14}$ W/cm^2. When the shift of the ion ground state is neglected and the shift of the atom ground state is taken into account explicitly, a light intensity of $1.2 \cdot 10^{14}$ W/cm^2 is found.

By measuring the ion signal it was established that at this laser intensity only a small fraction (less than 5%) of ionization events leads to Xe^{2+} production. Apparently, the used light intensity is less than the saturation intensity for the used wavelength, pulse duration and target atom. For longer pulses but otherwise similar conditions it is known [3] that the saturation intensity is lower. A tentative conclusion, drawn in ref [5] is that the difference is due to the resonance after absorption of six photons. In the limit of short light pulses, this resonance will not increase the ionization rate, in the limit of long pulses this resonance is indeed increasing the ionization rate.

[1] P. Agostini, J. Kupersztych, L. A. Lompré, G. Petite and F. Yergeau, Phys. Rev. A.

[2] T. S. Luk, T. Graber, J. Hara, U. Johann, K. Boyer, and C. K. Rhodes, J. Opt. Soc. Am. **B4** 847 (1987).

[3] R. R. Freeman, P. H. Bucksbaum, H. Milchberg, S. Darack, D. Schumacher and G. M. Geusic, Phys. Rev. Lett. **59** 847 (1987).

[4] P. Kruit and F. H. Read, J. Phys. E **16** 313 (1983).

[5] H. G. Muller, H. B. van Linden van den Heuvell, P. Agostini, G. Petite, A. Antonetti, M. Franco and A. Migus, Phys. Rev. Lett. **60** 565 (1988)

Two-Photon Absorption Sampling Spectroscopy for Fast Transient Luminescence Measurements

Y. Takagi and K. Yoshihara

Institute for Molecular Science, Myodaiji, Okazaki 444, Japan

Recent development of femtosecond lasers has stimulated interest in high time-resolution spectroscopic methods using nonlinear optical processes. In order to make sampling of fast transient luminescence, a nonlinear polarization is induced in a suitable material by an intense optical field giving rise to modulation of the refractive index or sum (difference)-frequency generation as in the optical Kerr shutter and up (down)-conversion, respectively.

In this report a new method of sampling spectroscopy based on two-photon absorption (TPA) is proposed. Two-photon resonant transition takes place in an atomic vapor using one photon from a pump wave and one photon from the luminescence to be examined. Detection is by the observation of fluorescences from appropriate excited energy levels, or of ionization due to further excitation following the two-photon transition. Time resolution is limited only by the pump duration time and wavelength selection is determined by the two-photon transition frequency minus pump-wave frequency. Two advantages of this method are that time resolution is not limited by dispersion in the medium and that luminescence in the deep UV can be detected. Also by scanning the pump-wave frequency the time-resolved luminescence spectrum is obtained.

Two-photon transition rate is roughly estimated [1] as $W=\sigma_1 F_1 \tau \sigma_2 F_2$, where σ_i and F_i are the absorption cross section and photon flux of the incident beams (pump and luminescence), respectively. τ is a duration time in a virtual excited state. Taking $\sigma_i = 10^{-16}\,cm^2$ (square of a typical atomic dimension), $\tau=\omega^{-1}=10^{-15}\,s$ (intermediate states are assumed to be far from one-photon resonance), and $F_1=10^{29}$ photons $cm^{-2}s^{-1}$ (easily available from a flashlamp-pumped mode-locked Nd:YAG laser), the transition rate is $W=10^{-18}F_2$. An absorption coefficient for the luminescence under the irradiation of the pump pulse is then $10^{-2}\,cm^{-1}$ assuming an alkali vapor of 1 torr. This value indicates the applicability of this method, if other undesirable linear or nonlinear processes are negligible.

In order to have some idea of the detectability, TPA spectra were taken for sodium (3S→5S, 6S, 4D, and 5D transitions), cadmium ($5^1S\to5^1D$), and mercury ($6^1S\to6^1D$) vapors using an excimer laser-pumped dye laser. The lowest pump energy required for fluorescence detection using a photomultiplier without data manipulation was about 1 μJ for sodium 3S→4D, while for an excitation very close to a one-photon resonance the signal enhancement was 10^3. Ionization detection was also made by inserting a copper-plate electrode (bias voltage 100-300 V) into the atomic vapor oven. In most cases the photon energy of the pump wave was enough for ionization from the two-photon excited state. The signal intensity was about one order of magnitude higher than in the fluorescence detection, but higher pump powers gave a background signal, which was proportional to the pump power.

Springer Series in Chemical Physics, Vol. 48: **Ultrafast Phenomena VI**
Editors: T. Yajima · K. Yoshihara · C.B. Harris · S. Shionoya

We have investigated the effect of an intermediate state on the time resolution and signal intensity. Two sets of optical parametric generators (OPG) with 0.1 mJ energy, 30 ps pulse width, and 10 cm^{-1} bandwidth pumped by a third harmonic of a mode-locked Nd:YAG laser were operated such that the sum-frequency of their outputs was resonant with the 3S→6S two-photon transition in sodium. First, we tuned a single OPG to this transition at 549.5 nm to check the signal size in the picosecond experiment. A pump energy of 1 μJ was necessary for the fluorescence detection, which is accidentally equal to the detectability in the nanosecond experiment. This is reasonable because the effective pump power is 10^2 times higher in pulse width but 10^{-2} times lower in bandwidth. Figure 1 shows fluorescence intensities at 285 nm (5P→3S) as a function of relative delay time between the two OPG outputs. OPG wavelengths were scanned around the 3S→3P resonance. When one of the incident beams was tuned to 3S→3P, the intensity increased and a long tail appeared. However when the detuning (Δ) was increased the long tail diminished and ultimately disappeared, so that only the cross correlation was observed. The time constant of the tail at Δ=0 was about 150 ps, which is two orders of magnitude shorter than the lifetime of the 3P state. Considering the D_1 and D_2 linewidths (Doppler width) of about 0.1 cm^{-1}, this tail probably gives a decay of a 3S→3P coherent superposed state. An additional longer decay at Δ=0 is likely to be due to the population decay in 3P state. We conclude that for the best time resolution intermediate state should be avoided.

We have applied the TPA sampling method to measurements of the fluorescence decay of dye solutions. A second (or fourth) harmonic of a mode-locked Nd:YAG laser was used for excitation of a dye solution and for TPA through a variable delay element. The pump and fluorescence beams were mixed by a dichroic mirror and focused into the vapor. Typical examples of the fluorescence decay are shown in Fig.2a and b for the fluorescence and ionization detections, respectively. In Fig.2a the fast decay probably results from amplified spontaneous emission at the sampled wavelength (462 nm). From the data in Fig.2b with a nonosecond decay (sampled at 634 nm) we can estimate the energy of the dye fluorescence sampled by the 10 ps pump pulse to be less than 1 nJ, assuming the fluorescence spectral width of 100 cm^{-1}. This

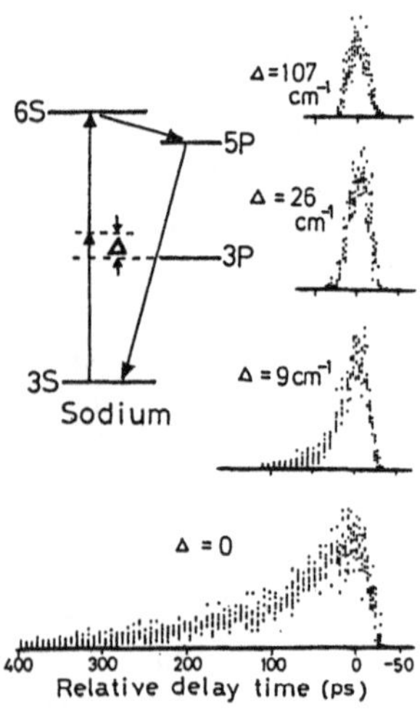

Fig.1 Fluorescence intensities of a sodium vapor as a function of relative delay time between two OPG

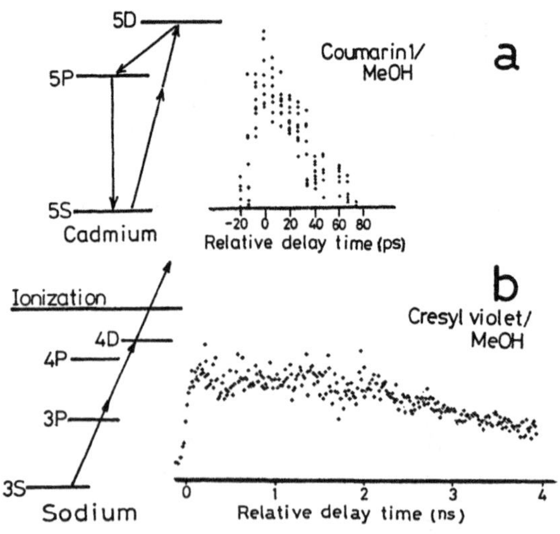

Fig.2 Fluorescence decays of dye solutions sampled by TPA. Pump wavelengths are (a) 266 nm and (b) 532 nm

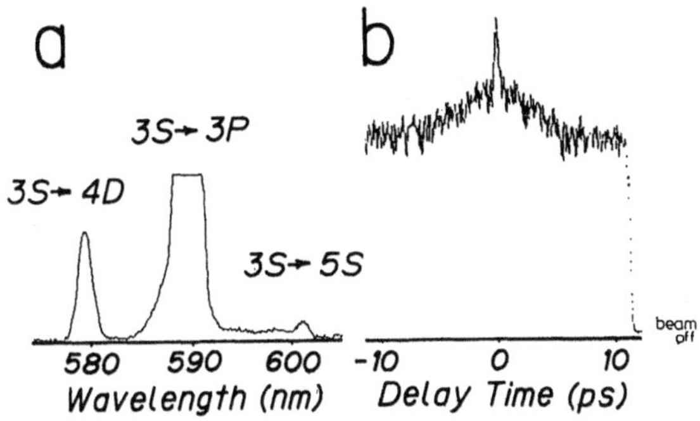

Fig.3 (a) TPA spectrum in sodium vapor and (b) autocorrelation trace using TPA at 578 nm

result is consistent with the detectability measurement described above because the product of the effective incident powers, to which the TPA probability is proportional, are comparable.

We have checked the applicability of our method in spectroscopic measurements using a synchronously pumped mode-locked dye laser. The sync-pump dye laser generates pulses at 76 MHz but the peak power is typically 10^{-3} times lower than the required pump energy of 1 μJ, leading to 10^{-6} times lower efficiency. Signal integration over 10^6 pulses is necessary for a comparable signal intensity. Wavelength of the sync-pump dye laser was scanned by moving a slit behind a dispersive prism [2] in the cavity. Figure 3a shows a TPA spectrum for sodium vapor taken by ionization detection. Peaks at 578 and 602 nm correspond to 3S→4D and 3S→5S transitions, respectively. The intense central peak originates from D-line excitation followed by a two photon ionization. We have measured an autocorrelation profile of the output pulses using an autocorrelator, in which a second harmonic crystal was replaced by sodium vapor. Figure 3b shows a trace of an autocorrelation signal at 578 nm. It gave the same shape as that obtained by second harmonic generation except for a large offset due to TPA coming from an uninteresting region where the two incident beams are not overlapped. This offset can be eliminated by using an atomic beam instead of the vapor, or by taking an autocorrelation at 602 nm using circularly polarized beams with opposite senses, where TPA with $\Delta m = \pm 2$ for sublevel selection-rule is not allowed [3]. It is noted that TPA sampling method gives a time resolution as high as the second harmonic generation. TPA sampling would give a high time-resolution since TPA is free from group-velocity dispersion effects. The TPA linewidth (0.1 cm^{-1}) of atomic vapor is much narrower than our laser bandwidth. Nevertheless, the autocorrelation width in Fig.3b, which is shorter than the reciprocal of 0.1 cm^{-1}, indicates that a fraction in the laser bandwidth larger than the TPA linewidth can contribute to the transition.

A more sensitive measurement using an electron multiplier detector in the atomic beam condition is now undertaken. TPA sampling would be a useful method for high time-resolution spectroscopy particularly in the deep UV region, where existing nonlinear sampling techniques are not applicable.

1. A. Gold, "Two Photon Spectroscopy" in the course of the Enrico Fermi School (Varenna 1967).
2. M.D. Dawson, T.F. Boggess, D.W. Garvey and A.L. Smirl, Opt. Commu. 60, 79 (1986).
3. M.D. Levenson and N. Bloembergen, Phys. Rev. Lett. 32, 645 (1974).

Application of the Time Characteristics of Synchrotron Radiation to Transient Spectroscopy

T. Mitani[1], *H. Okamoto*[1], *Y. Takagi*[1], *I. Yamazaki*[1], *M. Watanabe*[1], *K. Fukui*[1], *S. Koshihara*[2], and *C. Ito*[3]

[1]Institute for Molecular Science, Myodaiji, Okazaki 444, Japan
[2]Faculty of Science, University of Tokyo, Bunkyo-ku, Tokyo 113, Japan
[3]Faculty of Science, Nagoya University, Chigusa-ku, Nagoya 464, Japan

Synchrotron Radiation (SR) from an electron storage-ring can be utilized as a light source of pulses in a wide spectral region from the far-infrared to the soft X-ray. A high repetition rate of the pulse structure of the SR, whose frequency is 90 MHz (the multi-bunch mode) or 5.7 MHz (the single-bunch mode) with a duration time of 460 ps generated at the SR facility of the Institute for Molecular Science as shown in Fig. 1, is quite suitable for the time-resolved fluorescence and excitation spectroscopies in the subnanosecond time domain. In addition, a high stability of the time structure allows us to synchronize a CW mode-locked Nd:YAG laser (Quantronix 416, 10W) with the frequency of the SR pulses (90 MHz). A new system of the transient spectroscopy has been constructed for the first time by using a delay-time modulation technique. This system opens a unique possibility of expanding a measurable region of transient spectroscopy to infrared and far-infrared regions, where fast optical devices for detection have not been developed as yet.

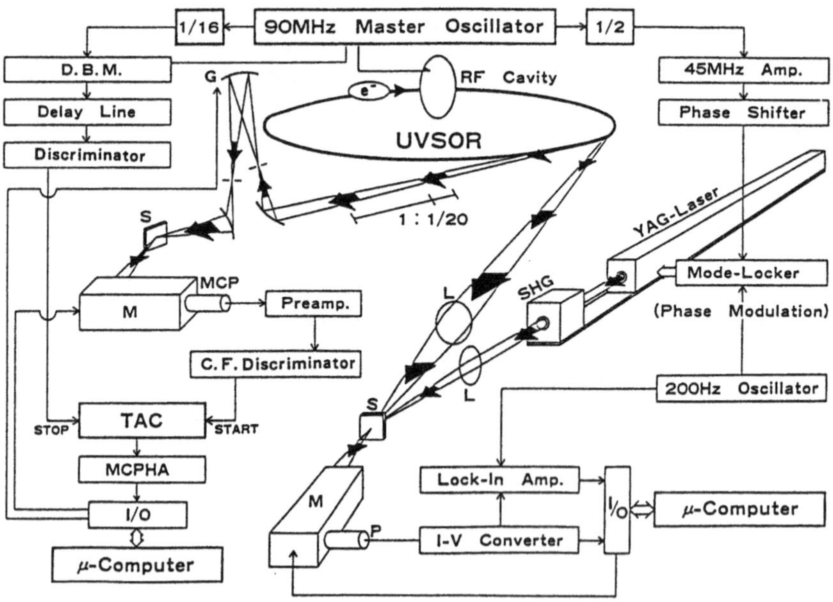

Fig. 1 Schematic diagram of the experimental setup for transient SR spectroscopy

Springer Series in Chemical Physics, Vol. 48: **Ultrafast Phenomena VI**
Editors: T. Yajima · K. Yoshihara · C.B. Harris · S. Shionoya

The experimental instrumentation, a) for the time-resolved fluorescence and excitation spectroscopies and b) for the transient absorption measurements is schematically illustrated on the right and left sides of Fig. 2, respectively. A typical number of monochromatic incident photons of the SR is 10^{2-3} photons/pulse with a spectral resolution of 0.1 %.

a) The time-resolved synchrotron spectroscopy with the aid of a time-correlated single-photon counting method has been successfully applied to anthracene crystals in the excitation energy region of 4-35 eV.[1] The time decay of the singlet exciton (S_1) fluorescence is significantly dependent on exciting photon energy, as shown in Fig. 2. Corresponding to the characteristic behavior of these time decays, the time-resolved S_1 fluorescence spectra presented in Fig. 3 show a drastic change within a time period less than 100 ps, particularly in a high energy region, which is nearly in resonance with the S_1 absorption peak. Furthermore, this short-time fluorescence has a high quantum efficiency when the high exciton bands are resonantly excited by the polarized light along the b-axis. Possible interpretations of the time correlation between excitation and fluorescence observed in the subnanosecond time period have been made on the basis of the exciton polariton model.[1] These experimental results demonstrate the capability of time-correlated spectroscopy in providing useful information on the dynamics of the electronic state in organic crystals.

b) Using the tuning method of the laser pulses as shown in Fig. 1, a delay time of the laser pulses, t_d, measured from the peak position of the SR pulses can be modulated by applying an AC voltage (e.g. 200 Hz) with an amplitude of Δt as shown in Fig. 4. A change of the transmitted SR light

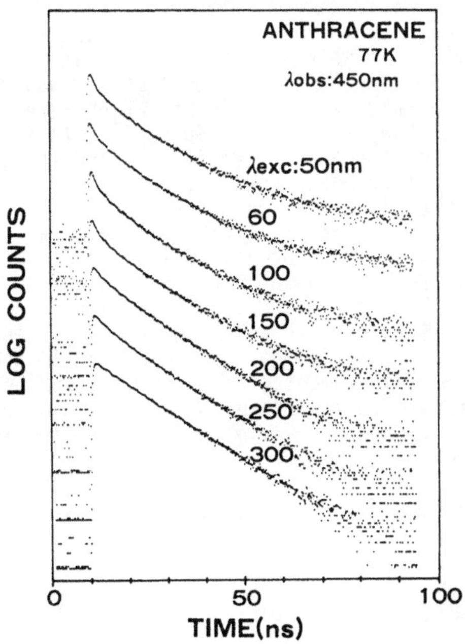

Fig. 2 The decay curves of the S_1 fluorescence of anthracene excited by the b-polarized light at various wavelengths at 77K

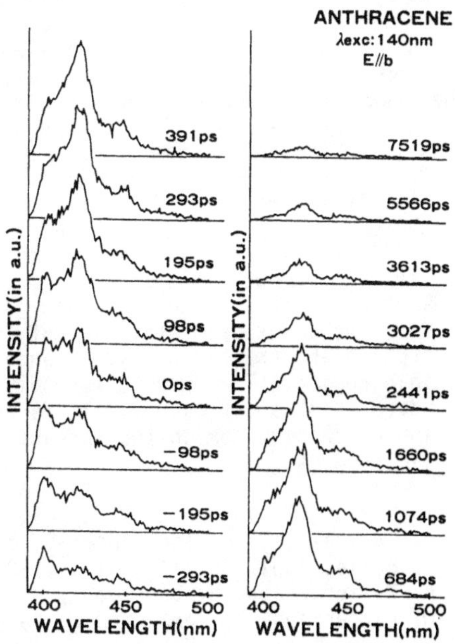

Fig. 3 The time-resolved fluorescence of anthracene excited at 140nm at 77K. The origin of time (0 ps) corresponds to the time of peak position of the exciting pulse

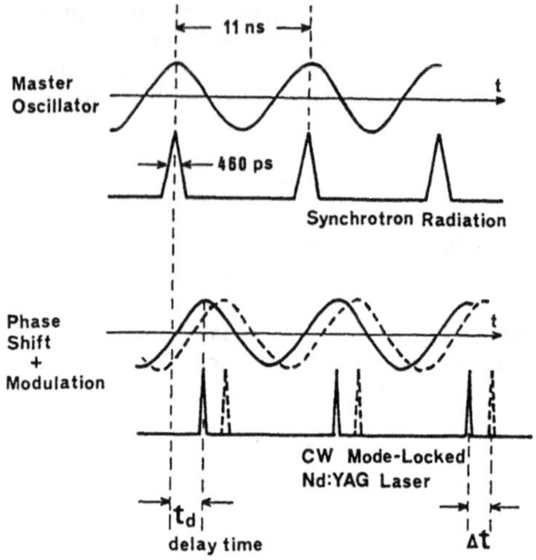

Fig. 4 The delay-time modulation technique in the synchronized SR and Nd:YAG laser system

induced by the modulated irradiation of the laser beam, which is sensitively dependent on t_d, is detected by a phase-sensitive lock-in amplifier as a modulation signal of the absorption coefficient.

The resolution of time in this transient spectroscopy is limited by both of the rise time of the SR pulses and the pulse width of the laser, which was checked by the single-photon counting system to be within 50 ps. These time characteristics of this system are in contrast to those obtained by the transient spectroscopy using a N_2-laser[2], in which the SR was used as a continuous light source, and by the two-photon absorption spectroscopy using the Q-switch Nd:YAG laser pulses with the pulse width of 15 ns, which were synchronized with one of the SR pulses at the frequency of 30 Hz[3]. In this sense, the present system is considered to have a high sensitivity and to be suitable for the two-step transient absorption measurements under a relatively low density of excitations in a short time period (0.05 - 11 ns). Furthermore, a spurious signal due to a thermal change of the sample by a shot of intensive laser pulse, which occasionally causes a serious problem in low frequency modulation spectroscopy in solids, may be removed from the modulation signals, since it would not follow the high repetition rate of pulses. As an experimental test of the ability of this system, the transient absorption spectra of R6G dye in an aqueous solution has been measured in the visible region. The results were almost in agreement with those obtained by standard laser spectroscopy. It is possible to expand the measurable energy region of this system to the infrared or vacuum ultra-violet region, which is now in progress.

1. T. Mitani, T. Yamanaka, M. Suzui, T. Horigome, K. Hayakawa, and I. Yamazaki, J. Luminescence 39, 313 (1988).
2. V. Saile, Appl. Opt. 19, 4115 (1980).
3. R. Pizzoferrato, M. Casalboni, R. Francini, U.M. Grassano, F. Amtonangeli, M. Piacentini, N. Zema, and F. Bassani, Europhys. Lett. 2, 571 (1986).

Part VIII

Dynamics on Surfaces and at Interfaces

Femtosecond Laser Photoionization Mass Spectrometry of Molecules on Surfaces

S.V. Chekalin, V.V. Golovlev, A.A. Kozlov, V.S. Letokhov, Y.A. Matveetz, and A.P. Yartsev

Institute of Spectroscopy, USSR Academy of Sciences,
Troitzk, 142092 Moscow Region, USSR

1. Introduction

The exit point of a photoelectron or a photoion is localized with an indeterminacy much lower than the laser wavelength. Indeed, the principal localization limit is determined by the Heisenberg indeterminacy principle, i.e., by the de Broglie wavelength for the photoelectron (photoion), which at thermal energies comes to 10 - 0.1 Å.

This makes it possible to utilize the resonant photoionization of molecules for their imaging with a spatial resolution of a few ängström units in a device the type of a photoion (or photoelectron) microscope proposed by Letokhov [1,2]. The highest spatial resolution should be inherent in a photoion projector based on the photodetachment of certain chromophores. This seems to be of special interest in revealing the sequence of bases in nucleic acids or the sequence of amino acids in proteins.

To develop a photoion microscope, it is necessary to learn how to detach with a laser pulse a chromophore from a large molecule adsorbed on the projector tip surface. Apparently ultrafast laser excitation is in principle capable of depositing through multiple-photon processes a substantial amount of energy in the chromophore, which gives reason to expect that it will be ionized and detached from the molecule. As no such process was known, we have launched a few years ago investigations into the photoionization of surface molecules.

When irradiating molecular crystals with nanosecond UV laser pulses [3] and picosecond visible laser pulses [4] falling within the electronic absorption band of the crystals, the resulting molecular ions were found to be produced without perceptible fragmentation. The mechanism of this phenomenon (thermal desorption of molecules, followed by their resonant stepwise photoionization in vacuo) was suggested and verified by experiment in [5] for UV laser pulses of moderate intensity. The experiments [6] on irradiating a complex molecule containing an aromatic chromophore with UV laser pulses 5 ps in duration revealed a rise of the chromophore peak in the mass spectrum. But to examine the formation of ions directly on the sample surface, with the desorption of neutral molecules being excluded, use should be made of subpicosecond laser pulses, for in that case one can guarantee that no neutral molecule will have enough time to move away from the surface for more than 0.3 nm during the laser pulse [5]. The experiments [7,8] on the femtosecond desorption of ions showed the me-

Springer Series in Chemical Physics, Vol. 48: **Ultrafast Phenomena VI**
Editors: T. Yajima · K. Yoshihara · C.B. Harris · S. Shionoya

chanism of production and detachment of ions in this case to be substantially different from the one indicated above.

In this paper, we present the results of our recent experiments on the femtosecond photoionization of large molecules, particularly tryptophan and tryptophan-containing proteins, which point to the real possibility of detaching a chromophore from a surface molecule.

2. Experimental Technique

The experiments were conducted with the aid of a standard photoionization time-of-flight mass spectrometer like, for example, the one described in [3,5]. The vacuum in the instrument's chamber was kept at 10^{-7}- 10^{-8} Torr. The laser radiation (300-fs pulses 10^{-4}J in energy at 308 nm and up to 10^{-3}J in energy at 615 nm) was incident upon the sample at a grazing angle of 80° and was focused upon it with a lens 10 or 20 cm in focal length. In the experiments, we observed the mass spectra of the ions produced upon irradiation of the sample and measured the threshold energy densities at which these spectra appeared. The detection limit of our time-of-flight mass spectrometer, at which mass spectra could still be registered reliably, corresponded to the arrival of some 10^2 - 10^3 ions at the detector (a secondary electron multiplier). The kinetic energy distribution of the ions was found from the shape of the respective mass peaks. The data obtained were compared with the same spectra resulting from irradiation of the same samples and with the same geometry of experiment with 15-ns pulses at 308 ns from a pulsed XeCl laser or with 10-ns pulses at 630 nm from a pulsed Q-switched laser.

Figure 1 shows schematic diagrams of irradiation of molecules in the photoionization mass spectrometer chamber with femtosecond and nanosecond laser pulses and their combinations.

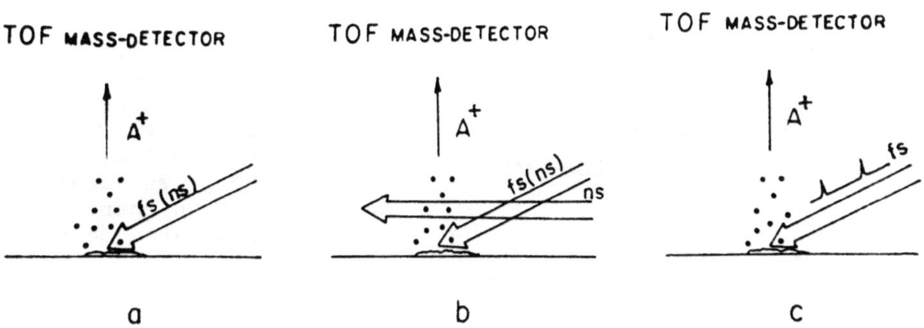

Fig. 1. Schematic diagrams of sample surface irradiation in the photoionization mass spectrometer chamber: a) Irradiation with a single femtosecond or nanosecond pulse; b) Desorption of molecules from the sample surface with a femtosecond or nanosecond pulse, followed by their photoionization with a second pulse of nanosecond duration; c) Irradiation with two femtosecond pulses. The second pulse is delayed relative to the first by a few picoseconds.

3. Experiments with Try and Try-Containing Tripeptide

The experiments revealed some specific features which are especially evident from comparison between the mass spectra resulting from irradiation with femtosecond and nanosecond laser pulses in the UV (308 nm) and visible (620 nm) regions of the spectrum.

The degree of fragmentation depends on both the laser pulse duration and wavelength. Figure 2 shows the mass spectra obtained upon irradiation of a pure tryptophan powder and indicates the corresponding laser pulse parameters. The UV pulse effects the multiple-photon ionization of Trp through the resonant S_1 state, while the visible pulse, the multiple-photon ionization and two-photon excitation of the S_1 state. As can be seen from comparison between the mass spectra of Figs. 2a and c, the ionization of the Trp chromophore is predominant in the latter case.

Of particular interest is the pulse-duration dependence of the threshold energy density E_{thr} for the appearance of mass spectra (Fig. 3) in the case of the multiple-photon UV ionization of the Trp Gly Ala tripeptide. It is clearly seen that the

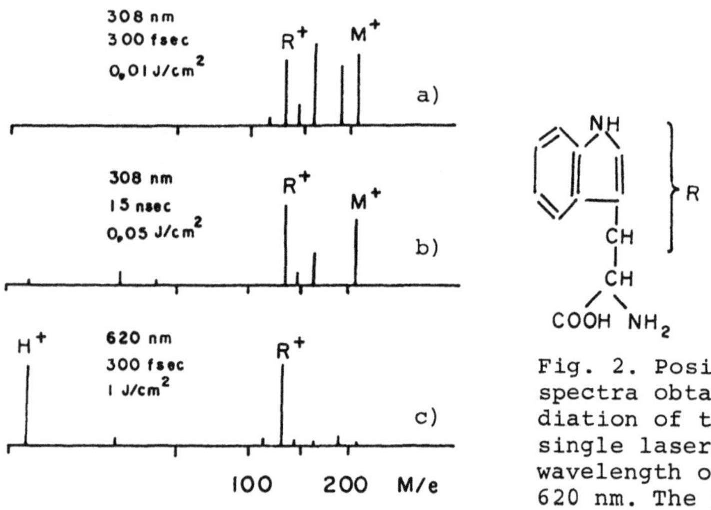

Fig. 2. Positive ion mass spectra obtained upon irradiation of tryptophan with single laser pulses at a wavelength of 308 nm and 620 nm. The pulse duration and irradiation energy density are indicated on the left.

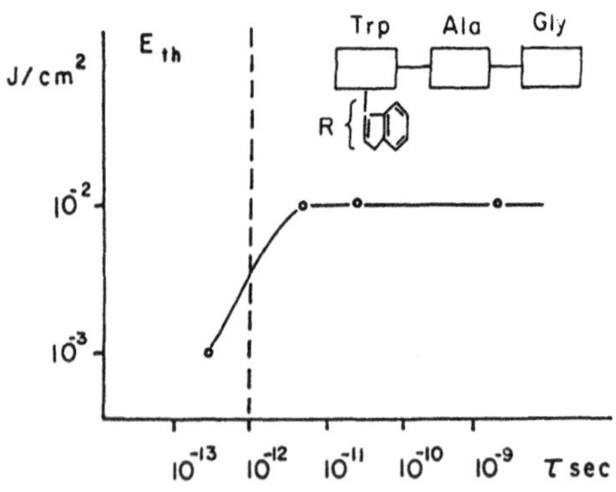

Fig. 3. Threshold energy density for the appearance of the tripeptide mass spectrum as a function of the UV laser pulse duration

threshold energy fluence is reduced by a factor of 10 in the case of irradiation with femtosecond pulses. The same effect is observed with the pure tryptophan powder. The threshold energy density for the appearance of the 300-fs pulse mass spectrum $E_{thr}^{fs} = 10^{-3}$ J/cm^2 (3 x 10^9 W/cm^2), is by an order of magnitude lower than that of the 15-ns pulse mass spectrum, $E_{thr}^{ns} = 10^{-2}$J/cm^2 (6 x 10^5 W/cm^2).

It should be noted that the observed mass spectra are of photoionization origin and have no relation to the formation of plasma due to a strong heating of the irradiated surface. The formation of plasma at high energy densities (observed only upon irradiation with nanosecond pulses) becomes clearly manifest in the character of the mass spectrum: it loses its sharply defined structure. Besides, the surface heating was estimated (10-100 deg). This value is insufficient for plasma formation.

The kinetic energies of the ions observed upon irradiation with femtosecond pulses are much higher than in the case of irradiation with nanosecond pulses, the distribution width does not exceed 1 eV (the spread function of the instrument), while for 300-fs pulses, this width amounts to 25 eV. What is more, the distribution maximum in the case of excitation with the femtosecond pulses is shifted a few tens of electron-volts towards the high-energy side, the shift reaching its maximum for the lightest mass (R^+) and diminishing with increasing mass.

The experimental results listed above suggest that the mechanisms by which ions are produced upon irradiation with nanosecond and femtosecond pulses are different. We ran a series of additional experiments which allowed us to reveal the role of local electric fields in the desorption and acceleration of the ions.

First, it was found that the velocity of the neutral fragments desorbed by the 300-fs laser pulses did not exceed the thermal value. For this purpose, an excimer laser beam was passed 0.5mm above and parallel to the sample surface, which ionized the fragments in the gas phase (Fig. 1b). The excimer laser pulse was delayed relative to the 300-fs pulse for a time from a few tens of nanoseconds (delay accuracy) to several tens of microseconds. The experimental conditions were selected so that no pulse taken separately yielded any ion signal. The width of the ion peaks observed in that case corresponded to ion energies of no more than 1 eV.

Second, we observed the change of the kinetic energy distribution of the desorbed chromophore ions upon irradiation (using the geometry of Fig. 1c) of the sample with two identical femtosecond laser pulses with a spacing of $\Delta t = 7$ ps. To generate the femtosecond double pulse, one half of the laser beam cross section was shut off with a plane-parallel quartz plate 4.2 mm in thickness. In that case the proportion of low-energy peaks in the kinetic energy distribution of the ions increased. In the case of double-pulse irradiation, some ions apparently form from the neutrals that have moved within 7 ps a few tens of ängström units away from the sample surface, these ions acquiring much lower velocities than those formed on the surface.

The fields required to accelerate the ions (around $E(M^+)/eR$ 0.5 V/Å) are fairly strong and can substantially facilitate their desorption from the sample surface by compensating for the effect of polarization of the medium.

The role of local electric fields in the acceleration of ions also agrees with the results of the experiments [8] on the femtosecond laser pulse desorption of ions formed on the surface of polycrystalline metal-free phthalocyanine crystals. In that case, the kinetic energies of the ions did not exceed 1 eV, which could be explained by the induced photoconductivity in the crystals in conditions of intense irradiation that prevented the formation of electric microfields. Further investigations are necessary to reveal the mechanism responsible for the formation of so strong surface electric field microinhomogeneities.

4. Tryptophan-Containing Proteins. Possible Spatial Effects

The above experiments have demonstrated that the femtosecond pulses are capable of detaching the chromophore ion from a molecule located directly on the sample surface. The mass spectra themselves might be used as a source of information on the location and surroundings of chromophores when studying more complex molecules. For example, if a tryptophan molecule is located on the surface of a protein globule, an ion can be desorbed. But if it is inside the globule, the ion formed will apparently not be able to escape. Thus, the ion signal may depend not only on the concentration of the chromophore being excited, but also on its location in the globule.

To investigate these possibilities, we ran preliminary experiments on the photoionization mass spectroscopy of samples containing relatively small amounts of Trp. Taken as the objects of study were crystalline proteins Ig Bovine, Ig G Rabbit, and Albumine Human. The absorption of the protein in the spectral range 280-310 nm is determined mainly by tryptophan. The photoionization mass spectra of all samples featured mass peaks typical of tryptophan. The intensity distribution of these peaks is approximately the same as the one obtained for pure tryptophan under similar conditions. The main difference between the spectra for the 300-fs UV or visible pulses is that the kinetic energy of the ions from IgB, for example, does not exceed 3 eV. Also, the kinetic energy distribution of the ions is substantially narrower than in the case of pure tryptophan or tripeptide. This apparently points to the absence of strong electric microfields at the surface of the IgB sample.

5. Conclusion

The experiments briefly described in the present paper demonstrate that femtosecond laser pulses offer new opportunities for the resonant multiple-photon ionization mass spectrometry of polyatomic surface molecules.

First, the fast femtosecond excitation makes it possible to produce molecular and fragmentation ions directly on the surface being irradiated. The kinetic energies of the ions can help one to judge about the electric fields on the surface. Of considerable interest is the femtosecond double-pulse irradiation technique whereby the first pulse desorbs neutral molecules from the surface and the second effects their resonant photoionization at some controllable distance from the surface. This technique extends the experimental studies started in [10] and opens up new possibilities for investigation into the molecular ion-surface interaction.

Second, two-photon excitation with an intense femtosecond pulse makes it possible to improve the selectivity of ionization of the chromophore (tryptophan in our case) in large molecules. This is of interest in the photoionization mass spectrometry of bioorganic chain molecules.

And, finally, the femtosecond excitation and ionization of chromophores in large globular molecules are sensitive to whether the chromophores are located on the surface or inside the molecules. This allows one to use the femtosecond photoionization mass spectrometry for studying the three-dimensional structure of bioorganic molecules.

References

1. V.S.Letokhov: Kvantovaya Elektronika (in Russian), 2, 930(1975); Comm.Atom.Molec.Phys., 11, 1(1981)
2. V.S.Letokhov: Laser Photoionization Spectroscopy (Academic Press, Orlando, San Diego, 1987), Ch.12
3. V.S.Antonov, V.S.Letokhov, A.N.Shibanov, Pis'ma Zh.Eksp.Teor. Fiz., 31, 471 (1980) (in Russian)
4. V.S.Letokhov, V.G.Movshev, S.V.Chekalin, Zh.Eksp.Teor.Fiz., 81, 480 (1981) (in Russian)
5. S.E.Yegorov, V.S.Letokhov, A.N.Shibanov, Kvantovaya Elektronika, 11, 1393 (1984) (in Russian)
6. V.S.Antonov, S.E.Yegorov, V.S.Letokhov, Yu.A.Matveyets, A.N. Shibanov, Pis'ma Zh.Eksp.Teor.Fiz., 36, 29 (1982) (in Russian)
7. S.V.Chekalin, E.F.Gavrilov, A.A.Kozlov, V.S.Letokhov, Yu.A. Matveyets, A.P.Yartsev, Proc. of IV Intern.Conf. Ultrafast Phenomena in Spectroscopy, Vilnius (1987)(World Publ.Co.,Singapore, 1988
8. A.A.Kozlov, V.S.Letokhov, Yu.A.Matveyets, S.V.Chekalin, A.P. Yartsev, Pis'ma Zh.Eksp.Teor.Fiz., 47, 294 (1988) (in Russian)
9. S.V.Chekalin, V.V.Golovlev, A.A.Kozlov, Yu.A.Matveyets, A.P. Yartsev, V.S.Letokhov, Journ.Phys.Chem.(submitted)
10. V.S.Antonov, V.S.Letokhov, Yu.A.Matveyets, A.N.Shibanov, Laser Chemistry, 1, 37 (1982).

Picosecond Photoionization Mass Spectroscopy and Optical Spectroscopy of Hot Semiconductor Surfaces

D. von der Linde, B. Danielzik, K. Sokolowski-Tinten, and P. Harten

Institut für Laser- und Plasmaphysik, Universität-GHS Essen, D-4300 Essen 1, Fed. Rep. of Germany

1. Introduction

When intense picosecond laser pulses are incident on strongly absorbing semiconductors the surface is heated up in a time comparable with the pulse duration. The rapid rise of the surface temperature is accompanied by ultrafast sublimation and desorption of atoms and molecules from the surface. For sufficiently high pulse energy, melting and vaporization of the material may occur. With picosecond laser heating these processes occur at unprecedented high rates. Interesting phenomena such as highly superheated transient phases can be observed and new distinct surface phenomena may be expected. Here, we report measurements of picosecond time-resolved desorption of atoms and molecular radicals from laser-heated surfaces and measurements of the ultrafast melting kinetics in Si and GaAs.

2. Picosecond Laser Mass Spectroscopy

We demonstrate the extension of laser photoionization mass spectroscopy to the picosecond time regime enabling the observation of ultrafast evaporation and desorption. A pump-probe scheme is used in which the surface is heated up by a pump pulse causing desorption of particles. An intense UV probe pulse following after some delay time photoionizes the desorbed neutral particles via multiphoton processes. The photoions are detected with a time-of-flight mass spectrometer (TOFMS).

Figure 1 shows a typical TOF mass spectrum obtained from a GaAs (100) surface after exposure to a single 25 ps heating pulse at 532 nm followed by a 266 nm photoionization probe pulse after 75 ps. The energy fluence of the heating pulse in our laser desorption experiments was much less than the threshold fluence for melting (45 mJ/cm^2). From the measured flight times we readily identify $^{69}Ga^+$ and $^{71}Ga^+$, the two common isotopes of gallium, and molecular arsenic, As_2^+, As_3^+, and As_4^+. The atomic As^+-signal was very weak. The group of line near 156 amu is attributed to Ga_2O^+-radicals from desorption of surface oxides. The strong alkali lines presumably originate from the chemical agents used during wafer surface preparation.

By varying the delay time of the photoionization pulse (probe pulse) the desorption kinetics of each species can be measured. As an example the measured Ga_2O^+-signal is shown as a function of delay time in Fig. 2. A monotonic step-like increase of the signal to an approximately constant level is observed. The risetime is about 50 ps. The relatively large scatter of the data is due to spatial surface inhomogenieties which cause changes of the signal when the surface is raster-scanned during the measurement.

The data in Fig. 2 are consistent with simple first order thermal desorption of neutrals followed by probe pulse photoionization. We have calculated the surface temperature using a detailed numerical code of laser surface heating. The dashed curve in Fig. 2 shows the calculated number of desorbed particles as a function of time convoluted with the probe pulse to

Springer Series in Chemical Physics, Vol. 48: **Ultrafast Phenomena VI**
Editors: T. Yajima · K. Yoshihara · C.B. Harris · S. Shionoya

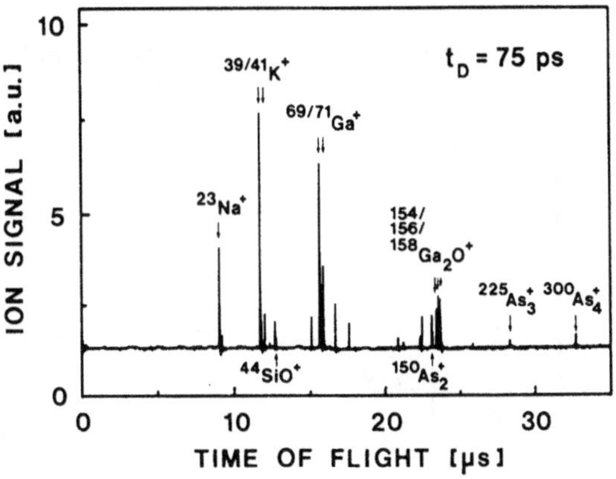

FIG. 1: TOF mass spectrum from a GaAs surface

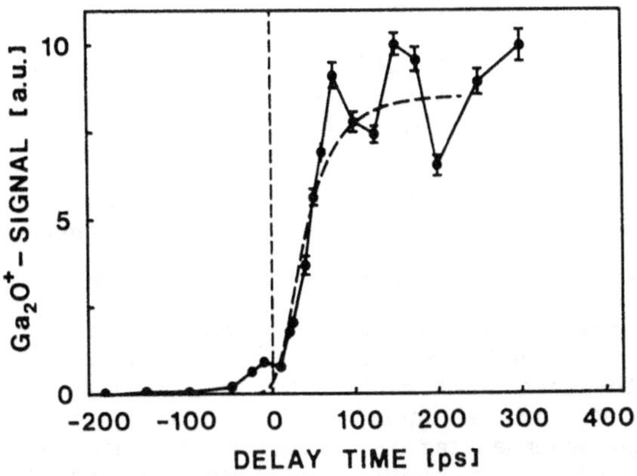

FIG. 2: Ga_2O^+-desorption signal vs. delay time

take into account photoionization. A good fit of the experimental data is obtained with an activation energy A = 1.2 eV.

3. Optical Reflectivity Measurements

Increase of the pump pulse energy fluence eventually leads to melting of the surface. In Si and GaAs melting is associated with a strong increase of the optical reflectivity due to the development of a metallic liquid surface layer. Detailed information about the kinetics of the solid-liquid phase transformation can be obtained from time-resolved reflectivity measurements. However, the reflectivity increase due to melting often competes with reflectivity changes caused by other processes. For instance, lattice heating leads to an increase of the reflectivity of Si and GaAs in the visible spectrum, whereas the excitation of a dense electron-hole plasma by the pump pulse causes a decrease of the reflectivity in the infrared [1].

Figure 3 shows two examples of the measured reflectivity change of Si at 532 nm during and after photoexcitation by a 20 ps UV pulse (266 nm): (a) and (c) correspond to an energy fluence of 45 mJ/cm² slightly above the melting threshold and to 80 mJ/cm², respectively. The experimental data are compared with two different melting models [2] (solid curves): (a,c) interface motion limited by the kinetics of atomic rearrangement; (b,d) heat-flow controlled interface motion. The first allows for dynamic superheating of the solid, whereas in the second the temperature of the solid never exceeds the melting point. In the superheating model the reflectivity rise

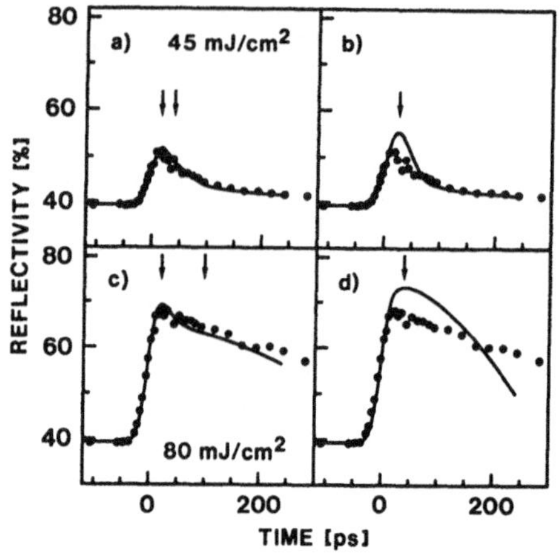

FIG. 3: Reflectivity of Si vs. delay time for two fluences

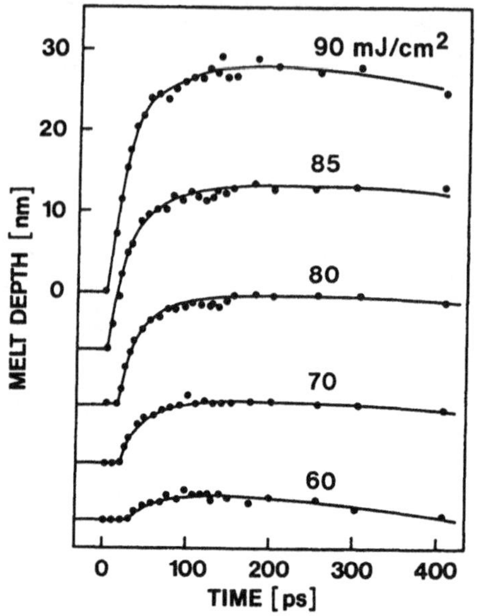

FIG. 4: Melt depth in GaAs vs. delay time for rising fluence

is dominated by the increase of the reflectivity of the superheated solid, and the reflectivity maximum coincides with the maximum of the surface temperature (left arrow), whereas the melt depth maximum (right arrow) occurs at later times. On the other hand, the reflectivity rise is dominated by the developing liquid layer when superheating is neglected. In this case the reflectivity maximum corresponds to the maximum melt depth (b and d). Note that the two models predict rather different time behavior of the reflectivity. The comparison with the experimental data indicates that a much better agreement is obtained when superheating effects are included.

The contributions to the pump-induced reflectivity changes from lattice heating are substantially reduced in the spectral region in the near infrared. In fact it can be shown that in Si and GaAs the reflectivity changes are to a very good approximation determined by the thickness of the molten layer when p-polarized probe pulses at 1064 nm incident at the principal angle are used [3]. In this case the instantaneous melt depth and the velocity of the liquid-solid interface can be obtained from the time-resolved reflectivity measurements. Examples of melt depths for GaAs are shown in Fig.4 for different values of the energy fluence of the 532 nm heating pulses. At the beginning of melting the thickness of the liquid layer increases very rapidly. The melt depth reaches a maximum 100 to 200 ps after the heating pulse. The melting velocities are given by the slope of the curves. We find that the maximum melt-in velocity increases from about 70 m/s near threshold to almost 800 m/s for twice the threshold fluence. Somewhat higher velocities have been observed in silicon [3].

4. Conclusions

A method of ps-photoionization mass spectroscopy has been introduced and applied to studies of time-resolved desorption from laser-heated GaAs surfaces. Optical reflectivity measurements give clear indication of superheating of the solid phase and have enabled measurements of melt-in velocities approaching 1000 m/s.

1. H. Kurz and N. Bloembergen: Mat. Res. Soc. Symp. Proc. 35, 3 (1985)
2. F. Spaepen and D. Turnbull: In Laser Annealing of Semiconductors, ed. by J.M. Poate and J.W. Mayer, (Academic, New York 1982) p.15
3. B. Danielzik, P. Harten, K. Sokolowski-Tinten and D. von der Linde: Mat. Res. Soc. Symp. Proc. 100, 471 (1988)

Picosecond Surface Reaction Dynamics and Carrier Processes at Semiconductor Interfaces

R.J. Dwayne Miller, J.J. Kasinski, L.A. Gomez-Jahn, and L. Min

Department of Chemistry, University of Rochester, Rochester, NY 14627, USA

One of the most fundamental steps in many surface reaction processes is that of interfacial electron transfer. The exact mechanism of electron transfer at surfaces needs to be established to realize rational control of the surface chemistry. To study the dynamics of surface electron transfer processes, three, highly surface specific, optical probes have been developed in conjunction with semiconductor liquid junctions. These are surface restricted transient gratings, optically generated surface acoustic waves, and in-situ electro-optic sampling of surface space charge fields. These three optical probes have been employed to measure interfacial carrier population dynamics, interface structure and carrier thermalization, and surface transport respectively at TiO_2(001) and GaAs(100) interfaces.

The laser system used for these studies consisted of fiber optic compressed 3 psec pulses from a Q-switched and mode locked YAG laser. Above band gap excitation for the TiO_2 (GaAs) studies was at 355 nm (532 nm) and the below gap probe at 532 nm (1.064μm). The GaAs studies and surface acoustic studies were conducted with ~100 psec pulses from the uncompressed YAG output. Surfaces were chemically etched and characterized using standard electrochemical procedures.

Grating studies of TiO_2(001) surfaces were conducted both in air and in the presence of an aqueous liquid junction of PH=13.5. The in air studies serve as a control to determine the surface induced electron-hole pair recombination under flat band or zero field conditions. The in-air decays are non-exponential and are well described by the ambipolar diffusion controlled kinetic model of HOFFMAN et al./1/ (see insert to Fig. 1). For highly doped samples ($10^{19}/cm^3$) in the presence of 1eV of band bending, 500A wide space charge field, (determined relative to an SCE reference electrode) the decays showed a 40% component with a decay time of 460 psec. In the presence of the space charge field the electron-hole pairs separate and are prevented from recombination. At the lowest excitation level ($3x10^{18}/cm^3$), the space charge field is screened by ~30%. Nevertheless the transport of minority hole carriers, generated within the space charge field, is increased by the electric field. Transit times for these carriers to the surface are on a 10 psec time scale. The slow 460 psec decay observed indicates that the holes accumulate at the surface, and thermalize before interfacial transfer. The observed dynamics in connection with the high quantum yield for interfacial hole transfer (>80%) are consistent with OH^-_{aq} as the initial hole acceptor. However, chemisorbed OH_s cannot be conclusively ruled out as the initial acceptor.

Optical excitation of surface acoustics (SAWs) by non-radiative carrier relaxation was also employed to study the TiO_2 interface. The SAWs are observable as oscillations in the diffracted signal that arises from surface corrugation effects accompanying SAW propagation. The increase in the SAW amplitude and rapid decay in the electron-hole pair phase grating demonstrate that Auger

Springer Series in Chemical Physics, Vol. 48: **Ultrafast Phenomena VI**
Editors: T. Yajima · K. Yoshihara · C.B. Harris · S. Shionoya

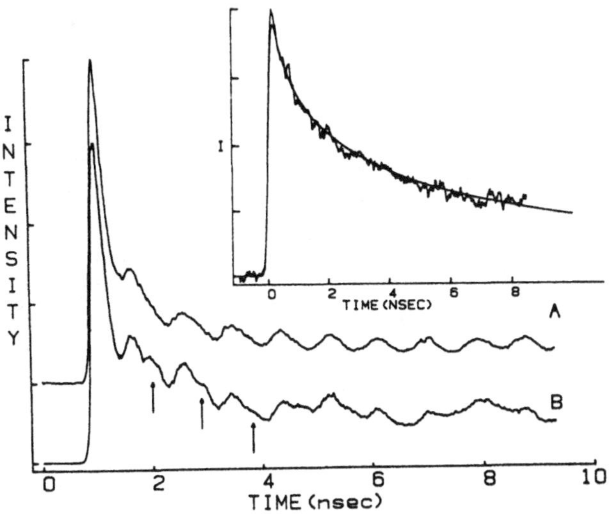

Fig. 1. TiO_2 grating studies Curve A: TiO_2 (001) in air, excitation level 10^{20} photocarriers/cm^3.
Curve B: TiO_2/H_2O, excitation level 10^{20} photocarriers/cm^3.
Inset: TiO_2 in air (small SAW modulations subtracted) low excitation conditions, $3x10^{18}$/ cm^3. Theoretical fit is the solid line using a surface recombination velocity (S) of 10^4cm/ sec.

recombination processes start to dominate at excitation levels above $3x10^{19}$/ cm^3. (TiO_2 in air, curve A, Fig. 1). A study of the bounded TiO_2/H_2O interface found the appearance of an anomolous high velocity interfacial acoustic mode travelling in the water half space (Fig. 1 curve B - indicated by arrows). The speed of sound of this mode ($4.4 \pm .2 \times 10^5$ cm/sec) corresponds more closely to ice than bulk phase water and indicates a phase restructuring at the TiO_2/H_2O interface. This result demonstrates that H_2O is strongly perturbed by the polar TiO_2 surface and bulk dielectric properties cannot be used at the interface in describing the dielectric relaxation processes of electron transfer.

Similar studies were conducted on Si doped GaAs (100) surfaces. In contrast to TiO_2, GaAs (100) in air does not illustrate flat band recombination dynamics except at high carrier injection levels of 10^{19}/cm^3 or greater (Fig. 2 curve A). As the excitation level is decreased, the signal intensity at t>3 nsec is found to be saturated for over two orders of magnitude (Fig. 2 curves B and C). Proper signal scaling is not observed until excitation conditions correspond to less than 10^{12} photons/cm^2 - which corresponds to typical surface state densities. The observed saturated signal is attributed to long lived surface state traps. To further confirm that the signal arises from surface states, the surface was photowashed using 1-2 W/cm^2 from an Ar^+ laser.[2] Photowashing the first few atomic layers off the surface eliminated the signal and flat band dynamics were observed. It is noteworthy, that the diffraction efficiency from the surface states is significantly higher than the free carriers. The enhanced phase grating component suggests a highly delocalized (excitonic) electronic state for the filled surface states of GaAs rather than a localized one. However, photorefractive effects may be responsible for the surface state detection.[3]

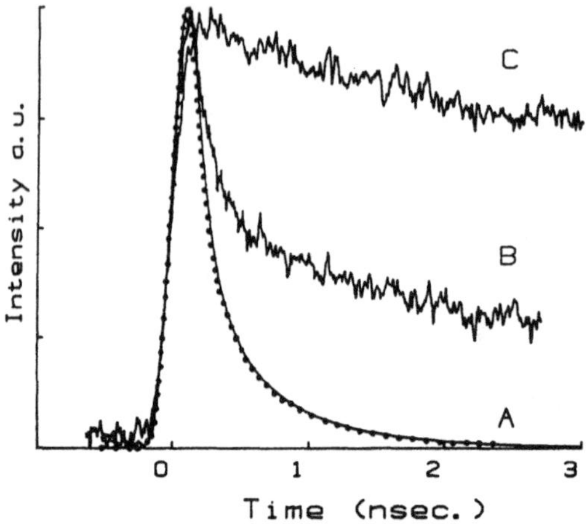

Fig. 2. Grating studies of GaAs (100) doping $7x10^{17}/cm^3$ in air. Curve A: excitation level $1.8x10^{19}/cm^3$, dots are the data points, solid curve is a theoretical fit with $S = 3x10^6$ cm/sec.
Curve B: excitation level $2.8x10^{17}/cm^3$
Curve C: excitation level $7.2x10^{16}/cm^3$ similar results found for other doping levels

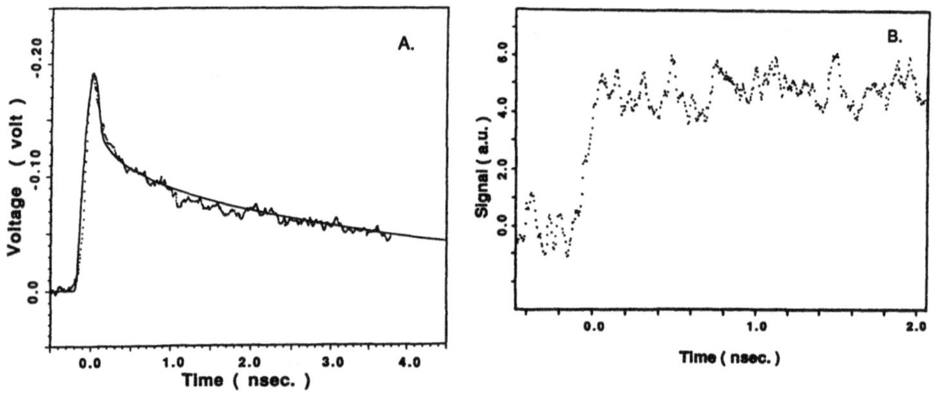

Fig. 3. A: Electro-optic sampling studies of n-GaAs (100) in air, excitation level $3x10^{19}/cm^3$. The solid line is a theoretical fit with $S=3x10^6$ cm/sec
B: excitation level $1x10^{17}/cm^3$

In-situ electro-optic sampling (EOS) studies were also conducted of the space charge field. At high carrier levels, the intrinsic space charge field is completely screened. The field that develops is well described by the photo-Dember effect with surface state recombination included (Fig. 3A). At low excitation conditions, the signal remains saturated until levels less than 10^{12} photons/cm^2 as in the grating studies. The space charge field induced carrier separation dominates the signal at these levels. Minority carrier transport to the surface, as determined from the signal rise time, occurs in less than 20 psec as expected. Femtosecond studies are in progress to fully resolve the space charge transport.

Both the EOS and surface grating methods provide new optical probes of surface state trapping dynamics at GaAs surfaces with 10^{-3}-10^{-4} monolayer sensitivity. The grating studies give evidence for surface state trapping on a picosecond timescale. The intrinsic surface state trapping rate constants need to be separated from carrier transport dynamics which can be sorted out with EOS. The results demonstrate the importance of the surface state trapping dynamics in understanding surface state mediated chemistry at this surface.

1. C.A. Hoffman, K. Jarasiunas, H.J. Gerritsen, and A.V. Nurmikko, Appl. Phys. Lett. 33, 536 (1978).
2. S.M Beck and J.E. Wessel, Appl. Phys. Lett. 50, 149, (1987) references therein.
3. A.L. Smirl, G.C. Valley, K.M. Bohnert and T.F. Boggess, IEEE, J. Quant. Electron. 24, 289 (1988).

Direct Observation of Photodynamics in Opaque Organic Microcrystals: A Picosecond Diffuse Reflectance Laser Photolysis Study

N. Ikeda[1,*], *M. Koshioka*[1], *H. Masuhara*[1], *N. Nakashima*[2,**], *and K. Yoshihara*[2]

[1]Department of Polymer Science and Engineering, Kyoto Institute of Technology, Matsugasaki, Kyoto 606, Japan
[2]Institute for Molecular Science, Myodaiji, Okazaki 444, Japan
*Present address: Chemistry Department, College of General Education, Osaka University, Toyonaka, Osaka 560, Japan
**Present address: Institute of Laser Engineering, Osaka University, Suita, Osaka 565, Japan

Laser photolysis method is now recognized as an indispensable technique to study photoprimary processes in chemistry, physics, biology, and in a wide field of application. However, this conventional (transmission type) technique is restricted to transparent samples. In recent years, a diffuse reflectance laser photolysis method has been developed.[1] This method is powerful because it gives absorption spectra of transient species in opaque and scattering materials such as organic microcrystals, semiconductors as well as insoluble polymer powders, dyed fabrics, molecules adsorbed on silica gels, etc. Since photophysical and photochemical processes are very fast because of strong interchromophoric interactions in these systems, we have extended this method to picosecond time domain,[2] and studied transient absorption spectra and its excited state dynamics of opaque organic microcrystals.

A picosecond diffuse reflectance laser photolysis system with a mode-locked Nd^{3+}:YAG laser was set up, in which a double-beam optical arrangement was adopted.[2] Samples, contained in a cell with 2 mm thickness, were excited with a single 355 nm pulse (17 ps, 1 mJ), and monitored by a picosecond continuum as a wide-band analyzing light. The picosecond diffuse reflected light from the sample was detected by a multichannel photodiode array (MCPD1) through a polychromator. A part of the continuum beam was detected by another polychromator with MCPD2, and was used as the reference for a spectral shape of the analyzing light and shot-to-shot variations. The spectral data were obtained by averaging over several measurements. Since the diffuse reflected light is relatively weak, it is important to correct the contribution of sample emission. All experiments of the powder sample were performed under aerated conditions at room temperature.

The transient absorption intensity in diffuse reflectance was displayed as % absorption.[1] Figure 1 shows transient absorption spectra of microcrystalline benzophenone and p-terphenyl at 100 ps after excitation. Those spectra are assigned to be T_n-T_1 and S_n-S_1 absorptions (in exciton state), respectively, and are also in agreement with the spectra obtained by nanosecond transmittance laser photolysis of its single crystal. Figure 2 shows a series of time-resolved absorption spectra of benzil in microcrystal and in solution. Although the obtained spectra were broader

Springer Series in Chemical Physics, Vol. 48: **Ultrafast Phenomena VI**
Editors: T. Yajima · K. Yoshihara · C.B. Harris · S. Shionoya

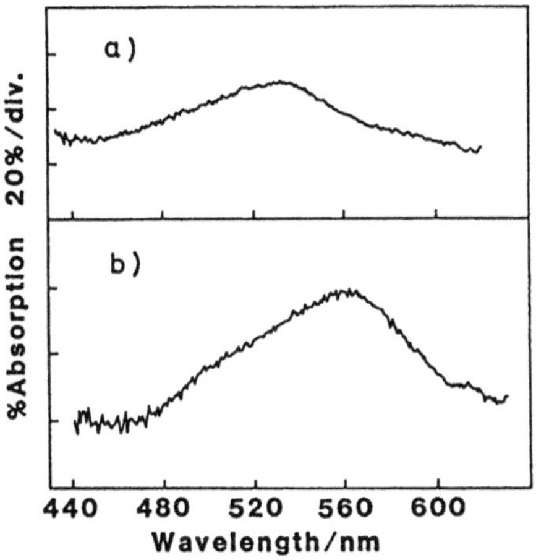

Fig. 1
Transient absorption spectra of organic microcrystals at 100 ps after excitation;
a) benzophenone
b) p-terphenyl

and red shifted in microcrystal than in solution, a similar spectral change occurred but with a different time constant. Those fast and slow transients can be assigned to be S_1 and T_1 states, respectively. The geometry of the excited states is reported to be skewed in crystal and trans planar in solution. As can be seen from Fig. 2, the intersystem crossing rate of benzil in microcrystal is faster than in solution. However, in the case of benzophenone there was no appreciable difference of the triplet rise curves between solid and solution phases.[2]

In molecular crystal phase, a high density laser excitation will usually bring about the efficient singlet-singlet annihilation. Our success in measuring the excited singlet absorption seems to be due to a low quantum

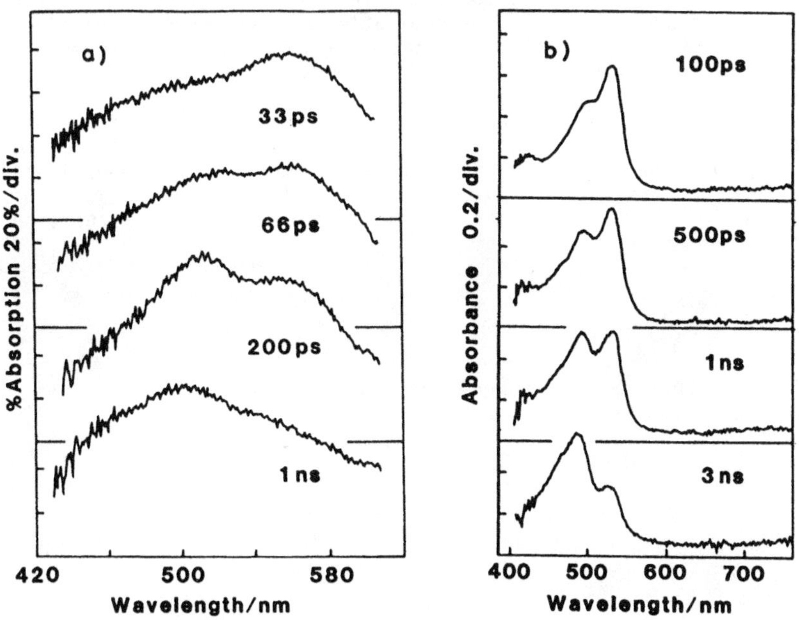

Fig. 2. Transient absorption spectra of benzil in microcrystal (a) and in cyclohexane solution (b)

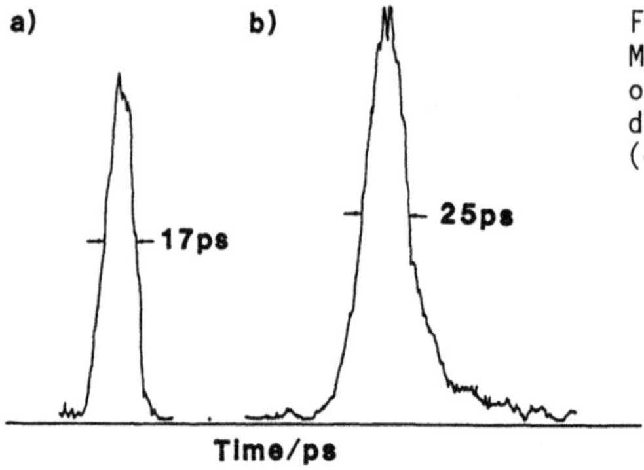

Fig. 3
Measurement of the pulse width of incident ps pulse (a) and diffuse reflected pulse (b) (diffuser; silica gel powder)

yield of the excited singlet benzil and a small molar extinction coefficient of p-terphenyl at 355 nm. Both conditions lead to a relative low exciton density.

For applications in picosecond time scale of diffuse reflectance laser spectroscopy to opaque and optically thick samples, the following point should be considered. As both exciting and monitoring pulses undergo numerous reflections, refractions, and diffractions in the microcrystals, the time resolution of this method might be lower than that of the transmittance laser photolysis. Actually, according to our preliminary measurements using a streak camera, the pulse width of picosecond diffuse reflected light is found to be ca. 1.5 times longer than that of the incident one (Fig. 3). In this sense, a response time in the powder system could be longer than in solution system, and this depends on the scattering coefficient and both absorption coefficients of the ground state and of transients. For more quantitative analysis, the following points are being examined in this group i)precise determination of the response function, ii)broadening effect upon absorption spectra, iii)multiphoton processes, iv)effect of defects or surface states of microcrystals.

In conclusion, we have performed for the first time a picosecond transient absorption spectral and kinetic measurement of singlet excited states of organic microcrystals by the diffuse reflectance laser photolysis method. This is especially powerful for the understanding of photoreactivity in crystalline environments since many of reaction intermediates might be nonluminescent.

References

1. F.Wilkinson, J. Chem. Soc. Faraday Trans. 2, 82 (1986), 2073, and references sited in.
2. N.Ikeda, K.Imagi, H.Masuhara, N.Nakashima, and K.Yoshihara, Chem. Phys. Lett., 140 (1987), 281.

Sequential Excitation Energy Transport in Stacking Multilayers: A Comparative Study Between Photosynthetic Antenna and Langmuir-Blodgett Multilayers

I. Yamazaki[1], *N. Tamai*[2], *and T. Yamazaki*[1]

[1]Faculty of Engineering, Hokkaido University, Sapporo 060, Japan

[2]Institute for Molecular Science, Myodaiji, Okazaki 444, Japan

1. Introduction

The sequential excitation energy transport has been studied with a biological antenna and artificial Langmuir-Blodgett (LB) multilayers by means of a picosecond time-resolved fluorescence spectrophotometer.[1] We will start with a general discussion on the energy relaxation of electronically excited molecules incorporated in organized molecular assemblies. The chromophore distribution, fractal or fractal-like structure, might play an important role in relaxation pathways. The kinetic behaviors of the sequential excitation transport in stacking multilayered architectures are compared between biological and artificial antennas.

2. Excitation Energy Relaxation in Molecular Assemblies

Electronically excited molecules incorporated in molecular assemblies may undergo energy relaxation in various pathways summarized as follows:

$$M^* \longrightarrow M + h\nu \qquad \text{fluorescence} \qquad (1)$$

$$M^* + Q \longrightarrow M + Q^* \qquad \text{energy trapping} \qquad (2)$$

$$M^* + M \longrightarrow M + M^* \qquad \text{energy migration between donors} \qquad (3)$$

$$M^* \longrightarrow M_m^* \longrightarrow M_n^* \dashrightarrow \qquad \text{energy migration to lower energy sites} \qquad (4)$$

$$M_m^* \searrow h\nu_m, \qquad M_n^* \searrow h\nu_n$$

$$M^* + A \longrightarrow M + A^* \qquad \text{energy transfer to adjacent layer} \qquad (5)$$

Processes 1-4 occur in a single monolayer and are important particularly in LB films. Process 4 is observed as a red shift of the fluorescence as time going on in the picosecond region. The fluorescence decay curve cannot fit the Förster equation based on the assumption of random and uniform distribution of chromophores, but fit the decay function based on the irregular distribution of fractal or fractal-like structure.[2,3] Process 5 represents the interlayer energy transfer between donor and acceptor layers. The interlayer energy transfer in artificial systems should compete with the energy dissipation of processes 2-4 taking place in respective layers.[4] On the other hand, the energy transport in biological antennas takes place such that the interlayer transport is a dominant process, and any other paths of energy loss is negligible.[4,5]

Springer Series in Chemical Physics, Vol. 48: **Ultrafast Phenomena VI**
Editors: T. Yajima · K. Yoshihara · C.B. Harris · S. Shionoya

3. Sequential Excitation Transport in Stacking Multilayer Systems

In the phycobilin-chlorophyll system of red and blue-green algae (Fig. 1a), the photoexcitation energy is transferred from the outer surface to the inner core of the reaction centers with the efficiency of near unity. [5] Previous study [5,6] has demonstrated that, following picosecond laser excitation, the fluorescence spectrum changes with time ranging 0-400 ps, indicating that the

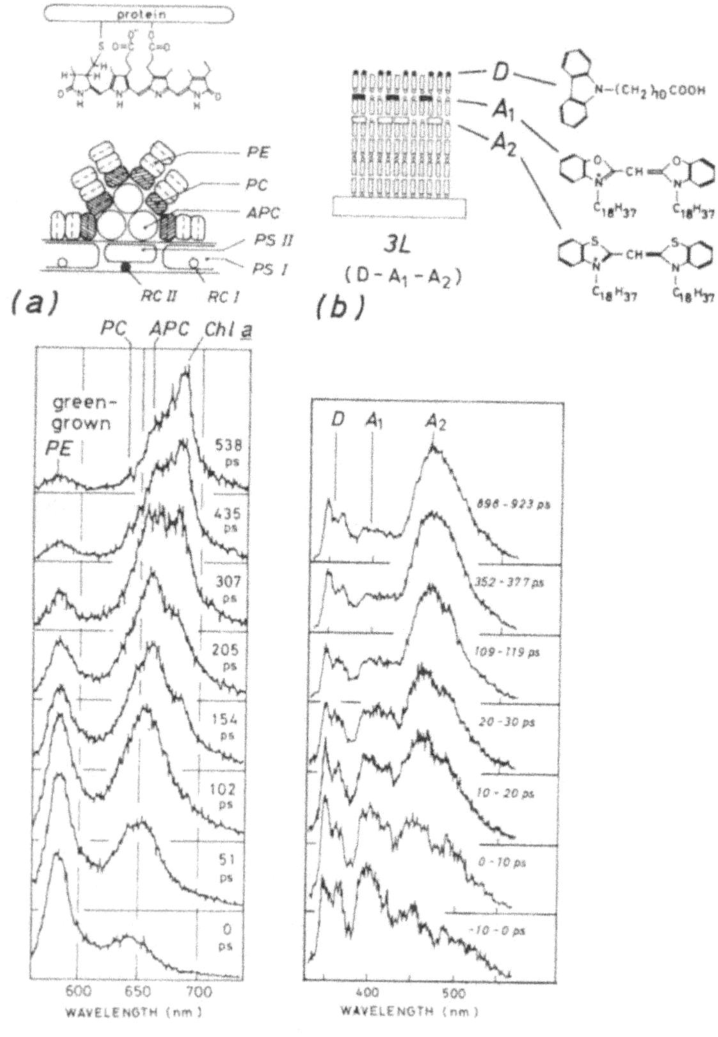

Figure 1. Schematic representation of multilayered structures of a biological antenna of algae (a), and artificial LB films (b). Pigment chromophores are open-ring tetrapyrroles in algae; in LB films, carbazole with a long hydrocarbon chain (D) and different kinds of cyanine (A_1 and A_2) are used in this study.

excitation energy is transported successively through respective pigment proteins along the phocobilisome rod; i.e., PE-PC-APC-Chl a. In this study, the LB multilayer as an artificial analogue of the biological antenna was investigated. The LB films containing carbazole and three types of cyanine dyes as energy donor and acceptor (Fig. 1b) were prepared.[4] The fluorescence bands of respective pigments are emitted successively during 500 ps, as the excitation energy is transferred from the outer surface to the inner layer.

Phenomenological aspects of the sequential energy transfer are parallel between the biological and the artificial systems. Commonly to these two cases, the fluorescence rise and decay curves are characterized by a rapid rise in the acceptor but much slower decay in the donor. Such a sequential energy transport can be fitted to the equation of $\exp(-2kt^{1/2})$ type rather than $\exp(-3kt^{1/3})$ type. The kinetic parameter values, e.g., fluorescence rise and decay times and rate constant of the energy transfer are close with each other. However, the transfer efficiency is higher in the biological antenna than in the LB multilayer; the efficiency in each step is 0.9 in phycobilins whereas it is 0.5 in LB films. The transfer efficiency is determined predominantly by the branching ratio between the energy transfer to adjacent layer to the energy trapping within a layer. Lower efficiencies in LB multilayers are a consequence of relatively large density of traps in monolayers due to dimer and/or higher aggregates of chromophores. Recent study on LB films [2,7,8] has demonstrated that the distribution of pigment chromophores is not uniform nor random in LB monolayers, and form an irregular structure, i.e., island structure. Such an irregular distribution yields a large number of trap sites with their density being much higher than estimated from the actual concentration of chromophores.[8]

Contrary to the artificial LB multilayers, the biological antenna system has a regular architechture of chromophore distribution, in which tetrapyrroles are distributed with a specific array in polypeptide networks. In this study, two kinds of phycobilisomes in which the stacking structure is different were investigated. It was found that, in the inner core of phycobilisome rod, there exists a special channel for the excitation transport straightforward to the reaction center.

References

1. I. Yamazaki, N. Tamai, H. Kume, H. Tsuchiya, K. Oba, Rev. Sci. Instrum. 56, 1187 (1985).
2. N. Tamai, T. Yamazaki and I. Yamazaki, Chem. Phys. Lett., 147, 25 (1988).
3. N. Tamai, T. Yamazaki, I. Yamazaki and N. Mataga, In Ultrafast Phenomena V, ed. by G. R. Fleming and A. E. Siegman, Springer Ser. Chem. Phys., Vol. 49 (Springer, Berlin, Heidelberg 1986) p.449.
4. I. Yamazaki, N. Tamai, T. Yamazaki, A. Murakami, M. Mimuro and Y. Fujita, J. Phys. Chem. (1988) in press.
5. I. Yamazaki, M. Mimuro, T. Murao, T. Yamazaki, K. Yoshihara and Y. Fujita, Photochem. Photobiol. 39, 233 (1984).
6. I. Yamazaki, N. Tamai, T. Yamazaki, M. Mimuro and Y. Fujita, In Ultrafast Phenomena IV, ed. by D. H. Auston and K. B. Eisenthal, Springer Ser. Chem. Phys., Vol. 38 (Springer, Berlin, Heidelberg 1984) p.490.
7. I. Yamazaki, N. Tamai and T. Yamazaki, In Ultrafast Phenomena V, ed. by G. R. Fleming and A. E. Siegman, Springer Ser. Chem. Phys. Vol. 46 (Springer, Berlin, Heidelberg 1986) P.444.
8. I. Yamazaki, N. Tamai, T. Yamazaki, J. Phys. Chem. 91, 3572 (1987).

Fluorescence Lifetime of Dye Molecules Near a Metal Surface

F.R. Aussenegg, A. Leitner, M.E. Lippitsch, and H. Reinisch

Institut für Experimentalphysik, Karl-Franzens-Universität, Graz, Austria

1. Introduction

Ten years ago it was shown theoretically by CHANCE, PROCK and SILBEY [1], that the lifetime of an excited dye molecule is shortened when approaching a metal surface (CPS theory). This shortening can be explained by coupling of near-field components of the molecular transition dipole to the electron gas in the metal [2]. The metal then acts as an additional antenna for the emitted radiation, and simultaneously as a lossy material in the near-field of the oscillating molecular dipole and hence a source for damping by dissipation. Both effects cause additional deexcitation pathways with a rate large compared with that of the free molecule's radiation into the far field. Up to now this phenomenon has been investigated directly only for luminescence from long-lived states [2, 3], but never for fluorescence from the first excited singlet state. To fill this gap, we have performed fluorescence decay studies with picosecond time-resolution on the dye rhodamine 6G near a silver surface.

2. Experimental

Rhodamine 6G molecules were positioned in a well-defined distance from a thermally evaporated silver film (substrate at room temperature, evaporation rate 0.1 to 0.5 nm s^{-1}). Continuous bulk-like silver films of 200 nm mass thickness were used in comparison with islands-films of 4 nm mass thickness. The distance between the silver surface and the dye was controlled by a spacer layer of evaporated silicon monoxide (SiO_x). Sub-monolayers of the dye with an average intermolecular distance of 4 nm, which is high enough to prevent significant intermolecular dipole-dipole interaction, were used. A more detailed account of the preparation is given in [4].

For excitation, pulses of 6 ps duration with a wavelength of 527 nm from a passively mode-locked and frequency-doubled Nd-phosphate glass laser system were applied. Fluorescence emission was detected spectrally integrated and time-resolved by a streak-camera.

3. Results and discussion

On the solid silver film the observed fluorescence decay curves (Fig. 1 a) could be readily fitted by a single exponential. Figure 2 shows the resulting lifetimes *vs* spacer layer thickness *d* (circles). A comparison with results from the CPS theory gives good agreement with respect to the lifetime - distance dependence, when the

Springer Series in Chemical Physics, Vol. 48: **Ultrafast Phenomena VI**
Editors: T. Yajima · K. Yoshihara · C.B. Harris · S. Shionoya

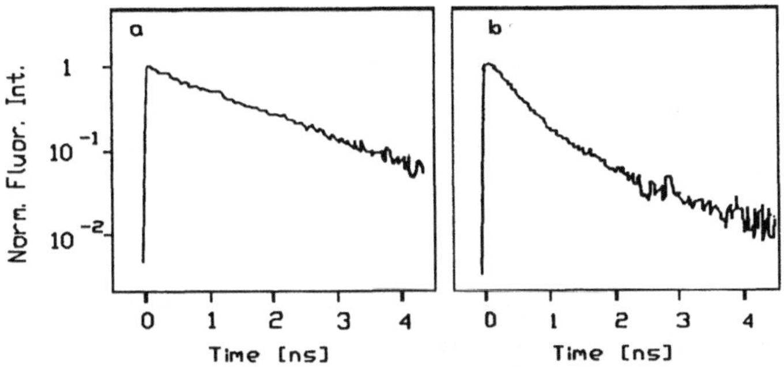

Fig. 1: Fluorescence decay curves for rhodamine 6G, 20 nm above a solid silver film (*a*) and a silver islands film (*b*)

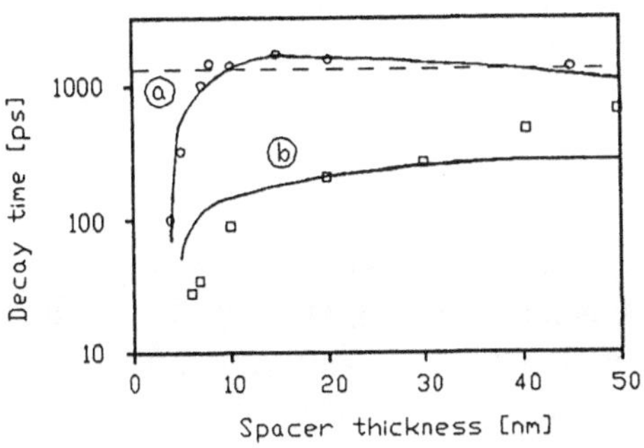

Fig. 2: Fluorescence lifetime of rhodamine 6G on silver *vs* spacer layer thickness *d*. *Circles*, experimental values for compact silver films. *Squares*, experimental values for island films. *Dashed line*, lifetime for the molecule on the substrate without metal. Curve *a*, result from CPS theory for molecular orientation parallel to metal surface. Curve *b*, the same for perpendicular orientation

orientation of the molecular dipole is assumed to be parallel to the surface (curve a in Fig. 2). This demonstrates that the CPSmodel is applicable also to fluorescence decays proceeding on a sub-nanosecond time scale and not only for microsecond luminescence as shown so far [2, 3]. In addition the expectation to have the molecules lying flat on the surface is verified.

As is well known, submicroscopic metal structures are an appropriate tool for enhancing interactions between molecules and light fields [5]. Therefore the influence of such structures on the lifetime was also studied. In this case (molecules on silver islands films, as shown in Fig. 3) the observed decay curves (Fig. 1 b) could no longer be fitted by a single exponential [4]. Hence for comparison an 'effective' lifetime had to be defined as the mean of the time constants involved weighted by their respective amplitudes. The dependence of this 'effective' lifetime on the distance between dye submonolayer and islands film is also shown in Fig. 2 (squares). Now, agreement with the CPS results cannot be obtained, neither for parallel nor perpendicular orientation of the molecule. Since in an islands film the surface is not smooth, parallel or perpendicular orientation can barely be defined, however, as is visualized in Fig. 3. In addition, as is also clear from the figure, there is no well-defined distance between dye molecules and metal. Already from these arguments a disagreement with the CPS results seems reasonable and a non-exponential decay would be expected. In any case, the quantitative knowledge of

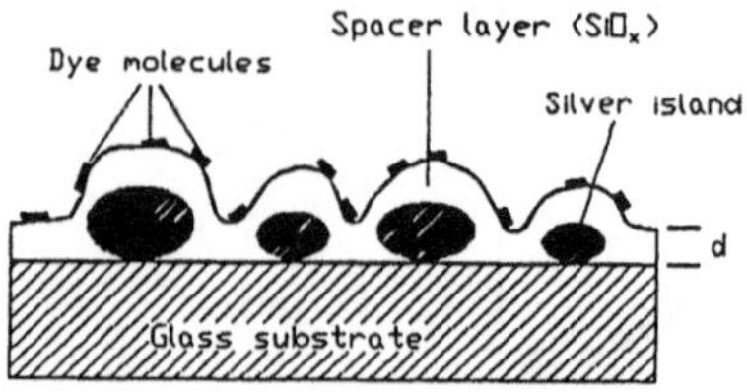

Fig. 3: Schematic sketch of dye molecules on a metal islands film

lifetime shortening is important for an intended application of submicroscopic metal structures for enhancing interactions between molecules and electromagnetic light fields.

4. Acknowledgement

Support of this work by the Austrian *Fonds zur Förderung der wissenschaftlichen Forschung,* grant no. 5628, and the *Erwin Schrödinger-Gesellschaft für Mikrowissenschaften* is gratefully acknowledged.

5. References

1. R. R. Chance, A. Prock, and R. Silbey, *Adv. Chem. Phys.* **37**, 1 (1978)
2. P. M. Whitmore, H. J. Robota, and C. B. Harris, *J. Chem. Phys.* **77**, 1560 (1982)
3. H. Kuhn, Spectroscopy and applications of monolayer assemblies, pt. I: Principles and applications. *In:* Techniques of chemistry, A. Weissenberger and B. W. Rossiter, eds., Wiley, New York 1972
4. F. R. Aussenegg, A. Leitner, M. E. Lippitsch, and H. Reinisch, *Surf. Sci.* **189/190**, 935 (1987)
5. A. Wokaun, Surface-enhanced electromagnetic processes. *In:* Solid state physics, vol. 38, eds. H. Ehrenrein and D. Turnbull. Academic Press, Orlando 1984

Molecular Aspects of Fast Fluorescence Dynamics in Amorphous Poly(*N*-Vinylcarbazole) Films

H. Sakai, A. Itaya, and H. Masuhara

Kyoto Institute of Technology, Department of Polymer Science and Engineering, Matsugasaki, Kyoto 606, Japan

Poly(N-vinylcarbazole) (abbreviated hereafter as PVCz) film is a representative system whose photophysics and photoconductivity have been studied in detail. It is amorphous, however, its local geometrical structure is governed by polymer configuration and conformation [1]. They give two spectrally distinct intramolecular excimer fluorescence. One excimer with the maximum at 420 nm is due to the normal sandwich structure of neighboring carbazolyl groups, while the other around 370 nm is to a partial overlap structure. The former and the latter structures are formed in isotactic and syndiotactic sequences, respectively. This important molecular aspect is considered here, since several works on this polymer by picosecond kinetic spectroscopy have never examined them.

PVCz(r) and PVCz(c) are the polymers prepared by radical and cationic polymerizations, respectively. Both are stereoblock polymers, and the former and the latter consist of the ratio of syndiotactic to isotactic sequence, 3:1 and 1:1, respectively. Film sample was prepared by casting benzene solution with slow evaporation rate under a N_2 atmosphere in dark. The concentration of their tetrahydrofuran solution was adjusted to be around 5×10^{-5} mol dm^{-3} where any interpolymer interactions are neglected. All the samples were completely degassed. Fluorescence rise as well as decay curves and time-resolved spectra were measured with a picosecond time-correlated single-photon counting system with a microchannel-plate photomultiplier [2]. Since this system does not require an intense excitation pulse and has both a wide dynamic range and a high time-resolution, accurate fluorescence spectral and kinetic data can be obtained without complicated photophysical behavior due to S_1-S_1 annihilation.

Time-resolved fluorescence spectra are shown in Fig. 1. For all the systems at early delay times, the fluorescence maximum was observed at 370 nm and no appreciable sandwich excimer fluorescence in the long wavelength region was detected. It should be noticed that the shoulder was observed at 350 nm and it was clearly detected in solution. This can be ascribed to the carbazole monomer fluorescence. With time on going from picosecond and nanosecond time regions, the monomer fluorescence disappeared quickly, partial overlap excimer followed, and the sandwich excimer became the main component.

Namely, excitation energy migration is confirmed to be sequential.

Springer Series in Chemical Physics, Vol. 48: **Ultrafast Phenomena VI**
Editors: T. Yajima · K. Yoshihara · C.B. Harris · S. Shionoya

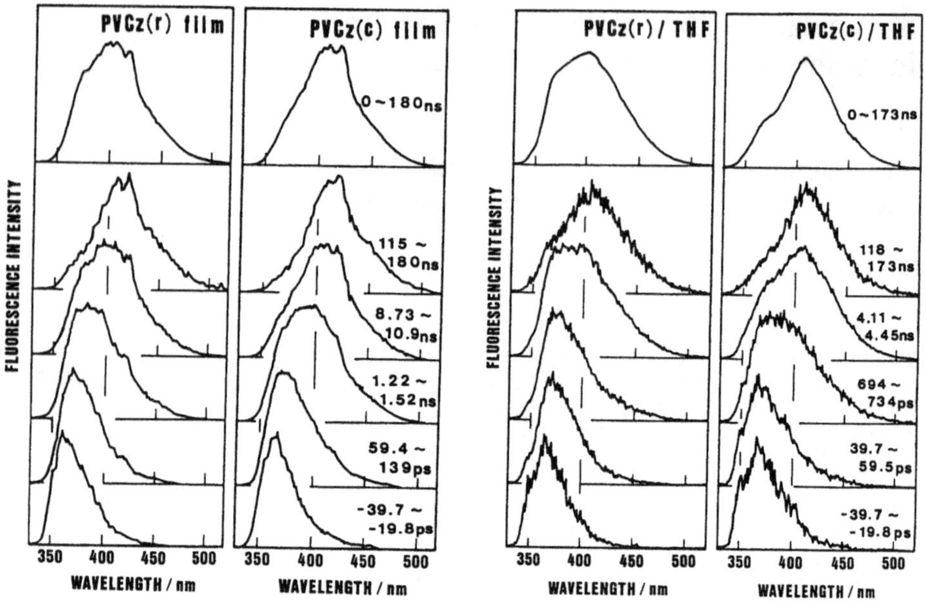

Fig. 1. Normalized picosecond time-resolved fluorescence spectra.

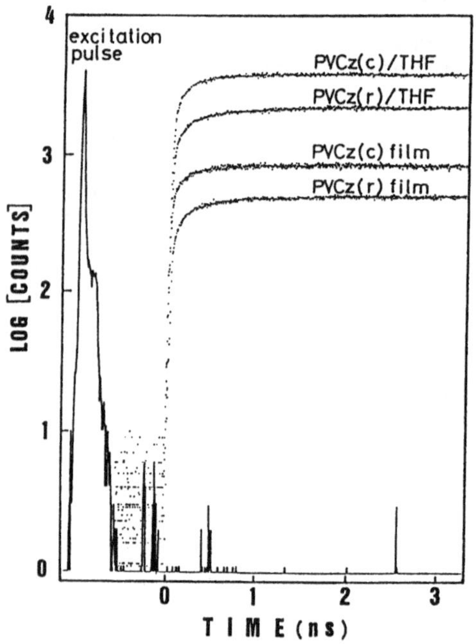

Fig. 2. Fluorescence rise curves of sandwich excimer (475 nm) in solution and film systems.

The fluorescence rise and decay curves of both PVCz in film were monitored at 350 nm and 380 nm where the monomer fluorescence and the partial overlap excimer are respective main components. The dynamics of the sandwich excimer in both solution and solid film were monitored at 475 nm, as shown in Fig. 2. Multi-exponential functions were used for their

Table 1. Multi-exponential analysis of fluorescence rise and decay curves.

	350 nm		380 nm		475 nm	
	τ_i ns	A_i %	τ_i ns	A_i %	τ_i ns	A_i
PVCz(r) film	0.012	83.2	0.035	39.1	0.023	-711(50.7%)
	0.200	11.0	0.354	27.3	0.236	-298(21.3)
	1.16	3.6	1.52	17.1	1.27	-392(28.0)
	6.05	1.6	7.06	10.9		
	35 fix	0.5	35 fix	5.7	35 fix	1431
PVCz(c) film	0.012	79.3	0.013	66.0	0.033	-372(61.1%)
	0.146	15.0	0.255	17.7	0.276	-112(18.4)
	1.03	3.7	1.43	7.6	1.55	-125(20.5)
	6.37	1.4	7.14	5.0		
	35 fix	0.6	35 fix	3.6	35 fix	634
PVCz(r) in THF	0.012	83.4	0.020	53.2	0.025	-338(45.7%)
	0.187	7.9	0.454	15.3	0.325	-203(27.5)
	1.14	5.1	1.69	16.9	1.95	-198(26.8)
	5.42	2.0	6.63	7.1		
	35 fix	1.5	34.7	7.6	35 fix	829
PVCz(c) in THF	0.017	80.2	0.020	62.0	0.032	-447(64.6%)
	0.178	12.2	0.292	19.5	0.306	-226(30.6)
	0.927	4.5	1.59	8.7	2.44	- 35(4.7)
	5.38	1.6	7.65	3.4		
	35 fix	1.6	33.9	6.4	35 fix	809

analysis, since at least three molecular species were identified in the time-resolved spectra. The results are summarized in Table 1. All rise and decay time constants are irrespective of observation wavelength and common to PVCz(r) and PVCz(c), and only amplitude factors are different from each other.

It is worth noting that the time constants are also common to film and solution systems. We consider that local geometrical structures in syndiotactic and isotactic sequences of the polymer determine fluorescence dynamics of amorphous polymer films. This suggests an energy relaxation mechanism of PVCz film, not involving efficient energy migration, which is being studied in detail.

REFERENCE

1. A. Itaya, H. Sakai, and H. Masuhara, Chem. Phys. Lett., 138, 231 (1987).
2. I. Yamazaki, N. Tamai, H. Kume, H. Tsuchiya, and K. Oka, Rev. Sci. Instr., 56, 1187 (1985).

Part IX

Energy Transfer and Relaxation

Picosecond and Femtosecond Infrared Spectroscopy with CW Diode Lasers

P. Anfinrud, C. Han, P.A. Hansen, J.N. Moore, and R.M. Hochstrasser

Department of Chemistry, University of Pennsylvania,
Philadelphia, PA 19104, USA

1. Introduction

From a chemical standpoint it is important to find methods of studying vibrational spectra on extremely short time scales. Many photophysical, biological, and photochemical processes have very short lived intermediates whose structures require accurate determination. Vibrational spectra contain valuable information on molecular structure that is not usually available from the optical spectrum, particularly for condensed systems.

Raman spectroscopy, which has now been applied to chemical and biochemical problems on the ps [1-4] and fs [5] time scales has, in addition to its advantages, some disadvantages compared with IR spectroscopy. For example, ultrafast Raman scattering experiments have so far required a resonance effect. This means that only those vibrational transitions can be studied that can somehow be resonance enhanced. Also, the Raman response is a tensor of polarizability derivatives that for complex molecules is difficult to predict theoretically and so the Raman anisotropy is not a sharp indicator of molecular structure. The IR spectrum on the other hand is a property of all polyatomic molecules and the anisotropy is determined by a vector property in the molecular frame so that it relates directly in many cases to geometric properties of the molecule. This paper describes recent studies aimed at studying fast processes in molecular systems using infrared spectroscopy.

In a recent paper a new method of obtaining IR spectra with ca. 15 ps resolution was described [6]. The technique was readily adaptable by using polarized beams to measurements of geometrical properties of molecules in solutions [7] and in proteins [8]. In this paper new results on the photophysics and chemistry of metallocarbonyls in solutions as well as in hemoglobin and myoglobin are presented. Furthermore the earlier method has now been extended into the femtosecond regime and new results on these carbonyl systems were obtained recently.

Considerable progress toward a versatile ultrafast IR spectroscopic method was made recently [9] using either parametric generators or down conversion. These methods and the diode laser approach reported here can be employed with many different types of lasers or dual laser systems and will soon lead to high sensitivity, high repetition rate and continuously tunable IR spectroscopic approaches.

2. Experimental

The infrared spectroscopic technique incorporates the output of a tunable cw diode laser (Spectra Physics, Laser Analytics Division, Bedford, MA) as

Springer Series in Chemical Physics, Vol. 48: **Ultrafast Phenomena VI**
Editors: T. Yajima · K. Yoshihara · C.B. Harris · S. Shionoya

the IR probe. The IR beam is passed through the sample and a change in the IR absorbance is induced by a pulse of visible or ultraviolet radiation. Time-resolved detection of the transmitted IR is achieved by sum-frequency mixing the IR with a second pulse of visible radiation in a nonlinear crystal which in this case is lithium iodate. This mixing upconverts the IR into the visible region and allows detection with a photomultiplier tube. The photolyzing and upconverting pulses may be variably delayed at a fixed IR frequency to obtain kinetics, the IR frequency may be scanned at a set delay time between the visible pulses to obtain time resolved IR spectra, and the relative polarizations of the photolyzing and upconverting pulses may be varied to observe transient anisotropy in the IR absorbance.

Picosecond time-resolved IR spectroscopy has been achieved using this technique in which the pulses of visible radiation are obtained from two dye lasers, each synchronously pumped by the frequency doubled output of a single mode-locked and Q-switched Nd:YAG laser [6]. Single pulses (ca. 560 nm, 20 ps FWHM, 1-10 μJ/pulse) may be selected from these dye lasers at a repetition rate of up to 1.2 kHz, and used as the photolyzing and upconverting pulses without amplification. This has allowed infrared spectra to be obtained with spectral and temporal resolutions of 5 cm^{-1} and 15 ps, respectively.

Femtosecond time resolution has been achieved using the apparatus shown in Fig. 1. The frequency doubled output of a cw mode-locked Nd:YAG laser is used to synchronously pump a cavity dumped rhodamine 6G dye laser. The output pulses are shortened to 300 fs (autocorrelation FWHM) using an optical fiber compressor and subsequently amplified. A new approach to high repetition rate multipass amplification has been developed where the frequency doubled output of a Q-switched cw Nd:YAG laser (65ns FWHM, 900μJ, 1kHz) is used to pump rhodamine B in ethylene glycol in a 1 mm flow cell. Six passes through the gain medium, which is oriented at Brewster's angle relative to the pump beam, are sufficient to amplify a 0.7 nJ input pulse to 2.5 μJ at 580 nm with no apparent pulse broadening. Significantly, the integrated nanosecond amplified spontaneous emission is typically less than 3% of the amplified pulse energy, thus

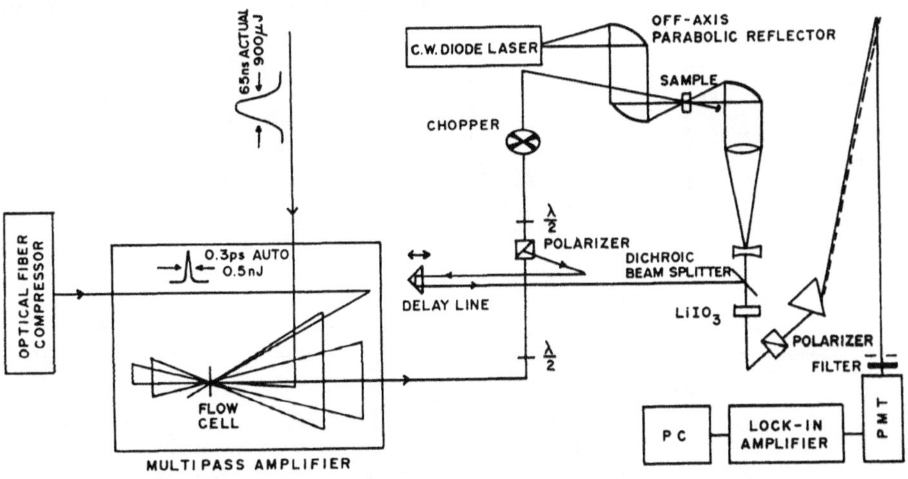

Fig. 1. Experimental apparatus for femtosecond transient IR spectroscopy

obviating a saturable absorber. The amplified output is variably split in a Glan Laser prism to derive the photolyzing and upconverting beams. Off-axis parabolic reflectors bring the cw IR (ca. 10 mW) to a common focus (ca. 120 μm spot size) with the photolyzing beam which is synchronously chopped at half the amplifier repetition rate. The transmitted IR is recollimated and telescoped down to about 3mm before frequency summing with a variably delayed upconverting pulse in $LiIO_3$. The polarizer, dispersing prism, and interference filter isolate the upconverted light from the transmitted upconverting beam. The differential transmitted IR intensity (photolyzed vs. unphotolyzed) is recovered by demodulating the PMT response at half the amplifier repetition rate with a lock-in detector.

3. Femtosecond HbCO photobleaching kinetics

A 15 mM solution of carboxy-hemoglobin (HbCO) in D_2O was prepared and flowed through a 56 μm thick sample cell. The diode laser was tuned to the peak of the bound CO stretch of ground state HbCO. The time dependent change in IR transmission due to photolysis at 580 nm is depicted in Fig. 2 (dots). These data are fitted to an instantaneous response (dashed line) using the instrument response function (300 fs FWHM autocorrelation, solid line). As evidenced by the visual quality of the fit, the photobleaching transient is rapid on the 300 fs time scale with no observable recovery of the bleach. This shows that geminate recombination of vibrationally unexcited CO does not occur on this timescale. Petrich et al. [10] recently reported a 2.5 ps partial recovery (14%) of the Soret band bleach which was attributed to geminate recombination. This discrepancy possibly arises from the complicated dynamics in the excited state hemoglobin optical absorption which is not detected in the IR experiment.

The transient IR spectrum at 10 ps time delay was obtained by plotting the normalized differential transmitted IR intensity at various IR wavelengths (Fig. 3). The absorption linewidth is fairly well approximated by the convolution of the diode laser bandwidth (ca. 10 cm^{-1} due to lasing on multiple longitudinal modes) with the linewidth for the bound CO absorption (ca. 8 cm^{-1}). This clearly demonstrates that the spectral resolution of this technique is determined by the cw diode laser bandwidth

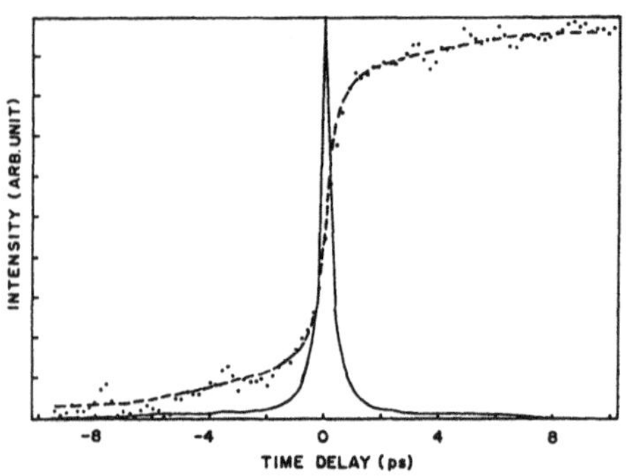

Fig. 2. Photobleaching transient of HbCO (dots); generated fit (dashed line); instrument response function (solid line)

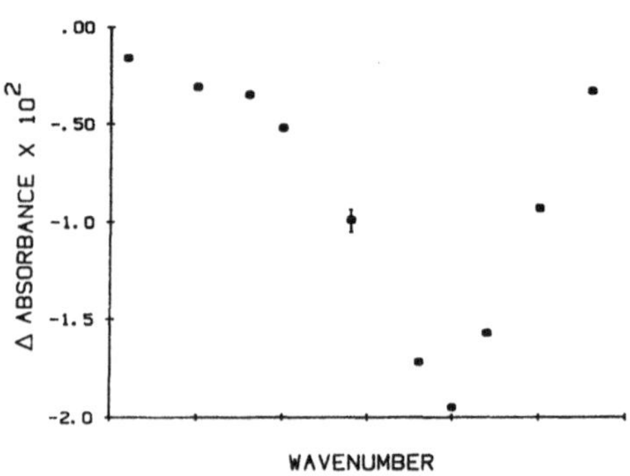

Fig. 3. Transient IR spectrum of HbCO at 10 ps time delay. The wavenumber axis tick marks occur at 5 cm^{-1} intervals

rather than the bandwidth of the femtosecond pulse (nearly transform limited at ~70 cm^{-1}), in contrast to IR pulse excitation methods.

The spectrum in Fig. 3 is virtually indistinguishable from one obtained at 0 ps delay. Hence, we have found no evidence (within our time resolution and S/N) for an excited state HbCO absorption within a 30 cm^{-1} region about the peak of the ground state CO absorption. The absence of a detectable absorption implies that the linewidth of any bound CO transition in this region is broad, reflecting its transient nature. This enables us to place an upper limit on its dissociative lifetime of ~200 fs, which is faster than the appearance of a deoxylike species in the optical spectrum. This is consistent with the conclusions by Petrich et al. [10] that the CO dissociates in 50 fs.

4. Iron-carbonyl bond geometry determination

Picosecond measurements of the anisotropic changes in IR absorbance upon photodissociation of CO from the caboxy- derivatives of hemoglobin, myoglobin and protoheme have allowed the equilibrium iron-carbonyl bond geometries to be determined for these systems in ambient temperature solution [7,8]. The results may be presented as either a photoselection anisotropy, r, or a polarization ratio, R, as given in Table 1. These values may be used to deduce the FeCO geometries because the measurements are made with a time resolution which is short compared with the overall rotational reorientation of the molecules in solution. Two simple models are those in which the distribution of CO configurations is described either by a delta function, yielding a single value α for the angle between the CO bond axis and the heme plane normal, or by a model in which the CO bond can adopt with equal probability any angle within a cone of semiangle β. The measurements reveal nothing of the azimuthal orientation of the CO group in the molecular frame. In the case of the heme proteins, the delta function is an appropriate model because of the uniform environment and narrow distribution for the CO location in any given conformation. Each conformation is characterized by a specific IR band, and if the X-ray

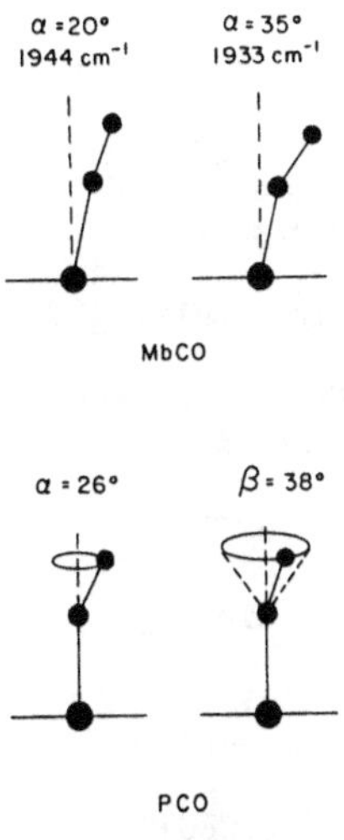

Fig. 4. FeCO configurations of MbCO. Models for PCO(90%G) structures. —, heme planes

Table 1. Polarization ratios and anisotropies used to obtain angles α and cone semi angles β for heme proteins and free heme in solution. The IR band positions are given in cm^{-1}. The aqueous solvent mixtures (by volume) were of glycerol (G), ethylene glycol (EG) and methanol (M).

	ν(CO)	R	r	α	β
MbCO	1944	1.74	-0.165	20°	-
MbCO	1933	1.38	-0.101	35°	-
HbCO	1951	1.78	-0.171	18°	-
PCO(90%G)	1962	1.59	-0.141	26°	38°
PCO(60%EG)	1962	1.69	-0.158	22°	32°
PCO(70%M)	1966	1.80	-0.174	17°	25°

diffraction data [11] is transferable to solution, the results presented here allow a sharp definition of the ligand geometries, as illustrated in Fig. 4 for MbCO. The crystallographic data are not necessarily transferrable to solution: indeed the preferred FeCO bond geometries [11] are not consistent with the solution structures determined here. For free hemes in solution the distribution is unlikely to be described well by either of the simple models. However, both illustrate that the FeCO unit experiences considerable distortion by the solvent since the unit is known to be linear and perpendicular to the heme plane in crystals [12,13]. The degree of distortion was found to vary with solvent, the greatest distortion being observed for PCO in 90%G, as illustrated in Fig. 4.

5. Transition metal carbonyls

IR spectroscopy has been used extensively to study the photochemistry of transition metal carbonyls through the observation of both matrix isolated and reacting transient species [14]. The technique reported here has extended the time resolution available for such studies from > 100 ns into the ultrafast regime. Photolysis of $[CpFe(CO)_2]_2$ in solution has been found to result in both bleaching of the trans isomer terminal CO stretch at 1960 cm^{-1} and the appearance of new IR absorption bands on the ps timescale. These preliminary results demonstrate the applicability of the technique to the study of ultrafast inorganic photoreactions.

Acknowledgements

Technical assistance from Laser Analytics is gratefully acknowledged. This research was supported by grants from the NSF (CHE-8603672) and the NIH (GM12592).

References

1. J. Terner, J.D. Stong, T.G. Spiro, M. Nagumo, M.F. Nicol and M.A. El-Sayed: Proc. Natl. Acad. Sci. USA 78, 1313 (1981)
2. T.L. Gustaffson, D.M. Roberts and D.A. Chernoff: J. Chem. Phys. 79, 1559 (1983)
3. C.K. Johnson, G.A. Dalickas, S.A. Payne and R.M. Hochstrasser: Pure & App. Chem. 57, 195 (1985)
4. See references in: C.K. Johnson and R.M. Hochstrasser: In Ultrafast Spectroscopy, ed. by W. Kaiser (Springer, Berlin, Heidelberg 1988)
5. J.W. Petrich, J.L. Martin, D. Hovde, C. Poyart and A. Orszag: Biochemistry 26, 7914 (1987)
6. J.N. Moore, P.A. Hansen and R.M. Hochstrasser: Chem. Phys. Lett. 138, 110 (1987)
7. P.A. Hansen, J.N. Moore and R.M. Hochstrasser: Chem. Phys., in press.
8. J.N. Moore, P.A. Hansen and R.M. Hochstrasser: Proc. Natl. Acad. Sci. USA 85, in press.
9. T. Elsaesser and W. Kaiser: Chem. Phys. Lett. 128, 231 (1986)
10. J.W. Petrich, C. Poyart and J.L. Martin: Biochemistry 27, 4049 (1988)
11. J. Kuriyan, S. Wilz, M. Karplus and G.A. Petsko: J. Mol. Biol. 192, 133 (1986)
12. S.M. Peng and J.A. Ibers: J. Am. Chem. Soc. 98, 8032 (1976)
13. W.R. Scheidt, K.J. Haller, M. Fons, T. Mashiko and C.A. Reed: Biochemistry 20, 3653 (1981)
14. M. Poliakoff and E. Weitz: Adv. Organomet. Chem. 25, 277 (1986)

Vibrational Relaxation Measurements of Carbon Monoxide on Metal Clusters

E.J. Heilweil, R.R. Cavanagh, and J.C. Stephenson

National Bureau of Standards, Center for Chemical Physics, Room B-268, Building 221, Gaithersburg, MD20899, USA

Measurements using tunable picosecond infrared pulses of CO(v=1) vibrational population relaxation for carbon monoxide bound to transition metal cluster compounds and supported metal particles give the first time-resolved evidence for vibrational-to-electronic energy damping of adsorbates on metals.

1. Introduction

There exists an ever-increasing need to directly measure microscopic energy dynamics of adsorbates on chemically active metal surfaces. It is the mechanisms and rates of electronic and vibrational excitation migration within an adsorbate, or from the adsorbate to the substrate, which play key roles in such fundamental processes as surface diffusion, desorption and chemical reaction. Until recently, damping rates (T_1^{-1}) of adsorbate vibrational modes on surfaces could only be inferred from bandwidth measurements and theory. However, ultrafast time domain techniques have now reached a level of sophistication that in situ measurements of energy transfer on reactive metal surfaces can now be made. Here, we present the first direct measurements of vibrational energy relaxation for carbon-monoxide ligands of metal-carbonyl cluster compounds and CO chemisorbed on colloidal (≥30 Å diameter) metal particles dispersed on silica.

2. Experimental

The method employed in this work is based on a tunable picosecond infrared pump-probe saturation-recovery approach. This technique was previously developed to obtain T_1 lifetimes for OH and other strongly IR-allowed modes of adsorbates on dielectric substrates [1]. For the current application, however, ultrashort tunable infrared pulses in the 5 micron (ca. 2000 cm^{-1}) region are required. These pulses are obtained by difference frequency generation in a nonlinear $LiIO_3$ crystal from second harmonic (532 nm) and tunable dye laser (≈600 nm) pulses derived from a 10 Hz Nd^{+3}:YAG laser system. IR pulses produced in this fashion are routinely characterized as having a 15 ps FWHM duration, 10-15 μJ energy (±20% shot fluctuation) and 4 cm^{-1} FWHM bandwidth [2].

In the experiment, a single IR pump pulse ($I \leq 10$ mJ/cm^2) directly excites population to the v=1 level of a CO-stretching mode of the sample. This excitation produces a transient increase in sample transmission T (at the pump frequency) which recovers exponentially in time (with decay time T_1) to the unexcited sample transmission level (T_o). A time-delayed weak pulse at the same frequency probes the sample transmission recovery, and for the CO-containing systems studied here, yields a measure of the vibrational population relaxation of that fundamental mode.

Springer Series in Chemical Physics, Vol. 48: **Ultrafast Phenomena VI**
Editors: T. Yajima · K. Yoshihara · C.B. Harris · S. Shionoya

The above method has been extended from our previous studies of mono-metal carbonyl compounds [2] to the investigation of vibrational relaxation of CO bound to metal clusters of increasing nucleus size [3]. These systems include the stable cluster compounds $Rh_2(CO)_4Cl_2$, $Rh_4(CO)_{12}$, $Co_4(CO)_{12}$, and $Rh_6(CO)_{16}$ in dilute, room temperature $CHCl_3$ solution, some of these same molecules adsorbed on pressed fumed SiO_2 supports, and now CO adsorbed on much larger Rh and Pt particles (supported on silica). The main thrust of this investigation has been to examine the effect of increasing metal nucleus size on the CO(v=1) T_1 lifetime and thereby determine whether new relaxation mechanisms become operative when "free-electron" behavior for sufficiently large metal particles sets in.

3. Results

The most pertinent results are summarized below:

• The CO-stretching vibrations for CO in small metal cluster compounds (≤6 metal atoms) in room temperature $CHCl_3$ solution give $400 \leq T_1 \leq 750$ ps. These long lifetimes are attributed to the large energy gap between the excited CO(v=1) mode (ca. 2000 cm^{-1}) and the next highest metal-CO and metal-metal vibrations (200 - 600 cm^{-1}) of these "molecular" systems.

• A novel bi-exponential decay has been observed for $Co_4(CO)_{12}$ in $CHCl_3$ solution. This response is thought to arise from initially rapid (ca. 50 ps) CO-stretch energy equilibration followed by a much slower (350 ps) combined relaxation into the lower vibrational modes of the system.

• For $Rh_2(CO)_4Cl_2$, $Co_4(CO)_{12}$ and $Rh_4(CO)_{12}$ supported on silica in vacuum and at room temperature, a reduction of T_1 of a factor of four occurs compared to T_1 in the solution phase (see Fig. 1a).

• Induced transient absorption signals have been detected when the IR frequency is tuned to the peak or low energy side of an absorption band. This behavior is concurrent with multiple photon absorption and most likely arises from CO(v=2) and higher overtone excitation.

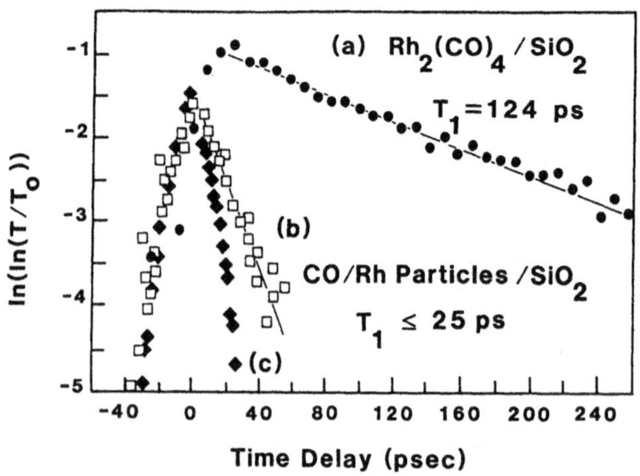

Fig. 1 Transient T_1 vibrational population decays of CO(v=1) for (a) isolated $Rh_2(CO)_4/SiO_2$ surface species and (b) CO adsorbed on large (ca. 35 Å) rhodium particles supported on SiO_2 disks. Figure 1 (c) shows the 22 ps FWHM second harmonic infrared pulse autocorrelation function.

• Samples prepared by reduction of metal salts ($RhCl_3 \cdot 3H_2O$ or PtH_2Cl_6) on silica exhibit absorption features from CO adsorbed on large (ca. 35 Å, 10^3 atoms) Rh and Pt metal particles and for rhodium, isolated metal sites e.g., $-Rh(CO)_2$. These absorptions give rise to short (≤ 25 ps) and long (ca. 125 ps) lifetimes, respectively, in the <u>same</u> inhomogeneously distributed sample. T_1 relaxation decays for one $CO/Rh/SiO_2$ sample are shown in Fig. 1. The decreased magnitude of T_1 for CO on the large particles compared to the smaller cluster compounds and isolated metal sites suggests the participation of a fast relaxation channel, presumably damping by electron-hole pair formation or other electronic properties of the metal.

It is anticipated that further measurements of this type on other adsorbate-metal systems will help unravel the detailed mechanistic, kinetic and microscopic structural characteristics of chemically reactive metal surfaces.

References

[1] E. J. Heilweil, M. P. Casassa, R. R. Cavanagh and J. C. Stephenson in Springer Series in Chemical Physics, Vol. 46, eds. G. R. Fleming and A. E. Seigman (Ultrafast Phenomena, 1986) page 465.

[2] E. J. Heilweil, R. R. Cavanagh and J. C. Stephenson, Chem. Phys. Letters, **134**, 181 (1987).

[3] R. R. Cavanagh, E. J. Heilweil and J. C. Stephenson, J. Electron Spec. and Related Phenom., **45**, 31 (1987).

Vibrational Relaxation Pathways of the N-H Stretch of Pyrrole in Liquids

J.R. Ambroseo and R.M. Hochstrasser

Department of Chemistry, University of Pennsylvania, Philadelphia, PA 19104, USA

A series of experiments have been conducted in order to measure the rate and to elucidate the mechanism of vibrational energy (T_1) relaxation of the symmetric N-H stretch (E_V=3400cm^{-1}) in liquid and solution phase pyrrole. The impetus for these studies lies in the need to expand the understanding of the dynamics of vibrational relaxation and to provide the first conclusive investigation of N-H relaxation.

The work in this field was pioneered by Kaiser and coworkers (1). They generated a wealth of T_1 information particularly on the relaxation of C-H stretches and coupling of these vibrations to internal motions in both solute and solvent. Further T_1 measurements in liquids (1,2) and solids (3) were made, but little knowledge of specific relaxation pathways was obtained. Recent studies in solids (4) and liquids (2), including the current work, show that far from being limited to a few picoseconds T_1 lifetimes often lie in the 100's ps or even ns (4) regime. Moreover, the nature of vibrational energy transfer in pyrrole allows for the separation of contributions from intramolecular and intermolecular pathways. The effects upon T_1 rates and mechanisms of various solvents, isotopic substitution, and temperature have also been examined.

In the experiments reported here, the N-H stretch is resonantly excited by an ca. 4ps infrared pulse (Δv=10cm^{-1} FWHM) obtained by difference frequency mixing (in $LiIO_3$) a fundamental pulse (1054nm) from a Nd^{3+}:phosphate glass laser and a 775nm pulse from an amplified continuum. The resulting populations in the N-H stretch and other vibrational modes as well as their time evolution were determined by spontaneous anti-Stokes Raman scattering from a 527nm pulse. The signal-to-noise was adequate to enable studies of dilute solutions of pyrrole in ethers and benzenes.

The vibrational population lifetime of the N-H stretch (Fig. 1a) at 298K was 42±3ps for neat pyrrole. It was also found to be temperature independent to 353K. The relaxation pathway was studied by measuring the population in the four C-H stretches around 3100cm^{-1} and the symmetric C=C stretch at 1144cm^{-1} following pumping of the N-H stretch. In neat pyrrole, less than 1% of the initial N-H excitation flows into C-H motion whereas ca. 32% is transferred to the C=C stretch (T_1~4ps, Fig. 1b). Thus, the system demonstrates a remarkable propensity for coupling the N-H and C=C stretches. Upon deuteration of the C-H bonds, the efficiency of N-H to C=C (E_V=1094cm^{-1} in CD_4NH) decreases to ca. 18% while the N-H lifetime rises to 64±8ps. This relative insensitivity to isotopic substitution suggests the C-H modes are not involved in the N-H relaxation. Further, the lack of sensitivity to the internal mode shifts resulting from deuteration would be consistent with an intermolecular mechanism.

Springer Series in Chemical Physics, Vol. 48: Ultrafast Phenomena VI
Editors: T. Yajima · K. Yoshihara · C.B. Harris · S. Shionoya

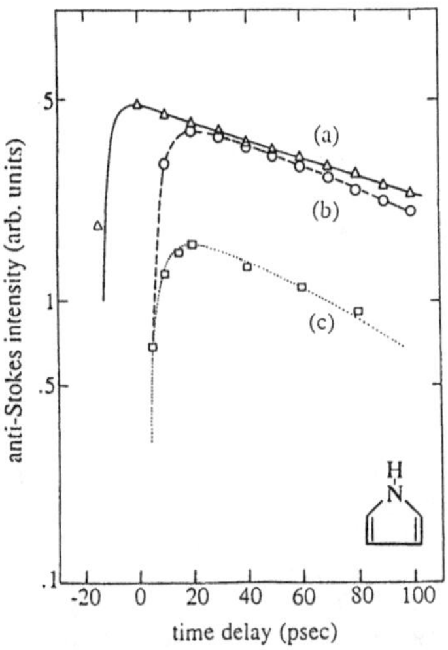

Figure 1. Population evolution following N-H excitation in (a) the N-H stretch, (b) the ring mode in neat pyrrole, and (c) the 991 cm^{-1} mode in benzene.

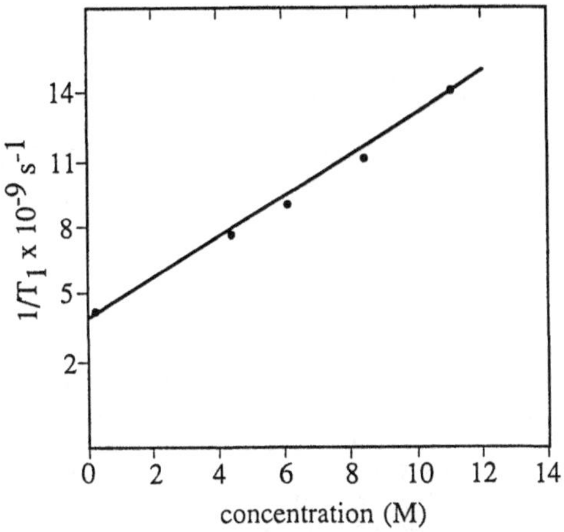

Figure 2. Concentration dependence of the vibrational population rate of relaxation of solvated N-H stretches in ether solutions.

The intermolecular nature of the relaxation pathway is strongly supported by experimental observations for solvated samples. In a 50% solution of pyrrole in protobenzene, ca. 5% of the N-H population (T_1=79±9ps) of solvated pyrrole molecules was transferred to the 991cm^{-1} C=C stretch (T_1~12ps) in benzene (Fig. 1c). Deuteration of the solvent yielded no change in the quantum efficiency and a negligible change in the lifetime (T_1=90±11ps). These results indicate the N-H stretch of a pyrrole molecule is coupled to the C=C stretch in a neighboring molecule, either pyrrole or benzene, via the π-H hydrogen bond.

Additional evidence of these conclusions is given by the T_1 studies of the N-H stretch in ether solutions (Fig. 2). In these samples, lifetimes extending to 220±20ps were observed and no C=C stretch population was detected. The concentration dependence was found to obey Stern-Volmer kinetics of the form $1/T_1 = 1/T_1(i) + n_0\gamma_{01}$. Here, $T_1(i)$ is an intramolecular relaxation time, n_0 is the population of pyrrole molecules in the v=0 level of the N-H stretch and γ_{01} is the bimolecular rate constant governing the relaxation of molecules excited in v=1 during collisions with those in v=0. The values: $T_1(i)$ = 240ps; and $\gamma_{01} = 8.2 \times 10^8$ $lmole^{-1}s^{-1}$, the latter being in the range expected for a diffusion controlled reaction. Thus for neat pyrrole, the intramolecular relaxation is almost entirely suppressed in favor of an intermolecular process while in the dilute solutions the opposite situation prevails.

This research was supported by the NSF and the NIH.

References

1. A. Laubereau and W. Kaiser, Rev. Mod. Phys., 50 (3), 607 (1978).
2. E.J. Heilweil, M.P. Cassassa, R.R. Cavanagh and J.C. Stephenson, J. Vac. Sci. Technol., B3 (5), 1471 (1985).
3. S. Velsko and R.M. Hochstrasser, J. Phys. Chem. 89, 2240 (1985).
4. T.J. Trout, S. Velsko, R. Bozio, P.L. DeCola and R.M. Hochstrasser, J. Chem. Phys., 81 (11), 4746 (1984).

Picosecond Infrared Spectroscopy of Semiconductors and Molecules

W. Kaiser, R.J. Bäuerle, T. Elsaesser, H.-J. Hübner, and A. Seilmeier

Physik Department E11, Technische Universität München,
Arcisstraße 21, D-8000 München 2, Fed. Rep. of Germany

In recent years, the generation of tunable picosecond infrared pulses has made substantial progress. The wavelength range from 1 μm to 10 μm is covered continuously combining different generation techniques. Infrared dye lasers pumped in the traveling-wave geometry give single pulses which are tunable between 1.1 μm and 1.45 μm with a narrow bandwidth of 7 cm^{-1} /1/. High peak intensities of up to 50 MW/cm^2 and pulse durations between 4 ps and 20 ps may be generated. Nonlinear frequency converters based on parametric amplification and difference frequency mixing produce tunable pulses of narrow bandwidth between 1.2 μm und 10 μm /2/.

In this paper we report on two spectroscopic applications of these infrared pulses: the nonlinear absorption of free carriers in InAs and the vibrational redistribution in the electronic ground state of large organic dye molecules.

The stationary free-carrier absorption of polar semiconductors has been studied in a variety of p- and n-doped samples to determine the wavelength dependence and the absolute value of the absorption coefficient /3/. The experimental results have been analyzed by the simple Drude formula or by more detailed quantum-mechanical models, which demonstrate the important role of phonons in the absorption process. There is, however, little information on the change of the absorption coefficient with the thermal energy of the carriers, especially for the case when the temperatures of the lattice and the carrier system are different. Here we report on a picosecond investigation of the transient absorption of hot carriers which demonstrates the influence of carrier temperature on the absorption cross-section.

Undoped as well as n-doped InAs samples are used in the present experiments. The carrier concentration in the undoped crystal is less than 2×10^{16} cm^{-3}; the n-doped sample contains 1.5×10^{18} electrons per cm^3. The thickness of both crystals amounts to approximately 300 μm. At room temperature, the absorption edge of the undoped and n-doped samples is located around 0.345 and 0.390 eV, respectively. The pump and probe technique is applied in the experiments. The medium infrared pulses are split in an intense excitation pulse and a probe pulse of weak intensity. The polarization of the pump pulse is chosen perpendicular to that of the probe pulse to suppress the coherent artifact occurring around zero delay time for parallel electric field vectors of the two pulses. The experiments are performed with sample temperatures between 70K and 300K.

In the first experiment, the intrinsic sample is studied at room temperature. A hot electron-hole (e-h) plasma of a maximum carrier temperature $T_c \approx 450K$ is created by two-photon absorption of the intense excitation pulse at 0.25 eV (wavelength 5 μm). The probe pulse monitors the absorption change at 5 μm, i.e. below the band gap. The result of this measurement for an excitation density of approximately 10^{18} e-h pairs per cm^3 is presented in Fig.1. The normalized change of transmission, $\Delta T/T_0$, is plotted versus the delay time

Springer Series in Chemical Physics, Vol. 48: Ultrafast Phenomena VI
Editors: T. Yajima · K. Yoshihara · C.B. Harris · S. Shionoya

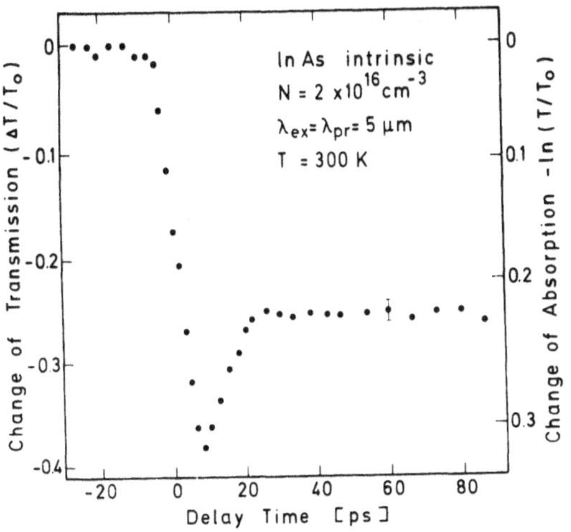

Fig. 1 Transient free-carrier absorption of a hot electron-hole plasma generated by two-photon excitation in intrinsic InAs.

between excitation and probe pulses; ΔT is given by $\Delta T = T - T_0$, where T_0 and T are the sample transmissions prior to and after excitation. We observe a fast decrease of transmission, which relaxes within 30 ps to a residual value constant up to delay times of several hundreds of picoseconds.

The fast kinetics disappears completely when electron-hole pairs are generated with negligible excess energy, i.e. for excitation pulses of a photon energy of $E_G/2$ (E_G: bandgap energy). In this case, only the long-lived, temporally integrating absorption change is observed. These findings demonstrate that the enhancement of absorption at early delay times is indicative of the hot carriers, whereas the second slow contribution represents the absorption of the thermally relaxed electrons and holes. This second component is mainly due to holes because of their larger infrared absorption cross-section as compared to the electron absorption. The long-lived signal disappears by carrier recombination on a time scale of several hundred nanoseconds.

In the second experiment, the n-doped sample is excited with pulses at 0.19 eV (6.5 μm), i.e. the energy of two photons is not sufficient to generate electron-hole pairs. The carrier density is constant and the absorbed photons heat only the electron gas. The experimental data for a sample temperature of 70K are shown in Fig.2. The density of absorbed photons amounts to approximately 5×10^{17} cm^{-3} heating the electrons to a maximum temperature $T_c \approx 550K$. The absorption of the hot electrons rises quickly and decays within 70 ps to the initial value reflecting the cooling of the carriers. In contrast to Fig.1 a long-lived signal component corresponding to excess electron-hole pairs is not observed. At room temperature, a similar behavior is found. However, the absolute value and the decay time of the absorption change are considerably smaller.

The quantum-mechanical theory of free carrier absorption depends upon two factors relevant to our experiments /4/: First, the Fermi distribution function of the electrons in the conduction band changes with increasing carrier temperature; second, the absorption cross-section is related to the density of phonons which couple to the carriers in the absorption process. The

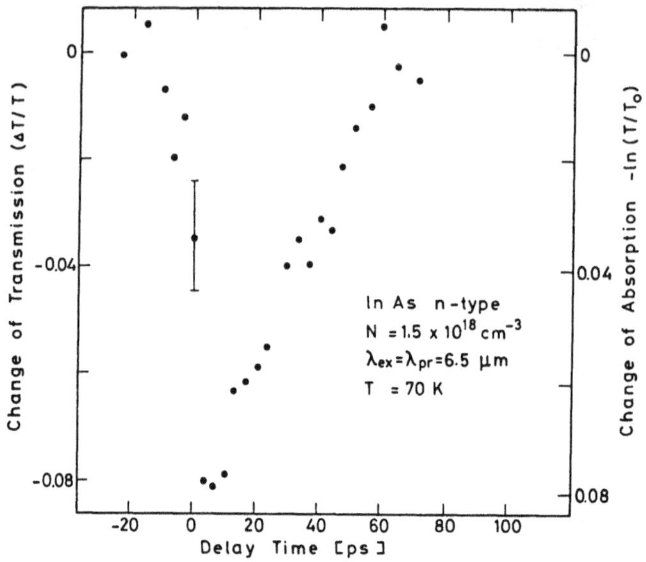

Fig. 2 Transient absorption of hot free electrons in n-doped InAs at a wavelength of 6.5 μm.

phonon density increases by the cooling of hot carriers. The quantitative analysis shows that the electron heating leads to changes of the absorption coefficient of less than 2 percent in the range of carrier temperatures occurring in our experiments. According to Fig.2 we observe larger changes of absorption. This finding suggests a transient nonequilibrium phonon distribution in the sample. More detailed model calculations of the absorption and phonon kinetics will be presented elsewhere.

As a second application of tunable infrared pulses we discuss the lifetimes of vibrational modes of large molecules in solution. There exists a considerable amount of data for excitation frequencies around 3000 cm^{-1}, i.e. for C-H stretching modes. Here a systematic study of vibrational states of energies between 1000 cm^{-1} and 2000 cm^{-1} is presented for the first time.

In our experiments, different vibrational modes are populated via resonant absorption of infrared light pulses. The excess population is monitored by a second visible light pulse promoting the excited molecules to the electronic S_1 state. The emitted fluorescence light is a measure of the instantaneous population of the investigated vibrational state.

Now we work with a Nd:glass laser as a pumping source for the infrared pulse generators, the IR pulse duration is reduced to a value of 2 ps.

Experiments are performed with a solution of coumarin 6 in C_2Cl_4. Fig.3 shows the infrared spectrum of coumarin 6 between 1200 cm^{-1} and 1800 cm^{-1}. A series of distinct absorption lines, mostly skeletal modes, is observed. The strongest lines around 1600 cm^{-1} are due to C-C stretching modes of the chromophore. The well separated line at 1728 cm^{-1} represents the C=0 stretching mode of the molecule. In the following we present time-resolved data measured at two frequency positions at 1594 cm^{-1}, the strongest line in Fig.3 (C-C stretching mode) and at 1728 cm^{-1} (C=0 stretching mode).

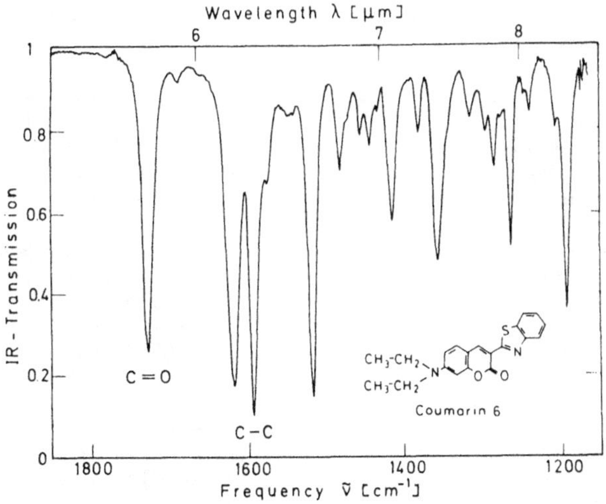

Fig. 3 Part of the infrared spectrum of coumarin 6 in C_2Cl_4 (c = 2.5×10^{-3}M, ℓ = 2mm). Inset: Molecular structure of coumarin 6.

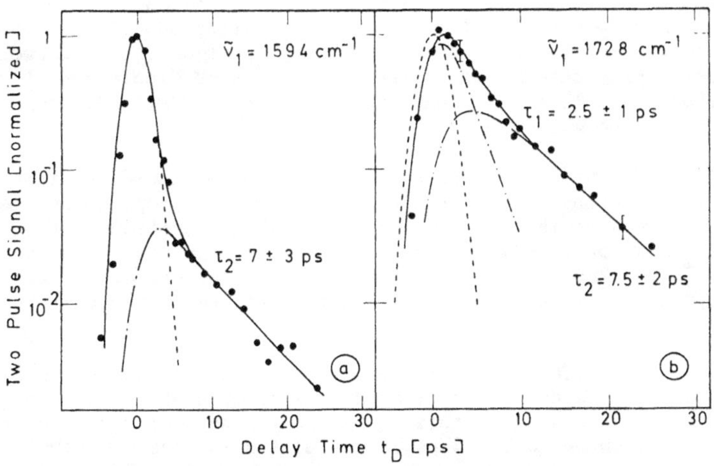

Fig. 4 Fluorescence signal as a function of time after resonant excitation of (a) a C-C stretching mode and of (b) the C=0 stretching mode. Broken line: cross-correlation curve; dash-dotted lines: the fluorescence components.

Information on the vibrational relaxation is obtained by the observation of the fluorescence as a function of the delay time. In Fig.4 experimental results are shown after excitation of (a) a C-C stretching mode and of (b) the C=0 stretching mode. In both cases the fluorescence signal consists of two components. At short delay times a rapidly rising and decaying signal is found. In Fig.4a it closely follows the cross-correlation curve (broken line). In Fig.4b it decays with a time constant of 2.5 ps. This component represents the population and the depopulation of the initially excited vibrational mode. In other words, we find a T_1 lifetime of 0.8±0.4 ps for the C-C skeletal mode and of 2.5 ps for the C=0 stretching mode. The skeletal modes around 1600 cm^{-1}

exhibit very large Franck-Condon factors. Consequently, the fluorescence signals are strong and the decay may be observed over three orders of magnitude.

Subsequently, the excitation energy is intramolecularly redistributed within the vibrational manifold. A new energy distribution is established which is well described by a thermal distribution with an increased internal temperature. A rise of the internal temperature of approximately 70K is estimated from the energy supplied to the molecule and the specific heat of the molecule.

The vibrationally hot molecules give rise to a second fluorescence component which is clearly observed at longer delay times in Fig.4a and 4b. The decay of the longer-lived component represents cooling of the vibrationally hot molecules. In Fig.4 we find a time constant of 7 ps for the intermolecular transfer of vibrational energy to the solvent. This value is independent of the initially excited mode.

Eight vibrational modes between 1400 cm^{-1} and 1800 cm^{-1} were investigated. In most cases the initially excited mode decays with a time constant faster than the time resolution of the system (similar to Fig.4a). In all cases, however, the vibrational cooling of the molecules occurs with a time constant of approximately 7 ps.

More detailed information on the very fast population lifetimes is obtained by comparison of the absolute signal values with calculations based on a rate equation model. Our experimental results show that the magnitude of the fluorescence signal at longer delay times ($t \geq 10$ ps) is proportional to the amount of absorbed infrared energy determined by the known infrared dipole moment of the initially excited mode. This fact indicates that the longer-lived component represents a heating of the molecule.

The situation is different for the fluorescence signal around delay time zero. The transient population of the initially excited state depends on the infrared dipole moment and on the lifetime of the excited vibrational mode. The fluorescence signal is proportional to the instantaneous population and to the Franck-Condon factor. The knowledge of the Franck-Condon factors allows an estimate of the population lifetime based on the absolute peak values of the fluorescence intensities.

We calculate population lifetimes of the excited vibrational modes taking into account Franck-Condon factors deduced from Shpolskii spectra of coumarin 6 /5/. The lifetimes determined according to this procedure depend on the excited mode. For example, a time constant of 0.8 ± 0.3 ps is calculated for the C-C stretching mode (Fig.4a).

The two examples discussed here demonstrate the potential of our generator of tunable infrared pulses. Several additional applications are discussed elsewhere, e.g. the investigation of intersubband relaxation in multiple quantum-well structures /6/, coherent pulse propagation effects in air /7/, and proton transfer processes in large molecules /8/.

References

1 T. Elsaesser, H.J. Polland, A. Seilmeier, W. Kaiser: IEEE J. Quant. Electron. QE-20, 191 (1984)

2 T. Elsaesser, H. Lobentanzer, A. Seilmeier: Optics Commun. 52, 355 (1985)

3 W.G. Spitzer, H.Y. Fan: Phys. Rev. 106, 882 (1957)

4 see e.g. K. Seeger: Semiconductor Physics, Springer Series in Solid State Sciences, Vol. 40, Springer (Berlin 1985)

5 T.A. Tamm, P.M. Saari, Zurnal prikladnoj spektroskopie 32, 150 (1980)
6 A. Seilmeier, H.-J. Hübner, G. Abstreiter, G. Weimann, W. Schlapp, Phys. Rev. Lett. 59, 1345 (1987); R.J. Bäuerle, T. Elsaesser, W. Kaiser, H. Lobentanzer, W. Stolz, K. Ploog, Phys. Rev. Lett. B, in press (1988)
7 A. Seilmeier, M. Wörner, H.-J. Hübner, W. Kaiser, to be published (1988)
8 T. Elsaesser, W. Kaiser, Chem. Phys. Lett. 128, 231 (1986)

Time- and Frequency-Resolved Infrared Spectroscopy with Picosecond Pulses

H. Graener, T.-Q. Ye, R. Dohlus, and A. Laubereau

Physikalisches Institut, Universität Bayreuth,
D-8580 Bayreuth, Fed. Rep. of Germany

Recently we demonstrated a new picosecond double resonance technique for the determination of vibrational relaxation processes /1/, which is based on nonlinear infrared transmission spectroscopy /2/. A strong pump pulse excites a vibrational state and depletes the ground states with correspondingly rapid transmission changes at various probe frequencies. A careful analysis of the time- and frequency-resolved measurements provides detailed insight in the molecular processes. In the following we discuss dynamics of hydrogen-bonded molecules, which is studied by our technique for the first time on the ps time scale.

The strong pump pulse and the weak probe pulse are generated by separate multiple-step parametric devices. The two synchronized parametric pulses are derived from single pulses of an active-passively modelocked Nd-YAG laser. The pulse duration is approximately 12 ps and the bandwidth 11 cm^{-1}. Tuning range is 2700 to 4000 cm^{-1}. The probe transmission change $\ell n(T/T_o)$ is measured as a function of delay time t_D and frequency position ν of the probe pulse. $T_o(\nu)$ represents the equilibrium transmission value (pump beam blocked).

Two examples of our transient IR-probing spectra are presented in Fig. 1. Ethanol dissolved in CCl_4 (0.42 mol/liter) is investigated at room temperature. The pump pulse is tuned to 3320 cm^{-1} near the center of the broad O-H stretching absorption band of the sample. Fig. 1a shows the picosecond probe spectrum for t_D = 10 ps, which differs distinctly from the conventional IR spectrum. An increase of transmission (positive $\ell n\ T/T_o$) is observed for $3100 < \nu < 3430$ cm^{-1}, in contrast to a transmission decrease for $3430 < \nu < 3640$ cm^{-1}. Fig. 1b displays the probe spectrum for the same excitation conditions but at t_D = 100 ps with only little change apart from an amplitude decrease. The probe spectrum remains nearly constant for later times up to 4 ns.

In Fig. 2 two examples of time-resolved measurements are depicted for the same sample and excitation frequency as for Fig. 1. For the probing frequency of 3320 cm^{-1} (Fig. 2a) one observes a rapid increase of transmission at $t_D \approx 0$, followed by a decrease with a time constant of 20 $\pm$ 5 ps. For $t_D > 100$ ps an almost constant transmission change occurs. Tuning the probe frequency to 3500 cm^{-1} yields the strikingly different signal transient of Fig. 2b. At this frequency the transmission displays a somewhat delayed decrease. A time constant of 5 $\pm$ 3 ps

Springer Series in Chemical Physics, Vol. 48: **Ultrafast Phenomena VI**
Editors: T. Yajima · K. Yoshihara · C.B. Harris · S. Shionoya

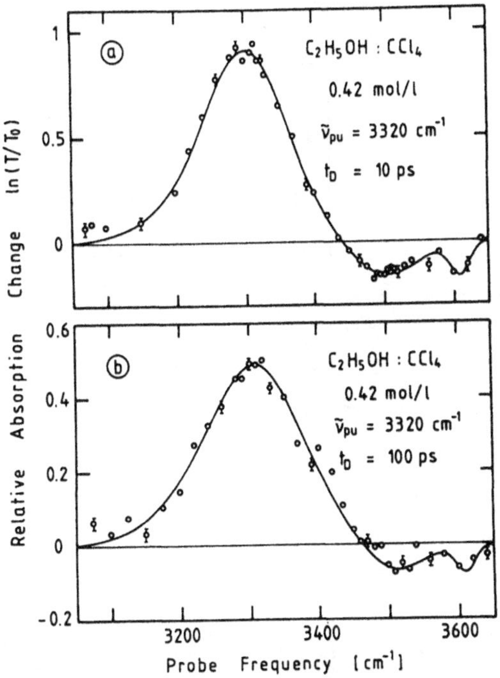

Fig. 1

Transient IR-absorption spectra of 0.42 mol/l ethanol in CCl_4; excitation frequency 3320 cm^{-1}
a) delay time 10 ps
b) delay time 100 ps

is inferred from the delayed minimum position. The transmission change then relaxes with time constant 20 $\pm$ 5 ps to an asymptotic value.

For the interpretation of these results it is recalled that the absorption around 3300 cm^{-1} is caused by hydrogen-bonded ethanol chains (oligomers), whereas the absorption bands of monomeric and dimeric ethanol molecules are situated at 3630 cm^{-1} and 3500 cm^{-1}, respectively /3/. So the transient bleaching around 3300 cm^{-1} may be caused by two mechanisms: either population changes of the O-H stretching vibration or breaking of hydrogen bonds, the binding energy (approximately 5.5 kcal/mol) of which corresponds to a frequency of about 2000 cm^{-1}. The transmission decrease around 3500 cm^{-1}, on the other hand, can only be explained by an increase of the number of dimeric or monomeric ethanol molecules; here, excited state absorption of the pumped OH vibration can be excluded since the corresponding frequency position lies on the low-frequency side of the excitation. As a result, convincing evidence for dissociation of H-bonds is presented. The rapid signal changes as compared with the long population lifetime $T_1 \simeq 70$ ps /2/ of ethanol monomers suggest that the breaking of hydrogen bonds occurs with the measured time constant of 5 $\pm$ 3 ps, i.e. is the dominant relaxation channel for vibrationally excited oligomers. Shortly after, the ethanol molecules may re-associate; the corresponding time constant is 20 $\pm$ 5 ps. For longer times (t_D > 100 ps) the hydrogen bonds are in a new thermal equilibrium. From the measured temperature dependence of conventional IR spectra we estimate a temperature rise of less than 10 K for the observed asymptotic probe transmission changes, consistent with energy arguments.

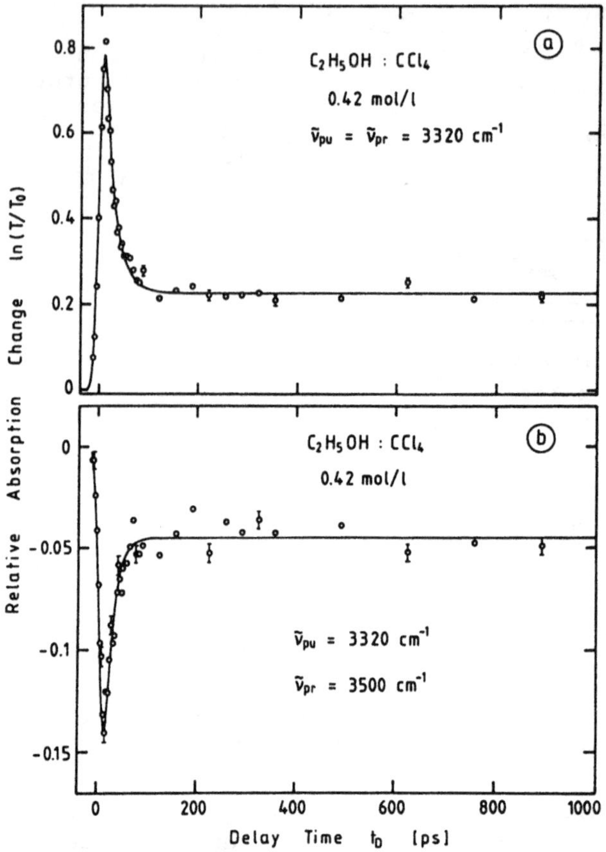

Fig. 2

Relative transmission change versus delay of 0.42 mol/l ethanol in CCl_4. Pump frequency 3320 cm^{-1}

a) probe frequency 3320 cm^{-1}

b) probe frequency 3500 cm^{-1}

In conclusion, it is pointed out that we have observed the dynamics of hydrogen bridges in ethanol solutions at room temperature using a novel ps spectroscopy in the IR. The dominant relaxation channel of vibrationally excited oligomers is made up by ultrafast breaking of H-bonds.

1. H. Graener, R. Dohlus and A. Laubereau, Chem. Phys. Lett. 140, 306 (1987)
2. E.J. Heilweil, M.P. Cassassa, R.R. Cavanagh and J.C. Stephenson, J. Chem. Phys. 81, 2856 (1984)
3. See, for example, G. Geiseler and H. Seidel, 'Die Wasserstoffbrückenbindung', (Vieweg, Braunschweig, 1977)

Femtosecond Dephasing Processes of Molecular Vibrations

W. Zinth, W. Holzapfel, and R. Leonhardt

Physik Department E11, Technische Universität München, Arcisstraße 21, D-8000 München 2, Fed. Rep. of Germany

Dynamical and static interactions in molecular liquids may be deduced from Raman and infrared spectra. In a limited number of cases the shape of the spectral band is understood by a definite line-broadening mechanism. Many realistic liquids, on the other hand, show complicated structures with non-Lorentzian asymmetric and partially overlapping bands. In addition, the spectral bandwidth may be considerably broader than 10 cm^{-1} corresponding to molecular time constants of less than 1 ps. The high time resolution now available in femstosecond coherent anti-Stokes Raman spectroscopy /1-3/ gives the possibility to study various line-broadening mechanisms on the time scale of the molecular interaction.

In the present experiments we applied a femtosecond laser system allowing tunable excitation of vibrational modes between 300 cm^{-1} and 3000 cm^{-1}. The laser pulses derived from a femtosecond unidirectional ring dye laser (pulse duration t_P = 80 fs) serve as the exciting and probing laser pulses at frequencies ω_L /4/. A standard synchronized tunable picosecond dye laser supplies the Stokes pulses at ω_S. Coherently scattered anti-Stokes light at $\omega_{AS} = 2\omega_L - \omega_S$ is detected as a function of the time delay between probing and exciting pulses (of perpendicular polarization). The high time resolution of the experimental system is readily demonstrated in Fig. 1 showing the coherent anti-Stokes signal obtained from the OD-stretching mode of liquid D_2O. The full circles are experimental data taken at a sample temperature of 65°C. These data points are close to the experimental response function determined

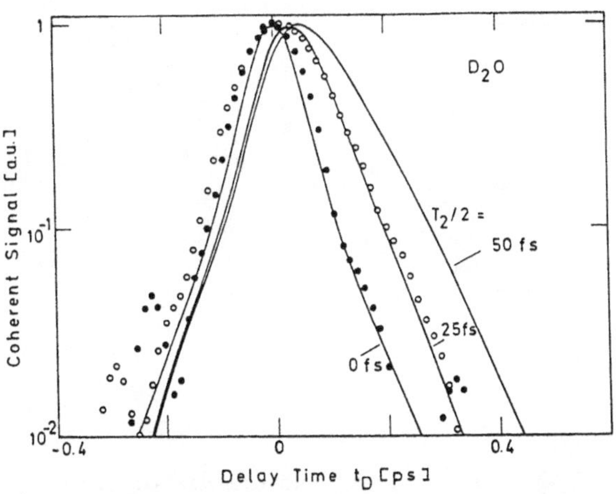

Fig. 1 Time-resolved coherent Raman signal from the OD vibration of liquid D_2O at 21°C (open circles) and 65°C (closed circles). The solid curves are calculated using a decay time $T_2/2$ of 0fs, 25fs, or 50fs.

Springer Series in Chemical Physics, Vol. 48: **Ultrafast Phenomena VI**
Editors: T. Yajima · K. Yoshihara · C.B. Harris · S. Shionoya

by a cross-correlation experiment. A different signal is obtained when the sample is held at a lower temperature of 21°C (open circles). Now the decay of the signal is delayed. The solid curves are calculated using the experimental response function and decay times of $T_2/2$ = 0, 25 fs, and 50 fs. We interpret the experimental data as follows: At low temperatures the coherent experiment studies the hydrogen bonded D_2O molecules. These complexes have a somewhat lower vibrational frequency than the non-bonded D_2O. We tentatively assign the dephasing time $T_2/2$ = 25 fs to these D_2O complexes. When the sample is heated the hydrogen-bonded complexes break up and the experiment probes the liquid of "monomeric" molecules. A very rapid decay of the signal is observed which is not resolved in our experiment.

Interesting information is obtained when studying the CH_3-stretching modes of liquid methanol. The spontaneous Raman spectrum of methanol shows two strong bands at 2835 cm^{-1} and 2943 cm^{-1}. In addition there are large shoulders around the two major lines. Time-resolved coherent Raman scattering gives the result presented in Fig.2 (solid points). The signal exhibits a pronounced beating structure superimposed to an exponential decay. The solid curve of Fig.2 was calculated using homogeneous dephasing times for the two Raman-active modes. This curve fits the experimental points quite well. In order to take into account the possibility of inhomogeneous broadening, the calculation was repeated with an inhomogeneous contribution of 25% and 50% to the Raman linewidth (see broken curves of Fig.2). These curves noticeably deviate from the experimental data points. We conclude that the CH-stretching modes of methanol are predominantly homogeneously broadened.

Results on the molecular interactions are obtained when measuring the CH-stretching modes of liquid acetonitrile. In Fig.3a the coherent signal decay is shown over three orders of magnitudes. At later delay times the signal follows the exponential decay representing homogeneous dephasing with T_2 = 1.74 ps. At earlier delay times the experimental data deviate from the exponential slope. This fact is apparent from Fig.3b where the signal is multiplied by $\exp(2t_D/1.74ps)$. An overshoot of the signal is seen at early delay times of up to 1.5 ps. Our first attempt was to fit the experimental decay curve by a Kubo line-shape model with an exponential decay of the frequency correlation function /5/. A ready agreement between theory and experiment was

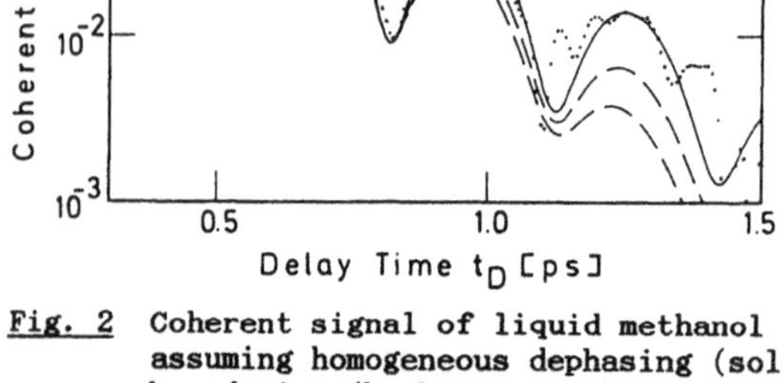

Fig. 2 Coherent signal of liquid methanol (points) and calculated curves assuming homogeneous dephasing (solid curve) and inhomogeneous line-broadening (broken curves) for 25% and 50% inhomogeneous contributions.

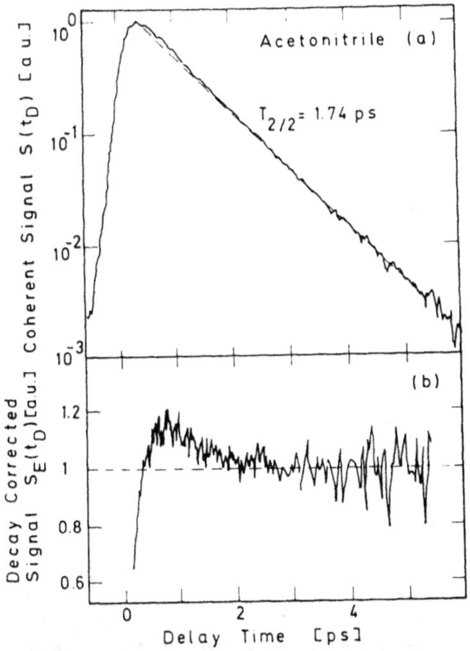

Fig. 3 Coherent signal from liquid acetonitrile. Intermolecular interactions show up by the overshoot of the signal at early delay times.

only obtained with a more elaborate theory, taking into account the close packing of the molecules in the liquid and the attractive part of the molecular interaction /6/.

Acknowledgement. The authors acknowledge valuable discussion with Professors W.Kaiser, S.F.Fischer, and E.W.Knapp.

References

1 R. Leonhardt, W. Holzapfel, W. Zinth, W. Kaiser: Chem. Phys. Lett. 133, 373 (1987)

2 R. Leonhardt, W. Holzapfel, W. Zinth, W. Kaiser, Rev. Phys. Appl. 22, 1735 (1987)

3 W. Zinth, R. Leonhardt, W. Holzapfel, W. Kaiser: IEEE J. Quant. Electron. QE-24, 455 (1988)

4 J. Dobler, H.H. Schultz, W. Zinth: Optics Commun. 57, 407 (1986)

5 D.W. Oxtoby, Adv. Chem. Phys. 40, 1 (1979)

6 E.W. Knapp, S.F. Fischer: J. Chem. Phys. 74, 89 (1981)

Effects of Coherence Transfer on Time-Resolved Coherent Anti-Stokes Raman Scattering and Transient Response of Resonant Light Scattering from Molecules

M. Hayashi [1], *Y. Nomura* [1], *Y. Fujimura* [1], *and Y. Ohtsuki* [2]

[1] Department of Chemistry, Faculty of Science, Tohoku University, Sendai 980, Japan

[2] College of General Education, Tohoku University, Sendai 980, Japan

1. Introduction

Many experimental and theoretical studies have been reported on the effects of vibronic dephasings (population decay and/or pure dephasing) on the time-resolved coherent anti-Stokes Raman scattering (CARS) and resonant light scattering from molecules in the heat bath. On the other hand, relatively little is known about the effects of bath-assisted dynamics such as coherence transfer between two vibrational transitions and feeding processes in the resonant states. Recently, femto-second quantum beats in time-resolved CARS spectra of polyatomic molecules in liquids have been reported by LEONHARDT et al. [1]. It is very interesting to analyze quantum beats appearing in the CARS spectra of molecular liquids in such an ultra-short time domain because one can obtain information on the dynamical behaviors in liquids. The first purpose of this paper is to show that the effects of the coherence transfer induced by the molecule-heat bath interactions are reflected directly on the time-resolved CARS spectra. The second purpose is to investigate the feeding effects on the time-resolved resonant light scattering from multilevel molecules.

2. Coherence transfer between Raman active vibrational transitions

We notice that in ultra-short time domains, two Raman active transitions $a \rightarrow c$ and $a' \rightarrow c'$ of nearby molecules in liquids correlate each other, and the coherence transfer between two transitions may take place through the bath modes. An expression for the time-resolved CARS intensity $I(\tau)$ as a function of the delay time τ between the pumping pulses and the probing pulse is derived by using the perturbative density matrix method [2]. The molecule-heat bath interaction effects on $I(\tau)$ are incorporated with aids of the Liouville space Feynman diagram in which the time evolution of the coherence between the Raman active transitions of two molecules can be easily seen. The coherence transfer matrix element expressing the time evolution of the coherence between two transitions can be evaluated by introducing coherence transfer constant Γ, and by solving the Master equations of the off-diagonal density matrix elements relevant to two transitions. In the case in which the frequency difference between two transitions $\Delta\omega$ is larger than the magnitude of the coherence transfer constant, the nonresonant time-resolved CARS intensity can be approximated to

$$I(\tau) \simeq C \exp[-\{\gamma_{ca} + \gamma_{c'a'} + i(\Delta\omega - \frac{2\Gamma^2}{\Delta\omega})\}\tau] \,, \tag{2-1}$$

where C is a constant associated with the transition moments and photon fields, and γ are the dephasing constants relevant to the Raman transitions.

Springer Series in Chemical Physics, Vol. 48: **Ultrafast Phenomena VI**
Editors: T. Yajima · K. Yoshihara · C.B. Harris · S. Shionoya

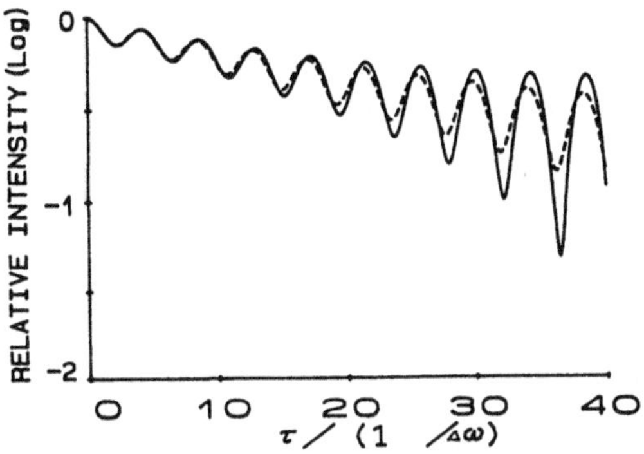

Fig. 1. Calculated time-resolved CARS intensities via coherently excited two Raman transitions of molecules in liquids in a four spherical layer model. The solid line represents the CARS intensity in a long range coherence case in which magnitudes of Γ^2 in the first, second, third and fourth layers are taken to be 0.009, 0.008, 0.007 and 0.006 in units of $\Delta\omega^2$, frequency difference between two Raman transitions. The dashed line represents the CARS intensity in a short range coherence case in which the magnitude of Γ^2 in the first layer is 0.009. The dephasing constants related to two Raman transitions are taken to be 0.03 and 0.01 in units of $\Delta\omega$ in both cases.

Equation(2-1) indicates that the frequency shifts of the quantum beats appear in the time-resolved CARS spectra of molecules in liquids as a result of the coherence transfer. In the case in which there is no coherence transfer, i.e. each molecule interacts with the heat bath independently, the quantum beats of the time-resolved CARS decay with dephasing constants γ.

We consider a multi-spherical layer model to deal with the effects of the coherence transfer of molecules in liquids. In this model the coherence transfer constants are characterized by the same magnitude within each layer. In Fig. 1, model calculations of the coherence transfer effects on $I(\tau)$ are shown using a four spherical layer model. Temporal separation τ is taken in units of $(\Delta\omega)^{-1}$. The time dependent behaviors of the amplitudes of the quantum beats in Fig. 1 originate from the coherence transfer effects.

3. Feeding effects on time-resolved resonant light scattering

We present a theoretical result of the feeding effects (bath-induced fluorescence) on the RLS from a displaced harmonic oscillator molecule embedded in the heat bath consisting of an ensemble of low frequency harmonic oscillators with the Boltzmann distribution. The feeding processes as well as the ordinary dephasing ones are taken into account in the double (cap and tilde) space representation [3]. The damping operator γ_{tot} is expressed as

$$\gamma_{tot} = \hat{\gamma} + \tilde{\gamma} - \gamma^{mix} , \tag{3-1}$$

where $\hat{\gamma}(\tilde{\gamma})$ refer to the system operators representing the population decays in the cap (tilde) space. Operators mixing the cap space with the tilde space, γ^{mix} consist of both pure vibronic dephasing and feeding operators. The feeding operator γ^f is defined by

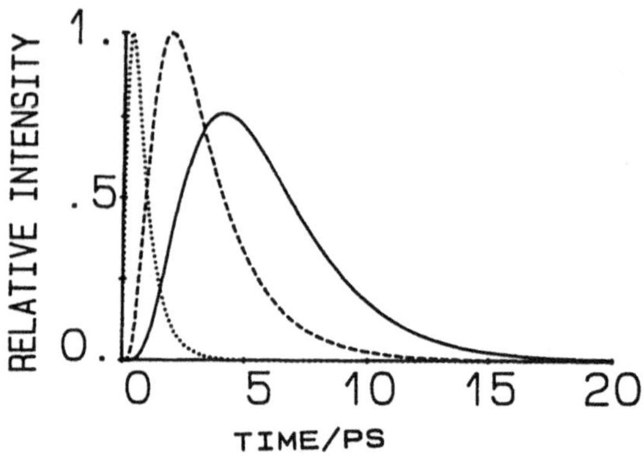

Fig. 2. Calculated time-dependent RLS. The excitation is assumed to the first excited vibrational level in the resonant state by a pulse denoted by a dotted curve. The solid and dashed curves represent the bath-induced fluorescence and resonance fluorescence, respectively. The parameter set used are ω= 300 cm^{-1} as the vibratiional frequency with potential displacement Δ = 0.5, $g(\omega)$ = 0.9 cm^{-1} and $g(-\omega)$ = 0.2 cm^{-1}, and $\gamma_{GE:GE}$ = 1.4 cm^{-1} as the magnitude of the electronic pure dephasing constant.

$$\gamma^{f} = [2g(\omega)\hat{b}\tilde{b} + 2g(-\omega)\hat{b}^{+}\tilde{b}^{+}]\hat{c}^{+}\hat{c}\tilde{c}^{+}\tilde{c} , \tag{3-2}$$

where $\hat{b}(\tilde{b})$ and $\hat{c}(\tilde{c})$ are the boson and fermion annihilation operators in the cap (tilde) space, respectively, and g denote the fourier components of the bath correlation functions. In Fig. 2, calculated time-dependent RLS spectra are drawn. We can see from Fig. 2 that the bath induced fluorescence gives a significant contribution as well as the resonance fluorescence and Raman scattering from the level just excited. Different time-dependent behaviors between the feeding and other mechanisms are identified.

4. Acknowledgement

This work was supported by the Grant-in-Aid for Scientific Research from the Japanese Government (63540322).

5. References

1. R. Leonhardt, W. Holzapfdel, W. Zinth, W. Kaiser: Chem. Phys. Lett. 133, 373 (1987)
2. M. Hayashi, Y. Nomura, Y. Fujimura; J. Chem. Phys. 89, 34 (1988)
3. Y. Ohtsuki, Y. Fujimura, to be published

Vibrational Dynamics in the S_1 and S_0 States of Dye Molecules, Studied Separately by Femtosecond Polarization Spectroscopy

G. Angel, R. Gagel, and A. Laubereau

Physikalisches Institut, Universität Bayreuth,
D-8580 Bayreuth, Fed. Rep. of Germany

In this summary induced dichroism of femtosecond pulses is discussed for the study of the subpicosecond population dynamics of several dye molecules in the S_1. A new version of time-resolved polarization spectroscopy is also demonstrated which provides novel information on Franck-Condon states of the S_0 manifold. Apart from rotational relaxation, which is negligible on the subpicosecond time scale, the information supplied by the two techniques is closely related to two-pulse transmission measurements and self-diffraction off induced population gratings.

The energy level diagram considered for S_1 relaxation dynamics is depicted in Fig. 1a. The excitation pulse promotes a small fraction of molecules to a distribution of Franck-Condon states (2) with large density of vibrational states ($\nu_{vib} \approx 10^3$ cm^{-1}). Anharmonic coupling leads to rapid energy-redistribution (time constant τ_{red}) with a subensemble of approximately isoenergetic vibrational states (3) ("dark states"). Energy transfer to the other vibrational degrees of freedom (4) is described by the vibronic decay time τ_{vib}. A novel technique exploiting polarization dependent amplification is depicted in Fig. 1b, which gives information on vibrational relaxation in the S_0. The required ini-

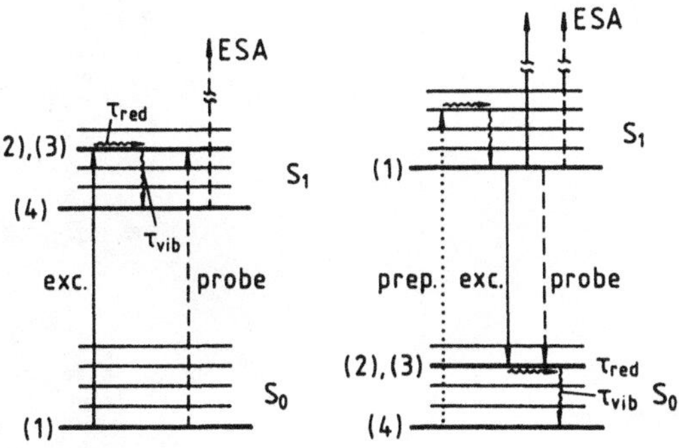

Fig. 1
Schematic of the energy levels and pump/probe schemes of transient polarization spectroscopy a) for vibrational dynamics in the S_1 state, b) for vibrational dynamics in the S_0 necessitating initial population of the bottom of the S_1 (dotted arrow). Data analysis includes population redistribution (τ_{red}), population decay (τ_{vib}) and dephasing (T_2) of vibrational level distributions.

Springer Series in Chemical Physics, Vol. 48: **Ultrafast Phenomena VI**
Editors: T. Yajima · K. Yoshihara · C.B. Harris · S. Shionoya

tial population of the upper electronic state is generated by an intense preparation pulse /1/. Again, rapid IVR among a vibrational subensemble (3) in the S_o represented by time constant τ_{red} is taken into account. A distribution of vibronic transition frequencies (1)-(2) with Gaussian shape and finite dephasing time T_2 of the individual transition is considered.

The measuring set-up is similar to recent ps experiments /2/, the probe pulse of same frequency being generated by a 10 % beam splitter. Our special femtosecond dye laser (Rh 6G/DQOCI) with non-cw operation has been described recently /3/. Single pulses of 115 fs duration and 10 μJ are generated, the exponential pulse wings extending over more than 3 orders of magnitude. The emission wavelength of 572 nm allows to excite higher vibrational levels. The laser system also generates a synchronized green (527 nm) preparation pulse of 1.3 ps required for the S_o technique of Fig. 1b.

An example of the observed S_1 dynamics is depicted in Fig. 2 for DQOCI in ethanol. The measured probe pulse energy transmission (full circles) behind a crossed polarizer is plotted versus time delay.

As expected theoretically the signal curve consists of three parts: (i) a coherence peak around t_D=0, (ii) a rapidly decaying signal overshoot for 0.1 ps < t_D < 2 ps related to vibrational population dynamics and (iii) an asymptotic decay for t_D > 2 ps. The decay time τ_{vib} = 390 ± 100 fs is directly obtained from the time-resolved signal decrease (ii). The modest vibrational signal contribution of ≈ 30 % at t_D = 300 fs represents strong evidence for rapid energy redistribution. Fitting the calculated curves

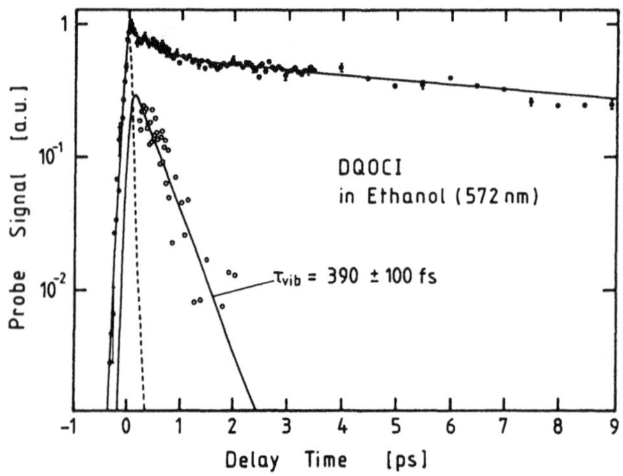

Fig. 2
Induced dichroism of the fs-probe pulse for the study of vibrational dynamics in the S_1 of DQOCI; full points, experimental; lines, calculated; subtracting the exponential signal asymptote the vibrational part (open circles) of the signal is calculated. At t_D=0 a coherence peak is noted which allows one to determine the vibronic dephasing time T_2 /4/.

(solid lines) to the data we find τ_{red} = 45 ± 5 fs and also T_2 = 90 ± 20 fs.

Our results are compiled in Table 1. The data on T_2 and τ_{red} are of semi-quantitative nature since they originate from the specific theoretical model. The values of τ_{vib} are directly observed and model-independent.

Table 1	ν_{vib} cm^{-1}	T_2 fs	τ_{red} fs	τ_{vib} ps
DQOCI/Eth	1100	90 ± 20	45 ± 5	0.39 ± 0.1
DODCI/Eth	1100	30 ± 20	5 ± 20	0.95 ± 0.3
Rh 101/Meth	700	60 ± 20	30 ± 10	0.95 ± 0.35
Rh 6G/HFIP	1570	≤ 40	≤ 20	1.5 ± 1
Rh 6G/CH_2Cl_2	1380	≤ 40	≤ 20	1.7 ± 0.8
Rh 6G/Eth	1030	≤ 40	≤ 20	2.4 ± 0.5
Rh 6G/DMSO	530	50 ± 20	30 ± 10	3.1 ± 0.6

The rapid energy redistribution may be explained by anharmonic coupling among the vibrational subensemble /5/. The time constant τ_{vib} represents the effective population decay time of levels (2), (3); two mechanisms appear to be important: (i) V-V energy transfer to the other vibrational degrees of freedom of the dye molecule; (ii) intermolecular energy transfer to solvent molecules. Recent investigations of the latter process have indicated time constants around 5-10 ps /6/. As a consequence process (i) appears to be dominant. The population decay of levels (2),(3) is believed to generate vibrationally heated molecules with a subsequent slower cooling via mechanism (ii). Estimates show that observation of the solvent interaction (ii) is below the limit of detectibility of our present measurements.

In conclusion we point out that subpicosecond (S_1) to picosecond (S_0) vibrational decay times have been observed for several dyes at 293 K. A single vibrational relaxation time cannot account for our observations. Clear evidence for rapid energy redistribution of a vibrational subensemble is obtained from a comparison with model calculations. Analysis of a coherence peak at t_D=0 necessitates the introduction of the finite vibronic dephasing time T_2. Values in the range ≤ 90 fs are suggested for the investigated dye molecules, the major dephasing mechanism being made up by vibrational redistribution, $T_2 \approx 2\,\tau_{red}$.

1. D. Ricard and J. Ducuing, J. Chem. Phys. 62, 3616 (1975)
2. D. Reiser and A. Laubereau, Appl. Phys. B 27, 115 (1982)
3. G. Angel, R. Gagel and A. Laubereau, Optics Commun. 63, 259 (1987)
4. M.W. Balk and G.R. Fleming, J. Chem. Phys. 83, 4300 (1985)
5. A. Amirov, J. Jortner, S. Okajima and E.C. Lim, Chem. Phys. Lett. 126, 487 (1986)
6. F. Wondrazek, A. Seilmeier and W. Kaiser, Chem. Phys. Lett. 104, 121 (1984)

Femtosecond Time and Frequency Resolved Fluorescence Spectroscopy of a Dye Molecule

A. Mokhtari[1], *J. Chesnoy*[1], *and A. Laubereau*[2]

[1]Laboratoire d'Optique Appliquée, Ecole Polytechnique, F-91128 Palaiseau, France
[2]Physikalisches Institut, Universität Bayreuth, D-8580 Bayreuth, Fed. Rep. of Germany

The electronic and vibrational relaxation dynamics of a dye molecule following immediately the absorption process have not yet been fully investigated. An experimental key to this problem is femtosecond fluorescence spectroscopy, which is favoured by several advantages : (i) only the population dynamics of one electronic state (S1) are involved , (ii) the molecule relaxes freely without perturbation by a probing pulse and (iii) spectral information is available without tuning the laser pulse.

We have improved the sensitivity and time resolution of the parametric up-conversion method for transient fluorescence spectroscopy of a dye molecule[1] in the femtosecond domain. For pumping and time gating, a cw modelocked dye laser, pumped by a frequency doubled YAG laser, is used (pulse duration adjustable down to 55 fs, average power 30 mwatts, repetition rate 50 MHz) [2].

The fluorescence signal is collected from a dye jet and focused into a short urea crystal (100 μm thick) which serves as a femtosecond light gate. The continuous up-converted signal is frequency analysed by a monochromator placed in front of a photon counting system. Optimising phase matching conditions, group velocity dispersion and geometrical factors lead to an overall time resolution of 130 fs for a signal level of 1000 counts per second in a detected bandwidth close to the spectral width of the laser pulse.

We have studied the fluorescence dynamics of nile blue in methanol (293K), which displays a weak but simple vibronic structure in the conventional fluorescence spectrum.with a vibronic modulation at 230 cm^{-1}. The time evolution of the fluorescence at different spectral intervals is displayed in Fig.1.

The rapid rise of the signal curves is limited by the time resolution of the experiment. After this rapid rise, we observe a slower evolution on a time scale of several picoseconds (in the range of 5ps) that resembles qualitatively previous observations [1]. In addition, we note a subpicosecond fluorescence decay particularly evident in the high frequency wing of the fluorescence band (Fig.1a). In Fig. 2 we show two examples of the transient fluorescence spectra together with the conventional steady-state spectrum.

Transient spectral red shifting, bandwidth narrowing and displacement of the vibrational bumps are clearly illustrated. The vibronic line positions give a direct measurement of the electronic frequency shift, while the band position is an integrated quantity determined by a summation over Franck-Condon factors. We find that the time evolution of the lines and of the band center are not simply correlated.

Springer Series in Chemical Physics, Vol. 48: **Ultrafast Phenomena VI**
Editors: T. Yajima • K. Yoshihara • C.B. Harris • S. Shionoya

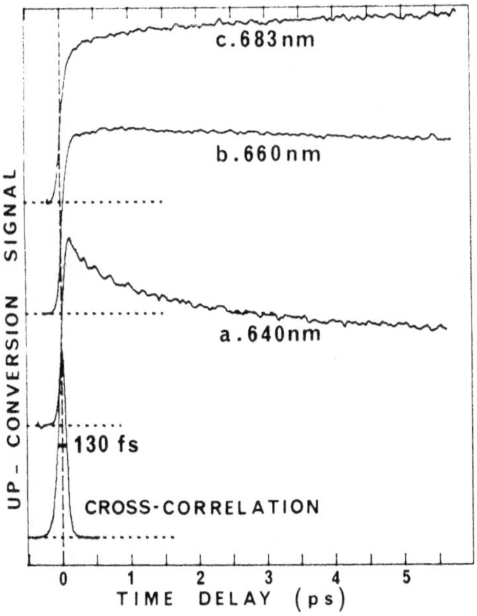

Fig.1
Time resolved fluorescence signal of nile blue dissolved in methanol, collected at the "magic angle" (rotation-free). The up-converted signal is traced at the wavelengths indicated in Fig.2. The time resolution is given by the width of the intensity cross-correlation function obtained by up-converting the light scattered by the sample. The detected spectral bandwidth is 12 nm.

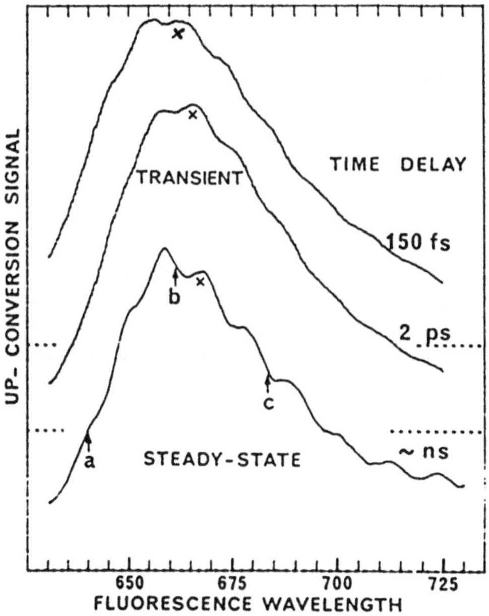

Fig.2
Transient fluorescence spectra of nile blue taken for two time delays after excitation. The laser pulses are deliberately lengthened to around 180 fs to resolve the vibrational structure with a 5nm resolution. The crosses show the evolution of a given line. Arrows a,b,c indicate the spectral position corresponding to the time resolved results in Fig.1.

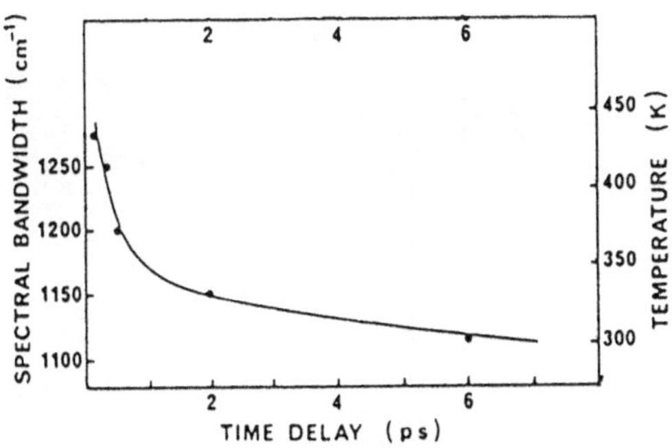

Fig. 3
Bandwidth (FWHM) of the fluorescence band associated to the relaxation of the vibrational temperature of the dominant 230 cm^{-1} active mode.

From the time resolved frequency spectrum, the full bandwidth is obtained and plotted in Fig. 3. The bandwidth is related to the calculated vibrational temperature of the active 230 cm^{-1} mode which is given as a secondary scale. We notice an approximately 400 fs initial decay close to the value obtained in a different pump-probe experiment [3] and that can be clearly attributed to vibrational relaxation, and then a subsequent decay on a ~ 4 ps time scale associated with molecular cooling having the same origin as for the molecular thermometer of Scherer et al.[4].

References

1- E.W. Castner, J.R. Maroncelli and G.R. Fleming : J. Chem. Phys. 86, 1090 (1987)
2- J. Chesnoy and L. Fini : Optics Letters 11, 635 (1986)
3- A.M.Weiner and E.P. Ippen : Chem. Phys. Lett. 114, 456 (1985)
4- P.O. Scherer, A. Seilmeier and W. Kaiser : J. Chem. Phys. 83, 3948 (1985)

Supercontinuum Spectroscopy of Ethyl Violet Using a Simple Pulse Compression Technique

M.M. Martin, F. Nesa, E. Breheret, and Y.H. Meyer

Laboratoire de Photophysique Moléculaire du CNRS, Bât. 213, Université Paris-Sud, F-91405 Orsay Cedex, France

Time resolved absorption spectroscopy of dyes is not straightforward because of the overlap of several electronic transitions: transient gain analysis in a wide spectral range is required. We report here a simple method of producing short pulses and supercontinuum from UV to IR and its use for analysis of the relaxation path of a triphenylmethane dye, Ethyl Violet. A delayed bleaching is observed at a particular wavelength range and is interpreted as the formation of an absorbing transient state possibly of charge-transfer nature.

1. Pulse Compression Techniques: from 10 ns to sub-picosecond

Time shape modification of a smooth laser pulse by saturable amplification was theoretically demonstrated in 1963 [1]. We extend this effect to the case of a spiked pulse in order to selectively amplify the first spike, in a multipass amplifier, and isolate it by means of a single pass saturable absorber. A pulse with spikes of short duration is produced at 610nm from a 30µm cavity dye oscillator filled with Rhodamine 640 and pumped by a 100ps, 560nm, 10µJ pulse. The latter is obtained, with the spectro-temporal selection method previously reported [2], from a R6G oscillator pumped with 10ns, 532nm, 10Hz, Q-switched Nd:Yag laser pulses. An 8 pass amplifier (Fig.1) pumped by 1mJ of the 532nm pulses leads not only to a high energy amplification of about 10^3 (ASE < 10% without saturable absorber) but to a strong distortion of the spiked 100ps input pulse shape as shown in Fig.2 obtained with a streak camera (ARP). Such an alteration is ensured by the saturation of the gain by the pulse leading edge. When the fluence within the pulse reaches, at a given pass, saturation value for the amplifier medium, the gain for the following spikes is reduced. During the next passes the saturation will be reached earlier in the pulse. After six

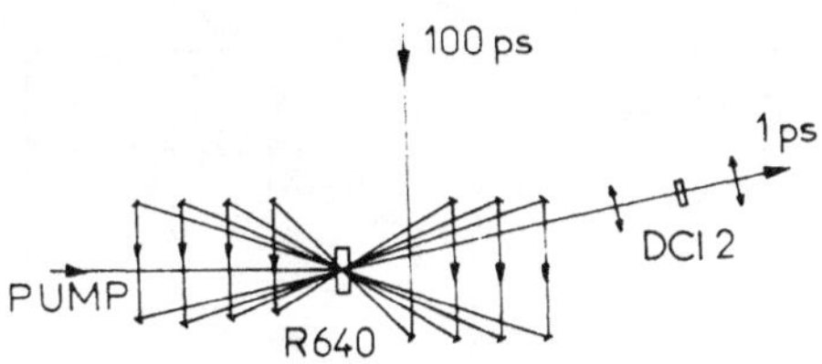

Fig.1 Multipass amplifier compressor

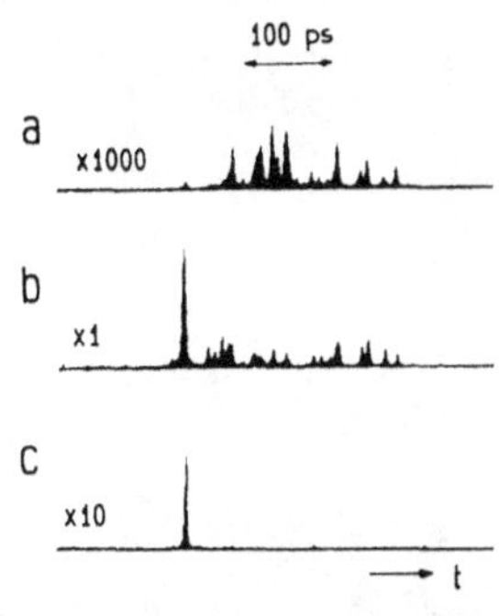

Fig.2 a: input pulse; b: after saturated amplification; c: after saturable absorber (instrument limited)

Springer Series in Chemical Physics, Vol. 48: **Ultrafast Phenomena VI**
Editors: T. Yajima · K. Yoshihara · C.B. Harris · S. Shionoya

passes in our set-up the first spike is largely predominant. Its measured fluence is equal to the saturation value F_s=0.5 mJ/cm^2 estimated from $F_s = 1/\sigma_e$ with $\sigma_e = 2.10^{-16}$ cm^2 for R640 at 610nm. The pump power is adjusted so that the gain recovery time is greater than the pulse duration. Focusing the output of the multipass amplifier in a 1mm cell filled with a solution of saturable absorber (DCI2 in ethanol, ground state absorption recovery time: 20ps) and adjusting the small signal transmission to $\simeq 10^{-4}$, we obtain a single ps pulse (Fig.2c). Subsequent high power amplifiers separated by saturable absorbers (DCI2) produce 0.6ps, 0.3mJ, around 610nm pulses. An intensity modulation (period 340 fs) arises from mode beating due to the 30µm cavity used [2b]. These pulses, when focused in a 1cm cell filled with water, generate a wide super-continuum from UV to IR from which narrow bands can be selected with interference filters. Thus sub-ps pulses at any wavelength are cheaply obtained from a single standard Q-switched Nd:YAG laser (non mode-locked).

2. Excited State Relaxation of Ethyl Violet (Fig. 3)

Relaxation dynamics in TPM dyes have been extensively studied in both steady state and time-resolved experiments from ns to fs (see ref. in [3-6]). The fast deactivation of the excited state was described by a barrierless torsional relaxation, but the existence of an intermediate electronic state in the relaxation process is much debated [3-6]. One difficulty in analysing transients in dyes is due to overlap of photoinduced absorption, ground state bleaching and gain. Figure 4 shows the "initial" variation of optical density ΔD measured at the end of the pump pulse with a polychromator and an OMA by the usual pump-probe method, from 350 to 850nm for Ethyl Violet in hexanol, at room temperature. This spectrum was deduced from measurements of the transmitted probe spectra before and just after pumping. The contribution of the excited state to the initial small gain coefficient ($\gamma = -2.3\ D/l$) for the probe light through the excited sample of length l was estimated from $\Delta\gamma$ and the plateau value of the quantity $\Delta\gamma/\gamma_0$ (γ_0: with no pumping). This contribution (thick line in Fig. 4 in optical density units) is found to be positive and negative above and below 570 nm, respectively.

Time dependent transmission curves in various solvents were directly recorded in a two-beam set-up successively tuned at 450, 550 and 670 nm by filtering the supercontinuum probe beam. In contrast with the sharp rise of the net gain at 670nm and of the net transient absorption at 450nm, a

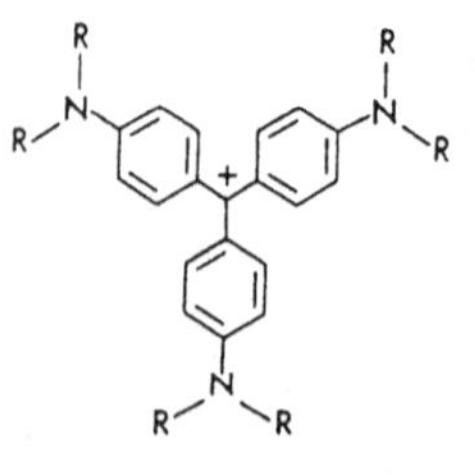

Fig.3 Ethyl Violet ($R = C_2H_5$)

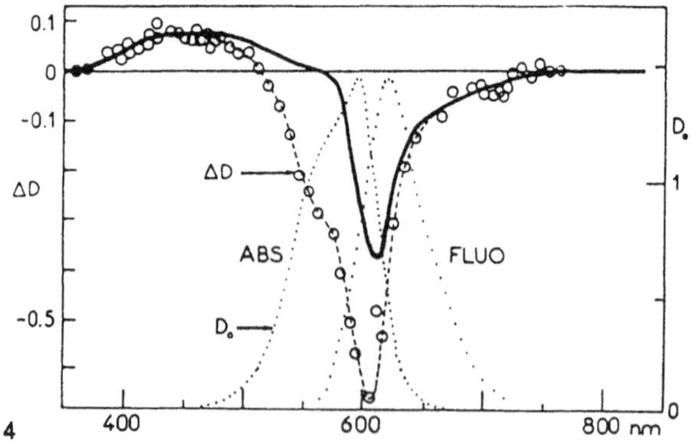

Fig.4

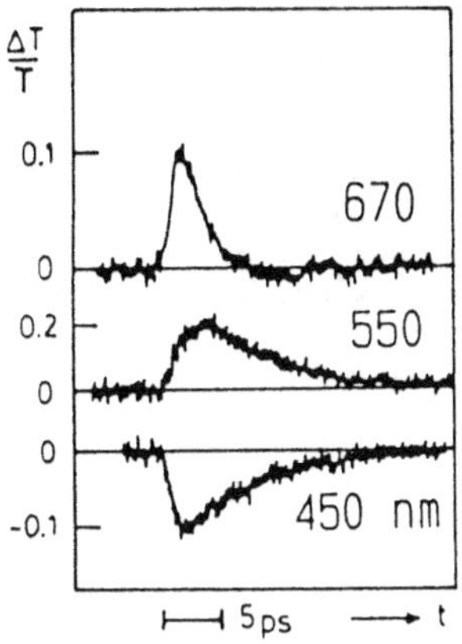

Fig.5

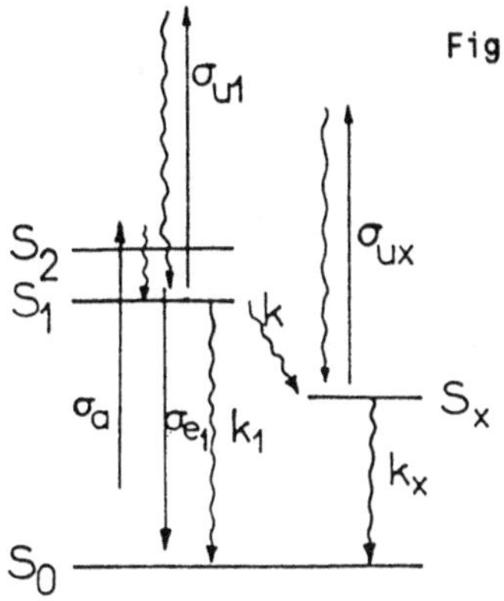

Fig.6

delayed absorption bleaching is observed around 550nm. In Fig.5, a delay of about 2ps can be seen for Ethyl Violet in tetrahydrofuran at room temperature. The delayed bleaching can be explained with a molecular relaxation model involving the formation of an intermediate state S_x as shown on Fig.6. We neglect the stimulated emission from S_x assuming that it may be either a distorted form of the excited state with a small σ_e or a distorted form of the ground state. The shape of the observed transmission curves for the probe beam through the excited sample can be explained with

$$T(t) = \exp\left[-\sigma_a N_0(t) + (\sigma_{e1} - \sigma_{u1})N_1(t) - \sigma_{ux}N_x(t)\right] l$$

A good fit can be obtained by adjusting the quantities $\sigma_{e1} - \sigma_{u1}$ and σ_{ux}. It shows that both S_1 and S_x states contribute to the observed decays at 450, 550nm and 670nm. At 550 nm, where $\sigma_{e1} \simeq 0$, σ_{u1} and σ_x are significant but smaller than σ_a. The rapid relaxation from S_1 to S_x with $\sigma_{ux} < \sigma_{u1}$ explains the delayed bleaching.

Fast relaxation involving the formation of a TICT state due to the electron donor character of the anilino groups was suggested in fluorescence quenching experiments [6]. We compared the S_x formation and relaxation rates in solvents of high and low dielectric constant of comparable viscosity: ethanol and dioxane, acetonitrile and tetrahydrofuran. No specific rôle of the solvent dielectric constant could be observed which is not a priori in favor of a TICT of the type defined by GRABOWSKI et al [7]. However formation of TICT states governed by molecular structure and by molecular subunits electron donor-acceptor properties were reported in non polar solvents and in gas phase [8]. This may also be the case for TPM dyes since the rotation of the anilino groups is obviously concomitant with a positive charge either on the central carbon atom or on one nitrogen atom: S_x possibly has a charge transfer nature.

1. L.M. Frantz, J.S. Nodvik: J. Appl. Phys. **34**, 2346 (1963)
2. a) Y.H. Meyer, M.M. Martin, E. Bréhéret, O. Benoist d'Azy: "Ultrafast Phenomena V" Ed. Fleming and Siegman: Springer Series in Chem. Phys. **46**,

89 (1986). b) Y.H. Meyer, M.M. Martin, F. Nesa, E. Bréhéret: J.Phys. **49** C7-397 (1987)
3. V. Sundström, T. Gillbro: J. Chem. Phys.**81**: 3463 (1984)
4. R. Mentzel, C.W. Hoganson M.W. Windsor: Chem.Phys.Letters **120**, 29 (1985)
5. D. Ben-Amotz, C.B. Harris: J. Chem. Phys. **86**, 485 , 5433 (1987)
6. M. Vogel, W. Rettig: Ber. Bunsenges Phys. Chem. **89**, 962 (1985)
7. Z.R. Grabowski, K. Rotkiewicz, A. Siemiarczuk, D.J. Cowley, W. Baumann: Nouv.J.Chim. **3**, 443 (1979)
8. W. Rettig: Angew. Chem. Int. Ed. Engl. **25**, 971 (1986)

The Effect of Overlapping Electronic Excited States on the Subpicosecond Fluorescence Anisotropy Decay Behavior of Tryptophan in Water

A. Ruggiero, D. Todd, and G.R. Fleming

Department of Chemistry and the James Franck Institute,
The University of Chicago, 5735 S. Ellis Avenue,
Chicago, IL 60637, USA

Introduction. The fluorescence properties of tryptophan and its derivatives in solution have been the subject of intense study for several years [1]. This research effort has been motivated primarily by its importance as an intrinsic fluorescent probe of protein structure and conformational dynamics [2,3]. The use of frequency conversion optical gating in this work results in an increase in time resolution of two orders of magnitude over previous measurements of the time resolved fluorescence anisotropy of tryptophan. An important parameter in the evaluation of time resolved fluorescence depolarization data is the experimental value of the anisotropy at zero time, r(0). It is the intrinsic polarization anisotropy of the chromophore and is a function only of the orientation of the absorption and emission transition dipole moments. The theoretical value for a chromophore with parallel absorption and emission transition dipoles is 0.4. Observations of experimental values of r(0) less than the theoretically predicted value generally indicate the existence of relaxation processes occurring on a time scale shorter than the time resolution of the experiment. These processes can involve actual reorientational or librational motion of the molecule or purely electronic (vibronic) relaxation mechanisms [4,5,6]. Prior to the measurements presented here, the theoretical limiting value (r(0)=0.4) for tryptophan had never been observed. The spectroscopy of tryptophan is complicated by the existence of two overlapping electronic transitions with roughly perpendicular transition moments, designated the ^1L_a and ^1L_b bands [7,8,9]. Theoretical modeling of fluorescence anisotropy decay of tryptophan including the effects of ^1L_a, ^1L_b level kinetics has shown that the form of the decay and the value of the initial anisotropy can depend strongly on the ratio of excited states prepared and the observation wavelength of the fluorescence [10].

Experimental. The ultraviolet fluorescence upconversion spectrometer developed for this work is shown in Fig. 1. The second harmonic of a mode-locked Nd:YAG laser is used to synchronously pump a cavity dumped dual jet hybrid R6G dye laser with chirp compensation. The 1 nm bandwidth, output pulses are cavity dumped and amplified at a 12 kHz repetition rate in a copper vapor laser based 6 pass amplifier. The second harmonic of the amplified pulse is used to excite the sample. The fluorescence is collected and upconverted by type I mixing with the fundamental in a beta-barium borate crystal. The upconverted signal is detected by gated photon counting with a solar blind PMT. A 500 fs response function for the system generated by the cross-correlation of a 578 nm dye laser fundamental with its second harmonic and detected at 192 nm is shown in the inset on the right hand side of Fig. 1. The system response function varied between 500 fs and 1 ps depending on the experimental excitation wavelength. Since the upconversion signal is proportional to the cube of the laser power, small changes in power can have a dramatic effect on the signal level. To insure the integrity of the data, the count period is determined by the product of the intensities of the fluorescence and the fundamental and thereby compensates for laser power drifts over the course of data acquisition.

Springer Series in Chemical Physics, Vol.48: **Ultrafast Phenomena VI**
Editors: T. Yajima · K. Yoshihara · C.B. Harris · S. Shionoya

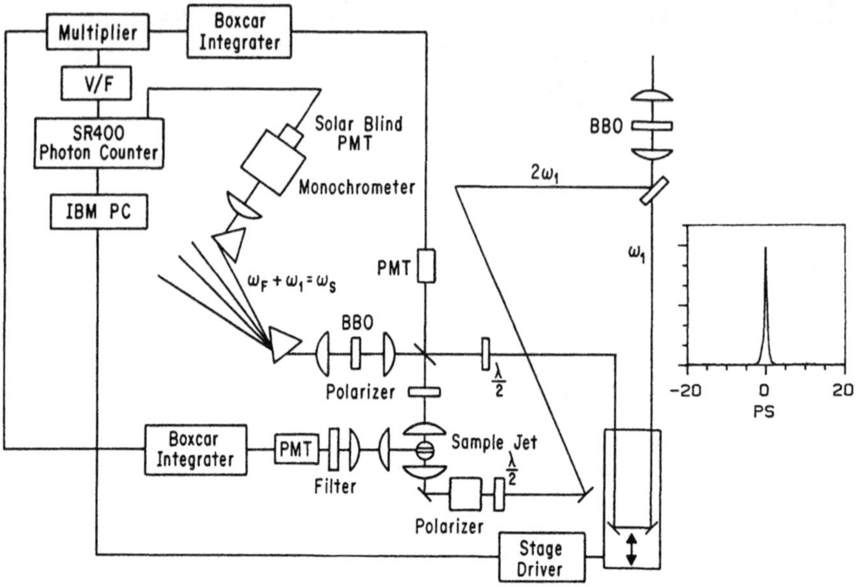

Fig. 1. Schematic for ultraviolet fluorescence upconversion spectrometer. Inset: 500 fs cross correlation of the 578 nm dye laser fundamental with its second harmonic. Sum frequency detection at 192 nm.

Results and Discussion. The raw parallel and perpendicular polarization decay components for tryptophan in water excited at 300 nm are shown in Fig. 2. Fluorescence at 330 nm (+/- 5 nm) on the short wavelength edge of the tryptophan emission was mixed with the 600 nm fundamental to produce the upconverted signal near 213 nm. A short decay component with a 2.6 picosecond lifetime is found in both the parallel polarization component of the emission and the total intensity decay of the fluorescence. The detection wavelength and time scale of the decay suggest that it is due to unrelaxed emission from the ^1L_b state. The raw anisotropy under these experimental conditions has an initial value of 0.4. The anisotropy decay consists of a 2.6 psec decay component and a slower decay component corresponding to the overall rotational diffusion of the molecule. Data simulated using the Cross-Fleming model for fluorescence depolarization including the effects of level kinetics are shown in the inset. The parameters used in these preliminary simulations were chosen to approximate the experimental conditions i.e. long wavelength mostly ^1L_a excitation and detection of short wavelength mostly ^1L_b emission. The rate constants used in the simulations were crude order of magnitude estimates based on literature values. The general form of the simulated decays are in good agreement with the raw experimental data. It is interesting to note that the unusual shape of the horizontal emission component is reproduced by this simple model. For purely orientational motion this polarization component would exhibit a simple risetime correlated with the decay of the parallel polarization component. Data collected at higher excitation energies (289 nm, 290 nm, 292 nm, 294 nm and 297 nm) under the same emission detection conditions, all have a short 1 to 4 psec lifetime component in the total fluorescence decay. The raw anisotropies constructed at these wavelengths, however, exhibit different initial anisotropies and short time decay behavior. The initial anisotropies vary from an r(0)=0.3 at 297 nm excitation to an r(0)= 0.18 at 289 nm. Evaluation of these data within the level kinetics model as

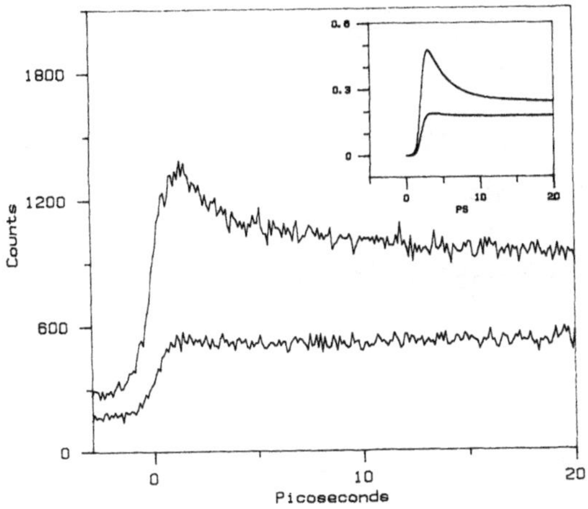

Fig. 2. Parallel (top curve) and perpendicular polarization components of the tryptophan emission. [300 nm excitation, 330 nm +/- 5 nm fluorescence, 600 nm gate pulse, upconverted signal detected at 213 nm.] Inset: Simulated data. See text.

well as acquisition of fluorescence anisotropy data on the long wavelength edge of the emission band is currently underway. Detailed simulations of the anisotropy along with refinement of the kinetic model to include the effects of solvent relaxation, will certainly be necessary to understand the complex polarization anisotropy of tryptophan at short times. In conclusion, the initial anisotropy and short time decay behavior of tryptophan are found to be complicated by non-rotational contributions to the anisotropy decay arising from the interaction of the two low lying ^1L_a and ^1L_b excited states on the one to two picosecond time scale. These initial results indicate that the use of time resolved fluorescence depolarization measurements to obtain information on short time librational motions of tryptophan sidechains in proteins may be severely complicated by contributions to the anisotropy from these electronic (vibronic) relaxation processes.

References.

1. D. Creed, Photochem. Photobiol. 39, 537 (1984).
2. I. Munro, I. Pecht, L. Stryer, Proc. Natl. Acad. Sci, U.S.A. 76, 56 (1979).
3. J.A. McCammon, M. Karplus, Acc. Chem. Res. 16, 187 (1983).
4. A. Szabo, J. Chem. Phys. 81, 150 (1984).
5. R.M. Hochstrasser in Probes of Biological Structure and Dynamics, Szabo and Masotti (eds.) Proceedings of NATO Advanced Study Institute, Acireale (1984).
6. A.J. Cross, D.H. Waldeck and G.R. Fleming, J. Chem. Phys. 78, 6455 (1983).
7. Y. Yamamoto, J. Tanaka, Bull. Chem. Soc., Japan 65, 1362 (1972).
8. N. Mataga, Y. Torihashi and K. Ezumi, Theoret. Chim. Acta 2, 158 (1964).
9. B. Valeur, G. Weber, Photochem. Photobiol. 25, 441 (1977).
10. Cross et al. (1983).

External Magnetic Field Effect on the Fluorescence of CS_2 Excited to the $V\ ^1B_2$ State

H. Abe[1], H. Hayashi[1], T. Imamura[2], and S. Nagakura[2]

[1]The Institute of Physical and Chemical Research, Wako, Saitama 351-01, Japan

[2]Institute for Molecular Science, Okazaki, Aichi 444, Japan

Since the first observation of the external magnetic field effect on the banded emission of CS_2 vapor excited by a nitrogen laser [1], several studies have been done to elucidate the origin of the effect[2,3]. In the nitrogen laser wavelength region, we have already shown that the integrated intensities of the excitation spectra for the banded emission are reduced by magnetic fields without changing the emission lifetime in the nanosecond time scale[4]. This indicates that certain intramolecular processes are accelerated by the fields and excludes the applicability of the Zeeman detuning model [3].

There remains, however, a contradiction in the term of " the intensity reduction without change of the lifetime". Therefore we have extended the investigation to the lowest allowed state; $^1B_2$ (Kleman's V state). Here, we have confirmed the generality of the phenomena by measuring the magnetic field effect on the intensities of fluorescence excitation spectra and the lifetime in the micro- to pico-second time scale.

Figure 1 shows fluorescence excitation spectra observed for the 6V band region of CS_2 in a supersonic free jet of 2 atm He by monitoring only banded emission. As can be seen from this figure, magnetic fields reduce the integrated intensity, S , with little Zeeman broadening. The magnetic quenching shows no systematic dependence on the rotational quantum number but the extent changes from line to line within one vibronic band. The integrated intensity of the whole region was found to decrease monotonically with the increase of the field. Up to 8.3 kG, no saturation for the magnetic quenching has been found. The same behavior could be observed for all other bands (9V, 10V, 13V, 15V).

The fluorescence decay curves observed in the absence and presence of magnetic fields are shown in Figure 2. CS_2 of 1.7 Torr was excited at the head of 6V band by using a synchronously pumped cavity-dumped dye laser[5]. Only banded emission was monitored. The fast decay component in the sub-nanosecond time scale can be recognized, which is more evident with increasing the field strength. The least square fit to a function of the form

$$I(t) = C_f \exp(-k_f t) + C_s \exp(-k_s t)$$

gives the lifetime of the fast decay component (t_s) to be 520 ps at 0 kG. Magnetic fields make the fast decay component conspic-

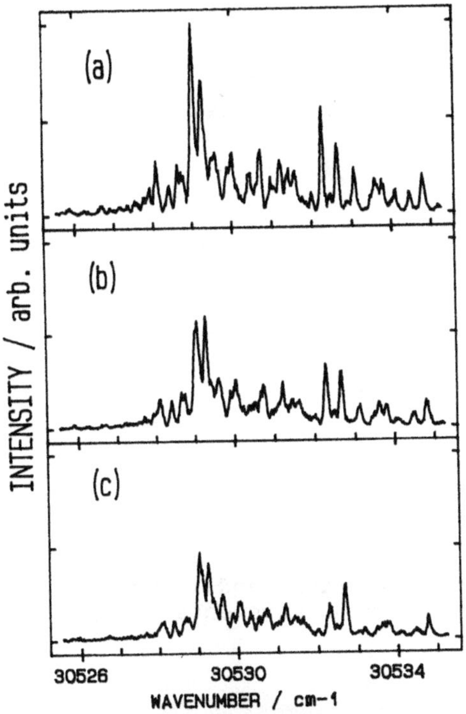

Figure 1. Fluorescence excitation spectra in the 6V band region of CS_2 in a supersonic jet of 2 atm He (a) at 0 kG, (b) at 4.9 kG, (c) at 8.3 kG

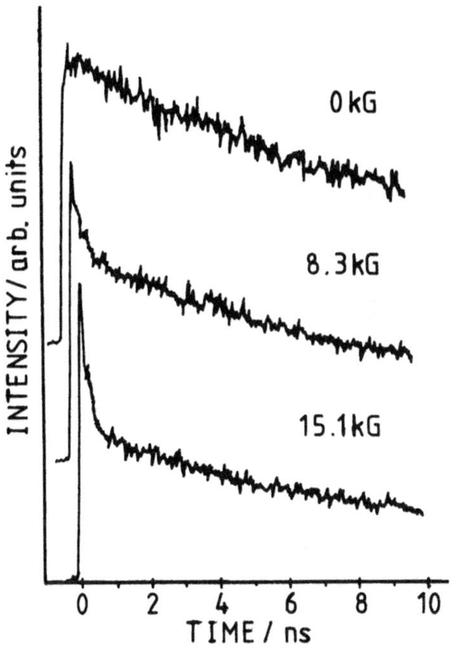

Figure 2. Magnetic field dependence of fluorescence decay of 1.7 Torr CS_2 excited to 6V band head. Only banded emission at 347 nm was monitored

uous and shorten t_s (250 ps at 8.3 kG and 190 ps at 15.1 kG) without changing slow decay.

These results indicate that CS_2 belongs to a so-called intermediate case molecule[6]. As seen in Figure 3, the dependence of pre-exponential facter ratio C_f/C_s on the field is almost linear, which is consistent with the theory[6,7]. No saturation however, was found up to 15.1 kG. This fact and that t_f is strongly dependent on the field strength as mentioned above are somewhat different from the case for usual organic molecule such as, for example, pyrazine[8] and suggest that some mechanism other than spin decoupling[7,8] would be operative in the case of CS_2.

Because of the proportional relation between C_f/C_s and the number of coupled states in the ISC process, the result of Fig. 3 indicates that the magnetic quenching of fluorescence can be explained in terms of the enhancement of the first dephasing process into triplet manifold due to the increase in the number of coupled states induced by magnetic fields.

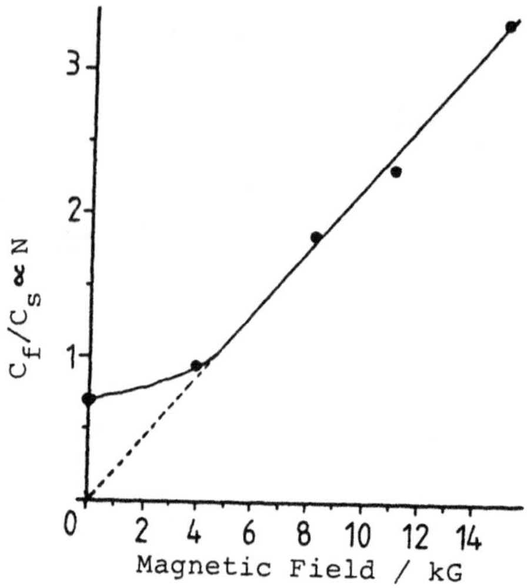

Figure 3
Plot of C_f/C_s as a function of magnetic field strength

The authors are indebted to Drs. N. Tamai and I. Yamazaki of IMS for their support during the decay measurements.

1. A. Matsuzaki, S. Nagakura: Chem. Lett. 81, 29 (1974) : Bull. Chem. Soc. Jpn 49, 359 (1976)
2. H. Orita, H. Morita, S. Nagakura: Chem. Phys. Lett. 81, 29 (1981)
3. S.J. Silvers, M.R. McKeever: Chem. Phys. 80, 177 (1983)
4. T. Imamure, S. Nagakura, H. Abe, Y. Fukuda, H. Hayashi: J. Phys. Chem. in press
5. I. Yamazaki, N. Tamai, H. Kume, H. Tsuchiya, K. Oba: Rev. Sci. Instrum. 56, 1187 (1985)
6. F. Lahmani, A. Tramer, C. Tric: J. Chem. Phys. 60, 4431 (1974)
7. P.R. Stannard: J. Chem. Phys. 68, 3932 (1978)
8. Y. Matsumoto, L.H. Spangler, D.W. Pratt; J. Chem. Phys. 80, 5539 (1984)

Picosecond Pulse Laser Photoelectron Spectra of Some Molecular Excited States

K. Kimura, K. Sato, K. Okuyama, and M. Takahashi

Institute for Molecular Science, Okazaki 444, Japan

Excited-state photoelectron spectra have been obtained successfully with a picosecond pulse laser at the Rydberg A state (v'=2) of ammonia and at different vibronic levels of the first singlet state of benzene in jets at 208.8, 228.7, and 252.7 nm. Some spectra are possibly assigned to resonant/non-resonant ionization before and after vibrational relaxation.

1. INTRODUCTION

It is quite attractive to employ a picosecond pulse UV laser in photoelectron spectroscopy to study the excited-state dynamics of molecules. Multiphoton resonant ionization makes it possible to provide a photoelectron spectrum of a specific excited state [1-5]. In general, the ionization of a resonant excited state is more or less in competition with deactivation processes such as electronic and vibrational relaxation [1-4], as shown schematically in Fig. 1.

Therefore, it is possible to study molecular excited-state dynamics by photoelectron spectroscopy. Such a photoelectron spectrum reflects the final ionic states as well as the resonant excited state, according to ionization selection rule. The rate of photoionization, k_i, is given by $k_i = I\sigma$, where I is the laser intensity and σ is ionization cross section. Therefore, if a mJ picosecond laser is used instead of a mJ nanosecond laser, we may expect that k_i increases by about 2-3 orders of magnitude.

In the present paper, we demonstrate several picosecond photoelectron spectra for some excited states of ammonia and benzene at 208.8, 228.7, and 252.7 nm.

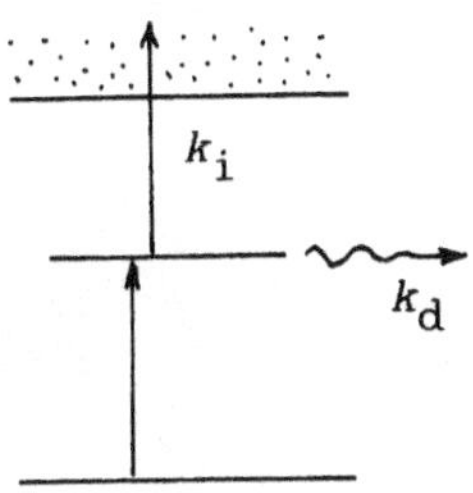

Fig. 1. Competition of ionization and deactivation at a resonant state. The rates of ionization and deactivation are shown by k_i and k_d, respectively

Springer Series in Chemical Physics, Vol. 48: **Ultrafast Phenomena VI**
Editors: T. Yajima · K. Yoshihara · C.B. Harris · S. Shionoya

2. EXPERIMENTAL

The picosecond pulse UV lasers of 208.8, 228.7, and 252.7 nm were produced though a hydrogen Raman shifter (10 atm), using the second harmonic generation (532 nm) of a mode-locked picosecond Nd-YAG laser (35 ps, 10 mJ, 10 Hz; Quantel Model 501DP). Photoelectron kinetic-energy spectra were measured with a molecular beam apparatus with a time-of-flight electron analyzer [1-3], by accumulating photoelectron signals for 5,000-10,000 times.

3. RESULTS AND DISCUSSION

3.1. Ammonia

At 208.8 nm, ammonia was ionized by two photon through the one-photon resonant Rydberg A state ($v_2'=2$), at which fast predissociation takes place. The resulting photoelectron spectrum is shown in Fig. 2, consisting of a vibrational progression, which can be explained in terms of ν_1^+ and ν_2^+.

The most interesting point is that although the picosecond spectrum (Fig. 2) is essentially the same as a nanosecond one in spectral pattern, the former intensity is much stronger. This means that the picosecond ionization is much larger than the nanosecond ionization.

3.2 Benzene

Photoelectron spectra obtained by (1+1) resonant ionization through the S_1 state of benzene at 252.7, 228.7. and 208.8 nm are shown by spectra (*a-c*) in Fig. 3 . Spectrum (*a*) originates from the S_1 6^11^1 vibronic level. Several peaks in spectrum (*a*) can be assigned to the vibrational levels of the ion, as indicated in a nanosecond photoelectron work [5]. Only ionization from 6^11^1 (or 6^11^2) is observed [3,5]. The lifetime of the 6^11^2 level has been reported to be 47 ns [6].

Spectrum (*b*) consists of a single peak at 1.5 eV and a broad band with several peaks in the lower energy region.

At 228.7 nm, benzene is initially excited to the 6^11^4 vibronic level which is 4200 cm^{-1} above the S_1 origin. The 1.5-eV peak corresponds to ($2h\nu$ - I_a), indicating that nonresonant ionization takes place, where I_a is the adiabatic

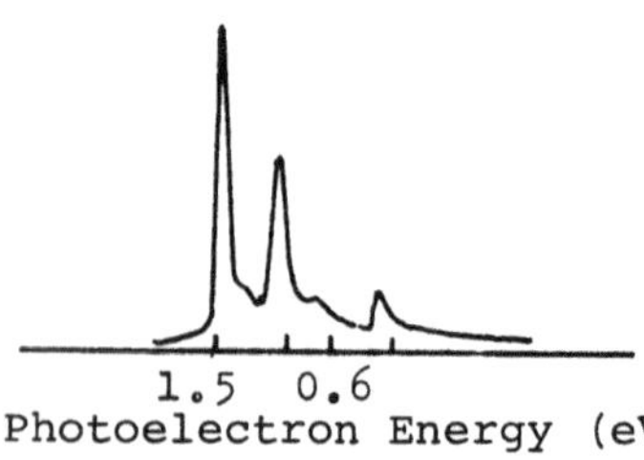

Fig. 2. The (1+1) photoelectron spectrum of ammonia at the Rydberg A state ($v_2'=2$)

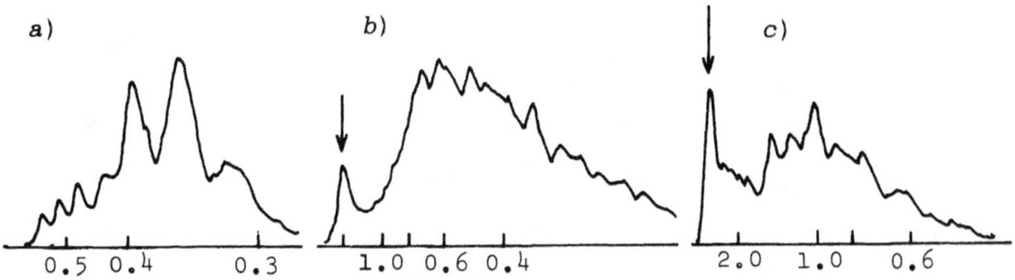

Fig. 3. The picosecond photoelectron spectra of the benzene S_1 state. (*a*) λ = 252.7 nm, the $6^1 1^1$ level; (*b*) λ = 228.7 nm, the $6^1 1^4$ level (4200 cm^{-1}); (*c*) λ = 208.8 nm, a higher vibronic level (8300 cm^{-1}). Arrows show 2-photon non-resonance

ionization energy (9.25 eV). Such nonresonant ionization was not detected with an ordinary nanosecond laser. The broad band in spectrum (*b*) is probably due to photoionization after IVR as previously suggested in the nanosecond work [3]. Several small peaks on the broad band may be from the optically prepared vibrational level.

Spectrum (*c*) shows a distinct peak at 2.7 eV which again corresponds to nonresonant ionization and a broad band with many peaks. Benzene is initially excited probably to a higher level (8300 cm^{-1} above the S_1 origin). The appearance of the several small peaks on the broad band suggests that ionization from vibrational levels takes place even in such high levels.

From the present picosecond laser work, we have found that the following three kinds of ionizations occur for the benzene S_1 state: (1) 2-photon nonresonant ionization given by ($2h\nu - I_a$); (2) 2-photon resonant ionization which reflects the optically prepared vibronic level; and (3) 2-photon two-step ionization in which the second step proceeds after IVR.

REFERENCES

1. K. Kimura, Adv. Chem. Phys. **60**, 161 (1985)
2. K. Kimura, Int. Rev. Phys. Chem. **6**, 195 (1987)
3. Y. Achiba, A. Hiraya and K. Kimura, J. Chem. Phys. **80**, 6047 (1984)
4. J. B. Pallix and S. D. Colson, J. Phys. Chem. **90**, 1499 (1986)
5. S. R. Long, J. T. Meek and J. P. Reilly, J. Chem. Phys. **79**, 3206 (1983)
6. M. Sumitani, D. V. O'Connor, Y. Takagi and K. Yoshihara, Chem. Phys. Lett. **108**, 11 (1984)

Simultaneous Analysis of the Fluorescence Decay Surface of Tryptophan as a Function of Temperature, pH, Quencher and Emission Wavelength

N. Boens, L.D. Janssens, and F.C. De Schryver

Department of Chemistry, Katholieke Universiteit Leuven, Celestijnenlaan 200 F, B-3030 Heverlee-Leuven, Belgium

Time-resolved fluorescence measurements [1,2] are a valuable tool in our understanding of the often complex photophysical behavior of biological systems. Tryptophan has been extensively used as intrinsic fluorescent probe for protein structure and dynamics. A detailed understanding of tryptophyl photophysics is needed for a proper interpretation of the experimental results. For reviews of tryptophyl photophysics, see refs. [3] and [4]. There is convincing evidence that tryptophan fluorescence decays at pH < 7 are biexponential with a subnanosecond component. Above pH 7, the decays are triple-exponential with the long component due to anionic tryptophan. All those results were obtained by conventional (single curve) data analysis in which the fluorescence decays were described individually in terms of exponential decay times τ_i and their associated scaling factors α_i. This analytical approach, although adequate in many cases, fails to take full advantage of relations that may exist between individual decays. The simultaneous (global) analysis of multiple curves [5,6] utilizes (and tests) those relationships by keeping some model parameters in common between different related experiments. This type of analysis imposes the model directly on the actual decay data. The advantages of the global analysis method over single curve analysis are the improved model testing capability and higher accuracy of the recovered parameters. In global analysis, the parameters kept common between related decays can be the decay times τ_i. However, in most instances, τ_i values will vary between analogous experiments. The set of α_i,τ_i are thus empirical descriptors of the fluorescence decays. The primary parameters of interest in simultaneous analysis are decay rate constants, activation energy, enthalpy and entropy.

Fluorescence decay curves of tryptophan were obtained using the 295 nm excitation of a mode-locked, cavity-dumped, synchronously pumped, frequency-doubled R6G dye laser with single photon timing detection [1,2]. Details of the fluorescence lifetime apparatus are given in ref. [7]. The tryptophyl decays were measured against monoexponential standards [8,9] as a function of pH, temperature, acrylamide concentration and emission wavelength. The decays (with 6000 to 10000 peak counts) collected in 1/2k channels were simultaneously analyzed in terms of various discrete decay rate constants, activation energies and frequency factors.

pH dependence

Fluorescence decays of tryptophan in phosphate buffer pH 6.0, 7.5 and 8.1, and in borate buffer pH 9.8 at 21°C were collected at analysis wavelengths ranging from 310 to 440 nm. Simultaneous analysis of 7 curves indicated that tryptophan at pH 6.0 decays biexponentially (Eq.(1), p=2) with decay times $\tau_1 = 760 \pm 11$ ps and $\tau_2 = 2.83 \pm 0.002$ ns. The two decay times were invariant (linked) between the decay curves. The preexponentials varied with wave-

Springer Series in Chemical Physics, Vol. 48: **Ultrafast Phenomena VI**
Editors: T. Yajima · K. Yoshihara · C.B. Harris · S. Shionoya

length : the short decay time τ_1 contributed more to the total fluorescence at shorter analysis wavelengths:

$$f_s(t) = \sum_{j=1}^{p} \alpha_j \exp[-t/\tau_j] \quad (1)$$

In Eq. (1) $f_s(t)$ is the fluorescence δ-response of the sample, p denotes the number of exponential terms considered in the decay, τ_j is the jth decay time and α_j its associated preexponential.

The global analysis of 5 tryptophyl decay curves at pH 7.5 gave a valid fit when all decays were analyzed simultaneously as a triple exponential (Eq. (1), p=3) with the three decay times invariant. The estimated decay times were $\tau_1 = 704 \pm 14$ ps, $\tau_2 = 2.85 \pm 0.01$ ns and $\tau_3 = 7.00 \pm 0.86$ ns. When the decays were analyzed one at a time, a biexponential model could in all cases specify the data. Global biexponential analysis of the curves, however, gave an inadequate fit. At pH 8.5 the fluorescence decays of tryptophan are triple exponential (Eq. (1), p=3, $\tau_1 = 700 \pm 19$ ps, $\tau_2 = 2.98 \pm 0.03$ ns, $\tau_3 = 7.84 \pm 0.13$ ns). Simultaneous biexponential analysis and single curve analysis also indicated that a third (long lived) decay component is present. Global analysis of 12 tryptophan fluorescence decays in borate buffer indicated that tryptophan at pH 9.8 decays triple-exponentially (Eg. (1), p=3) with decay times $\tau_1 = 628 \pm 28$ ps, $\tau_2 = 2.70 \pm 0.04$ ns and $\tau_3 = 8.38 \pm 0.05$ ns.

Temperature dependence

23 fluorescence decays of Trp in phosphate buffer pH 6.0 were collected at 320, 360, and 400 nm at temperatures ranging from 25.6 to 60.3°C. When the decays were analyzed globally according to Eq. (2) (p=3) with k_{ot}^1, A^1, E_a^1, k_{ot}^2, A^2 and E_a^2 linked between all curves, an adequate fit was obtained with $k_{ot}^1 = 1.3 \times 10^9$ s^{-1}, $A^1 = 3.5 \times 10^{13}$ s^{-1}, $E_a^1 = 93.3$ kJ/mol, $k_{ot}^2 = 4.3 \times 10^7$ s^{-1}, $A^2 = 8.8 \times 10^{12}$ s^{-1} and $E_a^2 = 25.3$ kJ/mol. The activation energy of the fast decay component is enormous. When the activation energies for the two decay components were linked ($E_a^1 = E_a^2$), a satisfactory fit was obtained with $k_{ot}^1 = 1.3 \pm 0.04 \times 10^9$ s^{-1}, $A^1 = 2.6 \times 10^5 \pm 1.6 \times 10^{12}$ s^{-1}, $k_{ot}^2 = 9.5 \pm 0.9 \times 10^7$ s^{-1}, $A^2 = 3.0 \pm 0.5 \times 10^{13}$ x^{-1}, and $E_a^1 = E_a^2 = 28.7 \pm 0.5$ kJ/mol. The frequency factor of the fast decay component is very small and has an extremely large standard deviation. These two global data analyses indicate that the short decay time τ_1 is independent of temperature. Indeed, the best fit was obtained when τ_1 was assumed to be independent of temperature while τ_2 was temperature dependent. The recovered decay parameters were : $k_{ot}^1 = 1.3 \pm 0.01 \times 10^9$ s^{-1}, $k_{ot}^2 = 9.6 \times 10^7$ s^{-1}, $A^2 = 3.2 \pm 10^{13}$ s^{-1}, and $E_a^2 = 28.9$ kJ/mol. The 23 decays were also analyzed individually as biexponentials (Eq. (1), p=2). Valid fits were obtained in all cases. τ_2 values could be described by Eq. (3), yielding the following values : $k_{ot}^2 = 1.3 \times 10^8$ s^{-1}, $A^2 = 6.1 \times 10^{13}$ s-1 and $E_a^2 = 30.7$ kJ/mol. The nonlinear analysis of the short decay times did not converge. A perfect fit was obtained when the 23 Trp decays were analyzed globally as biexponentials (Eq. (1), p=2) with τ_1 linked between all decays. The estimated τ_1 value was 717 ± 12 ps, equivalent to $k_{ot}^1 = 1.4 \times 10^9$ s^{-1}. Describing the long decay times τ_2 by Eq. (3) gave $k_{ot}^2 = 5.2 \times 10^7$ s^{-1}, $A^2 = 1.3 \times 10^{13}$ s^{-1} and $E_a^2 = 26.3$ kJ/mol.

$$f_s(t) = \sum_{j=1}^{p} \alpha_j \exp[-t\,(k_{ot}^j + A^j \exp(-E_a^j/RT))] \quad (2)$$

In Eq. (2) k_{ot}^j, E_a^j and A^j are respectively the temperature independent decay rate constant, the activation energy and the frequency factor for the jth exponential.

$$\tau_2^{-1} = k_{ot}^2 + A^2 \exp(-E_a^2/RT) \tag{3}$$

Quenching

14 fluorescence decays of Trp in phosphate buffer pH 6.0 at 22.5°C were collected at 320, 360, and 400 nm in the presence of acrylamide quencher concentrations ranging from 0 to 6×10^{-2} M. The decays were analyzed globally according to Eq. (4) with k_{oq}^1, k_q^1, k_{oq}^2, and k_q^2 linked between all curves. A satisfactory fit was obtained with $k_{oq}^1 = 1.24 \pm 0.03 \times 10^9\ s^{-1}$, $k_q^1 = 9.52 \pm 6.19 \times 10^8\ M^{-1}s^{-1}$, $k_{oq}^2 = 3.57 \pm 0.02 \times 10^8\ s^{-1}$, and $k_q^2 = 4.92 \pm 0.03 \times 10^9\ M^{-1}s^{-1}$. The large standard deviation of k_q^1 is noticeable. Analyzing the same data set globally with $k_q^1 = 0$ also gave a good fit with $k_{oq}^1 = 1.28 \pm 0.01 \times 10^9\ s^{-1}$, $k_{oq}^2 = 3.59 \pm 0.01 \times 10^8\ s^{-1}$ and $k_q^2 = 4.88 \pm 0.02 \times 10^9\ M^{-1}s^{-1}$.

$$f_s(t) = \sum_{j=1}^{p} \alpha_j \exp[-t\,(k_{oq}^j + k_q^j\,[Q])] \tag{4}$$

In Eq. (4) k_{oq}^j and k_q^j denote respectively the decay rate constant in the absence of quencher Q and the bimolecular quenching rate constant for the jth exponential term.

The simultaneous analysis of 42 tryptophan fluorescence decays measured at pH 6.0 as a function of temperature (20-60°C), quencher concentration (0-0.06 M) and analysis wavelength (31°-440 nm) indicated that the short decay component is constant under these experimental conditions.

The results obtained by the examination of the fluorescence decay surface of tryptophan demonstrate the power of this analysis approach in exploring complex photophysical behavior.

Literature

1. R.B. Cundall, R.E. Dale, eds.: Time-resolved fluorescence spectroscopy in biochemistry and biology, NATO ASI Series A : Life Sciences, Vol. 69 (Plenum Press, New York 1983)
2. D.V. O'Connor, D. Phillips : Time-correlated single photon counting (Academic Press, London 1984)
3. D. Creed : Photochem. Photobiol. 39, 537 (1984)
4. J.M. Beechem, L. Brand : Ann. Rev. Biochem. 54, 43 (1985)
5. J.R. Knutson, J.M. Beechem, L. Brand : Chem. Phys. Lett. 102, 501 (1983)
6. J.-E. Löfroth : Eur. Biophys. J. 13, 45 (1985)
7. N. Boens, M. Van den Zegel, F.C. De Schryver, G. Desie : In From photophysics to photobiology, ed. by A. Favre, R. Tyrrell, J. Cadet (Elsevier, Amsterdam 1987) p.93
8. M. Zuker, A.G. Szabo, L. Bramall, D.T. Krajcarski, B. Selinger : Rev. Sci. Instrum. 56, 14 (1985)
9. N. Boens, M. Ameloot, I. Yamazaki, F.C. De Schryver : Chem. Phys. 121, 73 (1988)

Picosecond Laser Photolysis of 1.8-Dibromoanthraquinone in Carbon Tetrachloride at Room Temperature

T. Nakayama, M. Ito, Y. Yuhara, K. Ushida, and K. Hamanoue

Department of Chemistry, Kyoto Institute of Technology,
Matsugasaki, Sakyo-ku, Kyoto 606, Japan

1. Introduction

From the measurements of phosphorescence and triplet-triplet ($T' \leftarrow T_1$) absorption spectra of α-chloro- and α-bromoanthraquinones at 77 K, we have concluded that the lowest triplet (T_1) states of these compounds are of mixed $n\pi^*$-$\pi\pi^*$ or $\pi\pi^*$ character, while the $n\pi^*$ triplet states are the lowest ones for 2-chloro- and 2-bromoanthraquinones [1,2]. Upon the picosecond laser photolysis of 1,8-dichloroanthraquinone (DCAQ) in solutions at room temperature, we have observed that the second triplet (T_2) state builds up and decays on the first picosecond time scale [3]. Since the abnormally long internal conversion time is obtained, this paper deals with the excited state dynamics of 1,8-dibromoanthraquinone (DBAQ) in carbon tetrachloride (CCl_4) at room temperature.

2. Experimental

The details of our picosecond transient absorption spectrometer have been given elsewhere [4]. The second harmonic (347.2 nm) from a mode-locked ruby laser was used for sample excitation. The mean pulse width (30 ps) and the time-zero point, t=0, were determined from the overlap of excitation and probing pulses by measuring the buildup of $T' \leftarrow T_1$ absorption of benzophenone (in n-heptane) at 530 nm. A double-beam optical arrangement was adopted, and absorption spectra in the 200-nm scanning region were recorded with two multichannel photodiode systems.

3. Results and discussion

Figure 1 shows the transient absorptions of DBAQ, observed by the picosecond laser photolysis in CCl_4 at room temperature. Clearly, Band A builds up and decays on the first picosecond time scale, while Bands B_1 and B_2 build up continously. Since the spectrum taken at 6 ns is identical with the $T' \leftarrow T_1$ absorption spectrum of DBAQ, it may be concluded that Band A is due to a precursor (the P state) of the T_1 state. When absorption bands of two transient species overlap each other, simple analysis of the time variation of them is impossible. As shown in Fig. 2, let the observed spectra be E and F at times t_2 and t_1 ($t_2 > t_1$), respectively, and let the absorbances of spectra E and F be $D_E(\lambda)$ and $D_F(\lambda)$, respectively. For spectrum E (the pure absorpion spectrum of the T_1 state), the ratio of the absorbances at wavelengths of λ_a and λ_b is given by

$$\frac{D_E(\lambda_a)}{D_E(\lambda_b)} = \frac{\epsilon_T(\lambda_a) C_T(t_2)}{\epsilon_T(\lambda_b) C_T(t_2)} = \frac{\epsilon_T(\lambda_a)}{\epsilon_T(\lambda_b)}, \tag{1}$$

Springer Series in Chemical Physics, Vol. 48: **Ultrafast Phenomena VI**
Editors: T. Yajima · K. Yoshihara · C.B. Harris · S. Shionoya

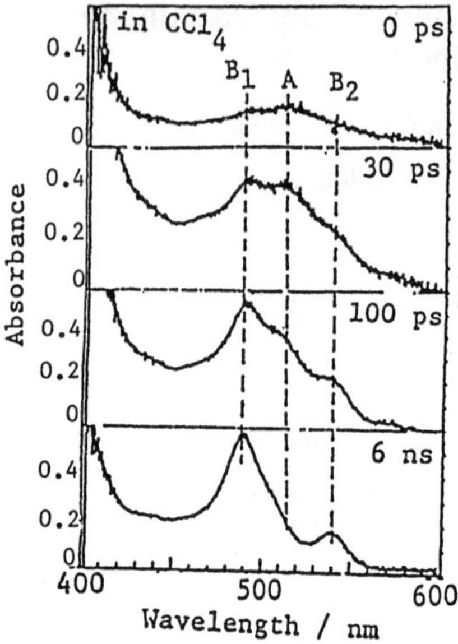

Fig. 1. Transient absorption spectra of DBAQ in CCl_4 at room temperature

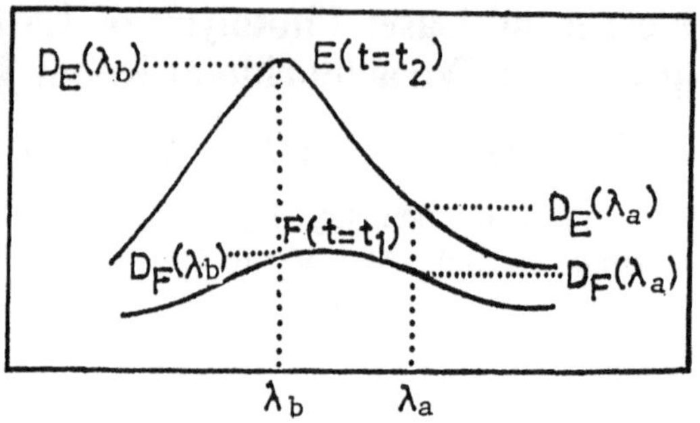

Fig. 2. A typical transient absorption spectrum due to the superpositions of the P state and T_1 state (F) and that of the T_1 state (E)

where ε_T and C_T denote the molar extinction coefficient and the concentration of the T_1 state, respectively. For spectrum F, $D_F(\lambda_a)$ and $D_F(\lambda_b)$ are given by

$$D_F(\lambda_a) = \epsilon_P(\lambda_a)C_P(t_1) + \epsilon_T(\lambda_a)C_T(t_1), \quad (2)$$

$$D_F(\lambda_b) = \epsilon_P(\lambda_b)C_P(t_1) + \epsilon_T(\lambda_b)C_T(t_1), \quad (3)$$

where ε_P and C_P denote the molar extinction coefficient and the concentration of the P state, respectively. Thus, one can get the following equation

$$D_F(\lambda_a) - \frac{D_E(\lambda_a)}{D_E(\lambda_b)} D_F(\lambda_b) = \alpha C_P(t_1). \quad (4)$$

Since $\alpha = \epsilon_P(\lambda_a) - \epsilon_T(\lambda_a)\epsilon_P(\lambda_b)/\epsilon_T(\lambda_b)$ is independent of time t , one can evaluate the relative time variation of C_P from the terms of the left-hand side of eq. (4). By this method, we analyzed the time variation of Band A in CCl_4 and the result is given in Fig. 3, where absorbances at λ_a=515 nm and λ_b=489 nm were chosen. The smooth curve corresponds to the theoretical best-fit absorbances calculated by a well-known convolution method, on an assumption of the following scheme: The lowest excited singlet (S_1) state $\xrightarrow{k_1}$ P state $\xrightarrow{k_2}$ T_1state. The most probable buildup (k_1^{-1}) and decay (k_2^{-1}) times were k_1^{-1}=20±10 ps and k_2^{-1}=100±20 ps, respectively. Similar results were also observed in toluene (k_1^{-1}=20±10 ps and k_2^{-1}=100±20 ps) and ethanol (k_1^{-1}= 30±10 ps and k_2^{-1}=80±20 ps).

The P state is not due to the S_1 state, because the time variation of the absorbance due to the P state can not be reproduced by the convoluted curve

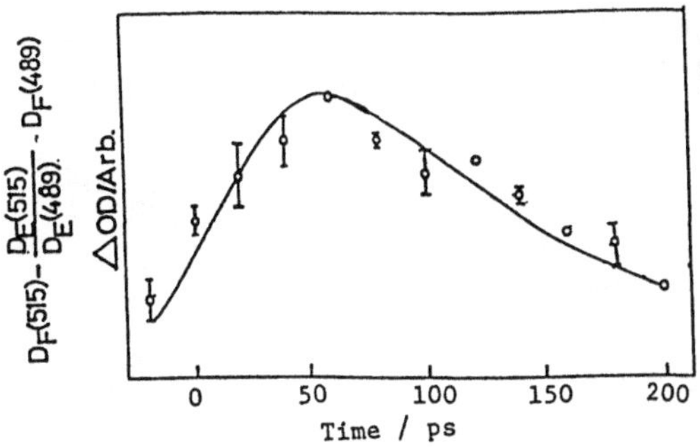

Fig. 3. Time variance of the relative concentration of the P state. The smooth curve corresponds to the theoretical best-fit absorbance

for the S_1 state. Based on the discussion for DCAQ in ref. (3), we conclude that the P state is also due to the T_2 state of an $n\pi^*$ character. Thus, the values of k_1 and k_2 reflect the rate constants for the intersystem crossing from the $S_1(n\pi^*)$ to $T_2(n\pi^*)$ state and the internal conversion from the $T_2(n\pi^*)$ to $T_1(\pi\pi^*)$ states, respectively. The value of k_1^{-1}(20±10 ps) for DBAQ is shorter than that for DCAQ (30-35 ps). This may be due to the internal heavy-atom effect of the bromine atom. The relatively longer lifetime of the T_2 state reflects that its $n\pi^*$ character may be different from a normal $n\pi^*$ one. We tentatively propose that the $T_2(n\pi^*)$ state has an intramolecular charge-transfer character between the oxygen and halogen atoms. This is based on the following conjecture: By the interaction of the carbonyl group with two halogen atoms, the geometrical molecular structure may be modified greatly. Since the electronegativity of the oxygen atom is larger than that of the halogen atoms, the effective electron transfer from the halogen to oxygen atoms may be possible. Since the electronegativity of the bromine atom is smaller that that of the chlorine atom, the charge-transfer character of the T_2 state of DBAQ may be larger than that of DCAQ. This may cause the smaller energy gap between the T_2 and T_1 states of DBAQ than that of DCAQ. Thus, the smaller value of k_2^{-1} for DBAQ (100±20 ps) than that for DCAQ (700-750 ps) is reasonable.

References

1. K. Hamanoue, T. Nakayama, Y. Kajiwara, T. Yamaguchi, and H. Teranishi, J. Chem. Phys., 86, 6654 (1987).
2. K. Hamanoue, Y. Kajiwara, T. Miyake, T. Nakayama, S. Hirase, and H. Teranishi, Chem. Phys. Lett., 94, 276 (1983).
3. K. Hamanoue, T. Nakayama, M. Shiozaki, Y. Funasaki, K. Nakajima, and H. Teranishi, J. Chem. Phys., 85, 5698 (1986).
4. T. Nakayama, S. Tai, K. Hamanoue, and H. Teranishi, Mem. Fac. Ind. Arts, Kyoto Tech. Univ. Sci. Technol., 29, 46 (1980); K. Hamanoue, T. Hidaka, T. Nakayama, and H. Teranishi, Chem. Phys. Lett., 82, 55 (1981).

Double- to Triple-Minima Change in the Adiabatic Potential Energy Curve and Exciton Relaxation Dynamics in α-Perylene Crystals

K. Mizuno[1], *M. Furukawa*[1], *A. Matsui*[1], *N. Tamai*[2], *and I. Yamazaki*[2,*]

[1]Department of Physics, Konan University, Okamoto, Kobe 658, Japan

[2]Institute for Molecular Science, Myodaiji, Okazaki 444, Japan

1 Introduction

Alpha-perylene crystal is one of the most interesting materials to study exciton-phonon interaction in solids. In alpha-perylene at room temperature free-exciton luminescence and excimer luminescence are simultaneously observed./1/ At low temperatures another luminescence which is referred to as Y luminescence appears with the photon energy in between free-exciton and excimer luminescence./2,3,4,5/ We have proposed a model that Y luminescence is a kind of the self-trapped exciton luminescence./6,7/ Since the Y-luminescence band is located between the free-exciton luminescence and excimer luminescence bands at low temperatures, the Y state which is the initial state of Y-luminescence is supposedly located in energy between the free-exciton and excimer states.

We have applied picosecond spectroscopy to study exciton dynamics in alpha-perylene crystals. We will discuss decaytimes of free-exciton luminescence in connection with the presence of the Y state, and then the barrier height between the free-exciton and self-trapped exciton states.

2 Results

Logarithmic luminescence intensity monitored at the position of the free-exciton luminescence band is plotted as a function of time in Fig.1. The bottom curve in the figure depicts time evolution of the 300.4 nm exciting light. The decaytimes which were obtained by a convolution technique applied to the luminescence decay curves in Fig.1 are plotted in Fig.2.

3 Discussions

3.1 Adiabatic Potential Curve at High Temperatures (297∿182 K)

It has been known that at high temperatures the free-exciton state is a metastable state and the excimer state is a stable state. Those excited states can be described by an adiabatic potential energy curve with two minima.

* present address:Faculty of Engineering, Hokkaido University, Kita, Sapporo, 060 JAPAN

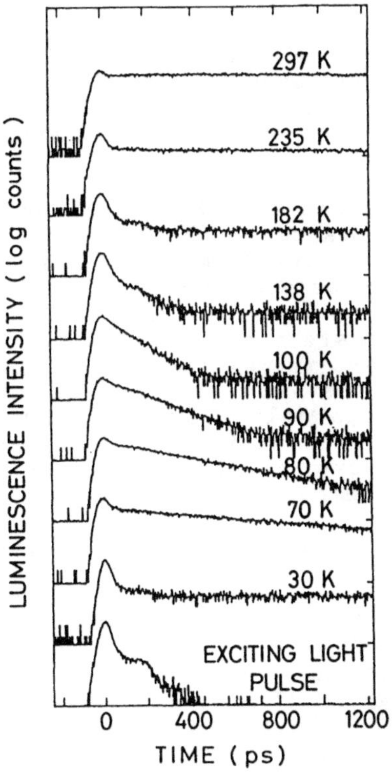

Fig.1 Free-exciton luminescence decay at several temperatures

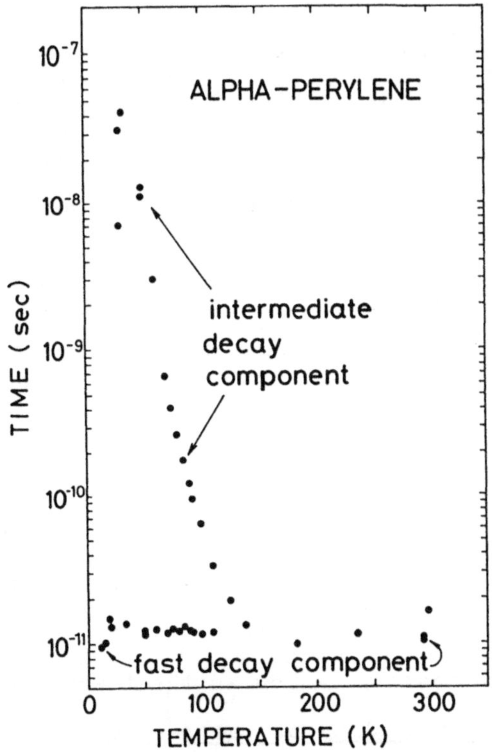

Fig.2 Decaytimes of the fast and intermediate components

In Fig.1 the luminescence decay curves at 297, 235, and 182 K are composed of a fast decay component (∿10 ps) and a slow decay component (75∿85 ns). The slow decay component is due to free excitons which are in thermal equilibrium with excimers, because the decaytime of the slow component is equal to the decaytime of excimer luminescence. The fast decay component is due to radiative annihilation of transient free excitons.

The decaytime τ for the transient component is generally expressed by an exponential function $1/\tau \propto \exp(-E_B/kT)$ where E_B is the self-trapping barrier height, k is the Boltzmann constant and T is the temperature./8/ In Fig.2 the decaytimes of the transient free-exciton luminescence are almost constant (∿10 ps) with temperature in the temperature range between 297 and 182 K. This means that the potential barrier height E_B between the free-exciton and excimer states is probably lower than the thermal energy at 182 K (126 cm^{-1})./7/

3.2 Adiabatic Potential Curve at Low Temperatures (below 138 K)

Due to the presence of the Y state the excited states at low temperatures are described by an adiabatic potential energy curve with three minima which correspond to the free-exciton, Y, and excimer states /7/. In Figs.1 and 2 an intermediate decay component appears in the free-exciton luminescence decay below 138 K. The intermediate decaytime was equal to the decaytime of Y luminescence. Therefore, it is concluded that the intermediate decay component is due to free excitons which are in thermal equilibrium with excitons in the Y state.

In the time-resolved luminescence spectra at 87 K (not shown here), two characteristic features appeared. (1) Free-exciton luminescence and Y luminescence were seen immediately after excitation, and soon after excimer luminescence was seen. (2) The risetime of excimer luminescence was equal to the decaytime of Y luminescence. These facts indicate that excitons always pass through the Y state to relax toward the excimer state and no direct relaxation of excitons to the excimer state occurs. Those two facts also suggest that the self-trapping barrier between the free-exciton and excimer states is high enough so that at 87 K excitons cannot surmount over the barrier existing between the free-exciton and excimer states.

The information on the self-trapping barrier height between the free-exciton and Y states is obtained from the fast decay component in Fig.2. The decaytimes of the fast components were constant below 138 K as well as above 138 K. This temperature independence of the decaytime of free-exciton luminescence means that barrier height between the free-exciton and Y states is supposedly as low as the thermal energy at 11 K (8 cm^{-1}).

3.3 Stability of the Y-state

In Fig.2 the intermediate decay component appears suddenly below about 138 K. This sudden uprise of the intermediate component suggests that a pronounced change in the adiabatic potential energy curve occurs with temperature. From the discussion in section 3 it is concluded that the Y state becomes stable below 138 K. At temperatures above 182 K, the intermediate decay component which is found to be related to Y luminescence is not observed. Therefore, the Y state seems to be located in energy above the free-exciton state, or, if the Y state is located in energy below the free-exciton state, the Y state may not form a clear minimum on the adiabatic potential energy curve.

4 References

1 H.Nishimura,A.Matsui and M.Iemura:J.Phys.Soc.Jpn.51,1341(1982)

2 J.Tanaka: Bull.Chem.Soc.Jpn. 36, 1237 (1963)

3 A.Inoue,K.Yoshihara,T.Kasuya and S.Nagakura: Bull.Chem.Soc. Jpn 45 (1972) 720

4 H.Auweter, D.Ramer, B.Kunze and H.C.Wolf: Chem.Phys.Letters, 85 (1982) 325

5 B.Walker, H.Port and H.C.Wolf: Chem.Phys. 92 (1985) 177

6 K,Mizuno and A.Matsui: J.Lumin. 38 323 (1987),

7 K.Mizuno and A.Matsui: J.Mol.Electronics 4 (1988) in press

8 A.S.Ioselevich and E.I.Rashba, JETP Letters: 40 1151 (1984)

Time-Resolved Exciton Self-Trapping in Pyrene Crystals

H. Port and R. Seyfang

3. Physikalisches Institut, Universität Stuttgart,
Pfaffenwaldring 57, D-7000 Stuttgart 80, Fed. Rep. of Germany

The dynamics of exciton self-trapping in organic molecular crystals is not well investigated up as yet. It is of interest, however, with respect to the interaction processes between electronic and phononic excitations leading to localized electronic states stabilized by lattice deformation. Pyrene represents a particular system, in which local self-trapping predominates due to the dimeric crystal structure and induces excimer formation.

Temperature dependent measurements of transient fluorescence spectra and fluorescence rise and decay times were performed in the high temperature phase of pyrene (113K - 300K). These prove the two-step excimer formation process via the precursor B-state discovered in our previous investigations [1].

In the present experiments ps-techniques have been applied using both time correlated Single Photon Counting and Streak camera detection (time resolution <15ps). In the SPC-experiment the synchronously pumped dye laser allows to vary the excitation wavelength in the crystal absorption region providing the system with variable excess energy.

The time resolved spectra reveal the blue fluorescence component B at the high energy side of the excimer emission which is not to be seen in steady state experiments (Fig.1a).

The B-fluorescence extends from the absorption origin at 376nm to about 450nm with its maximum at about 400nm. Its overall shape does not change with increasing temperature, whereas its intensity diminishes very fast.

From the Arrhenius plot of the B-fluorescence quantum flux (Fig.1b) a thermal activation energy of $E(Q) = (340 \pm 30)\,cm^{-1}$ is deduced.

In Fig.2a examples are given for the fluorescence transients of B- and excimer fluorescence at 115K measured at the respective intensity maxima 395nm and 460nm. Figure 2b shows a quantitative analysis of B-decay and excimer rise times as a function of temperature. The two time constants correspond to each other and shorten from 145ps at 115K to 20ps at 210K.

This result proves the precursor role of the B-state in the self-trapping process towards the excimer.

Springer Series in Chemical Physics, Vol. 48: Ultrafast Phenomena VI
Editors: T. Yajima · K. Yoshihara · C.B. Harris · S. Shionoya

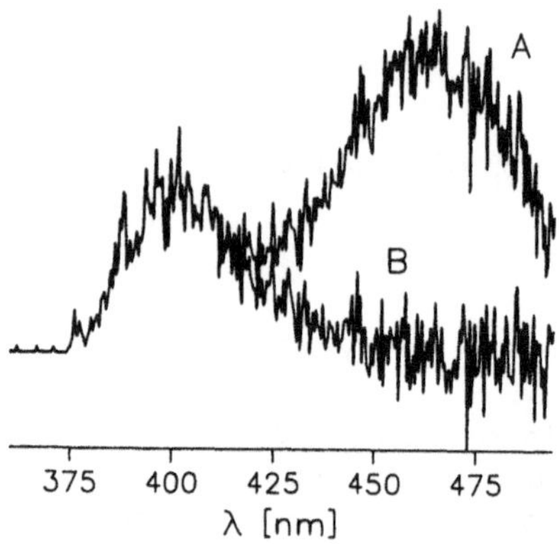

Fig.1a Time resolved fluorescence spectra at 115K. Curve A measured at time interval 0..500ps; Curve B obtained from A subtracting the excimer fluorescence contribution

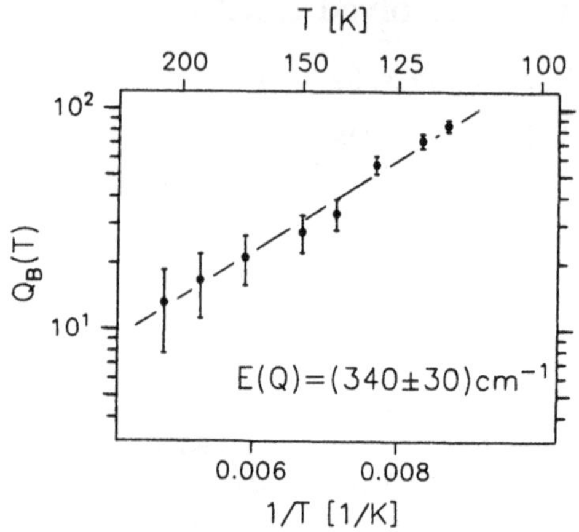

Fig.1b Temperature dependence of the B-fluorescence integral quantum flux Q_B

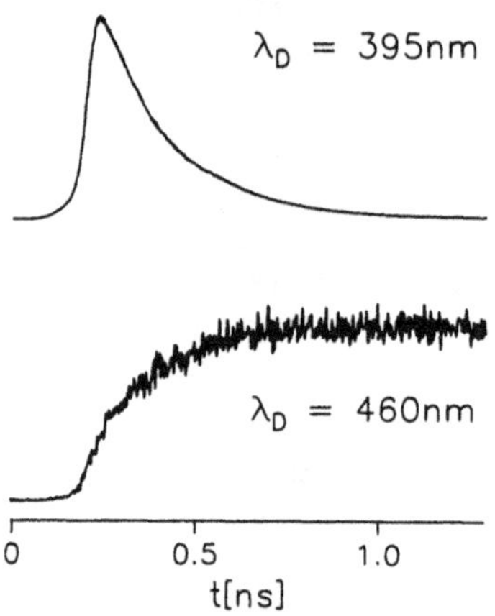

Fig.2a Fluorescence transients at 115K detected at 395nm (B-state) and at 460nm (excimer)

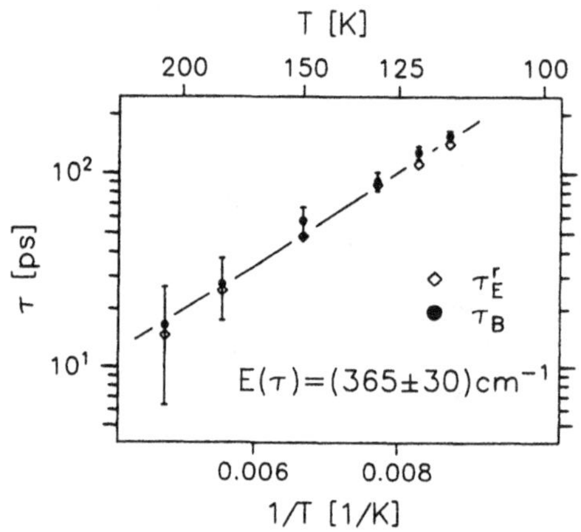

Fig.2b Temperature dependence of B-decay time τ_B and excimer rise time τ_E^r

The activation energy deduced from the time constants $E(\tau) = (365 \pm 30)cm^{-1}$ is in agreement with the value of E(Q) obtained above.

The same quantitative results are obtained for different excitation wavelength λ_{ex} between 328nm and the absorption origin at 376nm. Apparently the excimer dynamics in the high temperature phase of pyrene is independent of excess energy.

From a merely kinetic point of view the observed features fit into the scheme previously developed for the two-step excimer

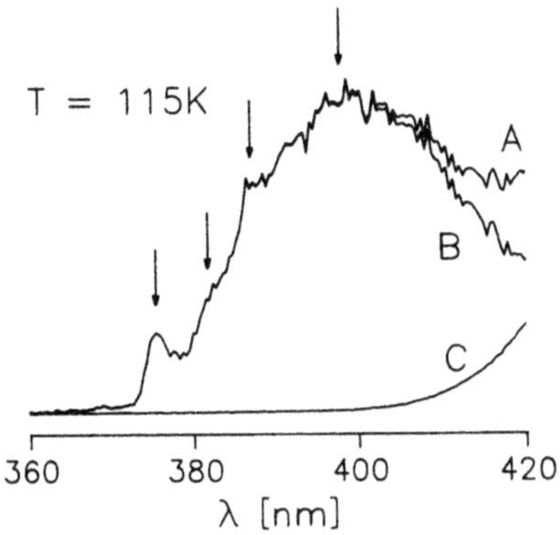

Fig.3 Short wavelength region of the time-resolved pyrene fluorescence spectra at 115K. Curves A and C measured at different time intervals 0..500ps and 1.9..6.3ns, respectively. Curve B, obtained by subtracting C from A, represents the transient B-fluorescence

formation process in α-perylene [2]. This kinetic scheme based on a system of rate equations has been adopted to the pyrene case successfully.

The kinetic model used does not specify the nature of the B-state. A more detailed analysis of the spectral characteristics of the B-fluorescence provides as essential features the clearly resolved zero-phonon line which coincides with the absorption line position [3,4] and the subsequent more intense broad fluorescence bands revealing slight substructure (Fig.3).

It seems obvious to explain the B-fluorescence structure as resulting from vibronic progression built-up on the zero-phonon lines and the concomitant phonon sidebands.

Based on this characterization we propose a modified scheme for self-trapping within the excited singlet state of pyrene in its high temperature phase. For further details see [5].

References

1. H. Port, R. Seyfang and H.C. Wolf, J. de Physique 46, C7-391 (1985)
2. B. Walker, H. Port and H.C. Wolf, Chem. Phys. 92, 177 (1985)
3. D. Fischer, G. Naundorf, W. Klöpffer, Z. Naturforsch. 28a, 973 (1973)
4. A. Matsui, H. Nishimura, J. Phys. Soc. Jap. 49, 657 (1980)
5. R. Seyfang, H. Port and H.C. Wolf, J. Lumin., in press

Part X

Chemical Reaction and Solvation Dynamics

Femtosecond Spectroscopy of Transition States in Reactions

A.H. Zewail

Arthur Amos Noyes Laboratory of Chemical Physics,
California Institute of Technology, Pasadena, CA91125, USA

Transition states of chemical reactions are the central configurations for the description of how reagents complete their trajectories to products. Over the past three decades both molecular beam scattering and photofragment product state distributions have been the main source of information for reaction dynamics. The information obtained is the *time-integrated* result of the interaction of fragments on the potential energy surface, PES. As pointed out by SMITH [1], this approach concentrates on the "before" and "after" of chemical events at the molecular level. The making of new bonds (and breaking of old ones) in elementary chemical reactions occurs in a very short time (picoseconds-femtoseconds), and to investigate chemical dynamics actually *during* the molecular processes that lead to chemistry requires new techniques with ultrashort laser pulses.

Such techniques have been developed in molecular beams to study the ultrafast chemistry of uni- and "oriented" bimolecular reactions [2]. These studies resulted in real-time viewing of energy redistribution, measurements of state-to-state rates and tests of microscopic statistical theories as prescribed by the PES. More recently, the technique was extended to the femtosecond time domain [3-6] to view the transition states of simple elementary reactions. The time scale for this femtochemistry [7] is now allowing for the observation of fragments during the collision or half-collision of a chemical reaction. Since the recoil speed of fragments is typically 1 km s^{-1}, the femtosecond time resolution is probing dynamics at fragments separation of < 0.5 Å.

In this report we will highlight some of the recent studies in ultrafast spectroscopy of chemical reactions and illustrate the applications to different classes of reactions. The apparatus for these femto- and picosecond experiments in molecular beams is described in previous publications from this group.

1. DIRECT BOND BREAKING

Here, we consider the elementary reaction:

$$ICN^* \rightarrow [I\cdots\cdot CN]^{\ddagger *} \rightarrow I + CN\ .$$

For this reaction we first used [3] subpicosecond pulses and later [4] much shorter pulses to probe the process of bond breaking (Fig.1).ICN was chosen to study because it offers several advantages. The excitation is to a repulsive state of the C-I coordinate, and the CN product can be detected using laser-induced fluorescence. There is also a large and ever-growing literature on the dynamics of the dissociation of ICN, including product state distributions, measurements of the anisotropy of product recoil, fragment alignment effects, etc.

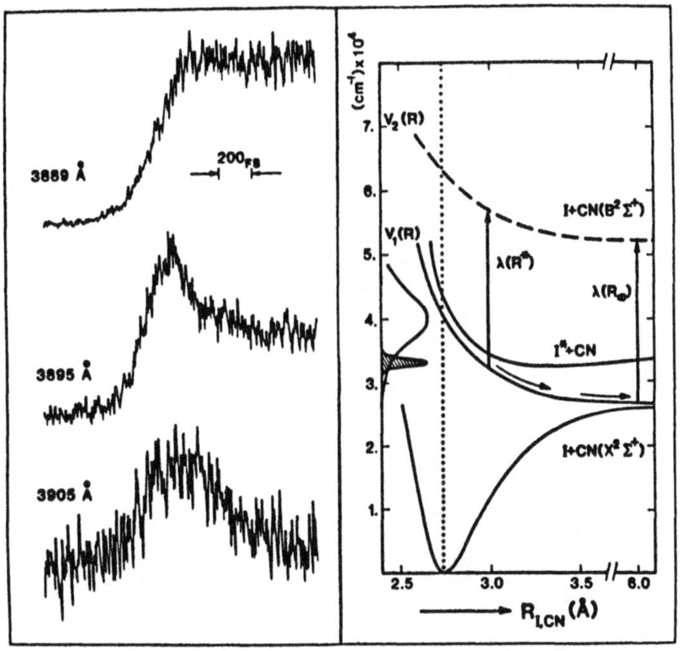

Fig. 1: Left: The fs transients observed for the reaction of ICN, detecting free fragments (top) and transition states (middle and bottom); Right: The PES involved [from Ref. 4]

From the delay and rise of the CN LIF signal we have measured the time it takes the CN and I fragments to separate on the repulsive surface (Fig. 2). To interpret these results, dissociation must be defined more precisely because the reaction is in continuous transition states from reactant to products. The question is: when in this transition does one consider the reaction to be over? The answer, for the experimental observations reported, is that the reaction is over when the CN is spectroscopically identifiable as the *free* CN radical, not as a CN radical perturbed by the close proximity of an I atom. Because the probe can be tuned to the free or perturbed CN transitions one can answer such questions, and measurements of transition state temporal behavior, which give direct information on the PES, can be made. The same technique was extended to studies of alkyl iodides where *direct* and *secondary* bond breaking was observed in reactions of R-I and I-R-I type systems [8].

2. EXTENSION TO BIMOLECULAR REACTIONS

The same techniques have been extended to studies of a special class of bimolecular reactions where the reagent molecules are oriented in close proximity. The reaction studied, in collaboration with R. B. Bernstein, is

$$H + OCO \rightarrow [HOCO]^{\ddagger} \rightarrow OH + CO \ .$$

The collision complex was observed (Fig. 3) as a delay of the OH fragment for a few picoseconds, indicating a large well involvement in the PES [9].

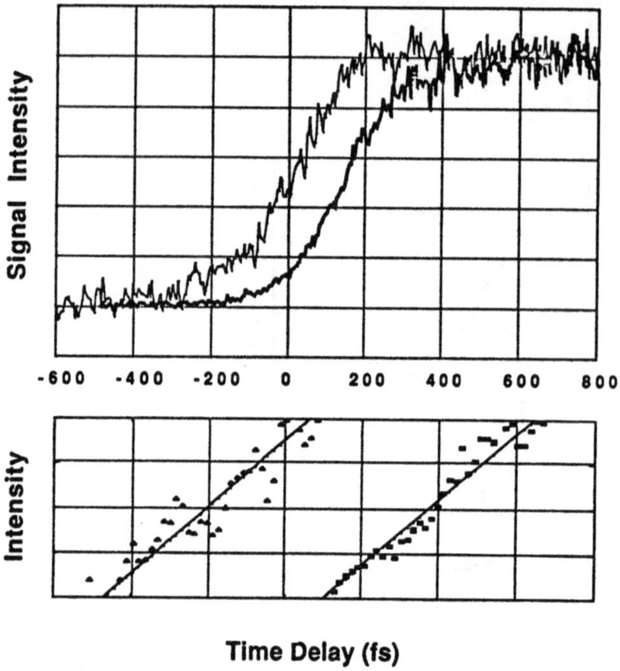

Fig. 2: The delay time observed (heavy trace) for the free fragments (CN) observed in the reaction of ICN. The light trace is for calibration of the zero of time [from Ref. 12]

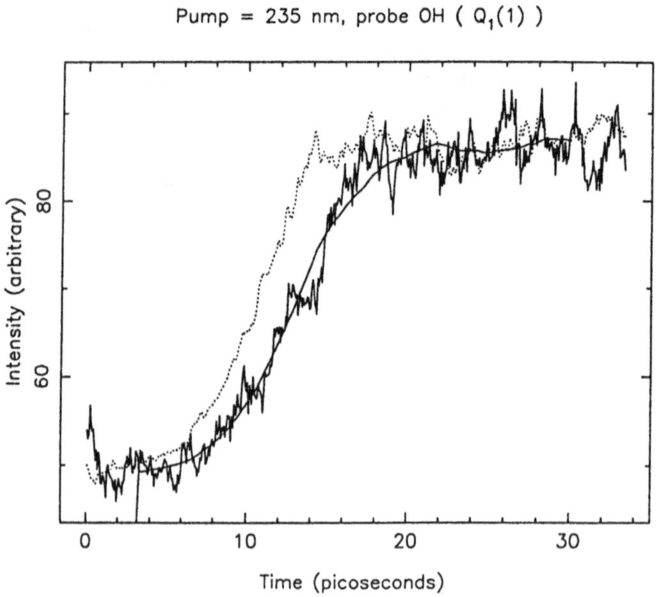

Fig. 3: The observation of OH fragment delay time in the reaction H + OCO. The dotted line is for calibration of the zero of time [from Refs. 9, 13]

3. RESONANCES IN DISSOCIATION

If, in the process of falling apart, more than one degree of freedom is involved (vibrational or electronic), the system might exhibit a quasi-bound state (or resonance). Manifestations of such a resonance in the real-time probing of fragment separation should be: 1) a delay in the appearance of free fragments, and possibly 2) the appearance of oscillations (reflecting the resonance frequency) of the wave packet of the dissociating molecules.

With these ideas in mind, we made attempts to obtain femtosecond transients of the reactions of alkali halides, where we expected such temporal features and there exists a wealth of cross-beam experiments. The work by BERRY and others [10] has provided the foundation for the descriptions of the surfaces. Because of the crossing between the *ionic* ground state and the *covalent* excited state of these salts, at fragment separation R_x, there are two channels for the reaction (Fig. 4). En route to products, the $[M\cdots X]^{\ddagger *}$ transition state molecules decide between these covalent and ionic channels. There are two limiting possibilities: either the packet is trapped on the *adiabatic* surface without crossing (resonance), or it crosses on the *diabatic* surface, as was the case for ICN. These two limits give rise to entirely different temporal behavior. If a resonance exists, the observed oscillation will provide the detailed nature of the surface and coupling.

Recently, we reported [5,6] on these observations of strong wave packet oscillations (strong resonance) in the dissociation reaction of NaI, and a much weaker resonance in the reaction of NaBr. These observations were made by exciting the salt to the covalent state(s) along the M-X coordinate using a femtosecond pump pulse. A fs probe pulse was used at different delay times and wavelengths (t and λ spectra) to detect the product free (on resonance with Na D-lines at 589 nm) or perturbed Na (off-resonance) atoms. These results (Fig. 5) are described in detail elsewhere [5,6].

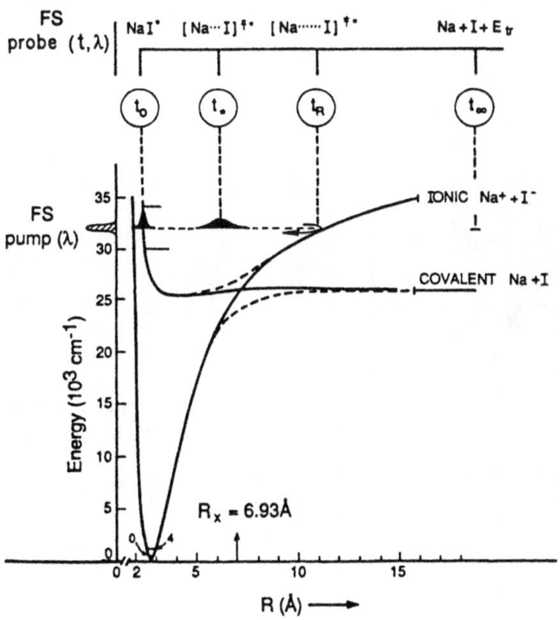

Fig. 4: A description of the femtosecond experiments for the detection of resonances in reactions of alkali halides [from Refs. 5,6]

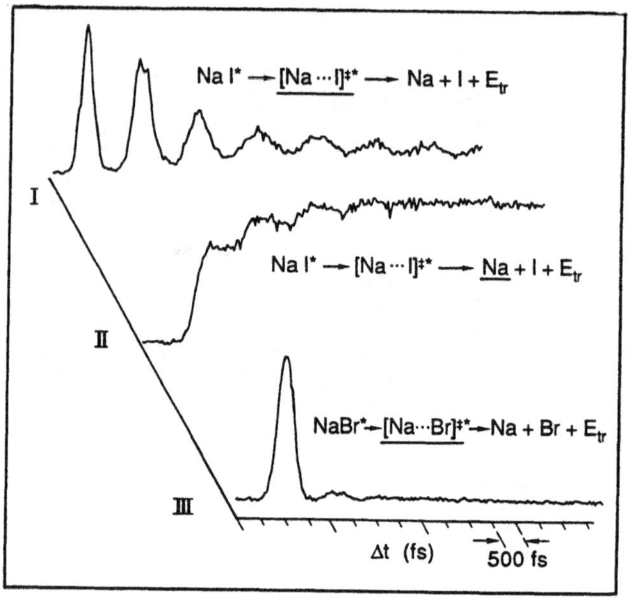

Fig. 5: Observed resonances in the dissociation reactions of Na I and Na Br [from Refs. 5,6]

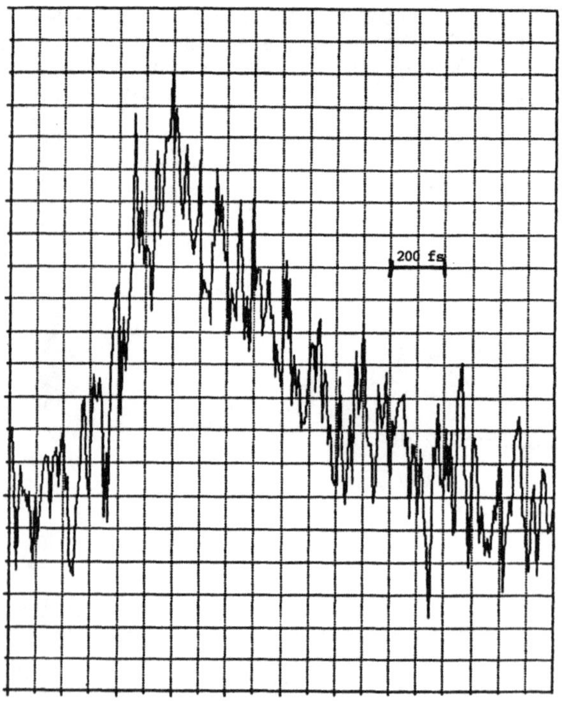

Fig. 6: The fs transients observed in the reaction of CH_3I, detecting the I atom in the process of separation [from Ref. 14]. Preliminary results

4. CONCLUSION

In conclusion, femtosecond spectroscopy of reactions promises to provide real-time viewing of the PES, of packet decay and resonances, and of transition states. Since the recoil velocity is typically 1 km s^{-1}, our fs time window gives a view of these dynamics with < 0.5 Å resolution for fragment separation. With the state-of-the-art in fs pulse width [11], this distance can be made < 0.1 Å, and many reactions, including R-I systems (Fig. 6), can be studied.

References

1. I.W.M. Smith: Nature 328, 760 (1987)
2. For recent review, see J.L. Knee and A.H. Zewail: Spectroscopy 3, 44 (1988)
3. N.F. Scherer, J.L. Knee, D.D. Smith, A.H. Zewail: J. Phys. Chem. 89, 5141 (1985)
4. M. Dantus, M.J. Rosker, and A.H. Zewail: J. Chem. Phys. 87, 2395 (1987)
5. T. Rose, M.J. Rosker, and A.H. Zewail: J. Chem. Phys. 88, 6672 (1988)
6. M.J. Rosker, T. Rose, and A.H. Zewail: Chem. Phys. Lett. 146, 175 (1988)
7. R.B. Bernstein and A.H. Zewail: Kaguku to Kogyo (Chemistry and Chemical Industry) 41, 108 (1988)
8. J.L. Knee, L.R. Khundkar, and A.H. Zewail: J. Chem. Phys. 83, 1996 (1985); L. Khundkar and A.H. Zewail, Chem. Phys. Lett. 142, 426 (1987)
9. N.F. Scherer, J.L. Khundkar, R.B. Bernstein, and A.H. Zewail: J. Chem. Phys. 87, 1451 (1987)
10. R.S. Berry: In Alkali Halide Vapors, eds. P. Davidorits and D. McFadden (Academic Press 1979) p. 78
11. See C.V. Shank's article and other articles in these Proceedings
12. M.J. Rosker, M. Dantus, and A.H. Zewail: submitted to Science
13. N.F. Scherer, C. Sipes, R.B. Bernstein, and A.H. Zewail: to be published
14. Work in progress in this group

This work was supported in part by the AFOSR (Grant No. 87-0071), and in part by the National Science Foundation.

Phase-Coherent Molecular Dynamics and Phase-Coherent Chemistry: Observation and Manipulation of Elementary Molecular Motion and Chemical Change

A.G. Joly, S. Ruhman, B. Kohler, and K.A. Nelson+*

Department of Chemistry, Massachusetts Institute of Technology, Cambridge, MA 02139, USA

Recent experimental and theoretical results have made it clear that phase coherent excitation is a very general feature of femtosecond spectroscopy. When a sufficiently short light pulse strikes any Raman-active or absorbing medium, i.e. nearly any molecular medium, coherent excitation is likely to occur. This is useful because, following its initiation, phase-coherent molecular motion can be monitored in real time by subsequent pulses to yield information about molecular dynamics and potential energy surfaces. In addition, some control may be exerted over molecular motion and even chemical change.

1. Femtosecond Phase-Coherent Excitation

Figure 1 illustrates the two main mechanisms through which phase coherent molecular motion is initated by a femtosecond pulse. Through impulsive stimulated Raman scattering, (ISRS), the electric field of a pulse exerts a sudden ("impulse") force on Raman-active modes.[1] Stimulated Raman scattering occurs through mixing among the frequency components of the ultrashort pulse. The impulse force initiates coherent vibrational oscillations which in general affect both the real and imaginary parts of the refractive index of the sample. Coherent oscillations can be observed by measurement of time-dependent absorption, birefringence, diffraction from transient gratings, etc. Time-resolved observations of coherent acoustic and optic phonons and molecular vibrations have been presented and reviewed.[1,2]

Figure 1b illustrates the results of absorption into an electronic excited state, e.g. S_1. In general, the S_0 and S_1 potential energy minima are displaced, e.g. an excited molecule may be "stretched" along vibrational coordinate Q. Following photoexcitation, an excited molecule undergoes vibrational oscillations about the S_1 minimum. If the excitation pulse is short in duration compared to the S_1 oscillation (i.e. the spectral width of the pulse exceeds the S_1 vibrational frequency), all the excited molecules oscillate in phase. Time-resolved observations of excited-state molecular vibrational oscillations [3], including oscillations in gas-phase NaI and NaBr which undergo photodissociation [4], have been reported.

Note that optical absorption of an ultrashort pulse also results in vibrational phase-coherence in the ground electronic state.[5] This has been explained [6] in terms of the "hole" which remains on one side of the S_0 po-

---

*Permanent Address: Department of Physical Chemistry, Hebrew University, Jerusalem 91904 Israel

+Presidential Young Investigator Awardee and Alfred P. Sloan Fellow

Springer Series in Chemical Physics, Vol. 48: Ultrafast Phenomena VI
Editors: T. Yajima · K. Yoshihara · C.B. Harris · S. Shionoya

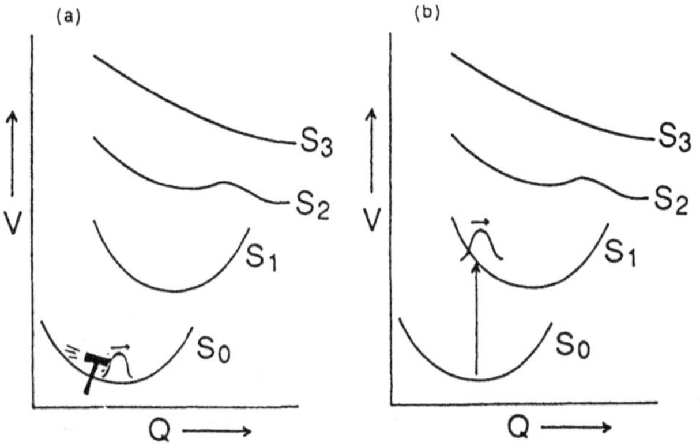

Figure 1. Schematic illustration of coherent wavepacket production with femtosecond pulses.
(a) Through impulsive stimulated Raman scattering, an ultrashort pulse exerts a sudden (impulse) driving force which initiates coherent vibrational motion in the ground electronic state.
(b) Optical absorption of an ultrashort pulse leads to coherent vibrational motion in the S_1 excited-state potential surface. Excitation into S_2 or S_3 leads to synchronized extension along coordinate Q with no return to the original geometry. This may correspond to molecular photodissociation or photoisomerization.
For either coherent excitation method to occur, the excitation pulse must be short in duration compared to the vibrational oscillation period, or in general compared to the time required for significant wavepacket propagation.

tential surface. The unexcited molecules are, on average, "compressed", i.e. on the left side of the S_0 potential. They continue their oscillatory motions which, although initiated thermally, become phase-coherent due to the selective removal of "stretched" molecules. Unlike the other effects discussed, this phenomenon depends on thermal excitation of excited vibrational levels and vanishes at low temperatures.

Of course, when resonant absorption of an ultrashort pulse occurs as in Fig. 1b, ISRS excitation of oscillations in S_0 also occurs and is resonantly enhanced. It has recently been shown that the two processes can be described in terms of the same nonlinear optical interaction.[6] For our purposes, resonance enhancement of ISRS excitation is important in that it may yield a substantial increase in the S_0 vibrational amplitude.

2. Liquid-Phase Molecular Dynamics and Reaction Dynamics

Spectroscopic exploitation of femtosecond phase-coherent excitation mechanisms has produced information about molecular dynamics in solid, liquid, and gas phases. We show two examples of data from molecular liquids. Figure 2 shows ISRS data from carbon disulphide liquid at 165K.[2,7] The impulsive torques exerted by the crossed excitation pulses through impulsive stimulated (rotational) Raman scattering initiate rotational motion which leads to net orientational alignment of the CS_2 molecules. This net alignment gives rise to birefringence and to signal. It is clear from the data that weakly oscillatory molecular librations occur in the liquid. The librational frequency yields an approximate value for the configuration-averaged torque due to intermolecular interactions, i.e. the average restoring force (librational "force constant") felt by molecules in the liquid. In addition, the rapid dephasing of the li-

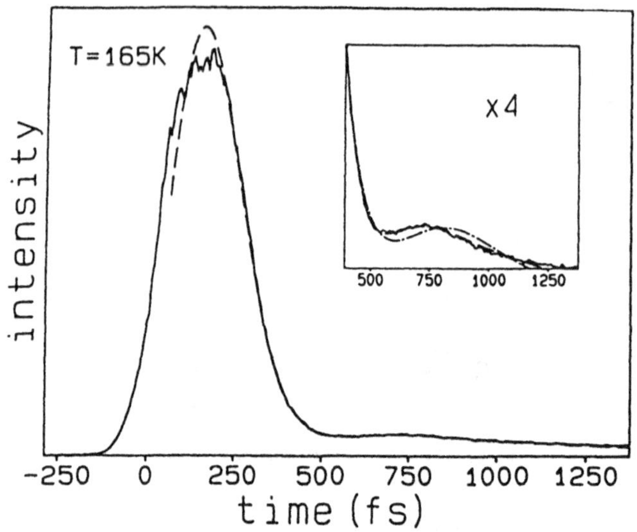

Figure 2. Impulsive stimulated scattering data (solid curve) from CS_2 liquid at 165 K. The data were recorded in a transient grating configuration. The crossed 65-fs excitation pulses initiate molecular rotational motion at t=0. The data show that orientational alignment increases after t=0 and is weakly oscillatory. The dashed-curve fit which lies under the data in the inset is based on a model of inhomogeneous librational dephasing. The broken-curve fit shows the inadequacy of a model of homogeneous dephasing.

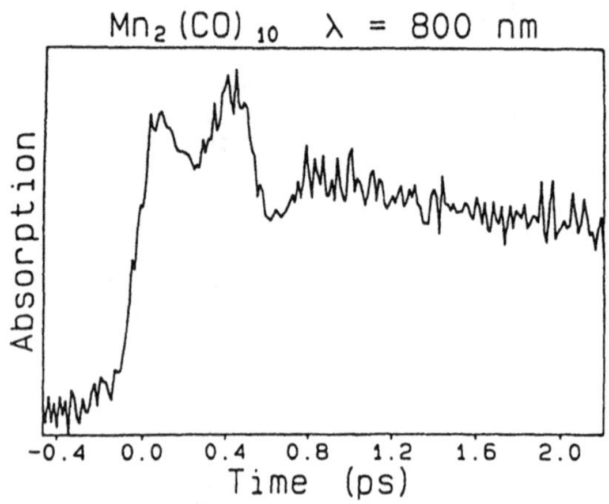

Figure 3. Transient 800-nm absorption data for $Mn_2(CO)_{10}$ in ethanol solution following 310-nm excitation. The $Mn(CO)_5$ photofragments absorb at 800nm. The oscillations in the data suggest that Mn-Mn stretching vibrations can occur, i.e. that the excited state is weakly bound. This is interpreted in terms of a barrier to dissociation produced by interactions with solvent molecules, which resist separation of the $Mn(CO)_5$ fragments.

brational oscillations is due primarily to inhomogeneity in the liquid (i.e. in the intermolecular torques), the extent of which is also determined approximately. Thus basic information about liquid-state molecular dynamics and intermolecular interactions is extracted from the data.[8]

Figure 3 shows transient absorption data recorded following pulsed 308-nm excitation of a solution of dimanganese decacarbonyl, $(CO)_5Mn\text{-}Mn(CO)_5$, in methanol.[9] The probe wavelength, 800 nm, was selected to monitor formation of $Mn(CO)_5$ fragments following photocleavage of the metal-metal (M-M) bond. The time-dependent absorption strength is oscillatory for $t < 1$ ps. Although further experimentation is necessary to determine unambiguously the source of the oscillations in the data, they may be due to excited-state M-M stretching vibrations, as suggested by Fig. 1b. This is interesting because the excited state of $Mn_2(CO)_{10}$ is believed to be unstable with respect to dissociation, as in level S_3 of Fig. 1. Neighboring solvent molecules may resist separation of the large photofragments, resulting in a weakly bound excited state in solution. Thus the oscillations in Fig. 3 may provide a measure of the intermolecular forces through which the solvent influences chemical reaction dynamics and yield.

3. Large-Amplitude ISRS Excitation and Optical Manipulation of Wavepacket Propagation: Controlled Phase-Coherent Chemistry

In the ISRS experiments carried out to date, typical coherent vibrational amplitudes have been in the 10^{-5} - 10^{-4} A range. The amplitude is linearly proportional to the excitation pulse integrated intensity and to the Raman-scattering differential polarizability. Sufficient laser pulse energy is available to drive amplitudes well over 0.1A.[10], but at the high intensities needed competing nonlinear optical processes would lead to white-light continuum generation and sample damage in most materials. It appears likely, however, that large-amplitude (i.e. > 0.1A) molecular and optic phonon vibrational amplitudes can be driven by ISRS excitation. To achieve large amplitudes, resonance enhancement (often by factors of over 100) of the differential polarizabilities can be exploited with appropriate excitation wavelengths. In addition, a selected vibrational mode can be driven repetitively with a sequence of evenly spaced excitation pulses. This is similar to repetitively pushing a child on a swing (rather than delivering a single, excessively hard blow) to reach large amplitudes. The desired sequences of many excitation pulses can be produced from one pulse with pulse-shaping techniques.[11]

The possibility of driving chemical or structural change through ISRS excittion in the 0.1 -1A range has been discussed.[10] A different approach to controlled wavepacket propagation involving absorption and stimulated emission of femtosecond pulses has also been proposed.[12] In Fig. 4, one of the simpler possibilities is illustrated. Figure 4a shows how a reactive excited-state potential surface could be elucidated through ISRS excitation of large-

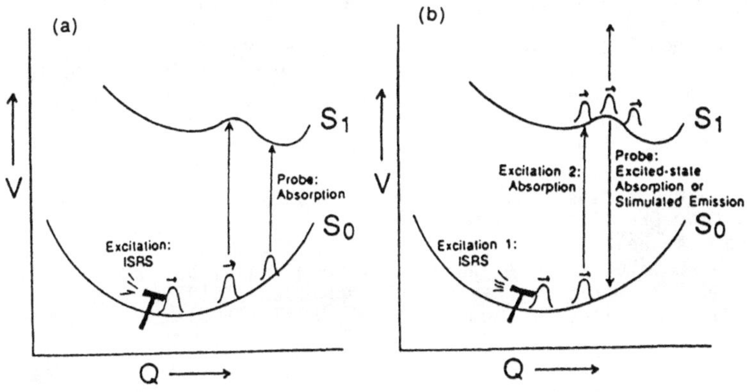

Figure 4. Schematic illustration of experiments involving large-amplitude ISRS excitation.
In (a), ISRS excitation followed by measurement of time-dependent absorption spectra yields a "map" of the reactive excited-state potential as a function of Q.
In (b), ISRS excitation followed by absorption produces an excited-state wavepacket whose initial position and momentum in S_1 can be controlled experimentally. In this manner sufficient momentum can be imparted such that a barrier to reaction in S_1 may be crossed. By varying the amplitude of ISRS excitation and the timing of the absorption into S_1, the dynamics on the S_1 potential surface can be manipulated.

amplitude vibrations in S_0 followed by measurement of absorption spectra of vibrationally distorted species. Figure 4b shows how large-amplitude ISRS excitation could be used to assist progress along an excited-state reaction coordinate. By varying the ISRS excitation amplitude and the relative timing of the two excitation pulses, the initial position and momentum of the excited-state wavepacket could be varied independently. In this manner the extent and effects of dissipation in the excited state due to coupling between the reactive mode and other (both) modes could be examined systematically. Investigation of the effects of different environments (e.g. different solvents) on barrier crossing dynamics and probabilities, and spectroscopic characterization of species at and near potential barriers will also be possible.

Many schemes other than that shown in Fig. 4b for coherent vibrationally assisted chemistry, including direct driving of some ground-state reactions[10] may be possible. Optical manipulation of coherent wavepacket propagation in reactive (and other) potential energy surfaces should play an important role in femtosecond molecular spectroscopy, especially with increased access to tunable high-energy pulses and pulse-shaping techniques.

This work was supported in part by NSF Grant No. DMR-8306701 and by NSF and contributions from NSF, AT&T Bell Labs, DuPont, Perkin-Elmer, and Chevron to a Presidential Young Investigator Award.

4. References

1. Y.-X. Yan, L.-T. Cheng, and K.A. Nelson: In Advances in Nonlinear Spectroscopy, ed. by R.J.H. Clark and R.E. Hester, Adv. in IR and Raman Spectroscopy Ser., v. 16 (Wiley, New York, 1988), p. 299,
2. S. Ruhman, A.G. Joly, B. Kohler, L.R. Williams, and K.A. Nelson: Revue Phys. Appl. 22, 1717 (1988).
3. M.J. Rosker, F.W. Wise, and C.L. Tang: Phys. Rev. Lett. 57, 321 (1986); K.A. Nelson and L.R. Williams: ibid, 58, 745 (1987); J.M.Y. Ha et al. ibid, 57, 3302 (1986).
4. T.S. Rose, M.J. Rosker, and A.H. Zewail: J. Chem. Phys. 88, 6672 (1988).
5. F.W. Wise, M.J. Rosker, and C.L. Tang: J. Chem. Phys. 86, 2827 (1987).
6. J. Chesnoy and A. Mokhtari: Revue Phys. Appl. 22, 1743 (1987); Phys. Rev. B, in press, and personal communication.
7. S. Ruhman, B. Kohler, A.G. Joly, and K.A. Nelson: Chem. Phys. Lett. 141, 16 (1987).
8. See also D. McMorrow, W.T. Lotshaw, and G.A. Kenney-Wallace: IEEE J. Quantum Electron. QE-24, 443 (1988).
9. A.G. Joly and K.A. Nelson: J. Phys. Chem., submitted.
10. Y.-X. Yan, E.B. Gamble, and K.A. Nelson: J. Chem. Phys. 83. 5391 (1985).
11. A.M. Weiner and J.P. Heritage, Revue Phys. Appl. 22, 1619 (1987).
12. D.J. Tannor and S.A. Rice: J. Chem. Phys. 83, 5013 (1986).

Ultrafast Laser Spectroscopy of Transient Ion Pair States in Solution

N. Mataga, H. Miyasaka, T. Asahi, S. Ojima, and T. Okada

Department of Chemistry, Faculty of Engineering Science, Osaka University, Toyonaka, Osaka 560, Japan

The photoinduced electron transfer (ET) and behaviors of produced transient ion pair (IP) states in liquids, rigid matrices, molecular assemblies and biological systems are the most fundamental and important central problems in photochemical primary processes in condensed phase [1-3]. The elucidation of factors underlying these processes is of crucial importance for the photochemical and photobiological reaction mechanisms [1-3] and is being much facilitated by the application of the picosecond-femtosecond laser photolysis method. We discuss here results of our ultrafast laser photolysis and time-resolved transient absorption spectral studies on the elementary processes of intra- and intermolecular transient IP formation due to the photoinduced charge separation (CS) in some typical systems.

Experimental

A microcomputer controlled ps laser photolysis system with mode-locked Nd^{3+}:YAG laser [4] was used for the transient absorption spectral measurements in 10 ps-ns region. For the measurements of time-resolved transient absorption spectra in the shorter time-region, a subps laser photolysis system was used. The output of a cw mode-locked Nd^{3+}:YAG laser operated at 82MHz rate was compressed by a fiber-grating system and the SHG of the compressed pulse synchronously pumped a pyridin-1 dye laser (300 fs fwhm, 1 nJ), the output of which was amplified to 0.4 mJ/pulse by 3-stage pyridine-1 dye amplifier pumped by a frequency doubled Q-switched Nd^{3+}:YAG laser operating at 10 Hz. The amplified pulse with several hundreds fs width was frequency doubled and used for exciting the sample. The rest of the 710 nm fundamental pulse was focused into D_2O to generate a white light probe pulse. Two sets of multichannel photodiode detectors were used to observe wide band transient absorption spectra.

Results and Discussion

(A) Photoinduced CS and CR of IP state in combined donor acceptor systems

We have examined the intramolecular photoinduced CS of p-$(CH_3)_2$N-ϕ-$(CH_2)_n$-(1-pyrenyl)(Pn) (n=1,2,3) and other related systems in acetonitrile solution. For example, we have measured the rate of photoinduced CS of P_1, P_2 and P_3 in acetonitrile by means of subps and ps laser photolysis and time-resolved transient absorption spectral measurements. The rise times of the intramolecular IP state have been determined to be ca. 1 ps, 8 ps and 20 ps for P_1, P_2 and P_3, respectively. As an example, the spectra of P_2 and rise curves of IP state of P_1 and P_2 are indicated in Fig. 1. On the other hand, the CR decay times of the intramolecular IP state have been determined to be ca. 10 ns, 3 ns and 1 ns for P_1, P_2 and P_3, respectively, which are much longer than the rise times of the CS state and in the reverse order with respect to the number n of the methylene chain.

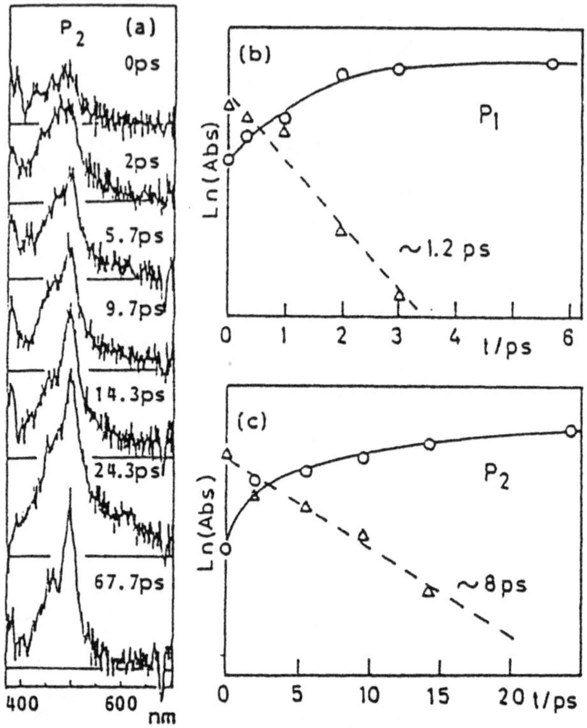

Fig. 1. Time-resolved transient absorption spectra of P_2 in acetonitrile solution at 22°C measured with a subps laser (355 nm, 800 fs) photolysis apparatus (a). Rise curves of the absorbance of pyrene anion band peak for P_1(b) and P_2(c). The spectra are not corrected for the chirping of the monitoring light pulse.

These results can be interpreted on the basis of the dependences of the ET rate upon the geometries and chain lengths as well as the relevant energy gap for the CS ($-\Delta G^\circ_{CS}$) and for the CR ($-\Delta G^\circ_{ip}$). For the chromophores of Pn, the $-\Delta G^\circ_{CS}$ values in acetonitrile are ca. 0.5 eV. Our theory [5] for the energy gap dependence of the CS rate constant k_{CS} assuming $\bar{A}=10^{12}\sim10^{13}\ s^{-1}$ gives $k_{CS}=10^{11}\sim10^{12}\ s^{-1}$ around $-\Delta G^\circ_{CS}\sim0.5$ eV, where $\bar{A}=2\pi|\langle i|\mathcal{H}'|f\rangle|^2/\hbar^2\langle\omega\rangle$ and $\langle i|\mathcal{H}'|f\rangle$ is the electronic matrix element responsible for ET. Since our theoretical calculations show a dull energy gap dependence of k_{CS} around $-\Delta G^\circ_{CS}\sim0.5$ eV, the large difference of k_{CS} between P_1 and P_2, P_3 cannot be ascribed to the small differences in $-\Delta G^\circ_{CS}$ but should be ascribed to the difference in $\bar{A}$. The $-\Delta G^\circ_{ip}$ value for these chromophores is estimated to be 2.86 eV in acetonitrile. Results of our detailed experimental [3,6] and theoretical [5] studies on the energy gap dependences of the CR of geminate IP's of various free donor and acceptor systems show that at $-\Delta G^\circ_{ip}\sim2.86$ eV, the CR is in the inverted region and the rate constant k_n is ca. $10^7\ s^{-1}$ assuming $\bar{A}=10^{12}\ s^{-1}$ (see B)). In the case of P_1, $\bar{A}$ seems to be larger and by taking $\bar{A}=10^{13}\ s^{-1}$ we obtain $k_n^{-1}=10^{-8}$ s in agreement with the observed recombination time $\tau_{CR}\sim10$ ns. In P_3, configurational change to sandwich type can take place immediately after ET which causes increase of $\bar{A}$ and the decrease of $-\Delta G^\circ_{ip}$, leading to the faster CR with $\tau_{CR} \sim 1$ ns. In P_2, the freedom for such configuration change is smaller than in P_3, which results in a little larger τ_{CR} value.

B) CR of geminate IP produced at encounter between fluorescer and quencher

The CR dynamics of geminate IP and dependence of its rate constant k_n upon $-\Delta G^\circ_{ip}$ are of crucial importance from various viewpoints such as mechanisms of organic photochemistry and biological photosynthesis. Contrary to the fact that only the result for the normal region but no result for the inverted region has been available for the CS in the fluorescence quenching reaction in polar solutions [2,3,7], only the result for the inverted region but no result for the normal region was available for the CR of IP produced by the fluorescence quenching reaction. By means of quantitative picosecond time-resolved absorption spectral measurements on the geminate IP's of various aromatic hydrocarbon-electron donating or accepting quencher systems, we have obtained experimental results of k_n not only for the inverted region but also for the normal region and have proved the bell-shaped energy gap dependence of k_n of the geminate IP produced by fluorescence quenching reaction for the first time [2,3,6]. The result in Fig. 2 is quite different from the dependence of the CS rate constant on $-\Delta G^\circ_{CS}$ [7], which cannot be interpreted satisfactorily by conventional theory but we need new theoretical interpretation [3,5,6].

(C) Relaxation processes of excited CT complexes

The electronic interaction between donor and acceptor in CT complexes will be considerably stronger than in the inter- and intramolecular exciplex type of systems with no direct interchromophore interactions in the ground state. The comparison of the relaxation processes of excited CT complexes

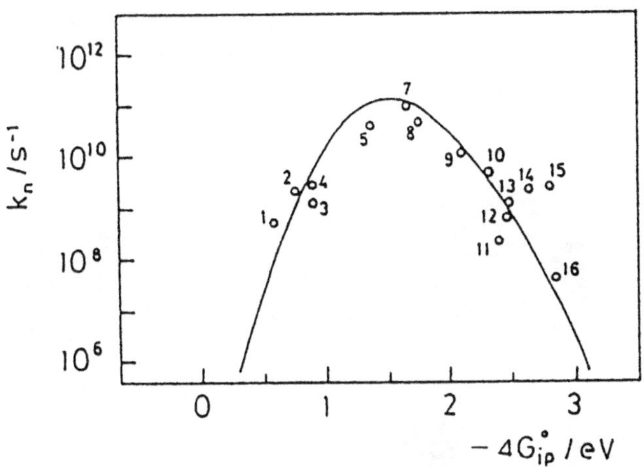

Fig. 2. The dependence of k_n of geminate IP produced by fluorescence quenching reaction on $-\Delta G^\circ_{ip}$ in acetonitrile solution (see ref. 6).
o: observed values for various aromatic hydrocarbon fluorescer-nitrile, acid anhydride and amine quenchers.
—: calculated by the method of ref. 5d with $\bar{A}=10^{12}$ s^{-1}.

from FC to the relaxed IP state with the photoinduced CS processes of those exciplex type of systems in (A) and (B), and also the comparison between the CR dynamics of their IP's will be an interesting problem with respect to the elucidation of photochemical electron transfer mechanisms.

The CT interaction in the 1,2,4,5-tetracyanobenzene (TCNB)-toluene complex is weak in the ground state, while it becomes stronger in the excited state causing relaxations to equilibrium IP state, being accompanied by conformational change and solvent reorientation. Figure 3 shows time-resolved transient absorption spectra of TCNB-toluene system measured with subps laser photolysis system exciting at 355 nm. The rapid rise of broad band nearly equal to the time response of apparatus was observed immediately after excitation. This broad absorption spectra is different from the sharp $TCNB^-$ like absorption band observed at long delay times.

With increase of the delay time, the sharp $TCNB^-$ band at 465 nm grows with the rise time of 33 ps. The time dependent spectral change from the broad band to the sharp $TCNB^-$ band seems to indicate the relaxation from a partial CT state to an almost complete IP state accompanied with rearrangements in the donor acceptor geometry and environment [8]. We have also made detailed transient absorption spectral studies of this system in 10 ps-ns regions. The intensity of the sharp band at 465 nm increases with increase of the delay time in 10 ps time region. In addition, a broad absorption in wavelength region longer than 650 nm was observed. This broad absorption also shows rise in 10 ps region and is very similar to the absorption of toluene dimer cation D_2^+. The rise curves of both absorption

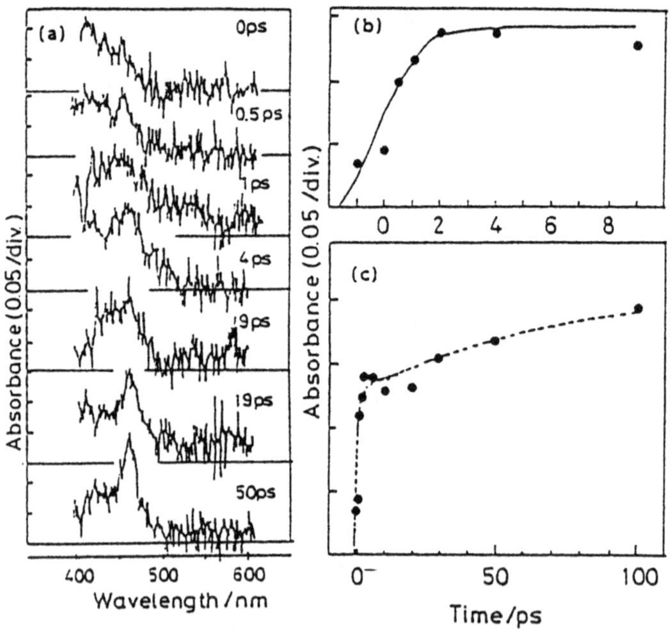

Fig. 3. Time-resolved transient absorption spectra of TCNB-toluene system excited with a 355 nm, 800 fs laser pulse (a) and time profiles of transient absorbance at 465 nm (b,c). The spectra are not corrected for the chirping of the monitoring light pulse.

coincide with each other with the same rise time as obtained by subps photolysis in Fig. 3. Therefore, we can conclude that it takes ca. 30 ps to produce the equilibrium excited state ($A^-D_2^+$) from the excited FC state in this system.

We have made preliminary time-resolved transient absorption spectral measurements on this complex in acetonitrile solution. The results indicate that it takes 10∿20 ps for the rise of the sharp $TCNB^-$ absorption band even in acetonitrile solution. The $-\Delta G^\circ_{IP}$ value of TCNB-toluene IP in acetonitrile is a little smaller than that of Py-DMA in acetonitrile [5d,9] and the energy difference between the FC state or partial CT state immediately after excitation and the relaxed IP state may be almost the same or larger than the $-\Delta G^\circ_{CS}$ value of the Py-DMA system. Nevertheless, the photoinduced IP formation of P_1 in acetonitrile, where the interchromophore interaction in the ground state is negligible, is much faster than that of TCNB-toluene, as discussed in (A). The complete CS to form the relaxed IP in toluene solution is assisted by the dimer cation formation, while, even in acetonitrile solution, the solvation does not make the IP formation so fast as in P_1. The latter fact might be ascribed to the much larger rearrangement of donor acceptor geometries and environments orientations necessary to realize complete CS in the more strongly interacting pairs. In the case of stronger donor and acceptor pairs such as TCNB-naphthalene, pyrene-pyromellitic dianhydride [9] etc., the IP state formation becomes very rapid.

Another important problem related to the excitation of the CT complex is the fact that the CR rate of the IP formed by excitation of the CT complex is different from that of the IP formed by CS at encounter between fluorescer and quencher. The lifetime of Py^+-$TCNE^-$ IP which is formed by encounter is ca. 200 ps in acetonitrile while the IP of the same pair formed by excitation of the CT complex undergoes ultrafast decay within 1 ps. Results of our studies on various systems show that the energy gap dependence of the CR decay of IP formed by excitation of the CT complex is not bell-shaped but the CR rate increases with decrease of $-\Delta G^\circ_{IP}$, which might mean that the CR decay in this case can be regarded as a nonradiative transition in the supermolecule of donor and acceptor.

(D) On the mechanism of hydrogen transfer and IP formation in ^{3}BP-amine systems

We have made accurate and quantitative laser photolysis studies over ps-ns regions by means of time-resolved wide band transient absorption spectral measurements on ^{3}BP-amine (DH) systems in various solvents. For example, ^{3}BP-Diphenylamine in nonpolar solvent shows triplet exciplex formation in competition with H transfer but no ketyl radical formation from the exciplex, while the same system in acetonitrile shows IP formation in competition with H transfer but no ketyl radical formation from the relaxed IP. We have obtained similar results in many other benzophenone amine systems [10]. These results show the possible participation of the non-relaxed CT state as the transition state of the H transfer where the relaxed IP or exciplex formation is competing with H transfer.

$$^3BP + DH \longrightarrow {}^3(BP^{-\delta}\cdots DH^{+\delta})^{\ne} \begin{array}{l} \nearrow BPH\cdot + D\cdot \\ \searrow BP^-\ldots DH^+ \text{ or } (BP^-\cdot DH^+) \end{array}$$

ACKNOWLEDGMENT

N. M. acknowledges the support by a Grant-in-Aid (No. 62065006) from the Japanese Ministry of Education, Science and Culture.

References

1. N. Mataga, Pure Appl. Chem. 56, 1255 (1984).
2. N. Mataga, Acta Phys. Polon. A71, 767(1987).
3. N. Mataga, in: Photochemical Solar Energy (Proc. 7th Internat. Conf. Photochemical Conversion and Storage of Solar Energy), ed. J. Norris, Elsevier, in press.
4. H. Miyasaka, H. Masuhara and N. Mataga, Laser Chem. 1, 357 (1983).
5. T. Kakitani and N. Mataga, (a) J. Phys. Chem. 89, 8 (1985), (b) Chem. Phys. 93, 381 (1985), (c) J. Phys. Chem. 89, (1985), (d) ibid. 90, 993 (1986), (e) ibid. 91, 6277 (1987).
6. N. Mataga, T. Asahi, Y. Kanda, T. Okada and T. Kakitani, submitted to Chem. Phys.
7. D. Rehm and A. Weller, Israel, J. Chem. 8, 259 (1970).
8 (a) S. Iwata, J. Tanaka and S. Nagakura, J. Am. Chem. Soc. 88, 894 (1966), (b) N. Mataga and Y. Murata, J. Am. Chem. Soc. 91, 3144 (1969), (c) H. Masuhara and N. Mataga, Z. Phys. Chem. N.F. 80, 113 (1972). (d) S. Nagakura, in: Excited State, Vol. 2, ed. E. C. Lim, Academic (1975), pp. 322-383.
9. N. Mataga, H. Shioyama and Y. Kanda, J. Phys. Chem. 91, 314 (1987).
10. H. Miyasaka, K, Kamada, K. Morita and N. Mataga, manuscript in preparation.

Geminate Recombination in Excited State Proton Transfer Reactions: Picosecond Dynamics in Electrolyte Solutions

E. Pines[1], *D. Huppert*[1], *and N. Agmon*[2]

[1]School of Chemistry, Sackler Faculty of Exact Sciences, Tel-Aviv University, Tel-Aviv 69978, Israel

[2]Department of Physical Chemistry, The Hebrew University, Jerusalem 91904, Israel

1. INTRODUCTION

Geminate recombination plays an important role in proton dissociation reactions [1]. We have recently shown [2] that excited-state acids dissociate via a number of reversible dissociation-geminate recombination cycles. The dissociation profiles of such excited acids are nonexponential, having a $t^{-3/2}$ dependence at long times. These nonexponential tails are affected by temperature and isotopic (H/D) substitution [3].

As a model molecule we have used [4] 8-hydroxy pyrene 1,3,6-trisulfonate (HPTS). Having a 4-fold negative charge after dissociation, this molecule recombines readily in the excited singlet state with its geminate proton. The intrinsic "chemical" dissociation rate, as determined from the slope of the decay curve, is initially $10^{10}s^{-1}$ in water. After 1nsec it is slowed down due to geminate recombination by an order of magnitude. As a result, the observed steady-state dissociation rate, namely the reciprocal area under the dissociation profile, is much slower than the true "intrinsic" chemical rate.

The reversibility of the excited state proton dissociation can be further demonstrated by carrying out the experiment in strong acidic media [4] or in volume limited environments [6]. In both cases a steady-state amplitude of the undissociated HPTS is created.

The dissociation profiles can be accurately reproduced by assuming that the dynamics of the separated ion-pair obeys the Debye-Smoluchowski equation (DSE) with reversible "back-reaction" boundary conditions. The two "intrinsic" rate-coefficients (κ_d, κ_r) for dissociation to and recombination from the contact distance, a, are taken as free parameters. For the actual numerical solution we have applied an efficient Chebyshev propagation algorithm which allows for large time-steps. Agreement with experiment deteriorates for supercooled water. We think that rotational diffusion of the anion, which becomes comparable to κ_r at low temperatures, leads to a "dynamic steric effect". At the higher temperatures, our treatment [5] is superior over the conventional two-state rate-equation models, such as the classical Eigen scheme [7], whose asymptotic temporal decay is exponential rather than a power-law.

Our model correctly predicts [6] the total (excited+ground-states) dissociation yields of HPTS [1] with no additional free parameters. From the rate-parameters of the transient measurements we estimate [4] the steady-state dissociation rate and pK^* value, with encouraging agreement to older fluorimetric (constant illumination) measurements [8].

In the present work we utilize the excited HPTS molecule as a microscopic probe for the study of electrolyte solutions in the time-domain, with a resolution of ca. 100psec. This timescale is short compared to the diffusional times but longer than the characteristic relaxation of the solvent (water). Therefore the DSE is valid over the entire experimental time-range.

Springer Series in Chemical Physics, Vol. 48: **Ultrafast Phenomena VI**
Editors: T. Yajima · K. Yoshihara · C.B. Harris · S. Shionoya

2. EXPERIMENTAL AND METHOD

Solutions of $5x10^{-5}$M HPTS in 2:1(v) water-methanol mixtures at 60°C were irradiated by the 3rd harmonic, 25 psec pulse, of a mode-locked Nd:Yag laser. The psec time resolved fluorescence of the acidic HPTS form was detected by a streak camera.

The composition of the solution was chosen to maximize the geminate recombination yield. The pH of the solutions was between 5.5 and 6.0 assuring that homogeneous recombination was insignificant in the experimental time-range. The temperature of the solution was raised to 60°C to eliminate the abovementioned "dynamic steric effect". Under present conditions the average number of geminate proton rebindings to the HPTS anion, before ultimate escape to the bulk, is calculated to be 2.7, twice as much as in pure water [4]. Hence geminate recombination is the main process which determines the apparent HPTS dissociation rate in this environment.

3. RESULTS AND DISCUSSION

Fig. 1 shows the effect of an added electrolyte ($NaNO_3$) on the transient dissociation profile of HPTS. The averaged streak camera records clearly show a suppression of the long-time fluorescence tail by the added salt. This is explained by a nonequilibrium screening effect which reduces the Coulombic attraction in the ion pair.

The relaxation time of the ionic atmosphere around a central ion suddenly formed in solution is [9] $2/D\kappa^2$ where D is the relative diffusion coefficient between the ions and $1/\kappa$ is the ionic atmosphere radius. In water at room temperature a solution of 150mM univalent electrolyte will have a relaxation time of 350psec. This is roughly an order of magnitude slower than the time needed for the proton to reach the nearest neighbor ion. The much faster proton can thus be pictured as moving through a largely unrelaxed ionic atmosphere which still has much of its predissociation characteristics. Only when the proton covers the nearest neighbor distance, R_{av}, would it be influenced by the screening of the other ions [10]. This model was found by us to be very simple to use and yet superior to the conventional Debye-Huckel approach, which fails in our case both because of the relatively high electrolyte concentrations (0.01-0.6M) and the nonrelaxed nature of the ionic atmosphere.

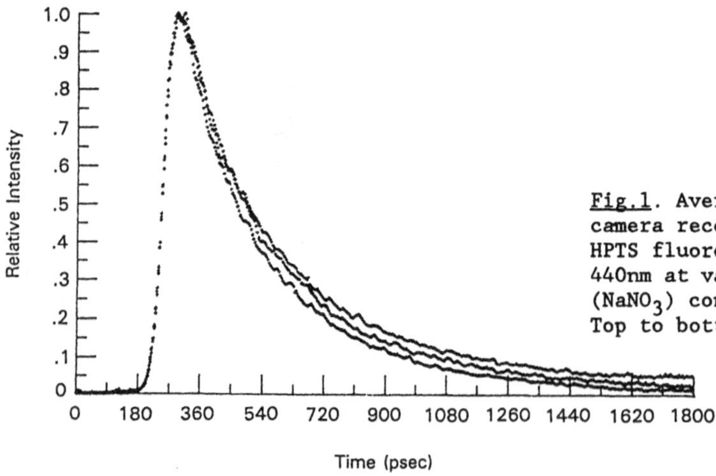

Fig.1. Averaged streak camera recordings of the HPTS fluorescence decay at 440nm at various salt ($NaNO_3$) concentrations Top to bottom: 0,30,150mM.

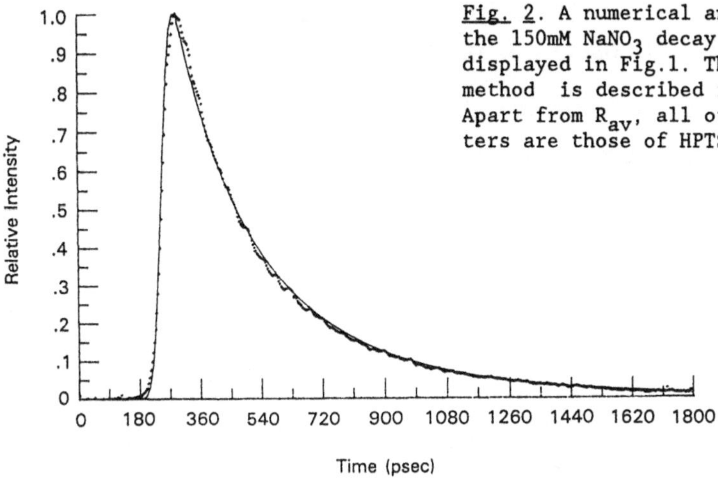

Fig. 2. A numerical analysis of the 150mM $NaNO_3$ decay curve displayed in Fig.1. The analysis method is described in the text. Apart from R_{av}, all other parameters are those of HPTS in pure water.

Fig.2 shows the analysis of the HPTS fluorescence decay in the presence of 150mM $NaNO_3$. The numerical calculation was carried out assuming that no screening of the Coulombic potential exists until the proton has reached the nearest neighbor distance $R_{av}^3 = 3000/4\pi cN_A$ where N_A is Avogadro's number and c is the concentration of the oppositely charged ion in moles-per-liter. At larger distances the Coulombic potential is assumed to be totally screened. Although a good fit could be achieved without any adjustable parameters, we allowed ourselves some freedom in order to achieve the best fit. In Fig.2 R_{av} is taken as 16Å compared to a calculated value of 14Å.

In conclusion, geminate recombination of protons in excited-state acid dissociation gives access to the dynamical properties of electrolyte solutions. Further work is in progress aiming to bring new experimental and theoretical insights into this much investigated yet largely unresolved research field.

REFERENCES

1. H.P. Haar, U.K.A. Klein and M. Hauser: Chem. Phys. Lett. 58, 525 (1978)
2. E. Pines and D. Huppert: J. Phys. Chem. 87, 4471 (1983)
   J. Chem. Phys. 84, 3576 (1986); Chem. Phys. Lett. 126, 88 (1986)
3. D. Huppert and E. Pines: In Ultrafast Phenomena V, edited by G.R. Fleming and A.E. Siegman (Springer, Berlin, 1986), p.315
4. E. Pines, D. Huppert and N. Agmon: J. Chem. Phys. 88, 5620 (1988)
5. N. Agmon, E. Pines and D. Huppert: J. Chem. Phys. 88, 5631 (1988)
6. N. Agmon: J. Chem. Phys. 88, 5639 (1988); ibid., in press.
7. M. Eigen, W. Kruse, G. Maas and L. De Maeyer: Prog. React. Kinet. 2, 287 (1964)
8. A. Weller: Prog. React. Kinet. 1, 189 (1961); J.F. Ireland and P.A.H. Wyatt: Adv. Phys. Org. Chem. 12, 139 (1976)
9. P. Debye and H. Falkenhagen: Phys. Z. 29, 401 (1928)
10. N. Agmon: Chem. Phys. Lett. 141, 122 (1988)

Barrierless Isomerization in Solution Studied by Pico- and Subpicosecond Spectroscopy

U. Åberg, E. Åkesson, H. Bergström, T. Gillbro, and V. Sundström

Department of Physical Chemistry, University of Umeå, S-901 87 Umeå, Sweden

1. INTRODUCTION

A photoisomerization reaction can be classified according to the shape of the potential surface characterizing the reaction. Two main cases can be distinguished. One case refers to the situation where initial and final states of the reaction are separated by a high potential barrier. A second type of reaction is characterized by a smooth potential surface without a barrier between the reactant and the product conformations. The most recent theory related to this latter situation is the Bagchi, Fleming, Oxtoby (BFO) theory (1) and extensions thereof (2).

TPM molecules have traditionally been considered to represent the barrierless case. In this work we have used picosecond and sub-picosecond pulses to study the isomerization dynamics of two other barrierless molecules, belonging to the cyanine dye family , pinacyanol and 1,1'-diethyl-4,4'-cyanine. The dynamics has been studied as a function of viscosity, temperature and excitation and analyzing wavelength.

2. RESULTS AND DISCUSSION

Figure 1 below summarizes the viscosity dependence observed in the n-alcohol series methanol-hexanol at constant temperature. The isomerization time constant is seen to increase linearly with viscosity within this viscosity interval. A linear viscosity dependence of the reaction rate is predicted by the BFO theory for a pinhole sink at the minimum of the excited-state potential. According to the BFO theory this sink function should also produce a non-exponential decay, the shape of which is excitation wavelength dependent. However, no deviations from single exponential decay were observed at the wavelengths and viscosities studied here. As is discussed below we observe a wavelength dependence of the reaction rate, but this dependence is related to the variation of the analyzing wavelength and not to the excitation light. Consequently, we believe that the effects giving rise to non-exponential decay and excitation wavelength dependence as discussed by the BFO theory may very well exist in the molecules studied here, but they occur on a timescale unaccessible with the presently

Springer Series in Chemical Physics, Vol. 48: **Ultrafast Phenomena VI**
Editors: T. Yajima · K. Yoshihara · C.B. Harris · S. Shionoya

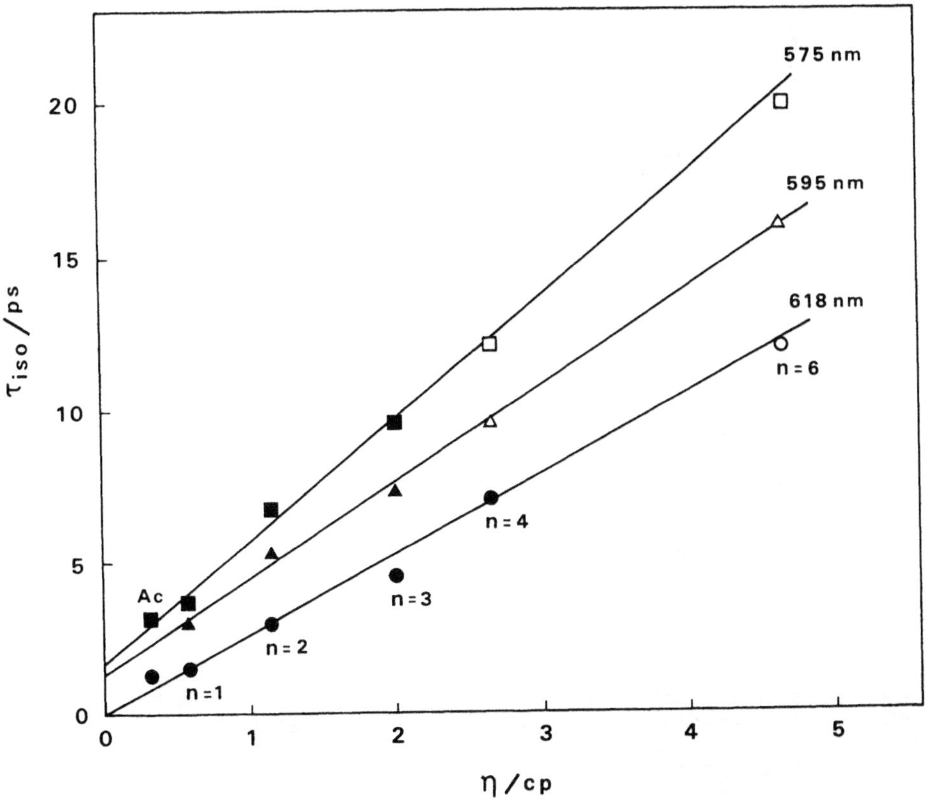

Fig. 1. Viscosity dependence of isomerization time constants

used time resolution. The fact that the isomerization time constant has a linear viscosity dependence down to the shortest value measured, 1.5 ps, at 0.5 cP also suggests that the rate of internal conversion from the sink is very high (probably higher than $10^{13}s^{-1}$) and that the sink is narrow.

Several of the previously studied TPM molecules have been shown to have wavelength dependent relaxation times. This wavelength dependence was shown to be associated with the variation of the probe wavelength in a pump and probe experiment (3,4), but there is no agreement as to the origin of this effect. The two cyanine molecules studied in this work also exhibit this puzzling wavelength dependence (5). In a previous publication (5) we tentatively assigned the effect to the excitation wavelength dependence of the reaction rate predicted by the BFO theory. However, those experiments were performed with a one-color pump-and-probe setup in which the excitation and probe wavelengths vary in tandem. In the present work we have used a setup allowing independent tuning of excitation and probing wavelengths. The wavelength dependence of the isomerization time constant of 1,1'-4,4'-C as measured with the two methods is displayed in Fig. 2. It is evident that the wavelength dependence of the

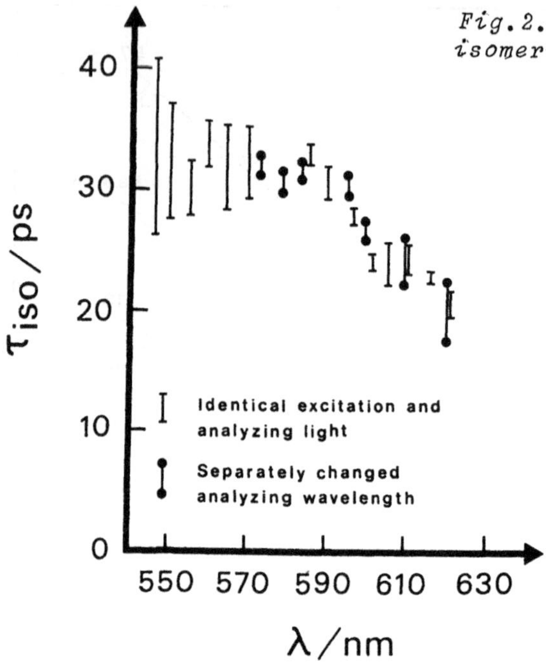

Fig.2. Wavelength dependence of isomerization time constants

isomerization rate is associated to the variation of the analyzing wavelength; tuning the excitation wavelength over most of the absorption spectrum has no effect within experimental error on the reaction rate.

At present we are considering the following possibilities to explain the observed wavelength dependence: A. More than one ground state species with slightly differing relaxation rates exist; B. An effect suggested by Ben-Amotz and Harris (4) to explain the wavelength dependence observed in TPM molecules; C. Ground state vibrational relaxation in a non-reactive mode coupled to the reactive mode; D. Solvent reorganization around the excited molecule. Considering all the experimental facts the possibilities C and D seem to be the most realistic explanations to the effect, but in order to distinguish between the two, more detailed experiments must be performed.

3. REFERENCES

1. B. Bagchi, G.R. Fleming and D.W. Oxtoby: 78, 7375(1983)
2. B. Bagchi, A. Singer and D.W. Oxtoby: Chem. Phys. Lett. 99, 225(1983); B. Bagchi, Chem. Phys. Lett. 115, 209(1985) Chem. Phys. Lett. 139, 119(1987)
3. V. Sundström and T. Gillbro: J. Chem. Phys. 81, 3463(1984)
4. D. Ben-Amotz and C.B. Harris: J. Chem. Phys. 86, 4856(1987); ibid. 86, 5433(1987)
5. E. Åkesson, H. Bergström, V. Sundström and T. Gillbro: Chem. Phys. Lett. 126, 385(1986)

Freezing of an Isomerization Reaction at Phase Transition

J. Korppi-Tommola[1], *A. Hakkarainen*[1], *T. Hukka*[1], *and J. Subbi*[2]

[1]Department of Chemistry, University of Jyväskylä, SF-40100 Jyväskylä, Finland

[2]Institute of Chemical Physics and Biophysics, Academy of Sciences of the Estonian SSR, Lenini Puistee 10, Tallinn 200105, USSR

Isomerization reactions are assumed to be responsible for the non-radiative decays observed in many dye solutions subject to laser excitation. In many of the systems studied isomerization has been taken as a single coordinate minimum energy path. Observed kinetics have been fitted to theoretical models including high and low barriers,random motion of the solvent,the viscosity of the solvent and recently also frequency dependent barrier crossing (1-3). Early developed Kramer's theory seems to work either in low-viscosity or in high-viscosity solutions for medium and high barrier cases.

We have studied the isomerization reaction of a cyanine dye diethyltetramethylindocarbocyanine iodide (DiIC2) dissolved in n-alcohols from methanol to dodecanol. Since the dye is known to have a long-lived photoisomer, absorbing at about 570 nm, the fluorescence spectra were recorded at several excitations (4). No emission, which could be assigned to the photoisomer, was detected.

Single photon counting and multichannel plate photomultiplier detection was used to measure fluorescence lifetimes at several temperatures in each solvent. Solutions were placed in a spinning cell and signals collected at right angles at magic angle polarization using low repetition rates of the synchronously pumped picosecond dye laser. Standing cell and high repetition rate results did not show any cumulative effects one would expect to see from a long-lived photosomer.The observed kinetics showed nice single exponential behaviour throughout solvents and temperatures studied, a strong indication of one emitting species. The observed lifetimes varied from less than 100 ps at low viscosities to about 2.6 ns, which was used as the radiative lifetime of DiIC2, at high viscosities. Some decay curves in butanol solution at various temperatures are shown in Fig. 1a.

In all solvents from methanol to hexanol the observed lifetimes became practically fixed at the freezing point of the solution. In dodecanol, decanol and octanol solutions a sudden jump of the observed lifetimes can be observed at freezing temperatures (Fig. 1b.). This can be taken as evidence of hindering of the isomerization reaction assumed to take place in the excited state (4). The observation can be correlated to the phase transitions of the solvents concerned. With less potential to self-associate long-chain alcohols crystallize more promptly than glass forming and strongly self-associating shorter alcohols.

The Arrhenius plots of the kinetic results are shown in Fig. 2a and the results plotted at constant viscosities are shown in Fig. 2b. The plots at constant viscosities indicate two important results 1) for solutions from methanol to butanol the barrier height E_o seems to depend only slightly on viscosity and an average barrier height of $8.0 kJmol^{-1}$ can be used for methanol through hexanol solutions 2) for solutions from hexanol to dodecanol the barrier height decreases from about $17 kJmol^{-1}$ at 1 cP to about $11 kJmol^{-1}$ at 20 cP. In the latter case E_o is obviously strongly viscosity dependent , a fact, which has to be included in theoretical models used to fit the data. Qualitatively similar results have been reported for the same dye recently (4).

We have tested three different theoretical models on photochemical isomerization of DiIC2 in solution. The observed kinetics of the dye in n-alcohols was fitted first to early Kramers'

Springer Series in Chemical Physics, Vol. 48: **Ultrafast Phenomena VI**
Editors: T. Yajima • K. Yoshihara • C.B. Harris • S. Shionoya

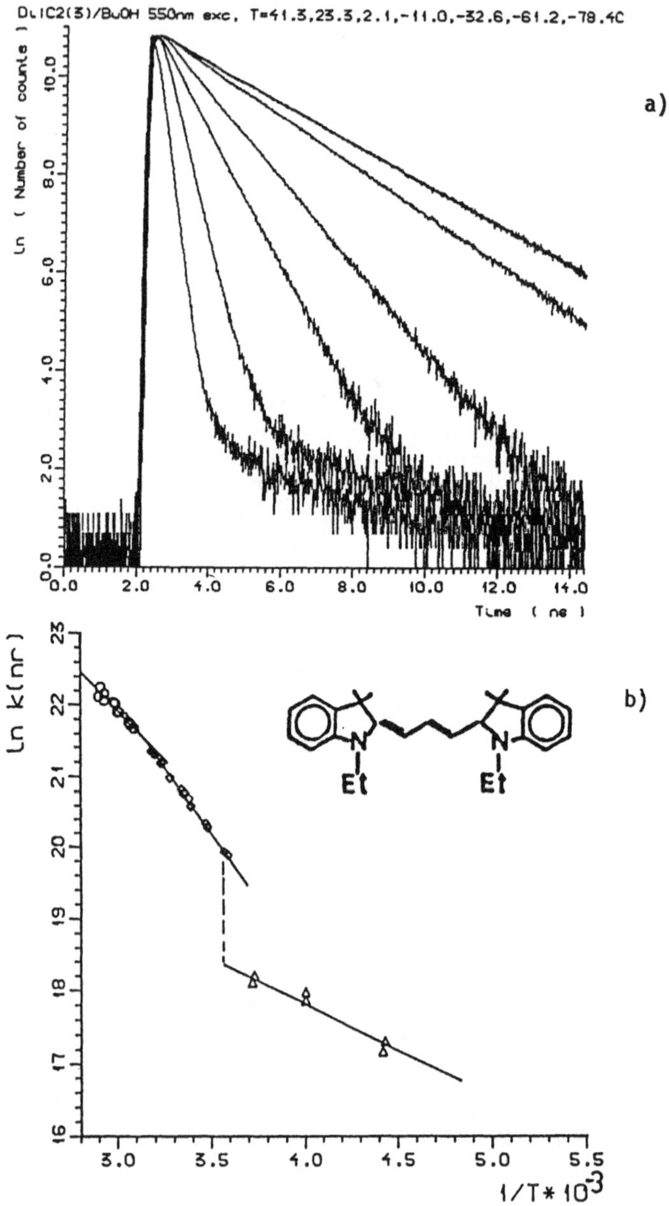

Fig. 1. a) Observed fluorescence lifetimes of DiIC2 in ethanol, b) Arrhenius plot of the non-radiative lifetime in dodecanol at various temperatures

theory, using all experimental points (except those obtained in dodecanol solutions) and viscosity dependent barrier heights, adequately described by a function

$$E' = E_o + \frac{A}{\eta} \tag{1}$$

Fig. 2. a) Arrhenius plots of the DiIC2 dye in n-alcohols

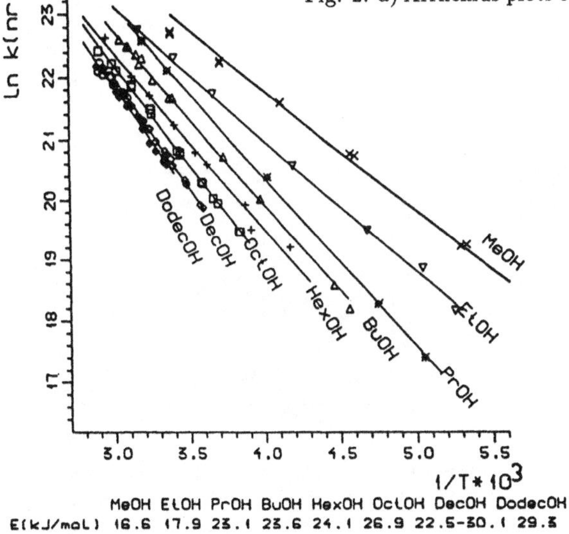

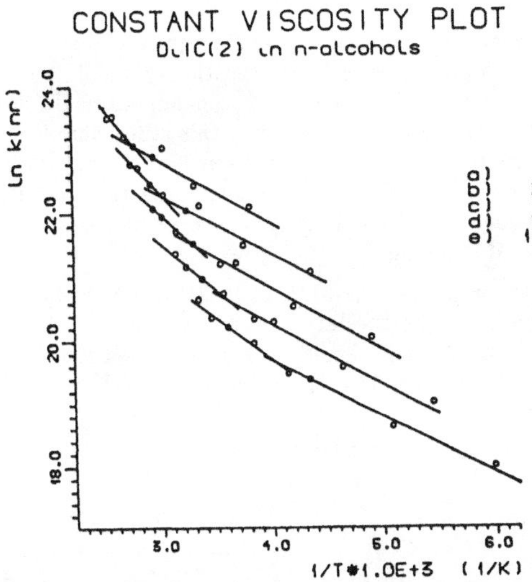

b) The non-radiative constants taken from Arrhenius plots at constant viscosities

where E_o is the lowest barrier height in the viscosity region concerned and A is a constant determined from a least square fit. As seen in Fig. 3a Kramers' theory only slightly underestimates the observed kinetics at high viscosities. Almost equally good fit is obtained using a viscosity independent 'average' barrier height of $8.0 kJmol^{-1}$. The kinetics of hexanol through dodecanol solutions can be nicely fitted to Kramers' equation provided that the strong viscosity dependence of the barrier height is taken into account using equation (1). Kramers' theory predicts nicely the reduced non-radiative rates of DiIC2 in n-alcohols from C6 to C12 (Fig. 3b). Using an average barrier height gives poor results in this case.

Grote and Hynes have brought up the idea of frequency dependent barrier crossing rate and derived a simple equation to account for non-Kramers' behaviour of isomerization reactions (2). Qualitatively, slower rates due to 'viscous drag' are obtained as compared to Kramers' theory

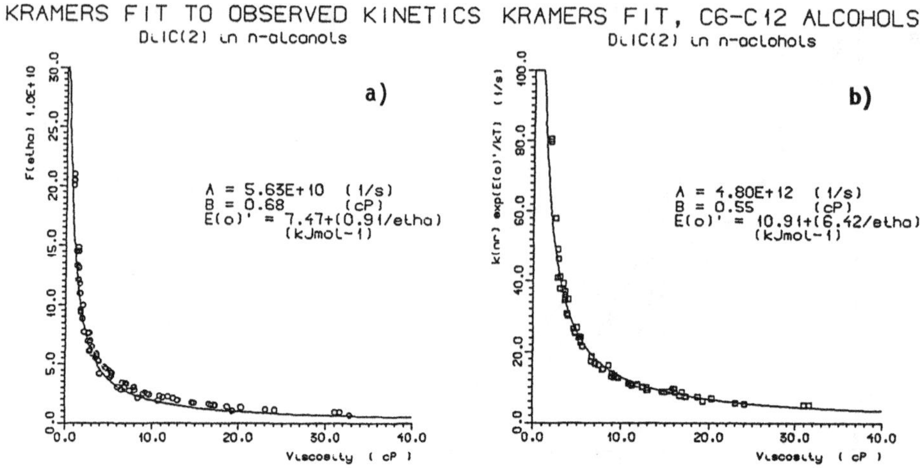

Fig. 3. a) Kramers' fit to the kinetics in all solvents except dodecanol b) Kramers' fit to the kinetics using viscosity dependent barrier heights in solvents from hexanol to dodecanol

at high viscosities. Application of this theory requires information about frequency dependent frictions, not easily available. However, in the high viscosity limit, the theory predicts η^{-1} dependence of the reduced rate constants (5). In the present case η^{-1} dependence could not be confirmed, though the viscosity range, form 1 cP to 35 cP, covered in this study should be representative. A 'modified Smoluchowski equation' $k_{red} = D * \eta^{-\alpha}$ often used to fit isomerization kinetics (5) gives good fits for DiIC2 dye in n-alcohols, as well. Including the viscosity dependence of the barrier height makes the exponent α approach unity. However, this empirical equation has not been given a reasonable physical meaning.

Recently, Robinson et al. (6) published an empirical equation to describe non-radiative kinetics of seven dyes in n-alcanols. In the present case the latter model, which we call the 'extended Kramers' equation' fails to reproduce the observed kinetics. The 'extended Kramers' equation' is not, even in theory, capable of predicting strong viscosity dependencies of the reduced rates obtained in this study.

Finnish Academy of Sciences is gratefully acknowledged for financial support.

References

1) J. Schroeder and J. Troe, in Annual Revue of Physical Chemistry, Eds. H.L. Strauss, G.T. Babcock and C.B. Moore, 38, 163 (Annual Reviews Inc., Palo Alto 1987)

2) R.F. Grote, J.T. Hynes, J.Phys.Chem., 73, 2715 (1980)

3) H.A. Kramers, Physica, 7, 284 (1940), J.L. Skinner and P.G. Wolynes, J. Chem.Phys. 72, 4913 (1980)

4) E. Åkesson, V. Sundström and T. Gillbro, Chem.Phys.Letters, 121, 513 (1985), E. Åkesson, V. Sundström and T. Gillbro, Chem. Phys. 106, 269 (1986)

5) S.P. Velsko, D. H. Waldeck and G.R. Fleming, J.Chem.Phys. 78, 249 (1983)

6) J. Lee, S.-B. Zhu and G.W. Robinson, J. Phys. Chem, 91, 4273 (1987)

Trapping and Solvation of Electrons in Aqueous Media

A. Migus, S. Pommeret, N. Yamada, A. Antonetti, and Y. Gauduel

Laboratoire d'Optique Appliquée, Unité INSERM 275,
Ecole Nationale Supérieure de Techniques Avancées,
Ecole Polytechnique, Batterie de l'Yvette,
F-91120 Palaiseau, France

Introduction

On the occasion of the first of these meetings ten years ago [1], when ultrafast was still picosecond, molecular dynamics of liquids was clearly identified as an important target of ultrafast spectroscopy [2]. The shortening of pulse duration down to the femtosecond time scale, combined with the escalation of computer simulation studies, was in fact required for such a development.

In this paper, we will center on the implication of the femtosecond spectroscopy in the monitoring of the fast elementary steps of electron solvation in aqueous media (homogeneous solutions and organized assemblies) at ambient temperature and we will describe the theoretical framework in which our results are discussed.

Experimental set-up

Our experimental set-up has been already described in detail [3]. It includes a colliding pulse mode locked dye laser with an intracavity four-prism arrangement which generates 60-80 fs duration pulses at a wavelength of 620 nm and a four stage amplifier pumped by the second harmonic of a 10 Hz repetition rate Q-switched neodymium-YAG laser. The amplified beam is separated into two parts. The first one is frequency doubled in a KDP crystal to produce 5-10 μJ pulses at 310 nm while the other one is focused into a 2 mm water cell to produce a white light continuum. This white beam is divided into two parts ; one (the probe) overlaps the pump in the sample while the other one (the reference) misses the sample. Both beams are collected into a monochromator and directed towards two photodiodes. Due to the large spectral region under investigation (typically from 0.4 to 1.3 μm) the chirp of the continuum induces an important shift of the zero (temporal overlap between the pump and the probe). Therefore careful experiments are simultaneously conducted at each wavelength to determine the zero time delay and the time response function of the apparatus [4]. Transient spectra are then reconstructed taking into account the amplitude of the signal and can then be represented in 3-D plots (Fig. 1).

Femtosecond generation of excess epithermic electrons

In the specific case of aqueous solutions at ambient temperature, apart from the discovery of a short-lived hydrated electron [5], very little is known about the mechanisms governing the electron-water molecule interactions during the solvation process. Up to now the great majority of electron solution studies in aqueous solutions have been realized by injecting electrons using pulse radiolysis methods, providing at best a 10 ps accuracy [6]. In this case, the appearance of the hydrated electron was always limited by the instrumental resolution. Flash photolysis permits the direct generation of electrons with low excess energy. In our case we have used 310 nm pulses to generate electrons through the ionization of ferrocyanide $Fe(CN_6)^{-4}$ ions in water solution (0.45 μM) or to induce charge separation in aromatic hydrocarbon molecules with

Springer Series in Chemical Physics, Vol. 48: **Ultrafast Phenomena VI**
Editors: T. Yajima · K. Yoshihara · C.B. Harris · S. Shionoya

a low ionization threshold, (such as phenothiazine), embedded in the hydrophobic core of micelles [7]. We found also that direct photoionization of water molecules was possible with 310 nm light even though water is transparent up to 190 nm. This excitation is likely due to two-photon absorption but in this case the global yield is much lower than in the previous ones, making the experiment quite difficult. However there are several advantages to photoionizing the neat solvent, avoiding for instance the perturbation of the microscopic environment by any solute molecule.

After the initial ionization event, the subsequent energy exchanges result in the formation of several intermediates states including quasi-free or dry electrons (e^-_{qf}), thermalized electrons (e^-_{th}), localized electrons (e^-_{loc}) and finally solvated electrons (e^-_{sol}).

$$\text{Donor(D)} \xrightarrow{h\nu} D^+ + e^-_{qf} \longrightarrow e^-_{th} \longrightarrow e^-_{loc} \longrightarrow e^-_{sol} \qquad (1)$$

Results in bulk liquid water and isotopes

In H_2O we find that a high energy tail of an infrared absorption band extending above 1250 nm begins to appear within the excitation but is fully developed only after a delay of 0.4 ps. This infrared band then disappears in a fraction of picosecond while a structureless absorption band rises. This asymmetric band which peaks around 1.7 eV (720 nm) with a high energy tail extending above 3 eV is clearly assigned to the hydrated electron. All this behavior can be easily followed on the 3D plot which has been reconstituted from the kinetics at each wavelength.

An analysis of our kinetics data (Figure 1) shows that they are mostly consistent with a model which assumes the existence of two species [8]. The first

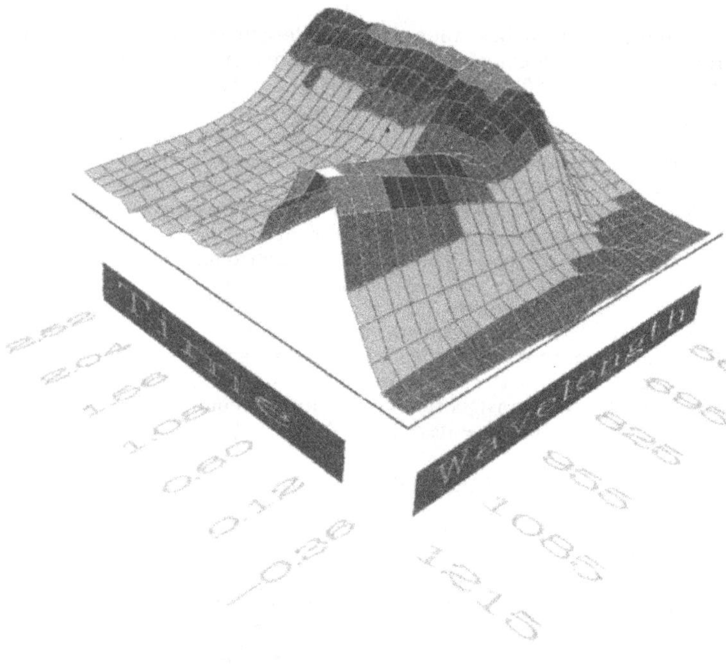

Fig. 1. Transient absorption spectra of excess electrons in pure liquid water excited at 310 nm. Wavelength is in nm and time in ps. This 3D plot is reconstituted from the experimental kinetics taken at different wavelengths from 500 nm up to 1250 nm.

one, absorbing in the infrared, does not appear instantaneously but with a characteristic time T_1 (110 $\pm$ 20 fs) and then relaxes with a constant time T_2 (240 $\pm$ 20 fs) towards the solvated state.

These dynamical investigations demonstrate that the electron solvation in pure liquid water proceeds through, at least, one intermediate state (presolvated state : e^-_{loc}) characterized by an infrared band and whose lifetime is in the femtosecond régime. These studies provide a unique experimental basis for testing recent theoretical approaches to excess electron coupling in liquid water. The measured initial trapping time of the electrons (1.1×10^{-13} s) is longer than the estimates of thermalization time in water [$(2.4 - 4) \times 10^{-14}$ s] [9]. This would suggest that if the early energy loss rate of the electrons in water is about 10^{13} eV/s, the electrons would get thermalized before being localized [9]. Important theoretical investigations on electron localization and solvation have been performed using computer simulation methods. Several groups have deduced information on the structural aspects of the hydrated electron by developing molecular dynamic simulations of water using Monte Carlo and Feynman path integral techniques [10-12]. In models based on electron-water pseudopotentials, the electron is described as a quantum particle and water molecules are treated classically using SPC, ST_2 or MCY models.

Computer simulations show a repartition of favorable sites for the initial electron localization in liquid water [12]. These sites are attributed to fluctuations of the electronic density and of solvent molecules' orientation, and correspond to a range of potential well depth extending from 0 to -32.2kcal/mol. Our femtosecond experiments cannot establish if the electron creates its own trapping site (self-trapping mechanism) or searches for pre-existing shallow traps identified as small structural fluctuations or miniclusters. However, in agreement with a recent plausible picture of electron trapping in water [9,12], our IR data are compatible with a distribution of pre-existing sites which behaves monotonically in energy down to - 1.4 eV. The efficient role of shallow traps in the appearance of a presolvated state (i.e. $V_o \simeq 0.58$ eV for a cross section around 20 Å^2) would correspond to spatially extended electron-traps of about 4 angströms.

Turning to the final transition to the fully solvated state (e^-_{sol}), it is assumed that the ultrafast relaxation of the infrared species (240 fs) corresponds to a contraction of the trap size associated with a modification of the charge distribution inside the cavity of the hydrated electron. Numerous theoretical studies have led to the conclusion that once the electron gets localized, a reorientation of the water molecules in the first solvation shell should intervene. Theories treating liquid water as a dielectric continuum predict that the shortest lifetime of a trapped electron is determined by the longitudinal relaxation time of the solvent τ_L [13,14]. From the fixed charge approximation, τ_L is related to the Debye relaxation time by the expression $\tau_L = \tau_D \epsilon_\infty / \epsilon_o$, in which ϵ_o and ϵ_∞ are the static and the high frequency dielectric constants respectively. A dielectric relaxation implies a continuous modification of the medium interacting with the electron. This should appear as a continuous spectral shift of the electron absorption band from the infrared towards the visible [14]. Even though there is an analogy between the time T_2 and the estimate of the longitudinal one (0.2 - 0.5 ps) the absence of a significant spectral shift between the two transitions of the electron (infrared and visible) suggests that the configurational changes of water molecules around the electron are limited to an extremely small number of movements and favours electronic transition of the electron during the final liquid relaxation.

The exhaustive identification and comprehension of localized and fully relaxed states of electrons in liquid water evidently need further experimental data on coupling of subexcitation electrons with water molecules and theoretical treatments of electron solvation as a non-adiabatic process.

Excess electrons in sequestered aqueous microdroplets

Reversed micellar systems (aerosol OT - hydrocarbon - water systems) offer the opportunity to investigate electron hydration in unique aqueous media where the microstructure and states of sequestered water bubbles is dependent on the water pool size defined as the $[H_2O]/[AOT]$ ratio (W) [15]. At low water content ($W \simeq 20$), strong ion-dipole interactions lead to a specific organization of water molecules around the sodium ion and provide favourable distribution of ionic clusters $[Na(H_2O)_n^+]$ within the inner aqueous core. In this section we will discuss only some of the recent data on electron solvation in large aqueous microdroplets (W = 50). The relevant results are summarized in Figure 2. In reversed micelles (RM) without a water pool (W = 0), the risetime of the induced absorption at 720 nm can be well fitted to a build-up time of 80 ± 10 fs (curve b). This time, taking into account the position of the zero time delay (curve a), is assigned to the formation of a trapped state of electrons in the non-polar phase of RM. The time-resolved spectroscopic data obtained following the photo-ionization of PTH in RM (W = 50) can be fitted to the convolution of the pulse profile and the signal rise which is made of a contribution of two components (curve c). The first component corresponds to the appearance of localized electrons in the hydrocarbon phase of RM. The second time constant has been determined from experiments performed in the infrared (Figure 2B). At 1000 nm, the kinetical model takes into account the existence of a transient state of the electron before solvation in the water bubble. The best fit of the infrared kinetics gives an appearance time T_1 of 140 fs and a lifetime of 270 fs [15]. These femtosecond investigations in RM demonstrate that the solvation of photo-electrons in water microdroplets does not proceed in a single step. The rise-time T_1 of the infrared tail includes both the electron ejection thermalization and localization inside the aqueous inner core of RM.

Once the electron gets trapped, the final stabilization towards a fully solvated state occurs in 270 fs. These data, showing the existence of a precursor of solvated electrons in sequestered aqueous microdroplets, demonstrate that the presence of counterions Na^+ does not provide more favourable spatial distribution of pre-existing deep traps in RM than in pure liquid water. It is interesting to notice that in large water droplets (W = 50) the lifetime of this presolvated state is similar to the value obtained in bulk water. Additional femtosecond investigations are in progress to establish if the mechanisms of electron localization and solvation in smaller water clusters are influenced by the properties of the sequestered interfacial water.

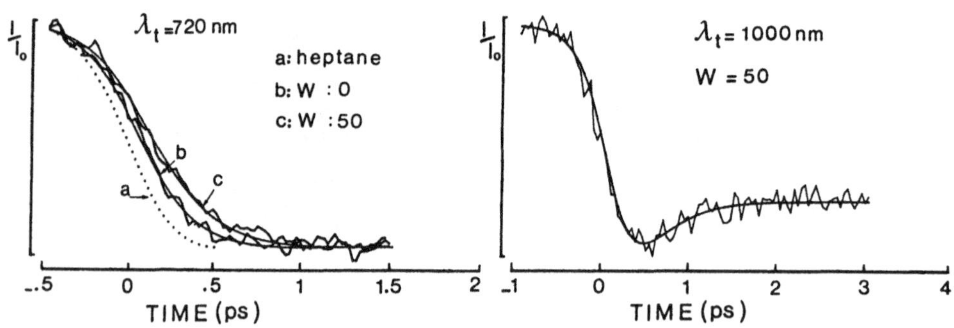

Fig. 2. Time-resolved induced absorption following femtosecond UV excitation of phenothiazine (4×10^{-4} M)in reversed micelles (AOT, 0.15 M, n-heptane).

Acknowledgments. The authors acknowledge Olivier Gallant for assistance in realizing 3D plots of transient spectra.

References

1. "Picosecond Phenomena", Eds. C.V. Shank, E.P. Ippen, S.L. Shapiro (Springer-Verlag, Berlin, Heidelberg, 1978).
2. G.A. Kenney-Wallace, ref. 1, page 208.
3. A. Migus, A. Antonetti, J. Etchepare, D. Hulin, A. Orszag, J. Opt. Soc. Am. B, **2**, 584, 1985.
4. Y. Gauduel, A. Migus, J.L. Martin, Y. Lecarpentier, A. Antonetti, Ber. Bunsenges. Phys. Chem., **89**, 218, 1985.
5. E.J. Hart, J.W. Boag, J. Am. Chem. Soc. **84**, 4090, 1962.
6. C. D. Jonah, J.R. Miller, M.S. Matheson, J. Phys. Chem., **81**, 1618, 1977.
7. Y. Gauduel et al., Revue Phys. Appl. **22**, 1755, 1987; Biochemistry **27**, 2509, 1988; Rad. Phys. Chem. (in press).
8. Y. Gauduel, J.L. Martin, A. Migus, N. Yamada, A. Antonetti : In "Ultrafast Phenomena V" Eds. G.R. Fleming and A.E. Siegman, (Springer-Verlag, Berlin, Heidelberg, p. 308, 1986).
9. A. Mozumder. Rad. Phys. Chem. (in press).
10. G.D. Jonah, C. Romero, A. Rahman, Chem. Phys. Lett., **123**, 209, 1986.
11. A. Wallqvist, D. Thirumalai, B.J. Berne, J. Chem. Phys., **183**, 1986.
    A. Wallqvist, G. Martyna, B.J. Berne, J. Phys. Chem., **92**, 1721, 1988.
12. J. Schnitker, P.J. Rossky, G. Kenney-Wallace, J. Chem. Phys. **85**, 2986, 1986 ; J. Schnitker and P.J. Rossky, J. Chem. Phys. **86**, 3462, 1987, P.J. Rossky and J. Schnitker, Phys. Chem. (in press).
13. D.F. Calef and P.G. Wolynes, J. Phys. Chem. **87**, 3387, 1983.
14. L.D. Zusman and A.B. Helman, Chem. Phys. Lett. **114**, 301, 1985.
15. Y. Gauduel, S. Pommeret, A. Antonetti : In "Structure and reactivity of reversed micelles", Pileni Ed. (Elsevier, Amsterdam), in press.

Dynamics of Polar Solvation

G.R. Fleming[1] *and M.P. Maroncelli*[2]

[1]Department of Chemistry and the James Franck Institute, The University of Chicago, 5735 S. Ellis Avenue, Chicago, IL 60637, USA

[2]Department of Chemistry, 152 Davey Laboratory, Penn State University, University Park, PA 16802, USA

The dynamics of polar solvation has become a topic of recent interest because of its importance in determining how a solvent will affect charge transfer reactions [1]. The results of experiments performed by us [2] and several other groups [3] have led to the conclusion that theoretical models that treat the solvent as a simple continuum fluid are inadequate to describe the actual dynamics. The experimental results deviate in three significant ways from the predictions of Debye-Onsager homogeneous continuum models. First, the average time scales observed are usually larger than τ_L, the longitudinal relaxation time, and lie somewhere intermediate between the τ_L and τ_D extremes. (Here τ_D is the measured dielectric response time). Second, in cases where the solvent relaxation could be followed over several decay times with a large signal to noise ratio, the decays were clearly nonexponential, resembling stretched exponential decays. Finally, at large values of the dielectric constant the response slows significantly. We have suggested that these three findings result from molecular aspects of the solvent response not included in the continuum description. As has been pointed out by several authors [4-6] for molecular scale solutes, nearest neighbor solvent molecules will not enjoy the full cooperativity implied by τ_L and will react on a time scale closer to τ_D. The response from the distant solvent regions will have the rapid τ_L-type response and so the overall response will therefore be slower and more complex than that predicted by the simple continuum response. Several years ago Calef and Wolynes [4] using a Smoluchowski-Vlasov equation discussed many of these ideas. More recently Wolynes [7], followed by Rips et al. [8,9] and Nichols and Calef [10] have developed the well-known mean spherical approximation (MSA) to describe the dynamics. The MSA is a treatment of a dipolar hard sphere solvent which obtains an approximation to the radial distribution function from first principles. It thus constitutes a molecular theory of solvation dynamics. In a parallel study we developed an inhomogeneous continuum model [11] in which the finite size of molecules is mimicked by a distance dependent dielectric function $\varepsilon(r)$. For particular choices of $\varepsilon(r)$ the MSA and inhomogeneous continuum models give essentially identical behavior. One significant aspect of both theories is that ions solvate more rapidly than dipoles whereas in the homogeneous continuum picture the solvation of both species has the same time constant. The origin of this behavior lies in the range of the interaction. Simply put, the ions reach out further and thus the overall response is more heavily biased by the τ_L-type contribution. A third approach is that of molecular dynamics simulation where the individual molecules are dealt with explicitly. In this paper we discuss our data with respect to the MSA model, the inhomogeneous continuum model and describe some of the simulation results.

Experimentally we exploit the time resolved fluorescence Stokes shift to construct a correlation function $C(t) = (\nu(t) - \nu(\infty))/(\nu(00 - \nu(\infty))$. Theoretical correlation functions are given the symbol $S(t)$. Figure 1 shows

Springer Series in Chemical Physics, Vol. 48: **Ultrafast Phenomena VI**
Editors: T. Yajima · K. Yoshihara · C.B. Harris · S. Shionoya

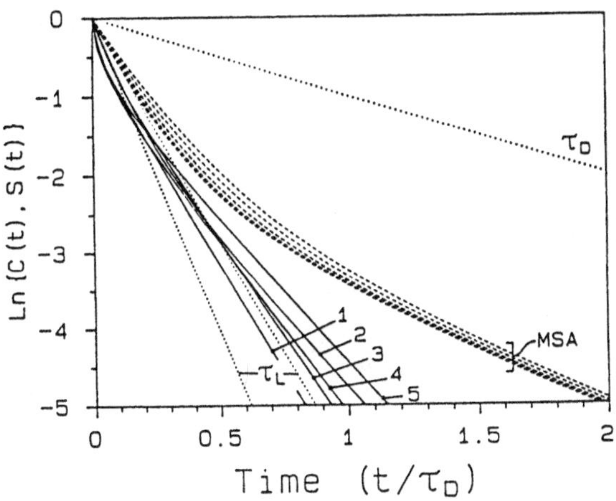

Fig. 1. Comparison of experimental C(t) response functions (solid curves) with the corresponding S(t) functions calculated from the MSA model (dashed curves). The experimental data are from Ref. 2 and the parameters used to calculate the S(t) curves are listed in Table I in Ref. 12. The curves shown correspond to the following solvents and temperatures. Experimental: n-propanol at 1=295 K. 2=273 K, 3=251 K, 4=232 K, and 5=221 K; calculated: 1,2,3,4,5 from top to bottom. For reference, the limiting exponential τ_D and τ_L decays are shown as dotted lines. τ_L varies slightly with solvent and temperature; the pair of lines shown here are the extreme values.

experimental data for Coumarin 153 in n-propanol over the temperature range 221 K-295 K. The time scale is given in units of t/τ_D. Thus if a simple continuum model applied all the curves would lie on top of each other and follow the line marked τ_L (in fact τ_L/τ_D varies slightly with temperature - the pair of dotted lines show the extreme values). The MSA model predictions are also very similar over the temperature range but clearly contain substantially more curvature than the real data. The MSA calculation in Fig. 1 is for an ion. A recent calculation by Rips et al. [9] for a dipole does not improve the quality of the description. In protic solvents Rips et al. [9] suggest that structure breaking in the vicinity of the solute is responsible for this failure.

Both the MSA (for dipoles) and the inhomogeneous continuum models do much better at reproducing the data in aprotic polar solvents. Figure 2 shows the inhomogeneous continuum calculation for Coumarin 153 in n-methyl proprionanide at 244 K and 273 K. Note that the relaxation is now very much slower than τ_L. The calculated curve uses an inverse exponential form for $\epsilon(r)$ with a single length parameter for both temperatures [11]. The same value of λ predicts the correct time dependence over a wide range of dielectric constants (80-300) and temperatures in different solvents. The scale of the perturbed solvent region for a dipole interaction corresponds to about two solvent shells.

We now turn to a discussion of the molecular dynamics results [12]. The systems investigated consisted of monatomic ions immersed in large spherical clusters of ST2 water. Relaxation of the solvation energy following step function jumps in the solute's charge, dipole moment and quadrupole moment were determined from equilibrium molecular dynamics simulations under the assumption of a linear solvent response. The effect of charge (+, 0, -) and size (small (S) with Lennard-Jones parameter σ = 3.1 Å and large (L) σ = 7.0 Å) was also investigated. With the exception of the S+ solute the ordering of the decay rate of the correlation functions is potential (charge) > field gradient (quadrupole) > field (dipole). In addition to the faster decay the potential correlation functions show pronounced librational oscillations that are much less pronounced in the other correlation functions. Comparing the simulated correlation functions with the MSA predic-

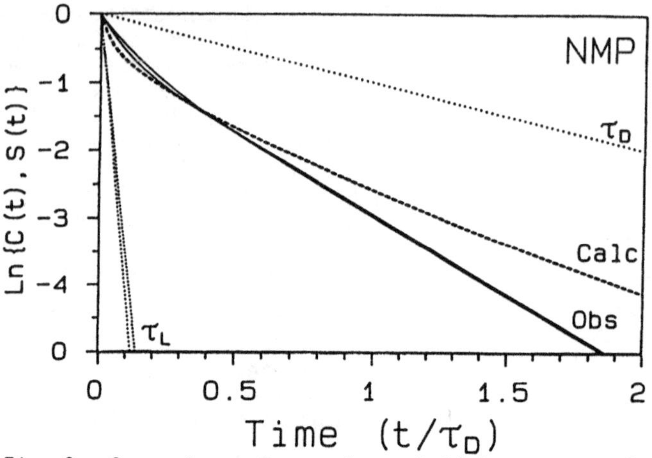

Fig. 2. Comparison of experimental C(t) responses (Eq. 1-5, data from Ref. 2) and S(t) curves calculated from the inverse exponential model of ionic solvation. The experimental data were obtained with Coumarin 153 as the probe molecule in N-methylpropionamide (244 and 273 K). The calculated curves are for ionic solvation using values $\varepsilon_1 = \varepsilon_\infty$ and λ = 1.83. The parameters used in single Debye form representations of $\varepsilon(\omega)$ of the experimental solvents are, in the order ε_B, ε_∞, τ_D (ps) at 244K:299, 6, 1890, and at 273K: 215,6,392 [13].

tions as with the experiments in alcohols leads to slower calculated than "observed" decays. A simple continuum model more closely approaches the simulated results but neither theory is able to account for the variations we see in potential correlation functions for different solutes. The main reason for the failure of the MSA approach is that the dynamics is dominated by the first shell as Fig. 3 shows. Also shown in Figure 3 is the single particle time correlation function for molecules in the first shell. The contribution of any one molecule to the potential tcf decays on this time scale. Thus the fast response clearly results from concerted motion of many molecules. The number of molecules actually required to achieve the

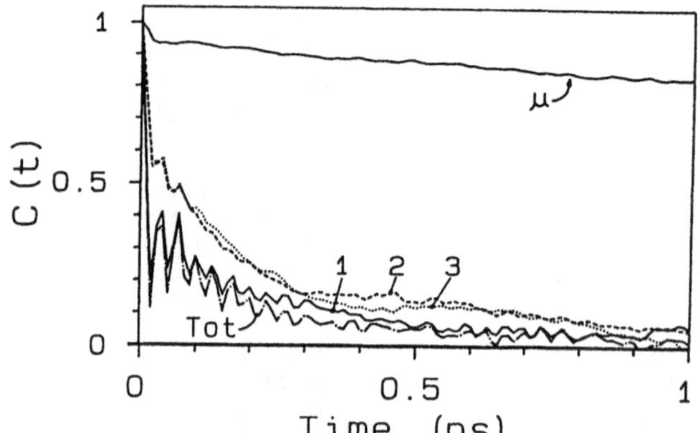

Fig. 3. Potential time correlation function for a SO solute decomposed into individual shell contributions. The solid curve is the total response and the curve marked μ is the single particle dipole tcf.

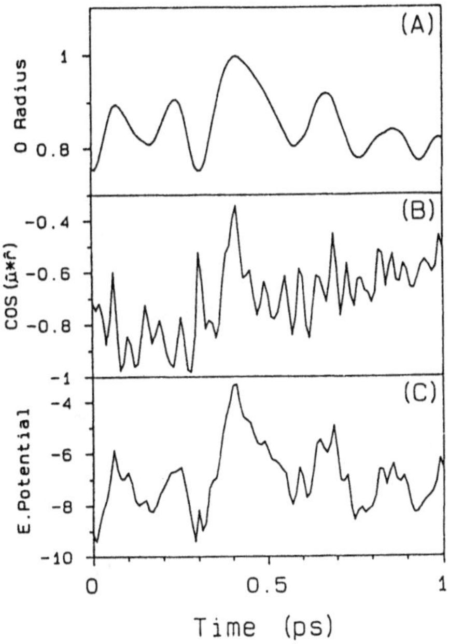

Fig. 4. Properties of the trajectory of a single solvent molecule in the first shell of an S+ solute. A. Solute-oxygen distance. B. cos(μ · r) where is the dipole direction and r is the solute-oxygen vector. C. The single molecule contribution to the oxygen potential.

full cooperativity is 10-40. The first shell dominates the response, contributing 85% of the zero time amplitude, and producing the large oscillations which appear to result from a collective librational motion of all the molecules in the first shell. A further significant point in the first shell dynamics is that the potential tcf differs from the polarization tcf indicating that the solute does not view solvent molecules as single point charges but rather interacts with their extended charge distribution.

The last aspect we discuss is the molecular basis for the differences observed between charged and uncharged species. Figure 4 shows the behavior of a first shell molecule for an S+ solute. Panels A and B give information on translation and rotation, respectively. For an uncharged solute (e.g. S0) the fluctuations in the electrical potential arise almost entirely from the rotational motion. However, as panel C shows for S+, translational motions play an important role in potential fluctuations. Waters in the first shell of S+ point one of their negative charges directly at the solute and are held rather tightly. This tight binding is manifest in the coherent vibrational character of the solute-oxygen distance shown in Fig. 4A. Thus while the spikes in Fig. 4C do come from motion that is rotational in character the overall contour of the potential trajectory closely matches excursions in the oxygen to solute distance. Thus the much slower potential tcf response for S+ as compared to S0 results from the importance of slower translational mechanisms in the former but not the latter solute.

Thus the results of this simulation do not lend support to the MSA picture of solvation dynamics. Of course water is one of the most highly structured solvents and its dynamics may well be more complex and solute specific than other polar solvents. Simulations are clearly needed on simpler liquids such as acetonilirile.

This work was supported by a grant from the National Science Foundation. We thank Edward Castner, Jr. and Biman Bagchi for their contributions to this work.

References.

1. See for example, I Rips and J. Jortner, J. Chem. Phys. 87, 2010 (1987), H. Sumi and R.A. Marcus, ibid 84, 4894 (1986) and references therein.
2. E.W. Castner, Jr., M.P. Maroncelli and G.R. Fleming, J. Chem. Phys. 86, 1090 (1987).
   M.P. Maroncelli and G.R. Fleming, ibid 86, 6221 (1987).
3. S.G. Su and J. Simon, J. Phys. Chem. 91, 2693 (1987). V. Nagarajan, A.M. Brearley, T.J. Kang and P.F. Barbara, J. Chem. Phys. 86, 3183 (1987).
4. D. Calef and P.G. Wolynes, J. Chem. Phys. 78, 4145 (1983).
5. V. Friedrich and D. Kivelson, J. Chem. Phys. 86, 6425 (1987).
6. R.F. Loring and S. Mukamel, J. Chem. Phys. 87, 1272 (1987).
7. P.G. Wolynes, J. Chem. Phys. 86, 5133 (1987).
8. I. Rips, J. Klafter and J. Jortner, J. Chem. Phys. 88, 3246 (1988).
9. I. Rips, J. Klafter and J. Jortner, J. Chem. Phys., in press.
10. A.L. Nichols and D. Calef, J. Chem. Phys, in press.
11. E.W. Castner, Jr., G.R. Fleming, B. Bagchi and M. Maroncelli, J. Chem. Phys., in press.
12. M.P. Maroncelli and G.R. Fleming, J. Chem. Phys., in press.
13. S.J. Bass, W.I. Nathan, R.M. Meigham and R.H. Cole, J. Phys. Chem. 68, 509 (1964).

Ultrafast Molecular Dynamics in Solvating Liquids

W.T. Lotshaw, D. McMorrow, C. Kalpouzos, and G.A. Kenney-Wallace

Lash Miller Laboratories, Department of Chemistry,
University of Toronto, Toronto M5S 1A1, Canada

The concepts underlying molecular motion, and the correlated dynamics involved in solvation phenomena, have undergone renewed scrutiny in the past decade as we have witnessed a transition from picosecond to femtosecond laser spectroscopy [1]. The ability to monitor *directly* time-ordered events makes femtosecond spectroscopic techniquies viable for attacking the fundamental problems of molecular motion in disordered, condensed phase systems which could be addressed previously only with frequency domain techniques. In this paper we present new optically heterodyned optical Kerr effect (OHD-OKE) data on several room temperature molecular liquids and discuss briefly the implications of these results on models of electron localization and solvation [2], and on the dynamic solvation of electronically excited states. The laser system and experimental details have been discussed at length elsewhere and will not be repeated here [3].

Figure 1 illustrates schematically the scattering of a quasi-free electron through a disordered medium, in which the fluctuations in free energy as a function of medium density are reflected in the extended array of potential energy wells of varying depth and shape. The model for electron localization proposes that the electron samples the statistical structure of the medium that is frozen on this time scale, with the distribution of potential wells formed by the van der Waals or dipole induced clustering of molecules providing a distribution of preexisting localization sites. All evidence indicates that a

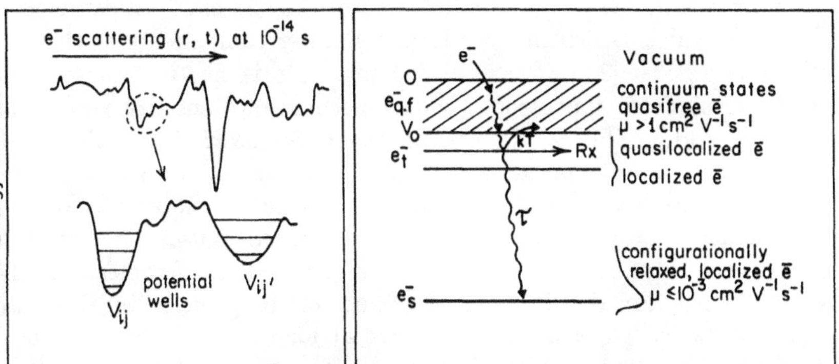

Figure 1. Schematic diagram for the trapping and solvating of a quasi-free electron

Springer Series in Chemical Physics, Vol. 48: **Ultrafast Phenomena VI**
Editors: T. Yajima • K. Yoshihara • C.B. Harris • S. Shionoya

minimum depth and cluster size is required in order to localize the electron below the conduction band of the medium (cf., fig. 1), and that residence time as a trapped electron must persist into the reorientation time of the molecules for solvation to occur. Hence, solvation about the initially trapped electron, e_t^-, can lead to a configurationally relaxed solvated electron, e_s^-. In alcohols, for example, this solvation time is typically 10^{-11} to 10^{-9} s [5]. It is thus relevant to ask, what are the dynamical structures present at the earliest times? How does one describe the response of the molecules to a sudden external perturbation, and how does the inhomogeneity of the liquid's local structure affect the measured response? Femtosecond spectroscopic techniques provide a powerful tool to address these issues.

In previous reports we have noted striking similarities in the dynamics of all liquids composed of anisotropic molecules. In particular, the OHD-OKE response of such liquids can be resolved into four dynamically distinct temporal regions, three of which originate in motions of the nucleii, with the fourth arising from the instantaneous distortion of the electronic polarizability to the applied optical field [3,4]. While the qualitative appearence of the data for different liquids may be quite different owing to different amplitudes of the different signal contributions, all can be resolved into these four dynamical regions and described in a qualitative fashion by a single, simple model [3,4]. Furthermore, it has been demonstrated for CS_2 that the exponentially decaying contribution with a lifetime of >1 ps behaves in a manner that is consistent with Debeye-Stokes-Einstein diffusive reorientation, while the subpicosecond nuclear relaxations appear to be intrinsically non-diffusive, arising from coherently-driven vibrational modes of the local fluid structure [3,4,6]. In liquids consisting of anisotropic molecules this intermolecular vibrational motion is dominated by oscillatory orientational motion (*molecular libration*) of the molecules in the local intermolecular potential wells.

Figures 2 gives the OHD-OKE responses for *neat*-pyridine and *neat*-bromoform at 298 K initiated with 65 fs pulses centered at ~633 nm. The bromoform data (fig. 2, lower) is dominated by a large electronic response centered at $\tau = 0$, and the coherent oscillation of the 154 cm^{-1} *intra*molecular vibrational mode which is driven by different Fourier components of the spectrally broad laser pulse [3,7]. Because this molecule is nearly isotropic, the signal contributions from anisotropic *inter*molecular motions are small. The OHD-OKE for pyridine (fig. 2, upper), on the other hand, is dominated by these intermolecular, anisotropic motions. The initial, rapid decay of the signal arises from the destructive interference of the different oscillators (molecular librators), all excited *in phase* by the ultrashort optical pulse. The rapid dephasing is a consequence of the inhomogeneous distribution of molecular environments, and hence the distribution of oscillation frequencies in these disordered liquid systems. The non-zero intercept at long times in both curves is indicative of a residual anisotropy of molecular orientations (an *orientational anisotropy*), the decay of which arises from diffusive reorientational motion.

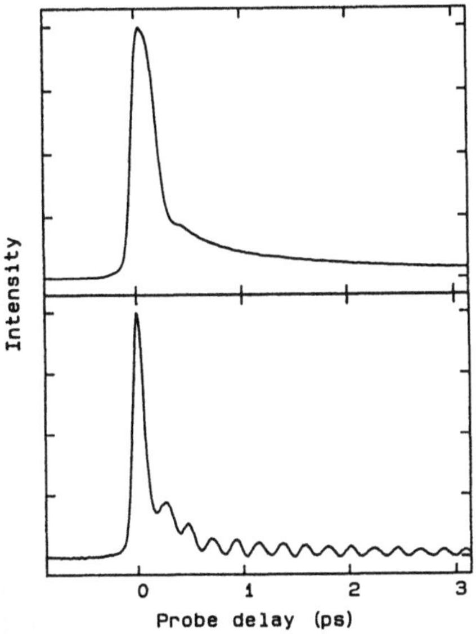

Figure 2. OHD-OKE data for pyridine (upper) and bromoform (lower) liquids at 298 K

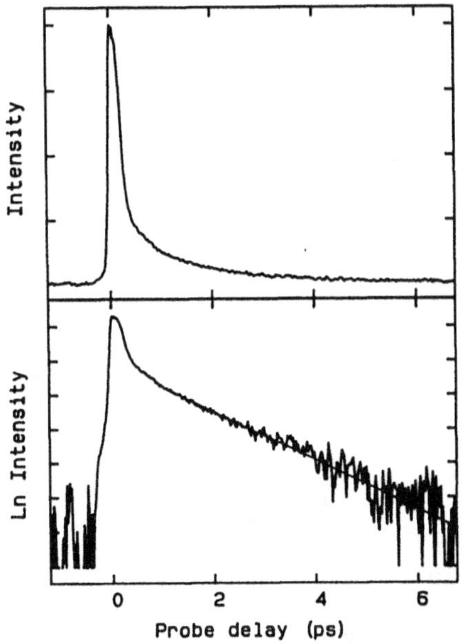

Figure 3. OHD-OKE data for acetonitrile liquid in linear and semi-log formats

Because of its importance in studies on dynamic solvation, we focus at this point on the ultrafast dynamics of acetonitrile. Figure 3 gives the raw OHD-OKE data for *neat*-acetonitrile, presented in linear and semi-log formats. Evident in this data, following the initial (instantaneous) rise of the electronic response, is the rapid signal decay associated with the inhomogeneously-broadened ensemble of coherently-excited molecular librators, followed by the slower decay of the orientational anisotropy. From a least-squares fit to the linear portion of the semi-log data ($\tau > 1$ ps, linear over > four lifetimes) a time constant of 1.4 ps is obtained for diffusive reorientation of acetonitrile.

The short-time dynamics of acetonitrile are investigated further in fig. 4. The upper panel presents again the raw OHD-OKE data. The lower panels present (in linear and semi-log formats) reduced data in which a tail-matched, exponentially decaying response generated to reproduce the 1.4 ps decay of the orientational anisotropy has been subtracted off [3]. The complexity of the dynamics at short times is evident, with the linear portion of the lower panel fitted to a straight line with a time constant of ~400 fs. This relaxation has been discussed [4] in terms of local density fluctuations which destroy the local structure of the liquid on a ~400-600 fs time scale. The observed dynamics in acetonitrile are consistent with those of numerous anisotropic molecular liquids investigated to date [3].

The residual OHD-OKE data of fig. 4 (middle and lower) contains contributions from both the instantaneous electronic response and the ensemble of

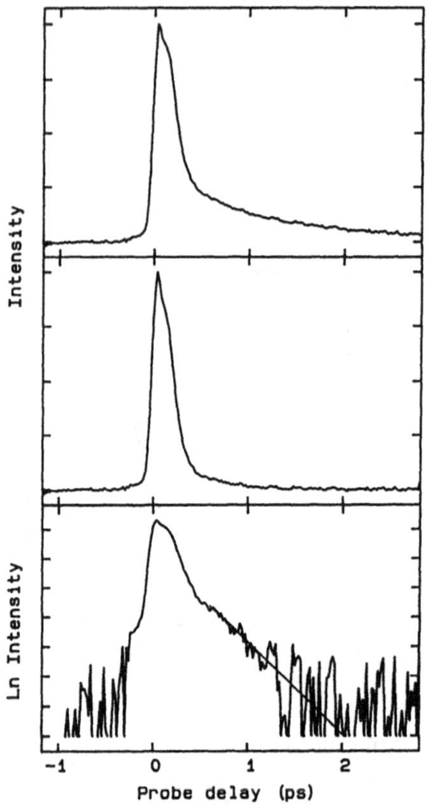

Figure 4. (upper) OHD-OKE data for neat acetonitrile at 298 K; (middle and lower) reduced data with tail-matched exponentially decaying 1.4 ps response subtracted off, presented in linear and semi-log formats. The straight line represents a ~400 fs decay

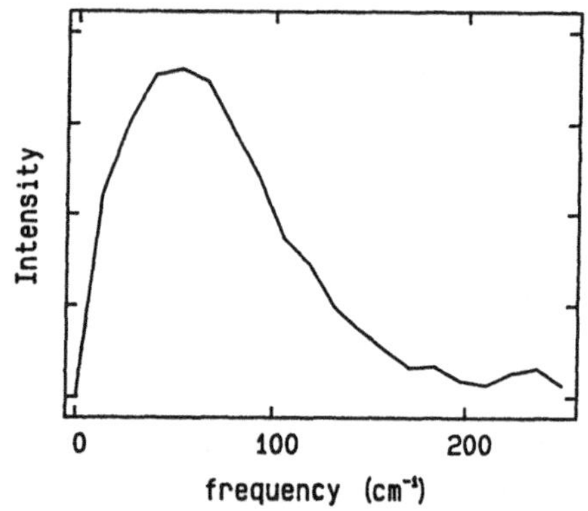

Figure 5. The coherently excited distribution of intermolecular oscillators, $D'(\omega)$, obtained from the raw OHD-OKE data for acetonitrile as is described in [7]

coherently-excited intermolecular oscillators, with this latter contribution containing information on the inhomogeneous distribution molecular environments. Elsewhere in this volume [8] we present a technique for obtaining directly the distribution function associated with this ensemble of oscillators. The resultant spectrum for acetonitrile is given in fig. 5. For the weakly polarizable, highly anisotropic acetonitrile molecule this skewed spectrum, which exhibits a maximum at $\sim$50 cm^{-1} and a width of $\sim$90 cm^{-1} FWHM, can be associated with the distribution of molecular librators in this liquid. The vibrational frequencies of the individual molecules reflect the site-to-site variations in the local intermolecular potentials. Hence, femtosecond optical Kerr measurements provide direct information on the inhomogeneity of the local structure in these disordered condensed-phase systems, with the distribution $D'(\omega)$ providing a direct experimental link to the models of pre-existing electron localization sites in liquids. Furthermore, as is evident from figs. 3-5, liquid acetonitrile exhibits high-frequency, non-diffusive orientational molecular motions extending well beyond 100 cm^{-1} ($\sim$3 THz), in addition to the slower dynamics of the diffusive reorientation process. These ultrafast motions, which are evident in all anisotropic liquids thus far investigated, should be of significance in the early time dynamics of the solvation process, whether it involves an electronically excited species or localized electrons. We gratefully acknowledge the U.S. Office of Naval Research and NSERC, Canada for financial support.

---

1. Special issue on Ultrafast Optics and Electronics, ed. by A.M. Johnson, IEEE J. Quant. Elect., QE-24 (1988).
2. G.A. Kenney-Wallace and C.D. Jonah: J. Phys. Chem. 86, 2572 (1982), and references therein.
3. D. McMorrow, W.T. Lotshaw, and G.A. Kenney-Wallace: IEEE J. Quant. Elect. QE-24, 443 (1988).
4. C. Kalpouzos, D. McMorrow, W.T. Lotshaw, and G.A. Kenney-Wallace: Chem. Phys. Lett., in press.
5. G.A. Kenney-Wallace: in Picosecond Phenomena, ed. by C.V. Shank, E. Ippen, and S.L. Shapiro (Springer-Verlag, New York, 1978) p. 208.
6. S. Ruhman, B. Kohler, A.G. Joly, and K.A Nelson: IEEE J. Quant. Elect. QE-24, 470 (1988).
7. D. McMorrow, W.T. Lotshaw, and G.A. Kenney-Wallace: Chem. Phys. Lett. 145, 309 (1988).
8. D. McMorrow, W.T. Lotshaw, and G.A. Kenney-Wallace: this volume.

Unified Theory of Solvation Dynamics in Nonlinear Optical Processes and Electron Transfer

S. Mukamel and Yi Jing Yan

Department of Chemistry, University of Rochester,
Rochester, NY 14627, USA

A novel semiclassical theory which provides a unified description of nonlinear optical lineshapes and molecular rate processes was developed recently.[1-4] Consider a molecular system with two electronic levels ($|a\rangle$ and $|b\rangle$) in a solvent, whose Hamiltonian is

$$H = |a\rangle H_a \langle a| + |b\rangle H_b \langle b| + H_{int} \tag{1}$$

where

$$H_{int} = V\,(|a\rangle\langle b| + |b\rangle\langle a|). \tag{2}$$

Here H_a and H_b represent nuclear degrees of freedom (both intramolecular and solvent). In nonlinear optical measurements $|a\rangle$ is the ground state and $|b\rangle$ is an electronically excited state and $V = -\mu\, E(r,t)$ is the dipole coupling with an external electromagnetic field. The same Hamiltonian may represent a rate process (such as electron transfer, or isomerization).[5] In that case V is the nonadiabatic coupling between the two reacting species. In order to express our results in a compact form we introduce two auxiliary quantities. The first is

$$\sigma_j\,(x) \equiv \langle \delta(x-U)\, \rho_j \rangle \tag{3}$$

where

$$\rho_j \equiv \exp(-H_j/kT)/\mathrm{Tr}\,\exp(-H_j/kT). \qquad j = a,b \tag{4}$$

$$U + E^o \equiv H_b - H_a \tag{5}$$

Here U is the solvation coordinate and $E^o = \hbar\omega_{ba}$ is the energy difference of the minima of H_a and H_b. $\sigma_j(x)$ is the probability distribution of the solvation coordinate U when the system is in the state j.

The second quantity is the conditional probability for the solvation coordinate to have the value x at time t, given that it had the value y at t=0 and that the system is in the state j, i.e.,

$$W_j(x,t;y) \equiv [\sigma_a(y)]^{-1} \langle \delta[x-U_j(t)]\, \delta(y-U)\rho_a \rangle\ , \qquad j=a,b \tag{6}$$

$$U_j(t) = \exp(iH_jt)\, U\, \exp(-iH_jt) \quad , \qquad j=a,b. \tag{7}$$

Note that by definition, $W_j(x,0;y) = \delta(x-y)$, and $W_j(x,\infty;y) = \sigma_j(x)$.

The absorption spectrum is simply given by $\sigma_a(\omega-\omega_{ba})$. Similarly, $\sigma_b(\omega-\omega_{ba})$ is the emission (fluorescence) spectrum in a steady state experiment. We next consider two time-resolved spectroscopic techniques which are commonly used in the studies of solvation: fluorescence and hole-burning (pump-probe) spectroscopies.[3,6] Both measurements are related to the nonlinear

Springer Series in Chemical Physics, Vol. 48: **Ultrafast Phenomena VI**
Editors: T. Yajima · K. Yoshihara · C.B. Harris · S. Shionoya

polarization $P^{(3)}$, and start with the application of a short pump pulse centered at t=0 with frequency ω_1. In a time-resolved fluorescence measurement, the solvation dynamics when the solute is in the excited electronic state |b> is probed by collecting a spontaneously emitted photon with frequency ω_2 at time t. The fluorescence signal is given by

$$S_{FL}(\omega_1,\omega_2,t) = \omega_1\omega_2^3\, W_b(\omega_2-\omega_{ba},t;\, \omega_1-\omega_{ba})\, \sigma_a(\omega_1-\omega_{ba}) \,. \tag{8}$$

In a time-resolved hole-burning measurement the absorption spectrum is measured with a probe pulse that is delayed relative to the pump pulse by time t. The hole-burning lineshape $S_{HB}(\omega_1,\omega_2,t)$ is defined as the difference between the absorption coefficient at ω_2 in the absence of a pump pulse and the absorption coefficient at ω_2 measured with a probe pulse that follows a pump pulse, and is given by

$$S_{HB}(\omega_1,\omega_2,t) = \omega_1\omega_2\, [W_b(\omega_2-\omega_{ba},t;\, \omega_1-\omega_{ba})+W_a(\omega_2-\omega_{ba},t;\, \omega_1-\omega_{ba})]\sigma_a(\omega_1-\omega_{ba}). \tag{9}$$

The rate of a chemical process (e.g., electron transfer) from |a> to |b> is given by[4]

$$K = \frac{2\pi\, (V^2/\hbar)\, \sigma_a(-E^o)}{1 + 2\pi\, (V^2/\hbar)\, \sigma_a(-E^o)\, (\tau_a+\tau_b)} \tag{10}$$

$$\tau_j = [\sigma_a(-E^o)]^{-1} \int_0^\infty dt\, [W_j(-E^o,t;-E^o) - W_j(-E^o,\infty;-E^o)]\,, \quad j=a,b. \tag{11}$$

Here τ_j is the relaxation time for a fluctuation of the solvation coordinate U at the transition state $U = -E^o$, when the system is in the state j. In the adiabatic limit (small V) the rate is proportional to V^2 and is given by the Fermi Golden Rule. In the opposite (adiabatic) limit the rate is given by the solvent relaxation time $(\tau_a + \tau_b)^{-1}$.

Eqs.(8-11) show that within the semiclassical approximation, the hole-burning and fluorescence measurements probe the same solvent quantities σ_j and W_j which enter into the calculation of rate processes. These equations provide a unique insight on the dynamics of optical and rate processes. The present derivation is very general and does not require the introdution of a specific model for solvation. Apart from considering a two state model and making a semiclassical approximation, we did not have to specify the model any further. We have derived explicit expressions for σ_j and W_j for polar solvents, which relate them to the dielectric function of the solvent $\varepsilon(\omega)$.[3,4]

We have calculated fluorescence and hole-burning lineshapes of a polar solute in ethanol at 247 K. For the dielectric function we assumed a Debye model[3] with a longitudinal relaxation timescale $\tau_L = 150$ ps. In Figure 1 we display the calculated time-resolved fluorescence spectrum of the retinal chormophore in bacteriorhodopsin, which has 29 optically active vibrational modes. The frames show fluorescence spectra measured at successively longer times. Each spectrum is labeled with the observation times in ps, measured from the time at which the sample interacts with the peak of the excitation pulse. In Figure 2 we show the hole-burning lineshapes for the same system.[3,6b]

The general theoretical framework described here provides a unified approach for calculating reaction rates and nonlinear optical lineshapes of polyatomic molecules in solution. Our theory of

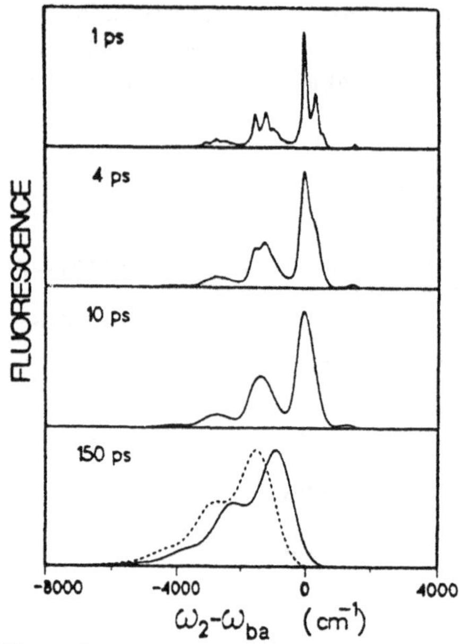

Figure 1

Time-resolved fluorescence spectra of a polyatomic solute in ethanol at 247 K following a 1 ps excitation pulse.[3] The pump frequency ω_1 is 1528 cm^{-1} above the 0-0 transition.

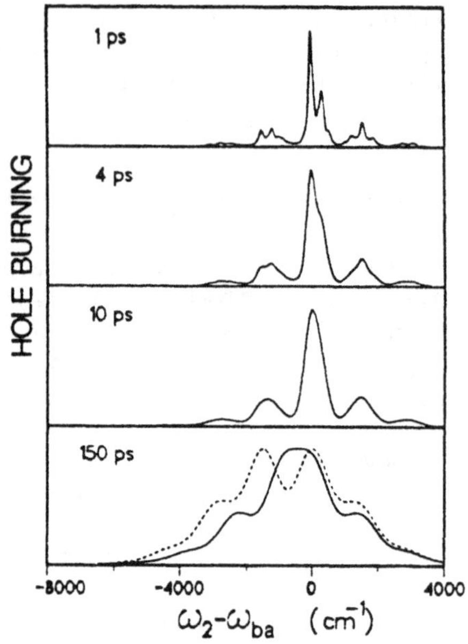

Figure 2

Hole-burning lineshapes for the same molecular model as Figure 1.

rate processes, such as electron transfer and isomerization in condensed phases, generalizes the Kramers and the Marcus theories and interpolates between the nonadiabatic and the adiabatic limits. We have demonstrated how solvation dynamics extracted from linear and nonlinear optical measurements (absorption, fluorescence, hole-burning, and $\chi^{(3)}$) may be used to predict electron transfer rates.

The support of the National Science Foundation, The Camille and Henry Dreyfus Foundations, the Office of Naval Research, the U.S. Army Research Office, and the Petroleum Research Fund, administered by the American Chemical Society, is gratefully acknowledged.

1. S. Mukamel, Phys. Rep. **93**, 1 (1982); Adv. Chem. Phys. **70**, Part I, 165 (1988); S. Mukamel and R.F. Loring, J. Opt. Soc. Am. B **3**, 595 (1986).
2. Y.J. Yan and S. Mukamel, J. Chem . Phys., **85**, 5908 (1986); ibid **86**, 6085 (1987).
3. R.F. Loring, Y.J. Yan, and S. Mukamel, J. Chem. Phys. **87**, 5840 (1987).
4. M. Sparpaglione and S. Mukamel, J. Chem. Phys. **88**, 3263, 4300 (1988); Y.J. Yan, M. Sparpaglione, and S. Mukamel, J. Phys. Chem. (in press).
5. M.W. Makinen, S.A. Schichman, S.C. Hill, and H.B. Gray, Science **222** , 929 (1983); G. L. Closs and J. R. Miller, Science **240**, 440 (1988); G. McLendon, Acc. Chem. Res. **21**,160 (1988); T. Gennet, D.F. Milner, and M.J. Weaver, J. Phys. Chem. **89**, 2787 (1985).
6. a) M. Maroncelli and G.R. Fleming, J. Chem. Phys. **86**, 6221 (1987);
   b) C.H. Brito Cruz, R.L. Fork, W.Knox, and C.V. Shank, Chem. Phys. Lett. **132**, 341 (1986); R.A. Mathies, C.H. Brito Cruz, W.T. Pollard, and C.V. Shank, Science **240**, 777 (1988).

Coherent Vibrational Motion in Liquids: The Inhomogeneously Broadened Distribution of Intermolecular Oscillators

D. McMorrow, W.T. Lotshaw, T.R. Dickson, and G.A. Kenney-Wallace

Lash Miller Laboratories, Department of Chemistry, University of Toronto, Toronto M5S 1A1, Canada

We present a simple and straightforward technique for obtaining the distribution function associated with the inhomogeneously broadened ensemble of oscillators excited in time-domain experiments with ultrashort laser pulses, and apply the technique to recent femtosecond optically heterodyned optical Kerr effect (OHD-OKE) data for two simple molecular liquids. This study provides new and detailed insight into the distribution of local intermolecular potentials in these disordered condensed-phase systems, and reveals the role of this distribution in determining the details of the ultrafast dynamics.

The low frequency, intermolecular vibrational motions of anisotropic molecules previously have been analyzed in terms of *assumed* functional forms which have been developed to *approximately* represent different physical models [1-3]. These analyses have provided an adequate description of the data in many cases, but are unable to reproduce the details of the temporal profile for liquids which exhibit more complex dynamics. Accordingly, the qualitative trends observed in the current study are consistent with those of previous analyses, however the distributions observed are significantly more complex than could be predicted by the simplified functional forms thus far utilized in analyzing the time-domain data. In particular, several liquids exhibit evidence for two and even three bands within the markedly asymmetric ~40-100 cm^{-1} FWHM spectrum (figure 2, *vide infra*). The method presented is quite general, and may be applied to obtain the frequency spectrum associated with any temporal signal that can be represented as a convolution of an effective driving function (or instrument function) with an impulse response function (cf., equation 1), if the driving function is a symmetric function of time. Furthermore, the result presented is exact to within the experimental limitations of the data.

In an OHD-OKE experiment the detected signal may be represented [1] as a convolution of the laser pulse intensity autocorrelation, $G_0^{(2)}(\tau)$, with the impulse response function of the induced birefringence, R(t). Assuming the separability of electronic and nuclear responses, $R(t) = \sigma(t) + r(t)$, and that the electronic response is instantaneous on the time scale of the ~65 fs laser pulses used in this study, the transmitted signal may be represented as

$$T(\tau) \propto G_0^{(2)}(\tau) + a\int_{-\infty}^{\infty} G_0^{(2)}(t-\tau)\, r(t)\, dt. \qquad (1)$$

Springer Series in Chemical Physics, Vol. 48: Ultrafast Phenomena VI
Editors: T. Yajima · K. Yoshihara · C.B. Harris · S. Shionoya

All of the information related to the nuclear response of the liquid to the applied optical field is contained in the impulse response function r(t). Because all nuclear motion is initiated *in phase* at $t = 0$, this impulse response may be represented as

$$r(t) = \int_0^\infty \sin(\omega t)\, D(\omega)\, d\omega. \tag{2}$$

It is straightforward to demonstrate that $D(\omega)$ may be obtained by *i*) *antisymmetrizing* the raw data, $T_a(\tau) = T(\tau) - T(-\tau)$, *ii*) performing a Fourier transform of the antisymmetrized response $T_a(\tau)$, and *iii*) dividing this result by the Fourier transform of the measured laser pulse autocorrelation. Thus,

$$D(\omega) \propto \frac{\mathcal{F}\{T_a(\tau)\}}{\mathcal{F}\{G_0^{(2)}(\tau)\}} \tag{3}$$

where $\mathcal{F}\{\cdots\}$ indicates the full exponential Fourier transform. Because the autocorrelation function $G_0^{(2)}(\tau)$ is symmetric with respect to $\tau = 0$, the purely electronic contribution (first term on the r.h.s. of (1)) vanishes in the antisymmetrization procedure. A similar antisymmetrization procedure has been utilized elsewhere in removing the coherent coupling contribution from time domain data [4]. The effects of the finite driving function are removed in the final operation, and it should be noted that this correction becomes less significant for shorter pulses and at lower frequencies. We note that since $T_a(\tau)$ and $G_0^{(2)}(\tau)$ are antisymmetric and symmetric with respect to $\tau = 0$, respectively, the numerator and denominator of (3) reduce to pure sine and cosine transforms. It is essential to maintain precisely the experimental $\tau = 0$ throughout this procedure [5].

Figure 1 gives the raw OHD-OKE data for *neat* CS_2 and *neat* bromobenzene at 298 K. The signal contribution associated with the imposition and decay of an orientational anisotropy ($\tau_{1/e} > 1$ ps) is not of primary interest here and may be removed from the data in a straightforward fashion [1] prior to the antisymmetrization procedure. In this case, the resultant distribution, $D'(\omega)$, represents the coherently-excited distribution of oscillators which, in general, is a function of both the equilibrium (thermal) oscillator distribution and the frequency spectrum of the exciting laser pulse. Figure 2 gives the distribution $D'(\omega)$ for the liquids of figure 1.

Previous analyses have assumed that $D'(\omega)$ could be represented by an approximately Gaussian distribution [2,3]. On inspection of the data of figure 2, it is not surprising that those analyses provide an adequate description of the short-time dynamics for the case of CS_2. On the other hand, bromobenzene and several other benzenes exhibit more complex temporal profiles that can not be reproduced with these simple representations. Accordingly, in figure 2, liquid bromobenzene exhibits a much more complex frequency distribution

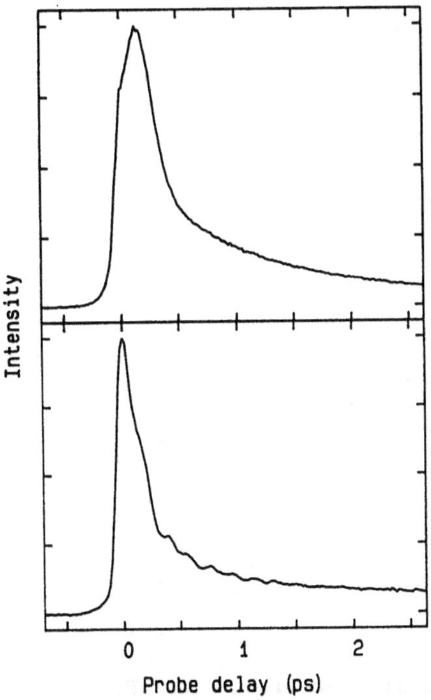

Figure 1. OHD-ODE data for CS_2 (upper) bromobenzene (lower) liquids at 298 K

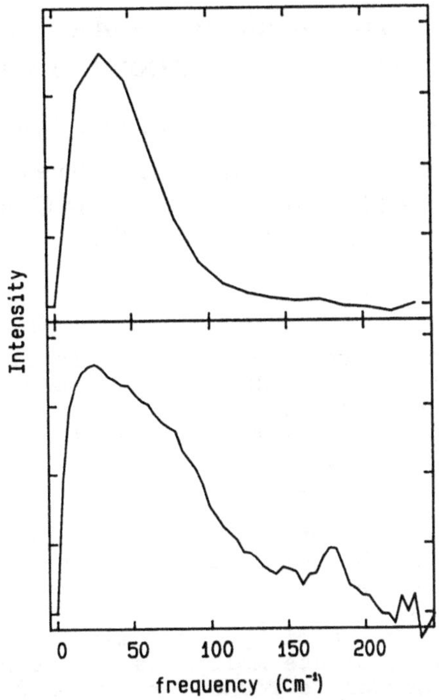

Figure 2. Distribution function $D'(\omega)$ for CS_2 (upper) and bromobenzene (lower)

than that of CS_2. A preliminary interpretation of the bromobenzene data involves librational motion about two disinct molecular axes, giving rise to the low frequency maximum at $\sim$25 cm^{-1}, and the higher frequency shoulder at $\sim$65 cm^{-1}. The data of figures 1 and 2 indicate clearly the role of the inhomogeneous distribution of molecular environments in shaping the ultrafast dynamics of these liquids.

---

1. D. McMorrow, W.T. Lotshaw, and G.A. Kenney-Wallace: IEEE J. Quant.Elect. QE-24, (1988) 443.
2. S. Ruhman, B. Kohler, A.G. Joly, and K.A. Nelson: IEEE J. Quant. Elect. QE-24, 470 (1988).
3. C. Kalpouzos, D. McMorrow, W.T. Lotshaw, and G.A. Kenney-Wallace: Chem.Phys. Lett., in press.
4. R.A. Engh, J.W. Petrich, and G.R. Fleming:J. Phys. Chem. 89, 618 (1985).

5. The authors wish to thank Robert Kitchenham for his assistance in the numerical computations.

Hydrodynamic and Molecular Contributions to Rotational Diffusion in Liquids

D. Ben-Amotz, T.W. Scott, and J.M. Drake

Corporate Research Science Laboratories,
Exxon Research and Engineering Company, Clinton Township,
Route 22 East, NJ 08801, USA

The large amplitude rotational motion of a solute molecule is a sensitive probe of local solvent structure and dynamics[1]. The experimentally measured rotational correlation function can be used to test microscopic friction models. Such models are of central importance in theories for chemical reaction dynamics in liquids[2].

The rotational anisotropy decay functions of several neutral solutes dissolved in both n-alkane and n-alcohol solvents have been measured using picosecond stimulated emission gain and fluorescence photon counting techniques. The results reveal a breakdown of hydrodynamic models for the rotational motion of small solute molecules. A turnover to near perfect hydrodynamic behavior is found for the largest solute studied (see Fig. 1).

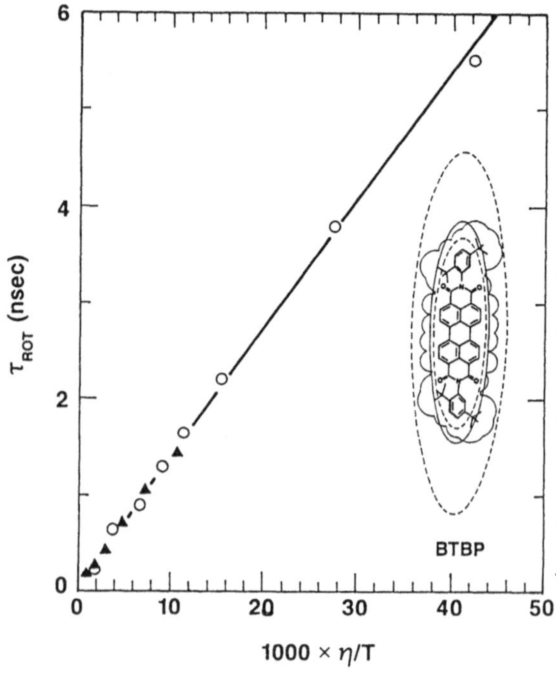

Figure 1. The rotational diffusion time of BTBP in various n-alkane (triangle) and n-alcohol (circle) solvents at 23°C as a function of solvent viscosity over temperature. The same behavior is observed when temperature dependent measurements are performed in n-hexanol and n-tetradecane. The solid ellipse in the inset is an outline of spheroid used to model BTBP. Spheroids the size of the small and large dashed ellipses would have to be assumed in order to get perfect agreement with stick and slip hydrodynamic predictions, respectively.

Springer Series in Chemical Physics, Vol. 48: **Ultrafast Phenomena VI**
Editors: T. Yajima · K. Yoshihara · C.B. Harris · S. Shionoya

The observed solute size dependence can be used to test various theoretical models for microscopic friction[3-5]. Zwanzig's zero frequency hydrodynamic model for the solute size effect[3] shows qualitatively the right turnover from stick to slip like behavior with decreasing solute size. This model does not, however, show either a sufficiently strong solute size dependence or the sub-slip friction often observed experimentally. We therefore suggest a simple generalized hydrodynamic model utilizing empirically derived frequency dependent shear viscosities of n-alkane solvents[6]. In this heuristic model the rotational anisotropy decay function is given by

$$R(t) = 2/5\ \mathrm{Exp}\{-3kTt/(3fV\eta(t)+I/t)\} \tag{1}$$

where the effective time dependent shear viscosity is

$$\eta(t) = A_r/(1+\tau_r/t) + A_s/(1+\tau_s/t). \tag{2}$$

The amplitudes A_r, A_s and time constants τ_r, τ_s depend on the solvent[5,6], V and I are the solute volume and moment of inertia respectively and f is a solute shape and boundary condition dependent factor[1,4,5]. The limiting behaviors of (1) at short and long times respectively are $R(t) = \mathrm{Exp}\{-kTt^2/I\}$ and $R(t) = \mathrm{Exp}\{-kTt/fV\eta_0\}$. These are the proper inertial (free rotor) and diffusive (Debye) limits, where $\eta_0 = A_r + A_s$ is the solvent's zero frequency shear viscosity. Crossover of the effective viscosity at intermediate times, as given by (2), is assumed to roughly represent the integral of the random force (torque) autocorrelation function experienced by the solute between time equal 0 and t.

Representative results of this model for the rotational diffusion of iodine molecules dissolved in n-alkane solvents are shown in Fig. 2 (solid lines). These results reveal substantial deviations from macroscopic hydrodynamic predictions (dashed lines). The triangles in Fig. 2a represent previously reported rotational diffusion times obtained from Raman line shape studies[7]. Calculations for larger molecules (e.g. p-terphenyl and BTBP) reveal relatively smaller deviations from macroscopic hydrodynamic predictions and very nearly exponential rotational anisotropy decays - again in qualitative agreement with experiment.

Three important predictions of our generalized hydrodynamic model are, i) sub-slip friction on small solute molecules, ii) non-zero intercepts in the zero viscosity limit, iii) non-exponential behavior (with both positive and negative curvature) in the rotational anisotropy decays of small solute molecules. Of these predictions the first two are born out by numerous previous studies[1,4,5,7]. The third prediction remains to be conclusively verified, although early time inertial non-exponentiality has been observed in some anisotropic Raman line shape studies and more recently in pump-probe time domain experiments[8].

Other molecularly based models[4,9,10] have been proposed to account for the breakdown of macroscopic hydrodynamics. All of these models undoubtedly contain features which are important for the proper description of microscopic frictional forces. The

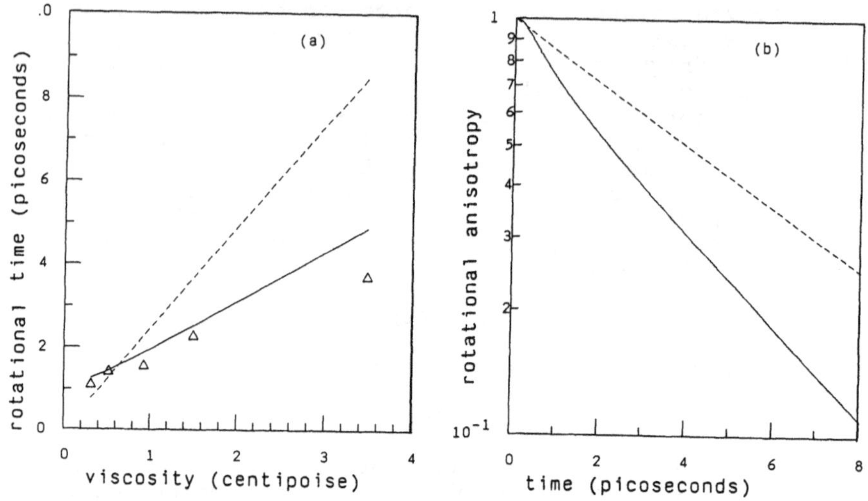

Figure 2. a) Predicted dependence of iodine's rotational time, $5/2 \int_0^\infty R(t)dt$, on the zero frequency viscosity of n-alkane solvents at 20°C. b) Predicted rotational anisotropy decay function, 5/2 R(t), for iodine in n-tetradecane at 20°C. Dashed lines are zero frequency slip hydrodynamic predictions, solid lines are calculated using our generalized hydrodynamic model and triangles are experimental points taken from [7].

substantial success of our very simple generalized hydrodynamic model suggests that the finite dynamic response time of real liquids represents a key molecular feature which strongly influences the motion of solute molecules on the picosecond time scale.

References

1. D. Ben-Amotz and T. W. Scott: J. Chem. Phys., 87, 3739, (1987).
2. J. T. Hynes, J. Stat. Phys.: 42, 149, (1986).
3. R. Zwanzig, J. Chem. Phys.: 68, 4325, (1978).
4. J. L. Dote, D. Kivelson and R. N. Schwatrz: J. Phys. Chem., 85, 2169, (1981).
5. D. Ben-Amotz and J. M. Drake: J. Chem. Phys, in press.
6. M. G. Sceats and J. M. Dawes: J. Chem. Phys., 83, 1298, (1985).
7. A. M. Goulay: J. Chem. Phys., 79, 1145, (1983).
8. A. B. Myers, M. A. Petreira, P. L. Holt and R. M. Hochstrasser: J. Chem. Phys., 86, 5146, (1987).
9. J. T. Hynes, R. Kapral and M. Weinberg: J. Chem. Phys., 69, 2725, (1978).
10. G. T. Evans: J. Chem. Phys., 88, 5035, (1988).

Influence of Functional Groups and Solvent on the Photoisomerization of Stilbenes

N.S. Park, N. Sivakumar, E.A. Hoburg, and D.H. Waldeck

Department of Chemistry, University of Pittsburgh, Pittsburgh, PA 15260, USA

The photoisomerization of t-stilbene is a model system for the investigation of solvent effects on chemical reactions. This study discusses the isomerization of stilbene and its 4,4'-disubstituted analogues in polar solvents. The investigation of substituted stilbene allows the influence of solute properties on the reaction dynamics to be evaluated. For the stilbenes in this study both the energy barrier height to isomerization and the reduced moments of inertia are changed by the substitution. The presence of different functional groups on the stilbene can also change the interaction with the solvent. Primarily the solute/solvent frictional coupling is expected to change, because the volume displaced upon isomerization changes and the nature of the solute/solvent coupling changes (i.e. stilbene couples via dispersion forces whereas 4,4'-dihydroxystilbene couples via dipole-dipole and/or hydrogen bonding interactions). The studies described here probe isomerization in two different homologous series of polar solvents. In the n-alkyl nitriles, which are non-associating solvents, it is possible to treat the isomerization as occurring on a solvated potential energy surface. However, in the n-alkyl alcohols which are associating solvents, it is not possible to extract a potential energy surface which is independent of the solvent dynamical response.

1. n-Alkyl Nitrile Studies

As shown previously [1-4], it is possible to exploit the general form of the expression for the rate constant to extract an activation barrier for the reaction. This form is

$$k_{iso} = F(\xi) \exp(-E/RT) \tag{1}$$

where E is the activation energy and $F(\xi)$ is a function of the solute properties and the solute/solvent friction. By performing Arrhenius type plots at constant solvent friction it is possible to obtain the energy barrier. Figure 1 shows such plots for t-stilbene in the n-alkyl nitriles where the solvent friction is modeled as being proportional to the solvent viscosity. Recent studies show that this method of obtaining a barrier is rather insensitive to the modeling of the solute/solvent friction [1,3]. The slopes of the isoviscosity plots appear to be insensitive to the changing solvation from propionitrile to decanenitrile [2]. The activation barriers for the three solutes studied are given in Table 1. For the case of stilbene and dimethoxystilbene it is possible to compare these barriers with the unperturbed barriers of 3.5 kcal/mol [3] and 5.7 kcal/mol [1], respectively. The barriers reported in Table 1 should be considered to be solvated barriers.

Shown in Fig. 2 is a fit of the preexponential factor, $F(\xi)$, to various models for the case of t-stilbene (qualitatively similar behavior is observed for the other solute probes). The dashed curve shows a best fit to a hydrodynamic form of the Kramers model [1-4], namely

Springer Series in Chemical Physics, Vol. 48: **Ultrafast Phenomena VI**
Editors: T. Yajima · K. Yoshihara · C.B. Harris · S. Shionoya

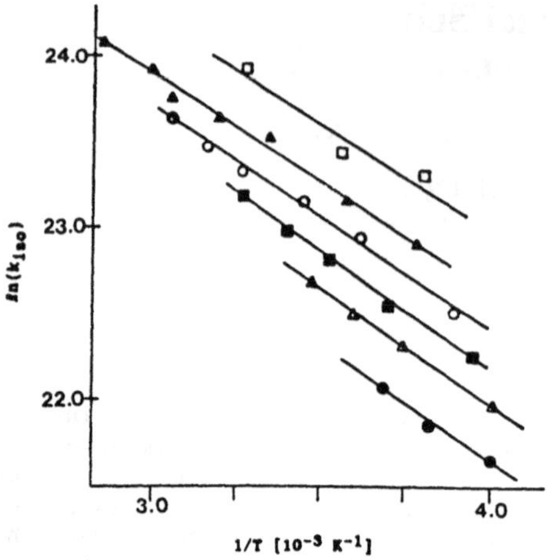

Fig.1 - Isoviscosity plots for t-stilbene in the n-alkyl nitriles. (●--5.0 cP; △--3.0 cP; ■--2.0 cP; o--1.3 cP; ▲--0.9 cP; □--0.5 cP).

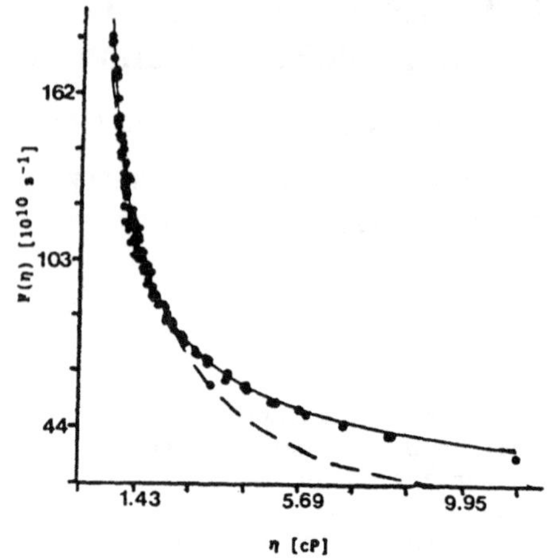

Fig.2 - Plot of reduced isomerization rate versus shear viscosity (see text).

Table 1: Parameters for n-alkyl nitrile studies (C_2H_5CN to $C_9H_{19}CN$)

Solute	E(kcal/mol)	B(10^{12})	a
t-stilbene	2.6	1.09	0.51
4,4'-dihydroxystilbene	4.3	1.67	0.44
4,4'-dimethoxystilbene	4.2	1.41	0.44

$$F(\eta) = \frac{A\eta}{B}\left((1 + (B/\eta)^2)^{\frac{1}{2}} - 1\right). \tag{2}$$

The failure of (2) to fit the data is similar to that observed in a wide variety of isomerization data. The solid line in Fig. 2 is a phenomenological form, B/η^a, which is known to fit isomerization rate data quite well. The trend in the parameters is similar to that reported earlier, that is as the barrier height increases the viscosity dependence becomes weaker. This trend has been interpreted in terms of the frequency dependence of the viscosity [4]. That is, as the timescale of the isomerization increases the ability of the zero frequency shear viscosity to model the friction fails. Various approaches to correcting this modeling of the friction have been discussed [1,3]. Although the dimethoxystilbene should displace approximately twice as much volume as stilbene upon isomerization, it has a weaker viscosity dependence, presumably caused by the increase barrier curvature.

2. n-Alkyl Alcohols

The isomerization dynamics in the n-alkyl alcohols is considerably more complicated than in the n-alkyl nitriles. Figure 3 shows isoviscosity plots for 4,4'-dihydroxystilbene, where the slope of the lines change for

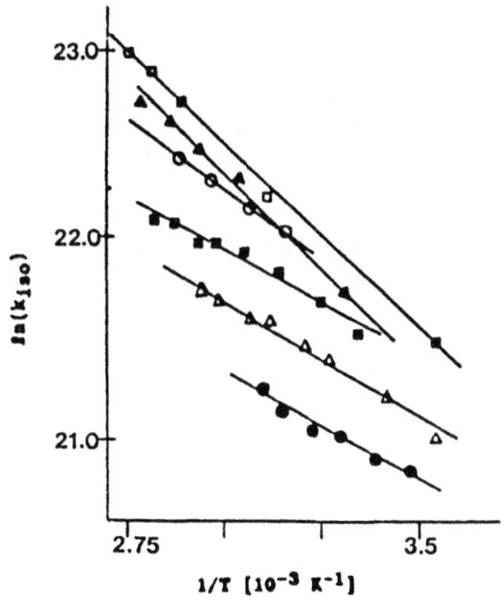

Fig.3 - Isoviscosity plots for dihydroxystilbene in the n-alkyl alcohols. (●--5.0 cP; △--3.0 cP; ■--2.0 cP; o--1.3 cP; ▲--1.0 cP; □--0.7 cP).

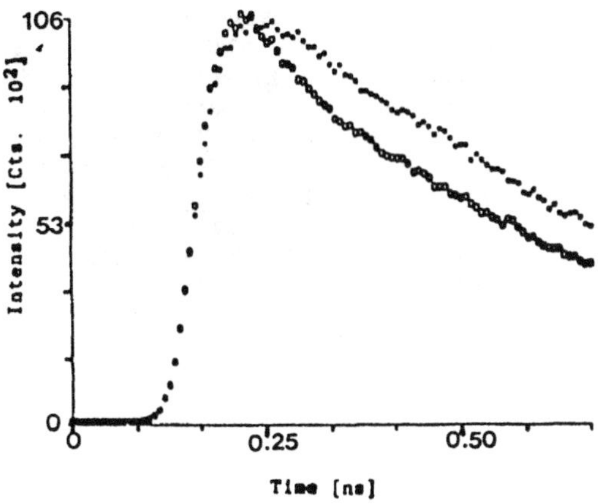

Fig.4 - Fluorescence decays of 4.4'-dihydroxystilbene in glycerol at 50°C showing the simultaneous Stokes shift and isomerization decay.

different viscosities. The separation of the barrier from the solvent effects has clearly failed. Similar behavior is observed for both t-stilbene [2] and 4,4'-dimethoxystilbene [1]. The explanation for this failure lies in the fact that the isomerization dynamics and the excited state solvation dynamics occur on similar timescales. Figure 4 shows the fluorescence decay profiles of 4,4'-dihydroxystilbene at two emission wavelengths which are distinctly nonexponential early time behavior - the long time behavior is single exponential with a wavelength independent decay constant.

A consistent picture of the isomerization dynamics in alcohols can be given by a consideration of three timescales. The first timescale is the "laboratory timescale" which is the timescale of the measured fluorescence decays - 1 nsec to 50 psecs. The second timescale is the "solvation timescale" which measures the time required for the solvent to respond to a new dipole size or orientation - 100 psec to 1 psec in n-alcohols. These relative timescales indicate that a fit of the decay curves at long time (when exponential) selects out of the prepared ensemble those solute molecules which are initially equilibrated with the solvent. The third timescale is the "reactive timescale" which is the characteristic time for the motion from reactant to product - as measured by $F(\xi)$ it is <1 psec. This last timescale suggests that the solute isomerizes while the solvent polarization field is out of equilibrium with the solute dipole. This view suggests that the reaction coordinate is intimately coupled to the time dependence of the solvents' polarization response.

These results underscore the importance of dielectric solvent effects, both static and dynamic, on chemical reactions.

References

1) D. M. Zeglinski and D. H. Waldeck: J. Phys. Chem. 92, 692 (1988).
2) a) J. Hicks, M. Vandersall, Z. Babarogic, K. B. Eisenthal: Chem. Phys. Lett. 116, 18 (1985); b) J. M. Hicks, M. T. Vandersall, E. V. Sitzmann, K. B. Eisenthal: Chem. Phys. Lett. 135, 413 (1987).
3) a) M. Lee, A. J. Bain, P. J. McCarthy, C. H. Han, J. N. Haseltine, A. B. Smith III, R. M. Hochstrasser: J. Chem. Phys. 85, 4341 (1986); b) S. H. Courtney, S. K. Kim, S. Canonica, G. R. Fleming: J. Chem. Soc. Faraday Trans. 82, 2065 (1986); c) S. K. Kim and G. R. Fleming: J. Phys. Chem. 92, 2168 (1988).
4) S. P. Velsko, D. H. Waldeck, G. R. Fleming: J. Chem. Phys. 78, 249 (1983).

Dynamics of Intramolecular Electron Transfer in Viscous Polar Solvents

T. Okada, K. Nakatani, M. Hagihara, and N. Mataga

Department of Chemistry, Faculty of Engineering Science, Osaka University, Toyonaka, Osaka 560, Japan

1. Introduction

Intramolecular electron transfer (ET) in the excited states of compounds with electron donor and acceptor groups combined by spacer have been studied extensively in order to elucidate the effect of the orientational motion of the surrounding solvent molecules on the ET process. The compounds listed below, except 35ATMA and ADMA, have no dipole moment due to intramolecular ET between two groups in the initial state of ET reaction, which is the excited state localized in pyrene or anthracene moiety. We measured decay curves of fluorescence from the locally excited (LE) state of pyrene or anthracene moiety of these compounds in viscous polar solvents. The observed decay time may correspond to the rate of intramolecular ET in solution. The compounds examined here should be ideal systems to study the role of polar solvent molecules in the course of ET reaction, since there is no ET interaction between two groups in the initial state immediately after excitation and the ET can take place very easily when the solvent orientation is favorable. Actually, the transient absorption spectra of these compounds show the change in the course of ET from the $S_n \leftarrow S_1$ band localized in pyrene or anthracene part to the superposition of donor cation and acceptor anion absorption bands [1]. In contrast to the system described above, some compounds show gradual change of electronic structure induced by solvation. For example, the electronic structure of the excited state of ADMA, which has a fairly large dipole moment in the excited state even in nonpolar solvent, changes gradually in the course of solvation through various states with different degrees of charge transfer and, accordingly, with different degrees of solvation and twisting angle between phenyl and anthryl moieties [1].

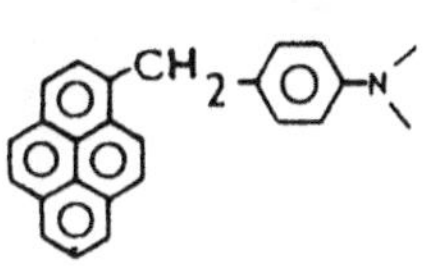

P1

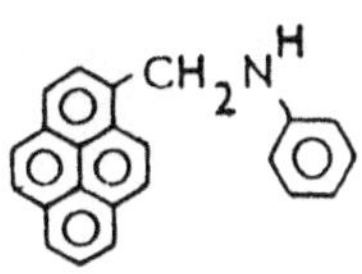

PNH1

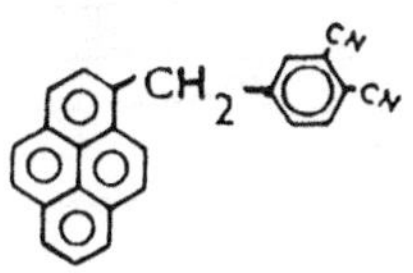

P1C

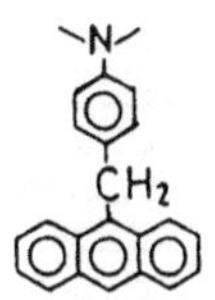

A_1

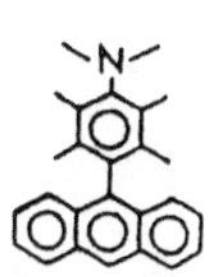

AHMA

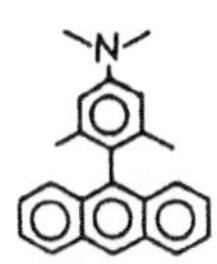

35ATM

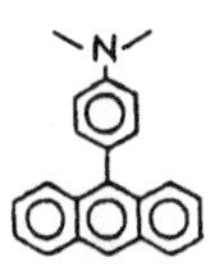

ADMA

Springer Series in Chemical Physics, Vol. 48: Ultrafast Phenomena VI
Editors: T. Yajima · K. Yoshihara · C.B. Harris · S. Shionoya

2. Experimental

Time-resolved fluorescence spectra and fluorescence decay curves were measured by 2-dimensional streak camera. The third harmonic of a picosecond pulse (355 nm, 25-ps duration) generated by a Q-switched mode-locked Nd:YAG laser was used for excitation of the sample. The preparations of the compounds are reported elsewhere [2]. All solutions for the measurements were deoxygenated by freeze-pump-thaw cycles or irrigating with a nitrogen gas stream.

3. Results and Discussion

Time-resolved fluorescence spectra of AHMA in 1-butanol is shown in Fig. 1. The LE fluorescence detected at about 410 nm was still observed at 6 ns after excitation. The LE fluorescence observed at ns time regime was detected also in the cases of A1 and P1. The LE fluorescence of AHMA in 1-butanol shows nonexponential decay as indicated in Fig. 2. The initial slope of the LE fluorescence, τ_{LE}, is listed in table 1. The temperature dependence of τ_{LE} of AHMA in 1-butanol has been measured. The observed activation energy of the decay rate is much smaller than that of viscosity of the solvent. The observed results were summarized as follows: (1) The decay time obtained from the initial slope of LE fluorescence is considerably shorter compared to the longitudinal dielectric relaxation time, $\tau_L = \tau_D \varepsilon_\infty / \varepsilon_0$, where τ_D is the Debye relaxation time, and (2) τ_{LE} is characteristic to each compound. (3) Observed decay curve of LE fluorescence was nonexponential.

A clear-cut interpretation for the above observations appears rather difficult at the present stage of investigation. It seems to be possible that small orientational fluctuations of the polar group of surrounding solvent molecules may induce intramolecular electron transfer since the electronic interaction between donor and acceptor in the excited state is reasonably strong. In fact, we have observed that ET takes place within a few ps for P1 in acetonitrile solution [3]. Nevertheless, the observed result suggests that the intramolecular ET dynamics of these compounds in

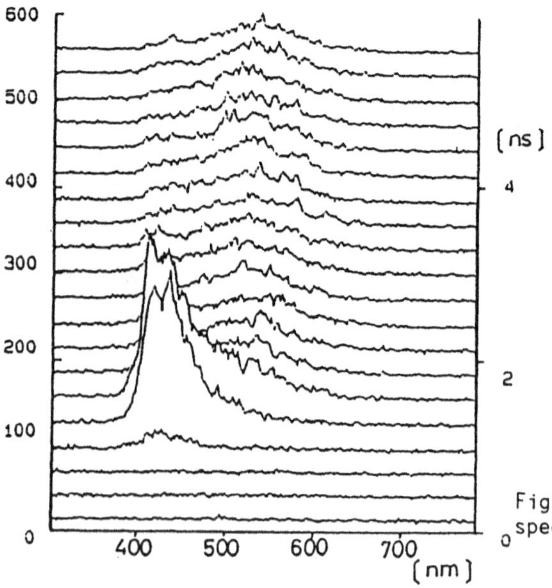

Fig. 1. Time-resolved fluorescence spectra of AHMA in 1-butanol.

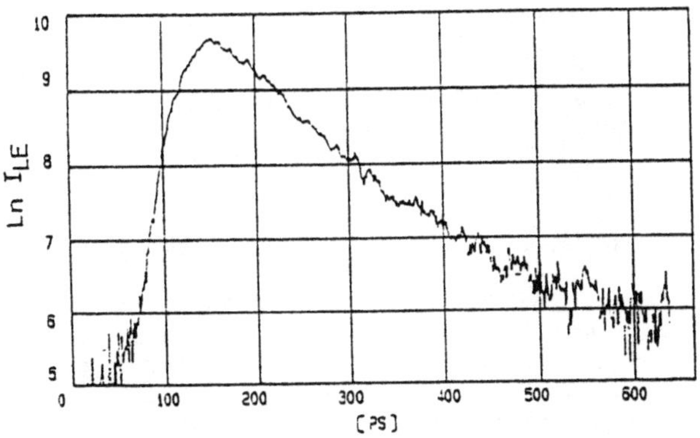

Fig. 2. Semi-logarithmic plot of the LE fluorescence intensity vs. delay time in the case of AHMA in 1-butanol.

Table 1. Initial slope of the LE fluorescence decay curve observed for the donor-acceptor combined systems.

Solvent	1-pentanol	1-butanol
A1	60 ps	50 ps
AHMA	140	100
P1	170	110
PNH1	115	80
P1C	115	100

viscous alcoholic solvents cannot simply be characterized by τ_L. For this result, we suggest here the following possible interpretations: (1) contribution of solvent relaxations with various τ_L values to ET process [4], (2) existence of presolvated state characteristic to each compound, (3)contribution of intramolecular vibrations to ET process [5], and (4) inhomogeneities in surrounding solvent structures which persist in ns time scale.

It should be noted here that there are many donor-acceptor combined systems with strong interactions between two groups for which the usual theoretical treatments are not necessarily valid. For example, the time-resolved fluorescence spectra of ADMA obtained by the present measurements showed gradual red shift with increase of the delay time after excitation, which might be ascribed simply to the stabilization of a large dipole of excited ADMA in the course of solvation. However, the electronic charge distribution between donor and acceptor groups in this molecule depends strongly upon the degrees of solvation as revealed by the time-resolved transient absorption spectra [1], leading to the gradual change of the dipole moment of the excited state in the course of solvation.

In view of the above results, it is of crucial importance to measure both the time-resolved transient absorption spectra and the time-resolved fluorescence spectra for the elucidation of the mechanisms of the intramolecular ET or charge transfer processes of the composite systems with combined electron donor and acceptor groups.

Acknowledgement

N. M. acknowledges the support by a Grant-in-Aid (No. 62065006) from the Japanese Ministry of Education, Science and Culture.

References

1. T. Okada, N. Mataga, W. Baumann, A. Siemiarczuk: J. Phys. Chem. 91, 4490 (1987).
2. M. Migita, T. Okada, N. Mataga, Y. Sakata, S. Misumi, N. Nakashima, K. Yoshihara: Bull. Chem. Soc. Jpn. 54, 3304 (1981); A. Siemiarczuk, Z. R. Grabowski, A. Krowczynski, M. Asher, M. Ottolenghi: Chem. Phys. Lett. 51, 315 (1977); N. Detzer, W. Baumann, B. Schwager, J.-C. Flohling, C. Brittinger: Z. Naturforsch., A: Phys., Phys. Chem., Kosmophys. 42, 395 (1987).
3. N. Mataga, H. Miyasaka, T. Asahi, S. Ojima, T. Okada: Ultrafast Phenomena VI (1988).
4. M. Sparpaglione, S. Mukamel: J. Chem. Phys. 88, 4300 (1988).
5. H. Sumi, R. A. Marcus: J. Chem. Phys. 84, 4894 (1986); W. Nadler, R. A. Marcus: ibid. 86, 3906 (1987).

Picosecond Investigation of Photoinduced Intramolecular Charge Transfer and Solvent Cage Relaxation Processes: Laser Dye DCM

J.C. Mialocq[1] *and M. Meyer*[1;2]

[1]CEA-IRDI-DESICP-DPC-SCM-UA 331 CNRS, CEN Saclay, F-91191 Gif-sur-Yvette Cedex, France

[2]Quantel S.A., F-91941 Les Ulis, Orsay Cedex, France

1. Introduction

The influence of the solvent on photoinduced intramolecular charge transfer and isomerization has attracted considerable interest during the last few years |1|. The trans-cis photoisomerization efficiency of styrenes depends largely upon the ability of the substituents to donate or accept an electron as well as on the solvent polarity |2|. We present recent picosecond fluorescence and absorption results which provide a detailed understanding of the photophysical and photochemical behaviour of the largely used laser dye 4-dicyanomethylene-2-methyl-6-p-dimethyl-amino-styryl-4H-pyran (DCM) |3|.

2. Results and discussion

Fluorescence lifetimes were measured via the time correlated single photon counting technique using as exciting light the tunable picosecond laser pulses from a cavity dumped dye laser synchronously pumped by the second harmonic of a cw mode-locked Nd-YAG laser. The picosecond absorption study was performed using 532 nm excitation by a single laser pulse (26 ps duration) and a continuum of white light as analyzing source.

The DCM spectral properties are remarkable as regards to the weak overlap between its absorption and emission spectra which both show important red shifts with increase of the solvent polarity. The Stokes shift $(\nu_A - \nu_F)$ of the fluorescence spectra with respect to the absorption spectra which is linearly related to the polar solvents Δf value according to the Lippert and Mataga theories indicates a large dipole moment increase ($\Delta\mu$ = 20.2 D) in the first excited singlet state and an increased intramolecular charge transfer. In order to evidence this intramolecular charge transfer, the resulting solvent cage relaxation and the possible formation of a twisted intramolecular charge transfer state (TICT) the fluorescence profiles of DCM in various polar solvents (dimethylsulfoxide, methanol, acetonitrile and chloroform) were analyzed. In dimethylsulfoxide, the fluorescence decay at 620 nm was very well fitted using a monoexponential law (τ=2.25±0.01ns) (Figure 1a). In methanol, a poor fit was obtained with a single exponential decay (Figure 1b) but fitting the decay curve with a sum of two exponential decays gave good χ^2 and Durbin-Watson (DW) parameter values. The long lifetime is τ_1=(1.37±0.03)ns, the short lifetime is τ_2=(0.35±0.17)ns (Figure 2) and the relative amplitudes are 93 and 7 percent respectively. A similar behaviour was observed in acetonitrile and chloroform.

Excitation of a 2 x 10^{-4}M DCM solution in dimethylsulfoxide gave a strong absorbance previously attributed to the first singlet excited state (λ_{max} = 450-460 nm) |3d|. The time profile of the 460 nm absorbance shows an initial increase which follows closely the integration of the pulse shape given by the 460 nm photobleaching and absorption recovery of a methanolic solution of pseudoisocyanine iodide (PIC) and a slow decay $\tau \sim$ 2.2 ns similar

Springer Series in Chemical Physics, Vol. 48: **Ultrafast Phenomena VI**
Editors: T. Yajima · K. Yoshihara · C.B. Harris · S. Shionoya

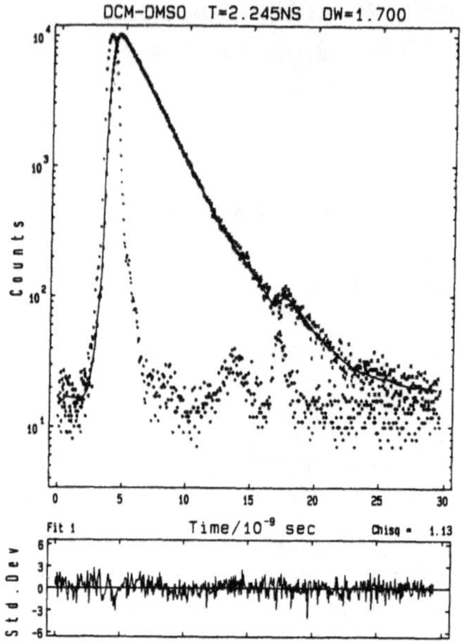

Fig. 1a : Fluorescence decay of DCM in dimethylsulfoxide.

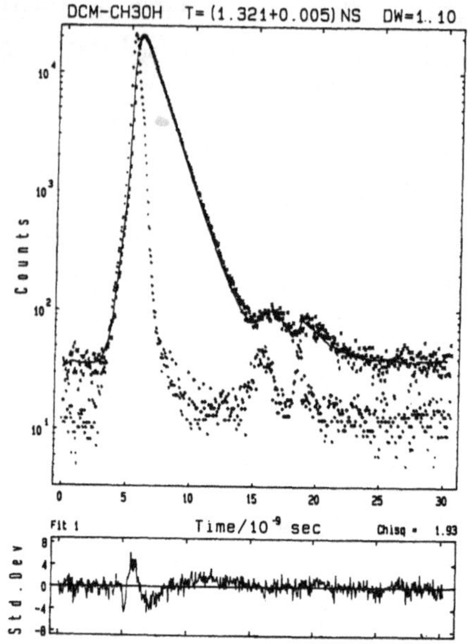

Fig. 1b : Fluorescence decay of DCM in methanol (single exponential fit).

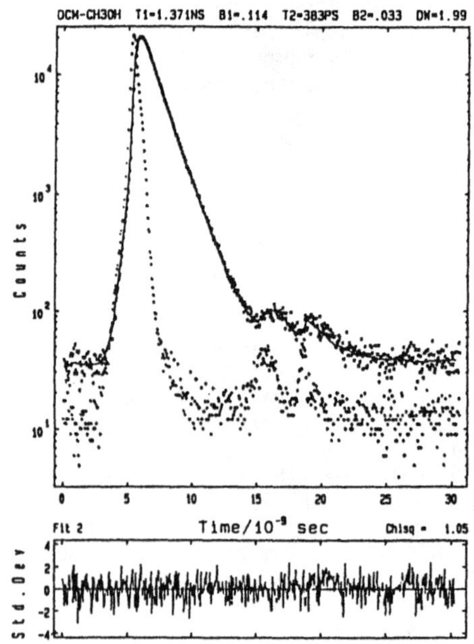

Fig. 2 : Fluorescence decay of DCM in methanol (fit with a sum of two exponentials)

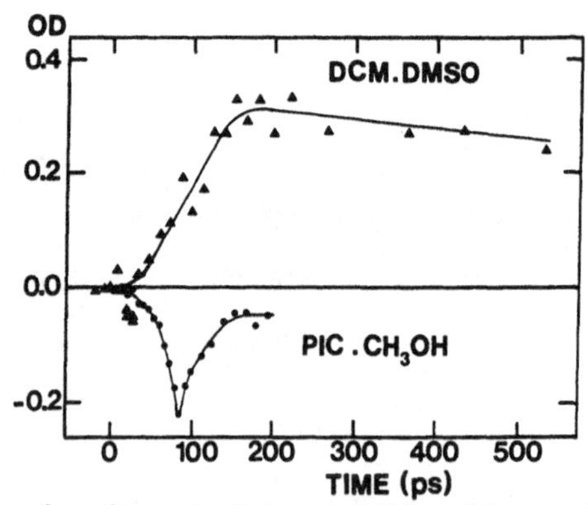

Fig. 3 : Ground state photobleaching and recovery of DCM in dimethylsulfoxide

to the fluorescence decay (Figure 3). The expected solvent relaxation following the fast intramolecular charge transfer is certainly much shorter than 50 ps and cannot be evidenced in the present experimental conditions.

The decay time measured in DMSO belongs therefore to a solvent relaxed planar fluorescent state. The two decays observed in methanol, acetonitrile and chloroform are attributed to the fluorescent states of the two DCM trans-cis isomers. The cis-isomer results from a trans-cis equilibrium in ambient light and has the shortest decay time as expected from the steric hindrance which inhibits the planeity and rigidity of the molecule favouring thereby internal conversion and intramolecular rotation. Our nanosecond laser, NMR and HPLC studies corroborate the existence of a trans-cis conversion more efficient in chloroform |3c|. The large dependence of the trans-isomer fluorescence lifetime in the non viscous polar solvents under study and the observed different photoisomerization efficiencies cannot be described by the Kramer's theory with a hydrodynamic model for the friction |1| and suggest that the solvent polarity which strongly affects the S_1 energy surface might induce an important modification of the barrier. Because the dipole moment of the trans-isomer is expected to be larger than that of the cis-isomer, due to the larger distance between the opposite charges, the excited trans-isomer is more stabilized in polar solvents than the excited cis-isomer. Moreover in polar solutions, when the torsional motion is accompanied by charge transfer, the potential energy surfaces in the ground and first excited states present local minima at the perpendicular geometry |4| which lowers the barrier between the fluorescent state and the perpendicular state. Besides the solvent polarity, specific interactions such as hydrogen bonding may be responsible for the change of the barrier energy. Indeed H-bonding in protic solvents is expected to induce some decrease in the amino lone pair electron delocalization, decreasing thereby the double bond character of the ethylenic bond of stilbenes substituted by amino-groups |5|. The absence of DCM photoisomerization in aprotic dimethylsulfoxide can be explained by a stronger double bond character inhibiting the torsional motion. Therefore, the observed lowering of the S_1 state in dimethylsulfoxide |3d|, and that of the perpendicular state are not decisive but the height of the energy barrier related to the double bond character and affected by H-bonding in protic solvents plays a major role in the trans-cis photoconversion. This property explains the more efficient laser action of DCM in dimethylsulfoxide.

References

1. a) S.P. Velsko, D.H. Waldeck and G.R. Fleming, J. Chem. Phys., 78, 249 (1983) ; b) E. Akesson, V. Sundström, T. Gillbro, Chem. Phys., 106, 269 (1986) ; B. Bagchi, Intern. Rev. Phys. Chem., 6, 1 (1987).
2. Y. Maeda, T. Okada and N. Mataga, J. Phys. Chem., 88, 2714 (1984).
3. a) P.R. Hammond, Opt. Comm. 29, 331 (1979) ; b) Z. H. Kang, M.R. Lan, N.E. Pin and G. Chu, J. Photochem. 29, 397 (1985) ; c) J.M. Drake, M.L. Lesiecki and D.M. Camaioni, Chem. Phys. Lett., 113, 530 (1985) ; d) M. Meyer and J.C. Mialocq, Opt. Comm., 64, 264 (1987) ; e) M. Meyer and J.C. Mialocq, J. Phys. 48, Suppl. N. 12, C7-541.
4. a) L. Salem and W.D. Stohrer, J.C.S. Chem. Comm., 140 (1975). b) Z. Smedartchina, F. Dietz, J. Photochem., 30, 13 (1985).
5. K.J. Smit, K.P. Ghiggino, Chem. Phys. Lett., 112, 369 (1985).

Picosecond Ultraviolet Multiphoton Ionization and Geminate Charge Recombination in Hydrocarbon Solvents

Y. Hirata and N. Mataga

Department of Chemistry, Faculty of Engineering Science, Osaka University, Toyonaka, Osaka 560, Japan

1 INTRODUCTION

Electrons produced by the ionization process in organic liquids usually form transient charge pairs. Parent molecular ions and their associated electrons are highly correlated and move randomly under the influence of mutual Coulomb attraction. The fate of the geminate ion pairs is considered to be determined by the diffusion in the Coulomb field, which is described by Smoluchowski equation. Many investigations have been done in radiation chemistry but the time resolution of those studies was rather poor and therefore the earliest stage of the charge recombination remains obscure. Although the photochemical studies in such field were few[1], they should provide important knowledge for the ultrafast geminate recombination phenomena because of their very high time resolution. In this respect, we have investigated the multiphoton ionization and subsequent charge recombination process for aminopyrene and pyrene in several hydrocarbon solvents.

2 AMINOPYRENE

Figure 1 shows the time resolved absorption spectra of aminopyrene in t-decalin. At 100 ps after the laser pulse excitation, the absorption band peaked at 520 nm was observed, which can be assigned to the $S_n \leftarrow S_1$ transition of aminopyrene. Only at quite short delay times was the additional transient absorption observed around 480 nm, which should be ascribed to the cation radical of aminopyrene. Reducing the excitation laser intensity by a factor of 4, the cation band became quite weak as shown by dot in Fig. 1. Therefore, we can conclude that the cation radical is formed by multiphoton ionization (probably two-photon ionization). The time dependence of transient absorbance measured at 480 nm is displayed in Fig. 2 (open circle). Rapid decrease of the transient absorbance is followed by a quite slow one. The latter is mainly due to the decay of the S_1 state, while the former should correspond to the geminate charge recombination. By subtracting the long lived component, the short lived one is shown by closed circles. Its decay is quite rapid and the lifetime seems to be less than the pulse width of the excitation laser (25 ps). Assuming that the distribution of the initial thermalization length of the ejected electron is given by δ-function at R_0, we have simulated the decay curve for several R_0s by using the time dependent Smoluchowski equation in a Coulomb potential, the analytical solution of which was given by Hong and Noolandi[2]. The decay curve strongly depends on the R_0 and we have obtained the value of 30-40 Å by simulation.

By adding benzonitrile as an electron scavenger, the decay time of the cation band increased. The yield of the cation radical increased with increasing concentration of benzonitrile. These results suggested that the electron was trapped by scavenger and this caused the drastic decrease of the mobility of the negative charge. In isooctane the yield of photoionization seemed high and we can estimate the R_0 from the dependence

Springer Series in Chemical Physics, Vol. 48: **Ultrafast Phenomena VI**
Editors: T. Yajima · K. Yoshihara · C.B. Harris · S. Shionoya

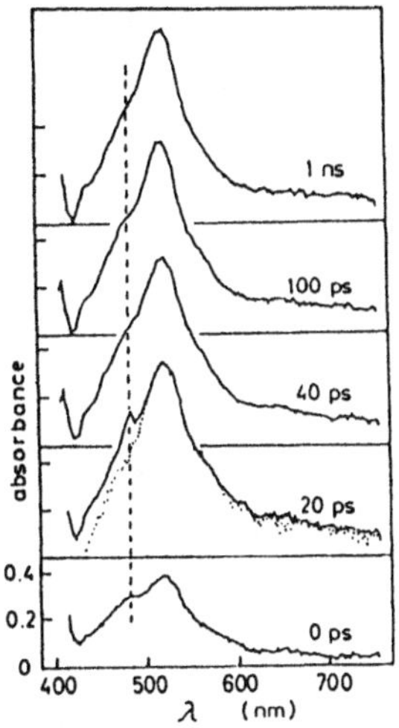

Fig. 1. Picosecond time resolved absorption spectra of aminopyrene in t-decalin. Delay times after the laser excitation are indicated

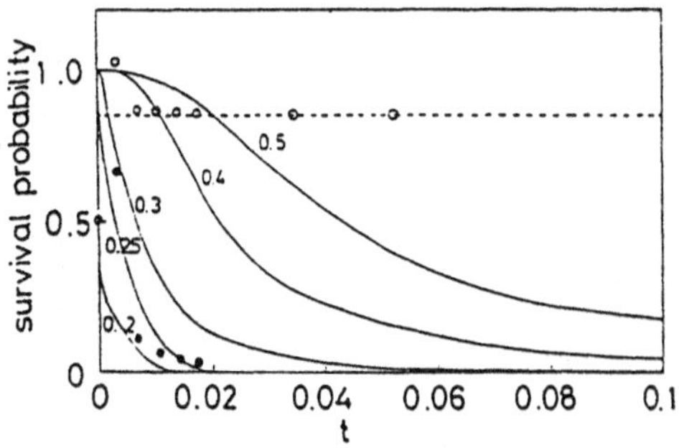

Fig. 2. Time dependence of the transient absorbance measured at 480 nm and expected decay curves of geminate ion pairs of several initial separations. The unit of time is $r_c^2/4D$

of the yield of trapped electron on the concentration of the scavenger[3]. The R_0 estimated in this manner was 40-50 Å which was slightly larger than that obtained from the decay curve measurements in t-decalin without electron scavenger. In radiation chemistry larger R_0s are obtained, although most of them are obtained for the anion-cation geminate pairs formed by the electron and hole transfer[4]. The shorter separation we obtained should be explained as follows. Since in the present case the excitation energy was much less than that of the electron pulse excitation used in radiation chemistry, the ejected electron should have less energy and be thermalized near the parent molecule. The measurements of the excitation energy dependence of R_0 are urgently desirable, which should give the detailed elucidation of the photoionization processes in solution.

The R_0 obtained from the scavenger experiments may be slightly larger than that from the direct measurements of electron-cation pairs because it is difficult to scavenge electrons formed in the proximity to the parent cation, which should recombine rapidly. The difference between the R_0 value of 30-40 Å and 40-50 Å may be explained in this manner, although these experiments were done in different solvents.

3 PYRENE

Figure 3 shows the picosecond time resolved absorption spectra of pyrene/benzonitrile in cyclohexane solution excited at 355 nm. The bands of the pyrene cation radical and $S_n \leftarrow S_1$ absorption of pyrene were observed. The measurements of the excitation intensity dependence clearly showed that the ionization was multiphoton process. Although in the spectra at short delay times the cation band was rather clear compared with those at long delay times, the contribution of the cation band was large even at delay times longer than 1 ns. The cation band showed rapid decay in hundreds of picosecond region, while the decay seemed in a few nanosecond.

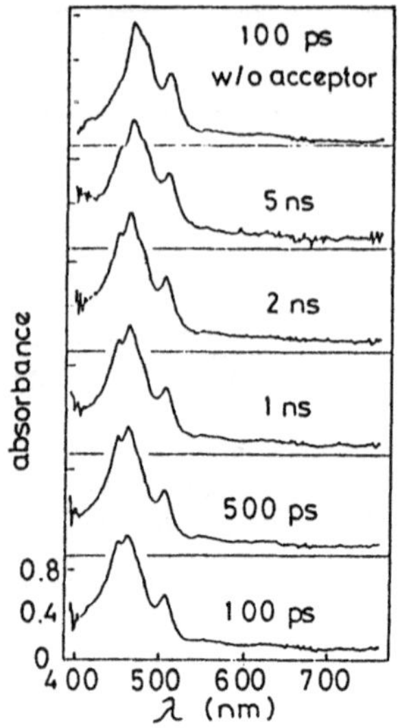

Fig. 3. Picosecond time resolved absorption spectra of the pyrene/benzonitrile in cyclohexane system. Delay times after the laser excitation are indicated in the figure

By using the mobility of typical molecular ions and the initial separation of 40-50 Å, we have estimated that the geminate recombination should occur in more than 10 ns. Therefore, the decay of the cation band should not be prominent in our measurements. If we use shorter separation the decay may occur in less than nanosecond, while the dissociation yield become quite low and leading to a quite low cation yield at 5 ns. These results strongly suggest that the photoionization process in solution is not so simple as it was believed. At least two mechanisms should be involved there. One may be quite similar to that observed in radiation chemistry, although the initial separation depends on the excitation energy. The other is the electron transfer from the highly excited neutral state to the electron acceptor, which will be observed only for the system of high concentration of electron scavenger because of very short lifetime of such excited state. The latter was actually observed by us for the pyrene/biphenyl/isooctane system[5]. In the present case the former correspond to the long lived component and the latter to the short lived one.

ACKNOWLEDGMENT

The support by a Grant-in-Aid(No. 62065006) from the Japanese Ministry of Education, Science and Culture is acknowledged.

REFERENCES

1. Y. Hirata, N. Mataga, Y. Sakata, and S. Misumi: J. Phys. Chem. 90, 6065 (1986), C. L. Braun and T. W. Scott: J. Phys. Chem. 91, 4436 (1987)
2. K. M. Hong and J. Noolandi: J. Chem. Phys. 68, 5163 (1987).
3. A. Hummel: In Advances in Radiation Chemistry, ed. by M. Burton and J. L. Magee, Vol. 4 (Wiley, 1974)
4. Y. Yoshida, S. Tagawa, and Y. Tabata: Radiation Phys. Chem. 28, 201 (1986).
5. Y. Hirata and N. Mataga: J. Phys. Chem. 89, 2439 (1985).

Picosecond Ketyl Radical Spectroscopy

N.A. Borisevich, N.A. Lysak, S.A. Tikhomirov, and G.B. Tolstorozhev

Institute of Physics, BSSR Academy of Sciences, 220602,
Leninsky Prospekt 70, Minsk, USSR

In the present work, by the methods of picosecond spectroscopy the influence has been studied of the solvent polarity, the type of the substituent in the BP molecule and the type of the hydrogen donor on the dynamics and mechanisms of charge transfer (CT) and ketyl radical formation. The results are presented of the investigations of intramolecular and intermolecular charge transfer for benzophenone and 4-dimethylaminobenzophenone. As hydrogen donors, aromatic (N,N-diethylaniline, DEA) and aliphatic (triethylamine, TEA) amines have been used.

In experiments an automated laser spectrometer was used [1]. The molecules being investigated were excited by third-harmonic single pulses from mode-locked Nd:glass laser (τ_p= 6 ps, λ_{exc} = 352 nm, E_p = 0.5 mJ). As the probe the wide-band picosecond continuum was used.

I. Fig.1 shows transient absorption spectra for the BP+TEA system in acetonitrile at different time delays θ between the exciting and probing pulses. At θ = 20 ps the absorption spectrum of the two-component system involved coincides with the spectrum of $T_1^* \rightarrow T_n^*$ absorption of pure BP. For θ = 500 ps the transient absorption spectrum of the system involved coincides with the spectrum of $D_0 \rightarrow D_1^*$ ketyl radical absorption. With θ from 20 ps to 500 ps no new absorption bands appear in the $\sim$660 - 710 nm region where the anion-radicals of BP absorp. This is direct evidence that in this case the hydrogen atom is transferred as a unit and the ketyl radical is formed directly in the encounter complex of triplet-state BP molecules and unexcited TEA molecules (one-stage mechanism).

For the BP+DEA system, along with the one-stage mechanism, new mechanisms of ketyl radical formation manifest themselves. As seen from Fig.2 the transient absorption spectrum of this binary system in acetonitrile (ε = 37) and hexane (ε= 1.4) at θ = 20 ps is a superposition of the absorption spectra of anion-radicals BP^- (λ_{max} = 700 nm), cation-radicals DEA^+ (λ_{max} = 480 nm) and $T_1^* \rightarrow T_n^*$ BP absorption. This points to the fact that for the system involved the photoreduction of BP has two stages, the electron transfer with the formation of an ion-radical pair taking place at the first stage. The characteristic rise-time of the optical density of ion-radical absorption

Springer Series in Chemical Physics, Vol.48: Ultrafast Phenomena VI
Editors: T. Yajima · K. Yoshihara · C.B. Harris · S. Shionoya

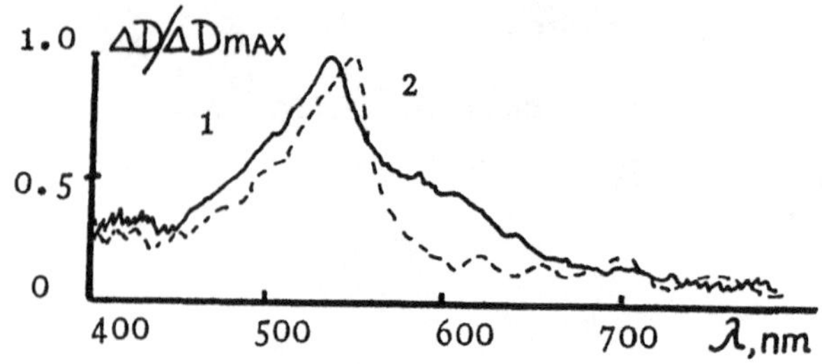

Fig.1. Transient absorption spectra of the BP ($C = 10^{-2}$ M)+TEA ($C = 1$ M) system in acetonitrile: 1- θ = = 20 ps; 2 - θ = 500 ps

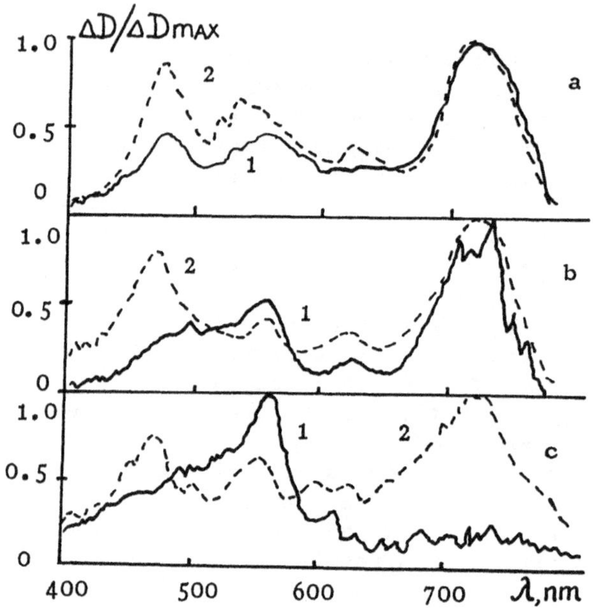

Fig.2. Transient absorption spectra of the BP+DEA ($C = 1$ M) system in hexane (1) and acetonitrile (2): a - θ = 20 ps; b - θ = 400 ps; c - θ = 1600 ps

in the spectral regions of λ_{max} = 480 and 720 nm turned out to be shorter than 5 ps. Such fast formation of CT states is surprising from the point of view of the commonly accepted ideas about the participation of triplet-state BP molecules (the characteristic time of intersystem crossing for BP in dences that at the DEA ($C = 1$ M) concentrations used the part of unexcited BP molecules forms a weak complex with DEA.

Further evolution of the present binary system in a polar and a nonpolar solvent proceeds differently. In a polar solvent acetonitrile the excited CT complex dissociates into ion--radicals. The ketyl radical formation in this case occurs in a time of about 1600 ps at collision of the anion-radical BP^- and the cation-radical DEA^+. By this time the integrated spectrum represents a superposition of the absorption bands of the ion-radicals BP^-, DEA^+ and the ketyl radical $\cdot$BP-H. In nonpolar hexane, a ketyl radical is formed at proton transfer in the excited CT complex (exciplex). The proton transfer rate in the excited complex estimated by the disappearance of the BP- ion--radical absorption turned out to be $K = 2.9 \cdot 10^9 s^{-1}$.

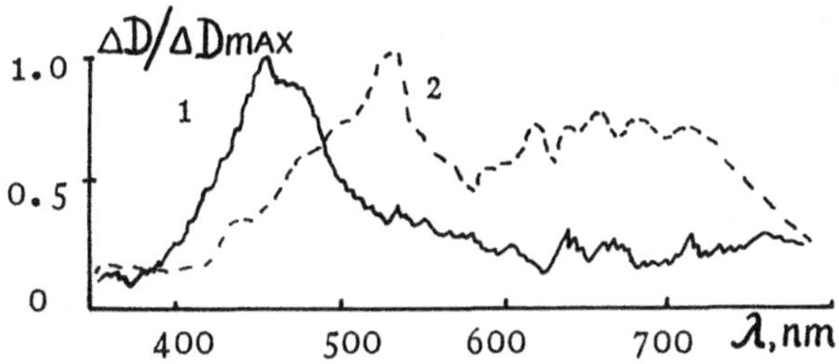

Fig.3. Transient absorption spectra of DMABP ($C = 10^{-2}$ M) in acetone (1 - θ =20 ps) and pentane (2 - θ = = 150 ps)

II. Earlier it was assumed that the attachment of aminoalcyl groups to 4 position of BP molecule must suppress the photoreduction of BP [2]. Fig.3. shows transient absorption spectra for 4-dimethylaminobenzophenone (DMABP) in pentane (ε = = 1.84) and acetone (ε = 20.7). In nonpolar solvent, the transient absorption spectrum is a superposition of the $T_1^* \rightarrow T_n^*$ absorption of BP (λ_{max} = 530 nm) and the absorption from the CT state (λ_{max} = 690 nm).

The transient absorption spectrum of DMABP in acetone is a relatively narrow band (λ_{max} = 460 nm) and corresponds to the absorption spectrum of twisted intramolecular charge transfer (TICT) states in polar solvents [3].

The differences between the two types of CT states of a DMABP molecule appear as differences in the formation and deactivation rates of these forms. The formation time of the TICT state is shorter than 5 ps, and the time of its deactivation depends on the solvent polarity and is equal to: τ =320 ps for acetonitrile, τ = 450 ps for acetone, τ = 890 ps for diethyl ether. In nonpolar solvent, the process of CT state formation is accomplished for 300 ps. The decrease in the rate of CT state formation in a nonpolar solution is likely to be due to the smaller probability of intramolecular detachment of an electron at $n \rightarrow \pi^*$ excitation as against the probability of TICT state formation at $\ell \rightarrow \pi^*$ excitation in polar solvents.

References

1. N.A.Lysak, S.V.Melnichuk, S.A.Tikhomirov, G.B.Tolstorozhev: Zh.Prikl.Spektrosk. 47, 267 (1987)
2. N.J.Turro: Molecular Photochemistry (W.A.Benjamin, Inc., New York, Amsterdam 1965)
3. C.Rulliere, Z.R.Grabowski, J.Dodkowski: Chem.Phys.Lett. 137, 408 (1987)

Electron Transfer Rates in Covalently Linked Donor–Acceptor Systems

S. Doraiswamy, G.B. Maiya, N. Periasamy, and B. Venkataraman

Chemical Physics Group, Tata Institute of Fundamental Research, Bombay, 400005, India

1. Introduction : Covalently linked energy/electron donor-acceptor systems are important components for directed energy/electron flow in molecular devices for harvesting solar energy. Several electron donor-acceptor (D-A) systems containing porphyrin/metalloporphyrin and poly-nitrobenzene were synthesized in which the D-A distance varies. The electron transfer rates when the donor is in the singlet excited state were determined in several solvents.

2. Experimental : The details of the synthesis of the covalently linked molecules are given elsewhere/1/. In all, twelve molecules were synthesized and these are labelled as follows : D-p-A, D-m-A and D-o-A where D is either tetraphenylporphyrin (TPP) or Zinc tetraphenylporphyrin (ZnTPP) and A is either 2,4-dinitrobenzene (DNB) or 2,4,6-trinitrobenzene (TNB). p,m,o refer to the point of attachment of A in the meso phenyl group of D, and the linkage between D and A is made through an oxygen atom (ether linkage). The synthesized molecules were purified by column chromatography. The solvents (toluene, $CHCl_3$, CH_3OH, CH_3CN and propylene carbonate (PC)) used were fluorescence-free, analar or spectrograde quality. Fluorescence decay profiles were obtained by picosecond laser excited time-correlated single photon counting method. A detailed description of the experimental set-up is given elsewhere/2/. The analysis of fluorescence decay data to fit a single or multi-exponential function was done by the non-linear least-squares method using iterative reconvolution and Marquardt procedure for the optimisation of parameters.

3 Results and Discussions : The fluorescence of ZnTPP (τ_o varying from 1.62 to 2.07 ns in different solvents) is efficiently quenched by 2,4,6-trinitro toluene (TNT) in intermolecular reactions. The quenching rate coefficients are 1.4×10^{10} $M^{-1}s^{-1}$ ($CHCl_3$) and 3.6×10^{9} $M^{-1}s^{-1}$ (PC), which are slightly higher than the diffusion-limited bimolecular rate constant, k_D (= 8 RT/3000η) at 25°C. The fluorescence quenching of ZnTPP by m-dinitrobenzene is slightly less efficient : 9.1×10^{9} $M^{-1}s^{-1}$ ($CHCl_3$) and 3.3×10^{9} $M^{-1}s^{-1}$ (PC). The fluorescence quenching of TPP (τ_o varying from 8.0 ns to 10.9 ns in the different solvents) by trinitrotoluene and m-dinitrobenzene are even less efficient :4.5×10^{9} $M^{-1}s^{-1}$ for TPP-TNT in $CHCl_3$ and 3.2×10^{9} $M^{-1}s^{-1}$ for TPP-DNB in $CHCl_3$. The above results are in agreement with the variation of free energy change, ΔG^o, calculated based on electron transfer mechanism using the redox potentials (Vs SCE) for ZnTPP (0.81 V), TPP (1.06 V), DNB (-0.93 V) and TNT (-0.85 V) measured in CH_2Cl_2/3/.

The fluorescence decay of the covalently linked molecules were fitted to one or two-exponential decay functions which are generally adequate for all cases, except a few. The observation of a short-lifetime component (in picoseconds) with a significant amplitude is associated with the existence of conformations of the covalently linked molecule which are favourable for electron transfer.

Springer Series in Chemical Physics, Vol. 48: **Ultrafast Phenomena VI**
Editors: T. Yajima · K. Yoshihara · C.B. Harris · S. Shionoya

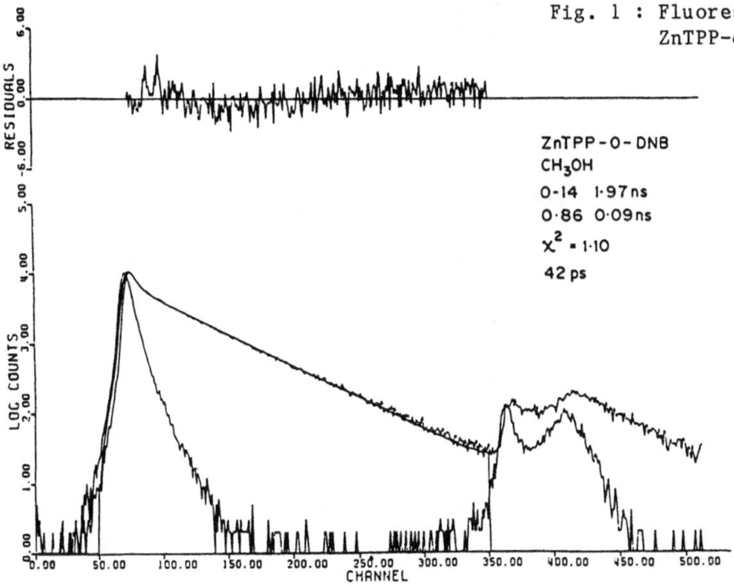

Fig. 1 : Fluorescence decay of ZnTPP-o-DNB in CH_3OH.

The fluorescence decay data of ZnTPP-p,m,o-DNB systems showed the presence of the short lifetime component in all the solvents. The lifetime of the short component varied from 90 ps (CH_3OH) to 150 ps (PC) for ZnTPP-o-DNB in polar solvents, whereas the lifetime of the long component (identified with the unreacted extended form of the molecule) is in nanoseconds (1.97 ns CH_3OH and 2.02 ns in PC). Figure 1 shows the fluorescence decay of ZnTPP-o-DNB in CH_3OH, along with the results of the fitting. The short lifetime is associated with the efficient electron transfer reaction and the electron transfer rate (k_{et}) was calculated using eqn. (1):

$$k_{et} = \tau_s^{-1} - \tau_1^{-1} \qquad (1)$$

where τ_1 is the lifetime of the long component and τ_s is that of the short component. The values of k_{et} are given Table 1 for ZnTPP systems in all the solvents.

Table 1 : Electron transfer rate ($10^{-9}k_{et}$, sec^{-1}) for ZnTPP-A in various solvents

Solvent / Molecule	Toluene	$CHCl_3$	CH_3OH	CH_3CN	PC
ZnTPP-p-DNB	0.5	2.7	2.9	1.7	2.4
ZnTPP-m-DNB	0.65	4.1	3.2	2.0	2.0
ZnTPP-o-DNB	3.2	8.5	11	9.5	6.2
ZnTPP-p-TNB	0.42	0.37	1.0	0.63	2.9
ZnTPP-m-TNB	0.27	0.30	0.34	0.60	0.56
ZnTPP-o-TNB	0.33	0.15	0.08	0.21	0.18

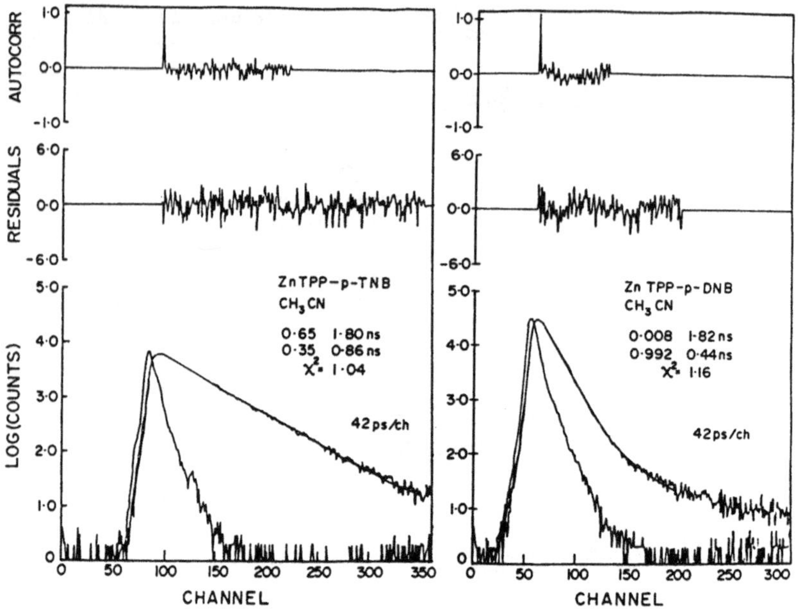

Fig. 2 : Fluorescence decay curves of ZnTPP-p-A (A = DNB, TNB) in CH_3CN.

The excited state electron transfer in ZnTPP-p,m,o-TNB systems was not as efficient as in the corresponding DNB system in all the solvents with the possible exception of ZnTPP-p-A in PC. Figure 2 shows the fluorescence decays of ZnTPP-p-A (A = DNB and TNB) system in CH_3CN. The k_{et} values for TNB systems are in general lower than those of the corresponding DNB systems (Table 1).

For TPP-p,m-A systems no short-lifetime component was observed in all the solvents indicating the absence of efficient electron transfer. However for TPP-o-A systems short-lifetime component (the lowest value in CH_3OH is 360 ps) was observed.

The values of k_{et} tabulated in Table 1 reveal two features. (1) The value of k_{et} is uniformly higher for ZnTPP-DNB system than that of the equivalent TNB system in all the solvents. (2) The values of k_{et} are seen to be in the order of -o> -m> -p for ZnTPP-DNB system whereas for ZnTPP-TNB system the trend is that -p> -m> -o. The trend observed for the ZnTPP-TNB systems is at variance with the general expectation that (i) k_{et} increases with decreasing distance between the donor and acceptor, and (ii) k_{et} increases with the decrease in the reduction potential of the acceptor except in the 'inverted' region.

The authors wish to thank the Department of Science and Technology of the Government of India for financial support to set up the Unit for Chemical Dynamics and Picosecond Spectroscopy.

References :

1. G. B. Maiya and V. Krishnan, J. Phys. Chem., 89, 5225 (1985)

2. N. Periasamy, S. Doraiswamy, G. B. Maiya and B. Venkataraman, J. Chem. Phys., 88, 1638 (1988)

3. G. B. Maiya and V. Krishnan, Unpublished Results

Excitation Transfer and Photo-Induced Electron Transfer in Conformationally Restricted Porphyrin Systems

A. Osuka[1], *K. Maruyama*[1], *I. Yamazaki*[2], *and N. Tamai*[2]

[1]Department of Chemistry, Faculty of Science,
Kyoto University, Kyoto 606, Japan
[2]Institute for Molecular Science, Myodaiji, Okazaki 444, Japan

Green-plants and bacterial photosynthetic organisms employ closely associated tetrapyrrole pigments as carriers of photons and electrons. The recent X-ray analysis of the photosynthetic reaction center from two photosynthetic bacteria have shown that photo-active pigments are held in carefully controlled arrangements that optimize the efficiency of photon or electron transfers.(1) A useful approach to study the primary events of natural photosynthesis is to synthesize model compounds with known distances and orientations and elucidate the geometrical dependence of energy or electron transfer process.

Now we wish to report here the synthesis of a series of conformationally restricted dimeric porphyrins bridged by naphthalene or spiro-biindane spacer. In these dimeric porphyrins, the two porphyrin rings are rigidly fixed in a certain geometry by the combined steric restrictions of the spacer and the flaking ethyl groups. Spiro-biindane spacers were employed with an aim to mimicking twisted relative orientations of chromophores such as hemes and chlorophylls in natural systems.

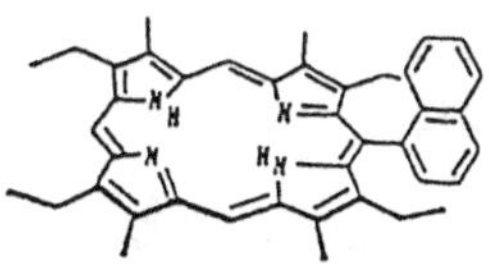

1-NP

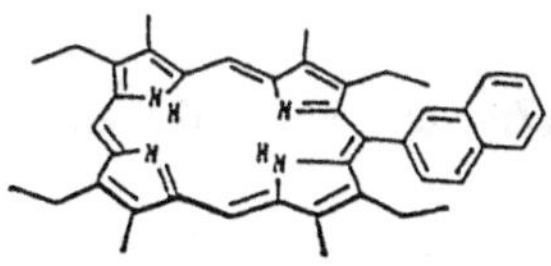

2-NP

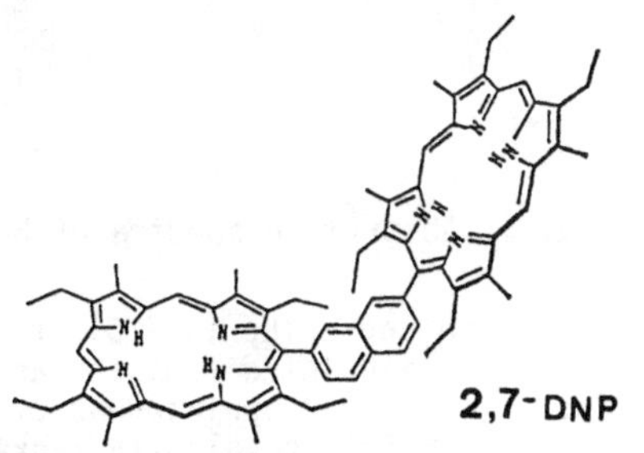

2,7-DNP

Diporphyrin model compounds were synthesized from the corresponding aryl dialdehydes by the method of Chang.(2) Bis-zinc and mono-zinc complexes of these diporphyrins were synthesized by the reaction with zinc acetate in dichloromethane and purified by flash column chromatography. The absorption spectra of 1-NP(Zn) and 2-NP(Zn) showed a Soret peak at 405.5 nm and Q-bands at 532.5 and 570 nm in dichloromethane. All bis-zinc complexes of naphthalene-bridged dimeric porphyrins (NDP) showed split Soret bands, but the shape and peak position of these dimers were unchanged relative to 1-NP(Zn), except for 1,3-NDP(Zn2) and 1,4-NDP(Zn2), which showed Q-bands at 538 and 573 nm. While the head-to-tail dimer 2,6-NDP(Zn2) displayed a mainly red-shifted Soret band at 412 nm, the face-to-face dimer 1,8-BDP(Zn2) displayed a blue-shifted Soret band at 388 nm. Interestingly, the relative intensities of the split Soret bands changed depending upon the dihedral angle of the two porphyrin rings. (Fig. 1) A plot of the dipole-dipole exciton energy (ΔE) vs. the geometrical parameter, $2(\cos\alpha + 3\cos\theta_1\cos\theta_2)/r^3$, showed a good linear correlation between the two quantities. (Fig. 2) From the slope, the magnitude of the transition dipole moment of the monomer effective for the exciton coupling was determined to be 7.9 D. This is in good agree-

Springer Series in Chemical Physics, Vol. 48: **Ultrafast Phenomena VI**
Editors: T. Yajima · K. Yoshihara · C.B. Harris · S. Shionoya

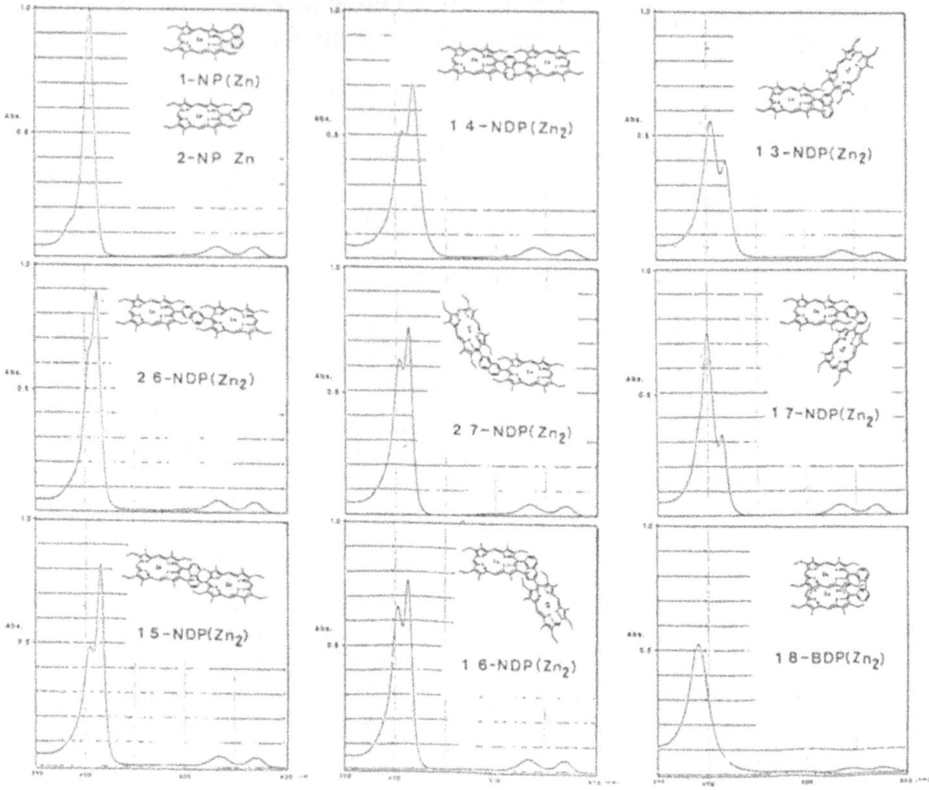

Fig. 1. Absorption spectra of NP(Zn) and NDP(Zn2) in CH_2Cl_2.

ment with the value (9.7 D) of the transition dipole moment of the monomer, which is calculated from the absorption spectrum of 1-NP(Zn). These results indicate the appropriateness of the treatment of the simple exciton theory for this covalently linked system in solution.(3)

As was reported in zinc-porphyrins (Zn-P) covalently bound to metal-free porphyrins (H2-P) with more flexible linkages (4), the intramolecular singlet excitation energy transfer from Zn-P to H2-P was observed in mono-zinc complexes of these dimeric porphyrins. The rate of the energy transfer (ke) were determined by measuring fluorescence decays (λem 580 nm) of the Zn-P as well as fluorescence rises (λem 690 nm) of the H2-P, using the pico-second time-resolved fluorescence spectroscopy.(5) Notably, all these mono-zinc complexes exhibited a single decaying component (4.9 - 197 ps), indicating that the singlet excited states of the Zn-P are quenched by the H2-P through single conformation. This situation permits us to analyze the geometry-dependence of the rapid intramolecular energy transfer process. Taking the radiative lifetime of 1-NP(Zn) (1.5 ns, τ_0), we can calculate the rate of energy transfer, ke, from the equation, $ke=1/\tau - 1/\tau_0$. These rates were in good agreement with the value calculated from the fluorescence rise of the H2-P. (Table 1).

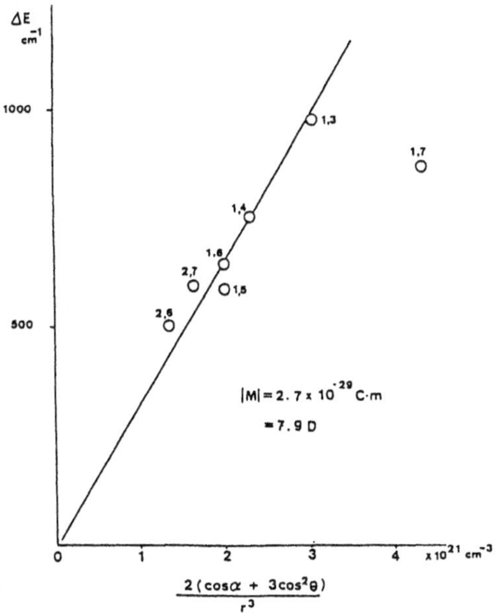

Fig. 2. Plot of exciton coupling energy ΔE vs geometrical parameter, $2(\cos\alpha + 3\cos\theta 1\cos\theta 2)/r^3$.

Table 1. Intramolecular Excitation Energy Transfer from Zn-P to H2-P.

porphyrin	τ_f(Zn-P)[a] (ps)	ke (x $10^{-10}s^{-1}$)	τ_r(H2-P)[b] (ps)	ke (x $10^{-10}s^{-1}$)
1,8-ADP	4.9	20	8.8	11
1,3-NDP	14	7.0	23	4.4
1,4-NDP	19	5.2	29	3.5
1,5-NDP	19	5.2	31	3.2
1,6-NDP	28	3.5	33	3.0
1,6-NDP'	22	4.5	33	3.0
1,7-NDP	14	7.0	28	3.5
2,6-NDP	45	2.2	45	2.2
2,7-NDP	36	2.7	47	2.1

a, Fluorescence lifetimes of Zn-P. b, Fluorescence rise times of H2-P.

c, Measured in air-saturated CH_2Cl_2 at 25°C.

1. J. Deisenhofer et al.,J. Mol. Biol., 180, 385 (1984).
2. C. K. Chang et al., J. Org. Chem., 48, 385 (1983).
3. A. Osuka and K. Maruyama, J. Am. Chem. Soc., 110, 4454 (1988).
4. R. L. Brookfield et al., J. Chem. Soc., Faraday Trans. 2, 82, 219 (1986), and references cited therein.
5. I. Yamazaki et al., Rev. Sci. Instrum., 56, 1187 (1985).

Part XI

Dynamics of Biological Processes

Ultrafast Spectroscopy of Biological Processes

J.W. Petrich[1], *J.L. Martin*[2], *and J. Breton*[3]

[1]As of January 1989: Department of Chemistry, Iowa State University, Ames, IA 50011, USA

[2]Laboratoire d'Optique Appliquée, Ecole Polytechnique, ENSTA, INSERM U275, F-91128 Palaiseau Cedex, France

[3]Service de Biophysique, Centre d'Etudes Nucléaires de Saclay, F-91191 Gif-sur-Yvette Cedex, France

1. Introduction

Ultrafast absorption and Raman spectroscopy are powerful means of obtaining information concerning protein dynamics. Here we discuss our experiments on photodissociation and rebinding of ligands to hemeproteins and on the primary charge separation in bacterial reaction centers. Even though the R to T transition in hemoglobin occurs in microseconds [1], studies on the ultrafast time scale provide a significant advantage in understanding such a process. Namely, the system has not had enough time to respond completely to the perturbation of ligand photodissociation. Hence the system is simplified as spectral changes can be more easily interpreted in terms of structural features near the chromophore and involvement of portions of the molecule far-removed from the chromophore need not necessarily be invoked. Furthermore, even if a protein reaction is very slow, the events that trigger the reaction may proceed very rapidly. In the reaction center, the primary electron transfer occurs in 2.8 ps at room temperature and a major problem is to understand how the protein is involved in rendering this reaction so efficient [2].

2. Vibrational Relaxation and Dissipation of Strain Energy in Photodissociated Carbonmonoxy Hemoglobin

The microscopic aspects of reversible ligand binding in hemoglobin (Hb) and myoglobin (Mb) are not yet understood. In particular, the cooperative mechanism by which successive ligands bind more easily to the Hb tetramer than the first ligand has yet to be unravelled. Time-resolved spectroscopic methods provide an excellent tool for investigating the rate and the mechanism by which information is transferred from one subunit of the Hb tetramer to another. Our subpicosecond Raman experiments [3] have yielded two important results bearing upon this question (Figs. 1, 2). Upon photodissociation of HbCO, a large downshift of the totally symmetric porphyrin mode ν_4 relaxes to a minimum in about 10 ps. This relaxation is attributed to vibrational cooling of the heme which has absorbed a photon of approximately 25 kcal/mol in excess of the bond dissociation energy. Such cooling has been predicted by steady-state measurements [4] and by molecular dynamics simulations [5]. After this rapid phase, ν_4, begins to downshift again with a time constant of about 30 ps. We have interpreted this second phase as the appearance of a T-like change at the intersubunit contacts. Due to tight coupling between the subunit interface and the F helix [3,6], this change accompanies a tilt of the proximal histidine which perturbs the porphyrin ring and is responsible for the downshift.

Springer Series in Chemical Physics, Vol. 48: **Ultrafast Phenomena VI**
Editors: T. Yajima · K. Yoshihara · C.B. Harris · S. Shionoya

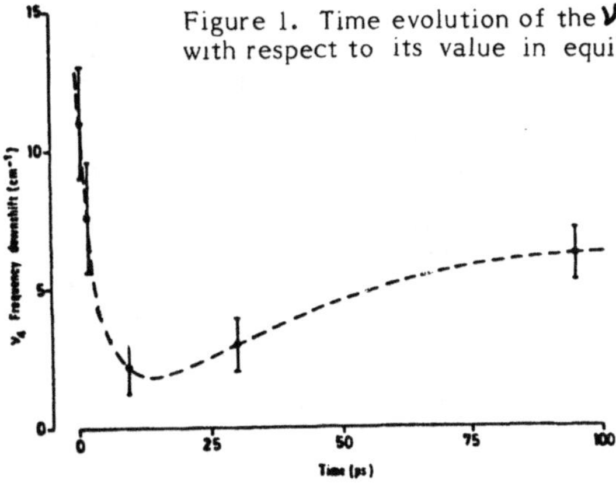

Figure 1. Time evolution of the ν_4 mode of the HbCO photoproduct with respect to its value in equilibrium unligated Hb.

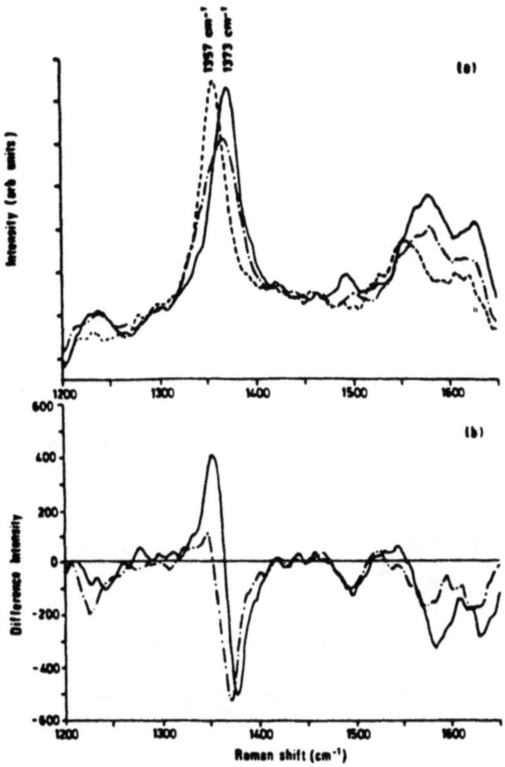

Figure 2. a) A totally symmetric stretching mode of the heme of the equilibrium unligated Hb and ligated HbCO is indicated at 1357 and 1373 cm^{-1}, respectively. The broad band is the 200-fs spectrum of this mode of 50% dissociated HbCO. b) The solid line is the difference spectrum between equilibrium unligated HbCO and unligated Hb. The broken line is the 200-fs spectrum of photodissociated HbCO from which the contribution from the ligated species has been subtracted [3].

3. Geminate Recombination of NO as a Probe of the Barrier to Ligand Rebinding

The barrier to ligand rebinding is a combination of electronic [7,8] and transiently evolving steric factors. Hence, the rate of recombination is sensitive to whether the ligand is CO, O_2, or NO [7]. The dramatic effect of NO observed in Fig. 3 is a result of the nature and relative population of the excited-state species created upon photodissociation and the presence of a rapid geminate recombination process (~10 ps) [7,8]. Because NO rebinding is so rapid, the geminate recombination of NO with hemeproteins provides a very sensitive probe of the heme pocket and parts

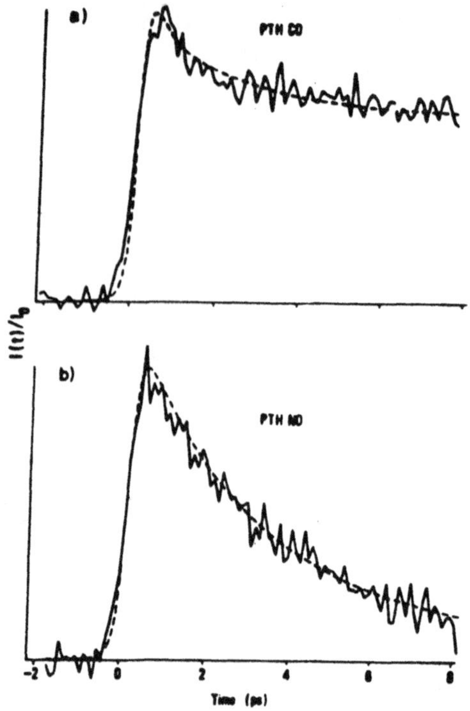

Figure 3. Transient bleaching of the Soret band of ligated PTH. a) PTHCO, 410-nm probe. The maximum absorption change is 0.17. b) PTHNO, 401-nm probe. The maximum absorption is 0.04. The 7-ps component represents geminate recombination of NO with PTH. The similarity among the bleaching kinetics of HbCO and PTH-CO and of HbNO and PTHNO are striking [7]. They suggest that purely electronic factors play a significant role in geminate recombination at room temperature.

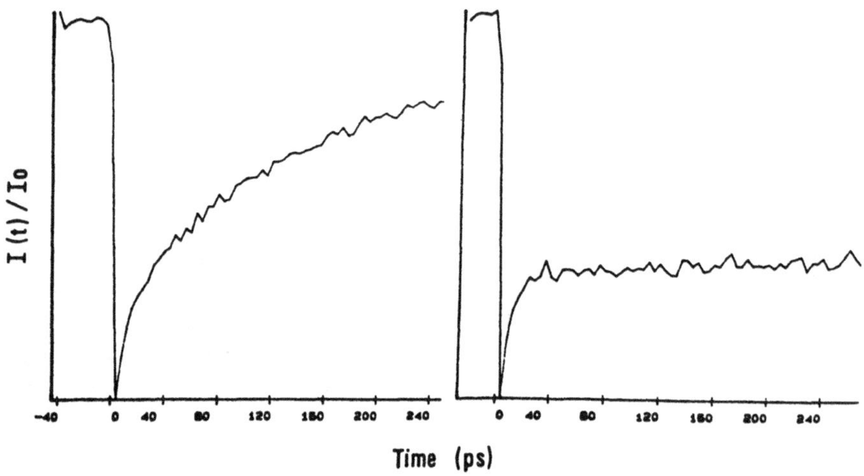

Figure 4. Geminate rebinding of NO to soybean Lb (left) and to Mb from Aplysia punctata (right) at room temperature. Even on a picosecond time scale at 300 K, the rebinding is well described by a power law function [8] indicative of a distribution of barriers to rebinding [11].

of the protein that are tightly coupled to it [8], thus providing a means of evaluating the contributions of the proximal and distal sides of the heme in the rebinding process. A striking result is that we find essentially identical rebinding kinetics of NO to horse heart Mb and to soybean leghemoglobin (Lb, a monomeric protein whose spacious hemepocket affords one of the highest known oxygen affinities [9]). Although the gross change in size of hemepockets between Mb and Lb does not alter the rebinding kinetics, it is nevertheless possible that specific alterations of the hemepocket may produce an effect. That this is the case is suggested by the rebinding of NO to Mb from Aplysia punctata (Fig. 4). Myoglobins from Aplysia are unusual in that they lack in general the distal histidine [10].

4. Charge Separation in Photosynthetic Bacterial Reaction Centers

The conversion of light energy into chemical free energy in the reaction center (RC) of photosynthetic purple bacteria is a highly efficient process that involves very fast reactions that compete efficiently with radiative lifetimes. The primary charge separation occurs between a bacteriochlorophyll dimer (P) and a bacteriopheophytin molecule (H_L) located on the L polypeptide subunit over a center-to-center distance of 17 Å. The structure of the RC, as solved by X-ray crystallography, shows that a monomeric bacteriochlorophyll (B_L) is located between P and H_L. The role of this molecule in the initial charge separation process is not yet understood and is the object of much experimental and theoretical scrutiny [2].

Direct excitation of the primary donor (P) leads to the formation of the radical pair $P^+H_L^-$ within 2.8 ps at 295 K in RCs from both Rb. sphaeroides and Rps.

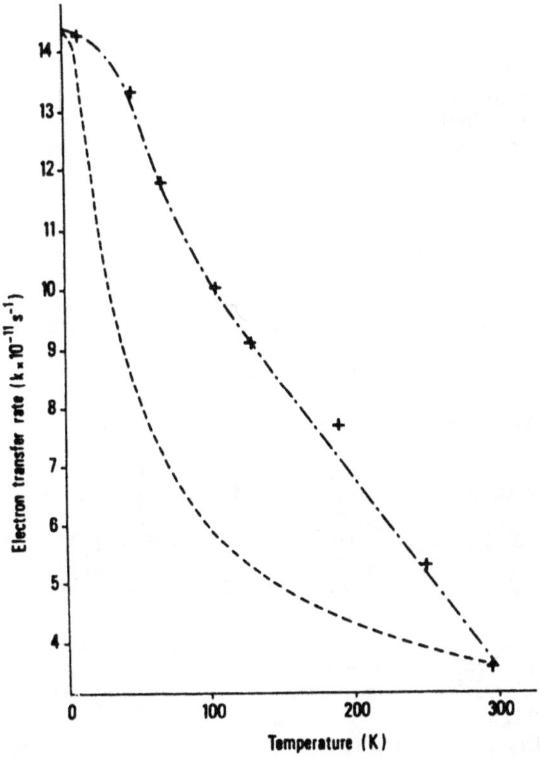

Figure 5. The variation of the electron transfer rate over the temperature range 10-300 K for RCs from Rps. viridis. The crosses represent the experimental points and the mixed line drawn through them serves only to aid the eye. The dashed line is a fit to the single mode approximation discussed in references 2 and 15. These data were obtained in collaboration with G.R. Fleming.

viridis. No evidence for a transient depletion of B_L as a result of the formation of a B_L^- or B_L^+ species is found. While this observation must be reconciled with the X-ray data, it is generally believed that B_L plays a role in the charge separation process. Several mechanisms for such a role have been suggested and are compatible with the absence of spectroscopic evidence for transient bleaching of the ground state absorption of B_L: 1) direct electron transfer from P* to H_L ($P^*B_LH_L \rightarrow P^+B_LH_L^-$) occurs via the virtual state $P^+B_L^-H_L$ in a "superexchange" mechanism; 2) the existence of B_L^- or B_L^+ has been postulated, but requires, in order to be consistent with femtosecond aborption spectroscopy, that the rate of depletion of this intermediate is much faster than its rate of formation; 3) an alternative to the two models above has been proposed [12] which uses a set of strongly coupled vibrational modes and is able to model hole burning and temperature dependence data [2]. See references 2, 13, and 14 for a summary and discussion of the possible mechanisms. The primary charge separation process is characterized by an apparent negative activation energy with an increase in rate by a factor of four for a decrease in temperature from 300 to 10 K (Fig. 4) and such behavior is expected for an activationless electron transfer process [14]. If we assume that the electron transfer is nonadiabatic, then the conventional expression for the rate can be used, $k = (2\pi/\hbar)|V|^2F$, where V is the electronic interaction matrix and F is a thermally averaged Franck-Condon factor. Assuming a coupling of the primary electron transfer to low frequency modes of the protein and using the single mode approximation [2,15], the electron transfer rate as a function of temperature for RCs from Rb. sphaeroides can be well fit using a 80 cm^{-1} mode [15]. The same approximation, however, gives a very poor fit to the data for the RCs from Rps. viridis and a 25 cm^{-1} mode is required to obtain the proper ratio of the rates at 10 and 300 K (Fig. 5). The difference in temperature dependence between the RCs of the two bacterial species may indicate that V decreases more strongly with decreasing temperature in Rps. viridis than in Rb. sphaeroides. It also, however, raises the question of whether the primary electron transfer is accurately described by a nonadiabatic process.

5. References

1. J. Hofrichter, J.H. Sommer, E.R. Henry, and W.A. Eaton, Proc. Natl. Acad. Sci. USA 80, 2235 (1983).
2. The Photosynthetic Bacterial Reaction Center, J. Breton and A. Veréglio, Eds. (NATO ASI Series, Plenum, New York).
3. J.W. Petrich, J.L. Martin, D. Houde, C. Poyart, and A. Orszag, Biochemistry 26, 7914 (1987).
4. S.A. Asher and J. Murtaugh, J. Am. Chem. Soc. 105, 7244 (1983).
5. E.R. Henry, W.A. Eaton, and R.M. Hochstrasser, Proc. Natl. Acad. Sci. USA 83, 8982 (1986).
6. B.R. Gelin, A W.-M. Lee, and M. Karplus, J. Mol. Biol. 171, 489 (1983).
7. J.W. Petrich, C. Poyart, and J.L. Martin, Biochemistry 27, 4049 (1988).
8. J.W. Petrich, J.C. Lambry, F. Stetzkowski, C. Poyart, and J.L. Martin. Submitted for publication.
9. E.G. Arutyunyan, Mol. Biol. 15, 27 (1981); J.B. Wittenberg, C.A. Appleby and B.A. Wittenberg, J. Biol. Chem. 247, 527 (1972).
10. L. Tentori, G. Vivaldi, S. Carta, M. Marinucci, A. Massa, E. Antonini, and M. Brunori, FEBS Lett 12, 181 (1971).
11. R.H. Austin, K.W. Beeson, L. Eisenstein, H. Frauenfelder, and I.C. Gunsalus, Biochemistry 24, 5355 (1975).
12. Y. Won and R.A. Friesner, Proc. Natl. Acad. Sci. USA 84, 5511 (1987).
13. A. Warshel, S. Creighton, and W.W. Parson, J. Phys. Chem., 92, 2696 (1988).
14. M. Bixon and J. Jortner, J. Phys. Chem. 90, 3795 (1986).
15. G.R. Fleming, J. Breton, and J.L. Martin, Nature 333, 190 (1988).

Femtosecond Excited-State Reaction Dynamics of Retinal-Containing Photosystems

W. Zinth, J. Dobler, K. Dressler, and W. Kaiser

Physik Department E11, Technische Universität München,
Arcisstraße 21, D-8000 München 2, Fed. Rep. of Germany

Bacteriorhodopsin (BR) is one of the best studied photoreactive systems. It functions as a light-driven proton pump building up a proton gradient across the cell membrane of halobacterium halobium /1/. The reaction scheme of BR may be understood as follows: In the primary step, light absorption initiates the isomerization of the retinal chromophore from the all-trans to the 13-cis configuration. This reaction establishes the starting condition for the subsequent much slower proton transport process /2-4/. It is the purpose of this paper to reveal the ultrafast molecular processes which proceed during the primary reaction. Due to the recent development of femtosecond techniques these ultrafast phenomena may now be directly studied /5,6,7/.

In our experiments we work with amplified pulses from a CPM dye-laser (Cu vapor pumped amplifier) for excitation (λ = 620 nm) and with probe pulses at different wavelengths selected from a femtosecond continuum. The change of transmission of the sample induced by the exciting pulses is measured with high precision as a function of time delay. The time resolution of the experiment depends on the width Δt of the cross-correlation function between exciting and probing pulses. Typical values of Δt = 90 fs are obtained in our apparatus permitting the investigation of dynamic processes faster than 50 fs.

The choice of the probing wavelength is of major importance for the interpretation of the observed absorption transients. (i) At short probing wavelengths in the region of the 0-0 transition of the molecule the absorption changes may be related to different processes, e.g. to cross relaxation of an inhomogeneous ground-state distribution, to excited-state processes, and to the formation of photoproducts. (ii) Working at longer wavelengths in the fluorescent region of the molecule, the ground-state processes may be neglected, i.e. a more straightforward interpretation of the experiment is possible /5,6/.

Time resolved changes of absorption observed on light-adapted Bacteriorhodopsin at room temperature are shown in Fig.1 for three probing wavelengths in the gain region of BR. At long probing wavelengths (λ = 850 nm, Fig.1c) a pronounced gain is found (the transmitted pulse is larger than the incident pulse). The gain decays at later time with a time constant of 500 fs. A slight 180 fs contribution is present at very early times. With decreasing probing wavelength the 500 fs contribution disappears. At 735 nm (Fig.1b) a 180 fs process dominates the decay of the gain. Fig.1a is taken at a still shorter probing wavelength of 660 nm, where the S_1-S_0 absorption of BR may still be neglected. At late times the build-up (with 500 fs) of an induced absorption of the intermediate J is seen. Around time zero we observe a very short-lived gain. A detailed analysis of the experimental data with a precise determination of time zero and of the cross-correlation function reveals that the gain around time zero is due to an additional intermediate with a lifetime of approximately 50 fs (50±25 fs).

Springer Series in Chemical Physics, Vol. 48: **Ultrafast Phenomena VI**
Editors: T. Yajima • K. Yoshihara • C.B. Harris • S. Shionoya

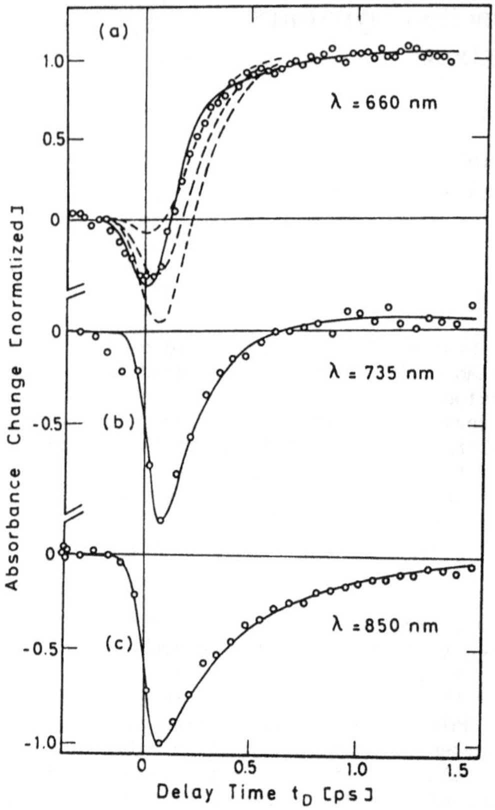

<u>Fig. 1</u> Time-resolved changes of absorption (negative values correspond to gain) induced by exciting femtosecond pulses at λ = 620 nm. The probing wavelengths are λ = 660 nm (a), λ = 735 nm (b), λ = 660 nm (c). The solid curves are calculated using the decay kinetics discussed in the text. The broken curves are calculated excluding the 50 fs kinetic.

The experimental data indicate that there exists an interesting rapid sequence of events in S_1, the excited electronic state, with time values of τ_1 = 50 ps, τ_2 = 180 ps, and τ_3 = 500 ps.

Taking into account the spectral properties of the transient signal and the known molecular data of retinal the following microscopic picture of the very early reactions is suggested (see Fig.2): The incident photons promote the retinal to the Franck-Condon state S_1** on the S_1 potential surface, where a number of vibrational modes are displaced relative to the S_1 equilibrium position /8/. Within 50 fs after light absorption an equilibration of vibrational modes occurs. During this first reaction the molecule remains practically unchanged along the coordinates of the low-frequency modes. The following slower reactive motion of the retinal is related to the 180 fs gain kinetics. In this process, part of the isomerization (presumably a rotation by 60 to 90 degree around the C13-C14 double bond) takes place and the molecules arrive at the bottom of the S_1 potential surface. From this energy position the isomerization continues to form the intermediate product J or the molecule returns via internal conversion to the original ground state with a time constant of 500 fs. Numerical estimates of the isomerization motion support the present interpretiations.

In conclusion it should be noted that additional experiments on other retinal containing systems gave evidence for a very similar hierarchy of events, indicating that the reaction mechanisms found in Bacteriorhodopsin is of a more general nature.

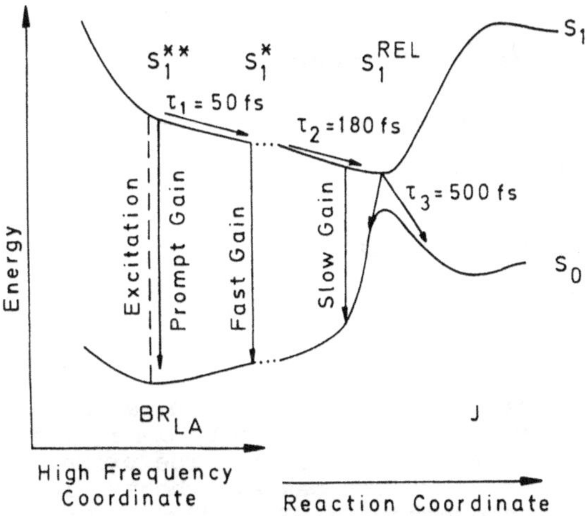

<u>Fig.2</u> Schematic of the ground-state (S_0) and excited-state (S_1) potential energy surfaces as a function of the high-frequency vibrational and the low-frequency reactive coordinates.

Acknowledgement: the experimental work was done in collaboration with Prof. D. Oesterhelt.

<u>References</u>

1. J.K. Lanyi: In <u>Bioenergetics</u>, ed. L. Ernster (Elsevier, Amsterdam 1984) p.314; W. Stockenius, R.A. Bogomolni, Ann. Rev. Biochem. <u>52</u>, 587 (1982)
2. M.C. Nuss, W. Zinth, W. Kaiser, E. Kölling, D. Oesterhelt: Chem. Phys. Lett. <u>117</u>, 1 (1985)
3. H.J. Polland, M.A. Franz, W. Zinth, W. Kaiser, E. Kölling, D. Oesterhelt: Biophys. J. <u>49</u>, 651 (1986);
H.J. Polland, W. Zinth, W. Kaiser: In <u>Ultrashort Phenomena</u>, Vol.4, eds. D.H. Auston, K.B. Eisenthal (Springer, Berlin 1984) p. 456
4. H.J. Polland, M.A. Franz, W. Zinth, W. Kaiser, E. Kölling, D. Oesterhelt: Biochem. Biophys. Acta <u>767</u>, 635 (1984)
5. J. Dobler, W. Zinth, W. Kaiser, D. Oesterhelt: Chem. Phys. Lett. <u>144</u>, 215
6. W. Zinth: Naturwiss. <u>75</u>, 173 (1988)
(1988)
7. R.A. Mathies, C.H. BritoCruz, W.T. Pollard, C.V. Shank: Science <u>240</u>, 777 (1988)
8. E.J. Heller, Acc. Chem. Res. <u>14</u>, 368 (1981)

Direct Observation of the Femtosecond Excited-State cis–trans Isomerization in Bacteriorhodopsin

R.A. Mathies[1], W.T. Pollard[1], C.H. Brito Cruz[2], and C.V. Shank[2]

[1]Chemistry Department, University of California, Berkeley, CA 94720, USA

[2] AT&T Bell Laboratories, Electronics Research Laboratory, Holmdel, NJ 07733, USA

1. Introduction

Femtosecond optical measurement techniques with a time resolution of ~6 fs [1] provide a means of directly observing the temporal evolution of photochemical reactions. The primary events in important photobiological systems can also be studied with these techniques. Bacteriorhodopsin (BR), the protein found in the purple membrane of *Halobacterium halobium*, is responsible for light-driven trans-membrane proton pumping. A variety of experiments have shown that light absorption drives a *trans* to *cis* isomerization of the bound all-*trans* retinal protonated Schiff base chromophore [2]. The primary photoproduct of BR, called J, is formed in ~500 fs and relaxes to K in 3 ps [3-5]. This report describes the use of femtosecond optical absorption spectroscopy to directly observe the excited state *trans* → *cis* torsional isomerization of the retinal prosthetic group in light-adapted bacteriorhodopsin, BR_{568} [6]. We also present a new theory for analyzing the femtosecond dynamic hole-burning spectra of BR_{568}.

2. Experimental Methods

To temporally resolve the excitation and subsequent photoisomerization of the BR_{568} chromophore, it is necessary to study this transformation with 6-fs optical pulses. Since the spectral width of a transform limited 6-fs pulse is extremely broad, these pulses present the unique opportunity to observe the full spectral response of the system on this rapid time scale. With this experimental approach, decay processes with very similar rates or with even a continuous distribution of rates can be identified by their spectral signatures. Such processes are very difficult to interpret with conventional single-wavelength probe data and first-order kinetic analysis.

The femtosecond pump and probe pulses are derived from the same amplified 60 fs optical pulse having an energy of 1 μJ and a center wavelength of 618 nm. The 6-fs probe pulse is formed by passing a portion of the initial pulse through a 12 mm length of optical fiber followed by a grating-pair compressor with prism compensation to accurately adjust the time position of the frequency components. The shortened probe pulse then passes through the excited sample to the spectrometer and diode array detector. The light-adapted purple membrane suspension was pumped through a nozzle at a sufficient velocity to ensure replacement of the sample between each laser shot (8 kHz). The integration time for a single spectrum was typically 30 sec.

Springer Series in Chemical Physics, Vol. 48: **Ultrafast Phenomena VI**
Editors: T. Yajima · K. Yoshihara · C.B. Harris · S. Shionoya

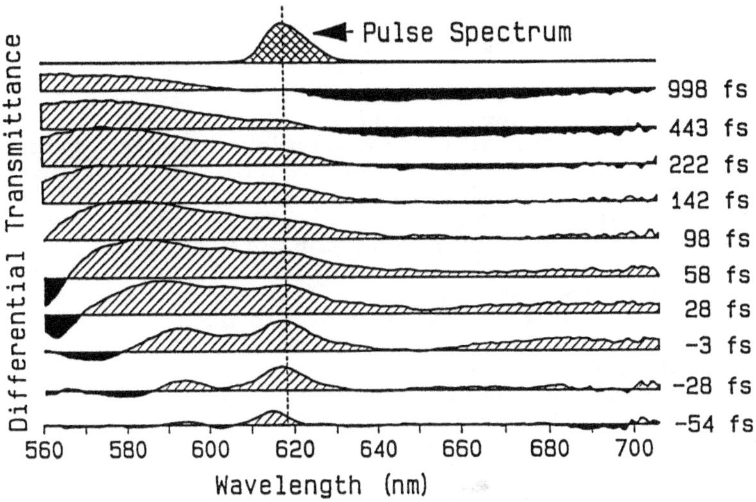

Fig. 1. Transient absorption spectra of light-adapted bacteriorhodopsin (BR_{568}) from -54 to 998 fs.

3. Experimental Results

Figure 1 presents the differential probe transmittance for various pump-probe time delays. We note four distinct contributions to these spectra. The dominant feature is the broad hole that grows in between 560 and 630 nm and reflects the bleaching of the BR_{568} absorption band. Competing with this at short times are a band of increased transmittance to the red of 650 nm and a strong band of increased absorption to the blue of 580 nm. The former we attribute to stimulated emission from the excited electronic state and the latter to excited state absorption from the same state. The rapid attenuation of the stimulated emission signal from -3 to 98 fs is best attributed to the ~100 fs torsional distortion of the molecule on the excited state potential surface which shifts the emission to higher wavelengths. The 580 $\rightarrow$ <560 nm blue shift of the excited state absorption features from -28 to 98 fs also argues that this is the time scale for excited state torsional isomerization. The amplitude of the bleach as indicated by the depth of the hole at ~580 nm is nearly constant from 28 to 222 fs, whereas the stimulated emission and excited state absorption signals evolve strongly from 0 to 150 fs. This supports the idea that the optical excitation is largely complete by the 28 fs spectrum and that _the temporal evolution of the spectra from 28 to 142 fs provides a direct observation of the excited state photoisomerization process in_ $\underline{BR_{568}}$. Finally, the appearance of a red absorption between 222 and 998 fs results from the formation of the ground state photoproduct J.

4. Theory of Femtosecond Dynamic Hole Burning

The dynamic hole-burning spectra can be analyzed in more detail to investigate excited state dephasing and energy relaxation processes. At early times the bleach has the width and approximate position of the pump pulse. At later times this feature evolves into the broad, featureless band seen after 58 fs. To understand these observations we have used a density matrix analysis to simulate the hole-burning spectra for BR_{568}.

The theory of dynamic hole-burning for 2-level systems has recently been described by Brito Cruz et al. [7]. We have generalized this theory for a system with one ground state vibronic level $|i\rangle$ and a multidimensional manifold of vibronic excited state levels $|k\rangle$ [8]. Eight contributions to the third order polarization are uncovered: four conventional level population terms, two "pump-polarization" terms, and two "perturbed free induction decay" (FID) terms. These expressions are difficult to evaluate for multilevel systems because each term contains two sums over the full vibrational subspace of the molecule. To make the evaluation of these expressions more tractible, we recognize that they may be recast in terms of a multidimensional excited state nuclear time correlation function denoted $\langle i|i(t)\rangle$. Here $|i\rangle$ is the initial ground state nuclear wavefunction and $|i(t)\rangle = e^{-iHt/h}|i(0)\rangle$ is the ground state wavefunction propagated on the <u>excited state</u> potential surface [9]. The Fourier transform of $\langle i|i(t)\rangle$ is essentially the homogeneous absorption lineshape. The final equation for the polarization to second order in the pump field and first order in the probe field is

$$P_{pr}^{(3)}(t) = \mu^4 \frac{i}{\hbar} \int_{-\infty}^{t} dt' \int_{-\infty}^{t'} dt'' \int_{-\infty}^{t''} dt''' \, e^{-(t-t')/T_2} e^{-(t'-t'')/T_1} e^{-(t''-t''')/T_2} \tag{1}$$

$$\Big\{ E_{pr}(t')E^*_{pu}(t'')E_{pu}(t''') \langle i|\, i(t-t')\rangle \langle i|\, i(t''-t''')\rangle \; +$$

$$E_{pr}(t')E_{pu}(t'')E^*_{pu}(t''') \langle i|\, i(t-t')\rangle \langle i|\, i(t''-t''')\rangle^* +$$

$$E_{pr}(t')E^*_{pu}(t'')E_{pu}(t''') \langle i|\, i(t-t''')\rangle \langle i|\, i(t'-t'')\rangle^* +$$

$$E_{pr}(t')E_{pu}(t'')E^*_{pu}(t''') \langle i|\, i(t-t'')\rangle \langle i|\, i(t'-t''')\rangle^* +$$

$$E_{pu}(t')E_{pr}(t'')E^*_{pu}(t''') \langle i|\, i(t-t')\rangle \langle i|\, i(t''-t''')\rangle^* +$$

$$E_{pu}(t')E_{pr}(t'')E^*_{pu}(t''') \langle i|\, i(t-t'')\rangle \langle i|\, i(t'-t''')\rangle^* +$$

$$E_{pu}(t')E^*_{pu}(t'')E_{pr}(t''') \langle i|\, i(t-t')\rangle \langle i|\, i(t''-t''')\rangle^* +$$

$$E_{pu}(t')E^*_{pu}(t'')E_{pr}(t''') \langle i|\, i(t-t''')\rangle \langle i|\, i(t'-t'')\rangle^* \Big\},$$

where T_1 and T_2 are the polarization relaxation and dephasing times, μ is the electronic transition moment between S_0 and S_1, and E(t) denotes the pump or probe field envelope. The first four terms with the pump $\rightarrow$ pump $\rightarrow$ probe sequence contribute to the level population changes. The 5th and 6th terms contribute to the pump-polarization signal while the last two terms give rise to the perturbed FID signal. When the pump and probe beams overlap in time, the coherent coupling between the two beams that is described by the pump-polarization and perturbed FID terms will give rise to rapid spectral and temporal modulation of the transmittance spectra.

The nuclear contribution to the polarization decay is contained in $\langle i|i(t)\rangle$ which can be obtained through a resonance Raman intensity analysis [10,11]. The relation between the resonance Raman cross section for a particular vibrational mode and $|i(t)\rangle$ is

$$\sigma_R \,\alpha \left| \int_0^\infty \langle f|\, i(t)\rangle\, e^{iEt/\hbar}\, e^{-\Gamma t}\, dt \right|^2 \qquad (2)$$

where $\langle f|$ is the final vibrational state in the Raman process, E is the incident laser energy, and Γ is a damping factor. The beauty of this relation is that by monitoring the intensity of a particular Raman transition, $\langle f|$ acts as a "filter" that analyzes that particular vibrational dimension of $|i(t)\rangle$. By measuring the intensities of all the allowed Raman transitions, $|i(t)\rangle$ may be determined. The excited state is characterized in terms of a ground$\rightarrow$excited state origin shift or equivalently in terms of the slope of the excited state potential surface evaluated with respect to the ground state normal coordinate. An analysis of the Raman intensities of BR_{568} has been performed by MYERS et al. [10,11]. They measured complete resonance Raman excitation profiles and developed a synthesis for $|i(t)\rangle$ which accurately described the intensities of the 29 Raman active modes. The homogeneous absorption band was found to be very broad (~3600 cm^{-1}) and unresolved due to strong Franck-Condon activity in a large number of modes and fast excited state relaxation.

5. Analysis of BR Hole-Burning Spectra

We have used this characterization of the S_1 excited state of BR_{568} to simulate the early time hole spectra [8]. The comparison of the experimental and calculated holes is given in Fig. 2. The calculated and experimental spectra evolve from a sharp feature at the pump wavelength at 0 fs to a broad band centered at 580 nm by 60 fs. The deviation of the calculated and experimental spectra on the extreme blue and red edges is due to the excited state absorption and stimulated emission which cannot be described by Eqn. 1. The sharp feature at 618 nm in the 0 fs spectrum is due to the pump-polarization and perturbed FID terms. These terms are expected to decay rapidly as time progresses and the pump and probe beams no longer overlap. Thus the evolution from a sharp spectrum at 0 fs to a broad hole at 60 fs is not an indication of spectral diffusion between sites in an inhomogeneously broadened band or excited state vibrational relaxation. The broad hole observed at 60 fs is completely consistent with the broad hole predicted from the resonance Raman intensity analysis. This provides a direct confirmation that the homogeneous lineshape in BR_{568} is very broad as a result of extensive Franck-Condon congestion. Stated alternatively, the excited state wavepacket leaves the Franck-Condon region rapidly (~20 fs) and permanently along multiple highly-displaced degrees of freedom. This rapid departure from the Franck-Condon region is an important element in producing the high efficiency of the isomerization process in BR_{568}.

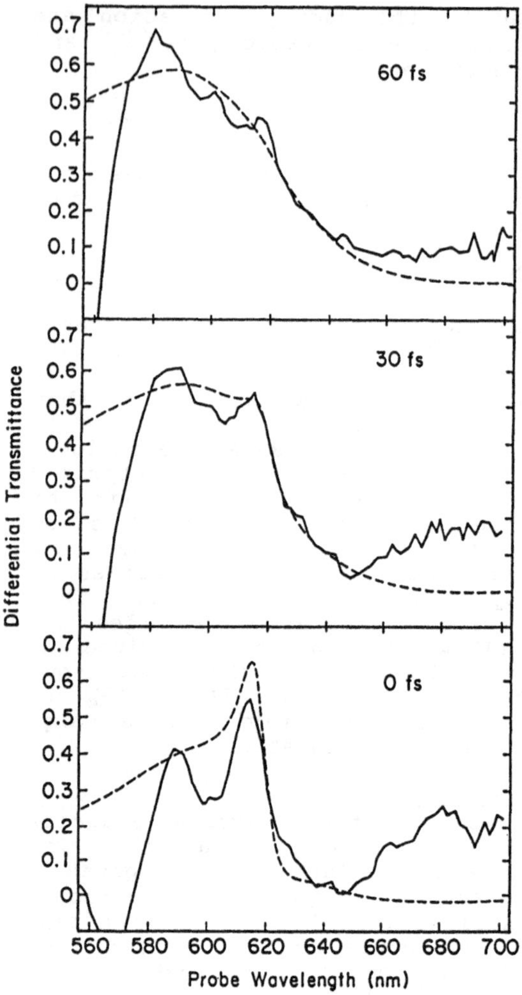

Fig. 2. Calculated and experimental lineshapes for BR_{568} at 0, 30 and 60 fs using excited state parameters from ref. [11]

This interpretation of the femtosecond spectra of BR_{568} was first presented at the Second European Conference on the Spectroscopy of Biological Molecules, Freiburg, FRG, September 6-10, 1987. This work was supported by the NSF (CHE 86-15093) and the NIH (GM 27057).

6. References

1. R.L. Fork, C.H. Brito Cruz, P.C. Becker, C.V. Shank: Opt. Lett. 12, 483 (1987).
2. R.R. Birge: Annu. Rev. Biophys. Bioeng. 10, 315 (1981).
3. A.V. Sharkov, A.V. Pakulev, S.V. Chekalin, Y.A. Matveetz: Biochim. Biophys. Acta 808, 94 (1985).
4. H.J. Polland, M.A. Franz, W. Zinth, W. Kaiser, E. Kolling, D. Oesterhelt: Biophys. J. 49, 651 (1986).
5. J.W. Petrich, J. Breton, J.L. Martin, A. Antonetti: Chem. Phys. Lett. 137, 369 (1987).

6. R.A. Mathies, C.H. Brito Cruz, W.T. Pollard, C.V. Shank: Science 240, 777 (1988).
7. C.H. Brito Cruz, J.P. Gordon, P.C. Becker, R.L. Fork, C.V. Shank: IEEE J. Quant. Elect. 24, 261 (1988).
8. W.T. Pollard, C.H. Brito Cruz, C.V. Shank, R.A. Mathies: J. Chem. Phys., submitted.
9. E.J. Heller, R.L. Sundberg, D. Tannor: J. Phys. Chem. 86, 1822 (1982).
10. A.B. Myers, R.A. Harris, R.A. Mathies: J. Chem. Phys. 79, 603 (1983).
11. A.B. Myers, R.A. Mathies: In Biological Applications of Raman Spectroscopy: Resonance Raman Spectra of Polyenes and Aromatics, ed. T.G. Spiro, Vol. 2 (J. Wiley & Sons, Inc., 1987) pp. 1-58.

Polarized Pump-Probe Spectroscopy of Exciton Transport in Bacteriochlorophyll *a*-Protein from *Prosthecochloris aestuarii*

T.P. Causgrove, S. Yang, and W.S. Struve

Department of Chemistry and Ames Laboratory-USDOE,
Iowa State University, Ames, IA 50011, USA

1. Introduction

Electronic excitation transport (EET) is the first process which follows light absorption by antenna pigments in green photosynthetic bacteria and plants. Such excitation migration is believed to proceed by an incoherent random-walk mechanism arising from Förster dipole-dipole coupling between chlorophyll molecules or aggregates. The EET dynamics are sensitive to the chromophore organization. This is unknown for any antenna system in green photosynthetic organisms, except for the BChl a-protein complex from the sulfur bacterium Prosthecochloris aestuarii. We have therefore studied EET in this BChl a-protein using polarized pump-probe spectroscopy, and analyzed the time-resolved depolarization in terms of this antenna system's well-characterized 3-dimensional structure [1].

The basic structural unit in BChl a-protein is a trimer of subunits containing 7 BChl a molecules each. Nearest-neighbor Mg-Mg separations for chromophores within a subunit range from 11.3 to 14.4 Å; there are no Mg-Mg separations shorter than 24 Å for chromophores belonging to different subunits. Hence, this system may present an example of Sauer's "pebble mosaic" model [2], in which EET (whose rate scales as R^{-6} for chromophores separated by R) is predicted to be far more rapid inside subunits than between subunits.

2. Experimental

A BChl a-protein solution in water (which contains BChl a-protein aggregates with a mean size of at least 30 trimers) was generously provided by R. E. Fenna. Samples housed in 800 μm cells were rotated at 12 Hz to minimize photooxidation by the laser.

A passively mode-locked rhodamine 590/DQOCI dye laser pumped by a multiline Ar^+ laser with 6W plasma tube provided linearly polarized 1.5 ps FWHM pulses between 598 and 609 nm with ~50 mW power at 125 MHz repetition rate. These wavelengths overlap the 603 nm BChl a-protein Q_x system. The pump-probe geometry of Ippen et. al. [3] was used with multiple modulation to evaluate the optical density transients $A_\parallel(t)$, $A_\perp(t)$ for probe pulses polarized parallel and perpendicular to the pump polarization.

3. Results

Figure 1 shows typical pump-probe transients $A_\parallel(t)$ and $A_\perp(t)$ for the BChl a-protein at 603 nm. They are dominated by coherent coupling artifacts during the first ~1.5 ps; they can be fitted for later times (continuous curves) with convolutions of the laser autocorrelation function with the expressions

Springer Series in Chemical Physics, Vol. 48: **Ultrafast Phenomena VI**
Editors: T. Yajima · K. Yoshihara · C.B. Harris · S. Shionoya

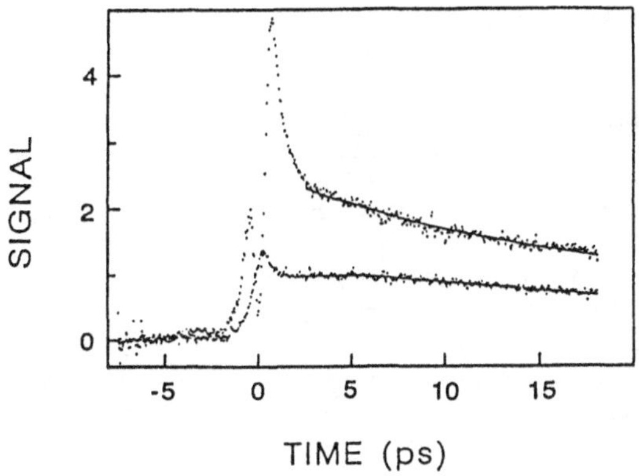

Fig.1

$$A_{\parallel}(t) = P(t)\{1 + 0.8[(1-a)\exp(-t/\tau) + a]\}$$
$$A_{\perp}(t) = P(t)\{1 - 0.4[(1-a)\exp(-t/\tau) + a]\}. \qquad (1)$$

P(t) is the isotropic decay function, evaluated from transients obtained with pump and probe polarizations separated by 54.7°. Nonzero $\underline{a}$ parameters are generally required to simulate the polarized photobleaching decays: the average value derived from six pairs of transients obtained at 603 nm (e.g. Fig. 1) is $\underline{a} = 0.477$. This is a consequence of the nonrandom chromophore organization. The average depolarization timescale τ at 603 nm is 4.78 ps.

The BChl $\underline{a}$-protein exists in at least two crystal modifications ($P6_3$ and $P6_1$), and the packing in aggregates formed in water solution has not been characterized. To model the EET depolarization, we consider for definiteness the $P6_3$ crystal structure shown in Fig. 2. Our model assumes Förster hopping between exciton states delocalized inside subunits containing 7 BChl $\underline{a}$ molecules. We arbitrarily restrict EET to migration between neighboring subunits, and distinguish two kinds of hopping. The first kind occurs with transition rate w_{AB} between subunits in the same trimer (e.g. A → B or D → E). The second kind occurs with rate w_{AD} between subunits in different trimers (e.g. A → D or B → H). The ratio w_{AB}/w_{AD} of transition rates is not known, because the exciton polarization which controls the dipole-dipole orientational factors in w_{AB} and w_{AD} has not been reported.

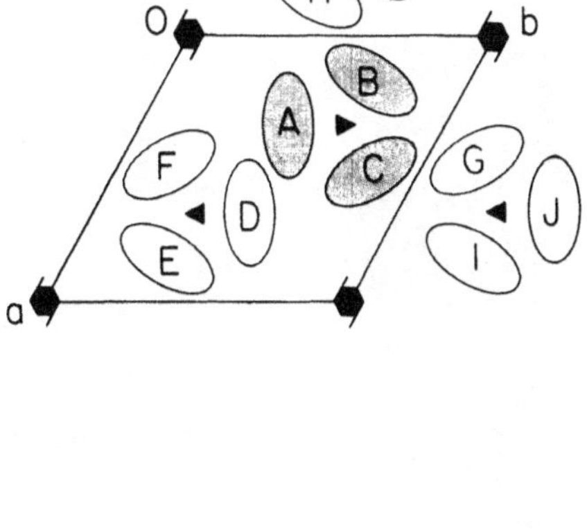

Fig.2

EET was simulated in $P6_3$ crystallites containing 27 trimers by solving the kinetic equations for the subunit excitation probabilities following initial excitation of subunit A. For a given orientation $\hat{d}_A \equiv (\alpha, \beta, \gamma)$ of the probed transition moment in subunit A, the corresponding orientations $\hat{d}_i$ for other subunits i were generated from crystal symmetry. These were weighted by the subunit excitation probabilities and averaged over the random crystallite orientations in solution to yield theoretical polarized transients $A_{\parallel}(t)$, $A_{\perp}(t)$. These computed transients depend on the assumed transition rates w_{AB}, w_{AD} and on the projection γ of the subunits' probed transition moment along the crystallite c-axis. Table I lists several combinations of w_{AB}, w_{AD} which yield theoretical anisotropy functions $r(t) = (A_{\parallel}-A_{\perp})/(A_{\parallel}+ 2A_{\perp})$ which are indistinguishable within experimental S/N from the phenomenological anisotropy function $r(t) = (1-a)\exp(-t/\tau) + a$, where $\tau = 4.78$ ps is the mean depolarization timescale and $a = 0.477$ is the mean residual anisotropy. (The latter parameter is related to γ through $a = (3\gamma^2 - 1)^2/4$). This Table emphasizes that a continuum of combinations of w_{AB} and w_{AD} can account for the observed decay, so that depolarization experiments alone cannot determine these transition rates separately. In our kinetic model (which assumes migration only occurs between proximate subunits), our data are consistent with the ranges $0 < w_{AB}, w_{AD} \lesssim (15\ \text{ps})^{-1}$ in combinations typified by Table I.

TABLE I. Förster transition rates from 27-trimer simulation of EET in BChl a-protein from P. aestuarii

w_{AB}/w_{AD}	w_{AB}^{-1}, ps	w_{AD}^{-1}, ps	$w_{AB} + w_{AD}$, ps^{-1}
∞	14.3	∞	0.070
3	20.2	60.7	0.066
1	28.5	28.5	0.070
1/3	66.0	22.0	0.061

1. Matthews, B. W., Fenna, R. E., Acc. Chem. Res. 1980, **13**, 309.
2. Philipson, K. D., Sauer, K., Biochem. 1972, **11**, 1880.
3. Ippen, E. P., Shank, C. V., Bergman, A., Chem. Phys. Lett. 1976, **38**, 611.

Picosecond Excitation Transport in Photosynthesis: Factors for Optimization of Light Harvesting

A. Freiberg, T. Pullerits, and K. Timpmann

Institute of Physics, Estonian SSR Academy of Sciences, 202400 Tartu, USSR

The present study has been inspired by the following motivations:

(i) From the theoretical point of view the main principles of the very efficient conversion of the solar energy into an electrochemical form by photosynthetic organisms are quite well understood (for a recent review, see [1]). However, the extent to which these principles are realized in one or another natural organism is still to be established;

(ii) During the past few years considerable advances have been made in explaining the composition and the structure of the photosynthetic unit (PSU) of purple and green bacteria (for a review, see [2]). Nevertheless, there is still only poor evidence about the pathways of the excitation energy migration between different pigment-protein complexes of PSU and related rate constants;

(iii) All pigments in photosynthetic organisms are bound to protein environment. Besides, these pigment-protein complexes are often embedded into the lipid membrane [2]. The effect of this specific environment on the spectral characteristics of pigments and dynamics of electronic excitation as well as charge transport may be essential and needs to be cleared up.

1. Spectral and Structural Heterogeneity of the Light-Harvesting Antenna

A number of purple photosynthetic bacteria, e.g. *Rhodobacter sphaeroides* and *Chromatium minutissimum*, contain light-harvesting pigment-proteins of three spectrally well-resolved types: B800, B850 and B875 (numbers indicate the bacteriochlorophyll *a* (BChl*a*) Qy absorption maxima in nm). The picosecond time-domain fluorescence kinetics of these bacteria in dependence on the emission wavelength have been studied under physiological conditions as well as at 77 K. It has been shown that the electronic excitation transport is directed from the short-wavelength light-harvesting antenna BChl *a* forms to the long-wavelength one, and from the latter to reaction centers (RC). As long as the RCs of purple bacteria are surrounded and interconnected by a constant number of B875 pigment-proteins [3,4] the energetically downhill transport of excitations is accompanied by their spatially directed transport towards RCs. As a result, most of the excitations are very quickly localized around RCs where, by means of charge separation, the solar energy is stored for further use. The observed macroscopic excitation trapping time by photoactive RCs, which is equal to about 60 ps at 77 K as well as at room temperature

Springer Series in Chemical Physics, Vol. 48: **Ultrafast Phenomena VI**
Editors: T. Yajima · K. Yoshihara · C.B. Harris · S. Shionoya

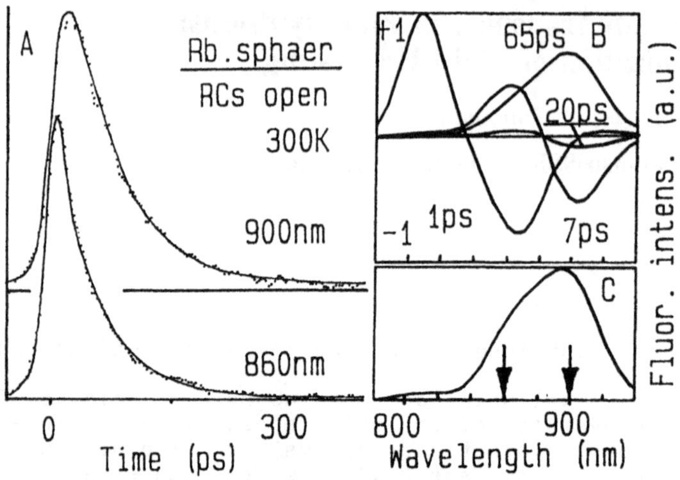

Fig. 1. A) Experimental fluorescence decay kinetics at the given wavelengths of *Rb. sphaeroides* chromatophores (dotted line) and corresponding model curves (continuous line). Excitation intensity at 796 nm about 0.01 W/(cm.cm); B) The calculated amplitude spectra for separate lifetimes; C) The calculated time-integrated fluorescence spetrum

in all the purple bacteria investigated (including *Rhodospirillum rubrum* where due to the lack of B800 and B850 antenna forms the number of BChl *a* molecules per RC is several times smaller), supports this conclusion. A comparison of experimental fluorescence kinetics with the results of model calculations based on coupled rate equations (see Fig. 1), reveals the rates and pathways of heterogeneous excitation transport in purple bacteria: excitations reach from B800 BChl *a* molecules to B850 ones, both situated on the same pair of pigment-binding polypeptides [2], in 1-2 ps. The excitation transport from B800-850 pigment-protein complexes to B875 complexes takes less than 10 ps for most (about 70%) excitations. The minor part of the B850 excitations is transferred to B875 in about 40-50 ps.

2. Long-Range Molecular Order

The light-harvesting antenna of green bacteria containing BChl c is by an order of magnitude larger than that of purple bacteria, which presents more rigorous requirements for the structure optimization. According to that all the antenna BChl c molecules are concentrated in extramembraneous bodies - the chlorosomes, whereas RCs and the most part of the antenna BChl *a* are associated with the cytoplasmic membrane. Selective excitation of BChl c Qy transition moments by picosecond laser pulses has revealed that the chlorosome excitations are transferred to membrane pigments with a characteristic time of 30-60 ps. Such a high transfer rate is somewhat unexpected for such a big antenna. Figure 2 shows that the BChl c fluorescence polarization is nearly constant during the excited state lifetime and attains a limiting high value ($p=0.42\mp0.02$). These results demonstrate that the

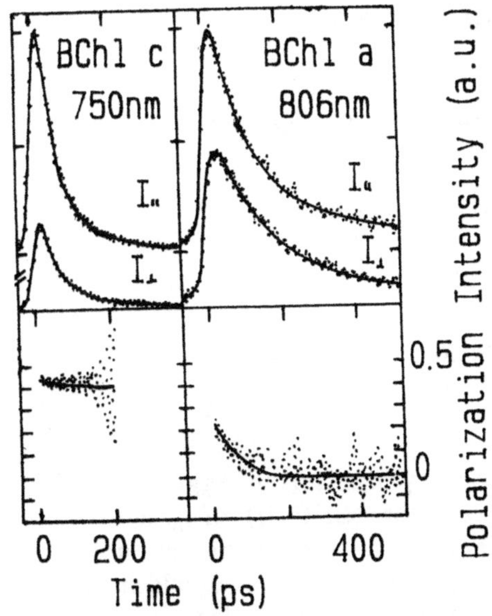

Fig. 2. Room-temperature polarized fluorescence kinetics of BChl *c* (detected at 750 nm) and BChl *a* (806 nm) and the time dependence of the corresponding calculated fluorescence polarization degree $p(t)=(I_{\parallel}-I_{\perp})/(I_{\parallel}+I_{\perp})$ of living cells of green bacterium *Chloroflexus aurantiacus*. Excitation intensity at 724 nm 0.2 W/(cm.cm)

excitation energy transfer in green bacteria is favoured by almost parallel transition moments of the chlorosome chromophores (or their coupled associates). It has been shown by the linear dichroism method that these transition moments are oriented along the long axis of the chlorosome [5].

3. Inhomogeneous Spectral Broadening

Optical transitions of molecules in solids are generally inhomogeneously broadened by virtue of random strains or by chromophore interactions. There is a strong evidence that pigment-protein bands of purple bacteria are inhomogeneously broadened as well. (A corresponding experiment on *R. rubrum*, see [6].) We have shown by means of computer simulation that in case the inhomogeneous bandwidth of B875 is larger than the homogeneous width, the conditions for an additional localization of excitations in B875-RC complexes and, thereby, for the shortening of the overall trapping time, are provided.

The authors thank Drs. Z.G.Fetisova and V.I.Godik (Laboratory of Molecular Biology and Bioorganic Chemistry, Moscow State University) for the fruitful collaboration.

References

1. Z.G.Fetisova, A.Yu.Borisov, M.V.Fok: J.theor.Biol. 112, 41 (1985)
2. G.Drews: Microbiol. Reviews 49, 59 (1985)
3. T.G.Monger,W.W.Parson: Biochim.Biophys.Acta 460, 393 (1977)
4. K.R.Miller: Nature 300, 53 (1982)
5. Z.G.Fetisova, S.G.Kharchenko, I.A.Abdourakhmanov: FEBS Letters, 199, 234 (1986)
6. V.I.Godik, K.E.Timpmann, A.M.Freiberg: Dokl. AN SSSR 298, 1469 (1988)

Excitation Energy Annihilation in Aggregates of Chlorophyll a/b Complexes

T. Gillbro [1], Å. Sandström [1], M. Spangfort [2], R. van Grondelle [3], and V. Sundström [1]

[1] Department of Physical Chemistry, University of Umeå, S-901 87 Umeå, Sweden
[2] Department of Biochemistry, University of Lund, S-221 00 Lund, Sweden
[3] Department of Biophysics, Physics Laboratory of the Free University, NL-1081 HV Amsterdam, The Netherlands

In this work we study the excitation energy transfer in aggregates of the light-harvesting chlorophyll a/b complex (LHC II) of green plants. The method we use is based on the excitation annihilation that can be observed if more than one excitation is created within a system of connected antenna pigment molecules, a so-called domain. By performing time-resolved absorption recovery measurements at different excitation light intensities one can obtain the annihilation rate, γ_2, and the rate of single excitation decay, γ_1. In general

$$\frac{dn}{dt} = -\gamma_1 \cdot n - \frac{1}{2}\gamma_2 \cdot n^2$$

where n stands for [Chla*]. In fig. 1 we have plotted 1/n(dn/dt) vs n and for two experiments with the excitation intensities $1.25 \cdot 10^{13}$ and $2.5 \cdot 10^{13}$ photons/cm^2, we obtain $\gamma_2 = 2.8 \cdot 10^9$ and $2.4 \cdot 10^9$ s^{-1}, respectively. From the absorption recovery traces at low excitation intensities, i.e. $< 10^{13}$ photons/cm^2, we find that $\gamma_1 \approx 3 \cdot 10^9$ s^{-1}.

There is another more common way to obtain γ_2 and that is to measure the fluorescence quantum yield. Paillotin et al. [1] have developed a theory, where a parameter $r = 2\gamma_1/\gamma_2$ is used as a measure of excitation annihilation. For large values of r, trapping and losses dominate, while for small r the annihilation is relatively strong. In this work we have obtained the quantum yield by integrating the normalized decay curves and plotted this value as a function of the number of absorbed photons/cm^2 (see fig 2). We obtain a good fit to theory (see dotted line in fig.2) if we choose r = 3. Putting $\gamma_1 = 3 \cdot 10^9$ s^{-1}, we obtain $\gamma_2 = 2 \cdot 10^9$ s^{-1}, in good aggreement with the results of fig. 1. Thus the two methods of calculating γ_2 give the same result within the experimental error. From fig. 2 it is also possible to calculate the number of interacting chlorophylls in a

Springer Series in Chemical Physics, Vol. 48: **Ultrafast Phenomena VI**
Editors: T. Yajima · K. Yoshihara · C.B. Harris · S. Shionoya

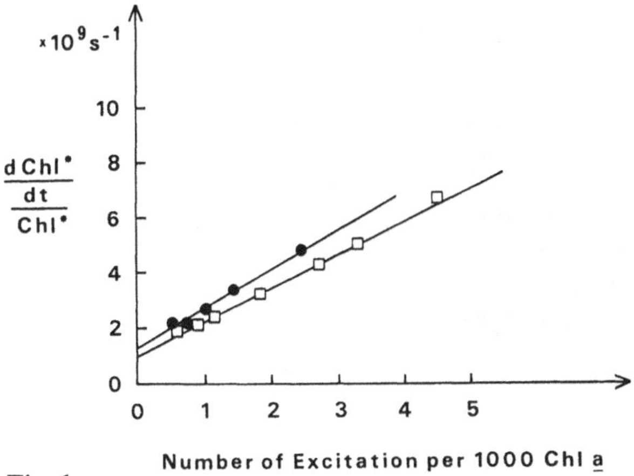

Fig 1
This figure shows (dCHL a*/dt)/Chl a*, as calculated from two different absorption recovery traces at 678 nm, vs number of excitations/1000 Chl a: 1.25 * 10 13 photons/cm2 and 2.5 * 10 13 photons/cm2. The slopes give γ_2 as described in the text.

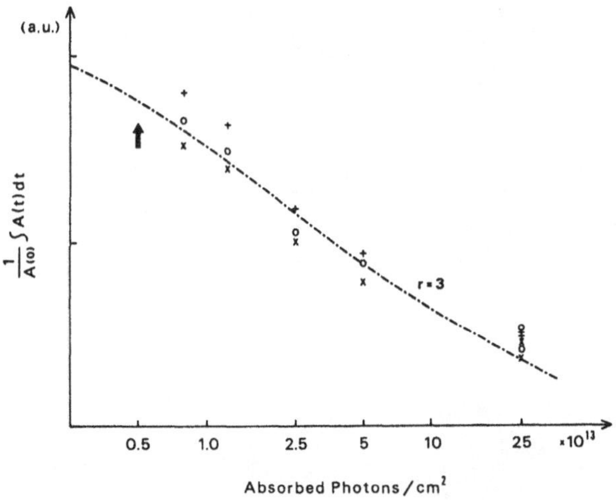

Fig. 2
Calculated normalized time-integrated absorption decays at 678 nm vs number of absorbed photons/cm2. The dashed line represents a fit to eq. 20 of ref. 1 using r=3. The point at which there is one excitation/domain is indicated by the arrow.

domain, N_D. The arrow in fig. 2 show the intensity at which 1 photon is absorbed per domain. This intensity corresponds to N_D= 1000. If the energy transfer occurs on a square lattice the following equation is valid [1]:

$$\frac{\gamma_2}{2} = (4\,k_h + \gamma_1)/(\frac{1}{\pi}\,N_D \ln N_D + 0.195\,N_D)\,.$$

From this the nearest-neighbour hopping rate, k_h, can be estimated to be about $10^{12} s^{-1}$. Since such a large domain size is obtained this hopping rate represents a very efficient energy transfer between chlorophyll molecules situated in both the same and different chlorophyll a/b complexes. In the future it will be of interest to see if such an efficient transfer of energy takes place over large distances between photosystem 2 units in thylakoid membranes.

Reference

1. G. Paillotin, C.E. Swenberg, J. Breton and N.E. Geacintov, Biophys. J. 25, 513 (1979).

Picosecond Studies of Dynamic Solvent Effects on a DNA Intercalator by a Synchroscan Streak Camera System

M. Ishikawa

Tsukuba Research Laboratory, Hamamatsu Photonics K.K., Tokodai 5-9-2, Tsukuba, Ibaraki 300-26, Japan

1. Introduction

Furocoumarins (psoralens) are well known not only as drugs for human skin diseases but also as molecular probes in DNA studies. The specific intercalation of furocoumarins between pyrimidine base pairs of double-stranded DNA, followed by UV induced covalent bond formation, is an essential aspect of the photobiology of furocoumarins. In order to obtain information from DNA molecules by using fluorescence of furocoumarins, it is indispensable to elucidate solvent effects on the fluorescence properties of furocoumarins. As a representative of furocoumarins, bergapten (5-methoxypsoralen) was chosen, because our preliminary results showed that dramatic changes in the fluorescence quantum yield were observed compared with those of furocoumarins in going from a non-protic solvent (cyclohexane) to a protic one (ethanol).

2. Experimental

Time-resolved fluorescence spectra were measured with a synchroscan streak camera system (Hamamatsu C1587), equipped with 82 MHz (M1955) and 4 MHz (M2171) synchroscan units. Picosecond UV pulses (ca.310 nm,) were generated by frequency-doubled synchronously-pumped and cavity-dumped dye lasers (Spectra-Physics). Fluorescence was detected at the magic angle (54.7°) for the vertically polarized excitation pulses.

3. Results and Discussion

In a solvent of cyclohexane-ethanol mixtures, two types of solvent effects must be taken into consideration: dipole-dipole interaction between ethanol and a fluorophore and a hydrogen bonding effect of ethanol on a fluorophore. Addition of ethanol into bergapten solution of cyclohexane caused changes having isosbestic points in absorption spectra (Fig.1), indicating hydrogen bonding between ethanol and a fluorophore in the ground state. The equilibrium constant for the hydrogen bonding was 1.0 M^{-1}, at 25 $^{\circ}C$; that is, only 5.8 % of fluorophors did not form hydrogen bonds even in 100 % ethanol in the ground state. However, the changes in fluorescence decay curves depended dramatically on ethanol concentration as is shown in Fig.2.

One possible mechanism for the ethanol concentration effect is the pseudo-Jahn-Teller effect [1],[2] induced by vibronic interactions between an $n\pi^*$ state and a $\pi\pi^*$ state on nonradiative processes. Interacting electronic states in furocoumarins are identified as the lowest $S_1(\pi,\pi^*)$ state and the nearby $S_2(n,\pi^*)$ state. Because the hydrogen bonding effect promotes the S_2 (n,π^*) state, and the dipolar effect of ethanol stabilizes the S_1 (π,π^*) state, the energy gap between these two states becomes wider with increasing ethanol concentration. Thus the rate of nonradiative tran-

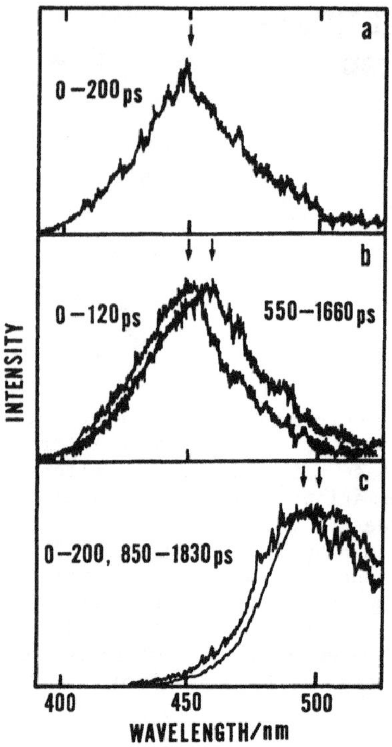

Fig. 1 Time-resolved fluorescence spectra of bergapten at 298 K
a:100% cyclohexane, b:2.5% ethanol
c:100% ethanol

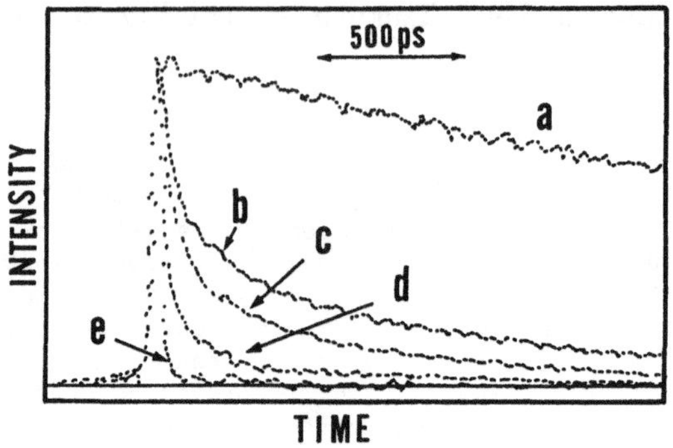

Fig. 2 Fluorescence decay curves of bergapten in various ethanol concentrations at 298 K
a:100%, b:5.0%, c:2.5%, d:1.0%
e:0%

sitions from the $S_1(\pi,\pi^*)$ state to the ground state is diminished due to the reduction of vibronic interactions.

Next, we studied a temperature effect on deactivation mechanisms of some furocoumarins in alcohols. Both fluorescence intensity and T-T absorption increased with decreasing temperature [2],[3]. We concluded that a nonradiative pathway which depends on temperature is not $S_1 \rightarrow T_1$ intersystem crossing but $S_1 \rightarrow S_o$ internal conversion, because the contribution of photochemical reactions to the nonradiative processes was negligibly small. Fluorescence intensity of other furocoumarins was also increased by adding ethanol to cyclohexane solutions. Moreover, the fluorescence intensity increased under pressure (3000 Kg/cm^2, as a max.value) at 298K.

All the above results are consistently explained by the pseudo-Jahn-Teller effect. Firstly, the solvent effects are explained by considering the following solvation equilibrium:

$$\text{Solvent(ethanol)} + \text{Fluorophore} \underset{k_2}{\overset{k_1}{\rightleftarrows}} \text{Solvated-Fluorophore}, \; K_s = k_1/k_2, \tag{1}$$

where K_s is the equilibrium constant for the solvation process. Secondly, the van't Hoff's eq. (2) can be applied to the above system in order to explain temperature and pressure effects on $S_1 \rightarrow S_o$ internal conversion:

$$d(\ln k_s)/dT = \Delta H/RT^2 \quad \text{and} \quad [\partial(\ln k_s)/\partial p]_T = -\Delta V/RT, \tag{2}$$

where ΔH is the enthalpy change in the solvation process, and ΔV is the change in the partial volume for the solvent-fluorophore system. Equation (2) immediately explains the experimental results obtained by changing

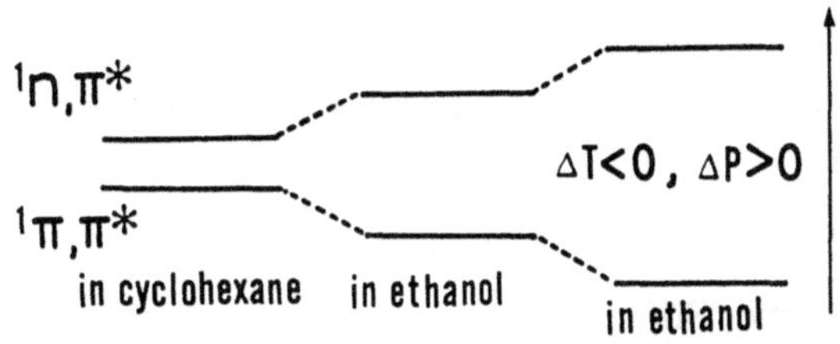

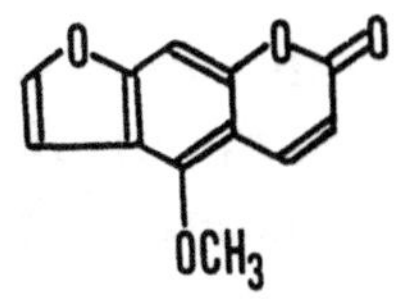

Bergapten

Fig.3 Schematic diagram of two interacting energy levels' separation changed by solvent, temperature and pressure

temperature and pressure, because ΔH is negative for the solvation process and ΔV is negative under pressure. Thus the results that fluorescence intensity increased with decreasing temperature and increasing pressure were due to the reduction of the vibronic interactions by the further solvation, that is, the further separation of the energy gap between interacting states as illustrated in Fig.3.

Furthermore, in addition to the nπ* - ππ* interaction, we must pay attention to the differences in dynamical behavior of a hydrogen bonding effect and a dipolar effect of ethanol to better understand the results indicated in Fig.1 and Fig.2. The rate of stabilization of the lowest S_1(π,π*) state by solvent dipolar reorientation is rather slow. On the other hand, a hydrogen bond is already formed in the ground state. Therefore, hydrogen bonding allows non-bonding electrons in the fluorophores to be affected instantaneously after absorption of light by movements of electrons within the solvent molecules. However, hydrogen bonds newly formed during the lifetime of the excited state probably do not contribute much to the nπ* - ππ* interaction, because the lifetime of hydrogen-bond-free species is very short. Figure 2 shows a time dependent spectral shift observed in 100 % ethanol, indicating temporal progress of the stabilization by the solvent dipolar relaxation. In the middle case (2.5 % of ethanol), a blue shift is observed in an early stage of relaxation. This blue-shifted spectrum coincides with the spectrum observed in 100 % of cyclohexane. Thus we can assign these time-resolved spectra as follows: the blue shifted spectrum is from hydrogen-bond-free species in the Franck-Condon state, and the red shifted spectrum is due to hydrogen-bonded fluorophors stabilized further by the dipolar interaction with ethanol depending on ethanol concentration.

4. Acknowledgment

The author is grateful to Dr.M.Yamashita (ETL) and his co-workers for providing the original version of the programs used in the fluorescence decay curve analysis.

References

[1] E.C.Lim: In Excited States, ed. by E.C.Lim, Vol.3 (Academic Press, New York 1977) p.305
[2] T.I.Lai, B.T.Lim and E.C.Lim (1982) J.Am.Chem.Soc.104,7631
[3] M.Ishikawa, K.Kikuchi and H.Kokubun, unpublished results

Picosecond Fluorescence and Absorbance Study of Charge Separation and Charge Stabilization Processes in Photosystem II Particles

A.R. Holzwarth, G.H. Schatz, H. Brock, and C.G. Colombano

Max-Planck-Institut für Strahlenchemie, Stiftstr. 34-36,
D-4330 Mülheim a.d. Ruhr, Fed. Rep. of Germany

1. Introduction

The primary process of excitation trapping, charge separation and charge stabilization in photosystem II (PS II) are still much less studied than those in bacterial reaction centers (RCs). Agreement has been arrived at on the nature of the primary (intermediate) electron acceptor to be pheophytin a. However, it is still poorly understood whether the radical pair is also formed in RCs in which the first quinone acceptor (Q) is reduced (closed RC). Early experiments with PS II particles have suggested efficient charge separation and consecutive charge recombination with a time constant of approx. 2 - 4 ns under such conditions (1). The results of the analysis of chlorophyll (Chl) fluorescence decay data (2) have led to conclusions which were in conflict with these studies.

2. Materials and Methods

Oxygen-evolving PS II particles with an antenna size of about 80 Chl/P680 have been isolated from thermophilic Synechococcus sp.. Picosecond fluorescence was measured with a single-photon timing apparatus. Picosecond absorbance changes were probed in the wavelength range from 620 to 700 nm. These measurements were carried out with a dual wavelength picosecond pump-probe spectrometer with high sensitivity which was developed in our laboratory. It consists of two synchronously emitting, independently tunable dye lasers as the source of excitation and probe pulses. An improved modulation technique and the use of advanced data analysis procedures enabled a signal to noise ratio close to the shot-noise limit, thus allowing for a drastic reduction of the required excitation densities. The instrument allows absorption measurements with absorption changes as low as $\Delta I/I \simeq 10^{-4}$ limited by the noise in the probe beam.

3. Results and Discussion

A correlated picosecond absorbance and fluorescence study using particles with open and closed RCs has been carried out (3). With open RCs excited state lifetimes of about 80 ps and 500 ps are observed in fluorescence. Closing the RCs by reduction of Q_A with dithionite (or 20μM DCMU plus light) results in an increase of the lifetimes by approximately a factor of 2 - 3 and in a decrease of the amplitude ratio α_1/α_2 in the fluorescence decays by a factor of 3 - 4. With open PS II the absorbance changes

Springer Series in Chemical Physics, Vol. 48: **Ultrafast Phenomena VI**
Editors: T. Yajima · K. Yoshihara · C.B. Harris · S. Shionoya

were described by a sum of three exponentials. The associated absorbance difference spectra were attributed to the exciton trapping and electron transfer processes (3,4). With closed RCs the absorbance changes were biexponential with lifetimes τ_1 = 170-260 ps and τ_2 = 1.6-1.75 ns. The resulting lifetimes show a good agreement with the corresponding fluorescence lifetimes (3).

A kinetic model which accounts for the described observations is proposed in Fig. 1 (see also ref. 3,4). It assumes that Chl^* decay in PS II is trap-limited. For both open and closed RCs the rapid decay phase reflects the equilibration of energy between the antenna and the radical pair, while the slow phase is associated with the overall decay of the resulting equilibrium by deactivation of the antenna and the forward reaction of the radical pair. The exciton equilibration between antenna Chl and reaction centers is assumed to occur on a much faster time scale, i.e., within $\leq$10 ps. The rate constants k_i of Fig. 1 have been calculated from the kinetic data and are given in Table 1. The electron transfer step $I^-Q \rightarrow IQ^-$ is characterized by a time constant ($1/k_2$) of 450 ps in open RCs. The latter process vanishes upon reduction of Q. The most striking change upon closing the RCs consists in a large reduction of the value of k_1 as compared to open RCs. From the comparative analysis of the kinetics of PS II with open and closed RCs we conclude that the reduction of the quinone acceptor Q_A has the following effects: i) the rate constant of charge separation is decreased by a factor of 6; ii) the yield of primary radical pair formation is reduced by a factor of 2 and the mean lifetime of excitons in the Chl antenna is increased by a factor of 3 to 4; iii) the stan-

Table 1. Rate constants k_j (ns^{-1}) according to the kinetic scheme in Fig. 1 calculated as described in the text using the results of fluorescence kinetics for open and closed RCs of PS II

	open	closed	open/closed-ratio
k_A	0.9	1.1	0.8
k_1	9.3	1.5	6.2
k_{-1}	2	2.4	0.8
k_2	2	-	
k'_2	-	0.3	

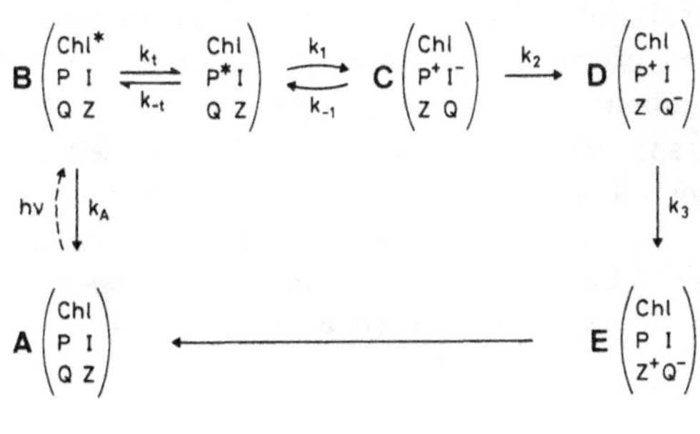

Fig. 1: Kinetic model of the primary reactions in open PS II. In this scheme the rate constants describe: $k_A = k_d + k_{rad}$, the radiationless plus the radiative decay of Chl^*; k_t and k_{-t}, the process of entering and leaving the trap, k_1, the apparent primary charge separation; k_{-1}, charge recombination to the excited state; k_2 and k_3, the process of charge stabilization associated with the relaxation of I^- and P^+ rereduction, respectively. For closed PS II the rate constant k_2 is replaced by k'_2 leading to as yet unspecified product(s) instead of D (4).

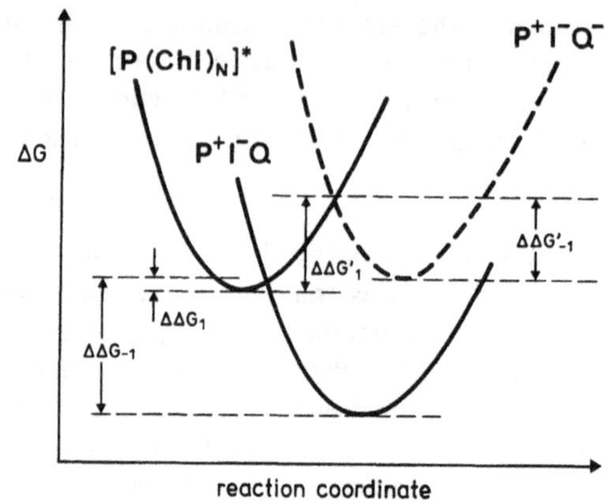

Fig. 2: Potential energy diagram for the primary charge separation and charge recombination processes in open and closed (Q^-) PS II reaction centers with an attached antenna of about 80 Chl/P680.

dard free energy of the state $P^+_{680}I^-Q^-_A$ is raised by about 50 meV as compared to the state $P^+_{680}I^-Q_A$, iv) charge separation in open RCs is exergonic and associated with a decrease in free energy of 38 meV, and v) in closed RCs charge separation is endergonic affording a free energy increase of 12 meV. In view of these findings we conclude that the long-lived (ns) chlorophyll fluorescence generally observed from closed PS II is not delayed charge recombination luminescence but prompt emission from the antenna chlorophylls, in contrast to the interpretation of ref. (1).

The control of P^+I^- formation by the redox state of Q is a consequence of the electrical field created by the negative charge on Q^-. This can be understood from the model shown in Fig. 2. In analogy to bacterial RCs we assume that in the case of open RCs the potential energy curves of the radical pair state P^+I^- and of the excited state P^* cross near the potential minimum of the latter state. This implies a basically activationless charge separation reaction for open RCs. The recombination reaction does require thermal activation. however. As discussed above, for closed RCs the standard free energy of the radical pair state is increased due to the interaction of P^+I^- with Q^-. Now the charge separation process is no longer activationless thus resulting in a large decrease in the corresponding rate constant k_1. In contrast, the activation energy for the recombination process is expected to be changed only to a small extent hence affecting the rate constant k_2 only to a small extent.

According to our kinetic scheme (see Fig. 1), the state B is a generalized excited state of any chlorophyll species, including all the antenna Chl^* as well as P^*. Hence, k_1 is the apparent rate constant of charge separation. It is different from the intrinsic rate constant, k_1^{int} by the factor of partitioning of the exciton between P and all Chl molecules in the core antenna (4). This partitioning can be described by the rate constants k_t and k_{-t} of entering and leaving the trap, respectively, and the degree of degeneracy of the state ($Chl^* \cdot P$). We calculate from our data an intrinsic rate constant of about 370 ns^{-1} (according to $k_1^{int} = 9.3\ ns^{-1} \cdot 80 \cdot 1/2$) corresponding to a time constant of about 2.7 ps. Interestingly, this is very close to the value of 2.8 ps which was measured directly for the primary charge separation in isolated bacterial reaction centers devoid of any antenna pigments.

References

(1) Shuvalov,V.A., Klimov,V.V., Dolan,E., Parson,W.W., and Ke,B., 1980. FEBS Lett. 118: 279-282.
(2) Schatz,G.H. and Holzwarth,A.R., 1986. Photosynth. Res. 10: 309-318.
(3) Schatz,G.H., Brock,H., and Holzwarth,A.R., 1987. Proc. Natl. Acad. Sci. USA 84: 8414-8418.
(4) Schatz,G.H., Brock,H., and Holzwarth,A.R., 1988. Biophys. J. 54: in print.

Picosecond Absorption Spectra of a Reaction Center from a Novel Thermophilic Photosynthetic Bacterium *Chromatium tepidum*

T. Nozawa [1], *M. Terauchi* [2], *T. Kobayashi* [2], *and M. Hatano* [3]

[1] Department of Molecular Chemistry and Engineering, Tohoku University, Aoba, Aramaki, Sendai 980, Japan
[2] Department of Physics, Faculty of Science, University of Tokyo, Hongo, Bunkyo-ku, Tokyo 113, Japan
[3] Chemical Research Institute of Non-aqueous Solutions, Tohoku University, Katahira, Sendai 980, Japan

Picosecond transient absorption spectroscopy has been applied to a new reaction center purified from a novel thermophilic photosynthetic bacterium *Chromatium tepidum*. Although the ground state spectrum of the reaction center shows significant differences from that of a mesophilic purple bacterium (*Rhodobacter sphaeroides* R-26), the primary charge separation kinetics in the reaction center of *C. tepidum* were essentially the same as those of *R. sphaeroides* R-26. The significant differences in the ground state spectra are reflected in the profile of the transient absorption spectra. Futhermore electronic transition assignments were discussed.

1. Introduction

Photosynthetic light energy conversion into chemical energy occurs in a membrane protein called the 'reaction center'. Generally a bacterial reaction center contains four molecules of bacteriochlorophyll (BChl), two of bacteriopheophytin (BPh), two quinones (Q), and a non-heme iron (Fe) [1]. Recent X-ray crystal analysis revealed that the pigments in the bacterial reaction center are organized in a transmembrane dye array [2]. In the periplasm (outside of membrane), a pair of BChl molecules are found in close contact with one another. They are called the 'special pair' ($BChl_2$ or P), because spectroscopy has shown that they are in contact and that their oxidation is the primary event in photosynthesis. Two alternative paths of electronically conducting chromophore extend from this special pair to the single non-heme iron atom which is positioned on the opposite (cytoplasmic) side of the membrane: each path consists of the sequence $BChl_2$-BChl(monomeric bacteriochlorophyll)-BPh-Q_a or Q_b-Fe. Only one of the two paths is actually utilized. In a photosynthetic reaction center a very fast forward electron transfer as compared with a backward reaction has been disclosed as the key to the high efficiency of the primary photosynthetic reaction (especially in the charge separation reaction) [3]. The time course of the primary electron transfer step has been studied [4,5] on several photosynthetic bacteria. A newly prepared reaction center from a novel thermophilic bacterium *Chromatium* (*C.*) *tepidum* showed distinct differences in the ground state absorption spectrum from those of mesophilic bacteria [6]. Since the mechanisms of the electron transfer process may have close correlation to the electronic structures of the special pair, we observed the picosecond (ps) transient absorption spectra to examine whether the differences in the ground state absorption spectra have any effect on the primary process of photosynthesis, and whether any information about the electronic transition assignment can be obtained from the picosecond absorption change in the near infrared region.

Springer Series in Chemical Physics, Vol. 48: **Ultrafast Phenomena VI**
Editors: T. Yajima · K. Yoshihara · C.B. Harris · S. Shionoya

2. Experimental

A reaction center of C. tepidum was prepared as described before [6]. The final purification by HPLC with a DEAE-analog column gave two types of reaction center samples with and without cytochrome c. The reaction center with cytochrome c was used in all the experiments unless otherwise stated. The reaction center of R. sphaeroides R-26 was prepared by the method reported in the literature [1]. The picosecond absorption spectra were obtained in 3 mm (or 2 mm) cells with OD_{800} = 5.5/cm in 700-900 nm, and OD_{800} = 16.5 in the 530-730 nm region. We used 20 mM Tris buffer with 0.1% dodecy-β,D-maltoside (or 0.1% Triton X-100) and 0.0125% of lauryl-dimethylamine-N-oxide. The details of the optical system used were described previously [8]. A mode-locked Nd-YAG laser (Quantel, Model 472) was used for both excitation and probe light. The second harmonic (532 nm, 50ps) attenuated to 100 μJ/pulse was used as excitation light. The excitation pulse causes photooxidation of about 50% of the reaction center on each flash. The repetition rate of the data aquisition was 1 Hz for C. tepidum, and 0.25 Hz for the R. sphaeroides R-26 reaction center.

3. Results and Discussions

In Fig. 1 the ground state absorption spectra are compared for reaction centers from C. tepidum and R. sphaeroides R-26. Reaction centers from purple photosynthetic bacteria show three distinct absorption bands in the near infrared region [4]. The longest band (around 860-900 nm) belongs to the special pair P because it bleaches when P is oxidized. The next longest band around 800 nm mainly corresponds to the monomeric BChl a, however, it has been proposed that the higher energy exciton band of P also exists around 800 nm. The 750 nm band is due to BPh bands. Figure 1 shows that the position of the special pair absorption band and intensity vary significantly, while the rest of the bands (i.e., the monomeric bacteriochlorophyll band (around 800 nm) and bacteriopheophytin band (around 750 nm) are kept unchanged among the purple photosynthetic bacterial reaction centers with BChl a. R. sphaeroides, R. capsulata, and R. rubrum have the special pair absorption band at around 860 nm at room temperature, while C. vinosum, and C. tepidum have the band at around 885 nm [4-7]. In addition to the difference in the position of the absorption maximum, the

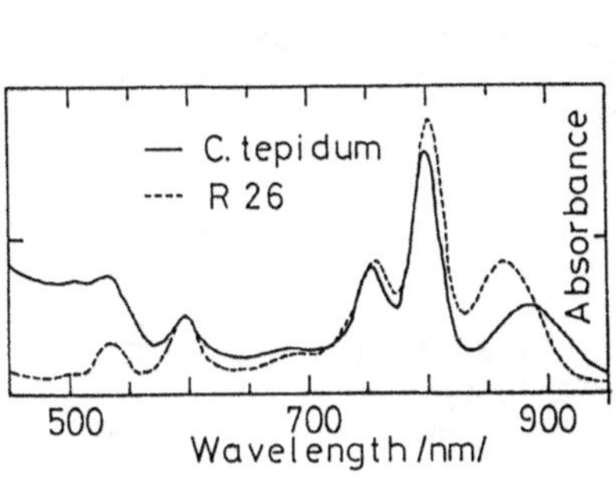

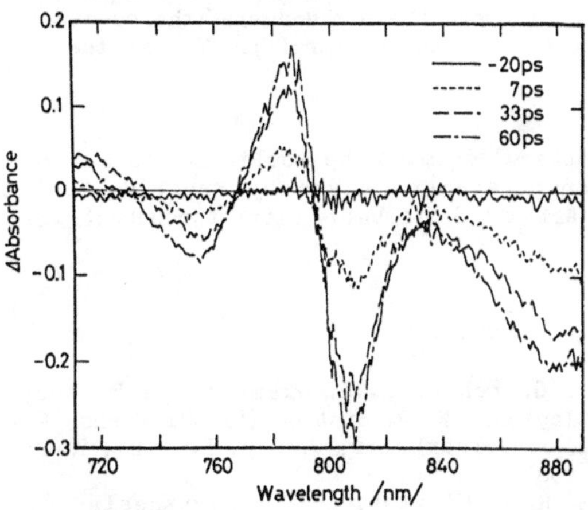

Fig. 1 & 2: Ground state and transient absorption spectra of reaction centers

intensity of the special pair bands for the latter group of reaction centers are smaller than those for the former by 60 %. The origins of this shift are not clear, but the presence of the tightly bound cytochrome c for the latter group has a critical effect [6]. The origins of the reduced intensity are still unrevealed.

The flash pulse at the wavelength of 532 nm excites the Qx transition of BPh, the Q band of cytochrome c and the longest wavelength band of carotenoid. Thereafter, the excitation energy is transferred to P. In Fig. 2 the transient absorption spectra are shown for the reaction center of C. tepidum. The formation of P^+ was indicated from bleaching at the 880 nm absorption band. This process occurred within the instrumental response time (20 ps). The increase and decrease of absorbance around 790 nm and 810 nm occurred also in the instrumental response time. These processes were observed in exactly the same way for the reaction centers from both C. tepidum and R-26 in the limit of the present experimental setup. The changes around 790 and 810 nm can be explained by merely a band shift of the 800 nm band caused by the ionization of the special pair, or by the existence of a special pair band with the same decay constant [4]. It should be noted that the profiles of the transient absorption spectra between C. tepidum and R-26 are fairly different, especially in the relative magnitudes of bleaching for the special pair band to that of the 800 nm band. Detailed studies with polarized light at low temperature using a shorter excitation pulse will give more definite conclusions about the correlation between electronic structures and functions of reaction centers.

Bleaching of the BPh band at 750 nm showed some delay from the special pair bleaching for both reaction centers. That is, the BPh bleaching occurred with a delay of less than 20 ps after the bleaching of P. This indicates that the rate constant of the electron flow from P to BPh is around 20 ps in the limit of the present instrumental response time. There was essentially no difference in this process for the reaction centers between C. tepidum and R-26. The recovery of the 755 nm bleaching to half its magnitude took place with a time constant of about 180 ps.

In conclusion, the data in the picosecond transient absorption spectra showed that the primary charge separation steps in reaction centers were not so sensitive to position and intensity of the special pair band in the ground state spectrum so as to give significant changes (more than the instrumental sensitivity (20 ps)) in the time constants. The interactions which gave these phenomena (the spectral shift and intensity decrease) did affect the spectral profile in the transient spectra.

Acknowledgement: We would like to acknowledge the kind hospitality and encouragement given by Prof. R. E. Blankenship and Dr. J. T. Trost (Arizona State University) throughout this study.

4. References

1. G. Feher, M. I. Okamura, In Photosynthetic Bacteria, ed. by R. L. Clayton, W. R. Sistrom (Plenum Press, New York 1968) p.349

2. J. Deisenhofer, O. Epp, K. Miki, R. Huber, H. Michel, Nature 318, 618 (1985)

3. R. K. Clayton, In Photosynthesis: Physical Mechanisms and Chemical Patterns (Cambridge University Press: Cambridge, 1980) p.1

4. W. W. Parson, Annu. Rev. Biophys. Bioeng. 11, 57 (1982)
5. D. Holten, C. Kirmaier, L. Levin, In Prog. in Photosynthesis Research, ed. by J. Biggins (Martius Nijhoff, The Hague 1987) p.169
6. T. Nozawa, J. T. Trost, T. Fukada, M. Hatano, J. D. McManus, R. E. Blankenship, Biochim. Biophys. Acta 894, 468 (1987)
7. L. Lin, J. P. Thornber, Photochem. Photobiol., 22, 37 (1975)
8. J. Iwai, M. Ikeuchi, Y. Inoue, T. Kobayashi, in Protochlorophyllide and Greening, ed. by C. Sironval, M. Brouers (Martinus Nijhoff, The Hague 1984) p.99

Time-Resolved Fluorescence Spectra of D-Amino Acid Oxidase: A New Fluorescent Species of the Coenzyme

F. Tanaka[1], N. Tamai[2], and I. Yamazaki[2]

[1]Mie Nursing College, 100 Torii-cho, Tsu 514, Japan
[2]Institute for Molecular Science, Okazaki 444, Japan

1. Introduction

D-Amino acid oxidase is a flavoenzyme with molecular weight of 40000 per monomer and contains flavin adenine dinucleotide (FAD) as coenzyme. The enzyme is in an equilibrium state among monomer, dimer and higher associated forms depending on its concentration [1]. A number of works have revealed that the physico-chemical properties of monomer, including the enzyme activity, are quite different from dimer or oligomers of the enzyme. The binding process of FAD has displayed a positive cooperativity [2]. The fluorescence lifetimes of FAD of D-amino acid oxidase in the picosecond region suggest that the microenvironment surrounding FAD in monomer is quite different from that in dimer or oligomer [3]. It has been also shown that the remarkable quenching of fluorescence of FAD in the protein is due to an electron transfer from tryptophan to the excited FAD [4].

In the present work we demonstrate that the normal fluorescence of FAD in D-amino acid oxidase appears after a meta-stable fluorescent state decays.

2. Experimental

D-Amino acid oxidase was purified from hog kidney according to the method previously reported. FAD was purchased from Nakarai Chemicals, LTD. (Kyoto) and purified by a column chromatography on DEAE-cellulose. The fluorescence decay curves and time-resolved fluorescence spectra were measured with a synchronously pumped, cavity-dumped dye laser and a picosecond time-correlated, single-photon counting apparatus [5]. Temperature of the samples was controlled with circulating water. Typical time width of the instrumental responce function was 30 ps. The observed decay curves were analyzed with multi-exponential decay functions by a non-linear least-squares iterative convolution method based on the Marquardt algorithm.

3. Results and Discussion

Fluorescence decay curves of FAD in D-amino acid oxidase were measured at 20 $^{\circ}$C and 5 $^{\circ}$C at eight different levels of the enzyme from 100 to 0.78 μM. The decay curves monitored at 530 nm were all non-exponential and analyzed with two-, three- and four-exponential decay functions, the results of which were compared. The best fit was obtained when the observed decay curves were analyzed with four-exponential decay functions. Figure 1 shows an observed decay curve obtained at 20 $^{\circ}$C (concentration, 0.78 μM in 0.017 mM pyrophosphate buffer at pH 8.3). The values of the lifetimes were almost constant within 15 % upon changing the concentration. Averaged values of the lifetimes were $\tau_0 = 26.6 \pm 0.7$ ps, $\tau_1 = 44.0 \pm 4.2$ ps, $\tau_2 = 177 \pm 11$ ps and $\tau_3 = 2.28 \pm 0.21$ ns at 20 $^{\circ}$C, and $\tau_0 = 25.2 \pm 3.0$ ps, $\tau_1 = 50.3 \pm 8.7$ ps, $\tau_2 = 228 \pm 27$ ps and $\tau_3 = 2.75 \pm 0.33$ ns at 5 $^{\circ}$C.

Springer Series in Chemical Physics, Vol. 48: **Ultrafast Phenomena VI**
Editors: T. Yajima · K. Yoshihara · C.B. Harris · S. Shionoya

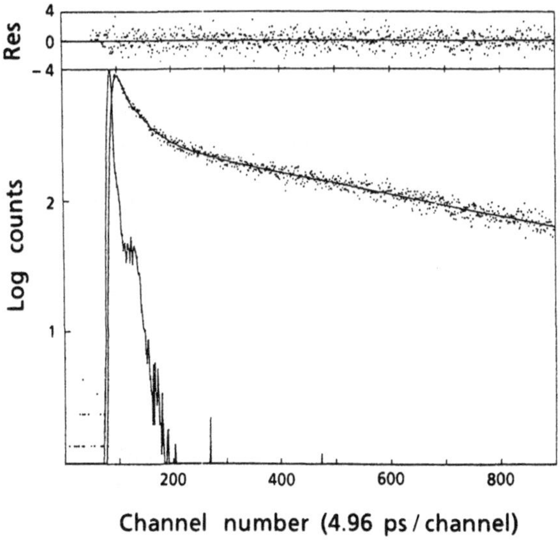

Fig. 1 Fluorescence intensities of FAD of D-amino acid oxidase at 20 °C are shown with dots. The emission was monitored at 530 nm upon excitation at 315 nm. The enzyme was dissolved in 0.017 M of pyrophosphate buffer at pH 8.3 (conc 0.78 μM). The calculated decay curve is indicated with a solid curve. Decay parameters at the best-fit were τ_0 = 26.5 ps (-1.013), τ_1 = 50.0 ps (0.797), τ_2 = 197 ps (0.127), τ_3 = 2.58 ns (0.077), where the values in the parentheses are component fractions.

The component fractions of the shortest lifetime were -0.998 ± 0.029 at 20 °C and -0.949 ± 0.084 at 5 °C. They were always negative and close to -1 within experimental accuracy. The fact suggests that the normal fluorescent state of FAD forms after a meta-stable state of the excited isoalloxazine decays. The rise-up component with negative component fraction was not observed in free FAD. The fluorescent species with lifetimes of positive pre-exponential factors, τ_1, τ_2 and τ_3 are assigned to be the associated forms of the enzyme including dimer, the monomer and free FAD dissociated from the enzyme, according to the previous work [3].

Time-resolved fluorescence spectra of FAD of D-amino acid oxidase were measured at 20 °C and 5 °C at the concentration of 100 μM. The spectra at 20 °C are shown in Fig. 2. The first spectra obtained by accumulating from -20 ps to 0 ps have the maxima at 584 nm at 20 °C and 557 nm at 5 °C. Successively, the emission spectra obtained accumulating for every 20 ps shifted toward shorter wavelength to 520 nm. The spectra became steady at the gated times later than 100 ps after the pulsed excitation, which is in good agreement with the spectrum obtained by steady-state excitation. The time-resolved spectra of free FAD were also examined. In free FAD the spectrum did not shift at all.

The remarkable blue shifts in the fluorescence spectra of the coenzyme of D-amino acid oxidase are consistent with the appearance of rise-up component in the decay curves. Since they were not observed in free FAD, these phenomena are considered to take place only in the protein. Such an ultrafast change in the spectra suggests that the electronic state of the excited flavin changes coupled with the dynamics of the protein structure. The blue shift in the time-resolved fluorescence spectra is rather unusual. The meta-stable fluorescent state may be related with one form of various types of hydrogen-bondings [6], or proton transferred complexes between isoalloxazine which contains four proton acceptors and one proton donor, and nearby amino acid residues.

Ackowledgments
We are grateful to Prof. Mataga and Prof. Nishimoto for stimulating discussions on the present work. We also thank the Computer Center, Institute for Molecular Science, Okazaki Research Institutes, for the use of HITAC M-680.

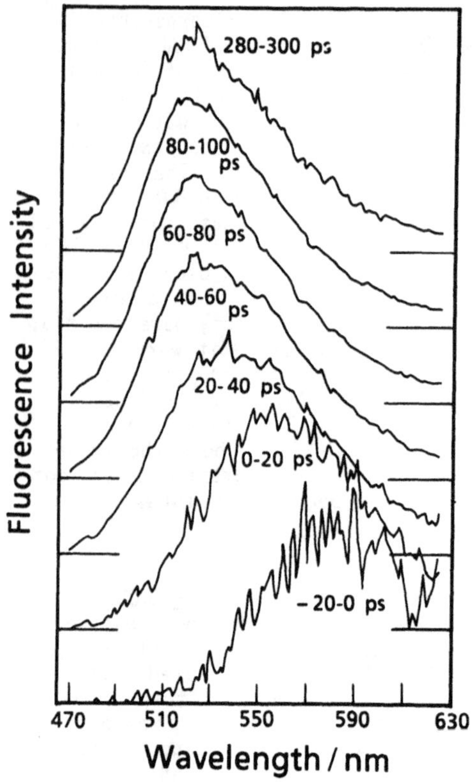

Fig. 2 Time-resolved fluorescence spectra of FAD of D-amino acid oxidase at 20 °C are shown at various gated times after pulsed excitation at 315 nm. The gated times for the accumulation are indicated beside the spectra. The maximum wavelength of the first spectrum (-20 - 0 ps of the gated time) was 584 nm. Successively, these were 556 nm for 0 - 20 ps, 539 nm for 20 - 40 ps, 526 nm for 40 - 60 ps, 522 nm for 60 - 80 ps. The maximum wavelength was 520 nm later than 100 ps of the gated time. The enzyme was dissolved in 0.017 M of pyrophosphate buffer at pH 8.3 (conc 100 μM).

References

1. H. Tojo, K. Horiike, K. Shiga, Y. Nishina, H. Watari, T. Yamano : J. Biol. Chem. 260, 12615 (1985)
2. F. Tanaka, K. Yagi : Biochemistry 18, 1531 (1979)
3. N. Nakashima, K. Yoshihara, F. Tanaka, K. Yagi : J. Biol. Chem. 255, 5261 (1980)
4. A. Karen, N. Ikeda, N. Mataga, F. Tanaka : Photochem. Photobiol. 37, 495 (1983)
5. I. Yamazaki, N. Tamai, H. Kume, H. Tsuchiya, K. Oba : Rev. Sci. Instrum. 56, 1187 (1985)
6. K. Nishimoto, Y. Watanabe, K. Yagi : Biochim. Biophys. Acta 526, 34 (1978)

Index of Contributors

S. Svanberg, W. Persson (Eds.)

Laser Spectroscopy VIII

Proceedings of the Eighth International Conference, Åre, Sweden, June 22–26, 1987

1987. 315 figures. XX, 474 pages. (Springer Series in Optical Sciences, Volume 55).
ISBN 3-540-18437-6

Contents: Applications of Laser Spectroscopy to Basic Physics. - Laser Cooling, Trapping and Manipulation of Atoms and Ions. - Quantum Jumps. - Quantum Optics, Squeezed States and Chaos. - Atomic Spectroscopy. - Molecular Spectroscopy. - Clusters, Surfaces and Solids. - Miscellaneous Laser Spectroscopy Experiments. - Laser Spectroscopic Diagnostics. - Spectroscopic Techniques. - Spectroscopic Sources. - VUV Spectroscopy. - Index of Contributors.

This book contains the proceedings of the Eighth International Conference on Laser Spectroscopy (EICOLS '87) held in Åre, Sweden, June 22–26, 1987. The more than 140 contributions record the talks and posters in which leading experts in the field presented the latest developments in laser spectroscopic sources, laser spectroscopic techniques and their applications to fundamental and applied physics and related fields.

The topics treated include:
- Applications of Laser Spectroscopy to Basic Physics
- Laser Cooling, Trapping and Manipulation of Atoms and Ions
- Quantum Jumps
- Quantum Optics, Squeezed States and Chaos
- Atomic Spectroscopy
- Molecular Spectroscopy
- Clusters, Surfaces and Solids
- Laser Spectroscopic Diagnostics
- VUV Spectroscopy

Springer-Verlag Berlin Heidelberg New York London Paris Tokyo Hong Kong

Springer Series in Chemical Physics

Editors: V. I. Goldanskii R. Gomer F. P. Schäfer J. P. Toennies

1 **Atomic Spectra and Radiative Transitions** By I. I. Sobelman
2 **Surface Crystallography by LEED** Theory, Computation and Structural Results. By M. A. Van Hove, S. Y. Tong
3 **Advances in Laser Chemistry** Editor: A. H. Zewail
4 **Picosecond Phenomena** Editors: C. V. Shank, E. P. Ippen, S. L. Shapiro
5 **Laser Spectroscopy** Basic Concepts and Instrumentation By W. Demtröder 2nd Printing
6 **Laser-Induced Processes in Molecules** Physics and Chemistry Editors: K. L. Kompa, S. D. Smith
7 **Excitation of Atoms and Broadening of Spectral Lines** By I. I. Sobelman, L. A. Vainshtein, E. A. Yukov
8 **Spin Exchange** Principles and Applications in Chemistry and Biology By Yu. N. Molin, K. M. Salikhov, K. I. Zamaraev
9 **Secondary Ion Mass Spectrometry SIMS II** Editors: A. Benninghoven, C. A. Evans, Jr., R. A. Powell, R. Shimizu, H. A. Storms
10 **Lasers and Chemical Change** By A. Ben-Shaul, Y. Haas, K. L. Kompa, R. D. Levine
11 **Liquid Crystals of One- and Two-Dimensional Order** Editors: W. Helfrich, G. Heppke
12 **Gasdynamic Laser** By S. A. Losev
13 **Atomic Many-Body Theory** By I. Lindgren, J. Morrison
14 **Picosecond Phenomena II** Editors: R. M. Hochstrasser, W. Kaiser, C. V. Shank
15 **Vibrational Spectroscopy of Adsorbates** Editor: R. F. Willis
16 **Spectroscopy of Molecular Excitions** By V. L. Broude, E. I. Rashba, E. F. Sheka
17 **Inelastic Particle-Surface Collisions** Editors: E. Taglauer, W. Heiland
18 **Modelling of Chemical Reaction Systems** Editors: K. H. Ebert, P. Deuflhard, W. Jäger
19 **Secondary Ion Mass Spectrometry SIMS III** Editors: A. Benninghoven, J. Giber, J. László, M. Riedel, H. W. Werner
20 **Chemistry and Physics of Solid Surfaces IV** Editors: R. Vanselow, R. Howe
21 **Dynamics of Gas-Surface Interaction** Editors: G. Benedek, U. Valbusa
22 **Nonlinear Laser Chemistry** Multiple-Photon Excitation By V. S. Letokhov
23 **Picosecond Phenomena III** Editors: K. B. Eisenthal, R. M. Hochstrasser, W. Kaiser, A. Laubereau
24 **Desorption Induced by Electronic Transitions DIET I** Editors: N. H. Tolk, M. M. Traum, J. C. Tully, T. E. Madey
25 **Ion Formation from Organic Solids** Editor: A. Benninghoven
26 **Semiclassical Theories of Molecular Scattering** By B. C. Eu
27 **EXAFS and Near Edge Structures** Editors: A. Bianconi, L. Incoccia, S. Stipcich
28 **Atoms in Strong Light Fields** By N. B. Delone, V. P. Krainov
29 **Gas Flow in Nozzles** By U. Pirumov, G. Roslyakov
30 **Theory of Slow Atomic Collisions** By E. E. Nikitin, S. Ya. Umanskii
31 **Reference Data on Atoms, Molecules, and Ions** By A. A. Radzig, B. M. Smirnov
32 **Adsorption Processes on Semiconductor and Dielectric Surfaces I** By V. F. Kiselev, O. V. Krylov
33 **Surface Studies with Lasers** Editors: F.R. Aussenegg, A. Leitner, M.E. Lippitsch
34 **Inert Gases** Potentials, Dynamics, and Energy Transfer in Doped Crystals. Editor: M. L. Klein
35 **Chemistry and Physics of Solid Surfaces V** Editors: R. Vanselow, R. Howe
36 **Secondary Ion Mass Spectrometry, SIMS IV** Editors: A. Benninghoven, J. Okano, R. Shimizu, H. W. Werner
37 **X-Ray Spectra and Chemical Binding** By A. Meisel, G. Leonhardt, R. Szargan
38 **Ultrafast Phenomena IV** By D. H. Auston, K. B. Eisenthal
39 **Laser Processing and Diagnostics** Editor: D. Bäuerle

www.ingramcontent.com/pod-product-compliance
Lightning Source LLC
Chambersburg PA
CBHW071926130926
51338CB00020B/40
9783642836466